CELLULAR SIGNAL PROCESSING

An Introduction to the Molecular Mechanisms of Signal Transduction

Friedrich Marks

Ursula Klingmüller

Karin Müller-Decker

Garland Science
Taylor & Francis Group

Vice President: Denise Schanck
Senior Editor: Janet Foltin
Assistant Editor: Katherine Ghezzi
Production Editor: Karin Henderson
Copyeditor: Heather Whirlow Cammarn
Illustration: Blink Studio Limited and Nigel Orme
Layout: Georgina Lucas
Cover Design: Matt McClements
Indexer: Liza Furnival

Friedrich Marks is Professor of Biochemistry at the University of Heidelberg and Emeritus Department Head, German Cancer Research Center. **Ursula Klingmüller** is a Lecturer at the University of Heidelberg and Head of the Systems Biology of Signal Transduction Division, German Cancer Research Center. **Karin Müller-Decker** is a Lecturer at the University of Heidelberg and Head of the Tumor Models Unit, German Cancer Research Center.

10-digit ISBN 0-8153-4215-2 (paperback)
13-digit ISBN 978-0-8153-4215-1 (paperback)

Library of Congress Cataloging-in-Publication Data

Marks, Friedrich, 1936-
  Cellular signal processing/Friedrich Marks, Ursula Klingmüller, Karin Müller-Decker
    p. ; cm.
  Includes bibliographical references and index.
  ISBN 978-0-8153-4215-1
    1. Cellular signal transduction. I. Klingmüller, Ursula. II. Müller-Decker, Karin. III. Title.
    [DNLM: 1. Signal Transduction—physiology. 2. Intracellular Signaling Peptides and Proteins—physiology. QU 375 M346c 2009]
  QP517.C45M37 2009
  571.7'4—dc22
                                2008029273

Published by Garland Science, Taylor and Francis Group, LLC, an informa business
270 Madison Avenue, New York, NY 10016, USA, and
2 Park Square, Milton Park, Abingdon, OX14 4RN, UK.

Printed in Great Britain by Ashford Colour Press Ltd

Visit our web site at http://www.garlandscience.com

"For Lilli, who opened my eyes to the wonderful world of binarily encoded patterns. F.M."

Preface

All life is problem solving
  Sir Karl R. Popper

Cellular signal processing, often called signal transduction, is a subject that ranks among the most rapidly developing fields in biomedical sciences. Diseases such as cancer, diabetes, dementia, psychoses, and cardiovascular disorders are caused in part by disturbances in cellular signal processing, and the majority of therapeutic drugs as well as many narcotics target corresponding cellular pathways.

This book is intended for undergraduate and graduate students in biology, biochemistry, bioinformatics, and pharmacology as well as for medical students. It emerged from our scientific work at the German Cancer Research Center and from a series of lectures we have given at the University of Heidelberg for many years. Although cell biology textbooks provide excellent introductions into the field of signal transduction, our students consistently requested a more comprehensive treatment of the subject that, in particular, presented evolutionary aspects and questions of medical interest. Moreover, we identified the need for a textbook that would offer a more complete overview of the field than that currently provided by available textbooks and monographs.

The most serious obstacle we faced in developing this book is the rapidly increasing amount of data that is still more or less in a preliminary form. In order to address this problem and keep the reader from losing the thread of the narrative, we decided to organize the content in a way that is different from that in other textbooks. To this end, the discussion is guided by three major principles that dominate cellular signal processing, namely the protein network, its energy supply, and its evolution.

These principles are introduced in the first three chapters. Chapter 1 deals with the ability of proteins to form a "neural" network that, together with the genome, resembles a cellular brain. In this network each protein acts as a "molecular neuron" that, according to the laws of chemical thermodynamics and mathematical logic, transforms input signals (environmental and systemic stimuli) into output signals (cellular behavior aiming at adaptation). According to the second law of thermodynamics, a data-processing apparatus needs to be supplied with energy. For the protein network, this energy is derived from a series of exergonic biochemical reactions such as redox processes, GTP and ATP hydrolysis, protein phosphorylation, and discharge of the membrane potential. This biochemical arsenal is explained in Chapter 2. An essential condition of survival in prokaryotes in an unpredictable environment is the ability to process signals, a function that must have evolved in parallel with metabolism and reproduction at the dawn of life. As discussed in Chapter 3, the biochemistry of cellular signal transduction that was initiated billions of years ago has been conserved during evolution, thus rivaling the conservatism of metabolic and genetic mechanisms. Chapters 4 through 16 then examine variations of these basic themes in terms of signal transduction in eukaryotic cells. The final chapter places cellular signal processing in context with systems biology and describes the formation and function of networks.

One of the hallmarks of the book is its presentation of a unifying concept of signal transduction that traces cell signaling to a few simple biochemical switching reactions that link protein–protein interactions with energy-delivering processes. To this end, the book devotes attention to important aspects that play a major role in these processes and that, in many textbooks, are either treated in a cursory manner or not at all. These topics include prokaryotic signal transduction, lipid signaling, hedgehog and Wnt signaling, Toll-like receptors and innate immunity, small G-proteins, steroid hormones and nuclear receptors, signals controlling mRNA translation, the role of ubiquitylation in signal processing, sensory signaling, ion channels, and neurotransmitter signaling. Presently the genome – considered to be the innate memory store of the "cellular brain," is emerging as a highly complex and dynamic system, which forms a regulatory network of its own. Since the exploration of this matter is just in the very beginning, we confined ourselves to a discussion of those interactions between the genome and the protein network that are sufficiently understood today.

The highly visual illustration program and features such as sidebars, which highlight complementary perspectives to the main discussion, help to strengthen the student's understanding of the essential concepts. The figures in the book are available in both PowerPoint® and JPEG formats on the book's Website, which can be accessed at www.garlandscience.com.

Academic lecturing is a process of giving and receiving. Therefore, we are grateful to our students for their attentive and critical interest and many helpful suggestions. We are particularly indebted to Ingrid Hofmann, Peter Krammer, Frank Rösl (all, German Cancer Research Center Heidelberg), Stefan Rose-John (University of Kiel), and Felix Wieland (University of Heidelberg) for reviewing individual chapters and providing critical comments. We would also like to express our appreciation to the following for their valuable critiques of the manuscript: Joseph Eichberg (University of Houston), Mark A. Lemmon (University of Pennsylvania School of Medicine), Shigeki Miyamoto (University of Wisconsin, Madison), Bradley J. Stith (University of Colorado, Denver), Wei-Jen Tang (University of Chicago), and Alexey Veraksa (University of Massachusetts). Thanks also to our colleague Martin Frank for his support in the preparation of figures showing three-dimensional protein structures. Finally, we would like to thank Heather Whirlow Cammarn, our copyeditor, and the editorial staff at Garland Science, in particular Janet Foltin and Katherine Ghezzi. Their never-ending patience, support, and encouragement were an invaluable experience and indispensable condition for the preparation of this book.

Factual errors are, of course, inexcusable but, unfortunately, not unavoidable. In this respect, we ask our readers for forbearance and welcome suggestions for improvement. We are aware of the fact that, in such a rapidly developing field, ideas, concepts, conclusions and the latest "facts" are highly perishable goods. Many of the questions raised today will certainly be answered in the near future, whereas other problems not yet recognized will become apparent. Nevertheless, we hope that the book serves as an indispensable resource for students and scientists who are exploring and working in one of the most exciting fields in the life sciences.

Friedrich Marks
Ursula Klingmüller
Karin Müller-Decker
Heidelberg

Contents

Contents

List of Headings

The "Brain of the Cell": Data Processing by Protein Networks

The term signal transduction has become established for the molecular mechanisms by which cells process information transmitted by exogenous or endogenous stimuli. The goal of signal transduction is to find the response that optimally safeguards survival. In this chapter we shall address the following questions:

- What is the nature of the cell's information-processing system?

- According to which principles does the system work, and what is it able to do?

- Is the computer a suitable metaphor for cellular signal processing, or would it be described better in neurological terms?

1.1 Metaphors and reductionism: temptations and limitations

Living beings are connected inseparably with their environment. Appropriately, apart from many other qualities, organisms are data-processing or cognitive systems. On the basis of a program of inherited and acquired information, they attribute a distinct "meaning" to sensory impressions and other exogenous stimuli that may be understood as "input signals." This enables them to calculate an "output signal," that is, to adjust their behavior, which includes genetic readout, metabolism, reproduction, shape, and motility, to the given environmental situation. This ability, which is a condition of survival, is an essential characteristic of life, from primeval cells to humans.

The processing of large and complex amounts of data requires a **network of switching devices** that, as a minimum, make Yes–No decisions according to the basic laws of mathematical logic and are able to adapt and learn. In biology, such a network is represented most impressively by the brain with its billions of interacting cells working as interconnected switching elements. However, the reception and interpretation of signals is not restricted to animals possessing a brain but is a general property of organisms, even of the most primitive bacteria. What is the data-processing equipment of a single cell or, so to speak, the "brain of the cell"? It must be a network of interacting molecules, each resembling a molecular "neuron," able to identify the meaning of an input signal and to "calculate" an output signal. This job is done primarily by proteins, interacting with other macromolecules such as nucleic acids of the genomic data bank. It is compelling to assume that as far as data processing is concerned both a network of cells (such as a brain) and a network of proteins (such as a cell) are working according to the same principles. After all, both have the same evolutionary background.

Whether an analogous conclusion can be drawn for the next higher level of complexity, for example, for populations of interacting individuals (Figure 1.1),

Figure 1.1 Data-processing networks at different levels of complexity Technical and biological data-processing (cognitive) networks exist at different levels of complexity. Even a single protein molecule may be understood as a data-processing network with the individual molecular domains that make up its three-dimensional conformation acting as switching elements. Meaningful data processing requires these switches to interact as directed by the laws of mathematical logic.

SOCIETY ⟶ network of individuals

ORGANISM ⟶ network of cells, e.g., neurons

CELL ⟶ network of macromolecules, e.g., proteins

PROTEIN ⟶ network of interacting domains

COMPUTER ⟶ network of microprocessors

remains a matter of worldview (*Weltanschauung*). For some people such a claim would have an unbearable biologistic taste, at least when it is extended to human society, whereas others may remain firm in their conviction of a unifying principle that—emerging in the course of evolution—dominates living nature.

Still another heuristically tempting metaphor is wandering through the popular and even the scientific literature: the data-processing network of microprocessors called a computer. Today it has become trendy to compare cells, brains, or even organisms with computers and, vice versa, to denote electronic networks capable of learning as "neural." However, this concept, although it has the advantage of clarity, rapidly leads into a cul-de-sac. Biological data processing—even if based on the same mathematical principles—is certainly more ingenious and sophisticated than today's computer technology, and it is still far from clear whether it can ever be interpreted and imitated by a technological approach. So one should treat the computer metaphor with caution, being aware of its narrow limits.

May not a similar objection be raised concerning the phrase "brain of the cell?" After all, we are still miles away from understanding the mode of operation of a real brain, which some people believe (perhaps in an exaggerated opinion of themselves) to be the most complex data-processing device in the universe. When compared with the computer metaphor, however, the "brain of the cell" concept has a clear advantage: it emphasizes the strong resemblance of biological data processing at levels of different complexity. As a consequence, the investigation of cellular data processing is expected to inspire brain research and vice versa.

Presently, both molecular brain research and molecular cell biology are still more or less in a phase of hunting and gathering. We are collecting and describing processes, trying to trace them back in a strictly reductionistic manner to elementary chemical and physical events. This approach has been extremely successful and is as indispensable as it is highly satisfying; it is also of tremendous importance for medical applications. Certainly the results obtained thus far provide an estimate of the complexity of biological data processing. However, the expectation that such a method will lead to a deeper understanding of cellular behavior appears to be as illusory as a neurobiologist's anticipation that detailed knowledge of molecular and cellular interactions in the brain would be sufficient to comprehend a systemic property such as consciousness or even mind, or the conviction of 19th century physicists that knowledge of the locations and impulses of all elementary particles would enable a complete prediction of the universe's future.

The reason for this is that complex systems such as living organisms are always much more than the sum of their parts. Life is the result of irreversible historical processes called evolution and development that continuously create new properties, which cannot be predicted just by counting and categorizing molecules. Only quite recently has molecular biology, hitherto the playground of reductionists, become aware of this problem. As a consequence, strong efforts are being made to extend molecular research into the area known as systems biology (see Chapter 17). By a combined approach of bioinformatics, biophysics, bioengineering, biochemistry, and molecular biology (including genome- and proteome-wide screening studies), one is trying to translate the language of molecular structures and interactions into the language of complex systems' behavior and the other way around. While there is no doubt that this methodology will dominate the bioscience field in the 21st century, we have to concede that presently such approaches are still in their infancy.

Thus, lacking a novel intellectual (or even mathematical) concept of the complexity of living systems, we have to be satisfied by reviewing the present results of our hunting and gathering of molecules and mechanisms, an endeavor that is not trivial. In fact, at present the flood of data is rising impressively and alarmingly. However, one should take care not to embrace new illusions; presently the physiological functions of no more than 10–20% of all signal-processing proteins are known. Taking as an example a special switching reaction such as protein phosphorylation, we know that at least one-third, or more than 10,000, of the cellular proteins undergo this chemical modification, but we understand the physiological relevance of less than 10% of the phosphorylations occurring *in vivo*, and this is still in a reductionistic rather than a systemic manner. Thus, our data collection is far from being complete, and many conclusions drawn today may have a rapid use-by date.

Nevertheless, certain basic principles and a couple of biochemical reactions upon which cellular data processing is based have emerged. They clearly show that the signal-transducing network primarily consists of proteins communicating with each other via quite a limited set of noncovalent interactions. These interactions are connected with a dozen or so biochemical reactions delivering the energy required for data processing, thus maintaining signal intensity and giving the flow of data a definite direction. The biochemistry of these reactions is surprisingly simple. In fact, complexity arises from an almost unlimited number of combinations of a few basic elements rather than from a huge number of complicated structures and reactions.

Summary

Data processing is a general property of organisms, being most efficiently developed in neuronal networks such as the brain. A network of proteins interacting with each other and with the genome is understood as the "brain" of a *single* cell. Both neuronal and molecular networks are assumed to work according to the same rules, enabling information processing on the basis of mathematical logic. Presently the investigation of biological data processing follows a strictly reductionistic approach of hunting and gathering, which results in the construction of (highly artificial) linear cascades of signal-transducing biochemical reactions. To arrive at a more in-depth understanding, this approach must be broadened by a comprehensive theory of (biological) complexity.

1.2 Information and signals

The ability to respond appropriately to environmental stimuli, called sensitivity (*Reizbarkeit*) in the older literature, belongs to the basic properties of life according to the maxim "the lower the degree of uncertainty, the higher the chance to survive." Therefore—and to repeat what has been stated above—organisms, from the simplest prokaryotes to humans, are cognitive, or data-processing and learning, systems.

On the molecular level, cognition is based on the unique ability of proteins to process data according to logical principles. Their tremendous functional versatility, their unsurpassed structural flexibility, and their incomparable capability to interact with each other and with other molecules make them efficient switching elements of an information-processing molecular network. Proteins are at the same time enzymes and constructive elements of the cell. In other words, the network of enzyme-catalyzed metabolic reactions, as well as the processes regulating cellular architecture and motility, are inseparably connected with the data-processing network. In a cell, nothing happens that is not under the control of this network. This system represents a state of extreme order that—according to the second law of thermodynamics—can be generated and maintained only by continuous energy consumption and on the basis of detailed information that has to be processed by the cell.

Information is understood as the content of news exchanged between partners—let us call them transmitters and receivers—aimed at reducing uncertainty and disorder. Such an exchange requires information to be represented by symbols such as letters and words, that can be transmitted via any physicochemical medium, for example, sound waves or printed paper (Figure 1.2). Physicochemical entities used for the exchange of information are called **signals**.

Information technology distinguishes between analog and digital signals. Analog signals are omnipresent in music and spoken language. They are characterized by a continuous encoding of information: within fixed limits, the signal may take on any value and may be of any length. In contrast, digital signals are generated by discontinuous encoding of information in the form of short pulses, the intensities of which take on only a few values. The simplest example of a digital signal is binary encoding with only two values. In addition, information can be encoded in the frequency, that is, the interval between the individual signal pulses. A classical example is provided by the Morse alphabet, which also clearly demonstrates the interchangeability of analog and digital signaling systems. Living systems continuously change between the two systems.

In information technology, signals are mostly electromagnetic impulses and waves, or sound waves, whereas inter- and intracellular communication preferentially makes use of molecules and ions (Table 1.1). As a consequence, cells process data primarily by use of biochemical reactions.

Frequently the symbolic systems are changed in the course of signal transduction: for instance, when a sensory impression is transformed into a electrochemical nerve impulse. Such transformations resemble the translation of languages. They require a **code**, that is, an instruction of translation by which one symbolic system is related to another one: for instance, the system of German language to the system of English language, the system of visual impressions to the system of nerve impulses, or the system of nucleic acid structure to the system of protein structure.

Information may be encoded in both the intensities and the temporal sequence of signals, called **amplitude- and frequency-dependent modulation**. As an example, a black-and-white sketch is based mainly on amplitude modulation, while a color picture is based on both amplitude- and frequency-modulated light signals. An analogous difference is that between noise and music. These exam-

Figure 1.2 Communication by signal transmission and signal transduction A transmitter encodes a message into a signal, which is transmitted to a receiver along a physicochemical medium. Having decoded the meaning of the signal, the receiver transmits a response signal provided a semantic agreement exists between both partners; that is, strictly speaking, both master the same language. In biological systems, receiver and transmitter may be individuals, cells, or protein molecules. The decoding of signals by a cell is usually called signal transduction.

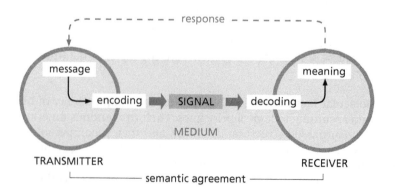

Table 1.1 Biological signals

Environmental stimuli: sound, light, temperature, touch, taste, and smell; molecules, xenobiotics, toxic substances, other stress factors

Between organisms: pheromones, gamones, sound, sight, touch, taste, smell

Between cells: systemic mediators such as hormones; local mediators such as tissue hormones, cytokines, lipid mediators, neurotransmitters, and nitrogen monoxide; cell surface proteins such as cell adhesion molecules

Within cells: second messengers such as cyclic nucleotides, diacylglycerol, inositol phosphates, phosphoinositides, Ca^{2+}; interaction domains of proteins

Within protein molecules: conformational changes

ples illustrate that, as compared with amplitude-dependent modulation, frequency-dependent modulation enables the transmission of much more information, a fact upon which information technology critically depends. Both types of modulation are also found in biological systems. The prototype of a frequency-modulated signal is the nerve impulse. However, many (perhaps most) biochemical reactions are oscillating as well and may generate or be regulated by frequency-modulated signals. Thus, frequency-dependent modulation of endogenous signals is expected to be the rule rather than the exception, and like the brain, a cell may be understood as an ever-oscillating system.

While classical information theory deals primarily with the techniques of encoding and transmission of news, in biology the **meaning** of signals has top priority. Since information can be encoded in any physicochemical medium, the meaning is medium-independent ("the medium is not the message"). Instead, it is read into the signal on the basis of an inherited and acquired pre-information or **program**. By convention, this process of signal decoding by the recipient is called **signal transduction**, although signal processing would be a more appropriate term.

From the principle of medium-independent signaling it follows that signals resemble keywords and are principally ambiguous (Figure 1.3): for different receivers, the same signal may have different meanings. This is an important point, since one can be tempted to seek the meaning of endogenous biological signals in their chemical structures, although it is exclusively assigned by the receiver. As an example, let us take the signaling molecule cyclic adenosine 3′,5′-monophosphate (cAMP). It is interpreted by many bacteria as a request to activate certain genes (details in Section 3.2.2), whereas the slime mold *Dictyostelium discoideum* responds to the same signal by aggregation into multicellular structures and development into a spore capsule. Numerous and completely different effects of cAMP occur in vertebrate cells. Another instructive example is provided by the vertebrate hormone estradiol that, depending on species and tissue, exhibits quite different effects (Figure 1.3). Of course, the reverse holds true as well: different signals may evoke the same response. For example, in the liver, the degradation of glycogen is induced by at least seven different hormones.

Although based mainly on biochemical interactions, inter- and intracellular signal transduction follows the same principles as language, where signals (words and sentences) contain conceptual information only for those who are proficient with the language or who possess the program of decoding. A comparison between language and chemical signal transduction immediately clarifies the analogy between the systems (Figure 1.4). Moreover, it tells us that there is a direct causal relationship between the cellular and the social levels of complexity: On one hand language is the result of elementary molecular events occurring in neurons (bottom-up effect), on the other hand emotionally charged sentences

Figure 1.3 Signals are ambiguous (the medium is not the message) As the scent of a medium-sized animal has a deterrent effect on small animals but attracts a larger animal, biological signals have completely different meanings depending on the species, the tissue type, and the physiological context, that is, on the program of signal decoding.

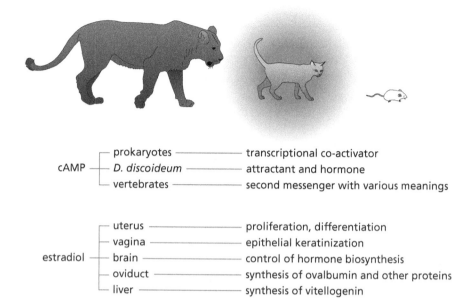

inevitably have an impact on the biochemistry of cells, including the activity of genes (top-down effect). Such a reciprocal resemblance is a characteristic feature of living systems. It will be discussed in more detail in Chapter 17.

Summary

Organisms are cognitive or information-processing systems. On the basis of inherited and acquired knowledge (the program), they relate a certain meaning to an environmental stimulus that is transmitted as a signal, or a distinct modulation of a physicochemical medium. Signals per se are meaningless. Their interpretation depends solely on the receiver's program and the physiological context. Therefore, a signal is ambiguous: it may have different meanings depending on the recipients and the given situation.

1.3 Proteins are binary switches

Information-processing systems are made up of interconnected switching elements or elementary calculators that transform an input signal into an output signal according to the laws of mathematical logic. In the simplest form, a switch represents the binary values 0 (OFF) and 1 (ON) (Figure 1.5). It is a special property of proteins to exist *at least* in two different states: an enzyme, for example, is

Figure 1.4 Resemblance between acoustic and chemical signal transmission In both cases the speaker's data bank (memory) constructs an image of its environment that is equipped with a meaning and transmitted as a signal (molecule or word) to a listener, who decodes the signal, interpreting its meaning by means of his signal-processing apparatus, that is, asking his memory store.

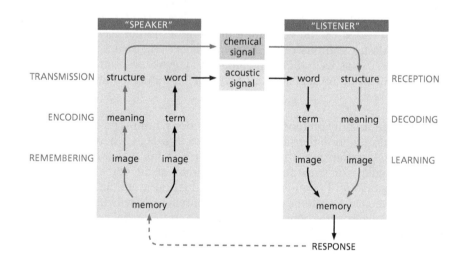

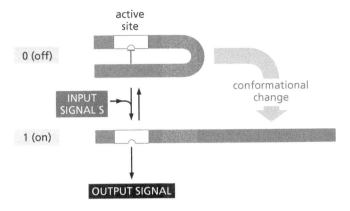

Figure 1.5 Proteins as binary switches Most proteins exist in at least two different functional states (here called states 0 and 1) that are represented by different conformations. In state 0 the protein is assumed to be inactive (in the closed or off position) because in this conformation the active site is blocked by an intramolecular interaction. An "input signal" is any influence that transforms or switches one state into the other. The conformational change thus induced releases the protein from autoinhibition, resulting in an "output signal" generated by the active site (another input signal may trigger the reverse 1→0 switch). This elementary data-processing unit is an oversimplification because in the cell most proteins exist in more than two different states or conformations.

either inactive (state 0) or active (state 1). The states differ in their three-dimensional structure or conformation. Any protein molecule may be considered, therefore, to be an elementary signal-processing unit. Input signals are influences that change the equilibrium between 0 and 1 and thus the functional state. As a rule such changes are caused by chemical ligands, but depending on the type of protein, changes may also be caused by other stimuli such as pressure, light, and temperature. The output signal is the effect resulting from the altered function of the protein, which in turn is a result of an altered conformation. For instance, an enzyme transforms the input signal "substrate concentration" into the output signal "product concentration."

For most enzymes, substrate concentration is not the only input signal but their activities are controlled and fine-tuned by numerous activators and inhibitors that act as additional input signals. This regulatory principle holds true also for other proteins. How such additional input signals manipulate the protein switch is explained by the theory of **allostery**. It is based on the concept of regulatory protein domains, which are not identical with the active center but, nevertheless, influence protein function via long-range effects on the protein's three-dimensional structure, the conformation. In its most concise form (as originally formulated by Monod, Wyman, and Changeux in the early 1960s), the theory assumes a protein to exist in two different conformations, being in equilibrium and differing in their activities as well as their affinities for regulatory factors, which we may call input signals S (Figure 1.6). To make clear that the allosteric effect is a binary switching event, we shall replace the historical terms T (tense) and R (relaxed) by 0 (=T) and 1 (=R).

Let us assume S to be a stimulatory signal or, in pharmacological terms, an **agonist**. In the absence of S, the protein may be assumed to be mostly in conformation 0: the switch would be in the OFF position. In the terminology of chemical thermodynamics, conformation 0 is said to have a low standard free energy. This means that the transition from 0 to 1 is an endergonic reaction that does not

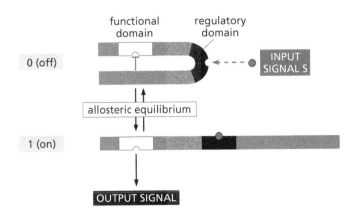

Figure 1.6 Allosteric switching of protein function A protein with spatially separated regulatory and functional (catalytic) domains is thought to exist in at least two conformations, 0 and 1, that are in an allosteric equilibrium. An agonistic input signal stabilizes conformation 1, thus shifting the equilibrium downward. In contrast, an antagonistic input signal would stabilize conformation 0, shifting the equilibrium upward (not shown).

occur spontaneously but requires a supply of energy. When S interacts with the regulatory domain of the protein, the equilibrium is shifted to conformation 1: the switch is turned into the ON position, provided the signal molecule S has a higher affinity for 1 than for 0 and the energy released upon binding of S is sufficient to push the conformational change.

The theory also explains the effect of an inhibitory signal or **antagonist**. It has a higher affinity for 0 than for 1 and arrests the switch in the OFF position. Frequently one is dealing with mixed agonists–antagonists that, depending on the concentration, act either as activators or as inhibitors.

Agonistic and antagonistic signaling factors are either small molecules and environmental stimuli or proteins that interact with their target protein via specific domains. These domains are amino acid sequences that are able to recognize and to bind complementary sequences of their own (homotypically) or of a different structure (heterotypically). They will be discussed in more detail in Section 1.8. Such protein–protein interactions occur according to the principles of allosteric regulation.

The interaction of a signaling molecule with a protein may be either covalent or noncovalent. Protein phosphorylation and other post-translational modifications represent covalent binding, while noncovalent binding is typical for the interaction of a hormone with its receptor or for most protein–protein interactions,

To return the switch back to 0, the agonistic signal S has to be removed from the binding equilibrium, a process known as signal extinction. A dilution or chemical inactivation of S is sufficient to reverse a noncovalent interaction, whereas a covalent modification requires a chemical cleavage, which as a rule has to be catalyzed by a specific enzyme (Figure 1.7).

When a protein consists of several subunits, the theory of allostery predicts **cooperativity**. Accordingly, the interaction of one subunit with the signaling molecule either facilitates (positive cooperativity) or hinders (negative cooperativity) the binding of additional signaling molecules, resulting in conformational changes of the other subunits. A cooperative effect is indicated by a sigmoidal curve when the activity of the protein, or the output signal (for instance, meas-

Figure 1.7 Interactions of signaling molecules with proteins (A) Binding of cyclic AMP (cAMP) as an example of a noncovalent interaction. The cAMP signal is extinguished by dilution and by enzymatic degradation, driving the reaction in the reverse direction. In an analogous manner, proteins interact with each other. (B) Phosphorylation as an example of a covalent interaction or post-translational modification. Both the forward and reverse reactions require enzymatic catalysis.

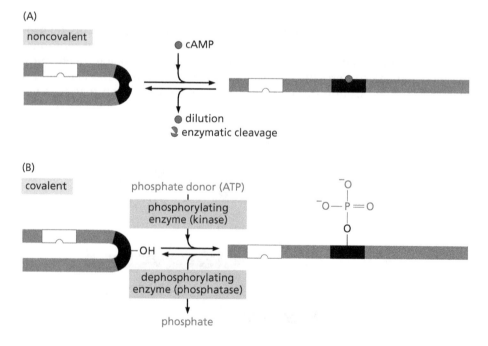

8

ured as enzymatic reaction rate), is drawn in relation to the ligand concentration, or the input signal S. Signal-transducing systems with a sigmoidal characteristic are resistant to low irregular signal intensities called **noise**, which is feared as a disruptive factor in information technology. In other words, cooperative proteins possess a built-in noise filter (Figure 1.8; for other molecular noise filters, see Figure 1.18). In contrast, proteins without quaternary structure mostly exhibit linear or hyperbolic kinetics, and thus they lack a noise filter. This may be one major reason that signal-transducing proteins mostly consist of several subunits forming oligomeric structures. Another reason is that a system with a sigmoidal characteristic operates like an on-off switch, whereas systems with linear or hyperbolic kinetics rather work as accelerators or brakes (Section 1.9).

Most signaling proteins form not only oligomers but, moreover, are expressed in several subtypes or **isoforms** that result from individual genes, from genes with alternative promoters, and from post-transcriptional modifications such as alternative splicing of pre-mRNA. Such isoforms may be restricted to individual tissues and cell types and may exhibit different functions. In addition, they provide the signal-processing apparatus with a high degree of **redundancy**, which in technology is considered to be a feature of superior design since it considerably diminishes the susceptibility to faults (see also Chapter 17).

The theory of allostery was originally developed to explain the cooperative behavior of proteins with quaternary structures. However, it is not restricted to such proteins (as is frequently stated in textbooks) but applies to any protein with separated regulatory and catalytic domains, though in the absence of a quaternary structure the allosteric interaction is not easily recognized from the kinetics but has to be proven by costly structural analyses.

Summary

Proteins may be looked upon as switching elements of a data-processing molecular network. Switching is due to a conformational change induced by an input signal. As a result, the protein function representing the output signal is modulated. The conformational change results from an intra- or intermolecular allosteric interaction between a regulatory domain (receiving the input signal) and a functional domain (transmitting the output signal). In proteins with a quaternary structure, cooperative allosteric interactions provide a noise filter.

1.4 Signal-transducing proteins are "nanoneurons"

The cells of the brain are specialized for information processing. The same holds true for the proteins of cellular signal transduction. They differ from metabolic enzymes or architectural proteins as neurons differ from liver or skin cells. For such signal-transducing proteins, the concept of working as simple binary switches describes a borderline case, since most of them are controlled by much more than one input signal.

In fact, in their capability to convert numerous input signals (or allosteric effectors) into an output signal or even digital into analog signals, such proteins—albeit on a lower level of complexity—resemble neurons, where a large number of dendritic synapses serve as signaling input terminals and the axonal synapse acts as an output terminal. In signal-transducing proteins, such synaptic contacts are replaced by structural domains. As a rule, the input terminals are binding sites or interaction domains for allosteric regulators, whereas the functional domain (for instance, the catalytic center of an enzyme) serves as the signaling output terminal (Figure 1.9).

Since function strictly depends on the three-dimensional structure of a protein, the conversion of input signals into the output signal occurs through (mostly allosteric) conformational changes acting in either an additive, synergistic (more

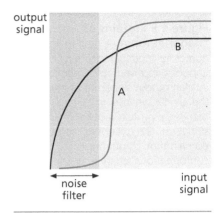

Figure 1.8 Noise suppression by a sigmoidal characteristic of signal transmission With sigmoidal kinetics (curve A), a significant output signal is produced only when the input signal surpasses a certain value. Such a system has a built-in noise filter and resembles a switch with first trigger pressure. In contrast, a system with hyperbolic characteristics (curve B) responds to very weak input signals including accidental noise. It resembles a rotary switch or an accelerator pedal.

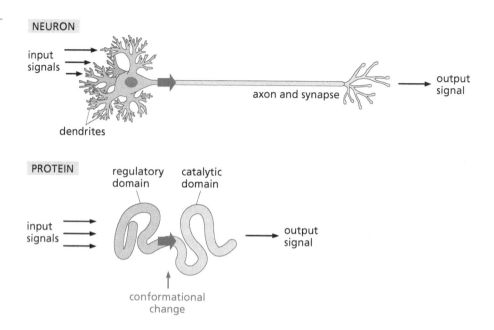

Figure 1.9 Proteins as "nanoneurons" A neuron converts numerous input signals (nerve impulses) received at the dendrites into an output signal transmitted at the axonal synapse. In an analogous manner, a protein receives input signals at its regulatory or interaction domain that, through a conformational change, become converted into an output signal released by a catalytic or functional domain.

than additive), or antagonistic manner. As will be explained below (Section 1.8) regulatory protein domains react not only with input signals from the outside but also with each other. These intramolecular effects are an important condition of signal conversion because they determine the algorithm of the protein calculator. Their elucidation ranks among the most important and most difficult tasks of current biochemistry and structural biology.

Summary

In general, signal-transducing proteins have more than one receiver site for regulatory input signals. In their ability to integrate various input signals and, to calculate an output signal, they resemble neurons, albeit at a lower scale of complexity.

1.5 Signal-transducing proteins are components of logical gates and neural networks

Switches—even simple binary ones—may be used to carry out fundamental logical operations of the type NOT, AND, OR, and NOR according to the rules of **Boolean algebra**. These operations are sufficient to process *any* kind of logical information. The corresponding circuits, called logical gates, are the elementary units of technical data-processing networks that, as already mentioned, resemble the nervous system to a certain extent. Because these technical devices are able to adapt and to learn, they are called neural networks, with the idea that some day they may imitate living systems, including intelligent behavior (artificial intelligence). Such a concept was formulated initially in the 1940s, when it was shown theoretically that a switching device of a few simplified and strongly idealized neurons, called "McCulloch–Pitts neurons," could carry out logical operations. When they are understood as cellular "nanoneurons", proteins are also expected to operate as logical gates and, like molecular McCulloch–Pitts neurons, assemble to form a data-processing network, or the "brain of the cell" (Figure 1.10).

It should be noted that like the concept of binary protein switches the rigid Yes–No decisions of Boolean logic represent a borderline case that cannot fully describe biological data processing. Instead, so-called **fuzzy logic**, based on the mathematical theory of fuzzy sets, is more suited to explain decision and regulatory

Figure 1.10 Proteins as logical gates The active protein (producing an output signal) is depicted in red (state 1), and the inactive protein is shown in gray (state 0). A protein working as a NOT gate does not transmit an output when it receives an input signal S. An AND gate produces an output signal only when it receives two input signals S1 and S2 simultaneously (so-called coincidence detector). An OR gate responds either to S1 or to S2. To produce an output signal, a NOR gate must receive only signal S1 but not signal S2.

processes occurring in response to the incomplete, imprecise, and even contradictory information an organism typically receives from its environment.

For the brain, the actual state of function of its neuronal network resembles what is called random access memory (RAM) in computer technology. The corresponding memory store of a cell is the functional state of its protein network. Like a brain, it uses its genetic and acquired program to construct an image of the surroundings (for a unicellular organism or for a sensory cell, it is the environment, and for a tissue cell, it would be the internal milieu of the body).This enables the cell to calculate a response necessary for survival. Therefore, the idea of a protein network representing something like the "brain of the cell" appears to be not too far-fetched, though one should refrain as much as possible from anthropomorphic metaphors, keeping in mind that a single cell does not "think." Moreover, the images constructed by the protein network are necessarily incomplete, because only those signals for which the network is phylogenetically and ontogenetically programmed are recognized and processed (though this holds true also for our brain!).

A more questionable or even dangerous metaphor is that of the "computer of the cell." As mentioned above, it certainly has the advantage of clarity and for this reason alone it is used at times in this book: for instance, when referring to switching elements or random access memories. Although information technology tries very hard to imitate biological data processing by neural networks, the results are still rather meager since certain principles of biological systems either are not fully understood or are realized only with difficulty. Thus the hardware of protein networks, in contrast to customary electronic networks, is not irreversibly wired but self-organizes continuously into ever-changing patterns depending on the task and the physiological context. This self-patterning is a direct consequence of a characteristic property of proteins, which is the ability to interact with each other and with other signaling molecules at any place in the cell. Thus, data-processing protein networks are continuously reorganized both spatially and temporally.

Summary

Interacting proteins form logical NOT, AND, OR, and NOR gates, enabling the cell to process any kind of information. When the extreme flexibility and complexity of cellular data processing is considered, the computer metaphor appears to be questionable. It would be more appropriate to speak of the "brain of the cell."

1.6 Privacy versus publicity

Signals that transmit news may have a private character, where they address a discrete partner, or they may be directed at the public (Figure 1.11). This principle

Figure 1.11 Private versus public signaling Adapted from J.J. Sempé, Volltreffer. München: Deutscher Taschenbuch Verlag, 1968.

also holds true for inter- and intracellular signal transduction. As shown in Figure 1.12, signaling between cells may range from highly public to highly private. Thus, the endocrine system works via public long-range effects, while the nervous system operates mainly via private short-range (para- and autocrine) effects. The highest level of privacy is reached when cells communicate via membrane-bound signaling molecules (juxtacrine signaling).

In an analogous way, proteins may interact with each other in a widespread or in a strictly targeted manner. Targeted short-range effects that maintain privacy are due to interaction domains, working like electronic contacts (see below), while for widespread long-range effects, an enzymatically active protein produces a large amount of diffusible messenger molecules that interact, either covalently or noncovalently, with a high number of remote partner proteins (publicity). A characteristic example of long-range effects is provided by second messengers such as cAMP or calcium ions, which are released in large amounts and may diffuse throughout the cytoplasm upon stimulation of a few transmembrane receptors (see Chapter 3).

Depending on the context, signaling through second messengers may have also a more private character. In fact, short-range interactions in subcellular networks (spatial privacy) are facilitated by potent mechanisms of signal extinction that restrict a signal to a very limited area of the cell. Moreover, certain proteins that are specialized to act as adaptors or scaffolds may bring interacting proteins into contact or dock them only at specific cellular sites (see Section 1.8; for some illustrative examples see Sections 4.6.1 and 11.5).

Another hallmark of biological signaling is temporal **transitoriness.** A neuronal signal, for instance, is short-lived in order to avoid overstraining the system. Long-lasting effects require, therefore, an oscillating signal that, at the same time, enables frequency-dependent modulation. This is a typical feature of signaling in the nervous system. Intracellular signal processing by protein networks is dominated by the same rule. Here a wide variety of inactivating mechanisms guarantees transitoriness. Moreover, signal-generating and signal-extinguishing proteins frequently are coupled to modular complexes that fulfill the conditions

Figure 1.12 Different pathways of signal transmission between cells and molecules Long-range signaling between cells is called endocrine when the signal molecule is transported by the bloodstream (typical for hormones), paracrine when the signal diffuses between neighboring cells across the extracellular matrix (typical for neurotransmitters and many so-called tissue hormones or local mediators), and autocrine when the signal re-acts on the transmitter cell (typical for neurotransmission and developmental processes). Short-range signaling through a direct contact between cells via membrane-bound signaling proteins is called juxtacrine (typical for developmental processes). While endocrine signaling has a public character, para-, auto-, and juxtacrine signals are more private. Endocrine signals find their intracellular counterparts in signal molecules (such as second messengers), which are produced by an enzymatically active protein providing intermolecular long-range signaling by diffusing through the cytoplasm. The more private para- and autocrine signaling resembles interactions that depend on a close encounter between the proteins. Protein phosphorylation provides an illustrative example. A direct signaling contact between proteins resembles juxtacrine signaling. The signals are symbolized by red triangles.

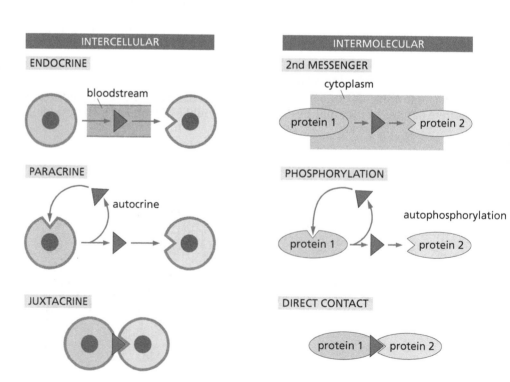

of molecular oscillators (see below). Therefore, a wavelike pattern and a frequency-dependent modulation of intracellular signals may be assumed to be the rule rather than the exception (for details see Sections 4.6.1, 11.7, and 14.5.4).

Summary

By interacting with each other and with other cellular constituents, proteins form a neural network of logical gates. Network formation is based on the property of proteins to communicate with each other by both short- and long-range interactions (thus resembling organisms). Short-range effects are due to direct contacts (using specific contact or binding domains), while long-range interactions are transmitted by signaling molecules. Thus, subcellular signal processing resembles both widespread hormonal signaling and targeted neuronal signaling. Both neuronal and subcellular signal processing are dominated by the same principles: spatial privacy, temporal transitoriness, redundancy of the components, and frequency-dependent modulation of the signals.

1.7 Cross talking and network formation

The nervous system is an extremely complex network of cells communicating via a large variety of para-, auto-, and juxtacrine interactions. This organization, being mandatory for a data-processing system, finds its counterpart at the subcellular level in the network of signal-transducing proteins, albeit with a lower degree of complexity.

Previously, only linear pathways of signal transduction connecting the cellular periphery with the metabolic and genetic apparatus could be analyzed. It became clear, however, that the picture thus obtained resembled an extreme oversimplification because in reality such pathways were found to be interconnected by a phenomenon called "signaling cross talk." In fact, this term is nothing but a synonym for signal processing by a protein network. It is now generally accepted that any signal received by a cell activates a substantial part of the data-processing protein network rather than stimulating just a simple sequence of a few biochemical reactions, a concept that, for an in-depth investigation, imperatively requires novel approaches such as that of systems biology (see Chapter 17). The situation closely resembles data processing in the brain, where a single sensory input induces an extended (diffuse) activation pattern rather than stimulating just a small group of specialized neurons (Figure 1.13).

Figure 1.13 Diffuse signal processing by the human brain
In the brain, the processing of even rather simple sensory inputs results in extended and diffuse excitation patterns involving a large number of neurons and developing on the basis of apparently chaotic background activity. On the molecular level, an analogous situation has to be assumed for cellular signal processing. The excitation patterns were visualized by a method measuring the blood circulation. (Adapted with modifications from G.D. Fischbach, *Sci. Am.* 267, 48–57, September 1992.)

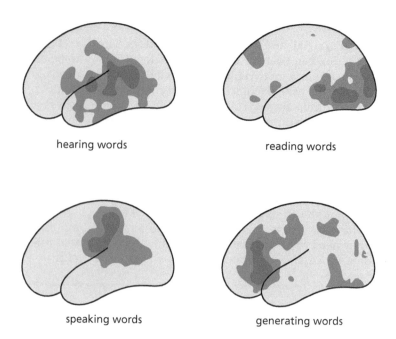

hearing words

reading words

speaking words

generating words

Impressive microscopic photographs and three-dimensional models showing the beauty and complexity of the neuronal network in the brain have given a vivid impression of the cross talk between nerve cells. For molecules, such a visualization is not yet possible. Instead we have to fall back on more indirect evidence. We know, for instance, that cytoskeletal and membrane structures play a key role in support of the signal-processing protein network. Thus, the complexity of such hardware, or scaffolding, may reflect the complexity of the "cellular brain" to a certain extent (Figure 1.14).

As far as teaching is concerned, signaling cross talk creates a problem because for the sake of clarity it is impossible to depict and describe all the interactions that occur when a cell receives a signal. So in textbooks—including this one—still more or less linear, artificial signaling cascades are presented. However, the reader should always keep in mind that this is an oversimplification, illustrating mechanistic principles rather than the complex reality.

1.7.1 Network biology

A new discipline of theoretical biology known as network biology is emerging that deals with the rules and laws dominating the formation and function of

Figure 1.14 The cytoskeleton: hardware support of the data-processing protein network The photographs show the extremely complex structure of cytoskeletal fibers in the neighborhood of a mitochondrion (A) and in the border zone between cytoplasm and nucleus (B). (A, adapted from S. Penman, *Proc. Natl Acad. Sci. U.S.A.* 92, 5251–5253, 1995. B, adapted from J.E. Darnell, H.F. Lodish, and D. Baltimore, Molecular Cell Biology, 2nd ed. New York: Scientific American Books, 1990.)

(A)

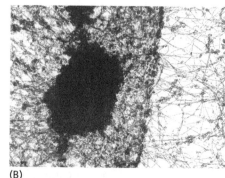

(B)

molecular networks. Combining molecular biology, biophysics, and mathematics, this research has focused on five types of networks: metabolic, genetic, transcription factor, protein interaction, and protein phosphorylation. These approaches are strongly facilitated by novel high-throughput techniques that use cDNA microarrays, protein chips, and other automated methods to allow genome- and proteome-wide screening. Moreover, a conceptual framework has been developed to characterize different types of networks (see Chapter 17).

Using simple organisms such as yeast, one can generate computerized graphs that show an illustrative, albeit rough, survey of intermolecular cross talk and allow the discovery of hitherto-unknown interactions, at least on paper. There is no doubt that a more systematic understanding of these networks is urgently needed. However, network biology is still in its infancy, particularly because of the difficulty in uncovering the physiological role of the interactions described. Therefore, one must wait and see whether this new discipline will significantly broaden our understanding of cellular phenomena, especially of signal processing, or—remaining on a descriptive level—if it will only provide a new mathematical formalism. In the following section, as well as in Chapter 17, the reader will be able to glimpse the problems network biology is facing.

1.7.2 Some remarks on the complexity of the "cellular brain"

When compared with the size of a genome, the number of genes encoding signaling proteins is rather modest. Thus, analysis of the human genome has shown that, of the approximately 30,000 genes, about 1500 encode signal receptors, 500 encode protein kinases, 150 encode protein phosphatases, and 1500 encode transcription factors, just to mention some of the most common switching elements of the data-processing protein network. Most of these genes contain several (up to 200) exons, which means that transcribed pre-mRNA probably undergoes alternative splicing. As a rule, the proteins encoded by such splice variants exhibit different properties such as ligand affinities, intracellular distributions, enzymatic activities, and stabilities (see Section 8.9). Arbitrarily assuming at least three different splice variants are produced per gene, one arrives at 4500 different signal receptors, 1500 protein kinases, 450 phosphatases, and at least 4500 transcription factors.

Each of these proteins is expected to undergo covalent post-translational modifications and so its properties are changed further. Let us again assume that three different modifications per protein occur on average (which certainly is an understatement). These modifications produce $2^3 = 8$ different states, since each modification exists in two states, either present or absent. This calculation increases the number of receptor variants to 36,000, kinase variants to 12,000, phosphatase variants to 3600, and transcription factor variants to 36,000, thus greatly outnumbering the total number of genes. The degree of complexity is furthermore enlarged dramatically by the understanding that most signal-processing proteins exist as homo- and hetero-oligomers rather than as monomers and that the number of functional states of a network increases exponentially with the number of the components. In fact, in a single cell the number of dynamic states generated by the interactions between signal-processing proteins must be astronomically high.

It must be emphasized, however, that in a given tissue cell only a fraction of these variants becomes realized. The pattern of active signaling proteins thus generated changes continuously, depending on the state of differentiation and the actual physiological context. This is what we have called the rearrangement of the hardware components and actual state of the random access memory. The overwhelming complexity of the "cellular brain" indicates that our efforts to understand life are still in their infancy.

Summary

The data-processing network of the cell is based on chemical interactions or cross talking between the components. Therefore, a single signal received by a cell evokes a diffuse excitation pattern rather than activating one or a few signaling pathways. As yet, the extreme complexity and flexibility of such networks prevents a more in-depth analysis of their physiological functions.

1.8 Interaction domains: how the network is plugged together

Proteins are assembled into highly organized signal-transducing complexes and data-processing networks. This occurs on demand in a fully reversible manner. How is this achieved? The original concept of proteins as freely diffusible molecules encountering each other more or less by chance is considered to be outdated today. Due to the extremely high concentration of macromolecules (known as macromolecular crowding), the interior of a cell needs to be viewed as a semi-solid medium in which free diffusion would be far too slow for life processes. In addition, neither a precise arrangement nor a controlled interaction of the components would be possible. However, both are mandatory for cellular data processing. Thus, protein interactions must be brought about by other more specific and more targeted processes.

1.8.1 Protein modules

Complex formation is a direct consequence of a characteristic structural feature known as the **modularity of proteins**. Proteins are composed of defined amino acid sequences or modules according to a unit construction system. Generally one may distinguish between functional and interaction modules. A functional module is, for example, the catalytic domain of an enzyme, whereas interaction modules serve as contact sites for all kinds of allosteric regulators—including proteins with complementary contact modules, diffusible signaling molecules such as second messengers, and systemic and environmental factors—as well as for cellular structures. Interaction modules guarantee relatively stable (though mostly noncovalent) binding between the partner molecules. In signal-transducing proteins, the interaction modules are identical with the regulatory domains that serve as terminals of input signals, while functional modules are the catalytic domains that generate output signals (see Figures 1.5, 1.6, and 1.8). As a rule, signal-transducing proteins have much more interaction modules than other proteins such as metabolic enzymes. In the course of evolution, the protein modules were recombined into new patterns by genetic rearrangements. Accordingly, innovations in signal processing (resulting in novel properties of the organism) are brought about by unique combinations of existing building blocks rather than by invention of entirely new protein structures. This mechanism of evolvability is a major driving force of complexity. By means of such a combinatorial strategy, cells establish the astronomically large number of protein contacts by using relatively few interaction domains.

The pattern of protein domains represents something like a signaling syntax, and the highly organized multimodular structures and multiprotein complexes thus formed resemble microprocessors to a certain extent. Nevertheless, and in contrast to a computer, proteins are not firmly interconnected but are assembled and disassembled upon demand by energy-consuming reactions. This reversible wiring is promoted, in particular, by covalent post-translational modifications of protein modules, generating ad hoc new sites for inter- and intramolecular contacts that immediately become untied when they are no longer needed. By this means, new interaction patterns are continuously formed between proteins, just as one might expect for an adaptable and educable neural network. The cooperation of the individual domains in such patterns lends protein interactions the necessary precision.

The principles of domain interactions are depicted in Figure 1.15, while in Table 1.2, selected interaction domains are summarized and arranged according to the recognition sites of their interaction partners. The first group includes domains that recognize short peptide sequences. The domains of the second group interact with peptide sequences containing covalent post-translational modifications due for instance to phosphorylation, acetylation, methylation, and ubiquitylation. Interaction domains not listed in the table include binding sites for extra- and intracellular signal molecules, found in receptors and in many other signal-transducing proteins, as well as for nucleotide sequences, found in transcription factors and other DNA- or RNA-binding proteins.

Domains interacting with post-translational modifications (Figure 1.15) recognize the modified amino acid together with a relatively short amino acid sequence in the immediate neighborhood. These flanking areas may differ from protein to protein, lending the interactions additional specificity, especially since the cognate interaction domains are also variable. Thus, a protein with a distinct recognition site does not simply interact with any other protein possessing the corresponding interaction domain but strictly selects its partner molecule.

This selectivity of interactions is impressively demonstrated by receptor tyrosine kinases. Upon ligand binding, these dimeric transmembrane proteins undergo trans-autophosphorylation of several Tyr residues in their cytoplasmic domains, thus generating contact sites for proteins with SH2 and PTB domains (see Table 1.2). Each of those proteins has a special SH2 or PTB variant that recognizes only one of the various phospho-Tyr residues on the receptor. In other words, each interacting protein finds its correct place on the receptor protein (a detailed discussion of this subject is found in Section 7.1.2). Conversely, the same recognition site may react with several *different* interaction domains. For instance, SH2 domains can interact with phospho-Tyr residues and sometimes also with SH3 domains, and PDZ domains may interact with each other (homotypically) as well as with carboxy-terminal amino acid sequences (see Table 1.2 and Section 7.1.4).

Figure 1.15 distinguishes between domains that interact with nonrelated domains (heterotypically) and those that interact with their own type (homotypically). Examples of domains interacting exclusively in a homotypical manner are the death domains DD and DED. They play a key role in apoptotic signaling (Section 13.2.3). Another group of interaction domains recognizes membrane components such as phospholipids. Such interactions enable reversible binding

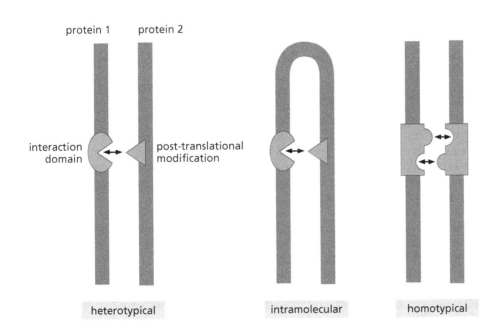

protein 1   protein 2

interaction domain   post-translational modification

heterotypical   intramolecular   homotypical

Figure 1.15 Interaction domains
The figure shows interaction domains of proteins that are binding sites for other protein molecules carrying a specific recognition signal. To the left, an interaction domain is depicted that recognizes a heterotypical signal, for instance, a post-translational protein modification. In the middle, a heterotypical interaction is shown occurring in the same protein molecule. The scheme to the right shows a homotypical interaction between two identical interaction domains. Other interaction domains recognize low molecular weight messenger molecules, specific nucleic acid sequences, and membrane lipids. Receptors exhibit an enormous variety of interaction domains for intercellular signals and environmental stimuli.

Table 1.2 A selection of interaction domains

Interaction partner	Interaction domain
Group 1: the recognition site is a short amino acid sequence	
Pro-x-x-Pro	SH3
Arg-x-x-Lys	SH3
Pro-Pro-x-Tyr	WW
x-y-z-Val-COOH or -Leu-COOH	PDZ
Group 2: the recognition site is a short amino acid sequence containing a covalent modification	
phospho-Tyr	SH2, PTB
phospho-Ser	14-3-3, WW, MH2, etc.
phospho-Thr	FHA, WD 40
acetyl-Lys	bromo
methyl-Lys	chromo
ubiquityl-Lys, polyubiquityl-Lys	UIM, UBA, CUE, UEV, PAZ, NZF, etc.
Group 3: the recognition site is a membrane component	
diacylglycerol	C1
phosphatidylserine	C2
phosphatidic acid	C2
phosphoinositides, Gβγ subunits	PH
3-phosphorylated phosphoinositides	FYVE
3-phosphorylated phosphoinositides	PX
Group 4: the interaction domain is identical with the recognition domain (homotypical interaction)	
death domain (DD)	DD
death effector domain (DED)	DED
CARD	CARD
PDZ	PDZ
SAM	SAM

of proteins to membranes and generation of signal-processing protein complexes in the neighborhood of membrane receptors. Many membrane lipids are, in addition, signaling molecules or second messengers that control the function of the interacting protein.

1.8.2 Intramolecular interactions

Interactions between regulatory domains may also occur within the same protein molecule, with various consequences for protein function. Thus, one interaction domain may either suppress or promote the function of another, like an antagonist or an agonist, or two different interaction domains may become occupied sequentially to induce a specific function. It is easily conceivable that such interactions provide logical gates.

Frequently, intramolecular interactions stabilize special protein conformations. This provides a general mechanism to keep a signaling protein in an inactive conformation (or vice versa in an active one). Such an intrasteric blockade is broken up by an input signal that either competes with one of the domains or removes the interacting residue (if it is a covalent modification such as a phosphate group) or induces a conformational change by binding to a third domain (Figure 1.16). One immediately recognizes that the lifting of an intramolecular

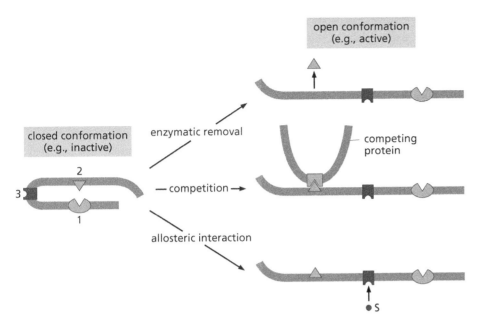

open conformation
(e.g., active)

enzymatic removal

closed conformation
(e.g., inactive)

competing
protein

— competition →

allosteric interaction

● S

Figure 1.16 Conformational change by lifting an intramolecular interaction Due to an interaction between domains 1 and 2, a protein is kept in a closed conformation, which in this example is assumed to be equivalent to an inactive state (domain 1 represents an active center). There are at least three ways (representing input signals) in which the closed conformation may become unfolded so that the protein becomes activated: enzymatic removal of a critical residue, generally a covalent amino acid modification, in domain 2; interaction with a competing protein exhibiting a binding domain for domain 2; or covalent or noncovalent allosteric interaction of a signaling molecule S with a regulatory domain 3. S may be a low molecular weight factor such as a second messenger, a membrane lipid, a protein, a nucleic acid, or a covalent modification such as phosphorylation.

blockade resembles a reversible switching process such as those depicted in Figures 1.5 and 1.6. Examples of this central theme of cellular signal processing are found throughout this book (see, for instance, Sections 2.6.1 and 7.1.2).

1.8.3 Docking domains

A dogma of biochemistry states that to become recognized by an enzyme, a substrate has to bind stereospecifically to the catalytic center. However, in contrast to metabolic enzymes, most signal-transducing proteins are able to interact with a wide variety of substrates. This is typical, for example, for the majority of protein kinases. For such proteins the conventional mechanism of substrate recognition would create a problem of specificity. Therefore, they frequently exhibit additional substrate binding sites that with a high degree of selectivity recognize a docking motif of the substrate protein that is distinct from the site interacting with the catalytic center. Such docking domains not only confer specificity but also help the protein to find its substrate. In other words, they may have a stimulatory effect on the signaling event. An example is discussed in Section 11.2.2.

Like other interaction motifs, certain docking domains may be generated as needed by covalent post-translational modifications of amino acid residues. Such a mechanism renders the process of enzyme–substrate encounter as an adjustable target of input signals and logical AND gates. Examples are provided by the so-called priming of protein kinases and substrates of ubiquitin ligases by phosphorylation (see Sections 4.6.4, 9.4.4, and 12.16).

1.8.4 Adaptor and scaffold proteins

Since the assembly of multiprotein complexes at specific cellular sites would require proteins with many different docking domains, transferring the job of targeting and assembling to a third party would dramatically increase the system's degree of freedom. This might have been the evolutionary pressure for the development of adaptor and scaffold proteins, which dominate signal processing, particularly in eukaryotes (Figure 1.17). Adaptor proteins are composed mainly of interaction domains and frequently lack functional domains. As "molecular plugs," they fit together other proteins to form functional units. By use of a single adaptor, a cell may interconnect several receptor types with the same intracellular effector mechanism (see, for example, Sections 3.3.4 and 6.3).

From adaptor proteins it is only a short step to the scaffold(ing) proteins that fulfill the function of a platform in that they arrange other proteins into specific

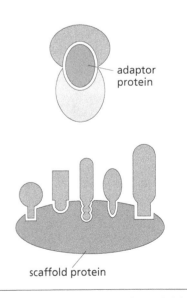

adaptor
protein

scaffold protein

Figure 1.17 Adaptor and scaffold proteins By means of interaction domains, adaptor and scaffold proteins plug together other proteins to form functional units and multiprotein complexes.

19

switching patterns, thus enabling highly selective interactions. These multiadaptors can be compared most closely with the sockets of electronic devices. They are extremely important for signal processing since they establish new but reversible connections between signaling pathways, at the same time giving them temporal and spatial specificity. Examples of scaffold proteins are found throughout the book (see, for instance, Section 11.5).

1.8.5 The complexity conundrum

The modular composition of proteins is a fundamental feature of biological evolution. Many interaction domains are, therefore, found already in prokaryotes, while others have developed later (such as, for example, the SH2 domains simultaneously with the appearance of tyrosine phosphorylation in multicellular animals). In the post-genomic era, this evolutionary conservatism provides a basis for a census of signaling proteins in that by computerized genome-wide analysis the total number of proteins with related domains and putatively related functions (such as protein kinases, G-proteins, AAA+ proteins, transmembrane proteins) can be determined.

Such studies demonstrate that the collection of eukaryotic interaction domains is predetermined in simple uni- and multicellular organisms such as yeast and the nematode *Caenorhabditis elegans* and has been enlarged only marginally in humans. Dramatically enlarged, however, is the number of *combinations* of such domains in a protein molecule, that is, the complexity of modular protein structure. For example, sequences encoding the SH3 domain have been found in 31 yeast, 132 nematode (*C. elegans)*, and 273 insect (*Drosophila*) genes but in 894 human genes. In fact, the principle of (almost) free mixing of protein modules enables biological evolution to constantly construct new data-processing circuits, thus climbing the ladder of complexity, although the number of basic units remains rather limited. Moreover, in the course of evolution, an interaction domain may have split into a wide variety of isoforms that differ in their properties such as binding specificity.

This tendency of combinatorial evolution, which also becomes apparent in other processes such as the alternative splicing of pre-mRNA (Section 8.9), may explain what has been called the complexity conundrum: why the size of the human genome (30,000 genes) has about the same size as that of the mouse and differs so unexpectedly (not to say disappointingly) little from that of the tiny and apparently primitive worm *C. elegans* (19,000 genes), the small insect *Drosophila melanogaster* (14,000 genes), or even a weed such as *Arabidopsis thaliana* (26,000 genes).

Summary

The hardware basis of cellular data processing is the modularity of protein structure, consisting of specific amino acid sequences that fulfill the role of communicative and functional domains. In the course of evolution, these modules have been combined in patterns of ever-increasing complexity. Signal-transducing proteins differ from metabolic enzymes and structural proteins because they have a particularly high number of interaction domains that receive input signals or enable other kinds of specific interactions (such as substrate and cellular targeting and intramolecular interactions). Interaction domains are either intrinsic to protein structure or formed on demand by post-translational modifications of amino acid residues. Adaptor and scaffold proteins are specialized for assembling proteins and other factors into signal-transducing complexes.

1.9 Generation of signaling patterns

In this section we examine some mechanisms used by signal-processing proteins to generate a wide variety of signaling patterns with important consequences, particularly for the type and pattern of the output signal and the stability of the signaling system. The output signal may resemble a pulse or it gradually increases or decreases like a sound growing louder or becoming softer, and exhibits either a continuous or an oscillatory pattern. Stability is understood as resistance to perturbations and as the ability to respond over a prolonged period of time without running the risk of becoming desensitized by overstraining. In technology such properties are considered to be a feature of high performance. For biological systems they are a matter of survival.

Perturbation insensitivity or **robustness** can be improved by adaptation, noise filtering, redundant components and mechanisms (see also Chapter 17), and temporal stability of signaling by oscillatory behavior. While the benefits of redundancy have already been addressed in Section 1.3, noise filtering, adaptation, and oscillatory signaling require a somewhat more detailed treatment.

Switches and regulators. Mathematical analysis of biochemical reactions, carried out on the basis of the fundamental laws of chemical kinetics (see Chapter 17), supplies a series of simple network components or circuits, the functions of which are described by a set of nonlinear differential equations. Examples of these circuits are depicted in Figure 1.18. In cells such elementary units are combined into highly sophisticated devices that, if one likes such a metaphor, may be looked upon as biochemical microprocessors.

Provided a protein exists in two states ON (1) and OFF (0), which are converted by a signal S (an allosteric effector), the relationship between signal intensity and response is linear or hyperbolic for a monomeric protein and sigmoidal for a protein consisting of allosterically cooperating subunits (Figure 1.18A). Sigmoidal behavior resembles that of a switch that requires some pressure to overcome its built-in resistance to accidental triggering. This resistance is equivalent to a **noise filter**. In contrast, linear or hyperbolic behavior resembles that of a sliding regulator such as a throttle.

Switch-like devices with noise filters are also established by **positive feedback**, for instance, when the signal-receiving protein activates itself or stimulates an activating factor. In the beginning the autostimulation increases slowly and a rise of signal intensity has no significant consequence. However, when the signal intensity surpasses a critical value, the response suddenly reaches its maximal value and the protein activity (that is, the output signal) becomes independent

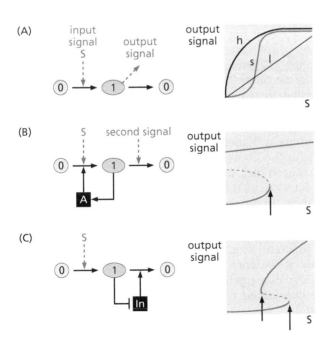

Figure 1.18 Simple biochemical switching elements An input signal S transfers a protein from an inactive state 0 (gray circles) into an active state 1 (red ellipses) that transmits an output signal and returns to state 0. (A) Simple switch with—depending on the structure of the protein and the reaction type—linear (l), hyperbolic (h), or sigmoidal (s) characteristics. (B) State 1 activates an activator A that promotes the activation of 0 by S. This switch has a discontinuous characteristic and the protein remains active unless it is inactivated by a second signal. (C) State 1 inactivates an inhibitor In that promotes the re-switching of 1 to 0. Again the characteristic of this switch is discontinuous, but in contrast to case B, it can re-switch to 0 by its own.

of the input signal S. The result is a discontinuous signal–response curve (Figure 1.18B). In principle, the process is irreversible unless an additional signal switches the protein back to the 0 state. Autophosphorylation of the insulin receptor serves as an example. It is induced by the insulin signal and results in an insulin-independent long-term activation of the receptor until it is abolished by a phosphoprotein phosphatase activated by a second signal (details in Chapter 7). Without countercontrol, positive feedback systems tend to develop in an explosive and catastrophic manner.

A discontinuous function resembling a (reversible) toggle switch is obtained when the positive feedback is due to the abolition of an inhibitory effect, that is, when the signal-activated protein inhibits or destroys an inhibitor (Figure 1.18C). Such a case, resembling the arithmetical instruction **"minus times minus is equal to plus"** (double negation), is particularly abundant in biological systems because it works with preformed components and, therefore, guarantees a particularly quick and robust response. Moreover, it again provides a highly efficient noise filter. Examples of this regulatory principle are found throughout this book (for a particularly clear one, see Section 5.9).

Oscillators. Oscillating systems open up the possibility of frequency-dependent modulation of signals. When the input signal S not only activates the protein but also stimulates a factor that sets the protein switch back to 0, the signal–response curve shows a **dampened oscillation** resembling an **adaptation by negative feedback** (Figure 1.19A). In Section 3.3.4 a classical example, the chemotactic reaction of bacteria is described. Another prominent case is the adaptation of sensory organs (see Chapter 15).

Adaptation contributes to robustness. Moreover, it is a fundamental prerequisite to learning and a critical condition for a signal-receiving device to respond over a wide range of signal intensity with constant sensitivity, thus being able to track down and pursue the source of the signal.

To generate an **undampened oscillation**, a combination of positive and negative feedback (or a twofold negative feedback) is required. To this end, at least three proteins have to interact (Figure 1.19B).

Such systems convert into oscillators at intermediate signal intensities. They are suited best for long-term signaling. An example is the oscillating cAMP signal that induces cell aggregation and differentiation of the slime mold *Dictyostelium discoideum*. cAMP stimulates, via a cellular receptor, its own production (positive feedback), while at higher concentrations it inactivates its receptor and simultaneously promotes its own degradation by stimulating a cAMP-phosphodiesterase (negative feedback). Another example is provided by the oscillations of intracellular Ca^{2+} signals, representing one of the most impressive cases of an oscillating signal (details in Section 14.5.4). Here the positive component is a Ca^{2+}-induced Ca^{2+} release from the endoplasmic reticulum, whereas the negative component consists of an inhibition of this release and an activation of Ca^{2+} pumps at higher Ca^{2+} concentrations. When such systems consist of many units (in the above examples, slime mold cells or Ca^{2+}-transporting membrane channels) and the inhibitory field is considerably more extended than the stimulatory field, the output signal may develop into a time-independent **spatial pattern**. In the reverse case, when the inhibitory field is the smaller one, the output signal spreads temporally and spatially as a **wave**. It is easily conceivable that such systems play an essential role in morphogenesis, with *Dictyostelium* differentiation being a well-known example. Moreover, a steadily increasing body of evidence indicates that wavelike patterning and frequency modulation of signals are general hallmarks of subcellular data processing. Thus a cell may be considered to represent a "musical" rather than only a trivial "chemical" system!

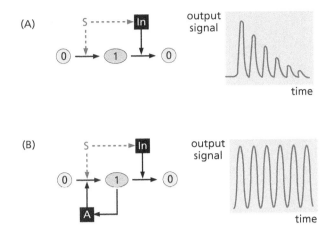

Figure 1.19 Two biochemical oscillators As in Figure 1.18, the activation (0 → 1) and deactivation (1 → 0) of a signal-controlled protein is depicted. Shown are two systems generating either a dampened or an undampened oscillation.
(A) The signal S simultaneously activates an inhibitor In that re-switches 1 to 0. This results in a dampened oscillation of the output signal. (B) The protein activated by the signal S stimulates both its activation, via an activator A (positive feedback), and its deactivation, via an inhibitor In (negative feedback). As a result an undampened oscillation of the output signal may develop within a certain intensity range of the input signal S.

Summary

By combining negative and positive feedback loops, interacting proteins may self-organize into complexes resembling technical devices that transform input signals into special patterns of output signals. Circuits resembling double negation according to the rule "minus times minus is equal to plus" are particularly abundant. Of special interest are oscillating systems, since they enable frequency-dependent encoding of signals.

Further reading

Arkin A & Ross J (1994) Computational functions in biochemical reaction networks. *Biophys. J.* 67, 560–578.

Barabasi AL & Oltvai Z (2004) Network biology: understanding the cell's functional organization. *Nat. Rev. Genet.* 5, 111–113.

Bhattacharyya RP, Remenyi A, Yeh BJ, et al. (2006) Domains, motifs, and scaffolds: the role of molecular interactions in evolution and wiring of cell signaling circuits. *Annu. Rev. Biochem.* 75, 655–680.

Bray D (1995) Protein molecules as computational elements in living cells. *Nature* 376, 307–313.

Bruggeman FJ & Westerhoff HV (2007) The nature of systems biology. *Trends Microbiol.* 15, 45-50.

Changeux JP & Edelstein SJ (2005) Allosteric mechanisms of signal transduction. *Science* 308, 1424–1428.

Ferrell JE Jr (1996) Tripping the switch fantastic: how a protein kinase cascade can convert graded inputs into switch-like outputs. *Trends Biochem. Sci.* 21, 460–466.

Ferrell JE Jr (2002) Self-perpetuating states in signal transduction: positive feedback, double-negative feedback and bistability. *Curr. Opin. Cell Biol.* 6, 140–148.

Flynn DG (2001) Adaptor proteins. *Oncogene* 20, 6270–6272.

Jordan JD, Landau EM & Iyengar R (2000) Signaling networks: the origins of cellular multitasking. *Cell* 103, 193–200.

Lim WA (2002) The molecular logic of signaling proteins: building allosteric switches from simple binding domains. *Curr. Opin. Struct. Biol.* 12, 61–68.

Maniatis T & Reed R (2002) An extensive network of coupling among gene expression machines. *Nature* 416, 499–506.

Papin JA, Hunter T, Palsson BO, et al. (2005) Reconstruction of cellular signaling networks and analysis of their properties. *Nat. Rev. Mol. Cell Biol.* 6, 99–111.

Pawson T & Nash P (2003) Assembly of cell regulatory systems through protein interaction domains. *Science* 300, 445–452.

Pufall MA & Graves BJ (2002) Autoinhibitory domains: modular effectors of cellular regulation. *Annu. Rev. Cell Dev. Biol.* 18, 422–458.

Seet BT, Dikic I, Zhou MM, et al. (2006) Reading protein modifications with interaction domains. *Nat. Rev. Mol. Cell Biol.* 7, 473–483.

Tyson JJ, Chen KC & Novak B (2003) Sniffers, buzzers, toggles and blinkers: dynamics of regulatory and signaling pathways in the cell. *Curr. Opin. Cell Biol.* 15, 221–231.

Zhu X, Gerstein M & Snyder M (2007) Getting connected: analysis and principles of biological networks. *Genes Dev.* 21, 1010–1024.

Supplying the Network with Energy: Basic Biochemistry of Signal Transduction

As discussed in Chapter 1, cellular signal transduction is first of all due to biochemical reactions that mediate the interactions within the protein network while at the same time supplying the energy required for data processing. This chapter describes the most abundant and important switching reactions, and will provide the reader with an "Ariadne's thread" to help find a path through the maze of details discussed throughout the rest of the book.

2.1 No order without work

The goal of data processing is to minimize the level of uncertainty. This is synonymous with creating order or a higher level of organization, which thermodynamically resembles a lowering of entropy in the system (see Sidebar 2.1). According to the second law of thermodynamics, such a task requires energy. Data-processing systems, thus, depend on work, or on a permanent energy supply (Figure 2.1): a computer has to be continuously supplied with electricity and a brain with metabolic energy. The same condition applies to the signal-processing protein network of the cell. In fact, the overall theory is that signal transduction reflects a fundamental principle of life: namely, that a process of self-organization (the emergence of order) must always be coupled with another process that provides energy by producing disorganization or chaos. But what is going on at the molecular level?

We return to the concept of a protein resembling a simple binary switch that can be switched on or off (being aware of the fact that this is a borderline case) (Figure 2.2). Such a switching process resembles a chemical reaction that is not in equilibrium. This means that when the forward reaction is exergonic (releasing free energy and thus occurring spontaneously), the reverse reaction necessarily must

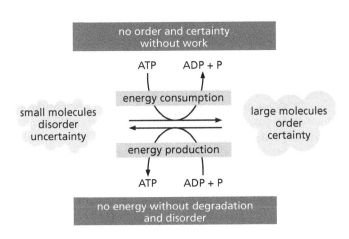

Figure 2.1 Energy processing by living systems Biological energy processing is a direct consequence of the second law of thermodynamics. In order to synthesize large molecules (such as sugars, lipids, proteins, and nucleic acids), maintain a highly ordered state, and minimize the degree of uncertainty, an organism consumes energy, first and foremost by hydrolyzing ATP. ATP is produced (energy is stored) by degrading large molecules (taken up as food) and generating disorder in the environment. Therefore, organisms are called "dissipative systems."

Sidebar 2.1 A living cell resembles a charged battery
The second law of thermodynamics states that, in a *closed* system (such as the universe), disorder measured as entropy increases continuously. In contrast, cells and organisms are *open* systems that, for a certain period of time (the lifespan), are able to struggle against this verdict, albeit at the expense of their environment. This phenomenon is explained by chemical thermodynamics. A chemical reaction either delivers or requires energy. Only a certain portion of this energy, traditionally called "free energy," can be used for work—transformed into chemical, mechanical, electrical, or electromagnetical energy—whereas the rest is lost as heat, increasing the disorder. The energetic turnover of a reaction occurring at constant temperature and constant pressure (as in cells) is described by the **Gibbs-Helmholtz equation:**

$$\Delta H = \Delta G + T\Delta S$$

where ΔH is the change of total energy, ΔG is the change of free energy, ΔS is the change of entropy, and T is the absolute temperature. The term $T\Delta S$ represents the "bound" energy that unavoidably is processed as heat. According to $T\Delta S = \Delta H - \Delta G$, a reaction occurs spontaneously at negative values of ΔG (indicating a loss of free energy by the system), because then the entropy and thus the degree of disorder increases. A positive value of ΔG means, on the other hand, that the reaction needs the supply of free energy because the entropy decreases. Such a reaction does not occur spontaneously.

Since the dimension of G and H is kilocalories per mol,

the energy status of a reaction depends on the concentration of the molecules involved. Thus the ΔG value of ATP hydrolysis is about −12 kcal/mol under normal cellular conditions (low ADP concentration) but becomes much less negative, even approaching 0, when the ratio of ATP to ADP concentrations decreases in extreme stress situations, exhausting the cell's energy store.

In fact, depending on the conditions, ATP hydrolysis and any other reaction has the tendency to reach a thermodynamic equilibrium. In this situation ΔG is equal to 0. This means that the reaction has suffered from "heat death" and has become useless as an energy-delivering process. Therefore, the biochemical status of a living cell represents a system far out of thermodynamic equilibrium, thus disposing of a large amount of free energy that is used to maintain a state of high order. Such a state requires a continuous supply of free energy from the environment. This free energy is hidden in chemical bonds (or, in plants, derived from sunlight), for instance in food, where it is released by metabolism and stored in energy-rich cellular compounds such as ATP and macromolecules. In other words, a living cell resembles a battery that continuously has to be recharged at the expense of its environment, whereas in a dead cell the battery has been irreversibly discharged. Systems that maintain a high degree of organization by producing disorder in their environment are called "dissipative."

For his pioneering work on non-equilibrium thermodynamics and dissipative systems, Ilya Prigogine was awarded the 1977 Nobel Prize in Chemistry.

be endergonic (needing an energy supply), or vice versa. Therefore, a protein switch must be coupled with an energy-delivering reaction, because otherwise it could be switched into one direction only but not back. Moreover, in a network of switching units, the input signal becomes dispersed rapidly and fades away; therefore, it has to be permanently amplified. Finally, the targeting of the signal requires energy derived from irreversible, strongly exergonic reactions. We have to conclude, therefore, that all cellular switching processes are coupled with biochemical reactions delivering a large amount of energy. These are, in particular, redox reactions, the hydrolysis of nucleoside triphosphates (NTPs) and other energy-rich compounds, the degradation of macromolecules, and ion flows along electrochemical gradients across membranes (Figure 2.3).

2.1.1 Post-translational modifications

Frequently the energy-supplying biochemical reactions coupled with signal transduction generate covalent changes of the signal-transducing protein. Such changes are called post-translational modifications. They are fully reversible with the forward and the reverse reactions catalyzed by enzymes. As summarized in Figure 2.3, the most frequent ones are:

- oxidations and nitrosylations of thiol groups, and glutathionylation

- phosphorylations, catalyzed by protein kinases

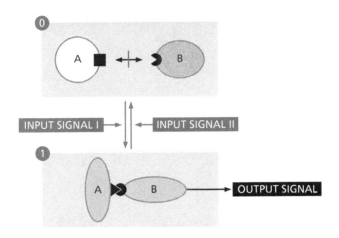

Figure 2.2 Switching element consisting of two interacting proteins An interaction between the proteins A and B takes place only in state 1. Input signals enable the switching process (in both directions), while the activity of the AB complex resembles the output signal. According to the rules of chemical thermodynamics, one of the two reactions is exergonic with the reverse reaction being necessarily endergonic (requiring an energy supply). As a rule, exergonic reactions pass off spontaneously, though frequently very slowly, due to a high activation energy. In such a case the input signal has a catalytic effect; that is, it lowers the activation energy. When the reaction is endergonic, the input signaling is coupled with an energy-delivering process.

- ubiquitylations, catalyzed by ubiquitin transferases
- acetylations, catalyzed by acetyltransferases
- ADP-ribosylations, catalyzed by ADP-ribosyltransferases
- partial proteolysis, catalyzed by proteases

The biochemical and functional principles of these modifications are explained in the following sections. Additional covalent reactions of proteins involved in signaling include methylations of glutamate/aspartate carboxyl groups and of lysine amino groups, hydroxylations, and reversible binding of *N*-acetylglucosamine. They are discussed in Sections 3.3.4, 8.2.4, 8.7, and 9.4.3, respectively.

As far as signal transduction is concerned, covalent modifications of proteins serve at least two purposes (Figure 2.4).

(1) The conformational and functional change caused by the switching process in the effector protein (protein B in Figure 2.2) is "frozen in" to a certain extent, until the modification is reversed by corresponding enzymes (such as oxidoreductases, phosphoprotein phosphatases, ubiquitin proteases, deacetylases, or ADP-ribosylhydrolases) or sometimes nonenzymatically. Such enzymes are also targets of input signals. Thus, as compared with a noncovalent conformational change, covalent post-translational modifications provide additional possibilities to feed input signals into the switching process and, in particular, to control this process temporally.

(2) The covalent change provides a contact site for noncovalent interaction with other proteins and non-protein molecules, provided these partners possess an appropriate interaction domain, like a lock fitting a key. Such interactions

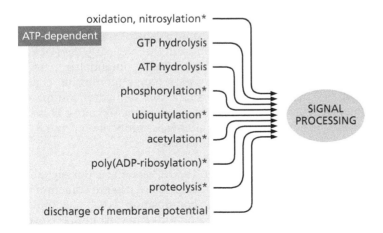

Figure 2.3 Most frequent noncovalent and covalent switching reactions of cellular signal processing The reactions summarized here lead to noncovalent changes or covalent post-translational modifications (marked by asterisks) of signal-transducing proteins. They depend on or are coupled with energy-supplying processes such as ATP hydrolysis (gray box) or redox reactions.

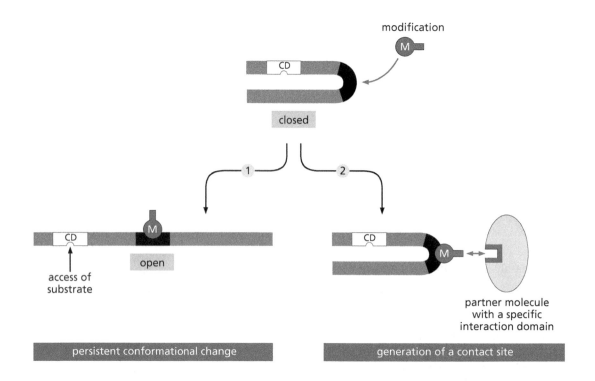

Figure 2.4 Two consequences of covalent post-translational modification The modification M may change the protein conformation in such a way that, in the case of an enzyme, a catalytic domain (CD) becomes uncovered, allowing the access of a substrate (pathway 1), or generates a contact site for a partner molecule with a corresponding interaction domain (pathway 2). This mechanism plays a key role in the formation of signal-transducing protein complexes and for the intracellular localization of proteins. Frequently a single modification has both effects. By the same token, a post-translational modification may have an inhibitory (antagonistic) effect when it stabilizes the closed conformation of an enzyme (not shown). The scheme may be generalized. Apart from the catalytic domain of an enzyme, any interaction domain (for instance, a binding site for an allosteric effector or a cytoskeletal component) may be either uncovered or hidden by the conformational change triggered by a post-translational modification.

are required for the formation and intracellular targeting of signal-processing complexes and networks of macromolecules. Thus, covalent modification also controls a switching process spatially.

Summary

Because it is an entropy-lowering process, cellular signal processing depends on a permanent energy supply. The energy is derived from metabolic redox reactions, the hydrolysis of nucleoside triphosphates (ATP, GTP) and other energy-rich molecules such as NAD^+, controlled protein degradation, and a discharge of the membrane potential. The enzymes catalyzing these processes are targets of input signals. Every signaling event is coupled with at least one of these energy-supplying biochemical reactions. Frequently this coupling results in a reversible covalent change (post-translational modification) of the signaling protein.

2.2 Redox and nitrosylation switches: a balance on a narrow ridge

We begin our examination of energy-supplying reactions with redox processes because they provide the primary energy source of an aerobic cell. Ironically, their role beyond metabolism has been recognized only recently: the redox potential, defined as the (electrochemically measured) energy released either by oxidation or upon withdrawal of electrons or separation of hydrogen, appears to have been used for signal transduction from the very early days of evolution. In fact, proteins respond to changes of the redox potential by conformational rearrangements, provided they possess structures monitoring changes of the redox potential. These redox sensors are the sulfhydryl groups of cysteine residues that are oxidized to disulfide and sulfenic acid groups. This simple reaction, which may occur even nonenzymatically, probably represents one of the most ancient post-translational modifications.

A second group of redox sensors involved in bacterial redox signaling are iron ions bound in so-called sulfur clusters. These are protein cofactors that consist of inorganic iron sulfide entities of the composition $[Fe_{2n}S_{2n}]$ complexed by Cys (and His) residues of the protein. The iron ions can exist in the redox states Fe^{2+}

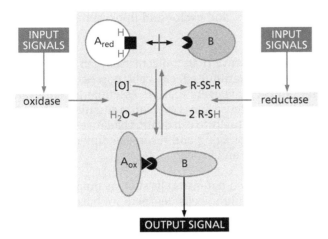

Figure 2.5 Redox switch Protein A interacts with and changes the conformation and function of protein B only when its SH groups have first been oxidized. Oxidants are reactive oxygen species [O] generated as intermediates during the enzymatic or nonenzymatic reduction of molecular oxygen. The corresponding enzymes (oxidases and oxygenases) are controlled by input signals. The same holds true for reductases that set the switch back to 0 by using peptides with thiol groups (R-SH) as reductants. The switching energy is derived from the oxidations. Like any protein switch, this one may also work in the opposite direction, that is, when A interacts with B only in a reduced form.

and Fe^{3+}. Such clusters are found in bacterial transcription factors controlling the production of enzymes that deal with the potentially hazardous effects of oxygen (see Section 3.5.5), but also in metabolic redox enzymes including ferredoxin (an electron carrier in photosynthesis), bacterial nitrogenase (Section 3.3.2), xanthine oxidase and the cytochrome bc_1 complex. While sulfhydryl groups are oxidized via a two-electron transition, iron-sulfur clusters are specialized on one-electron transitions as they are induced by free radicals.

To become a binary switch, the redox reaction must be forced under the control of enzymes. Moreover, it has to be reversible: the oxidation must be reversed by a reduction. Such a redox switch is shown in Figure 2.5. Here a signal-transducing interaction between two proteins occurs only when one of the partners is reversibly oxidized. The oxidases and reductases catalyzing these reactions are the sensors of input signals.

The oxidants generated by oxidases are **reactive oxygen species (ROS)**, whereas the reductases use peptides and proteins with thiol groups as well as reduced cofactors such as NADH and NADPH as reductants. ROS is a collective term for intermediates arising during the reduction of molecular oxygen to water. These intermediates are generated by one-electron transitions and are superoxide anion radicals, hydrogen peroxide, and hydroxyl radicals:

$$O_2$$
$$\downarrow + e^-$$
superoxide anion radical $O_2^{-\bullet}$
$$\downarrow + e^-$$
hydrogen peroxide H_2O_2
$$\downarrow + e^-$$
hydroxyl radical $OH^{\bullet}$
$$\downarrow + e^-$$
water H_2O

An additional ROS is singlet oxygen, $^1O_2$, generated during photosynthesis and as a byproduct of enzymatic oxidations such as those of unsaturated fatty acids. One-electron transitions are typically catalyzed by oxidases and oxygenases containing flavin nucleotides, heme, or non-heme iron as prosthetic groups. These families of enzymes include major ROS generators such as the NADPH oxidase of the plasma membrane and the xanthine oxidase of the Golgi apparatus, the respiratory-chain complexes of the mitochondrial matrix, as well as sulfhydryl oxidases, and cyclo- and lipoxygenases of unsaturated fatty acid metabolism.

Since ROS are more aggressive than molecular oxygen, their use is a risky venture. To neutralize their harmful effects, cells possess an arsenal of enzymes, reductants, and radical scavengers that are essential for survival in an oxygen atmosphere. When this defense line is overrun, a situation called **oxidative stress** emerges, which leads eventually to cell death. Oxidative stress is considered to be one of the major causes of aging and diseases connected with aging, such as dementia, atherosclerosis, and cancer.

ROS are used by specialized defense cells such as phagocytes to kill microbes. Moreover, early in evolution, cells learned to harness ROS generated in subtoxic concentrations for a controlled post-translational modification of proteins and thus for signal processing. As mentioned above, this modification affects the sulfhydryl groups of cysteine residues, which are reversibly oxidized to disulfide and sulfenic acid groups. Disulfide formation may occur intramolecularly or with other partners, in particular, glutathione and thioredoxin (Figure 2.6). In the latter case the protein is said to be glutathionylated or thioredoxinylated.

The ROS that typically oxidizes sulfhydryl groups is **hydrogen peroxide**. It is the most abundant and most important messenger in redox signaling. In contrast, superoxide anion radicals are less suited for signaling: their diffusion across membranes is limited by their negative charge, their intracellular lifetime is much shorter than that of H_2O_2, and they preferentially interact with iron sulfur clusters rather than with sulfhydryl groups. Finally, the extremely dangerous hydroxyl radicals are not used as messengers but destroyed immediately by radical scavengers and antioxidants. Whether an SH group is oxidized by H_2O_2 depends on its sterical accessibility and its acidity, which is determined by electrostatic effects of neighboring amino acid residues. As expected for a signaling event, the oxidation is highly selective. In fact, frequently only one of the many cysteine residues of a protein is attacked by the oxidant.

The reductases that reset the sulfhydryl-disulfide switch to 0 mainly use glutathione and thioredoxin as reductants. **Glutathione** is a Glu-Cys-Gly tripeptide found in cells in millimolar concentrations. It is rather resistant to ROS but is oxidized to a dimeric disulfide by signaling proteins with disulfide groups. This oxidation is catalyzed by the enzyme glutaredoxin, again via a disulfide intermediate, whereas glutathione reductases are responsible for the reduction of oxidized glutathione by NADPH. The much more effective **thioredoxins** are ubiquitous 12 kDa dithiol-oxidoreductases. Although mechanistically resembling glutaredoxins, they do not catalyze glutathione oxidation but catalyze their

Figure 2.6 Biochemistry of redox signaling: the forward reaction In the course of various enzymatic or nonenzymatic reactions, reactive oxygen species (ROS) are generated that oxidize cysteinyl thiol groups of proteins to sulfenic acid groups or to intra- and intermolecular (mixed, here with glutathione) disulfide groups.

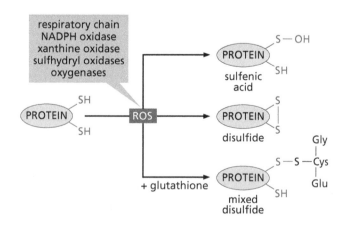

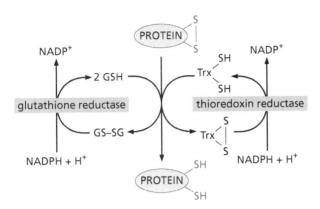

Figure 2.7 Biochemistry of redox signaling: the reverse reaction The protein modified by the redox signal (here with an intramolecular disulfide bridge) is reduced by nicotinamide adenine dinucleotide phosphate (NADPH) with glutathione (GSH) or thioredoxin (Trx) acting as an electron transmitter. As a rule, the reaction with glutathione occurs spontaneously and relatively slowly, whereas thioredoxin is a dithiol-oxidoreductase catalyzing a fast redox reaction. The flavoprotein enzymes glutathione reductase and thioredoxin reductase regenerate the reduced forms of glutathione and thioredoxin at the expense of NADPH.

own oxidation to intramolecular disulfides. These disulfides are reduced again by specific reductases, with NADPH acting as an electron donor (Figure 2.7).

Redox signaling has been found in both prokaryotes and eukaryotes and is probably more abundant than hitherto assumed. The proteins modified by this reaction include major signal transducers such as transcription factors (see, for instance, Section 3.5.5), protein kinases, and protein phosphatases (see Sections 2.6 and 7.1). A modification of redox signaling is the **nitrosylation** of cysteinyl–SH groups in proteins. In this case nitric oxide (NO) is the reactive agent generated by an enzymatic oxidation of arginine (for details see Section 16.5.1). NO is both an inter- and intracellular signal molecule. Downstream, it activates cytoplasmic guanylate cyclases by noncovalent binding and modifies other proteins by covalent nitrosylation forming cysteinyl–S–NO groups. When NO is pathologically overproduced, the nitrosylation occurs nonspecifically. The **nitrosative stress** thus evoked is as harmful for the cell as oxidative stress. Under normal conditions, however, NO production is strictly controlled and nitrosylation is exclusively used for signal-controlled post-translational modifications of proteins. Today more than 100 proteins are known, the functions of which are selectively regulated along this pathway. Even though nitrosylation is, strictly speaking, a nonenzymatic reaction, it specifically attacks only selected cysteine residues. The reason is the same as that for the selectivity of redox signals: intramolecular interactions modifying the nucleophilic reactivity (acidity) as well as the sterical accessibility of an SH group. An additional cause of selectivity is targeted interactions of NO synthases with the substrate proteins of nitrosylation.

Enzymes catalyzing denitrosylation set the nitrosylation switch to 0. Again glutathione and thioredoxin play a central role in this process. Thioredoxin transfers NO from the substrate protein to one of its cysteine residues and from there to glutathione where subsequently it is reduced and released by thioredoxin–S–NO or glutathione–S–NO reductase. This trans-nitrosylation of one protein by another seems to work in the reverse direction as well, with nitrosylated thioredoxin and glutathione functioning as NO donors for protein nitrosylation.

Summary

An ancient source of energy required for signal processing is redox reactions. Redox signaling is mediated by reactive oxygen species (ROS) generated along various metabolic pathways (or nonenzymatically). By oxidizing sulfhydryl groups, ROS, in particular H_2O_2, post-translationally modulate protein structure and function. The process is reversed by reductases that use peptides (glutathione) or proteins (thioredoxins) with sulfhydryl groups as reductants and NADPH as an electron donor. A related post-transcriptional protein modification with signaling character is the nitrosylation by NO of sulfhydryl groups. An overactivation of these signaling reactions results in oxidative or nitrosative stress with manifold pathological consequences.

2.3 Switches operated by enzymes that hydrolyze energy-rich compounds

From the very beginning, cells have learned to store the energy obtained from oxidative reactions and other elementary metabolic processes in metabolites containing energy-rich chemical bonds, first and foremost in the phosphoric acid anhydride bond of NTPs (see Sidebar 2.2). Most *targeted* cellular processes that result in a lowering of entropy—that is, those that lead to a state of higher order—are linked to the hydrolysis of NTPs such as ATP and GTP. These include the biosynthesis of macromolecules, the generation of ion gradients across cell membranes (resulting in the membrane potential), transport reactions and movements, as well as the great majority of signal-transducing reactions.

The phosphoric acid anhydride bond of NTPs is thermodynamically labile but, for the reasons explained in Sidebar 2.2, kinetically stable; that is, under the conditions of a living cell, its hydrolytic cleavage, although strongly exergonic, does not occur spontaneously but has to be catalyzed. This is achieved by two types of enzymes: nucleoside-triphosphatases (NTPases) and kinases. NTPases transfer the γ-phosphate residue of NTP to water, producing free inorganic phosphate (P_i). Kinases transfer the γ-phosphate residue to nucleophilic residues (XH) of other molecules, which become phosphorylated. In a second step, catalyzed by a phosphatase or proceeding spontaneously, the phosphate residue is transferred to water, yielding phosphoric acid:

$$\text{NTPase reaction:} \quad NTP + H_2O \rightarrow NDP + P_i$$

$$\text{kinase reaction:} \quad NTP + RXH \rightarrow NDP + RXP$$

$$\text{phosphatase reaction:} \quad RXP + H_2O \rightarrow RXH + P_i$$

Guanosine and adenosine triphosphatases (GTPases and ATPases) and protein kinases/phosphatases are particularly abundant switching devices of the signal-processing protein network. While the NTPases induce noncovalent conformational changes, the kinases/phosphatases catalyze covalent phosphorylations/dephosphorylations of the corresponding effector protein. The activities of these enzymes are regulated by an enormous variety of noncovalent interactions with other proteins, low molecular weight factors, and environmental stimuli as well as by post-translational modifications. These are the input signals driving the switching process either from 0 to 1 or from 1 to 0.

Summary

Most of the energy driving cellular signal processing is derived from the hydrolysis of ATP and GTP. These reactions are catalyzed by nucleoside-triphosphatases (NTPases) and protein kinases/phosphatases. The energy thus obtained is used for reversible conformational changes or post-translational modifications of signaling proteins.

2.4 GTPase or G-protein switch

One of the most simple and most abundant signal-transducing reactions is based on the hydrolysis of GTP catalyzed by GTPases. This reaction supplies as much free energy as ATP hydrolysis. In the late 1960s it was observed that the activation of the cellular adrenaline receptor required GTP (Section 4.3). The corresponding GTP-binding protein was identified as a GTPase interacting with the receptor. Since then numerous regulatory GTPases or G-proteins have been found to belong to the standard equipment of the data-processing protein network. As prototypes of switching elements using the energy derived from NTP hydrolysis, G-proteins are involved in almost all signal-processing mechanisms of eukaryotic cells. Evolutionarily, they originate in prokaryotic precursors (see Section 3.5.4).

Sidebar 2.2 The unique phosphoric acid In the data-processing protein network, most switching reactions are linked directly or indirectly to the strongly exergonic hydrolysis of NTPs and the resulting phosphorylation of proteins. What is the distinctive feature of phosphoric acid that makes it so indispensable for life processes and essentially irreplaceable by any other structure?

The phosphoric ester bond was most likely present in living systems at the very beginning of evolution. The discovery of ribozymes, RNA with enzymatic activity (the 1989 Nobel Prize in Chemistry awarded to Sidney Altman and Thomas R. Cech), led to the hypothesis that the most ancient biomolecules are ribonucleic acids rather than proteins as previously postulated. In fact, RNA exhibits several properties essential for life such as catalytic activity, the ability to copy itself, and the capability to store information in its linear structure. The latter requires sufficient chemical stability as the information would otherwise be lost rapidly and could not be transferred reliably to descendants.

For these purposes the phosphoric ester bond is particularly suitable, due to its additional negative charge, which a comparable sulfuric acid ester would be lacking. This charge renders RNA an extremely hydrophilic substance that cannot be lost across cellular membranes and protects the ester bond by electrostatic repulsion from nucleophilic attack such as hydrolysis (Figure 2.8).

Nevertheless, RNA is still relatively sensitive because hydrolysis proceeds via a cyclic 2′,3′-phosphate as an energetically favored intermediate. This fact may have provided the selective pressure for the development of DNA, which cannot form such an intermediate and, as a data carrier, is about 100 times more stable than RNA. In parallel with the evolution of DNA, RNAs were replaced as enzymes by proteins, probably because protein structures are much more flexible and variable than nucleic acid structures.

The synthesis of a macromolecule such as RNA is accompanied by a lowering of entropy, thus requiring an energy supply. This may be the reason for the evolutionary "invention" of NTPs. Their phosphoric acid anhydride bond is energy-rich and thermodynamically unstable. Nevertheless, it is kinetically stable since the hydrolysis requires a large amount of activation energy. This is again due to the negative charge, which protects not only the ester but also the anhydride bond from nucleophilic attack. Under the conditions of a cell, the hydrolysis of NTPs has to be enzymatically catalyzed. This peculiarity makes the NTPs, first and foremost, ATP, the universally applicable energy suppliers for the majority of endergonic biochemical reactions. As a rule such reactions proceed via *short-lived* intermediates where the enzyme is phosphorylated at suitable amino acid residues such as His, Asp, and Cys, with ATP as a phosphate donor. From this point, it was probably only a small evolutionary step toward the application of prokaryotic protein phosphorylation as a more persistent modulation of protein structure and function and thus as an energy-dependent switching reaction for cellular data processing.

Simultaneously with the development of protein phosphorylation cells may have used ATPase and GTPase reactions, the membrane potential, redox processes, and degradation of macromolecules to obtain the energy for data processing. Since the energy stored in the membrane potential and in macromolecules ultimately stems from the hydrolysis of NTPs, it may be stated without exaggeration that cellular data processing is mainly based on the unique chemical properties of phosphoric acid.

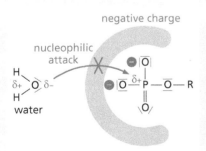

Figure 2.8 Phosphate group: self-protection against hydrolysis Through electrostatic repulsion, a cloud of negative charge at neutral pH shields the phosphate group from nucleophilic attack by polarized water molecules. Therefore, the hydrolytic reaction, although exergonic, is blocked since it requires a high activation energy. To occur under the conditions of a living cell, it must be sped up by enzymatic catalysis. R symbolizes an organic residue in the case of a phosphate ester group (as in nucleic acids) or a second phosphate group in the case of a phosphoric acid anhydride group (as in ATP).

A distinction needs to be made between "small" and "large" G-proteins. Small G-proteins are monomers of about 200 amino acids. According to their prototype, the p21 Ras protein, they are grouped into the so-called Ras superfamily. Large G-proteins are heterotrimeric $\alpha\beta\gamma$ complexes with the Ras-related α-subunit harboring the GTPase function. Other G-proteins include the ribosomal elongation factors, the signal-recognition particles of the endoplasmic reticulum, and the dynamins. The different G-protein families are discussed in detail in Chapters 4 and 10.

Regulatory GTPases have many functions that, at a first glance, seem to have almost nothing to do with each other. However, all these effects target cellular processes by making them irreversible through entropy reduction brought about by GTP hydrolysis.

2.4.1 G-proteins: amplifiers, rectifiers, and organizers

G-proteins are signal-transmitting and signal-transforming switches. They induce reversible conformational changes rather than posttranslational modifications of proteins. In an exemplary fashion G-proteins demonstrate two basic features of signal transduction: one, connection of signaling processes with an energy-supplying biochemical reaction and two, spatial and temporal control of signaling. A G-protein has bound GDP in state 0 and GTP in state 1, in a 1:1 ratio. In the GTP-loaded state the G-protein interacts with downstream effector proteins, changing their conformation and function in a time-dependent manner (Figure 2.9) or acting as an organizer, targeting them to defined sites such as the cell membrane where they come in contact with other proteins forming signal-transducing complexes (Figure 2.10). These functional changes and the altered intracellular localization of the effector proteins resemble the output signals of the G-protein switch.

Activation, or the exchange of GDP for GTP, has to be accelerated enzymatically and is thus an input terminal for signals. The same holds true for the reverse reaction, GTP hydrolysis. Being GTPases, G-proteins are able to switch back to position 0 on their own. However, their GTPase activity (and with that the duration of the active state) is modulated by other input signals within wide limits. Thus G-proteins are not only amplifiers and rectifiers but also adjustable timers.

2.4.2 Switching mechanism of G-proteins

All regulatory GTPases share an evolutionary ancient core of about 200 amino acids. It is called the **G-domain** and resembles the small G-protein Ras in size and basic structure. The G-domain distinguishes G-proteins from other GTP-hydrolyzing enzymes such as β-tubulin or phosphoenolpyruvate carboxykinase. By means of X-ray analysis and directed mutations, the architecture and mode of function of the G-domain could be elucidated for several G-proteins, beginning with the prokaryotic ribosomal translation factor EF-Tu.

The basic structural elements of the domain are six anti-parallel β-sheets with five α-helices grouped in between. Five of the β-sheets are placed in a plain surrounded by the helices. The catalytic center is located in the N-terminal area of this structure, containing a highly specific and high-affinity binding site for GTP/GDP. A characteristic structural element of this site is the P-loop between β-sheet 1 and helix 1. Here the negatively charged phosphate residues (present

Figure 2.9 GTPase switch as a controlled timer When it is loaded with GDP, the G-protein or GTPase (G) cannot interact with an effector protein. Only when GDP is exchanged for GTP (catalyzed by guanine nucleotide exchange factor, GEF) is the conformation changed allosterically, now enabling contact with a corresponding binding domain of the effector protein, which becomes altered structurally and functionally. The switch is set back to 0 by the endogenous GTPase activity of G (stimulated by the regulatory GTPase-activating protein, GAP). Input signals control the activity of GEF or GAP, whereas the functional change of the effector protein as induced by G resembles the output signal. Another effect of G is to bring its effector protein in contact with other components of the signal-processing network (see Figure 2.11).

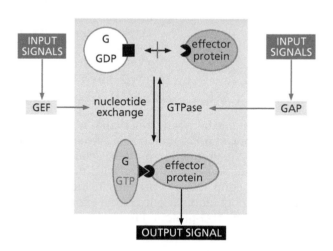

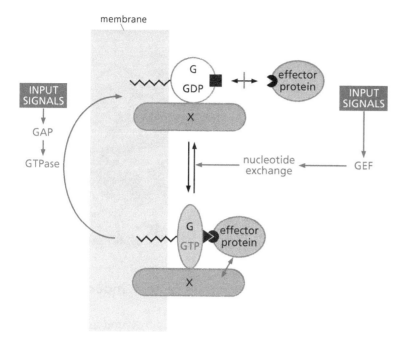

Figure 2.10 GTPase switch as a spatial organizer Upon GDP–GTP exchange, the GTPase G anchored to a cellular structure (for instance, to the plasma membrane through a lipid anchor, shown as a zigzag line) undergoes a conformational change allowing the binding of the effector protein, which is thus brought in contact with another protein X (red). When the interaction of G does not induce an energy-dependent conformational change of the effector protein, the activated GTPase continuously recruits effector proteins and passes them on to other proteins until it inactivates itself by the GAP-stimulated GTPase activity. The energy required for this organizing process is supplied by interactions between the effector proteins and protein X. A defect of the GTPase function, for instance due to a gene mutation, resembles a switch that cannot set back to 0. It may have fatal consequences, as will be explained later for the GTPase Ras.

as Mg^{2+} complexes) of the nucleotide are fixed by circularly arranged amino acids via electrostatic and hydrogen bonds (PM-motif) while the guanine base fits an adjacent hydrophobic pocket, the G-motif. Together PM- and G-motifs build up the catalytic center. The water molecule executing the nucleophilic attack on the phosphoric acid anhydride bond is polarized by a highly conserved Thr or Gln residue and placed into position in such a way that it can split off the terminal phosphate residue of GTP in the course of an S_N2 reaction.

Both GDP–GTP exchange and GTP hydrolysis are accompanied by far-reaching conformational changes of the G-protein, affecting in particular two structural elements of the G-domain, switches 1 and 2 (Figure 2.11). These elements contain highly conserved amino acid residues (Thr and Gly) forming hydrogen bridges with the γ-phosphate residue of GTP. When these bridges are pulled down in the course of the GTPase reaction, the switches fold down into another conformation similar to a relaxing string (Figure 2.12). These conformational changes of the switch regions are transferred onto the effector protein.

Figure 2.11 Structure of the small G-protein Ras as compared with the α-subunit of the trimeric G-protein transducin Ras exhibits the typical architecture of a G-domain. In the transducin subunit an additional helix bundle is inserted into the G-domain.

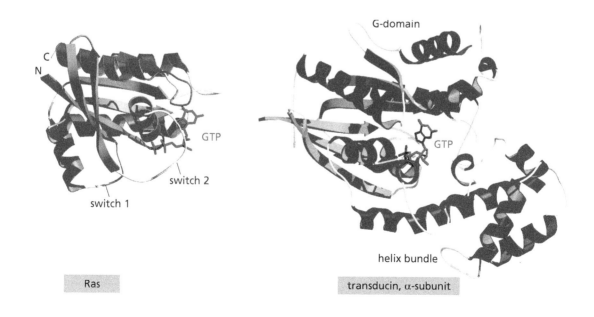

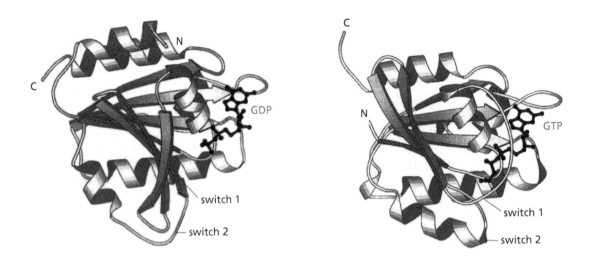

Figure 2.12 Effect of GDP–GTP exchange on structure of the G-domain The small G-protein Arf is shown as a GDP complex (left) or a GTP complex (right). The nucleotide exchange results in a considerable rearrangement of the switch regions.

2.4.3 Guanine nucleotide exchange factors: inducers of G-protein signaling

At G-proteins the exchange of GDP for GTP is thermodynamically favored by high GTP excess in the cell. Nevertheless, due to the extremely strong binding of GDP (and GTP) to the G-domain, it requires so much activation energy that it takes hours to occur spontaneously. Therefore, it has to be catalyzed enzymatically (for the same reason, G-proteins do not exist in cells in a guanine nucleotide-free form). The corresponding enzymes are called guanine nucleotide exchange factors (GEFs). Each G-protein has one or more GEFs (the human genome encodes more than 1000 GEFs). With the nucleotide-liganded G-protein, the GEF forms a ternary transition complex where the binding between nucleotide and G-domain is up to 10^6 times weaker as in the absence of GEF, resulting in a dramatic lowering of the activation energy (Figure 2.13). In this state the GDP–GTP exchange can proceed in a fraction of a second with the surplus GTP displacing the GEF from the G-protein.

GEFs weaken the affinity of G-proteins to GDP/GTP by deforming the nucleotide binding site. This catalytic mechanism has been studied in detail for mSOS, a GEF of the small G-protein Ras. mSOS blocks the phosphate binding site in switch 2 of the G-protein and simultaneously pushes aside switch 1 by an extended hairpin loop, resulting in the release of GDP from the binding pocket. Since GEFs have different structures, this mechanism may be representative in principle but certainly not in detail. Thus the GEFs of the large trimeric G-proteins are receptors with seven transmembrane domains and are unable for steric reasons to activate the GDP–GTP exchange at the GTPase-active α-subunit. Instead the exchange reaction may be assisted by the β-subunit. For G-proteins GEFs are the major transducers of input signals. To this end they contain,

Figure 2.13 Catalytic effect of GDP–GTP exchange factor GEF (schematically) Though it is exergonic, the transition from G-GDP to G-GTP is kinetically blocked by the high activation energy. The latter is lowered by GEF through the formation of a transition complex with G, GDP, and GTP; that is, GEF has the characteristic catalytic effect of an enzyme.

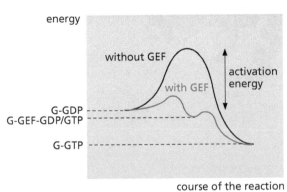

in addition to the catalytic domain, a wide variety of interaction domains that control substrate recognition, cellular localization (for instance membrane binding or transport into the nucleus) and cross talk with other signaling pathways.

2.4.4 GTPase-activating proteins: terminators of G-protein signaling

The GTPase reaction terminates G-protein signaling. It critically depends on a basic amino acid residue (mostly Arg) in the catalytic center of the G-protein, neutralizing the negative charge of the phosphate residue and thus facilitating the nucleophilic attack of water on GTP. In addition, this residue brings the water-binding Thr or Gln residue in position for the hydrolytic reaction. Small G-proteins lack the critical Arg residue; therefore, their G-domain is incomplete and the GTPase activity is very low. The lacking Arg residue is supplied by another protein called GTPase-activating protein (GAP), where it is localized at the tip of a loop structure that, as an "arginine finger," thrusts into the catalytic center of the G-protein (Figure 2.14). Strictly speaking, small G-proteins are heterodimeric GTPases consisting of a catalytic subunit and a regulatory subunit, (GAP). In addition to GEFs, GAPs are targets of input signals controlling G-protein activity.

A reliable rule of biology is that there is no rule without exception. In fact, the GAPs of the small G-proteins of the Ran and Rap families supply an asparagine instead of an arginine residue to stabilize the correct orientation of water and GTP in the catalytic center of the G-protein.

In contrast to the small G-proteins, the heterotrimeric G-proteins have a complete G-domain. Due to several inserts, in particular a bundle of six helices containing the critical Arg residue (see Figure 2.14), this G-domain is about three times as large as that of a small G-protein. As a consequence, the α-subunits of large G-proteins exhibit up to 10^4-fold higher GTPase activity than small G-proteins. Notwithstanding, nature did not miss the chance to use the GTPase reaction as a terminal for input signals. In fact, for many large G-proteins, so-called regulators of G-protein signaling (RGS) have been found that stimulate the GTPase reaction in a signal-dependent manner, though by a mechanism differing from that of GAPs (details in Section 4.3.2).

Summary

G-proteins are GTPases characterized by a unique G-domain in the catalytic center. They function as molecular switches: binding and hydrolysis of GTP results in a conformational change that is transferred to interacting effector proteins. As a result, the activity of the effector protein is changed or the formation of signaling multiprotein complexes is triggered. There are two types of G-proteins: monomeric (small) and heterotrimeric (large). For both types, GTP loading is catalyzed by enzymes called GDP–GTP exchange factors (GEFs). While large G-proteins exhibit intrinsic GTPase activity, small G-proteins require interaction with regulatory subunits called GTPase-activating proteins (GAPs). The GTPase activity of many large G-proteins is also controlled by accessory proteins. These as well as GEFs and GAPs are receptors of input signals.

2.5 ATPase switches

Similar to G-proteins, two families of ATPases are specialized to induce conformational changes in other proteins and to regulate the association and dissociation of (signal-transducing) protein complexes. These families include the chaperones and the AAA+ proteins. Both types are responsible for protein quality control and for targeted cell physiological processes. Other ATPases work as membrane transporters and nanotechnical motor proteins, causing mechanical alterations of the cell structure.

large G-protein α-subunit

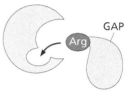

small G-protein

Figure 2.14 Mechanism of action of GTPase-activating protein GAP
GAP functions as a subunit of small G-proteins in that it provides an Arg residue required for the GTPase reaction. In the G-domain of the α-subunits of large G-proteins, this Arg residue is already present.

2.5.1 Molecular chaperones: how to make proteins fit for work

Molecular chaperones represent a large group of mostly unrelated proteins that assist with protein folding and protein complex assembly. They are essential for the maintenance of the highly ordered state of a living cell. Current theories of protein folding start from the assumption that the *native* state of a protein represents the most favorable thermodynamic conformation of the polypeptide chain, which arranges spontaneously from the genetically determined primary structure (Anfinsen theory; the 1972 Nobel Prize in Chemistry awarded to Christian B. Anfinsen, Stanford Moore, and William H. Stein). In an aqueous environment, the energy for the native folding stems from the tendency of proteins to hide the hydrophobic side chains of the amino acids in their interior; that is, to adopt a low-energy conformation. However, the native state is by no means always identical with the *functional* state. Moreover, a spontaneous folding into the native conformation has been observed only for small proteins under nonphysiological conditions. Larger proteins tend to form stable, incompletely folded intermediates during polypeptide biosynthesis at the ribosome and to nonspecifically aggregate because of the very high protein concentration of 300–400 mg/ml (called macromolecular crowding), thus rendering the cytosol a semi-solid medium. Moreover, a wide variety of stress factors including elevated temperature and starvation cause protein denaturation and misassembly (for this reason some chaperones are known as heat-shock proteins). Disturbances of correct protein folding and assembly impair cellular functions and may have severe cytotoxic effects with fatal consequences (for example, Alzheimer's dementia; see Section 13.1.3) Therefore, it is essential to prevent such accidents. This is the task of molecular chaperones.

Since chaperones generate order, their effects depend on energy-supplying reactions, primarily ATP hydrolysis. To this end they either have intrinsic ATPase activity or couple with ATPases. [Here a narrow definition of chaperones as proteins that modify protein conformations in a noncovalent ATP-consuming reaction is preferred. Other authors also consider enzymes that act on the tertiary structure of proteins as chaperones; for example, protein disulfide-isomerases and peptidyl-prolyl *cis/trans* isomerases, which catalyze covalent modifications and do not exhibit ATPase activity.]

Two types of heat-shock proteins, **HSP70** and **HSP90**, belong to the basic equipment of all cells and rank among the most abundant cytosolic proteins (though variants are also found in mitochondria and in the endoplasmic reticulum). They are expressed in numerous isoforms and are essential for survival. These chaperones cooperate with each other and not only protect proteins from stress but also have important additional functions. The HSP70 chaperone (found in both pro- and eukaryotes) promotes the folding of growing polypeptide chains. In conjunction with HSP90, it plays a central role in cellular signal processing by inducing the formation of functional protein conformations that are thermodynamically unstable. Examples include the activation of steroid hormone receptors (this topic is discussed in more detail in Section 8.5) and the activation loop phosphorylation of protein kinases (Section 2.6.1). As is typical for most signaling proteins, the functional forms of HSP70 and 90 are dimers. The function of the chaperones culminates in the **chaperonins.** These are large cylindrical complexes made up by stacked rings of chaperone aggregates. A hole inside the complex called an "Anfinsen cage" protects the ATP-consuming folding process of the substrate or client protein from disturbing influences.

Chaperones recognize their client proteins by linear hydrophobic amino acid sequences that are typically exposed by immature or denatured proteins. For complete folding, the client protein undergoes several cycles of binding to and dissociation from the chaperone driven by ATP hydrolysis. Like G-proteins, chaperones may be looked upon as binary switches (Figure 2.15). In state 0 they are loaded with ADP. The switching event $0 \rightarrow 1$ involves the exchange of ADP for

(A)

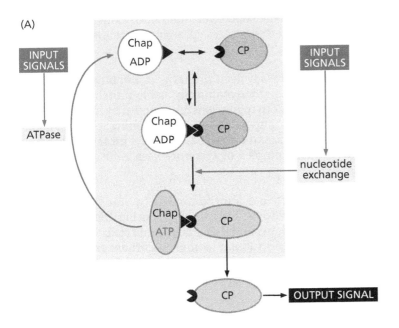

(B)

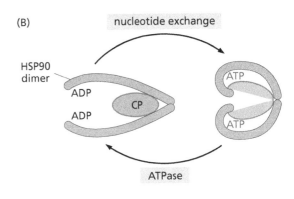

Figure 2.15 Chaperone switch (A) The ADP-loaded chaperone (Chap) interacts noncovalently with a client protein (CP). Following ADP–ATP exchange, the chaperone changes its conformation and induces a conformational change of the client protein. Upon ATP hydrolysis, the complex dissociates and the chaperone returns into the ADP-loaded conformation. Input signals affect both the nucleotide exchange and the ATPase reaction. The altered conformation (and function) of the client protein resembles the output signal. It should be remembered that not all proteins classified as chaperones are ATPases. (B) "Nutcracker" model: the chaperone dimer (in this example, heat shock protein HSP90) forms a clamp that, upon nucleotide exchange, changes into a more closed conformation, breaking up the conformation of the client protein.

ATP and the binding of the client protein. The latter activates the chaperone's ATPase function, thus releasing the energy required for the conformational change of the client protein. During step $1 \rightarrow 0$ the modified client protein dissociates from the chaperone, which returns to the initial state. Although various models have been developed (Figure 2.15), the precise mechanism of chaperone action still remains a mystery.

The output signal of the chaperone switch is the altered function of the client protein: for example, enzymatic activity or the tendency to correctly interact with signaling ligands. When the ligands are other proteins, the chaperone promotes the formation of functional complexes. In fact, the generation of multi-protein structures such as chaperonins, nucleosomes, and proteasomes is controlled by a series of highly specific chaperones.

Both the ADP–ATP exchange and the ATPase reaction are controlled by regulatory proteins or **co-chaperones** interacting with chaperones via specific domains. The ATPase activity of HSP70 chaperones is, for instance, stimulated by proteins called HSP40 and proteins with a J-domain (named after a bacterial co-chaperone DnaJ), while the ATPase activity of HSP90 is negatively regulated by Cdc37-type proteins (the name refers to a yeast protein involved in cell cycle control). Cdc37 and its homologs are multifunctional scaffolds that bring HSP90 in contact with a wide variety of client proteins.

The ADP–ATP exchange is catalyzed by a wide variety of nucleotide exchange factors that are under the control of signaling pathways. Moreover, it is inhibited by a HSP70-interacting protein (HIP), while an adaptor protein (HOP) brings about the contact between HSP70 and HSP90. Via such accessory proteins, input signals are transferred to the chaperone switch. This multi-stage regulation is strongly reminiscent of the control of G-protein function by GEFs and GAPs.

2.5.2 AAA+ proteins: formation and disruption of macromolecular complexes

AAA+ proteins (frequently referred to as triple-A plus proteins) are <u>A</u>TPases <u>A</u>ssociated with various cellular <u>A</u>ctivities. As indicated by the name, they constitute an extremely diverse family of pro- and eukaryotic proteins. The only structural feature these proteins have in common is an ATPase domain of about 240 amino acids called the AAA motif. This module contains AAA-specific sequences in addition to the so-called Walker A and B motifs that are typical for most

ATPases. As far as their functions are concerned, fluid transitions exist between AAA+ proteins and chaperones, and some AAA+ proteins are counted among the heat-shock proteins.

As "unfoldases," AAA+ proteins are involved primarily in the unfolding of proteins and in the dissociation of thermodynamically stable protein–protein and protein–nucleic acid complexes. Both require energy supplied by ATP hydrolysis. Therefore, AAA+ proteins play an important role in promoting the reversibility and repetition of cellular processes. An example is the AAA+ protein **NSF**, which dissociates the SNARE protein complex of secretory vesicle fusion, making it reusable (Section 10.4.2).

Other types of membrane fusion are also controlled by AAA+ proteins, for instance, in the course of peroxisome formation or during the post-mitotic reconstruction of the Golgi apparatus and the endoplasmic reticulum. An AAA+ protein, called **katanin**, dissolves tubulin complexes, thus depolymerizing microtubules. Katanin is concentrated in centrosomes, playing an important role in the separation of chromosomes during cell division.

Formation of the initiation complex of DNA replication is regulated by no less than 10 different AAA+ proteins, which play a similar role as G-proteins in mRNA translation (see Chapter 8). A major player in this process is the DNA helicase MCM that winds up the DNA double strand. Replication factor C, another hetero-oligomeric AAA+ protein, forms a complex with proliferating cell nuclear antigen (PCNA, an experimentally used indicator of cell division) and DNA polymerase δ at the replication fork. This complex displaces the polymerase α/primase complex, thus starting the elongation of the new DNA strand (for details on DNA replication, the reader is referred to textbooks of molecular biology).

An important cellular function is attributed to the **AAA+ proteases**. These enzymes contain, in addition to the AAA+ motif, a protease catalytic domain, located either on the same or on a separate polypeptide chain. Highly organized AAA+ proteases are the **proteasomes** and their prokaryotic predecessors. Proteasomes are barrel-like nanomachines specialized for protein degradation that takes place in a cavity inside the structure. They cooperate with chaperones in removing damaged proteins that can no longer be renatured and are potentially harmful to the cell. Moreover, they play a key role in signal transduction in that they terminate or promote signaling by eliminating stimulatory or inhibitory signaling proteins. Proteasomes consist of a catalytic 20S subunit enclosed by two regulatory 19S subunits. Both subunit types are multiprotein complexes exhibiting a mostly hexagonal structure with a central pore (for a schematic representation, see Figure 2.31). The AAA+ activity is located in the regulatory subunits that deliver the unfolded protein to the protease in the central cavity of the 20S subunit. Unfolding is required for threading the substrate protein into the narrow proteasome pore and facilitating proteolysis. Since the folded native protein represents a thermodynamically favored state, unfolding requires ATP energy.

The catalytic domains of the AAA+ unfoldase seem to be mobile polypeptide loops facing the central pore. By hydrolyzing one ATP molecule, they shift from an UP position to a DOWN position, thus working like a crowbar that breaks up the intramolecular interactions of the substrate polypeptide chain. Running such a machine is an expensive undertaking for the cell: depending on the stability of its conformation, the complete unfolding of an average protein may cost more than 500 ATP molecules! Nevertheless, the elimination of potentially harmful products as well as the precise termination of signaling seems to be worth the price.

For a protein to be degraded in proteasomes, it must be labeled by a polyubiquitin tag interacting with specific binding sites at the surface of the 19S proteasomal subunits. This tagging, catalyzed by E3 ubiquitin ligases, is the major target of signals controlling protein degradation (Section 2.8).

Proteasomes are molecular machines running on six (or more) cylinders, and the same holds true for most other AAA+ proteins catalyzing the unfolding and unwinding of polypeptide and DNA chains and the dissociation of complexes. Mechanical force is generated by ATP hydrolysis, which leads to the destruction of intramolecular interactions and drives motion. In fact, a particularly clear analogy to nanotechnology is provided by the AAA+ proteins of the **dynein** sub-family. These are motor proteins that under ATP consumption move along microtubule "tracks" of the cytoskeleton, transporting cargo such as organelles (see Section 10.3.5) and, during cell division, chromosomes. Dyneins are also components of eukaryotic flagellae and ciliae.

Like G-proteins and chaperones, the AAA+ proteins may be looked upon as binary switches with the effect on the protein or DNA substrate being the output signal (Figure 2.16). Again the input signals most probably control the ADP–ATP exchange as well as the ATPase reaction. How this occurs and which accessory proteins are involved is still widely unknown.

2.5.3 Motor proteins and membrane transporters: generation of motion

The dyneins are an example of motor proteins. Generally speaking, motor proteins are ATPases transforming the chemical energy derived from ATP hydrolysis into motion. By this means they change the cellular architecture and shape as well as the intracellular distribution of organelles and macromolecules and generate cell movements. To this end they interact with the actin and microtubule cytoskeleton that support the cell and serve as tracks for material to be transported.

During ATP hydrolysis, a motor protein undergoes a reversible conformational change, thereby bending and transferring mechanical energy to the cytoskeletal fiber. These single steps add up to a movement that results either in a shift of the interacting fiber or in a wandering of the motor protein along the fiber (for the purpose of transporting cargo) or in a pressure or tension between both partners (Figure 2.17). In addition to the dyneins, ATPases acting as motor proteins include the kinesins and the myosins. While kinesins, like dyneins, interact with microtubules, myosins prefer actin fibers. Motor proteins are major targets of signaling pathways.

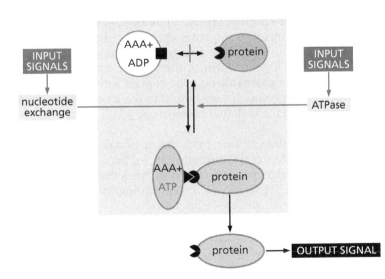

Figure 2.16 AAA+ protein switch
The switch widely resembles the GTPase switch (Figure 2.9). Again the altered conformation of the effector protein resembles the output signal.

Figure 2.17 Modes of operation of motor proteins (A) The motor protein (kinesin, myosin, or dynein) is anchored to a larger cell structure and moves the interacting cytoskeletal filament. (B) The cytoskeletal filament is anchored: the motor protein moves along the filament and may transport cargo. This mode of operation is typical for the interaction of kinesins and dyneins with microtubules. When both the motor protein and filament are anchored, a tension develops, resulting in either muscle contraction by an actin–myosin interaction or beating of flagella by a microtubule–dynein interaction. The gear-wheel drive serves as an illustration; eukaryotic motor proteins do not contain rotating structural elements but perform kinking and step-by-step movements (Figure 2.18).

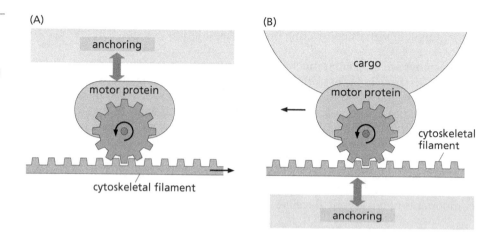

Kinesins

For the relatively large eukaryotic cell, an ATP-driven transport of cargo-filled vesicles and of macromolecular complexes is essential. The alternative would be free diffusion, which is neither targeted nor nearly fast enough when one considers, for instance, the enormous distances in nerve fibers. Therefore, the transport is carried out by kinesins, which are specialized in migrating along microtubule tracks and carrying cargo. For this purpose they hydrolyze one molecule of ATP per step; that is, the costs of intracellular transport are high. Kinesins play an important role in cellular data processing because they rearrange complexes of signaling proteins, thus adapting the "hardware construction" to the current demands. In addition, kinesins are components of the mitotic spindle apparatus.

Kinesins constitute a large and heterogeneous protein family including more than 100 members. They have in common an ATPase or motor domain in which the P-loops and switch regions 1 and 2 closely resemble the active centers of G-proteins and of myosins. These three protein families are assumed to have developed from an ancestral NTPase, with kinesins and myosins becoming molecular motors and G-proteins becoming molecular switches. In contrast, dyneins, like AAA+ proteins, belong to another phylogenetic tree.

Figure 2.18 depicts the basic structure of a conventional kinesin.* It is a tetrameric protein consisting of two heavy and two light chains each. A characteristic feature of the heavy chain is the N-terminal globular motor domain containing the binding site for microtubules, and a C-terminal globular tail domain with the binding sites for the light chain and other adaptor proteins. The central part of the molecule is an extended helical dimerization domain forming a coiled-coil structure with a partner protein of its kind. The helical regions of this domain are interrupted by hinge sequences providing the conformational elasticity required for the movement but also forming a rebend hairpin loop that structurally locks in the inactive form of the kinesin. The light kinesin chains contain interaction domains establishing the contact with the cargo, thus acting as adaptor proteins. Cargo vesicles are bound to the kinesin by so-called cargo receptors integrated into the vesicular membrane.

The stepwise moving pattern of kinesins is due to two reactions resembling the switching reactions of AAA+ and G-proteins: the exchange of ADP for ATP leads to a conformational change that is reversed by the subsequent ATP hydrolysis (Figure 2.18). Both these events as well as the loading with cargo are controlled

---

*In addition to these conventional kinesins, a nonconventional type exists that is characterized by the motor domain being placed either C-terminally or in the middle of the polypeptide chain. On microtubule tracks the nonconventional kinesins migrate opposite to the conventional kinesins, that is, toward the microtubule-organizing center (for details see textbooks of cell biology).

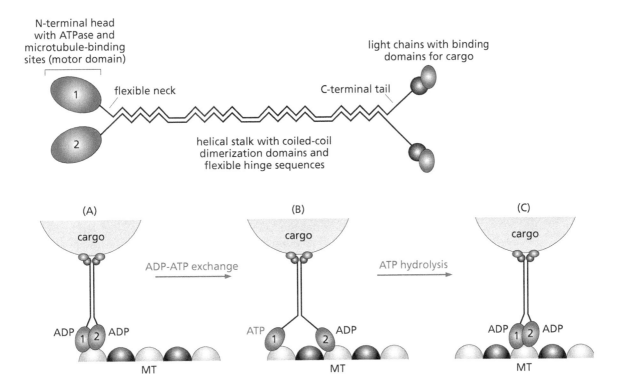

by input signals; however, these are still fairly unknown. An example of micro-tubule-directed cargo transport controlled by kinesin and dynamin phosphorylation is discussed in Section 10.3.5.

Myosins

Myosins constitute a protein family that in size and heterogeneity is comparable to the kinesin family. Together with their cytoskeletal partner F-actin they form contractile fibers enabling various types of movements such as muscle contraction, separation of cells during cytokinesis, and membrane extrusions and invaginations (for instance, amoeboid crawling, phago- and pinocytosis, and formation of microvilli). Whether or not myosins are also involved in cargo transport is an open question. In contrast to the kinesins they are unable to perform stepwise movements, although both types of motor proteins are structurally related, forming tetramers of two light chains and two heavy chains each. The heavy chains containing the ATPase- and actin-binding domains in the N-terminal region dimerize by an interaction of extended coiled-coil structures, whereas the myosin light chains have an autoregulatory function (see below).

Like G-proteins and kinesins, myosins undergo a two-step switching reaction: there is an exchange of ADP–ATP, which induces the conformational change required for the movement, followed by ATP hydrolysis, which returns the protein to the initial state. Both steps are stimulated by actin and blocked by regulatory proteins, the activities of which are controlled by input signals such as protein phosphorylation and Ca^{2+} ions (for details see Section 14.6.2). These regulatory proteins include the myosin light chains and, in the case of the conventional muscle myosin, tropomyosin, both of which hinder the interaction between myosin and F-actin.

In addition to the muscle myosins, a large number of nonconventional myosins have been found. They are regulated by light chains and also have numerous domains for interactions with other proteins and membrane constituents, so they are firmly linked into the signal processing protein network. The mechanisms under which these interactions work are unknown.

Figure 2.18 Basic structure and mode of movement of a kinesin tetramer The upper panel shows the basic structure of kinesin in a schematic manner. The motor protein consists of two heavy chains, held together by coiled-coil dimerization domains, and two light chains (red) noncovalently bound to the C-terminals of the heavy chains. In the lower panel the mode of movement is depicted schematically. (A) With its ADP-loaded head 1, the kinesin contacts a tubulin subunit of a microtubule MT. (B) Exchange of ADP for ATP causes a conformational change resulting in a spreading of the two heads at the neck region. This allows contact between head 2 and the next binding site on the microtubule, resembling a single step forward. (C) Subsequently ATP becomes hydrolyzed and head 1 catches up with head 2. For further details, see text.

Membrane transporters

Another large family of regulatory ATPases is specialized for transporting substances across membranes, mostly against a concentration gradient. These proteins are characterized by an A̲TP-B̲inding C̲assette and are called **ABC transporters**. Switching between an ATP- and an ADP-bound state, they mainly promote the excretion of molecules trapped in a conformational cage by the cell. ABC transporters are discussed in more detail in Section 3.2.1.

Summary

Various types of ATPases are involved in cellular signal processing. Chaperones help other proteins to acquire and maintain their functional conformations and to assemble functional complexes, in particular during *de novo* synthesis and signal transduction. In stress situations they may protect proteins from denaturation. Most chaperones are ATPases and use the energy derived from ATP hydrolysis for conformational changes. Some chaperones such as the heat-shock protein HSP90 play a particular role in inducing and stabilizing the active conformations of signaling proteins. Like G-proteins, chaperones are associated with regulatory proteins, modulating both ADP–ATP exchange and ATPase activity. Such proteins are receivers of input signals. AAA+ proteins are ATPases characterized by a unique ATP-binding motif. They are found in both pro- and eukaryotes, where they work as molecular machines. Their major functions include unfolding of proteins for degradation in proteasomes, disassembly of (signal-transducing) protein complexes, protein disaggregation, cellular movements, and modulation of cellular signal transduction. Prominent examples of AAA$^+$ proteins are proteasomes and dyneins. Motor proteins are a large and heterogeneous family of ATPases specialized for transforming, by conformational changes, the chemical energy derived from ATP hydrolysis into mechanical energy. By interacting with cytoskeletal anchor proteins they are able to move, thereby transporting cargo including signaling protein complexes, and to generate mechanical tension, resulting in changes of cell shape and cellular movements.

2.6 Protein phosphorylation

Reversible phosphorylation, the most abundant post-translational modification of protein structure, has been estimated to affect at least a third of all cellular proteins. Phosphorylation also represents the major switching reaction of cellular data processing, with step 0 → 1 catalyzed by protein kinases and step 1 → 0 catalyzed by phosphoprotein phosphatases (Figure 2.19). There are three phosphate donors and energy suppliers: ATP, GTP, and phosphoenolpyruvate, the latter used mainly by prokaryotes (Section 3.2.2). Targets of phosphorylation are the amino acid residues Ser, Thr, and Tyr (which deliver phosphoric acid esters), His (which delivers an energy-rich, unstable phosphoric acid amide), Asp (which

Figure 2.19 Protein phosphorylation switch The switch consists of a protein kinase (PK), a protein phosphatase, and a substrate protein. In the absence of ATP, the protein kinase is inactive (in analogy to other switching elements here it is formally shown as ADP-loaded). The active kinase transfers a phosphate residue (P) from ATP to the substrate protein, the conformation and function of which are thus changed, resembling the output signal. A phosphoprotein phosphatase sets the switch back to 0. Input signals control the activation of the kinase (summarized in Figure 2.26) and of the phosphatase.

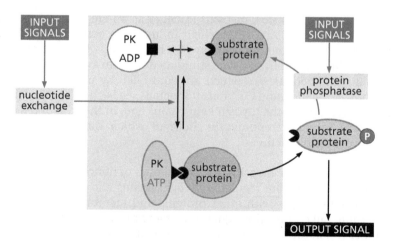

delivers a mixed acid anhydride), and Cys (which delivers a labile phosphoric acid thioester) (see Figure 2.20). Phosphorylation of Lys, Arg, and Glu residues may also be possible, but convincing proof of these processes is still lacking. Protein kinases do not phosphorylate free amino acids. Instead, the phosphorylatable amino acid residue (the phosphorylation site) has to be embedded in a short amino acid sequence of a protein. This so-called **consensus phosphorylation sequence** is recognized specifically by an individual protein kinase type. By convention the amino acid residues that are situated C-terminally of the phosphorylation site are numbered with a positive prefix and the residues that are situated N-terminally are numbered with a negative prefix. Depending on whether a consensus phosphorylation sequence is found only in a few or in many proteins, protein phosphorylation may be either a strongly selective event with a high degree of "privacy" or a more "public" reaction addressing a large number of different targets. However, even those protein kinases that are not highly selective because their cognate consensus phosphorylation sequence is abundant among proteins are able to select individual substrates by interacting with additional binding or docking sites, targeting subunits and scaffold proteins (see Sections 4.6. and 11.2 for more details).

The mechanisms of action of protein kinases and phosphatases as well as the impact of phosphorylation on protein structure are discussed in more detail below. Individual phosphorylations and questions regarding the evolution of protein phosphorylation are treated later.

2.6.1 Protein kinases: workhorses of signal transduction

Protein kinases constitute one of the largest enzyme families known. In the human genome, 518 kinase genes (1.7% of all genes, third-largest gene family) and more than 100 inactive pseudogenes have been found. The corresponding numbers for other species are 130 (2.1%, largest gene family) for the yeast *Saccharomyces cerevisiae*, 454 (2.4%, second-largest gene family) for the fly *Drosophila melanogaster*, and 1049 (4.1%, largest gene family) for the herb *Arabidopsis thaliana*. Also in prokaryotes, protein kinase genes cover a significant percentage of the genome.

There are two large kinase families: the mostly prokaryotic **histidine autokinases** and the so-called **eukaryotic protein kinases**. Eukaryotic kinases, which are also found in prokaryotes, catalyze the phosphorylation of either Ser and Thr

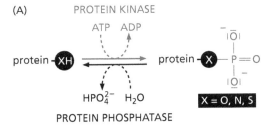

(A)

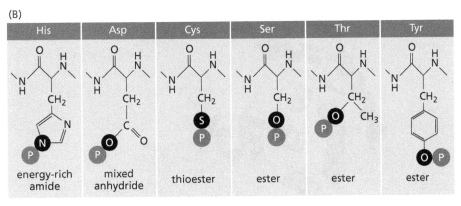

(B)

Figure 2.20 Biochemistry of protein phosphorylation (A) Reversible phosphorylation of a group XH on a protein as catalyzed by a protein kinase and a protein phosphatase. (B) Amino acid residues shown to become phosphorylated (black circles show the phosphate acceptor site X; red circles symbolize the phosphoryl group PO_3^{2-}, somewhat imprecisely biochemists call the transfer of this group between molecules "phospho=transfer reaction").

or of Tyr residues, or in a very few cases of all three amino acids. They are not related to the histidine autokinases. In addition, several minor families of protein kinases are known that have been called **atypical** because in their structures they differ considerably from the two major families.

Histidine autokinases: structural elements

The catalytic core of a histidine autokinase covers approximately 160 amino acids containing several conserved sequences that are characteristic for this kinase family. They form a binding site for the ATP–Mg^{2+}–substrate complex. This site is characterized by a typical structural element, the so-called Bergerat fold, and harbors the amino acid residues required for the phosphotransfer reaction. The conserved sequences are constituents of an α,β sandwich consisting of five planarly arranged anti-parallel β-sheets superimposed by a layer of three α-helices (Figure 2.21). This molecular architecture distinguishes the histidine autokinases from other protein kinases, relating them to chaperone HSP90, type 2 topoisomerase/gyrase (catalyzing the formation of superhelices from circular double-stranded DNA), and MutL recombinase (an enzyme for DNA repair). Together these enzymes constitute the **GHKL family** (from Gyrase, HSP90, Kinase, and MutL). They all generate states of higher order, thereby hydrolyzing ATP.

A special type of His autokinases is the EI enzymes of the bacterial phosphotransferase systems, which use phosphoenolpyruvate instead of ATP as a phosphate donor. The prokaryotic protein kinases are discussed in detail in Section 3.2.2.

Eukaryotic protein kinases: structural elements

Eukaryotic protein kinases are by far the largest and probably the most ancient kinase family. Their catalytic domain, covering about 250 amino acids, has another basic structure than that of His autokinases. Characteristic building blocks are 12 highly conserved subdomains, allowing an identification of the corresponding genes. These subdomains contain several invariant amino acid residues that are essential for the binding of the ATP–Mg^{2+} (or Mn^{2+}) complex and of the substrate protein, the phosphotransfer reaction, and the regulation of enzymatic activity. In the 1990s the three-dimensional structure was first elucidated by X-ray crystallography for cyclic adenosine 3′,5′-monophosphate (cAMP) -dependent protein kinase A (PKA). Further studies on numerous other kinases proved this structure to be representative for eukaryotic protein kinases (Figure 2.21). The predominant structural features are two lobes connected by a hingelike sequence, subdomain V, and forming the catalytic cleft. This structure opens and closes like a mouth with the ATP binding site localized inside and the binding site for the substrate protein outside the oral aperture.

Figure 2.21 Catalytic cores of protein kinases Shown are the catalytic centers of (A) a bacterial His autokinase (EnvZ of *E. coli*) and (B) a eukaryotic protein kinase (cAMP-dependent protein kinase of mice).

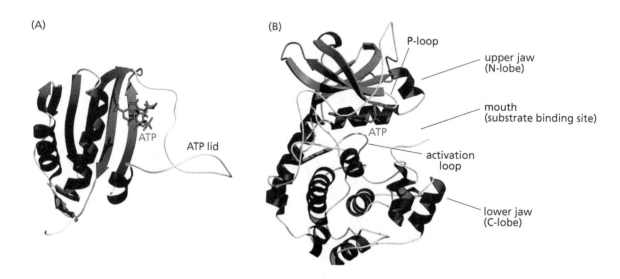

The N-terminal "upper jaw" (N-lobe) with subdomains I–IV is composed of five anti-parallel β-sheets and harbors the ATP binding site, whereas the larger C-terminal "lower jaw" (C-lobe) with subdomains VIA–XI consists mainly of α-helices that are responsible for binding the substrate protein and for the phosphotransfer reaction. Like GTP in G-proteins, ATP is fixed by hydrogen bridges and electrostatic bonds between its nontransferable α- and β-phosphate groups and corresponding amino acids (in subdomains I–IV) as well as by a hydrophobic pocket (in subdomain I) fitting the adenine residue. Subdomain I contains a glycine-rich sequence that is essential for binding the phosphoryl residues. In analogy to G-proteins and other phosphate-transferring enzymes, this sequence is called the P-loop.

In subdomain VIB (in the "lower jaw"), a catalytic loop structure is localized that is characterized by two invariant amino acid residues. One of them is an Asp residue essential for the phosphotransfer reaction; the other is a basic residue (Lys or Arg) neutralizing the negative charge of the γ-phosphate group of ATP, thus facilitating nucleophilic attack by the substrate protein. For the phosphotransfer reaction, the triphosphate group of ATP must be arranged linearly. This is brought about by a Mg^{2+} ion bound as a salt by acidic amino acid residues in subdomain VII. Subdomain VIII stabilizes the architecture of the "lower jaw" by means of a highly conserved Ala-Pro-Glu sequence. It also harbors a binding site for the substrate protein as well as an **activation loop** (also known as **T-loop** because it contains a strategic Thr residue).

The three-dimensional structure of the catalytic cleft is common to all eukaryotic protein kinases. The specificity for selective phosphorylation consensus sequences is due to subtle variations in the amino acid sequences of the substrate binding sites resulting in altered charge and hydrophobicity. Tyr-specific kinases differ from Ser/Thr-specific enzymes by a deeper catalytic cleft that is able to accommodate the more voluminous aromatic side chain of the tyrosine residue.

Different mechanisms of action

The different structures of His autokinases and eukaryotic protein kinases are mirrored by two fundamentally different mechanisms of action (Figure 2.22). Upon activation by an input signal, His autokinases autophosphorylate each other in a process called trans-autophosphorylation. The phosphorylation site (H box) is localized *outside* the catalytic center in a dimerization domain consisting

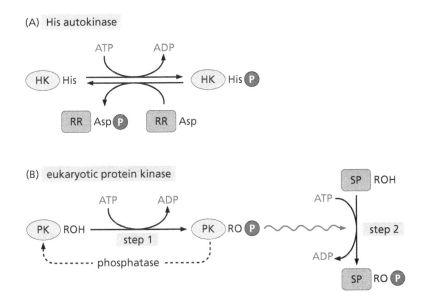

Figure 2.22 Different mechanisms of action for His autokinases and eukaryotic protein kinases (A) Upon autophosphorylation, His autokinases (HK) provide phosphate donors for an autophosphorylation of proteins called response regulators (RR) at Asp residues. (B) Eukaryotic protein kinases (PK) catalyze the phosphorylation of substrate proteins (SP) by ATP (step 2). To acquire full enzymatic activity, many of them have to undergo autophosphorylation first (or become phosphorylated by another kinase, step 1). However, in contrast to His autokinases, phosphorylation occurs at hydroxy-amino acids such as Ser, Thr, and Tyr (ROH). These phosphoryl groups (P) cannot be transferred to other proteins but instead change the conformation of the kinase, facilitating interaction with ATP and the substrate protein SP. To terminate eukaryotic protein phosphorylation, a protein phosphatase is needed to inactivate the (auto)phosphorylated kinase.

of two anti-parallel helices, known as the HPt domain (for details see Section 3.3). The amide bond of the phosphorylated histidine residue is unstable, that is, energy-rich. Therefore, in a kind of ping-pong mechanism, the phosphoryl group is immediately transferred to other proteins that are called response regulators. These are protein autokinases that use the autophosphorylated His kinases as phosphate donors for an autophosphorylation of aspartate residues. Thus, His autophosphorylation resembles a *short-lived transition state* of a phosphotransfer reaction between two proteins.

Eukaryotic protein kinases catalyze the phosphorylation of other proteins, though most of them also undergo (trans-)autophosphorylation of the activation loop *within* the catalytic center. This regulatory autophosphorylation is not a short-lived transition state of a phosphotransfer reaction; rather, it is a *stable post-translational modification* activating the enzyme by an allosteric conformational change.

In many protein kinases the activation loop closes the oral aperture and hinders contact between ATP and substrate protein (see Figure 2.21). When the kinase receives an input signal, the loop turns aside, allowing access to the catalytic center. For more sustained activation the loop has to be phosphorylated at one strategic Thr and/or Tyr residue. This is accomplished by either autophosphorylation or by another protein kinase (Figures 2.22 and 2.23). Most—but not all—kinases require such **activation loop phosphorylation** to become fully active.

Since most kinases are intrasterically blocked by inhibitory motifs, activation loop phosphorylation occurs only in the presence of input signals that release the enzyme from the intramolecular restraints and open the catalytic cleft between N- and C-lobe. But even then in many kinases the activation loop may not be accessible for phosphorylation but must be uncovered by an additional conformational change induced by a chaperone ATPase such as HSP90. In addition, this chaperone protects protein kinases from dephosphorylation, thus prolonging their active state.

Frequently the inhibitory motifs stabilizing the inactive kinase conformation are **pseudosubstrate domains.** These are polypeptide sequences that resemble the phosphorylation consensus sequence of a substrate protein and are bound by

Figure 2.23 Activation of protein kinases In an inactive protein kinase, the catalytic domain is blocked by an inhibitory peptide sequence. Interaction with an activator (a selection is shown in the pink box) induces a conformational change by which the inhibitory sequence becomes displaced and the catalytic domain is opened for the interaction with the substrate protein and ATP. To gain full activity, many protein kinases require, in addition, a phosphorylation of the activation loop in the catalytic center, which frequently requires the assistance of a chaperone such as HSP90 (see Figures 2.21 and 2.22). Such protein kinases are logical gates with at least three signaling inputs.

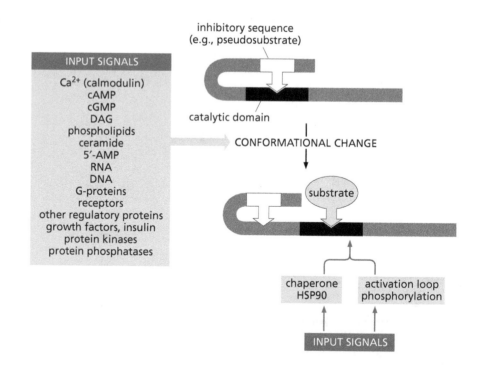

INPUT SIGNALS

Ca²⁺ (calmodulin)
cAMP
cGMP
DAG
phospholipids
ceramide
5′-AMP
RNA
DNA
G-proteins
receptors
other regulatory proteins
growth factors, insulin
protein kinases
protein phosphatases

inhibitory sequence
(e.g., pseudosubstrate)

catalytic domain

CONFORMATIONAL CHANGE

substrate

chaperone HSP90 — activation loop phosphorylation

INPUT SIGNALS

the catalytic center. However, since they lack the phosphoacceptor amino acid (which is replaced by a nonphosphorylatable residue, for instance, Ser by Ala), these sequences are not phosphorylated but act as competitive inhibitors. The interaction of the kinase with an input signal induces a conformational change displacing the inhibitory domain from the catalytic center (Figure 2.23). The pseudosubstrate sequence may be either a part of the kinase molecule (as in the case of protein kinase C, Section 4.6.3) or localized on separate subunits (as in the case of protein kinase A, Section 4.6.1).

Apart from pseudosubstrate sequences, other inhibitory structures may bring about intrasteric auto-inhibition. An example is provided by domains containing a phosphorylated tyrosine residue interacting specifically with certain phospho-Tyr binding domains. Activation of this kinase type requires an enzymatic dephosphorylation of the critical Tyr residue or a competitive interaction with a protein having a phospho-Tyr binding site. In most of those kinases, an additional phosphorylation of the activation loop is mandatory (for example, see tyrosine kinase Src, Section 7.1.2).

Kinases that require both the displacement of an auto-inhibitory domain from the catalytic center and an activation loop phosphorylation to be turned on have, so to speak, a double safety lock and two (or three if they require the assistance of a chaperone) separate input terminals for signaling. Thus, they represent logical AND gates or coincidence detectors of at least two different input signals. In addition, many protein kinases contain further input terminals, giving them the properties of more complex logical switches or molecular nanoneurons. Due to such safety locks, most protein kinases are inactive in the absence of input signals. There are, however, exceptions to this rule: for instance, glycogen synthase kinase 3, which is constitutively active and becomes *inhibited* by input signals (see Section 9.4.4). For a binary switching event it is insignificant, however, whether the event occurs by inhibition or by activation. What matters is that the switch has been turned.

2.6.2 Effects of phosphorylation on proteins

We shall now turn to the question of how phosphorylation changes the function of a protein. Exhibiting a pK value around 7, the phosphate group carries two negative charges in neutral medium, and, in addition, is able to form hydrogen bonds via its four oxygen atoms. We have to expect, therefore, rather pronounced effects on protein structure. Apart from this, the negative charge of the phosphate group could, solely by **electrostatic repulsion**, hinder negatively charged substrates from approaching the active center of an enzyme. This rather infrequent mechanism has been found for the bacterial enzyme isocitrate dehydrogenase kinase, which inhibits isocitrate dehydrogenase by phosphorylating a Ser residue in the catalytic core. A similar inhibition could be achieved experimentally by replacing this Ser residue by a negatively charged Glu or Asp residue but not by neutral residues, demonstrating an electrostatic mechanism. Moreover, X-ray crystallography did not provide any indication for a conformational change.

As a rule, however, phosphorylation leads to considerable **changes of protein conformation** (though it must not be overlooked that only a few proteins have been studied thoroughly in this respect). When, for instance, muscle glycogen-phosphorylase, a homodimeric enzyme, is phosphorylated by phosphorylase kinase at a single Ser residue, a long-range allosteric rearrangement (extending over a distance of 35 Å) takes place, opening the active center for the substrate. In the homologous yeast enzyme an allosteric inhibitor, glucose 6-phosphate, is displaced by such a rearrangement.

Apart from such allosteric long-range effects, a phosphate group causes conformational changes in the immediate vicinity as well. As an example, we have already discussed the effect on protein kinases of activation loop phosphorylation. Here electrostatic effects between the phosphate group and negatively charged amino acid residues help to open the active center for the substrates. Indeed, a comparable activation is achieved when the phosphate-acceptor amino acid is replaced by a negatively charged Glu or Asp residue. Some kinases are already equipped with such residues at the critical position in the activation loop and do not require phosphorylation to be activated. Sometimes phosphorylation does not exhibit any measurable effect on protein function, an occurrence that is known as **silent phosphorylation**.

Phosphoamino acid-binding domains: controlling complex formation, degradation, and sequestration

Many proteins contain contact domains that recognize and bind to amino acid sequences with phosphorylated Ser, Thr, or Tyr residues. Therefore, another important function of protein phosphorylation is to generate the complementary domains, thus bringing about **reversible contacts** between proteins (or between domains within the same protein molecule). Such contacts are essential for the *ad hoc* assembly of signal-processing complexes and networks (for more details see Sections 2.13 and 7.1.2). Binding sites for sequences with phosphorylated Tyr residues are the SH2 and the PTB domains, whereas phospho-Ser residues are recognized by 14-3-3, WW, and MH2 domains, and phospho-Thr by FHA and WD40 domains. These domains are found in a wide variety of proteins that function in a vast range of cellular events.

Leaving aside the formation of signaling multiprotein complexes, among these events protein degradation and sequestration are of uppermost importance for signal transduction. In both processes, protein phosphorylation plays a key role. Proteolytic protein degradation may influence signaling in both a positive and a negative way depending on whether it leads to release of an active fragment from an inactive precursor protein or to complete destruction. A major pathway of proteolysis makes use of ubiquitylation for targeting the protein to proteasomes. As explained in Section 2.8, ubiquitylation is catalyzed by three enzymes, with the **ubiquitin ligases** (enzymes E3) lending substrate specificity. Many ubiquitin ligases recognize their substrate proteins only when these have become phosphorylated (Section 12.16). Therefore, protein phosphorylation is a common trigger for ubiquitylation and degradation.

The same holds true for protein sequestration. The key players in this game are the **14-3-3 proteins**, which recognize their partner proteins by specific amino acid sequences carrying phosphorylated Ser residues. Like protein degradation, sequestration may have both negative and positive effects on signaling because it either reserves the protein for a certain period of time or facilitates its integration into a signaling complex. The 14-3-3 proteins are treated in more detail in Section 10.1.4. The major consequences of protein phosphorylation are summarized in Figure 2.24.

2.6.3 Protein kinase families of vertebrates: an overview

On the basis of structural and functional similarities, the eukaryotic protein kinases are subdivided into several families that can be arranged in a phylogenetic

Figure 2.24 Consequences of protein phosphorylation

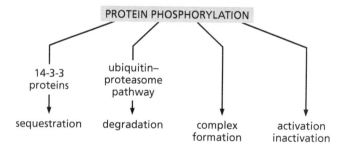

tree. The most abundant subfamilies of the 430 human Ser/Thr-specific protein kinases are discussed below.

AGC family

Because of strong homologies of their catalytic domains, 61 kinases are arranged in a group called the AGC family after their prototypical representatives: protein kinases A, G, and C. It includes the following enzymes:

- cAMP-activated protein kinases, PKA (Section 4.6.1)

- cyclic guanosine monophosphate (cGMP)-activated protein kinases G, PKG (Section 4.6.2)

- diacylglycerol- (and Ca^{2+}-) activated protein kinases cPKC and nPKC (Section 4.6.3)

- phosphoinositide-activated protein kinases PKB/Akt, PDK1, atypical PKC, and PKC-related PRK (Sections 4.6.3 and 4.6.6)

- Gβγ-activated protein kinases for G-protein-coupled receptors (GPCR kinases, see Sections 5.7 and 5.8)

- ribosomal S6 kinases S6K and RSK (Section 9.4.2)

As can be seen, many AGC kinases are activated by low molecular weight second messengers such as cAMP, cGMP, diacylglycerol, Ca^{2+} ions, and phosphoinositides as well as by Gβγ subunits of trimeric G-proteins.

Moreover, practically all of them require activation loop phosphorylation of a conserved Thr residue, which is catalyzed by the kinase itself (autophosphorylation) or by an upstream kinase, frequently the phosphoinositide-dependent kinase PDK1 (which activates itself by autophosphorylation). For maximal activation most AGC kinases have to be phosphorylated, in addition, on a Ser residue in a C-terminal hydrophobic sequence.

Ca²⁺/calmodulin-dependent kinases

For activation, these enzymes require Ca^{2+} ions that interact with a regulatory subunit, the Ca^{2+}-binding protein calmodulin (CaM). Among many other enzymes, the CaM kinase family includes the following:

- phosphorylase kinase of glycogen metabolism (Section 5.3.1)

- myosin light-chain kinase of the actomyosin system (Section 14.6.2)

- elongation factor 2 kinase (also known as CaM kinase III) of ribosomes (Section 9.3.4)

- multifunctional CaM kinases I, II, and IV (Section 14.6.3)

- distantly related CaM kinase kinases (Section 14.6.3)

The CaM kinase family also includes structurally and phylogenetically related kinases that do not contain calmodulin as a subunit. Examples include protein kinase D (Section 4.6.5) and 5′-AMP-activated protein kinase (Section 9.4.3). It should be noted that the conventional term CaM kinases is somewhat misleading because these enzymes do not phosphorylate their activator CaM.

CMGC kinases

These phylogenetically related protein kinases are proline-directed: they catalyze the phosphorylation of Ser and Thr residues in the neighborhood of Pro residues. The acronym CMGC is derived from prototypical members of this family:

- cyclin-dependent protein kinases CDK of cell cycle regulation (Section 12.4)
- MAP kinases (Section 11.2)
- glycogen synthase kinases GSK (Section 9.4.4)
- CDK-like kinases

Ste11 family

The kinases of this family are relatives of the yeast enzyme Ste11 (Section 11.4). Most of them are constituents of MAP kinase modules functioning as MAP2 and MAP3 kinases (Sections 11.2 and 11.3).

Further serine/threonine-specific kinase families of vertebrates

From the phylogenetic point of view, these enzymes cannot be classified as belonging to one of the major kinase families mentioned above. Among others, this group includes

- "casein kinases" CK1 and CK2 (Section 9.4.5)
- kinases WEE1 and MYT1 of cell cycle regulation (Section 12.4)
- kinases Nek/Nrk, Plk, and Aurora of mitosis control (Section 12.12)
- kinases IKK of the NFκB modules (Section 11.8)
- receptor kinases of the TGFβ/activin receptor family (Section 6.2)

The latter are found in a so-called Tyr kinase-like kinase family, the members of which are bona fide Ser/Thr kinases, albeit exhibiting a Tyr kinase-related structure. Other kinases belonging to this family are the MAP3 kinases Raf and MLK (Section 11.3), and the IRAKs (Section 6.3.2).

The **dual-specific Tyr/Thr kinases** of the MEK or MAP2K family, which are highly specific activators of MAP kinases, belong to the Ste11 family (Section 11.2). The 84 human **Tyr-specific protein kinases** are classified into cytoplasmic (soluble) and receptor tyrosine kinases. They will be discussed in detail in Sections 7.1 and 7.2. Some protein kinases are not related to the main group of eukaryotic Ser/Thr kinases and therefore are called **atypical kinases**. Representatives are, among others, the phosphatidylinositol 3-kinases (Section 4.4.3) and the related kinases mTOR (Section 9.4.1), the DNA-activated kinases DNA-PK and ATM/ATR (Section 12.9.1), and the His kinase-related pyruvate dehydrogenase kinases of mitochondria (Section 3.5.1).

2.6.4 Phosphoprotein phosphatases

Protein kinases switch their substrate proteins on or off by phosphorylation. To return the switch to the 0 position, the substrate protein must be dephosphorylated. This occurs by hydrolysis catalyzed by (phospho)protein phosphatases. Depending on structural parameters, the protein phosphatase family is classified in several nonrelated subfamilies:

- Ser/Thr-specific phosphoprotein phosphatases (PPP)
- Ser/Thr-specific phosphoprotein phosphatases requiring Mg^{2+} (PPM)
- conventional phosphotyrosine phosphatases (PTP)
- small phosphotyrosine phosphatases
- Thr/Tyr-specific Cdc25 phosphatases
- prokaryotic phosphohistidine and phosphoaspartate phosphatases
- prokaryotic protein kinases/phosphatases

PPP family

The enzymes of the PhosphoProtein Phosphatase family, including the sub-families PPP1, PP2A, PPP2B, and PPP5, have a catalytic core of about 280 amino acids that is characterized by three short sequences ranking among the most conserved structures of eukaryotic enzymes. The core consists of a β-sandwich formed by two planar β-sheets facing each other and being surrounded by several α-helices. In the upper part of the sandwich, two catalytically essential metal ions (either Fe^{3+}, Zn^{2+}, or Mn^{2+}) are bound to Asp and Glu residues. They polarize adjacent water molecules, thus activating them for a nucleophilic attack on the phosphoric ester bond of the substrate protein.

Except for the PPP5 subfamily, the PPP enzymes are composed according to a unit construction system in which a few catalytic subunits combine, mostly by noncovalent binding, with a wide variety of regulatory subunits targeting the enzymes to distinct cellular sites, inducing substrate specificity and modulating enzymatic activity (Figure 2.25). The eukaryotic phosphatase PP1 of the PPP1 subfamily has been estimated, for instance, to interact with no less than 50 different regulatory proteins. Prominent examples of regulatory subunits are those that render PP1 specific for either myosin light chains (Section 14.6.2) or, alternatively, enzymes of glycogen metabolism (Section 5.3.1). The subunit interaction is accomplished by a special interaction domain called RVxF motif.

From an evolutionary point of view, the PPP family is very old. Homologous enzymes are found in archaea and bacteria, perhaps playing a role in two-component signal transduction (Section 3.3). Interestingly, the prokaryotic enzymes also hydrolyze phospho-Tyr bonds, though these appear to be very rare in prokaryotes.

PPM family

These phosphoprotein phosphatases, for instance PP2C, need exogenous Mg^{2+} ions (therefore M). Although their amino acid sequences are quite different, the three-dimensional architecture of the catalytic core is similar to that of PPP enzymes. Obviously we are dealing with phylogenetic convergence. Whether or not the unit construction principle holds true also for PPM enzymes remains an open question. Like PPPs, PPMs also have a prokaryotic ancestry.

Tyrosine-directed and dual-specific protein phosphatases

Within this group, the **conventional phosphotyrosine phosphatases (PTPs)** constitute by far the most variable group of eukaryotic protein phosphatases (for more details see Section 7.4). In both structure and mechanism of action, they differ considerably from PPP and PPM enzymes. The catalytic core of PTPs covers 230 amino acids, forming a central group of seven anti-parallel β-sheets surrounded by α-helices. The phosphoric acid ester bond of the substrate protein is split by a nucleophilic attack of a cysteine residue, rather than by metal ion-activated water, and generates a labile phosphoric acid thioester as an intermediate that is hydrolyzed in a second step. The catalytically active cysteinyl–SH groups of these enzymes are oxidized quite easily to intramolecular or mixed disulfide or sulfenic acid groups, which are reduced again enzymatically. The activation of many PTPs is therefore assumed to be reversibly regulated by **redox signals** (Sections 2.2 and 7.1). Another important feature of PTPs is that, in contrast to the Ser/Thr-specific phosphatases, the regulatory domains are integrated into the enzyme molecule rather than being localized on separate subunits (Figure 2.25).

Some bacteria express conventional PTPs obtained from eukaryotes through horizontal gene transfer. These enzymes are potent virulence factors (Section 3.5.3). The **small PTPs** seem to be restricted mainly to eukaryotes. They have molecular masses around 18 kDa and are not related to conventional PTPs,

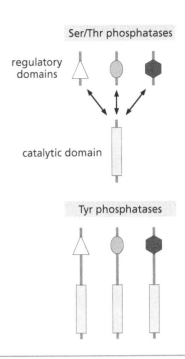

Figure 2.25 Different regulation of Ser/Thr phosphatases as compared with Tyr phosphatases In Ser/Thr phosphatases, a catalytic domain may combine noncovalently with various regulatory domains. In Tyr phosphatases, both catalytic and regulatory domains share the same polypeptide chain.

although they also hydrolyze substrates via a thioester intermediate. The dual-specific PTPs of the **Cdc25 family** are probably also restricted to eukaryotes. They are absolutely specific for the cell cycle-controlling protein kinases of the CDK family, activating these by dephosphorylation of two vicinal Tyr and Thr residues. Again the reaction involves a thioester intermediate, though no genetic relationship to other PTPs exists.

Prokaryotic histidine and aspartate kinases/phosphatases

These ancient enzymes are bifunctional autokinases/protein phosphatases undergoing auto-inactivation by auto-dephosphorylation. This mechanism again demonstrates that such prokaryotic protein phosphorylations represent transition states of phosphotransferase reactions. However, in some bacterial species, special phospho-His phosphatases such as SixA and phospho-Asp phosphatases such as CheZ and SpoOE have been found (see Section 3.5.3). These enzymes are highly specific. Since they have to be present in stoichiometric amounts and work only when the Asp-directed autophosphatase activity of their substrate kinases is intact, they are probably allosteric activators rather than genuine phosphatases, functioning in a manner analogous to that of GAPs in that they increase the endogenous phosphatase activities of His and Asp autokinases.

Other prokaryotic enzymes exhibiting both kinase and phosphatase activities are HPr kinase/phosphatase and isocitrate dehydrogenase kinase/phosphatase (Section 3.5.3). The combination of both functions in one enzyme appears to be an evolutionarily ancient model that was rejected by eukaryotes. In fact, the separation of kinase and phosphatase activities offers many more possibilities for regulation and network formation. A striking analogy exists between bifunctional kinases/phosphatases and G-proteins in that, via ATP or GTP hydrolysis, both modulate the function of substrate proteins and transfer the released phosphate group onto water.

For vertebrate cells the analysis of the overall protein phosphorylation status (measured by phosphoamino acid analysis upon labelling with ^{32}P, or by mass-spectrometry) yields 90% Ser, 10% Thr, and 0.05% Tyr phosphorylation, although the Ser:Thr:Tyr ratio in proteins is about 3:2:1 on average. These steady-state values are due to a combined action of protein kinases and protein phosphatases with the latter showing a bias towards phospho-Thr and in particular phospho-Tyr dephosphorylation.

2.6.5 Phosphorylation cascades and kinase modules

Phosphorylation reactions are often connected in series, with one kinase activating the next. Such reaction cascades or phosphorelays serve signal amplification, noise suppression, temporal sharpening of signals, frequency modulation, and cross talking with other signaling pathways. Frequently the kinases that form a cascade are assembled in modules stabilized by scaffolding proteins. Numerous examples of these cascades are presented in the subsequent chapters.

The most prominent and probably most versatile cascade type is the **MAP kinase cascade,** a central relay and distributor device of eukaryotic signal processing. It will be discussed in detail in Chapter 11. MAP kinase stands for Mitogen-Activated Protein kinase, a historical term because this kinase type was discovered by studies on mitogenic stimulation of cells. Since then it has become clear that MAP kinases are involved in many more cellular processes.

MAP kinase modules are assembled by three protein kinases connected in series. These are the MAP kinases proper; their regulator kinases are named MAP kinase kinases or MAP2 kinases, and the sensor kinases for input signals are called MAP kinase kinase kinases or MAP3 kinases (Figure 2.26). These elements of a MAP kinase module are assembled on scaffolding proteins in such a way that the analogy with integrated electronic circuits is quite striking.

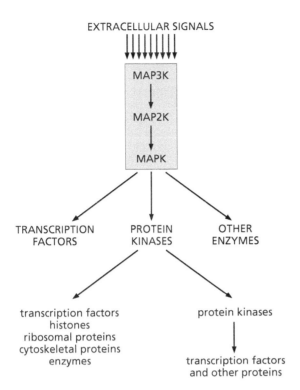

EXTRACELLULAR SIGNALS

MAP3K

MAP2K

MAPK

TRANSCRIPTION FACTORS   PROTEIN KINASES   OTHER ENZYMES

transcription factors
histones
ribosomal proteins
cytoskeletal proteins
enzymes

protein kinases

transcription factors
and other proteins

Figure 2.26 MAP kinase modules: universal relay stations Like a ganglion, a MAP kinase module (boxed) processes many different input signals and transmits a large number of output signals, thus influencing almost any cellular process.

MAP kinase modules are components of numerous signaling pathways connecting almost every receptor type with the metabolic machinery and the genome. The evolutionary success of the module is probably due to the fact that such a device works as a highly efficient noise filter, delivering an extremely sharp signal as a prerequisite for frequency modulation. Indeed, MAP kinase modules are integrated in positive and negative feedback loops in such a way that the conditions for generation of oscillating signals are fulfilled, though experimental proof is still lacking. A frequency-modulated code could easily explain why, depending on the conditions, a single MAP kinase module may exert contrary effects: for example, stimulation of cell proliferation and survival upon short-term activation versus an anti-mitotic effect or even induction of cell death upon long-term activation. Possibly the survival signal simply has another frequency than the death signal.

Summary

Phosphorylation ranks among the most abundant post-translational modifications of proteins. It is catalyzed by protein kinases that use ATP (or in a few cases GTP or phosphoenolpyruvate) as a phosphate donor. Protein phosphatases render the modification reversible. In addition to some minor families of atypical protein kinases, two unrelated superfamilies of protein kinases are known: His autokinases (predominantly found in prokaryotes) and kinases phosphorylating Ser, Thr, and Tyr residues. While Ser/Thr-specific kinases are found in both pro- and eukaryotes, Tyr kinases appear to be largely restricted to multicellular animals. Protein kinases are signal-operated switching elements. In the absence of input signals, most protein kinases are blocked by auto-inhibition. Many kinases require, in addition, phosphorylation of the activation loop (a structure in the catalytic center) to become active. Phosphorylation has three major effects on proteins: electrostatic repulsion of negatively charged partner molecules, conformational changes resulting in altered activity, and generation of contact sites for interacting proteins. Frequently several protein kinases are combined into multiprotein modules; the eukaryotic MAP kinase module being the most prominent example. Like kinases, protein phosphatases are also signal-operated switching elements. Ser/Thr-specific phosphatases exist in a few catalytic domains that noncovalently couple with a large number of regulatory subunits.

In Tyr-specific phosphatases, the regulatory and catalytic domains share the same polypeptide chain.

2.7 Protein acetylation: a central device of gene regulation

Apart from phosphorylation, the acetylation of proteins represents a further post-translational modification with signal character. It is catalyzed by acetyltransferases that transfer the acetyl residue from acetyl-coenzyme A to the ε-amino group of lysine or the guanidino group of arginine residues. Corresponding deacetylases hydrolyze the amide bond and thus reverse the switching reaction (Figure 2.27).

Both types of enzymes exist in many isoforms that, like protein kinases and phosphatases, differ from each other by their regulatory domains; that is, they respond to different input signals and exhibit different substrate specificities. The deacetylases are, moreover, subdivided into two families with quite different mechanisms of action: hydrolases, which transfer the acetyl residue onto water, and ADP-ribosyltransferases, which transfer the acetyl residue onto ADP-ribose derived from NAD$^+$ (more details are found in Section 8.2.3).

The acetylation of histones has been investigated most thoroughly since it plays a key role in gene activation, with histone acetyltransferases acting as transcriptional co-activators and histone deacetylases as co-repressors (for an in-depth discussion of this subject see Sections 8.2.2 and 8.2.3). However, protein acetylation is by no means restricted to histones but appears to be of similar importance as (the admittedly much better investigated) protein phosphorylation. This holds true for numerous transcription factors and related gene-regulatory proteins (for example, NFκB, Section 11.8, and p53, Section 12.9.3). Their intracellular distribution between cytoplasm and nucleus, their DNA-binding efficiency, and their interactions with other proteins are frequently modulated by acetylation, mostly in combination with phosphorylation. The catalyzing enzymes are called factor acetyltransferases. Their best-known representatives are CBP (CREB Binding Protein) and the homologous p300 (molecular mass of 300 kDa; CREB is the cyclic AMP response element binding protein). The **CBP/p300 acetyltransferases** are prototypes of transcriptional co-activators. Due to a large number of interactions (see the multidomain scheme in Figure 2.28), on chromatin they are able to form extended signal-transducing protein complexes including up to 20 different partner molecules, which bundle up various signaling pathways at distinct gene loci. Such sites are marked by gene-specific transcription factors that are components of the multiprotein complex recruited by CBP/p300. Transcription factors recognize their gene promoters by specific nucleotide sequences called response elements (see Chapter 8).

Figure 2.27 Acetylation switch The switching event consists of acetylation of a substrate protein on Lys or Arg residues. The acetate donor is acetyl-coenzyme A (Ac-CoA). The reaction is catalyzed by an acetyltransferase, whereas a deacetylase removes the acetyl residue from the substrate protein. Both enzymes are controlled by input signals. The conformational and functional change induced by acetylation in the substrate protein is the output signal. The switching energy is delivered by splitting the energy-rich thioester bond of Ac-CoA. Reloading of CoA with acetate is an ATP-consuming process.

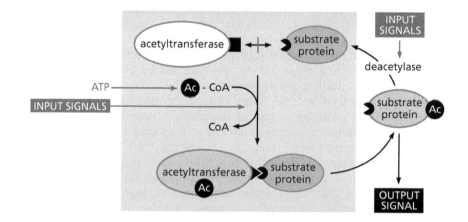

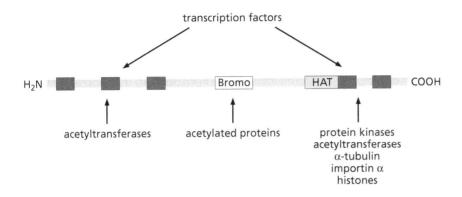

H_2N — Bromo — HAT — COOH

transcription factors

acetyltransferases

acetylated proteins

protein kinases
acetyltransferases
α-tubulin
importin α
histones

Figure 2.28 Acetyltransferase CBP/p300: an integrator of many signaling pathways The figure shows, in a simplified manner, the multidomain structure of the 2441 amino acid residue protein together with the interaction partners. HAT, histone acetyltransferase domain; Bromo, domain for interaction with acetylated proteins; black rectangles, other interaction domains. Further details are found in Sections 8.2.2 and 8.2.5. As in other figures throughout this book, the protein is depicted as a linear structure with the interaction domains symbolized as larger rectangles. Proteins are always shown with the amino terminus to the left and the carboxy terminus to the right.

From an evolutionary point of view, protein acetylation seems to be as ancient as phosphorylation since it is found in bacteria and archaea. For instance, the chemotaxis protein CheY of *Escherichia coli* (Section 3.3.4) is controlled by phosphorylation as well as by acetylation.

Summary

Reversible protein acetylation at Lys and Arg residues represents a widespread post-translational modification. By neutralizing the positive charge of histones, it provides a key reaction of transcriptional activation. Therefore, histone acetyltransferases and deacetylases are the standard co-activators and co-repressors of gene transcription.

2.8 Protein ubiquitylation: more than a signal of protein degradation

Post-translational modifications of proteins are achieved not only by addition of low-molecular weight groups such as phosphate or acetate but also by covalent binding of peptides. The prototype of such a peptide is ubiquitin, a highly conserved 8.5 kDa molecule containing 76 amino acids and found in all eukaryotic cells. Its binding to proteins represents an abundant signaling reaction comparable to phosphorylation and acetylation. In fact, in normally growing yeast cells up to 20% of all proteins have been found to be ubiquitin-conjugated and the same may hold true for animal cells. Since ubiquitylation is a highly dynamic process, this number represents a steady state rather than the absolute number of proteins ubiquitylated in any possible situation.

A ubiquitylating enzyme complex is a module consisting of three components: the enzymes E1 (ubiquitin-activating enzyme), E2 (ubiquitin-conjugating enzyme), and E3 (ubiquitin ligase). While E1 takes care of ubiquitin activation, E2 catalyzes the ubiquitylation of the substrate protein, which is specifically presented by E3 (Figure 2.29).

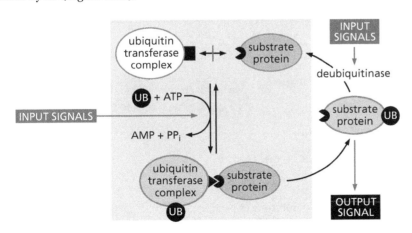

Figure 2.29 Ubiquitylation switch In principle the switch resembles the phosphorylation switch. In an ATP-consuming reaction, the ubiquitin transferase complex transfers ubiquitin (UB) to a Lys residue of a substrate protein (see Figure 2.30) where it may be removed again by a deubiquitinase. Both ubiquitylation and deubiquitylation are controlled by input signals. The conformational and functional change induced by ubiquitylation in the substrate protein is the output signal. The switching energy is delivered by the hydrolysis of ATP or, to be precise, of pyrophosphate (PP_i). This scheme applies to ubiquitin-like modifiers as well.

Ubiquitin—which if not immediately available may be proteolytically released from a precursor protein—is activated by binding via its C-terminal glycine residue to a Cys residue of E1. This formation of an energy-rich thioester bond requires ATP hydrolysis and provides the energy source of the ubiquitylation switch. From E1 the ubiquitin residue is taken over by E2—again transiently bound as a thioester—and transferred to a substrate protein where its C-terminal glycine residue forms a so-called isopeptide bond with the ε-amino group of a lysine residue. By bringing together E2 enzymes with selected target proteins and a wide variety of regulatory factors, the ligase E3 determines the substrate specificity of ubiquitylation. Moreover, E3 ligases are the signal-transducing components proper of the ubiquitin transferase complex, since they are the major receivers of input signals. While human E2s are encoded by about 40 genes, more than 500 different E3 genes have been found. The variety of ubiquitin ligases exhibiting different substrate specificities, control mechanisms, and tissue distributions thus rivals that of protein kinases. This aspect is discussed in more detail in Section 12.16.

Since the ubiquitin residue is removed from the substrate protein by specific proteases or isopeptidases, ubiquitylation is, like phosphorylation and acetylation, a reversible switching reaction. A substrate protein may be either **monoubiquitylated** by a single ubiquitin residue or **polyubiquitylated** by a polymerized ubiquitin chain (Figure 2.30). This chain consists of many ubiquitin residues connected by isopeptide bonds between the carboxy-terminal glycine residues and lysine residues. Since ubiquitin has seven lysine residues to be ubiquitylated, a wide variety of linear and branched polyubiquitin chains are constructed, rendering ubiquitylation the probably most versatile of all post-translational protein modifications. The mechanisms of ubiquitin chain assembly are still not known. Moreover, the reason a single protein can be monoubiquitylated in one case but polyubiquitylated in another case is not entirely clear. Possibly each reaction is catalyzed by a specific set of enzymes, or the activity of one enzyme is modulated by accessory factors. In fact, E3 ubiquitin ligases interact with several regulatory subunits that may be exchanged, depending on the input signals (see Section 12.16). For the discovery of ubiquitin-dependent protein degradation, Aaron Ciechanover, Avram Hershko, and Irwin Rose were honored by the 2004 Nobel Prize in Chemistry.

Figure 2.30 Ubiquitin transferase reaction Ubiquitin transferase is a complex of three cooperating enzymes, E1, E2, and E3. (A) Monoubiquitylation. In cells, ubiquitin (UB) is directly available or proteolytically released from a precursor protein. E1, the ubiquitin-activating enzyme, binds UB as an energy-rich thioester via a Gly–Cys bond. The reaction requires the hydrolysis of one molecule of ATP, generating a labile UB-AMP intermediate. From E1 the ubiquitin-conjugating enzyme E2 takes over the UB residue, binding it again as a high-energy thioester and transferring it to a substrate protein, where it forms an isopeptide bond between its C-terminal glycine and the ε-amino group of a Lys residue. The substrate protein is presented by the ubiquitin ligase E3 acting as an adaptor or scaffold protein. E3 plays a key role since it renders the ubiquitylation substrate-specific and is controlled by input signals (for details see Section 12.16). (B) Polyubiquitylation. Several UB residues become linked via isopeptide bonds. Depending on the linking site (Lys48 or Lys63), different downstream reactions result. Ubiquitin-like modifiers such as SUMO are activated and transferred in an analogous manner.

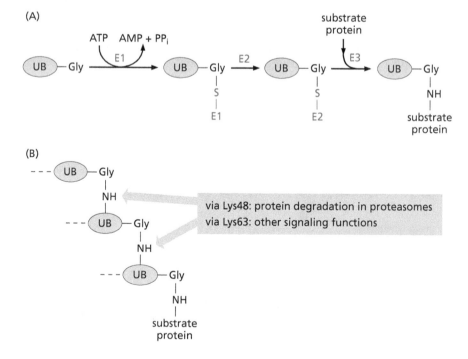

2.8.1 Ubiquitin-binding domains

The highly variable patterns of ubiquitylation have been compared with ZIP codes that are used to sort proteins to different destinations. To make sense, a ZIP code has to be deciphered. In fact, like phosphorylated proteins, ubiquitylated proteins make contact with other proteins by interacting with specific recognition and binding domains. More than 10 different types of such domains, ranging from 20 to 150 amino acids and exhibiting quite different structures, have been identified in a wide variety of proteins. Examples are provided by the ubiquitin-associating domain (UBA) and the ubiquitin-interacting motif (UIM) as well as the domain for coupling of ubiquitin conjugation to endoplasmic reticulum degradation (CUE). The binding of ubiquitylated proteins to such domains is rather weak and, therefore, readily reversible. It has been suggested that such low-affinity interactions enable the cell to quickly respond to input signals by assembling and disassembling a network of interacting proteins with built-in "dynamic instability." Nevertheless, the precise physiological role of most ubiquitin-binding domains is still poorly understood. Exceptions are the domains found in subunits of proteasomes that bring about the interaction with ubiquitylated proteins to be degraded (see below).

2.8.2 Cellular functions of ubiquitylation

The ubiquitylation of proteins has numerous and diverse consequences. Best known is its role in **proteolytic degradation of proteins in proteasomes.** To be delivered to the proteasome, proteins have to be labeled by a polyubiquitin tag linked via Lys48. Ubiquitin-binding domains showing some preference for this type of polyubiquitin chain are found in proteins of the 19S subunits of the proteasome and in adaptor proteins (Figure 2.31). These adaptors bind to proteasomes via ubiquitin-like domains. For the polypeptide chain to be threaded into the central hole of the proteasome, the ubiquitin chain has to be cut off by Lys48-specific deubiquitylating proteases that are components of the 19S regulatory subunits. The ubiquitin chain thus released is not further degraded but is recycled. After deubiquitylation, the protein to be degraded is unfolded by AAA+ proteins that are also found in the regulatory subunits of proteasomes (see

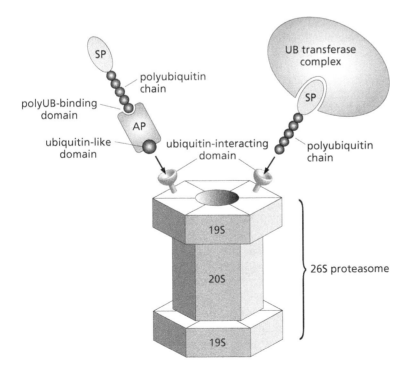

Figure 2.31 Delivery of polyubiquitylated proteins to the proteasome Substrate proteins (SP) to be degraded are labeled by polyubiquitylation and delivered to the 26S proteasome (consisting of a 20S catalytic subunit and two 19S regulatory subunits) either directly (right) or through an adaptor protein (AP, left).

Section 2.5.2). Frequently, the efficiency of ubiquitylation-induced proteolysis depends on phosphorylation (see above) or other post-translational modifications of the substrate protein. Thus, the combination of such modifications with ubiquitylation resembles a logical AND gate.

Ubiquitin-mediated protein degradation is one of the most important events in cellular signal processing, being involved in both signal generation—by inactivating inhibitory proteins (see, for example, Section 11.8)—and signal extinction. For instance, **transcription factors** are labeled for rapid degradation by ubiquitin ligases that are components of the transcription-inducing protein complex. Such a suicide mechanism prevents excessive and uncontrolled gene activation and, thus, allows a precise adaptation of the genetic readout to environmental conditions, albeit in quite a costly manner. In plants the ubiquitylation and proteolytic destruction of transcriptional repressors has been found to be even under direct hormonal control, in that E3 ubiquitin ligases are receptors of phytohormones such as auxin and gibberellic acid (Section 13.1).

Another example is provided by the **cell cycle**, each phase of which is controlled by CDKs. As discussed in Chapter 12, cyclins, which are the regulatory subunits of these kinases, are produced and degraded in a strictly defined order during the cell cycle, a process that is indispensable for the clockwork-like course of cell division. For degradation the cyclins have to be labeled by ubiquitylation. Polyubiquitylation plays a similar role in **DNA repair**. DNA damage leads to an interruption of the cell cycle to give the cell an opportunity to repair the defect or—if this turns out to be impossible—to commit suicide by apoptosis. Stop signals include both ubiquitin-mediated degradation of protein phosphatase Cdc25A, an enzyme that promotes the cell's transition into the phase of DNA replication (Section 12.4), and stabilization of the anti-proliferative transcription factor p53 by inhibition of ubiquitin-mediated proteolysis (Section 12.9.3).

A special case is the ubiquitin-controlled **processing of antigens** by antigen-presenting cells. Here specialized immune proteasomes dissect an antigenic protein into fragments of 8–10 amino acids which, together with the major histocompatibility complex, are exposed at the cell surface, stimulating cytotoxic T-lymphocytes.

In contrast to polyubiquitylation via Lys48, mono-ubiquitylation or polyubiquitylation via Lys residue 63 of ubiquitin does not label a protein for proteasome binding and degradation but, like phosphorylation, for interactions with other proteins. Examples such as ubiquitylation of histone 2B and the transcription factor NFκB are discussed in Sections 8.2.6 and 11.8.

In a substrate protein, several Lys residues may be monoubiquitylated. This type of multi-ubiquitylation plays an important role in **endocytosis**, as shown in exemplary fashion by the down-regulation of membrane receptors, through which cells protect themselves from overstimulation. Receptor down-regulation provides a major mechanism of adaptation to exogenous signals (see Section 7.1.1). In the course of down-regulation, the activated receptors are ubiquitylated and subsequently concentrated in endocytotic vesicles. These are split off from the membrane, transporting their cargo to endo- and lysosomes, where it is recycled or degraded (details in Section 5.8.1).

Disturbances of this mechanism may have serious consequences, as demonstrated by the proteins **BRCA1** (BReast CAncer protein 1) and **Cbl** (named after the Casitas B-cell lymphoma of mice). Both are E3 ubiquitin ligases, inducing the degradation of many signaling proteins that, when excessively expressed and activated, would promote tumor growth. BRCA1 is a tumor suppressor protein; its mono-allelic deletion causes a hereditary predisposition for breast cancer. [Another tumor suppressor protein, the genetic inactivation of which is causally related to familiar breast cancer, is BRCA2. Showing no homology to BRCA1,

BRCA2 is involved in the regulation of cell proliferation in response to DNA damage (Section 12.9).] In contrast, Cbl may become an oncoprotein through mutations destroying the ubiquitin ligase activity. Notwithstanding, the mutated Cbl still binds its substrate proteins, thus acting as a dominant-negative mutant that competitively inhibits the nonmutated Cbl (derived from the nonmutated allele). The major physiological function of Cbl is to mark growth factor receptors for down-modulation (see Section 7.1.1).

2.8.3 Ubiquitin-like modifiers

Ubiquitin stands for a larger family of protein-modifying polypeptides called ubiquitin-like modifiers (UBL). Their activation and binding mechanisms as well as their three-dimensional structures (but not necessarily their amino acid sequences) resemble those of ubiquitin. The most prominent representative is the protein **SUMO** (Small Ubiquitin-like MOdifier), which exists in several isoforms and is found in all eukaryotic cells. Several SUMO-specific E3 ubiquitin ligases [for instance, protein inhibitor of activated STAT; (PIAS; see Section 7.2.1) or Ran-binding protein RanBP2 (see Section 10.5)] and desumoylating proteases (called ubiquitin-like protein-specific proteases), as well as a rather large number of proteins the functions of which are regulated by sumoylation, are known. Substrates of sumoylation include transcription factors such as steroid hormone receptors, c-Jun, STAT, Myb, p53, and others, as well as the GAP of the small G-protein Ran controlling the transport of macromolecules across the nuclear membrane. The enzymes of sumoylation/desumoylation are components of the signal-processing protein network and are regulated, for instance, by protein phosphorylation, which may silence sumoylation signals by inhibiting SUMO ligases and stimulating SUMO proteases.

In contrast to ubiquitylation, sumoylation has never been found to mark a protein for proteolytic degradation but exclusively for interactions with other proteins. In fact, sumoylation may even *prevent* protein degradation by competing with ubiquitylation, thus prolonging the effectiveness of peptide and protein signals such as transcription factors. In addition, sumoylation has a direct effect on the activity of certain transcription factors such as the heat-shock factors HSF1 and 2 (not to be confused with heat-shock proteins), which are stimulated upon temperature stress.

Other ubiquitin-like modifiers are also involved in reactions to stress such as starvation and infections. One of those modifiers is the starvation-activated autophagocytosis* regulator **Aut7** that interacts with the membrane lipid phosphatidylethanolamine rather than with proteins. This provides another analogy to phosphorylation, which, under certain circumstances, may affect not only proteins but also phospholipids. Interestingly, ubiquitin-like modifiers are also used to control the ubiquitylation reaction proper. An example is the modification of certain E3 ubiquitin ligases by the modifier Nedd8, a reaction called **neddylation** (see Section 12.16).

2.8.4 Signal termination by deubiquitylating proteases

Among the 561 human protease genes, around 80 encode enzymes specialized for deubiquitylation of proteins, rendering ubiquitylation a reversible switching reaction. Additional proteases are responsible for the removal of ubiquitin-like modifiers such as SUMO and Nedd8. Considering the more than 500 E3 ubiquitin ligases, the enzymatic equipment of ubiquitylation clearly rivals that of protein phosphorylation, comprising in humans 518 kinase and 120 phosphatase genes.

---

*Autophagocytosis is a cellular function serving the degradation of intracellular material in lysosomes (or vacuoles). Such material may be organelles taken out of service, misfolded proteins, and invading microorganisms and viruses. Autophagocytosis plays a key role in programmed cell death and in insect metamorphosis. Moreover, in case of starvation, nonessential cellular components are degraded to provide energy for essential processes.

Deubiquitylating enzymes (DUBs) constitute a highly diverse group of (mainly) cysteine proteases with molecular weights from 188 to over 3000. Each of them has a limited set of substrates connecting it with distinct physiological functions. Specificity is related to the type of ubiquitin chain, the site of ubiquitylation, and the nature of the substrate protein. In contrast to many other proteases, DUBs are not produced as zymogens but become activated by binding the ubiquity-lated substrate. Such a mechanism conveys a high degree of selectivity and prevents accidental degradation of other proteins. In addition to signal termination, the protection of E3 ubiquitin ligases from spurious self-ubiquitylation is a major function of DUBs. For both reasons, they are frequently found to be associated with the ligases.

Summary

Ubiquitylation is an ATP-consuming post-translational modification that is restricted to eukaryotes. In its versatility and abundance, it rivals or even outdoes protein phosphorylation. It is catalyzed by a complex of three enzymes, E1, E2, and E3, with the ubiquitin ligases E3 determining the substrate specificity of the reaction. E3 ligases are signal-operated switching elements. Their genetic variability resembles that of protein kinases. Deubiquitylating proteases render the reaction reversible. Ubiquitylated proteins recognize their partners by specific ubiquitin-interacting domains. Cells make use of ubiquitylation for labeling proteins for degradation in proteasomes, for other protein–protein interactions (network formation), and for the control of protein activities. Proteasomal degradation is a major process of signal termination. It is induced by polyubiquitylation, mostly via Lys48 of ubiquitin, whereas the other effects depend on polyubiquitylation via other Lys residues or on monoubiquitylation. Ubiquitin is the prototype of a series of related peptides used for post-translational protein modification. In contrast to ubiquitin, these ubiquitin-like modifiers do not label proteins for proteolytic degradation but serve other purposes in signal transduction.

2.9 Mono- and poly(ADP-ribosylation)

Two types of post-translational modifications, mono(ADP-ribosylation) and poly(ADP-ribosylation), use the free energy released by the metabolic break-down of NAD$^+$ for signal transduction.

2.9.1 Mono(ADP-ribosylation)

This is a post-translational modification for which an ADP-ribosyl residue released from NAD$^+$ is transferred to a protein and linked with an Arg residue via a covalent *N-glycosidic bond* (Figure 2.32). The reaction is catalyzed by

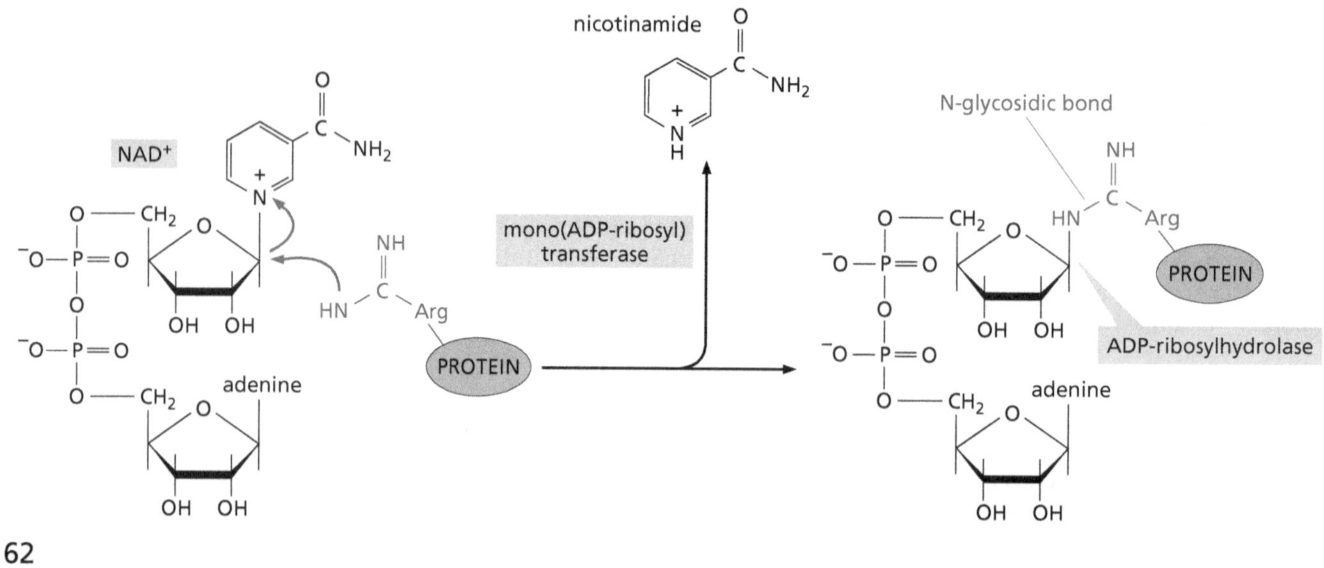

Figure 2.32 Mono(ADP-ribosylation) Via mono(ADP-ribosyl)transferase, the nicotinamide residue of the redox coenzyme nicotinamide adenine dinucleotide (NAD$^+$) is replaced by a protein, which becomes linked to the ADP-ribose residue via an Arg residue. The hydrolysis of this bond is catalyzed by an ADP-ribosylhydrolase. Both enzymes are controlled by input signals. The energy for the switching process is supplied by the hydrolysis of the N-glycosidic bond of NAD$^+$.

mono(ADP-ribosyl)transferases and reversed by ADP-ribosylhydrolases. These enzymes have been found in pro- and eukaryotes as well as in bacteriophages. In the human genome, the corresponding genes belong to the approximately 200 genes that were acquired from bacteria.

For some viruses and bacteria, ADP-ribosyltransferases serve as virulence factors. By this means bacteriophages reprogram the bacterial RNA polymerase in such a way that it preferentially transcribes viral genes. The ADP-ribosyltransferases of diphtheria, pertussis, and cholera bacteria are highly active toxins since they intervene strongly with G-protein signaling (more details in Section 4.3.3) or damage the cytoskeleton by ADP-ribosylation of actin. Other bacterial ADP-ribosyltransferases are probably not related to the toxins but control the activities of metabolic enzymes such as the nitrogenases of soil bacteria.

Eukaryotic ADP-ribosyltransferases are distributed over two nonrelated families, extracellular and intracellular enzymes. Extracellular transferases are bound to cell membranes by a glycosylphosphatidylinositol anchor (see Section 4.3.1). They are found particularly on cells of the immune system. Little is known about their substrate proteins and physiological role.

Intracellular ADP-ribosyltransferases are components of the signal-processing network, involved (among others) in the temporal limitation and extinction of signals. The $\beta\gamma$-subunits of trimeric G-proteins are inhibited (Section 4.3), and in photoreceptor cells, signal transduction from the trimeric G-protein transducin to cGMP-phosphodiesterase is blocked by ADP-ribosylation of the inhibitory subunit of the phosphodiesterase (Section 15.5). ADP-ribosylation suppresses, moreover, the polymerization of cytoskeletal proteins such as actin and desmin to filaments as well as the activity of certain metabolic enzymes such as glyceraldehyde-3-phosphate dehydrogenase.

A special type of ADP-ribosyltransferases, the **Sir enzymes** or sirtuins, are discussed in Section 8.2.3. These enzymes are both NAD$^+$-consuming histone deacetylases transferring the ADP-ribosyl residue to acetyl residues and are catalysts of ADP-ribosylation of proteins.

2.9.2 Poly(ADP-ribosylation)

Mono- and poly(ADP-ribosylation) are chemically closely related but biochemically different protein modifications with signaling character. For poly(ADP-ribosylation) the ADP-ribosyl residues derived from NAD$^+$ are linked via *ester-glycosidic bonds* to Asp and Glu residues of the substrate protein. Subsequently, additional ADP-ribosyl residues are added to the initial residue, forming linear and branched poly(ADP-ribosyl) chains of up to 200 units (Figure 2.33). The poly(ADP-ribosyl)polymerases (PARPs) catalyzing these reactions are highly conserved enzymes found in archaea and eukaryotes. Mammals express seven isoforms. Like mono(ADP-ribosylation), poly(ADP-ribosylation) is a reversible switching reaction: the ADP-ribosyl polymer is rapidly degraded by poly(ADP-ribose) glycohydrolases, and the initial residue is removed from the protein by ADP-ribosyl protein lyase. Substrate proteins of PARPs are the enzymes themselves (automodification), various transcription factors, proteins involved in DNA replication and mitosis, and histones.

Poly(ADP-ribosylation) of histones belongs to the reactions used by the cell to alter chromatin structure in such a way that gene transcription becomes possible (see Section 8.2.6). The histone-specific PARPs 1–3 have a DNA-binding domain with a two-fold zinc-finger structure (as found also in many transcription factors) and are strongly activated by damaged DNA, for instance, upon irradiation or contact with mutagenic chemicals. These and related observations (for example, in PARP-knockout mice) point to an important role of poly(ADP-ribosylation) in **DNA repair**. Other PARP isoforms have other regulatory domains

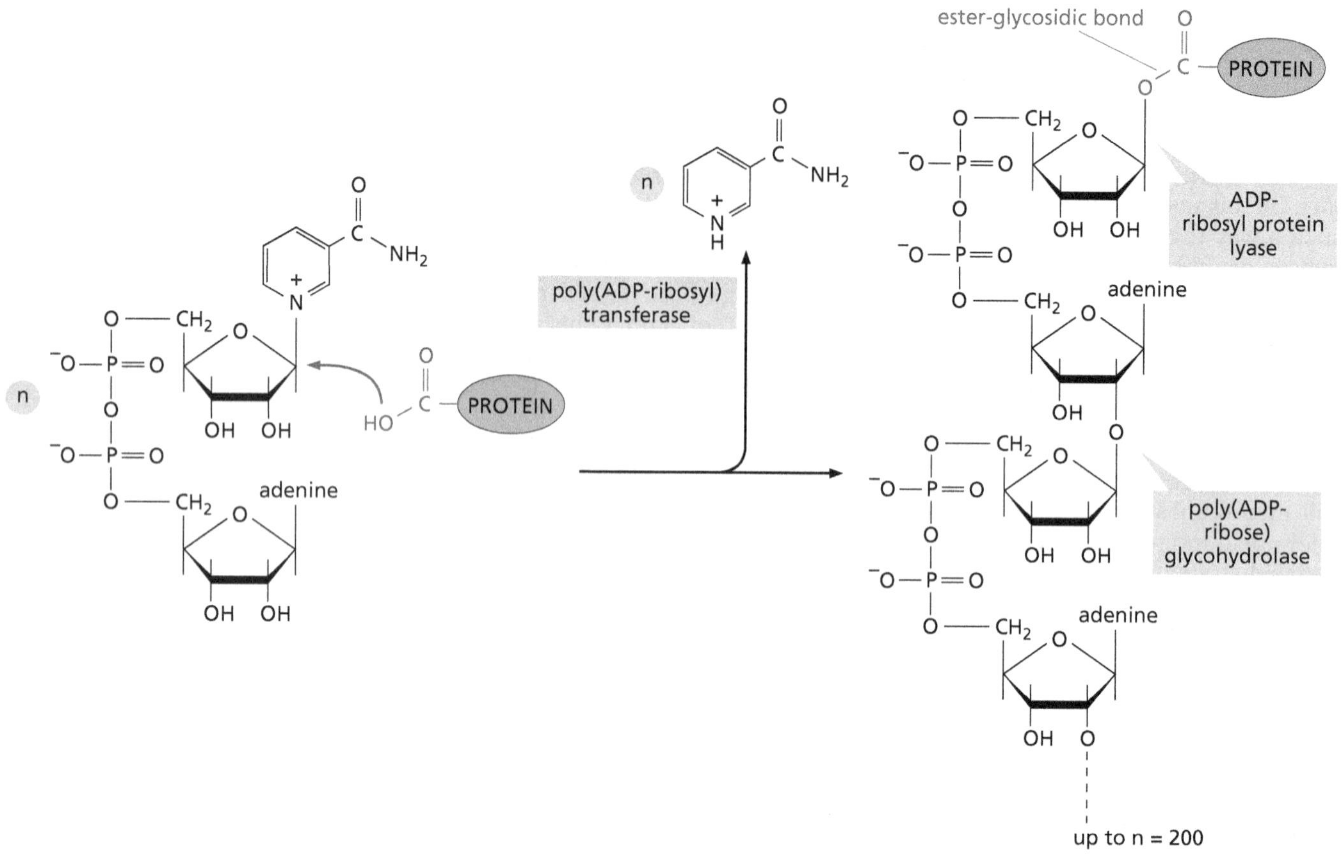

Figure 2.33 Poly(ADP-ribosylation)
For details see text.

rather than DNA-binding sites and a different subcellular distribution; therefore, they are assumed to fulfill functions different from those of histone-specific PARPs.

Summary

Mono- and poly(ADP-ribosylation) are two types of signal-operated post-translational protein modification deriving energy from NAD^+ hydrolysis. They are found in both pro- and eukaryotes and, although chemically closely related, are catalyzed by different enzymes. Bacterial ADP-ribosyltransferases are potent virulence factors. Poly(ADP-ribosylation) is used for histone modification in the course of eukaryotic transcription and becomes especially activated upon DNA damage. Mono- and poly(ADP-ribosylation) represent fields of research with many uncharted regions.

2.10 Ion channel switches: how to make use of electrical charge

Ion channel switches consist of signaling proteins that use the energy stored in the electric charge of cell membranes. This membrane potential is, in fact, an important energy source for signal processing, rivaling the hydrolysis of energy-rich bonds in NTPs, acetyl-coenzyme A, and NAD^+.

2.10.1 Membrane potential: a major energy provider for cellular data processing

The membrane potential is defined as the electrostatic charge caused by an imbalanced distribution of sodium, potassium, calcium, and chloride ions across cell membranes. This "capacitor" can be discharged, with its energy being used for signal transduction, and recharged at the expense of metabolic energy. In other words, cellular membranes function like electric batteries feeding a computer.

The key to understanding the membrane potential is the **semipermeability** of biomembranes. In aqueous solution, each ion binds several water molecules electrostatically: it is surrounded by a hydrate cover. For such hydrated ions, the lipid bilayer of cellular membranes constitutes a barrier that cannot be passed even under an extreme gradient of charge and concentration. Thanks to protein complexes working as pores, cell membranes are, nevertheless, permeable to ions. However, such pores allow only the small ions mentioned above to pass, whereas larger charge carriers such as amino acids and nucleotides, let alone electrically charged macromolecules such as proteins and nucleic acids, are held back. This semipermeability causes an imbalanced charge distribution.

To make this clear, one may imagine a model cell filled with a 100 mM solution of K^+ ions and containing potassium channels in its membrane. The positive charge is assumed to be matched by a corresponding amount of negatively charged macromolecules; in the beginning, the interior of the cell is electrically neutral. When such a cell is put into an aqueous solution, K^+ ions pass through the membrane channels along the concentration gradient while the negative counterions are prevented from leaving the cell. As a result, a charge difference (negative inside, positive outside) develops at the membrane, hindering the K^+ ion flow, which ends when the electrostatic power matches the chemical power of the ion gradient. This state or equilibrium, also called the Donnan equilibrium, is described by the **Nernst equation**:

$$P_K = \frac{RT}{zF} \ln \frac{[K^+]_o}{[K^+]_i}$$

where P_K is the equilibrium potential of K^+ ions in volts, R is the gas constant, T is the absolute temperature, z is the ion valence (1 for K^+), F is the Faraday constant, and $[K^+]_o$ and $[K^+]_i$ are the K^+ concentrations outside and inside the cell, respectively. If the cell contains additional ion species as well as the corresponding membrane channels, a Donnan potential may be calculated for each type. Under physiological conditions, the following outside:inside ion distributions are established: K^+ 2:1, Na^+ 2:1, and Cl^- 1:2. The resulting membrane potential is about –20 mV.

Despite this charge, a net flow of ions is impossible because the system is *equilibrated*. In other words, under these conditions the membrane potential cannot be used as an energy supplier for signal transduction and other purposes. Apart from this, the high osmotic pressure difference would cause an influx of water (through corresponding membrane pores), which may lead to a disruption of the cell membrane. To prevent this, the cell establishes a *disequilibrium* of the ion distribution to some extent approaching the following concentration ratios (again outside:inside): K^+ 20:1, Na^+ 1:7, and Cl^- 1:25 (actual concentrations are found in Figure 2.34). By this means, the membrane capacitor is charged up to a voltage of approximately –60 mV and can be used as an energy provider for coupled reactions, in particular signal transduction.

The charge energy is derived from ATP hydrolysis driving a membrane-bound ion transport system, the Na^+/K^+-dependent ATPase, which, for each hydrolyzed ATP molecule, pumps out three Na^+ ions and allows two K^+ ions to enter the cell. As a result one positive charge is removed from the cell for each pumping step. Consequently, the ATPase has been called an electrogenic ion pump. In addition, certain (not all) cell types possess chloride pumps to regulate the membrane potential, whereas Ca^{2+}-dependent ATPases or calcium pumps play only a minor role in charge control due to the comparably low Ca^{2+} concentration in cells.

To maintain the membrane potential, the ion pumps have to work permanently, since cell membranes are leaky for K^+ and Cl^- ions because corresponding ion channels remain open all the time. Therefore, inhibitors of the Na^+/K^+-dependent

Figure 2.34 Membrane potential as an energy supply of signaling reactions The figure shows the distribution of ions across the plasma membrane and the resulting resting potential as brought about by ATP-consuming ion pumps (Na^+/K^+- and Ca^{2+}-dependent ATPases) and ion exchangers. The numbers represent the average ion concentrations in millimoles per liter. The actual values differ between cell types. Moreover, depending on the resting potential and ion concentration, the gating of chloride channels may cause a charge-driven efflux rather than a concentration-driven influx (as shown here) of Cl^- ions. Signals alter the ion distribution by controlling ion channels (or ion pumps). This may result in either an increase (hyperpolarization) or decrease (depolarization) of the negative membrane potential. The resulting change of the membrane potential is recognized as a signal by voltage-controlled proteins, mostly ion channels. In contrast to K^+, Na^+, and Cl^-, Ca^{2+} entering the cell via the corresponding channels in most cells, does not alter the membrane potential to a significant extent but functions as a second messenger.

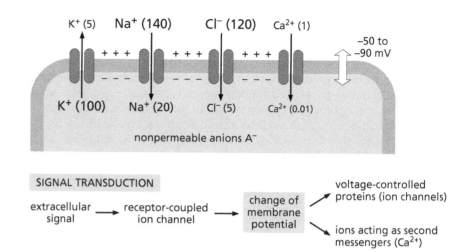

ATPase such as the active agents of foxglove (*Digitalis purpurea*) may cause a breakdown of the membrane potential, thus paralyzing the cell.

By controlling the activity of the ion pumps, the cell is able to alter specifically the membrane potential. This objective is also achieved by a controlled gating of ion channels, thus allowing an influx or efflux of ions depending on the concentration and charge gradients. As a rule, the membrane potential becomes more negative upon the opening of K^+ channels, an event called **hyperpolarization**. The gating of Na^+ channels has the opposite effect: because Na^+ ions leave the cell following the concentration gradient the interior of the cell becomes less negative. The cell is said to undergo **depolarization**. In contrast to sodium channels, chloride channels may have either a hyper- or a depolarizing effect depending on the concentration gradient of the ion and on the resting potential.* Thus, the effect of chloride channel gating depends on the cell type and the physiological context. Ion channels exist in a large variety of types that are operated either by changes of the membrane potential or by ligand binding (see Chapter 14).

2.10.2 Action potential: the standard signal of nerve and muscle cells

By acting on ion pumps and ion channels, hyper- or depolarizing signals either hinder or promote processes that depend on the membrane potential. A prototype of such a process is the action potential. Action potentials are waves of depolarization migrating with a speed of up to 120 m/s across cell membranes. These most rapid of all endogenous signals are produced by only a few cell types such as neurons, muscle cells, egg cells, and some others. Such cells have in common an **excitable plasma membrane**. In this membrane type, depolarization-operated Na^+ channels are arranged in such a way that they activate each other and stimulate cooperating K^+ channels which terminate the depolarization. Depending on the tissue, an action potential lasts between 0.1 (skeletal muscle) and 400 (heart muscle) ms.

Action potentials are generated according to an all-or-nothing rule, provided the depolarization of the cell membrane exceeds a threshold value, and they proceed, like the spark of a fuse, without dampening. The value of the resting potential (the membrane potential of the nonstimulated cell) in relationship to the threshold potential determines the probability of an action potential becoming

---

*The number of ions to be transported for depolarization or hyperpolarization is surprisingly small. According to the equation charge (in coulombs) = capacity (in microfarads per square centimeter) × voltage (in millivolts) and if a cell surface of 10^{-4} cm^2 is assumed, only 5×10^{-17} mol of K^+ ions has to be transferred to change the membrane potential to 50 mV. As compared with the intracellular K^+ content of 10^{-11} mol, this amount is practically negligible unless the ion flux lasts too long.

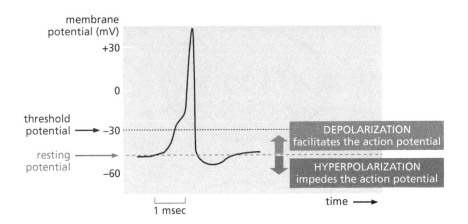

	active	inactive
Na$^+$ channel	D	H
K$^+$ channel	H	D
Cl$^-$ channel	H,D	H,D
Na$^+$/K$^+$ pump	H	D
Cl$^-$ pump	H,D	H,D

induced: the induction is impeded by hyperpolarization but facilitated by depolarization of the resting cell (Figure 2.35).

To trigger an action potential, depolarization is induced by the gating of Na$^+$ channels or nonspecific cation channels. Below the threshold potential, each channel responds independently. Above the threshold potential, the channels cooperate and they activate each other. The result is a strong influx of sodium ions, shifting the membrane potential to positive values for a short time (Figure 2.36). Simultaneously, leak channels for potassium ions are closed.

The Na$^+$ influx is terminated by a K$^+$ efflux. To this end, voltage-dependent K$^+$-channels are opened, though with some delay. To give the K$^+$ channels time for rapid repolarizatioin, sodium channels are desensitized, becoming reactive again only after a defined refractory period (see Sections 14.2 and 14.4.1). This sequence of events determines the frequency of successive action potentials and, at least in neurons, lets the signal spread in only one direction (for instance, along the axon towards the synapse) because during its refractory period a channel cannot respond to a signal transmitted by the next channel (Section 14.2). Again an analogy with a fuse that cannot burn down backwards is obvious (though in contrast to a fuse, the cell's membrane potential regenerates itself!).

Since action potentials follow an all-or-nothing rule, they cannot be modulated in their intensity but only in their frequency; thus they represent prototypes of oscillating signals. Depending on the tissue and species, this frequency varies between less than 1 (heart muscle) and more than 1000 (skeletal muscle) s^{-1}. The major effects of action potentials are muscle contraction and neurotransmitter secretion. In both cases Ca^{2+} ions are the intracellular key signals being released into the cytoplasm through Ca^{2+} channels that open upon depolarization (see Section 14.5).

2.10.3 Ion channels

Ion channels are switches driven by the energy of the membrane potential (Fig. 2.36). They consist of proteins that penetrate the cell membrane, forming a permeable pore that is lined by hydrophilic amino acid residues. For this purpose at least 4×2 transmembrane domains must be arranged in a circular pattern. Such transmembrane domains, mostly α-helices, are either parts of one polypeptide chain or distributed over several protein subunits. To open the pore—a process called gating—the transmembrane domains are assumed to twist like an iris aperture.

Depending on the type of input signals, a distinction is made between voltage- and ligand-controlled channels. Voltage-controlled channels contain transmembrane domains acting as voltage sensors because their conformation responds to changes of the membrane potential. Ligand-controlled channels harbor binding

Figure 2.35 Action potential When a depolarization exceeds a threshold potential (here assumed to be −30 mV) in excitable cells, an action potential of varying length may develop (here depicted for a nerve cell). Its induction is facilitated by rising (depolarizing) and impeded by lowering (hyperpolarizing) the resting potential. A few of the factors that change the resting potential when they are activated or inactivated (D, depolarizing effect; H, hyperpolarizing effect) are listed. Whether an influx or efflux of chloride ions induces hyper- or depolarization depends on the ratio between the resting potential and the intracellular chloride concentration.

Figure 2.36 Ion channel switch An input signal (a chemical ligand or a change of the membrane potential) controls the gating of an ion channel in the cell membrane. The resulting change of the cytoplasmic ion concentration induces a conformational and functional change of an effector protein, generating the output signal. Depending on the ion type, one distinguishes between pH-dependent effects (by protons), a direct modulation by ions of protein function (in particular by Ca^{2+}), and voltage-dependent alterations. The energy needed for signal transduction is supplied by ion gradients. These batteries become recharged by ATP-consuming ion pumps.

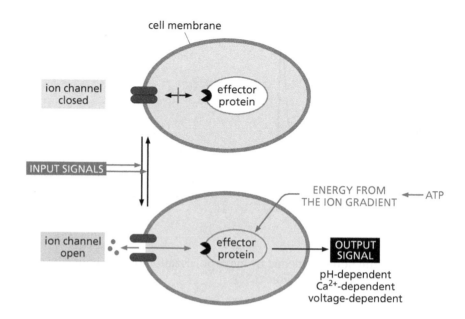

sites for extracellular or intracellular signal molecules. Some ion channels can be voltage- as well as ligand-controlled. A special type of ion channel responds to pressure.

In its simplest form, the gating of an ion channel resembles a binary switch in that the channel is either closed (value 0) or open (value 1). In reality, however, an adjustable time factor and a large number of interacting switches are added to allow transformation of the digital signal into an analog one. In addition, many channels possess a built-in switching-off device, turning the channel into a desensitized state upon prolonged activation. This protective mechanism strikingly resembles that of G-protein and protein phosphorylation switches, which also undergo auto-inhibition.

Ion channels let pass either Na^+, K^+, Ca^{2+}, or Cl^- ions. The **cation/anion selectivity** is determined by the electrical charge of the channel: positive for anions and negative for cations. The electrostatic interactions between channel and ion are rather weak since otherwise ion transport would be too slow (whereas in reality it approaches the rate of free diffusion). Therefore, highly charged amino acid residues such as carboxy and amino groups are avoided. Instead, the channel is lined by carbonyl groups (cation channel) or amide groups (anion channel) working as electrical dipoles. Depending on the ion type, the gating or closing of a channel has either a depolarizing or a hyperpolarizing effect. Cation channels may be either ion-specific or ion-nonspecific. Voltage-controlled channels are generally highly ion-specific. In addition to the transmembrane domains, they have a short P- (permeability) loop that dives into (but does not penetrate) the lipid bilayer of the membrane. [These P-loops must not be confused with the P (phosphorylation) loops of G-proteins and other phosphotransferases.] It contains several highly conserved amino acid residues and serves as a selectivity filter. Here the carbonyl or amide groups lining the pore compete with water for binding the ions. As a result the ions that are surrounded by water molecules are dehydrated since otherwise they could not pass the channel. To this end the interaction between ion and P-loop must supply a sufficient amount of energy to strip the hydrate cover. The more voluminous K^+ ions require less dehydration energy (80 kcal/mol) than the smaller Na^+ ions (98 kcal/mol). This difference together with the channel's diameter may explain the ion selectivity of voltage-controlled channels. Thus, the negative charge of the K^+ filter is sufficient to strip K^+ ions but not Na^+ ions, whereas a Na^+-selective channel, though it is able to dehydrate both ion types, is too narrow for K^+ ions to pass. In fact, voltage-dependent Na^+ channels are more than 10,000 times more permeable to Na^+

than to K$^+$, and vice versa. In contrast to highly specific voltage-controlled channels, most ligand-gated cation channels are permeable for K$^+$, Na$^+$, and Ca^{2+}.

Ion channel proteins constitute several families originating from ancient prokaryotic genes (details regarding the evolution of ion channels are found Section 3.5.7). The largest family, called the **S5–S6 family**, covers all passive and voltage-controlled ion channels, though some of them are controlled in addition by ligands. The term S5–S6 refers to a common structural feature, the transmembrane segments (or helices) 5 and 6 of the voltage-dependent K$^+$ channel of vertebrates (Section 14.4.1). This channel is composed of two subunits with six transmembrane helices (S1–S6) and a P-loop localized between S5 and S6. Helix S4 is the voltage sensor.

A simpler construction is shown by another K$^+$ channel of higher eukaryotes: the inwardly rectifying K$_{ir}$ channel, which consists of four subunits with two transmembrane helices each, resembling S5 and S6, and a P-loop in between (Section 14.4.2). This structure is quite similar to that of the prokaryotic K$^+$ channel KcsA. Still simpler are the S5–S6 channels lacking a P-loop, such as the epithelial Na$^+$ channel ENaC (Section 14.3), the ATP-receptor channel P2X (Section 16.6), the cAMP- and cGMP-controlled cation channels of sensory cells (Section 14.5.7), and various neuropeptide receptors of nonvertebrates. Starting from these simplest forms, the more complex ion channels are thought to have developed by gene doubling and gene enlargement. The subunits of the S5–S6 channels always form a single pore.

The **anion channels** that transport Cl$^-$ ions, in particular, belong to another family. They have a rather complex structure with at least 12 transmembrane domains and are dimers with two ion pores. A structurally unrelated chloride channel is CFTR, which belongs to the large family of ABC transporters. This ATP-dependent channel plays an important role in transepithelial electrolyte transport and is discussed together with other anion channels in detail in Section 14.7.2.

Other ion channel families cover the majority of **ligand-controlled channels**. By fusion of a S5–S6 channel protein with a prokaryotic glutamate binding protein, a special channel type, the ionotropic glutamate receptors, has developed. These are nonspecific cation channels that play a central role in the brain (Section 16.4). Receptor ion channels for other neurotransmitters constitute a separate family with the nicotinic acetylcholine receptor being the prototype (Sections 16.2 and 16.3). They are either nonspecific cation channels or anion channels made up of five subunits with four transmembrane domains each. A key role in cellular signal processing is played by the Ca^{2+} channels of the endoplasmic reticulum. They exist as tetramers, with six transmembrane domains and a P-loop-like structure each, and are under the control of intracellular signal molecules such as Ca^{2+}, inositol 1,4,5-trisphosphate, and others (Section 14.5.3).

Summary

The electrical charge of cell membranes (negative inside, positive outside) is due to semipermeability of the lipid bilayer and to ATP-driven ion pumps that generate disequilibrium of Na$^+$, K$^+$, Ca^{2+}, and Cl$^-$ ions. By coupling the influx and efflux of these ions through special ion channels with signal-transducing processes, cells use the energy stored in the "membrane capacitor" for signal processing. Excitable cells (particularly nerve and muscle cells) are able to generate action potentials, self-propagating waves of complete membrane depolarization that represent the fastest intercellular signal known. Action potentials are triggered when the membrane depolarization exceeds a certain threshold value. To generate an action potential, a particular arrangement of cooperating voltage-dependent ion channels is needed. Action potentials are frequency-modulated signals. Ion channels are membrane proteins with at least four transmembrane domains that form a central pore for the transport of either cations (Na^+, K^+, Ca^{2+}) or anions (mainly Cl^-). Most ion channels are either ligand- or voltage-operated (although some more specialized channels also respond to other stimuli such as pressure). In contrast to ligand-dependent channels, voltage-operated channels are ion-selective due to subtle differences in pore diameter and in the ability to dehydrate ions. All eukaryotic ion channels seem to have a prokaryotic ancestry. Major families are S5–S6 channels, including all voltage-dependent channels, and ligand-dependent channels related to the nicotinic acetylcholine receptor.

2.11 Proteolysis switch: protein degradation provides messenger molecules and energy

The energy required for signal transduction is supplied not only by the hydrolysis of energy-rich metabolites and by the membrane potential but also, in certain cases, by the hydrolytic degradation of macromolecules, in particular, proteins. Prominent examples include the proteolytic cascades regulating blood clotting, nonspecific defense (complement system), and programmed cell death (apoptosis). Such a mechanism kills two birds with one stone in that, besides energy, a functionally active fragment of the corresponding protein is released. This holds true also for signal transduction across cell membranes, which makes use of protein degradation to gain generation of both energy and signaling peptides that act as second messengers, resembling an output signal (Figure 2.37). An example of this type of signaling is the Notch system (Section 13.1). Of course, signal transduction by proteolysis is a borderline case and one may hesitate to call this kind of reaction a switch. After all, it is not easy to set such a switch back to 0, as this requires a costly and time-consuming de novo synthesis of proteins.

Summary

The energy stored in the peptide bond can be used to drive signal transduction, and the peptide fragment generated by limited proteolysis may function as a second messenger.

2.12 Receptors: how energy-supplying reactions are combined with signal transduction

Receptors are prototypical signal-transducing proteins. As such, they are subject to the same rules and laws as any other protein of the data-processing network, including allosteric regulation by noncovalent interactions, mainly with their ligands, and post-translational modifications for control and fine-tuning of receptor activity, oligomerization, and communication with other proteins via interaction domains. There is only one property that distinguishes receptors

Figure 2.37 Proteolysis switch Here the input signal activates a protease that releases an active fragment from an inactive precursor protein. The fragment produces the output signal. The switching energy stems from hydrolysis of the peptide bond.

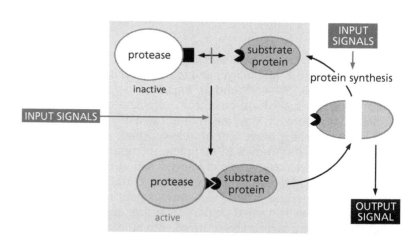

from other signaling proteins: they are specialized to recognize *extracellular* signals, acting as the sensory organs or, if one prefers a technical metaphor, the users' interface of the cell. Such signals may be endogenous messengers such as hormones, neurotransmitters, key metabolic compounds, or environmental stimuli. Like any signal-transducing process, receptor activity requires energy input. In fact, receptors provide a particularly instructive example of the combination of signal transduction with energy-supplying reactions.

The existence of receptors was proposed at the end of the 19th century. British physiologist J. N. Langley postulated that, in order to become effective, pharmacological drugs had to be bound in the organism by "receptive substances" (endogenous signals such as hormones and neurotransmitters were still unknown at this time). At about the same time, Paul Ehrlich proposed that antigens, but also other substances, were bound at the cell surface by receptors which he called "Seitenketten" (side chains), expecting them to recognize the *molecular shape* of the ligand according to the lock–key principle. This concept of a stereospecific structural complementarity between ligand and receptor is still the basis of modern receptor theory (Figure 2.38).

A receptor transforms an exogenous input signal into an intracellular signal that is fed into the data-processing protein network. Therefore, a receptor protein consists of a highly specific recognition or discriminator domain and an effector domain. The interaction of the recognition domain with the input signal is always noncovalent and thus fully reversible. It induces an allosteric activation of the effector domain, which then can interact with downstream signaling elements. Since most receptors are complex proteins with a quaternary structure, this effect is cooperative, with sigmoidal kinetics. In other words, receptors are switches with a built-in noise filter.

While recognizing the input signal with high selectivity, a receptor may interact with various downstream pathways that sometimes even have opposite effects (Figure 2.38). However, a single signaling pathway may also be controlled by various receptors (Figure 2.39). In addition, most receptors exist in several isoforms, combining a more or less identical discriminator domain with different effector domains. Taken together, this explains why the meaning of a signal is ambiguous and critically depends on the cell type and the physiological context (Figure 2.40).

The majority of drugs and toxins interact with cellular receptors of endogenous or environmental signals. They are called agonists when they imitate and antagonists when they counteract the effects of the natural ligands. In many cases, the dose of a drug or poison determines an agonistic or an antagonistic effect. Such effects are most easily explained by the theory of allostery, which predicts that receptor proteins exist in at least two conformations, an active one and an

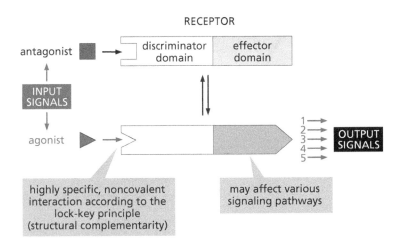

Figure 2.38 Principles of receptor function Like all signal-transducing proteins, receptors are composed of a receiver or discriminator domain, recognizing an input signal, and a transmitter or effector domain, generating an output signal. The receptor protein exists in an allosteric equilibrium between an inactive (upper) and an active (lower) conformation. Input signals stabilize either the active or the inactive form; that is, they exhibit either agonistic or antagonistic effects.

71

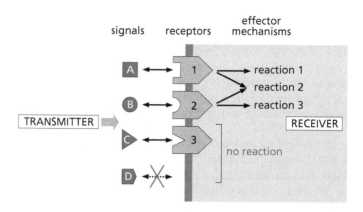

Figure 2.39 Lock–key principle of receptor function Signaling molecules fit the binding site of the receptor like a key fits a lock. Depending on the effector coupling of its receptor (pink), a signal (A or B) may induce different reactions (1 or 3) or two different signals (A and B) may evoke the same reaction (2). Receptors without effector coupling (3) and signals without a corresponding receptor (D) have no effect.

inactive one. Agonists stabilize the active form, pushing the allosteric equilibrium in the corresponding direction, whereas antagonists do the same with the inactive form (Figure 2.38). Upon overstimulation, most receptors reach a refractory state that does not depend any longer on the presence of a ligand. This desensitization represents a built-in mechanism of protection.

Receptors are coupled with all kinds of energy-supplying switching reactions. The **receptors of lipophilic ligands**, which can easily pass the cell membrane—for instance, the steroid hormones—are often found in the cytoplasm or nucleus. They constitute a family of ligand-controlled or nuclear transcription factors that derive their signaling energy from protein–DNA interactions and chaperone-catalyzed ATP hydrolysis (Section 8.3). Other cytoplasmic receptors such as the NO-binding soluble guanylate cyclases do not function as transcription factors (Section 4.5). Finally, plants express hormone receptors (for auxin and gibberellic acid) with a built-in E3 ubiquitin ligase activity that is stimulated upon ligand binding, triggering gene expression by inducing the proteolytic degradation of transcriptional repressors (Section 13.1.1).

The **receptors of hydrophilic ligands**, which are unable to penetrate the cell membrane, are transmembrane proteins, exposing their discriminator domains outside and their effector domains inside the cell. The effector domains are coupled with various switching proteins such as GTPases, protein kinases or phosphatases, ion channels, and others (Figure 2.41). Depending on the receptor type, this coupling is either covalent, with the effector protein being a domain of the receptor protein, or occurs via a noncovalent interaction between both partners.

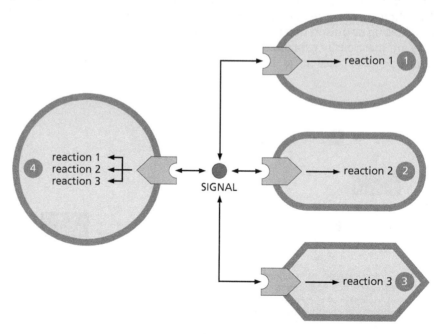

Figure 2.40 Ambiguity of signals As one lock and key may open the doors to different rooms, one signal may evoke different reactions in different cell types (cells 1, 2, and 3), provided its receptor (pink) exhibits different effector couplings. Different effector couplings may also exist within one cell (cell 4).

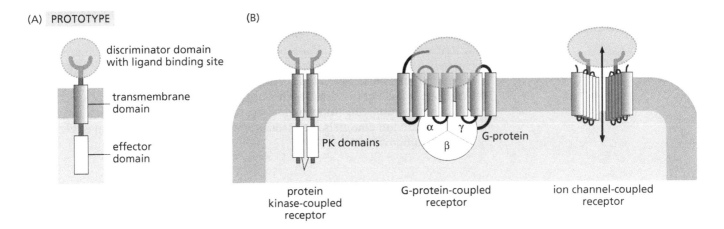

The three most abundant types of membrane receptors are shown in Figure 2.41 and are described below.

(1) **GTPase-coupled receptors** all have a characteristic structure of seven trans-membrane domains and activate via a large heterotrimeric G-protein enzymes that produce second messengers (Sections 4.3 and 4.4). Although the structural principle is of prokaryotic origin, these receptors are restricted to eukaryotes and constitute the largest receptor family in animals (not in plants; see Chapter 5). Receptor and G-protein are always separate.

(2) **Protein kinase or phosphatase-coupled receptors** exist at least as dimers, activating each other by trans-autophosphorylation or -dephosphorylation. There are three types of kinase-coupled receptors: histidine kinase-coupled receptors, found mainly in prokaryotes (Section 3.3); serine/threonine kinase-coupled receptors, abundant in all organisms (Chapter 6); and tyro-sine kinase-coupled receptors, found as yet only in animal cells (Chapter 7). Receptor and protein kinase are either separate proteins or parts of one polypeptide chain.

(3) **Ion channel-coupled receptors** have one or more ligand binding sites and are found in pro- as well as in eukaryotes (Section 3.5.7 and Chapter 16).

Special groups not shown in Figure 2.41 are **guanylate cyclase-coupled receptors**, which upon ligand interaction catalyze synthesis of the second messenger cGMP (Section 4.5), and **protease-** and **ubiquitin E3 ligase-coupled receptors**, which release peptide second messengers or trigger the degradation of signaling proteins (Section 13.1).

The human genome harbors approximately 1500 genes encoding putative recep-tor proteins. It has been estimated that receptors can occupy up to 25% of the cell surface which in an idealized average cell (10 µm radius) would amount to 300 µm². Given a radius of 5 nm, a receptor protein would occupy 75 nm² of this area. This means that a cell can expose at its surface 4 million receptor molecules at most, which is a value widely matched by experimental data.

From the 1500 receptor genes, only a few are expressed in a given tissue. Nonetheless, a cell possesses a surprisingly large number of possibilities to respond. Even if only 1% of the genes is active—that is, 15 different receptor types are expressed—a cell could theoretically discriminate between $2^{15} = 32,678$ different combinations of ligands (each receptor is assumed to exist only in two states, liganded or not liganded). This model calculation impressively demon-strates that the performance of a signal-processing apparatus depends on the combinations between, rather than on the actual number of, switching events. In the following chapters we shall explore additional examples of this principle.

Figure 2.41 Three most abundant types of transmembrane receptors (A) Prototype of a transmembrane receptor consisting of an extracellular discriminator domain, a transmembrane domain, and an intracellular effector domain. (B) The three most abundant types are the protein kinase (PK)-coupled receptor, forming dimers usually with two transmembrane domains; the G-protein-coupled receptor, with seven transmembrane domains and an associated trimeric G-protein ($\alpha\beta\gamma$) being the effector domain; and the ion channel-coupled receptor.

Summary

As sensory devices, receptors connect the environment with the data-processing protein network of the cell. Most receptors are transmembrane proteins, but special receptors are found also in the cytoplasm and in cellular organelles. Receptors specifically recognize exogenous signals according to the lock–key principle. To deliver the signal to the intracellular network, receptors are coupled with energy-supplying biochemical reactions. The corresponding enzymatic domains and ion channels are associated with the receptor protein either covalently or noncovalently.

2.13 Brief summary of experimental standard methods for investigation of signaling pathways: the classical approach

Signal transduction occurs in highly complex networks of interacting proteins. For the sake of clarity and to approach the problem experimentally, these networks are *artificially* dissected into individual linear cascades of interconnected partner molecules. The standard procedure to analyze such a signaling cascade is to manipulate a selected step with inhibitors or activators and to measure the effect on what are assumed to be the subsequent steps. This may be done by several experimental settings, as shown schematically in Figure 2.42 and with the help of an example in Figure 2.43.

The simplest approach is to employ chemical inhibitors. A large number of pharmacologically active drugs, as well as natural and synthetic toxins, affect signal-processing proteins. However, the specificities of such compounds do not always meet the expectations because they frequently interact with domains that, due to the modular protein structure, are found in more than one protein. A better, though more expensive, approach is to mutate the corresponding protein itself into a permanently active or an inactive form and then let this mutant compete with the cellular wild-type protein.

A constitutively active form is generally obtained by a genetic knock-out of a regulatory domain that exhibits an auto-inhibitory effect (as, for instance, the pseudosubstrate domain in protein kinases), whereas an inactivation of the effector domain results in a dominant-negative mutant that acts as a competitive inhibitor. By means of targeted (amino acid-specific) mutations, the effect measured may be related to structural parameters. Other procedures make use of anti-sense mRNA and siRNA (Section 8.8) as inhibitors. Mutated proteins and other

Figure 2.42 Analysis of a signaling cascade For this hypothetical reaction cascade starting with a receptor protein B is assumed to interact with downstream components C and D. To prove this, the effect on C and D of either inhibition or activation of B is measured. Methods to localize B in the cell or to prove a direct interaction between B and C are also mentioned. They will be discussed in more detail in Section 5.2.

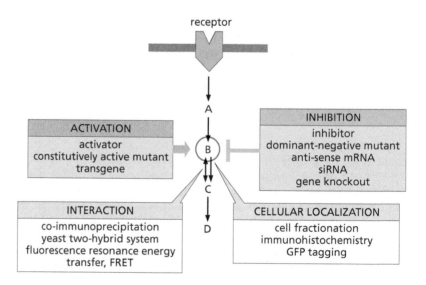

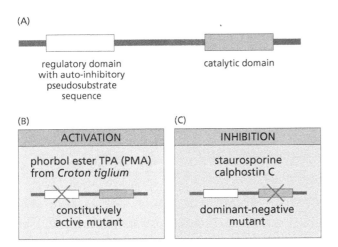

Figure 2.43 Experimental activation and inhibition of protein kinase C
(A) Diagram of the protein kinase C molecule with regulatory and catalytic domains. (B) Protein kinase C is activated relatively specifically by the phorbol ester TPA, a plant toxin (see Section 4.6.5) that binds to the regulatory domain mimicking the effect of the second messenger diacylglycerol. A constitutively active kinase mutant is obtained by inactivation or deletion of the auto-inhibitory pseudosubstrate domain C) The fungal toxin staurosporine (from actinomycetes) interacts with the catalytic domain but inhibits, in addition to protein kinase C, various other protein kinases. More specific is calphostin C (from *Cladosporium* species), which interacts with the regulatory domain. A dominant negative mutant is obtained by deletion of the catalytic domain.

inhibitors and activators mostly are tested on cell cultures. However, targeted mutations as well as the transgene and knockout procedures also allow studies on living model organisms.

When dominant-negative mutants are used, the investigator imitates nature. In fact, physiological protein variants resembling dominant-negative mutants are frequently employed by cells to control signaling pathways. As a rule, they are produced via alternative gene promoters or by alternative splicing of pre-mRNA. Inversely, constitutively active proteins may arise when auto-inhibitory domains (such as the pseudosubstrate sequences of protein kinases) are enzymatically removed. In contrast, constitutively active *gene* mutations are mostly pathological derailments that may cause diseases like cancer.

The experimental methods mentioned above are by no means free of problems and pitfalls. So, a transfection of cells with a large amount of manipulated protein may cause unpredictable disturbances of cellular signal processing that may not have anything to do with the process to be analyzed. In fact, when the complexity of cellular signal processing is considered, it is clear that a large amount of painstaking work is required to arrive at conclusive results.

The final goal of all such studies is to understand in detail the molecular mechanisms as well as the biological significance of a given process. For a mechanistic elucidation, a precise knowledge of the three-dimensional protein structure is mandatory. Here the methods of choice are X-ray crystallography and other sophisticated physical techniques such as mass spectrometry. The proof of physiological significance requires studies on appropriate model organisms and time-consuming clinical and epidemiological evaluations.

Summary

The standard procedure to analyze a sequence (cascade) of interconnected signaling reactions is to inhibit or stimulate a selected step and measure the effect on what are assumed to be the subsequent steps. This can be done by biochemical or pharmacological means or by genetic manipulation either *in vitro* or *in vivo*. Further mechanistic elucidation requires a detailed knowledge of molecular interactions and structures.

2.14 Model organisms for investigation of cellular signal processing

Although the complexity of cellular data processing has steadily increased during the course of evolution, the basic mechanisms entered the stage at the very beginning and have continued up to the present. Therefore, experiments with simpler organisms provide a key for understanding events going on in vertebrate cells and in humans. In addition to prokaryotes such as *Escherichia coli*, the following model organisms are most frequently used for the investigation of cellular signal transduction:

- yeast

- the slime mold *Dictyostelium discoideum*

- the nematode *Caenorhabditis elegans*

- the fruit fly *Drosophila melanogaster*

- the plant *Arabidopsis thaliana*

- several vertebrates such as the frog *Xenopus laevis*, the rat, and the mouse

As the simplest eukaryotes, **yeasts** are particularly easy to handle and to study and, most importantly, can be genetically manipulated almost as easily as bacteria. Yeast mutants have advanced the investigation of basic signaling mechanisms in a crucial manner, and upon searching for analogies and homologies, observations on yeast cells have led quite often to pioneering discoveries in vertebrates. Most studies were done using the bakers' yeast *Saccharomyces cerevisiae* and the fission yeast *Schizosaccharomyces pombe* (used in Africa for brewing the millet beer pombe). *S. cerevisiae* possesses about 6000 genes, and *S. pombe* has about 5000. Being unicellular organisms, yeasts are unsuited for investigation of problems related to multicellularity and tissue formation.

In this respect, ***Dictyostelium discoideum*** offers particularly attractive possibilities. Under favorable environmental conditions, the cells of this slime mold are living as free unicellular amoebae. Upon stress, however, they assemble into aggregates of up to 100,000 cells, which go for prey catching as a whole (therefore also called social amoebae) or, upon prolonged stress, differentiate into complex-structured spore carriers. This organism standing at the border between uni- and multicellularity is best suited for studies on chemotaxis, cell aggregation, cell motility, and cell differentiation especially since it can be genetically manipulated almost as easily as yeast.

The nematode ***Caenorhabditis elegans*** ranks among the simplest and, at the same time, best investigated metazoa. It is built up from exactly 959 somatic cells (plus a few thousand germ cells), the development and differentiation of which are precisely determined and can be studied in detail. Again the organism is easy to manipulate genetically, thus offering a significant experimental advantage. *C. elegans* is a standard model for studies on cell proliferation and programmed cell death and on intercellular interactions during embryonic development. In the course of such studies, surprising parallels to the situation in vertebrate cells have been revealed. With about 19,000 genes, the genome of the worm is already rather complex.

The tiny fly ***Drosophila melanogaster*** is considered a pet of geneticists and developmental biologists. It is easy to breed, has a very short generation time (of about 10 days), and is prone to gene mutations. The organization of the genetic material in giant chromosomes considerably facilitates cytogenetic and molecular genetic studies. So *Drosophila* is by far the best-investigated higher invertebrate. As far as the mechanisms of signal transduction are concerned, striking similarities to mammalian cells exist. In fact, many fundamental concepts on cellular data processing have been established based on studies on *Drosophila*. With about 14,000 genes, the genome of the fly is surprisingly small, at least when compared with that of the worm.

The plant ***Arabidopsis thaliana***, belonging to the family Cruciferaceae, is an abundant weed. The ease with which this plant can be bred in the laboratory has made it the botanical counterpart of *Drosophila*. Moreover, *Arabidopsis* is the first plant in which the genome has been completely sequenced. With 26,000 genes it is almost as large as the human genome. Studies on signaling have revealed, in addition to analogies, a number of unexpected differences between plants and animals.

The mechanisms of signal processing in **vertebrates,** especially in mammalian cells, differ from species to species only in minor variants. That means that studies on mice or rats are generally also meaningful for humans.

Cell cultures enjoy great popularity since they are simply organized, easy to manipulate, and usually inexpensive. However, cultivated cells behave differently than cells in an intact tissue, a reservation that holds true in particular for the very popular transformed cell lines, which are purely artificial products. Therefore, it is mandatory to verify the physiological relevance of a result obtained *in vitro* by studies on intact tissues or the whole organism. With this reservation, cell cultures are indispensable for the investigation of cellular data processing.

Summary

Because of the evolutionary conservation of cellular signal processing, the basic mechanisms can be investigated by use of simpler experimental models that provide the possibility of easy genetic manipulation. These include yeast, slime molds, nematodes, and insects as well as cell cultures. To prove the significance of the data thus obtained, *in vivo* experiments with vertebrates as well as (for humans) epidemiological and clinical studies are mandatory.

Further reading

Aguilar RC & Wendland B (2003) Ubiquitin: not just for proteasomes anymore. *Curr. Opin. Cell Biol.* 15, 184–190.

Bos JL, Rehmann H & Wittinghofer A (2007) GEFs and GAPs: critical elements in the control of small G proteins. *Cell* 129, 865–877.

Bukau B, Weissmann J & Horwich A (2006) Molecular chaperones and protein quality control. *Cell* 125, 443–451.

Corda D & Di Girolamo M (2003) Functional aspects of protein mono-ADP-ribosylation. *EMBO J.* 22, 1953–1958.

Davey MJ, Jeruzalmi D, Kuriyani J et al. (2002) Motors and switches: AAA+ machines within the replisome. *Nat. Rev. Mol. Cell Biol.* 3, 826–835.

DeMartino GN & Gillette TG (2007) Proteasomes: machines for all reasons. *Cell* 129, 659–662.

Doherty FJ, Dawson S & Mayer RJ (2002) The ubiquitin-proteasome pathway of intracellular proteolysis. *Essays Biochem.* 38, 51–63.

Ellis RJ (2006) Molecular chaperones: assisting assembly in addition to folding. *Trends Biochem. Sci.* 31, 395–401.

Fu M, Wang C, Wang J et al. (2002) Acetylation in hormone signaling and the cell cycle. *Cytokine Growth Factor Rev.* 13, 259–276.

Gallego M & Virshup DM (2005) Protein serine/threonine phosphatases: life, death, and sleeping. *Curr. Opin. Cell Biol.* 17, 197–202.

Geiss-Friedländer R & Melchior F (2007) Concepts of sumoylation: a decade on. *Nature Rev. Mol. Cell Biol.* 8, 947–956.

Goldstein LSB (2001) Molecular motors: from one motor many tails to one motor many tales. *Trends Cell Biol.* 11, 477–482.

Haddad JJ (2002) Antioxidant and prooxidant mechanisms in the regulation of redox(y)-sensitive transcription factors. *Cell. Signalling* 14, 879–897.

Haglund K, Di Fiore PP & Dikic I (2004) Distinct monoubiquitin signals in receptor endocytosis. *Trends Biochem. Sci.* 28, 598–604.

Hall A. (ed) (2000) GTPases. Oxford University Press.

Hanks SK (2003) Genomic analysis of the eukaryotic protein kinase superfamily: a perspective. *Genome Biol.* 4, 111.

Hanks SK & Hunter T (1995) The eukaryotic protein kinase superfamily: kinase (catalytic) domain structure and classification. *FASEB J.* 9, 576–595.

Hanson PI & Whiteheart SW (2005) AAA+ proteins: have engine, will work. *Nat. Rev. Mol. Cell Biol.* 6, 519–520.

Harper JW & Schulman BA (2006) Structural complexity in ubiquitin recognition. *Cell* 124, 1133–1136.

Harrison A & King SM (2000) The molecular anatomy of dynein. *Essays Biochem.* 35, 75–87.

Hicke L, Schubert HL & Hill CP (2005) Ubiquitin-binding domains. *Nat. Rev. Mol. Cell Biol.* 6, 610–621.

Hochstrasser M (2006) Lingering mysteries of ubiquitin-chain assembly. *Cell* 124, 27–34.

Huse M & Kuriyan J (2002) The conformational plasticity of protein kinases. *Cell* 109, 275–282.

Johnson ES (2004) Protein modification by SUMO. *Annu. Rev. Biochem.* 73, 355–382.

Johnson LN & Barford D (1993) The effects of phosphorylation on the structure and function of proteins. *Annu. Rev. Biophys. Biomol. Struct.* 22, 199–232.

Johnson LN & O'Reilly M (1996) Control by phosphorylation. *Curr. Opin. Struct. Biol.* 6, 762–769.

Kalhammer G & Bähler M (2000) Unconventional myosins. *Essays Biochem.* 35, 33–42.

Karcher RL, Deacon SW & Gelfand VI (2002) Motor-cargo interactions: the key to transport specificity. *Trends Cell Biol.* 12, 21–27.

Kennelly PJ (2001) Protein phosphatases—a phylogenetic perspective. *Chem. Rev.* 101, 2291–2312.

Kirkin V & Dikic I (2007) Role of ubiquitin- and Ubl-binding proteins in cell signaling. *Curr. Opin. Cell Biol.* 19, 199–205.

Kraus WL & Lis JT (2003) PARP goes transcription. *Cell* 113, 677–683.

Lambeth JD (2004) NOX enzymes and the biology of reactive oxygen. *Nat. Rev. Immunol.* 4, 181–189.

Manning G, Plowman GD, Hunter T et al. (2002) Evolution of protein kinase signaling from yeast to man. *Trends Biochem. Sci.* 27, 514–520.

Manning G, Whyte DB, Martinez R et al. (2002) The protein kinase complement of the human genome. *Science* 298, 1912–1934.

Minsky A, Shimoni E & Frenkiel-Krispin D (2002) Stress, order and survival. *Nat. Rev. Mol. Cell Biol.* 3, 50–60.

Nijman SMB, Luna-Vargas MPA, Velds A et al. (2005) A genomic and functional inventory of deubiquitylating enzymes. *Cell* 123, 773–786.

Pearl LH & Prodromou C (2006) Structure and mechanism of the Hsp90 molecular chaperone machinery. *Annu. Rev. Biochem.* 75, 271–294.

Pickart CM (2004) Back to the future with ubiquitin. *Cell* 116, 181–190.

Reichard P (2002) Ribonucleotide reductases: the evolution of allosteric regulation. *Arch. Biochem. Biophys.* 397, 149–155.

Rhee SG, Bae YS, Lee SR et al. (2000) Hydrogen peroxide: a key messenger that modulates protein phosphorylation through cysteine oxidation. *SciSTKE* 2000/53/pe1.

Saibil HR & Ranson NA (2002) The chaperonin folding machine. *Trends Biochem. Sci.* 27, 627–632.

Sauer RT, Bolon DN, Burton BM et al. (2004) Sculpting the proteome with AAA+ proteases and disassembly machines. *Cell* 119, 9–18.

Schrödinger E (1946) What is life? Cambridge University Press.

Schwartz DC & Hochstrasser M (2003) A superfamily of protein tags: ubiquitin, SUMO and related modifiers. *Trends Biochem. Sci.* 28, 321–328.

Smith S (2001) The world according to PARP. *Trends Biochem. Sci.* 26, 174–179.

Söti C, Pal C, Papp B et al. (2005) Molecular chaperones as regulatory elements of cellular networks. *Curr. Opin. Cell Biol.* 17, 210–215.

Sun L & Chen ZJ (2004) The novel functions of ubiquitylation in signaling. *Curr. Opin. Cell Biol.* 16, 119–126.

Toledano MB, Delaunay A, Monceau L et al. (2004) Microbial H_2O_2 sensors as archetypical redox signaling modules. *Trends Biochem. Sci.* 29, 351–357.

Ubersax JA & Ferrel JE Jr (2007) Mechanisms of specificity in protein phosphorylation. *Nat. Rev. Mol. Cell Biol.* 8, 530–541.

Verhey KJ & Rapoport TA (2001) Kinesin carries the signal. *Trends Biochem. Sci.* 26, 545–549.

Welchman RL, Gordon C & Mayer RJ (2005) Ubiquitin and ubiquitin-like proteins as multifunctional signals. *Nat. Rev. Mol. Cell Biol.* 6, 599–609.

Westheimer FH (1987) Why nature chose phosphates. *Science* 235, 1173–1178.

Woehlke G & Schliwa M (2000) Walking on two heads: the many talents of kinesin. *Nat. Rev. Mol. Cell Biol.* 1, 50–58.

Young JC, Barral JM & Hartl FU (2003) More than folding: localized functions of cytosolic chaperones. *Trends Biochem. Sci.* 28, 541–547.

Evolution of Cellular Data Processing

Being a condition of survival, the ability to adapt to an ever-changing environment has accompanied life from the very first day. This means that even the most ancient and primitive cell must have been able to recognize and to rate relevant environmental signals. As explained in Chapter 1, this is done by a data-processing protein network that, particularly for unicellular organisms such as prokaryotes, fully resembles a nervous system, justifying the metaphor of a cellular "nanobrain." Investigation of prokaryotic signal transduction has shown that, in its basic biochemical mechanisms, such a primeval network does not differ fundamentally from signal processing networks in eukaryotes and the mammalian nervous system, thus nourishing the concept of a common evolutionary ancestry that had entered the stage in the beginning of life. A closer look at today's prokaryotes gives us some idea of how this might have occurred in the past. It immediately makes clear that the data-processing network must have evolved in parallel to the metabolic network.

3.1 Evolution of biological signal processing

Organisms are classified in three domains: bacteria and archaea (or archaebacteria), together constituting the prokaryotes, and eukaryotes. Eukaryotes differ from prokaryotes by a nuclear membrane and an intracellular compartmentalization in structures such as endoplasmic reticulum, Golgi apparatus, mitochondria, and chloroplasts. Genetic studies have revealed that eukaryotes are related more closely to archaea than to bacteria (Figure 3.1). This was a somewhat surprising discovery because archaea were formerly thought to be the primary

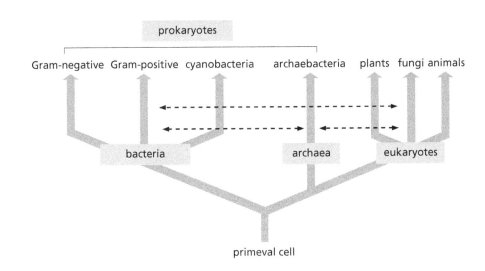

Figure 3.1 Phylogenetic three-domain tree Broken horizontal lines symbolize horizontal gene transfer. For other details see text.

organisms, in particular since they are settlers of extreme biotopes. The high degree of horizontal gene transfer between the three domains was another unexpected finding.

Once considered to be primitive, prokaryotes are now acknowledged as unexpectedly complex and highly organized forms of life. Their amazing ability to settle any possible ecological niche gives the reason for their overwhelming evolutionary success. Leaving aside adaptation by gene mutations, the cellular conditions of this ability are:

- rapid and reversible adaptation of gene expression and, thus, of the protein pattern (or "proteome") to the actual environmental conditions

- precise control of behavior such as motility, metabolism, and transport processes

- establishment of social interactions aimed at genetic recombination and division of labor in multicellular structures

To perform such adaptive responses, the cell must be a cognitive system. What are the primary environmental signals that a prokaryotic "nanobrain" has to decipher for the sake of survival? To exist in an unpredictable environment requires a permanent search for optimal conditions of life, or in other words, minimization of stress. It is conceivable, therefore, that the most elementary forms of signal processing are aimed at coping with stress situations such as starvation, disturbances of osmotic pressure, intoxication, dangerous radiation, and attack by predators. A cell unable to track down energy resources and blind to danger has no chance to survive. Thus, the processing of *environmental signals* stands at the very beginning (Figure 3.2). Social contacts or even the formation of multicellular colonies, including specialization of cells by differentiation aimed at a division of labor, may facilitate survival even more. Both require the evolution of an *intercellular* signaling system. It must be emphasized that, on the level of signal decoding, a cell does not make a distinction between environmental and intercellular signals: both are processed by the same network, provided the cell possesses the corresponding sensors or receptors.

A characteristic example of an ancient intracellular signaling reaction dealing with environmental stress is the **stringent response**. This is a primeval bacterial emergency reaction that saves energy. It leads to an almost complete stop of ribosome production and protein synthesis and, as a consequence, of cell proliferation. These are extremely expensive processes devouring almost 50% of the cellular energy. The stringent response is triggered by worsening life conditions, particularly starvation. Due to a lack of certain amino acids, uncharged tRNA accumulates at the ribosomes. As an immediate consequence, the ribosome-associated enzyme RelA, also known as stringent factor, becomes activated. RelA

Figure 3.2 Evolution of biological signal processing All cells are able to adjust their behavior and pattern of gene expression to exogenous (environmental and intercellular) signals provided these carry a meaning such as food, energy, danger, and partner cells. The meaning is decoded by the cellular apparatus of data processing. The intercellular signaling systems of multicellular organisms have evolved from intercellular communication that prokaryotes require for sexual, symbiotic, and parasitic interactions as well as for the formation of multicellular collectives.

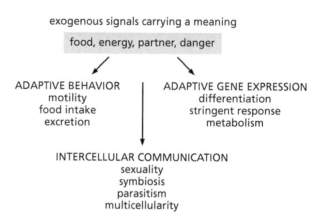

is a guanylate kinase that catalyzes the ATP-dependent pyrophosphorylation of GTP in position 3′, yielding ppGpppp, which subsequently is dephosphorylated to guanosine 3′,5′-bis(diphosphate) (ppGpp) by a specific phosphatase (Figure 3.3). ppGpp has the function of an intracellular emergency signal or "**alarmone**," binding to the catalytic β-subunit of RNA polymerase and inhibiting, in particular, the *de novo* synthesis of ribosomal RNA, ribosomal proteins, translation factors, and tRNA. Cells containing a defective *relA* gene (*relaxed* mutants) are unable to adapt, literally growing to death in deficiency situations. Evolutionarily advanced species such as myxobacteria use the stringent response for a more sophisticated survival strategy, the formation of multicellular spore capsules (see Section 3.4.2).

The enzymatic apparatus of ppGpp synthesis, including RelA-homologous enzymes, has also been found in **plant chloroplasts**, which are thought to have evolved from bacteria. Like the bacterial system, the plant system is stimulated under stress situations, here in particular by wounding. Eukaryotes possess, in addition, other highly efficient means to turn down translation upon amino acid deficiency, operating by mechanisms that differ entirely from the stringent response (Sections 9.3 and 9.4).

Summary

By processing a wide variety of exogenous stimuli prokaryotes easily adapt to environmental conditions. The most elementary forms of signal processing are aimed at coping with emergency situations (stress). An example is provided by the stringent response, an adaptation of cellular protein synthesis to environmental stress. The response is based on the generation of intracellular messengers (alarmones such as ppGpp) that suppress RNA polymerase activity. Social interactions, cell differentiation and formation of multicellular structures depend on the evolution of *intercellular* signaling systems. Environmental and intercellular signals are processed by the same protein-DNA network.

Figure 3.3 Signaling of the stringent response in *E. coli* Stress situations such as a shortage of certain amino acids cause an idling of translation, resulting in activation of the ribosomal kinase RelA. The latter transfers a pyrophosphoryl group from ATP to the 3′-position of 5′-GTP. ppGpppp thus generated becomes dephosphorylated enzymatically, yielding the "alarmone" ppGpp that acts as a transcriptional suppressor of genes encoding RNA and proteins required for translation. The result is an almost complete stop of protein synthesis. P, phosphate group.

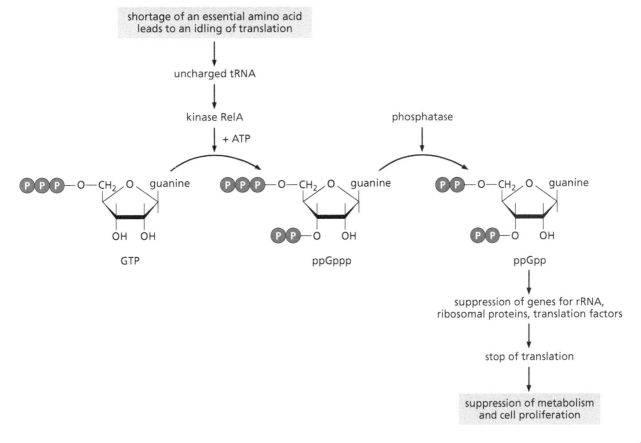

3.2 Signal-controlled transport across cell membranes: the ancient way to communicate

The controlled uptake of food and the excretion of (toxic) metabolic end products stand at the very beginning of any communication between an organism and its environment. Since the lipid bilayer of the plasma membrane is impermeable for most substances, even the most primeval cells were forced to develop regulative transport mechanisms based on the function of special membrane proteins.

There are four types of transport processes:

(1) **Passive transport**, *along* a concentration gradient, is made possible by membrane channels or carrier proteins (uniporters) and therefore has been called "facilitated diffusion." Examples of uniporters are the glycerol transporter of *Escherichia coli* and the glucose transporters GluT1 and GluT4 in liver and muscle or fat tissue, respectively (Section 7.1.3).

(2) **Primary active transport**, *against* a concentration gradient, requires energy derived from enzymatic hydrolysis of ATP; that is, membrane-bound transport proteins have intrinsic ATPase activity or associate with ATPases. These transporters are discussed in more detail in Section 3.2.1.

(3) **Secondary active transport**, *against* a concentration gradient, derives the energy from an ion flow occurring along a concentration gradient and coupled with the transport process. Examples are the lactose transporter LacY driven by a proton gradient in *E. coli* and the Na^+-Ca^{2+} exchanger powered by a Na^+-gradient (Section 14.5.1).

(4) **Group translocation** consists of a targeted transport process coupled with an energy-delivering chemical modification of the molecule to be transported. In most cases this is achieved by phosphorylation (the translocation of a phosphate group) requiring the hydrolysis of an energy-rich phosphoric acid derivative such as ATP or phosphoenolpyruvate (PEP). This process is discussed in detail in Section 3.2.2.

Even a very simple prokaryote such as *Mycoplasma genitalium* (harboring only 470 genes) expresses one passive transporter, 13 ATP-powered transporters, and one ion-gradient-powered transporter as well as two group translocation systems. For *E. coli*—although with about 4300 genes it is by no means the most complex prokaryote (the soil bacterium *Sinorhizobium meliloti* has 6300 genes, as many as yeast)—the corresponding numbers are 7, 194, 74, and 22, respectively.

3.2.1 ATP-powered membrane transporters

The proteins of this group are ATPases catalyzing the *targeted* transmembrane transport of ions or larger molecules, frequently against a concentration gradient.

Ion pumps

The bacterial **F-type transporters** use ATP hydrolysis to generate a proton gradient across the cell membrane, driving numerous active transport processes. The reverse reaction results in ATP production. In fact, the ATP synthases of mitochondria and chloroplasts belong to the family of F-type transporters. The **P-type transporters** represent another family of ATP-driven ion pumps. Examples are provided by the Na^+/K^+-dependent ATPases (sodium pumps) and the Ca^{2+}-dependent ATPases (calcium pumps). **V-type transporters** found in the membrane of intracellular vacuoles are restricted to archaea and eukaryotes.

ABC transporters

The largest and most versatile subfamily of ATP-powered membrane transporters is the ABC transporters, named after the ATP-Binding Cassette, a particular ATP binding site in the ATPase domain. Characteristic structural features of this highly conserved cassette are the ATP-binding Walker A motif or P-loop and the Mg^{2+}-binding Walker B motif (ATP is bound as a Mg^{2+} complex). Both motifs are common to nucleoside-triphosphatases and kinases (see Section 2.4).

ABC transporters are modular proteins consisting of four domains: two transmembrane domains, frequently with six transmembrane helices each, and two cytoplasmic domains with one ATP binding site each (Figure 3.4). These domains may exist as separate subunits or may be integrated in one or two polypeptide chains. The transport channel is established by a circular arrangement of the transmembrane helices.

In *E. coli* the ABC transporters constitute a large gene family (74 genes). They transport a wide variety of substances, from simple ions, amino acids, and sugars to peptides and proteins, and each transporter is specific for the uptake or excretion of a distinct substance. The uptake of material is facilitated by selective binding proteins mostly found in the periplasm of Gram-negative bacteria. They function as a kind of receptor and interact with the ABC transporters.

One of the best-studied ABC transporters is the **MsbA** protein of *E. coli*. It is the only transporter that is absolutely essential for the bacterial cell. MsbA is a multidrug resistance protein, which exists in more than 30 homologs. These proteins are responsible for the excretion of lipophilic and toxic (foreign) substances such as drugs. An adaptive overexpression of MsbA is one of the mechanisms that renders bacteria resistant to antibiotics. Resistant bacteria are assumed to be the reason for approximately 60% of iatrogenic diseases.

The MsbA protein is composed of two polypeptide chains with six transmembrane helices and one ATP binding site each. It probably functions as a "flippase," trapping the substrate in a cage at the inner side of the membrane and releasing it to the outside upon a conformational change brought about by ATP hydrolysis (Figure 3.4).

The human homolog of MsbA is the multi-drug resistance protein **MDR1**, also called P-glycoprotein. Its adaptive overexpression in tumor cells is one of the mechanisms leading to the feared resistance against cytostatics during cancer chemotherapy. MDR1 is the prototype of a large family of eukaryotic ABC transporters with several functions. Another human ABC protein, the cystic fibrosis transmembrane conductance regulator or **CFTR**, is strictly speaking not a transporter but a hormone-controlled chloride channel regulating the function of mucous membranes (for more details see Section 14.7.2). A defective *cftr* gene is the cause of cystic fibrosis or mucoviscidosis, which is one of the most common human hereditary diseases. In both MDR1 and CFTR, the four domains of an ABC transporter share a single polypeptide chain.

Figure 3.4 Bacterial ABC transporter of type MsbA The protein is composed of two identical subunits with six transmembrane domains (cylinders) and one intracellular ATP-binding site each. ATP hydrolysis causes a conformational change, promoting the excretion of toxic substances (red ball) trapped in a cage. The transporter becomes reloaded upon ADP–ATP exchange.

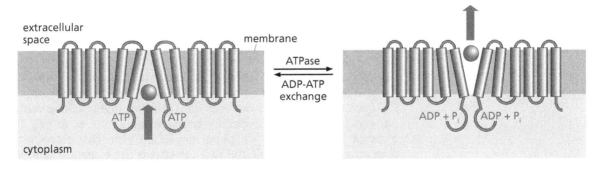

3.2.2 Group translocation systems: primeval forms of data-processing networks combining food uptake with gene regulation

The best-known representatives of this family of membrane transporters are the PEP (phosphoenol pyruvate)-powered **phosphotransferase systems**, abbreviated as PTS. They combine membrane transport with the regulation of gene transcription.

Prokaryotes are capable of switching their genes on and off in a specific and signal-controlled mode in order to adapt to the actual conditions of the biotope. As far as the molecular mechanism is concerned, there is no fundamental difference between prokaryotic and eukaryotic gene transcription. In both cases, an initiation complex containing RNA polymerase and a series of regulatory proteins assembles at the gene at sites that have been marked by transcription factors. This is the key event and primary target of signaling (for details see Chapter 8). A characteristic feature of prokaryotes is that genes controlling an individual metabolic process (structural genes) are combined into functional units known as operons. Operons are transcribed as a whole into a polycistronic RNA and are under the control of both stimulatory and inhibitory gene sequences called promoters, enhancers, and silencers. These are the binding sites of transcription factors. Stimulatory transcription factors facilitate and inhibitory factors (repressors) inhibit the assembly of an active RNA polymerase complex. The major signaling events controlling transcription factor and repressor activity in both pro- and eukaryotes are noncovalent ligand binding and covalent protein phosphorylation. Moreover, repressor activity is frequently regulated by corresponding metabolic products acting via feedback loops or by special signaling reactions. The active form of a transcription factor is always a homo- or heterodimer, enabling a cooperative effect with an intrinsic noise filter (Section 1.3). Transcriptional control supports the idea formulated in Chapter 1 that signal processing is based on combinatorial processes. A cell contains many more genes than transcription factors. Only 5% of the 4290 genes of *E. coli* encode gene-regulatory proteins. How, then, is *specific* gene regulation achieved? The answer is that the majority of bacterial gene promoters are regulated by a set of different transcription factors rather than by a single regulatory protein. Because of this combinatorial principle, a wide variety of environmental stimuli may affect a large number of genes by use of only a rather limited number of regulatory proteins. The cooperation between activators and repressors renders these systems logical gates (see Section 1.5).

PTSs manage the uptake of a variety of sugars and sugar alcohols by substrate phosphorylation, using PEP as the phosphate donor and energy provider (Figure 3.5). Phosphorylation is the starting reaction of substrate metabolism such as glycolysis. The process is embedded in a network of interacting proteins that adjust behavior and genetic readout to the food supply. PTSs, showing the inseparable connection between metabolism and signal processing, indicate that even on a rather simple level the cellular response to an exogenous stimulus has to be treated as a systemic answer of the whole organism rather than as an isolated biochemical reaction.

PTSs are abundant in bacteria but have not yet been found in archaebacteria and eukaryotes. Even among bacteria they are differentially distributed. *E. coli* expresses 22 PTSs with different substrate specificities, while the 10 times smaller genome of *Mycoplasma genitalium* contains only two PTS genes (for the transport of glucose and fructose) and *Mycobacterium tuberculosis* has none.

PTSs are multiprotein complexes consisting of highly conserved and variable parts. A prototype is the glucose-transporting PTS^{Glc} of *E. coli*. Its conserved parts are the two cytoplasmic enzymes EI and HPr (Histidine-containing phosphocarrier Protein), while the variable part is the transport complex EII. EII is

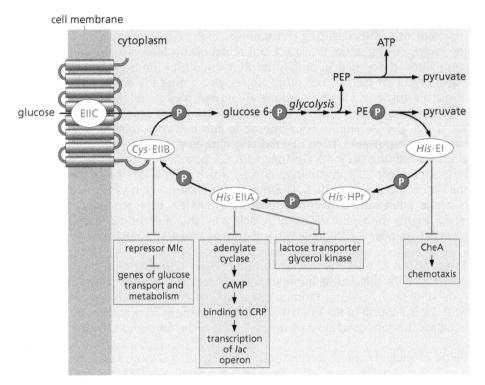

Figure 3.5 Phosphotransferase system (PTS) of *E. coli* The system couples the glycolytic pathway with glucose transport and regulatory reactions. Consisting of the cytoplasmic proteins EI, HPr, and EIIA and the membrane-bound channel complex EIIB/EIIC, it stepwise transfers a phosphate residue (P, red circle) from one of the two phosphoenolpyruvate (PEP) molecules derived from glucose 6-phosphate to glucose, which is taken up by the cell as glucose 6-phosphate. The second PEP molecule is used for ATP generation. When activated by glucose, the system changes cell motility and inhibits the uptake and metabolism of alternative nutrients (catabolite repression). The phosphorylated amino acids of the PTS proteins are depicted in italic type.

composed of three proteins: the membrane transporter EIIC and the associated proteins EIIA and EIIB (Figure 3.5). EI and HPr are separate cytoplasmic proteins, whereas the three EII subunits may either share one polypeptide chain (EIIABC; for instance, in the *N*-acetylglucosamine translocator of *E. coli*) or be partially separated (EIIA + EIIBC; for instance, in the glucose translocator of *E. coli*). EIIC contains eight transmembrane domains and as a dimer constitutes the trans-membrane channel.

The homodimeric enzyme EI catalyzes the hydrolysis of PEP, transiently binding the phosphate residue at a histidine residue (His189 in *E. coli*); thus it is acting as a histidine autokinase. From there the residue migrates via HPr, EIIA, and EIIB to glucose, which as glucose 6-phosphate passes the membrane channel EIIC. In both HPr and EIIA the phosphate residue is also bound to His residues; however, in EIIB, which is the glucose phosphorylase proper, the phosphate residue inter-acts with a cysteine residue.

Together these phosphorylations are thermodynamically unstable, short-lived transition states of phosphotransferase reactions. [Note that the phosphoamide bond of phospho-His contains so much energy that the reaction ATP + protein $\leftrightarrow$ ADP + protein-P is not shifted to the left only because the product (protein-P) becomes rapidly dephosphorylated by the subsequent reactions.] The more glu-cose available, the higher the rate of the cell's catabolic metabolism and the intracellular PEP level. Under such conditions the PTS machinery is running at top speed, phosphorylating glucose at a high rate and, thus, generating a rapid draining of phosphate. As a consequence, the steady-state concentrations of *phosphorylated* PTS proteins are low, whereas they rise upon glucose deficiency. The degree of phosphorylation of PTS components is used for a coupling of the transport process with accessory cellular functions. In fact, PTS represents the backbone of a simple but highly efficient signal-processing protein network enabling the cell to perfectly adapt to environmental conditions.

At a high glucose level, the now underphosphorylated EI inhibits the chemotaxis protein CheA, changing the motility pattern of the cell in such a way that it is able to approach the source of food (positive chemotaxis, see Section 3.3.4). Under the same conditions, underphosphorylated EIIB inhibits the transcriptional repressor Mlc, thus ensuring continuously high activity of the genes encoding the enzymes of glucose transport and metabolism. Moreover, as long as sufficient glucose is available, underphosphorylated EIIA binds and inactivates the transporters for lactose and other sugars, preventing the costly uptake of less attractive nutrients for which the metabolic enzymes are not available, since the corresponding genes, such as the *lac* operon, are also repressed by EIIA. Only when the glucose supply is beginning to run low does this inhibition, called **catabolite repression**, become gradually canceled, enabling the cell to exploit alternative sources of food. The trigger for this genetic and metabolic changeover is the increase in PTS phosphorylation, by which a signal is released that changes the transcriptional pattern. In Gram-negative bacteria such as *E. coli*, this signal is **cyclic adenosine 3′,5′-monophosphate (cAMP).** It is generated from ATP when highly phosphorylated EIIA activates the membrane-bound enzyme adenylate cyclase (see Sidebar 3.1 and also Section 4.4.1 for a reaction scheme).

The function of cAMP is that of an intracellular signaling molecule or second messenger. To this end it binds to a cAMP receptor protein, CRP (also called Catabolite gene Activator Protein, CAP), and the complex thus formed interacts with the α-subunit of RNA polymerase, functioning as a co-activator for a series of catabolite-repressed gene complexes such as the *lac* operon (Figure 3.5).

Being an ancient control mechanism, cAMP signaling is by no means restricted to *E. coli*-like bacteria; many other prokaryotes express a wide variety of adenylate cyclase isoforms. This holds true even for most primitive species such as mycobacteria that have been found to produce cAMP in unusually large amounts. The genome of *Mycobacterium tuberculosis*, for instance, encodes no less than 17 nucleotide cyclases as well as several cAMP phosphodiesterases and other cAMP binding proteins, in particular putative transcription factors. Evidence is accumulating that highly pathogenic *Mycobacteria* (causing tuberculosis, leprosy, and other diseases) use cAMP signaling for proliferation and subversion of host defense mechanisms. Therefore, the corresponding proteins are considered to provide potential targets for novel therapeutic approaches, in particular against bacterial strains that have become resistant to conventional drug therapy.

The PTS mechanism depicted in Figure 3.5 applies also to Gram-positive bacteria. However, for catabolite repression they pursue another strategy, because

---

Sidebar 3.1 Prokaryotic adenylate cyclases Adenylate or adenylyl cyclases catalyze the formation of cAMP from ATP according to

$$ATP \rightarrow \text{cyclic } 3′,5′\text{-AMP} + \text{pyrophosphate (PP}_i)$$

(the endergonic reaction is shifted to the right by the subsequent hydrolysis of pyrophosphate). Adenylate cyclases are ancient enzymes found in bacteria, archaea, and eukaryotes. Depending on the species, bacteria may express up to 40 different isoforms.

On the basis of primary sequence data, nucleotide cyclases are subdivided into six classes. Among these, three major families of prokaryotic adenylate cyclases are distinguished. Family 1 consists of adenylate cyclases of Gram-negative bacteria such as *E. coli* with a special function in catabolite repression. Family 2 contains adenylate cyclases of archaea and toxic adenylate cyclases of certain pathogenic bacteria, such as *Bacillus anthracis*, *Bordetella pertussis*, *Pseudomonas aeruginosa*, and *Bacillus bronchiseptica*. These enzymes are released together with other toxins into the host cell, causing a pathological overproduction of cAMP, for example, during anthrax and whooping cough. Family 3 consists of "eukaryotic" adenylate cyclases, some of which are found also in prokaryotes (for instance, in mycobacteria) but not in archaea (for details see Section 4.4.1).

Guanylate cyclases that transform GTP into cyclic 3′,5′-GMP seem to be entirely restricted to eukaryotes. Bacteria (but not archaea and eukaryotes) possess diguanylate cyclases that transform two GTP molecules into the second messenger cyclic di-GMP (see Section 3.4.2).

many of them—such as *Bacillus subtilis*—neither express a CRP-homologous protein nor produce cAMP. Instead, the catabolite-repressible genes are blocked by a transcriptional repressor that is removed upon glucose deficiency.

The key reaction is a phosphorylation of HPr at a serine (Ser) residue that competes with the EI-catalyzed His phosphorylation of HPr (which is still essential for the sugar uptake). The Ser phosphorylation requires ATP and is catalyzed by an HPr kinase/phosphatase. The latter is activated by a high intracellular level of ATP (and a low level of inorganic phosphate) as well as by intermediates of glycolysis; that is, under conditions of maximal glucose supply. In its structure and bifunctional activity (being both a kinase and a phosphatase), this primeval enzyme differs from all known protein kinases and phosphatases.

In contrast to the labile phospho-His bond, the ester bond of phospho-Ser is kinetically stable. Therefore, Ser-phosphorylated HPr is not a short-lived intermediate but can act as a transcriptional co-repressor by binding the so-called **catabolite control protein A (CcpA)**. This complex interacts with gene regulatory sequences named catabolite responsive elements, *cre*, and inhibits the transcription of catabolite-repressible operons (Figure 3.6). Upon glucose or ATP deficiency (or at a high level of intracellular inorganic phosphate), the kinase activity of the HPr kinase/phosphatase is turned down, whereas the phosphatase activity of the enzyme remains unchanged, now catalyzing the dephosphorylation of HPr. This leads to a dissociation of the HPr–CcpA complex and to an activation of the catabolite-repressed genes. Ser phosphorylation of HPr not only adjusts gene transcription to the food supply but also controls PTS by negative feedback, since the competing His phosphorylation, and thus the sugar uptake, is hindered.

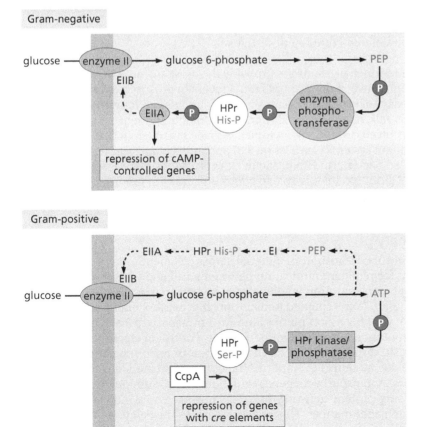

Gram-negative

glucose — enzyme II — glucose 6-phosphate — → — → PEP

EIIB

EIIA ← P ← HPr His-P ← P ← enzyme I phosphotransferase

repression of cAMP-controlled genes

Gram-positive

EIIA ← HPr His-P ← EI ← PEP

EIIB

glucose — enzyme II — glucose 6-phosphate — → — → ATP

HPr Ser-P ← P ← HPr kinase/phosphatase

CcpA →

repression of genes with *cre* elements

Figure 3.6 Catabolite repression in Gram-negative as compared with Gram-positive bacteria In Gram-negative species, the enzyme EIIA is underphosphorylated at high glucose concentration, inhibiting cAMP formation and thus the activation of cAMP-dependent (catabolite-repressed) genes. In Gram-positive bacteria, the PEP-dependent His phosphorylation competes with an ATP-dependent Ser phosphorylation of HPr. The Ser-phosphorylated HPr together with the protein CcpA inhibits transcription of catabolite-repressed genes.

Phosphotransferase system and evolution of signal transduction

PTS and catabolite repression are components of a data-processing protein network that regulates access to different nutrients in a highly economical manner. These components also cause the intake and metabolism of the particular carbon source to become adjusted at an optimal level that does not overtax the metabolic capacities of the cell. This enables bacteria to adapt perfectly to environmental conditions, albeit within their genetically programmed limits.

Of particular interest are the evolutionary aspects of the system. The network makes use of reversible protein phosphorylation, and the origin of this signaling reaction from short-lived enzymatic transition states becomes especially evident here. Moreover, with cAMP the prototype of a second messenger enters the stage for the first time. By transducing the effects of exogenous signals to the genome, this molecule has become an evolutionary model of success (though its function has changed from a cofactor of a gene-regulatory protein to an activator of protein kinases and ion channels that indirectly control transcription; see Sections 4.4.1 and 4.6.1). Further models of success are the two ways by which the activity of a transcription factor is controlled: either by ligand binding (cAMP–CRP) or by protein phosphorylation (Ser phosphorylation of HPr). These major principles of signaling to the genome have survived evolution up to today's most advanced organisms.

From one-component to multicomponent systems

The screening of domain libraries obtained by gene sequencing has shown that the majority of prokaryotic signal transducers combine just a single input with a single output domain, thus representing the simplest modular structure. Such proteins function as genuine binary switches. In order to distinguish them from the more complex two-component systems discussed below, these signaling devices are also called one-component systems. Most of them are transcriptional regulators with a DNA-binding and transactivating site as an output domain; some also have enzymatic functions. The input domains mostly recognize small intracellular molecules such as cAMP and metabolic products or are phosphorylated. The cAMP receptor protein CRP and the kinase substrate HPr provide examples of one-component systems.

One-component systems are probably the most ancient signaling proteins of all. The abundance of these simple transducers in prokaryotes indicates that, in the earlier days of evolution, linear signaling cascades with a rather moderate degree of cross talk may have predominated. One-component systems then seem to have evolved into more complex signal-transducing devices such as two-component systems (see Section 3.3) and multidomain proteins found in today's advanced bacteria and, in particular, in eukaryotes. Here a signaling protein with various input modules is able to communicate with numerous partner proteins, forming a multicomponent system and enabling intense cross talk between individual signaling cascades.

Summary

The controlled transport of material across cell membranes represents the most ancient form of communication between a cell and its environment. Transport against a concentration gradient must be powered by coupled ion flows or ATP hydrolysis. ATPase-coupled transporters that use ATP energy include the F, P, and V types, working as ion pumps, and ABC transporters (carrying an ATP-binding cassette) that are involved in food intake, control of electrolyte concentrations, and waste disposal, including the excretion of toxic substances. The phosphotransferase system (PTS) and the machinery of catabolite repression connect membrane transport with gene regulation. These components constitute a data-processing protein network that regulates access to individual nutrients in a highly efficient manner. This network enables the cell to adjust the uptake and metabolism of a particular carbon source to both the environmental situation and its metabolic capacities. In evolution, PTS provides an early example of signaling protein phosphorylation, whose origin from short-lived enzymatic transition states is obvious. Moreover, with cAMP the prototype of an endogenous signal (second messenger) enters the stage together with two other successful models of evolution: control of transcription factor activity by either ligand binding (cAMP and cAMP receptor protein, CRP) or protein phosphorylation (His-containing phosphocarrier protein, HPr). CRP and HPr are prototypes of one-component systems, the most ancient and most abundant signal-transducing devices in prokaryotes.

3.3 Sensor-dependent signal processing: two-component systems

Up to a certain extent, membrane-bound transport systems are the cellular organs for food intake and waste excretion. In PTS, food intake is coupled with a simple but highly efficient network of signal processing, allowing an adaptation to the particular source of food. Nevertheless, and despite their selectivity, transport processes of this type are "blind" to the environment. PTS controls the motility of the cell by modulating the activity of the chemotaxis protein CheA, but it does not enable a behavior of prey-seeking, a tracking down of optimal biotopes, or even more important, an escape reaction in the case of danger. In fact, such blind survival programs are sufficient only for very simple prokaryotes living in a constant biotope, whereas more complex and free-living species require specific **sensor proteins** working independently of food intake and metabolism. These proteins enable the cell to adjust both its genetic readout and its behavior not only to the food supply but also to many other environmental influences.

Since most of the environmental factors do not pass or must not be allowed to penetrate the cell membrane, they cannot be recognized by the cytoplasmic one-component systems. As a consequence, sensor proteins had to be placed at the cell surface. This was done by an evolutionary trick: namely, to distribute input and output domains onto two different proteins, with one taking on the role of a membrane-bound sensor, and the other, of a cytoplasmic effector. In other words, a two-component system was made from a one-component system. In addition, an effective mechanism of signal transduction between both partners was designed.

Two-component systems dominate the communication between bacteria and environment. Their sensor proteins are receptors for environmental stimuli, the meaning of which can be decoded by the data-processing network of the cell, an ability that is essential for survival. The effector proteins of the sensors are called **response regulators**. They transmit the signal from the sensors to the metabolic and genetic apparatus of the cell. With a few exceptions, sensor genes and their cognate response regulator genes are organized in operons.

Between sensor and response regulator, the signal is transduced by protein phosphorylation, in particular by His phosphorylation, which plays a similar key role as for PTS (Figure 3.7). In fact, all sensor proteins are **His autokinases** or at least associated with His autokinases. When the sensors are integrated in the plasma membrane, such two-component systems represent a prototype of a transmembrane signaling device operating by post-translational protein modification. But two-component systems are also used for intracellular signaling. In this case the sensor domain of a membrane-bound His autokinase is facing the cytoplasmic side (such as in the case of the energy sensor Aer; see Figure 3.13), or the His autokinase even lacks transmembrane domains (for an example, see Figure 3.10). In fact, about a third of all His autokinases known seem to be cytoplasmic rather than membrane-bound enzymes. They are assumed to process endogenous signals such as metabolite concentrations, redox status, and acidity.

Figure 3.7 Prototype of a two-component system with a transmembrane sensor The scheme shows a dimeric sensor protein (component I) with four transmembrane domains and two cytoplasmic His autokinase domains (red) as well as the corresponding response regulator protein (component II). Upon activation, the sensor undergoes trans-autophosphorylation on a cytoplasmic His residue with ATP as a phosphate donor. The phosphate residue (P) is transferred to an Asp residue in the receiver (or input) domain of the response regulator. By the resulting long-range conformational change, the effector (or output) domain of the response regulator becomes activated. Some response regulators thus control the behavior, while others stimulate the genetic readout (aiming at metabolic adaptation) of the cell. From the receiver domain, the phosphate group is removed either by auto-dephosphorylation, by the sensor protein (exhibiting both His kinase and Asp phosphatase activities), or by separate phosphatases. Here, *activation* by phosphorylation is shown. However, depending on the system, dephosphorylation may also stimulate downstream signaling (with the phosphorylated response regulator being the inactive form). Sensors with two transmembrane domains per monomer are particularly abundant. However, there are also sensor His kinases with no transmembrane helices or up to 20 transmembrane helices per monomer (for examples see Figures 3.10 and 3.11).

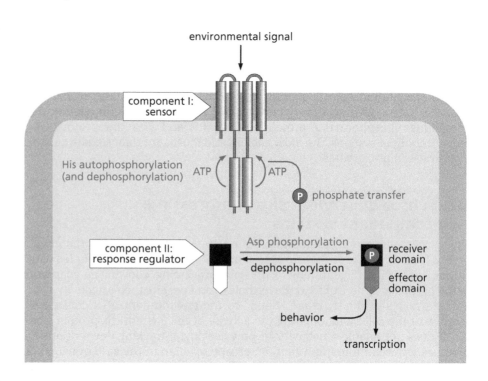

However, little is known about their role in metabolic regulation. We are much better informed about the functions of membrane-bound His kinases operating as sensors for environmental stimuli. Here signal transduction occurs as shown in Figure 3.7.

When the sensor receives an environmental signal, it undergoes His autophosphorylation at its cytoplasmic domain with ATP as a phosphate donor. Since sensors are homodimers, the autophosphorylation occurs in *trans*: that is, the monomers phosphorylate each other. As in the PTS, the His-phosphorylated sensor protein represents a short-lived energy-rich transition state with a very high turnover rate (it is estimated that less than 1% of the His autokinases of a cell are phosphorylated at a given time) that serves as a phosphate donor for the phosphorylation of the response regulator. This occurs at a highly conserved aspartate (Asp) residue and changes the conformation and the activity of the response regulator. Since the response regulator catalyzes this phosphorylation itself, it might be defined as an Asp autokinase. However, apart from the fact that the term kinase is linked to ATP (or GTP) serving as a phosphate donor, the response regulator protein does not contain the typical protein kinase domains.

Though thermodynamically labile, the acyl–phosphate bond of phospho-Asp becomes stabilized by interactions within the polypeptide structure since its energy is used for the conformational change of the response regulator. Therefore, the dephosphorylation of response regulators requires enzymatic catalysis, contributed either by the response regulator itself, by the corresponding sensor His kinase (being a bifunctional kinase/phosphatase), or by a special response regulator phosphatase such as CheZ in bacterial chemotaxis (Section 3.3.4) or SpoOE in spore formation (Section 3.5.3). Depending on the type of response regulator, the lifespan of the phosphorylated state may vary between seconds and hours. Stimuli acting as input signals may control both the His autokinase activity and the response regulator phosphatase activity of the sensor.

For the cell, two-component systems provide extremely efficient means to adjust the genetic activity to the environment. To this end, in their N-terminal (mostly extracellular) domains the sensor His kinases are equipped with receptor domains recognizing a huge variety of environmental stimuli. The structural variability of these receptor domains reflects the variability of the signals, whereas the cytoplasmic kinase domain is rather conserved.

Response regulators are signal-propagating proteins feeding the input signal into the data-processing protein–DNA network of the cell. Most of them are transcription factors exhibiting the characteristic architecture of a variable DNA-binding and transactivating domain combined with a conserved phospho-acceptor domain. Other response regulators may interact with proteins positioned downstream in a signaling cascade (see, for example, CheY in Figure 3.14), or the phospho-acceptor domains may be coupled with other signal-transducing domains, such as the catalytic centers of enzymes controlling the production of the intracellular messenger cyclic di-GMP (GGDEF and EAL; see Section 3.4.2) or post-translational protein modification by methylation (CheB; see Section 3.3.4). As far as their mode of function is concerned, response regulators seem to come close to digital (binary) switches (see Section 1.3). In the following sections, the mechanisms of action and cellular functions of two-component systems are explained and illustrated with selected examples.

3.3.1 To shrink or to burst: adaptation to osmotic pressure

For an organism living in an aqueous milieu, adaptation of cell volume and inside pressure to the osmolarity of the environment is a basic requirement of survival, since otherwise a cell would either burst or shrink and dry up. The osmoregulator of *E. coli* is the simplest of all two-component systems, demonstrating in an exemplary manner the characteristic features of these signal-transducing devices. It is composed of the osmosensor EnvZ (Env, envelope) and the response regulator OmpR (Omp, outer membrane protein), a transcription factor. The system controls the genes encoding the porins OmpF and OmpC (see Sidebar 3.2). These proteins constitute pores in the cell wall or outer membrane and regulate the transmembrane diffusion of hydrophilic and negatively charged molecules.

While the protein OmpF produces wide pores and is expressed at low osmotic pressure, OmpC forms narrow pores and is produced at high osmotic pressure, for instance, in the intestinal milieu, with its high concentration of dissolved substances of low molecular weight. Under such a condition of high osmolarity, the His autokinase activity of the sensor EnvZ is stimulated by a mechanism that is not yet fully understood. This results in an Asp phosphorylation of OmpR, which then induces transcription of the *ompC* gene while at the same time

---

Sidebar 3.2 Cell wall channels The family of **porin proteins** forms water-filled channels with a large diameter (1 nm) in the cell wall of Gram-negative bacteria as well as in the outer membrane of chloroplasts and mitochondria. In bacteria these channels serve the specific and unspecific exchange of material between environment and periplasm. The cell adjusts the expression of porins to the particular demands. For bacterial porins the exclusion limit of permeable molecules is at around 600 Da, with the transport following the concentration gradient passively. Though porins respond to changes of the membrane potential, the physiological role of this effect is still debated.

The pore of a porin consists of a β-barrel mostly composed of 16–22 anti-parallel β-sheets. A central bottle-neck formed by several loops connecting the β-sheets determines the substrate specificity of the channel. In the bacterial cell wall the porins are arranged as homotrimers. Parasitic bacteria may perforate the host cell membrane by means of porin-like toxins.

Another family of cell wall channels in Gram-negative bacteria is the **trans-periplasmic channel tunnels** of the type TolC. They have only one pore that again is formed by a β-barrel. Via a long tunnel domain they bridge the periplasm, directly interacting with transport proteins of the inner cell membrane. TolC channel tunnels are specialized for the export of proteins (for instance, bacterial toxins) and other molecules (for instance, substances toxic for the cell).

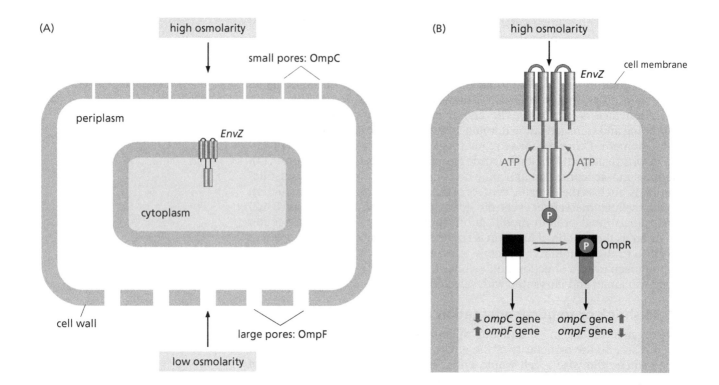

Figure 3.8 Two-component system of osmoregulation (*E. coli*) (A) Adaptation of pore size in the cell wall by osmolarity-controlled *de novo* synthesis of the pore proteins OmpC and OmpF. High osmolarity activates the sensor EnvZ in the cell membrane. (B) Control of *ompC/ompF* gene transcription by a two-component system consisting of the sensor His kinase EnvZ and the response regulator OmpR (P, phosphate). Dark gray vertical arrows symbolize inhibition or activation of gene transcription.

inhibiting *ompF* transcription (Figure 3.8). EnvZ is, thus, the prototype of a sensor responding to hyperosmotic stress. It is a protein of 450 amino acids exhibiting the characteristic architecture of a sensor His kinase, which always exists as a homodimer with one monomer phosphorylating the other one (trans-autophosphorylation, see Figure 3.9). Each monomer is anchored in the membrane by two N-terminal transmembrane helices flanking the signal recognition domain (this is the most abundant sensor type, but other forms with more than two transmembrane helices also occur; see Section 3.3.3).

In the cytoplasmic part of all sensor kinase dimers, a bundle of four helices constitutes a dimerization domain containing the His residues to be phosphorylated (so-called H-box) and the binding sites for the response regulator. The catalytic domain of the His kinase is localized in the C-terminal domain and oriented in such a way that trans-autophosphorylation of the subunits becomes possible (Figure 3.9). The dimerization domain of EnvZ also harbors the active center of an Asp-specific protein phosphatase that catalyzes the dephosphorylation of the response regulator, thus terminating signal transduction.

As EnvZ represents a typical sensor kinase, OmpR, a 27 kDa protein, is the prototype of a response regulator (and of a signal-transducing protein in general), consisting of an N-terminal regulatory or receiver domain and a C-terminal effector or transmitter domain. The conserved receiver domain contains the characteristic structural elements of a response regulator including the essential Asp residue. Its trans-autophosphorylation with His-phosphorylated EnvZ as a phosphate donor is again facilitated by a homodimerization of OmpR and induces a conformational change enabling the dimer to interact with the regulatory sequences of the *ompC* and *ompF* genes.

Like EnvZ, OmpR also exhibits Asp-specific protein phosphatase activity: it is able not only to activate but also to inactivate itself, thus, together with EnvZ, controlling the lifespan of the active state. So, the response regulator has the qualities of a switching element with a built-in timer. Such devices play an important role in the transformation of analog into digital signals and vice versa.

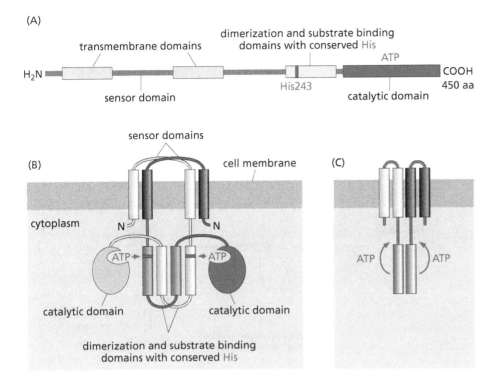

Figure 3.9 Sensor His kinase EnvZ of *E. coli* (A) Simplified domain structure of an EnvZ monomer with the strategic His residue 243 undergoing trans-autophosphorylation and the catalytic domain binding ATP. (B) Membrane anchoring and schematic folding geometry of an EnvZ homodimer. (C) Simplified scheme as used in other figures. The arrows symbolize His autophosphorylation.

3.3.2 Nitrogen fixation and hydrogen utilization

Enzymatic reduction of molecular nitrogen to ammonia represents a fundamental biochemical process essential for life on earth. It is managed by certain bacteria and archaea such as the rhizobia living as symbionts in leguminosae. Requiring the hydrolysis of 12 ATP per one N_2, the reduction of nitrogen ranks among the most expensive (that is, energy-consuming) biochemical reactions and is therefore under strict control:

$$N_2 + 6H^+ + 6e^- + 12ATP + 12H_2O \ \rightarrow \ 2NH_3 + 12ADP + 12P_i$$

The nitrogenases catalyzing the process are mostly molybdenum- (or vanadium-) containing iron–sulfur proteins with a complex quaternary structure. They become expressed only when nutritious nitrogen sources, such as amino acids and ammonia salts, are not available.

Nitrogenase expression is controlled by a two-component system consisting of the sensor NtrB and the response regulator NtrC (Ntr = nitrogen). NtrB is a prototype of a *cytoplasmic* sensor His kinase lacking transmembrane domains (other examples of sensor His autokinases without transmembrane domains, such as HoxJ, KinA, and CheA, are described later). The protein monitors the intracellular state of nitrogen metabolism. In the presence of good nitrogen sources, NtrB is blocked by an inhibitor protein in order to avoid an unnecessary and costly nitrogen reduction. With a deficiency of good nitrogen sources, NtrB phosphorylates the response regulator, which in turn stimulates the transcription of a gene encoding another transcription factor, NifA (Nif = nitrogen fixation). As a master key, NifA controls the whole collection of genes required for nitrogen fixation including that of nitrogenase (Figure 3.10).

Nitrogenases are extremely oxygen-sensitive enzymes. Nevertheless, the cell needs oxygen for ATP generation, which is then used as energy supply for nitrogen reduction. To adapt to an optimal oxygen concentration (micro-aerobic conditions), therefore, rhizobia possess an additional two-component system consisting of the sensor His kinase FixL and the response regulator FixJ [Fix =

Figure 3.10 Signaling pathways for the regulation of nitrogen fixation
The figure shows the cytoplasmic NtrB/NtrC two-component system inducing transcription of genes of nitrogen fixation (left), the membrane-bound FixL/FixJ two-component system controlling the adaptation to oxygen pressure (right), and the cytoplasmic NifL system expressed by some species to adjust gene transcription to the cellular redox state (bottom right). P, phosphate. For further details see text.

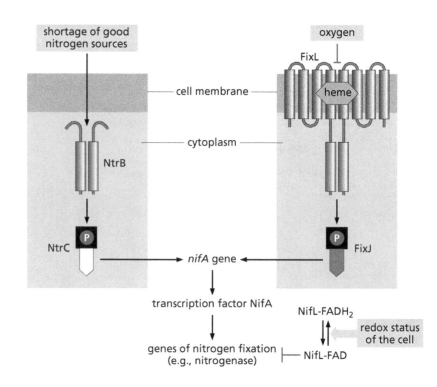

Sidebar 3.3 PAS and GAF domains As sensory domains recognizing a wide variety of input signals, these ubiquitous protein interaction motifs play a major role in both pro- and eukaryotic signal processing. Among the prokaryotic signal transducers, various His autokinases, adenylate cyclases, phosphodiesterases, ATPases, and diguanylate cyclases (Section 3.4.2) contain such domains functioning as regulatory elements.

PAS domains consisting of 100–300 amino acids are widely distributed among pro- and eukaryotes. The name PAS refers to three proteins where these domains were discovered originally:

- Period protein of *Drosophila*, involved, like homologous vertebrate proteins, in the regulation of genes controlling diurnal rhythms

- ARNT protein (Aryl hydrocarbon Receptor Nuclear Translocator) of vertebrates; among others, it controls (together with the aryl hydrocarbon receptor) genes encoding detoxifying enzymes (Section 8.4.3)

- Sim protein, a developmental regulator of *Drosophila*

In eukaryotes, PAS domains mainly establish homo- and heterotypical interactions between proteins. They are found in many signal-transducing proteins including protein kinases, receptors, and ion channels, as well as in phytochromes and other photoreceptor proteins of plants. The considerably smaller PAS domains of prokaryotes are distantly related to the eukaryotic domains. They are found mainly in signaling proteins such as the oxygen sensor FixL, the hydrogen sensor HoxJ, the sporulation-controlling sensor KinA, the aerotaxis-controlling sensor Aer, and the light sensor PYP.

GAF domains are named after the three different proteins in which they were first found: cGMP-phosphodiesterase, Adenylate cyclase (of the cyanobacterium *Anabaena*), and FhlA, a transcription factor of *E. coli*. They rank among the most ancient, widespread, and abundant small-molecule binding protein motifs and are found in organisms ranging from cyanobacteria to humans. Known ligands are the cyclic nucleotides cAMP and cGMP (that is not found in prokaryotes), while other ligands have yet to be identified.

GAF domains are distinct from cyclic nucleotide binding domains found in the bacterial transcription factor CRP and in eukaryotic effector proteins such as cAMP- and cGMP-activated protein kinases, cyclic nucleotide gated ion channels, and the Rap exchange factor EPAC (see Section 4.6.1). The regulatory domains of eukaryotic cGMP-controlled phosphodiesterases (Section 4.4.2) are GAF domains. The *Anabaena* adenylate cyclase is superactivated by positive feedback through an interaction of cAMP with GAF domains.

(nitrogen) fixation]. FixL represents a sensor with four transmembrane domains. It harbors a structurally unusual heme molecule bound by a PAS domain (see Sidebar 3.3). This heme serves as the oxygen sensor: at low oxygen concentrations, the system is active and the phosphorylated response regulator induces the transcription of the *nifA* gene, whereas at higher oxygen concentrations, oxygen occupies the Fe^{2+} of heme, thus inhibiting the His kinase activity of FixL (Figure 3.10). By this means an uneconomical synthesis of nitrogenase is avoided under conditions where the enzyme is running the risk of being inactivated by oxygen.

Another protective mechanism has been found, in particular in *Klebsiella* species. Here the operon encoding NifA contains an additional gene encoding the regulator protein NifL. By means of an associated flavin adenine dinucleotide (FAD), NifL monitors the redox state of the cell: in the presence of oxygen, the cofactor is oxidized to FAD and NifL represses transcription of the genes of nitrogen fixation (Figure 3.10). The reduced form $NifL-FADH_2$, indicating low oxygen pressure, does not exhibit this repressive effect. In contrast to the heme of FixL, here the redox cofactor is not a component of a sensor His kinase.

Hydrogen utilization

Many prokaryotes are able to cleave hydrogen, which accumulates as a byproduct of the nitrogenase reaction or is produced by anaerobic fermentation, into two protons and two electrons, resembling the oxidation of H_2 to water. This ancient reaction, which originally may have emerged in an oxygen-free primeval atmosphere, is catalyzed by hydrogenases and used as an energy supply in a wide variety of bacteria and archaea where the expression of hydrogenases is controlled by a two-component system. The sensor domain of the corresponding His kinase (for instance, the cytoplasmic HoxJ in *Ralstonia eutropha*, with Hox standing for H-oxidation) is via a PAS domain associated with a regulatory hydrogenase (HoxBC), an iron–nickel–sulfur protein acting as the H_2 sensor. The system's response regulator (here HoxA) represses transcription of the hydrogenase genes when phosphorylated. Hydrogen inhibits the His kinase of the sensor, thus promoting the accumulation of nonphosphorylated response regulator that induces gene transcription. This example demonstrates that the phosphorylation of a response regulator may be both a positive and a negative signal, a situation analogous to osmoregulation in yeast (see Section 3.5.1) and chemotaxis of *E. coli* (see Section 3.3.4).

3.3.3 Phosphorelays: devices for fine-tuning and cross talk

A rather baroque type of two-component system, transducing the signal along cascades of several interconnected phosphorylations, is represented by phosphorelays. In such relays each sensor His kinase delivers the phosphate residue to an Asp residue of a separate receiver domain, from where it migrates again to a His residue of a so-called HPt (His-containing Phosphotransfer) domain and finally to an Asp residue of a response regulator. HPt domains, which exhibit bundles of four helices, superficially resemble dimerization domains of sensor His kinases (Figure 3.9). The domains of a phosphorelay are either separate proteins or subdomains of one polypeptide chain. Like the PTS cascade, phosphorelays provide additional locations for cross talk with the signal-processing protein network, as well as probably more efficient noise suppression.

The phosphorelays most thoroughly studied are those controlling the sporulation of bacteria of the species *Bacillus* and *Clostridium*. Sporulation is triggered by environmental stress signals such as starvation. These signals are transduced by several sensor His kinases (for instance, the five membrane-bound or cytoplasmic sensor kinases KinA–E of *Bacillus subtilis*) to a highly conserved phosphorelay controlling the transcription of sporulation genes (Figure 3.11). The variety of stress signals is reflected by a corresponding variability of the sensors' extracellular receiver domains.

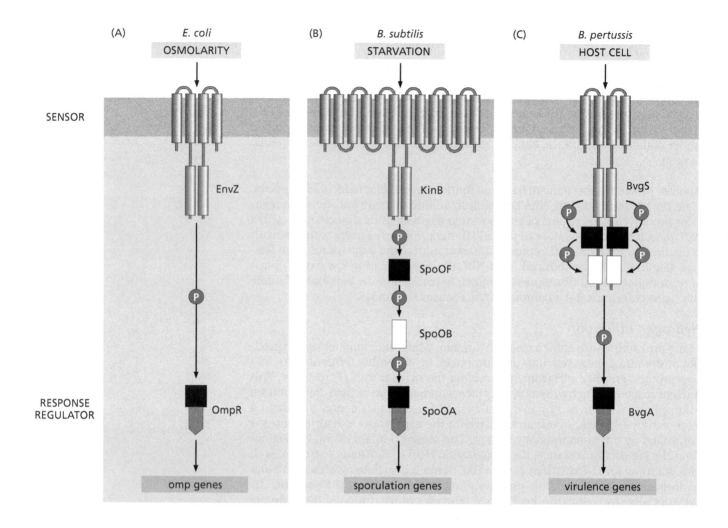

Figure 3.11 Selected two-component systems of bacteria
(A) Simple system of osmolarity control, shown for comparison (see also Figure 3.8). (B, C) More complex phosphorelay systems. Red cylinder, His kinase domain; black square, receiver domain (site of Asp phosphorylation) of a response regulator; gray wedge-shaped symbol, effector domain of a response regulator; white rectangle, HPt domain.

KinB, exhibiting 12 transmembrane domains per dimer, is a prototype of a group of sensors that functionally resemble the G-protein-coupled receptors of eukaryotes (Chapter 5); like those, they bind their ligands in the cavity formed by the gobletlike structure of multiple transmembrane helices. Such sensors may contain up to 40 transmembrane domains per dimer. Examples include the quorum sensors monitoring population signals in bacterial colonies (see Section 3.4.1). Another well-known phosphorelay regulates the transcription of the virulence genes of *Bordetella pertussis*. Here all components of the relay including the sensor are found on a single polypeptide chain.

3.3.4 The "nanobrain" of prokaryotes: learning by doing

A brain interprets sensory impressions and calculates a behavioral pattern aiming at adaptation. The data-processing protein network of bacteria does the same, albeit at a lower scale. The major type of prokaryotic behavior consists of controlled changes of motility. Thanks to the inspiring observations made by the German botanist Wilhelm Pfeffer at the end of the 19th century, it became clear that bacteria respond to beneficial and hazardous stimuli by seeking and flight movements, respectively. Such targeted movements are called taxis, and depending on the environmental stimulus, one distinguishes between chemotaxis, phototaxis, thermotaxis, osmotaxis, magnetotaxis, galvanotaxis (electricity), and thigmotaxis (touch). Representing a classical example of quasi-intelligent behavior, chemotaxis of *E. coli* ranks among the best-investigated physiological reactions of prokaryotes and is considered to be at present the most thoroughly studied mechanism in all biological data processing.

Chemotactic behavior

Prokaryotes qualified for taxis move by means of long flagella performing rotating or oscillating motions. In the absence of a stimulus, the cell executes a random walk, which in the case of *E. coli* consists of forward movements (swimming phases) that become interrupted after about 1 s each by a tumbling phase lasting for about 0.1 s. During each tumbling phase, the cell randomly orients into a new direction for swimming. This motility pattern results in an aimless three-dimensional zigzag course. For swimming the flagella are rotating counterclockwise, bundling to some kind of a propeller, whereas for tumbling the rotation is switched into a clockwise mode resulting in a disintegration of the flagellar bundle (in other species, the mode of rotation can be the other way around). Possessing only a forward and a neutral gear, the flagellar motor of *E. coli* is rather simple. Other species such as the soil bacterium *Sinorhizobium meliloti* are able to activate their flagella separately and stepwise, thus changing their speed, as well as to move in curves and to switch into a reverse gear.

The flagellar apparatus of bacteria is a complex device composed of many different proteins. Being an "electric motor," it obtains its energy from ion gradients across membranes. The rotation is brought about by sequential, possibly electrostatic interactions between rotor and stator proteins that are arranged in a circular pattern (Figure 3.12). The molecular mechanism for changing the direction of rotation is not yet fully understood.

When a bacterial cell picks up a stimulus, it changes the random walk into a targeted seeking or escape movement depending on whether the stimulus is an attractant (such as food) or a repellent (such as a poison). To this end, the swimming phases become prolonged when oriented correctly or shortened when oriented in the wrong direction. The cell is thus able to measure the concentration gradient of a substance and promptly adjust its behavior to the result obtained. A bacterium is too small to monitor a spatial concentration gradient, for example, between front and back. Instead, prokaryotes recognize *temporal* differences of concentration. To do this the cell must remember where it was before. In other words, targeted movements require learning and memory. Therefore, the data-processing protein apparatus of tactic behavior not only controls the flagellar machine but also must be able to learn and to adapt in a very short time, enabling the bacterial cell to make its decisions in a fraction of a second. As one may expect, bacterial chemotaxis can also be misused, for instance, by myxobacteria and myxamoebeae, which produce attractive signals to lure *E. coli*, their favored prey.

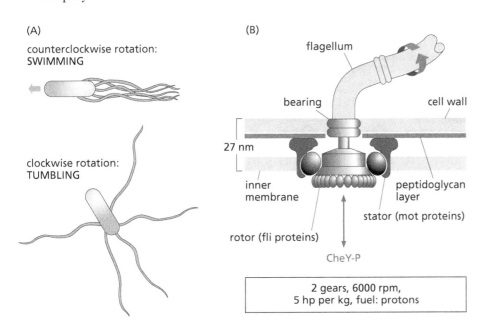

(A) counterclockwise rotation: SWIMMING

clockwise rotation: TUMBLING

(B) flagellum

bearing

cell wall

27 nm

inner membrane

rotor (fli proteins)

peptidoglycan layer

stator (mot proteins)

CheY-P

2 gears, 6000 rpm, 5 hp per kg, fuel: protons

Figure 3.12 Motility apparatus of *E. coli* (A) Two modes of movement. Depending on the direction of rotation, the 5–7 helical flagella either form a bundle, working like a propeller, or fall apart. (B) Schematic representation of the flagellar motor and its interaction with phospho-CheY (CheY-P).

Processing of chemotactic signals

Tactic signals are processed by a highly conserved two-component system that in its construction differs somewhat from the systems discussed above. The major difference is that sensor and His kinase are separate proteins and the His kinase interacts with two different response regulators that affect behavior rather than gene transcription by controlling the flagellar motor and the cellular memory. The separation of sensor protein and His kinase is highly economical since it allows a single signal-processing apparatus to be stimulated by a wide variety of environmental signals, thus enabling an integrated response of the cell.

Apart from the lack of intrinsic His kinase activity, the sensors of tactic signals resemble conventional sensor His kinases in that they exist as dimers with mostly two pairs of transmembrane helices with a variable signal recognition site at the outside (for Gram-negative bacteria, the periplasmic side) and a conserved cytoplasmic domain. These chemoreceptors are known as **methyl-accepting chemotaxis proteins (MCPs)** since they become methylated at several carboxyl groups in the course of a chemotactic response. The degree of change in methylation that occurs during signal processing and subsequently results in an inhibition of receptor activity resembles a bacterial short-term memory store. This feature is discussed in more detail later in the chapter.

E. coli expresses five different MCPs responding to certain sugars, amino acids, dipeptides, and oxygen as well as to changes in pH and temperature. To be recognized by an MCP, sugar molecules have to bind first to specific receptor proteins in the periplasm (Figure 3.13). The oxygen-sensing MCP called Aer binds the redox cofactor FAD at its cytoplasmic side via PAS domains. FAD is coupled to an electron transport chain and signals, by its redox state, either a surplus or a shortage of oxygen. Aer enables a more general **energy taxis** such as seeking for energy-rich food, the metabolism of which activates the cellular electron transport chains. By this means the cell may find a biotope with an optimal ratio between oxygen and oxidizable metabolic substrates. Other examples of energy-tactic sensors are the Ser receptor Tsr (Figure 3.13) and the stress sensor ArcB of *E. coli*. Tsr does not contain a redox cofactor but probably responds by a conformational change to pH fluctuations generated by redox reactions. ArcB contains a reactive dithiol group in a PAS domain, which under aerobic conditions becomes oxidized to an intramolecular disulfide group (Section 3.5.6).

While Aer represents an indirect oxygen sensor, other species such as *B. subtilis* and haloarchaea express direct oxygen sensors coupled with a heme moiety. All

Figure 3.13 Chemotaxis sensors of *E. coli* Depicted are the five membrane-integrated sensor proteins Tsr, Tar, Trg, Tap, and Aer, together with their preferred ligands. M, G, R, and P are periplasmic binding proteins for various sugars and peptides. The Che proteins, including His autokinases and response regulators, are the common downstream effectors of the sensors, controlling both rotation of the flagellar motor and, in a negative feedback loop, methylation of the sensor proteins. Ser, serine; Asp, aspartate; Gal, galactose; Mal, maltose; Rib, ribose; FAD, flavin adenine dinucleotide.

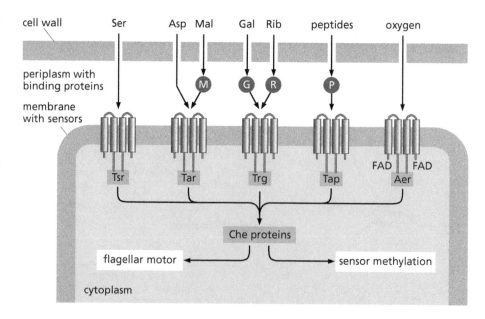

in all, the sensor array of *E. coli* appears to be rather meager, since there are bacteria that may express up to 10 times more different sensor proteins.

The apparatus of tactic data processing is highly conserved (Figure 3.14). It consists of the so-called chemotaxis proteins, two of which are response regulators:

- CheA, a cytoplasmic His autokinase

- CheW, an adaptor protein linking the sensor protein with CheA

- CheY, the response regulator controlling the flagellar motor

- CheB, the response regulator controlling adaptation, which is a demethylase for the sensor protein

- CheZ, an Asp-specific protein phosphatase for signal termination, supporting the weak autophosphatase activity of CheY

- CheR, a methyltransferase catalyzing methylation of the sensor protein by *S*-adenosylmethionine

It should be noted that the term "chemotaxis protein" is somewhat misleading, because the system processes not only chemotactic but all kinds of tactic signals. In fact, separation of the sensor domain from the His autokinase domain CheA has the advantage that a large amount of sensors can be coupled via the adaptor CheW with the same signal-processing network. We shall encounter this principle of domain separation again when discussing eukaryotic signal transduction.

Some models have been developed to explain the mechanism of signal transduction from the sensor to the His kinase CheA. They postulate either a rotation or a shift of the sensor's subunits. There is, for instance, experimental evidence indicating that a tiny pistonlike movement (of 0.16 nm) of the outer transmembrane segments of the sensor is sufficient to inhibit the His kinase activity of CheA (Figure 3.15). During such a minimal shift, the interactions between the side chains of neighboring helices would be preserved, providing the sensor with the properties of an elastic spring that automatically returns into the relaxed position upon termination of the signal. However, how a tolerable signal-to-noise ratio is maintained under such conditions remains an open question. Apart from

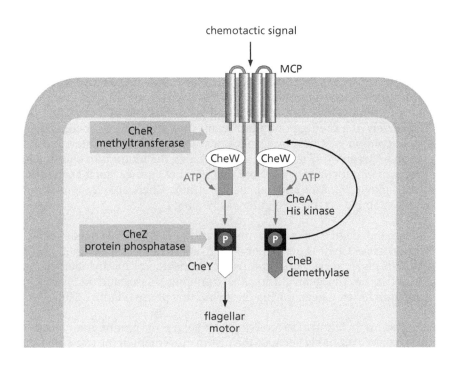

Figure 3.14 Chemotactic two-component system of *E. coli* The autophosphorylated His kinase CheA (red), coupled by the adaptor protein CheW to the membrane-bound sensor (MCP), is the phosphate (P) donor for Asp phosphorylation of the response regulators CheY (controlling the flagellar motor) and CheB (demethylating the sensor). Opponents are the CheY-specific protein phosphatase CheZ and the methyltransferase CheR. Both are constitutively active. The phosphorylated CheY switches the flagellar motor towards clockwise rotation. Red arrows indicate phosphotransfer reactions.

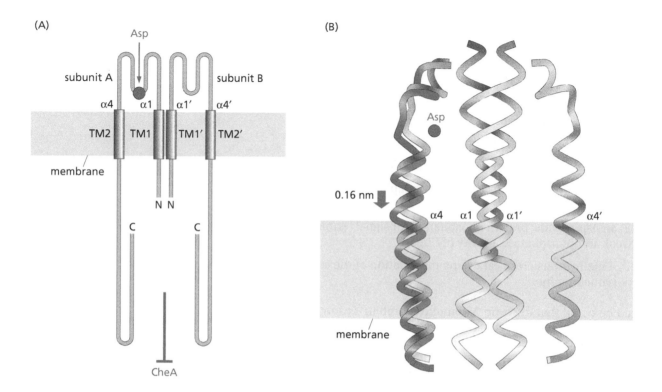

Figure 3.15 Model of transmembrane signaling by a chemotaxis sensor (*E. coli*) (A) Schematic representation of the aspartate sensor homodimer inhibiting the His kinase CheA upon binding of aspartate. (B) Vertical shift (red arrow) of the polypeptide chain α4 of the transmembrane domain TM2 induced by aspartate binding. (B, adapted from J.J. Falke et al., *Annu. Rev. Cell Dev. Biol.* 13, 457–512, 1997.)

this, it should be kept in mind that CheA is controlled not only by (chemo)taxis sensors but also by PTS (Figure 3.5).

The structure of His kinase CheA differs clearly from that of sensor His kinases such as EnvZ (Figures 3.16 and 3.17). Though CheA also exists as a dimer, the His phosphorylation occurs in N-terminal HPt domains rather than in the helix bundle of the dimerization domains. Moreover, each CheA monomer contains a regulatory domain for interaction with the adaptor protein CheW and the sensor as well as a substrate-binding domain for interaction with the response regulators.

Intelligent behavior by adaptation

In *E. coli* the activity of the His kinase CheA and thus the degree of phosphorylation of the Che proteins is high in the *absence* of attractants (or in the *presence* of repellents). In other words, the interaction of an MCP sensor with nutrients inhibits the Che signaling system. Due to the Asp phosphorylation, the response regulator CheY loses contact with CheA, instead interacting as an allosteric regulator with the switching protein FliM of the flagellar motor. This event causes the motor to rotate clockwise, into neutral gear, resulting in a tumbling of the cell. Conversely, the binding of an attractant inhibits CheA. As a consequence, dephosphorylation of CheY by CheZ gains the upper hand, interrupting the interaction between CheY and FliM and switching the motor into counterclockwise rotation, into forward gear. [Note that in other species (such as *B. subtilis*), attractants may *stimulate* and repellents *inhibit* CheA activity, providing an instructive example for the ambiguity of signaling.]

In parallel, the degree of sensor methylation becomes changed: while the methyltransferase CheR is permanently active, the demethylase CheB is stimulated by Asp phosphorylation. As a result the degree of methylation becomes diminished in neutral gear, during the tumbling phase, and increases upon switching into forward gear, during the swimming phase (Figure 3.18).

The stepwise methylation of several (four or five) methyl groups gradually desensitizes the sensor, probably because the piston movement of the transmembrane

(A)

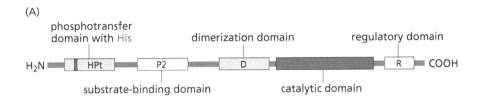

(B)

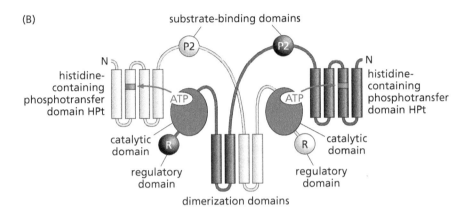

(C)

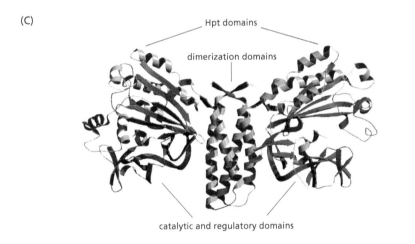

Figure 3.16 Histidine kinase CheA
(A) Domain structure of CheA.
(B) Schematic folding geometry of the homodimer. Autophosphorylated His residues are shown as red boxes. Catalytic domains with ATP-binding site are shaded red. (C) Molecular model of the CheA dimer based on X-ray crystallography.

segments becomes more and more blocked. This leads to reactivation of the His kinase CheA and thus to termination of the swimming phase. The system is switched back to position 0 and can now respond to the next higher attractant concentration, again with maximal sensitivity. In other words, the system has become adapted.

Our sensory organs are subject to similar adaptation, albeit following an entirely different mechanism. Such adaptive processes guarantee that the sensory apparatus remains fully sensitive over a wide range of signal intensity (at least 5 orders of magnitude). Thus, a targeted movement of bacteria is possible: without adaptation, the cell would become maximally excited at a minimal signal intensity, inevitably letting it go over the top.

Mainly due to the high phosphatase activity of CheZ, the degree of CheY phosphorylation changes within milliseconds, whereas sensor methylation is maintained for seconds. This difference is due to both the continuous activity of the methyltransferase CheR and the relatively slow auto-dephosphorylation of the phosphorylated demethylase CheB, which is *not* accelerated by CheZ. Due to this sluggishness, the sensor methylation plays the role of a molecular short-term

Figure 3.17 Active centers of some two-component proteins (Adapted from A.H. West and A.M. Stock, *Trends Biochem. Sci.* 26, 369–376, 2001.)

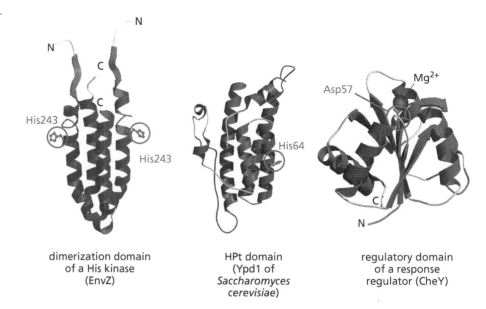

dimerization domain of a His kinase (EnvZ)

HPt domain (Ypd1 of *Saccharomyces cerevisiae*)

regulatory domain of a response regulator (CheY)

memory: the degree of methylation is high when the cell was in an area of high attractant concentration *immediately before*, and vice versa. It is easily understood that adaptation of the data-processing apparatus and learning or memory are two sides of the same coin; for example, I am eager to learn as long as my curiosity has not become satisfied or as long as I have not become adapted!

The prokaryotic "nanobrain"

Sensors and chemotaxis proteins are integrated into stable complexes (Figure 3.19). This fixed wiring guarantees a particularly rapid signal processing as well as a selective adaptation to individual stimuli. In the cell membrane, many thousands of such complexes are combined into large formations localized predominantly in the anterior pole of the cell. This sensor field may be formed by hexagonal lattices consisting of six triads of sensor dimers, each with associated CheW and CheA proteins (Figure 3.20). Such lattices are assumed to exhibit a high degree of cooperativity. This means that allosteric changes of conformation induced by the ligands in a few sensors spread across a larger distance. Such an arrangement is a potent signal amplifier and allows the transformation of digital input signals into analogous output signals. The result is a gradual behavior of the cell paired with high sensitivity. In principle, this situation resembles sensory signal processing in higher organisms. The (chemo)taxis system has been described, therefore, as a prokaryotic "nanobrain." Indeed, the phosphorylation and methylation pattern of the taxis proteins, or the excitation pattern of the

Figure 3.18 Chemotactic responses of *E. coli* Attractants inhibit and repellents stimulate the CheA-catalyzed phosphorylation of response regulators CheB and CheY, resulting in the opposite responses shown in the boxes. In the absence of CheA activity (lack of an attractant or contact with a repellent), CheB undergoes slow auto-dephosphorylation, whereas CheY is dephosphorylated rapidly by the phosphatase CheZ. As a consequence, the methylated state is more persistent than the phosphorylated state, providing a short-term memory store.

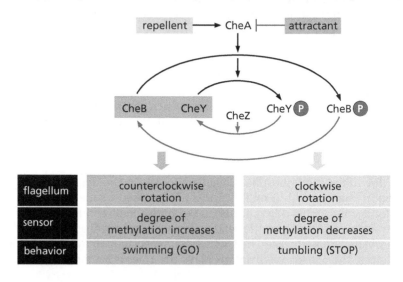

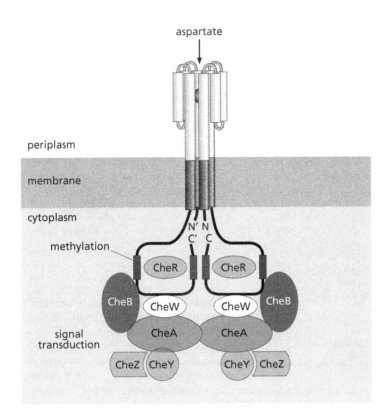

aspartate

periplasm

membrane

cytoplasm

methylation

signal
transduction

N' N
C' C

CheR CheR

CheB CheW CheW CheB

CheA CheA

CheZ CheY CheY CheZ

Figure 3.19 Chemotactic signaling complex of *E. coli* Schematic representation of the aspartate receptor Tar and the associated Che proteins. Helical domains of Tar are symbolized by cylinders. (Adapted with modifications from D.F. Blair, *Annu. Rev. Microbiol.* 49, 489–522, 1995.)

"nanoneurons", somehow portrays the actual environmental situation just as the excitation pattern of brain neurons does. It should be kept in mind, however, that the taxis system itself is embedded into a larger data-processing protein network, and thus it represents only a part of the bacterial "nanobrain."

Finally, it should be noted that although signal processing operates at the level of the single cell, chemotactic behavior is a systemic phenomenon involving a large number of cells that resemble a school of fishes. As we shall see in Section 3.4.1, the cells communicate by means of so-called quorum signals. These are hormone-like metabolites that also interact with sensors coupled to the chemotaxis system of signal transduction. Thus, in a particularly clear manner, bacterial chemotaxis illustrates a major challenge of today's research: to understand the function of a data-processing network and the emergence of new properties from complex systems, one has to proceed from studying single molecules or single cells to an investigation of large populations of interacting partners. That is, one must go beyond the reductionist approach (see Chapter 17).

Bacterial IQ

A closer look at the prokaryotic cell, with its clear conditions, strongly supports the metaphor introduced in Chapter 1 of the data-processing protein network as the "brain of the cell." Indeed, bacterial signal transduction widely fulfills basic requirements of a neural network, such as performing logical operations by parallel data processing using nontrivial algorithms that are able to adjust. Are bacteria really intelligent?

A partial affirmative answer to this question is given in the definition of the chemotactic response as quasi-intelligent behavior. But what is intelligence? Primarily it may be understood as an organism's potential to deal with environmental (and endogenous) challenges in a rational and flexible way. On the molecular level, the number of signaling proteins relative to the size of the genome may be taken as a rough measure of intelligence. This cellular IQ seems to be a matter of lifestyle: the higher the environmental challenges, the higher the IQ value. Analysis of a large number of prokaryotic genes has shown a clear

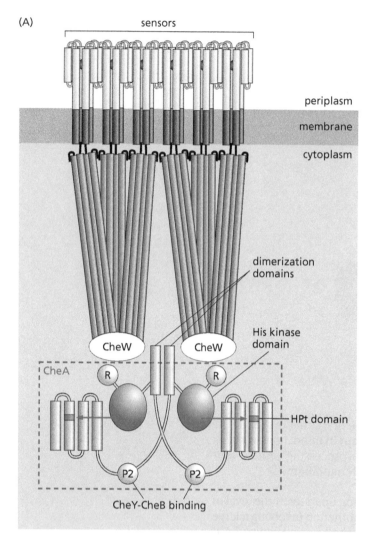

(A)

sensors

periplasm

membrane

cytoplasm

dimerization
domains

His kinase
domain

CheW   CheW

CheA

R   R

HPt domain

P2   P2

CheY-CheB binding

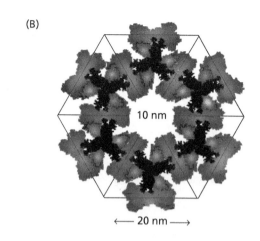

(B)

10 nm

←— 20 nm —→

Figure 3.20 Architecture of a sensor–CheA complex in the inner membrane of *E. coli* (A) CheA dimer interacting, via its regulatory domains (Figure 3.15), with two CheW adaptors. On average, three sensor molecules are bound to each CheW molecule via their extended (hairy) cytoplasmic domains. Several thousand such complexes constitute the bacterial nanobrain. (B) Model of a lattice of chemotactic receptor complexes, shown looking onto the plasma membrane (receptors are in black, CheW adaptor proteins are in gray, and CheA His kinases are in red). It is assumed that the sensory field of the cell membrane is formed by a large number of such hexagonal structures. (A, adapted with modifications from J. Stock and M. Levit, *Curr. Biol.* 10, R11–R14, 2000. B, adapted with modifications from T.S. Shimizu et al., *Nat. Cell Biol.* 2, 792, 2000.)

tendency for high IQ values in free-living microbes that encounter an ever-changing environment, whereas the price of living in luxury and idleness is stultification. Thus, the majority of symbiotic and parasitic microbes get by on an absolute minimum of signal transducers, frequently being devoid of two-component systems including that of chemotaxis. On this scale *E. coli*, the pet of microbiologists, turns out to be rather stupid. For instance, in comparison with free-living Gram-negative bacteria, which sometimes express more than 50 different chemotactic MCPs, the five MCPs of *E. coli* look rather poor, and the same holds true for other signal-transducing proteins (for example, 30 *E. coli* His autokinases versus up to 200 types in other species). This shows the limited value of studies relying just on one or a few "dumb" species.

It must be emphasized that our definition of intelligence is somewhat narrow since it does not include what is known as associative memory. This wonderful property enables us to extract common parameters from countless details that help us to recognize and categorize objects, subjects, and systems as related even when they are not identical. The absence of associative memory is a serious problem of today's computer technology, as demonstrated by daily experience. Whether prokaryotes in particular, or single cells in general, possess an equivalent of associative memory remains to be shown.

3.3.5 Phototaxis and the "invention" of rhodopsin

Phototaxis is defined as a targeted movement directed by light. It has been studied in more detail for **halobacteria**. These are a subclass of archaea inhabiting extreme biotopes, that are, for instance, high in salinity, such as the Dead Sea (salt concentration 4.3 M as compared to 0.6 M in ocean water). Halobacteria are able to utilize sunlight as an energy supply for ATP production. Light, particularly at short wavelengths, is also a deadly hazard. To deal with this problem, halobacteria have developed phototactic systems that recognize short-wavelength light as a repellent and long-wavelength light as an attractant. This enables the cells to actively find optimal environmental conditions.*

The molecular system decoding light signals resembles that of chemotaxis, though the sensors are, of course, quite different. Halobacteria express two types of light sensors or halobacterial transducers (Htrs): one (HtrI) for long-wavelength light and the other (HtrII) for short-wavelength light. Both resemble chemotaxis sensors as they interact with CheA and the other chemotaxis proteins upon illumination. However, being unable to respond to light on their own, the Htr proteins form complexes with two light-sensitive proteins, the sensory rhodopsins SRI and SRII. When the SRI/HtrI complex becomes stimulated, for instance by orange light, the cell responds with seeking behavior, whereas an escape reaction is induced by activation of the SRII/HtrII complex, for instance by blue light. In other words, prokaryotes equipped with such a signal-processing system are able to "see" colors.

Rhodopsins are proteins with seven transmembrane domains arranged in a goblet-shaped pattern. At the outside, this structure has bound retinal (vitamin A aldehyde) as an imide (Schiff base) at the ε-amino group of a lysine residue. In the absence of light, the retinal is locked in the *all-trans* configuration and the nitrogen of the imide bond is protonated, carrying a positive charge. Upon illumination, retinal isomerizes into the 13-*cis* form, resulting in an allosteric rearrangement of the transducer protein Htr that causes an activation of the phototactic system.

In addition to the phototactic sensor rhodopsins (SR type), other microbial rhodopsins are known. These participate in the transformation of light into metabolic energy. In fact, these bacterio- and halorhodopsins (BR and HR) play the role of light-powered ion pumps serving the generation of ATP and the control of the cellular pH value. Upon illumination, BR produces a proton gradient across the cell membrane by transferring, through a conformational rearrangement, the proton bound to the imide nitrogen from the inside to the outside of the cell. By powering a membrane-bound ATP synthase, the proton gradient then transforms light energy into chemical energy (Figures 3.21 and 3.22). HR uses the positive charge of the protonated Schiff base to bind chloride ions electrostatically. Here the proton does not dissociate from the illuminated protein but, again due to a light-dependent conformational rearrangement, the N–H dipole is folded down, releasing the chloride ion into the interior of the cell. This transport process regulates intracellular acidity.

Halobacteria perfectly adjust to environmental conditions. When there is ample food and oxygen (that is, the oxygen signal is received by heme-containing aerotaxis sensor HtrVIII and transduced to the Che complex), only light protection by the HtrII/SRII system is needed. Consistently, the energy-supplying HRs and BRs become expressed only upon oxygen shortage in the water, as caused, for instance, by an increase of the salt concentration due to evaporation.

Evolution of rhodopsins

Rhodopsins are universal light sensors that are classified in two families: microbial rhodopsins (type 1) and rhodopsins of the animal eye (type 2). Whether both

---

*Simpler systems of light protection are found in many prokaryotes. **Photoactive yellow protein** (**PYP**) has been investigated in more detail since in biophysics it provides a favored model for studies of the effects of light on protein conformation. PYP is a cytoplasmic protein containing *p*-hydroxycinnamic acid as a light-sensitive pigment. Short-wavelength light, via a molecular rearrangement of the pigment, induces a conformational change that evokes an escape reflex.

Figure 3.21 Reaction cycle of archaebacterial rhodopsin In the inset the reactions occurring in the cell membrane are shown: light-dependent *cis–trans* isomerization of retinal bound as a Schiff base at a lysine residue of rhodopsin. As a consequence, the conformation of the membrane-bound rhodopsin is changed in such a way that a proton bound to the Schiff base is released into the extracellular space. The reflux of the protons following an electrical charge gradient powers a membrane-bound ATP synthase. In halorhodopsin, the protons are not released but facilitate the influx of chloride ions.

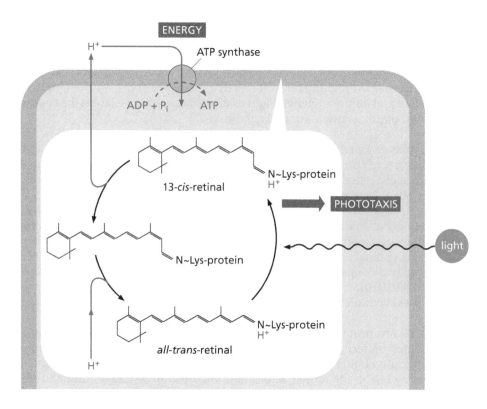

types developed independently or are the result of lateral gene transfer is still a matter of debate. Microbial type 1 rhodopsins are by no means restricted to haloarchaea but have been found in a wide variety of prokaryotes as well as in unicellular eukaryotes, green algae, and fungi. Thus, the type 1 rhodopsin CSR (<u>C</u>hlamydomonas <u>S</u>ensory <u>R</u>hodopsin) is found in the light-sensitive organelles of *Chlamydomonas* algae (Figure 3.23).

The molecular architecture of rhodopsin, with its seven transmembrane helices and an extracellular ligand binding site, is one of the most successful standard models of evolution. It serves as a receptor for retinal and light as well as for a wide variety of other signaling molecules, though the receptors of higher eukaryotes are probably not descendants of the microbial rhodopsins (see Section 5.1).

Light sensors of cyanobacteria

Other prokaryotes using light as an energy supply are the cyanobacteria or blue-green algae. They are considered to be the evolutionary "inventors" of

Figure 3.22 Phototaxis of haloarchaea Different rhodopsins (red) are shown integrated into the plasma membrane and serving energy-delivering ion transport (center) and controlling, as complexes with corresponding sensors, phototactic movements. Note that in the situation depicted by the right figure seeking behaviour is triggered by the HtrI/SRI complex, whereas an escape reaction is induced by the HtrII/SRII complex. For further details see text.

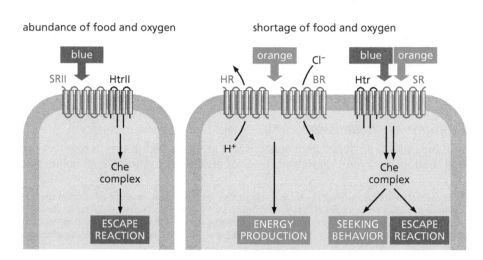

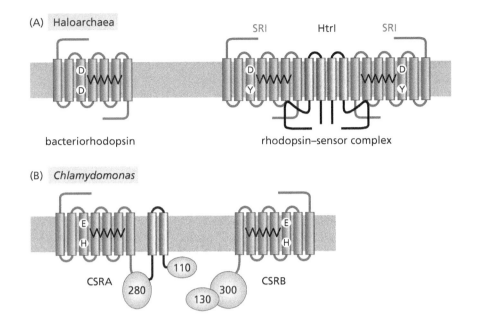

(A) Haloarchaea

SRI   HtrI   SRI

D
D

bacteriorhodopsin

D
Y

rhodopsin–sensor complex

D
Y

(B) *Chlamydomonas*

E
H

CSRA

280   110   130   300

CSRB

E
H

Figure 3.23 Rhodopsins of the microbial type (A) Two haloarchaeal rhodopsins are depicted: ion-transporting bacteriorhodopsin (left) and the rhodopsin–sensor complex of phototactic signal transduction (right). Retinal is symbolized by the zigzag line. The rhodopsin–sensor complex differs from the ion-transporting bacteriorhodopsin in that a negatively charged aspartate residue (D) in the third transmembrane helix, essential for ion translocation, is replaced by a neutral tyrosine residue (Y). Each sensor homodimer (HtrI) is shown to interact via its cytoplasmic domains with two rhodopsin molecules. Upon illumination, the transmembrane helices of the rhodopsins are shifted and interact with the transmembrane helices of the sensor, thus inducing signal transduction along the Che pathway. (B) Two sensoric rhodopsins (CSRA and CSRB) found in the light-sensitive structures of the unicellular eukaryote *Chlamydomonas*. They provide examples of light receptors that are not coupled to the chemotaxis complex. Instead, in both molecules rhodopsin is covalently bound to large proteins (characterized by the molecular weights given in kDa) that evoke a light-dependent influx of Ca^{2+} ions by functioning as transmembrane ion channels or by interacting with separate channel proteins. The Ca^{2+} current regulates the phototactic motility of the cell.

chlorophyll-dependent photosynthesis, being the ancestors of plant chloroplasts. In cyanobacteria, phototaxis is controlled by the phycochromes phycoerythrin and phycocyanin rather than by rhodopsins. Moreover, by means of these light-sensitive proteins, the cell is able to utilize light with wavelengths for photosynthesis to which chlorophyll is relatively insensitive.

The light-sensitive pigments of phycochromes are two tetrapyrrole derivatives, phycoerythrobilin (for short-wavelength light) and phycocyanobilin (for long-wavelength light), which are structurally related to bile pigments. Light induces an isomerization of the pigment molecules, acting as an input signal for the phototaxis system (Figure 3.24). In addition, as in rhodopsin, a proton is released and fed into the photosynthetic complex. Apart from the Che apparatus, another two-component system is used by the cell to adjust the expression of

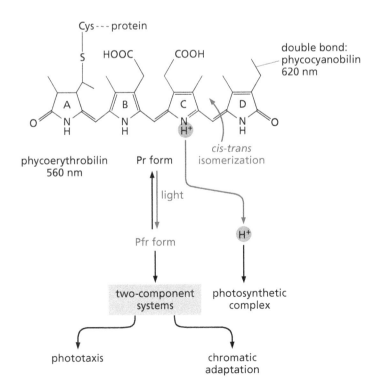

Cys --- protein

HOOC   COOH

double bond: phycocyanobilin 620 nm

A   B   C   D

phycoerythrobilin 560 nm

Pr form

cis-trans isomerization

light

Pfr form

H+

two-component systems

photosynthetic complex

phototaxis   chromatic adaptation

Figure 3.24 Light-sensitive reaction of cyanobacteria Light-dependent isomerization of the tetrapyrrole pigments in the phycochrome proteins is shown. The light-activated Pfr form activates two-component systems of phototaxis and chromatic adaptation. Absorption maxima of the pigments are given.

phycochromes to the light conditions, a response called chromatic adaptation. The phycochromes are the evolutionary ancestors of the phytochromes serving light adaptation in plants (Section 3.5.1).

Bacterial signal transduction: more elaborate than previously assumed

The sequencing of almost 200 prokaryotic genomes has provided a wealth of data showing that prokaryotic signal processing is a theme with many variations. Thus, depending on the species, putative sensor domains of transmembrane proteins are fused with His autokinases, adenylate cyclases, serine/threonine (Ser/Thr)-specific protein kinases, protein phosphatases, diguanylate cyclases (a novel type of second-messenger generating enzymes; see Sidebar 3.4), and DNA-binding domains. These interactions indicate the existence of transmembrane signaling mechanisms that go beyond the conventional two-component system, albeit in most cases the extracellular input signals as well as the downstream effector proteins of such receptorlike proteins are not known. Moreover, such enzymatically active domains are integrated in various cytoplasmic proteins, which by the presence of typical protein and DNA interaction domains are characterized as potential transducers in an intracellular signaling network. These data, which are derived from genomic screening, need to be validated by biochemical studies. Nevertheless, they show that the prokaryotic cell has mastered the task of constructing new protein species by recombining a limited set of modules in a virtuoso manner, thus laying the foundations for most of the signaling processes found in mammalian cells.

Summary

Two-component systems provide the standard model of receptor-coupled signal processing in prokaryotes. The sensor expresses His autokinase activity or couples with His autokinase. The effector protein, the response regulator, usually is a transcription factor. About two-thirds of all sensors are transmembrane proteins processing environmental stimuli; the rest are found in the cytoplasm processing intracellular signals. Upon activation by a signal, the sensor undergoes ATP-dependent His autophosphorylation, immediately transmitting the phosphate residue to a regulatory Asp residue of the response regulator. Phosphatase activities render the signaling process reversible. Adaptation of *E. coli*-like bacteria to osmotic pressure provides a simple example of the function of a two-component system. Depending on the phosphorylation status, the corresponding response regulator induces transcription of genes encoding either small or large cell wall pores. Genes encoding the enzymatic machinery of bacterial nitrogen fixation are under the control of several two-component systems that adjust transcription to both the availability of nitrogen sources and the toxic potential of oxygen. Oxygen-sensing systems are coupled with heme or redox cofactors. Expression of the enzymes used for the primeval energy-supplying oxidation of hydrogen is controlled by two-component systems containing specific H_2-sensor proteins. Phosphorelays resemble two-component systems but consist of more than two partners. Such phosphorylation cascades offer additional possibilities for control, network formation (cross talk), and signal fine-tuning. By means of a universal two-component system consisting of individual sensors and a common apparatus of signal transduction (Che complex), prokaryotes are able to adjust their movement pattern to a wide variety of chemical and physical stimuli acting as either attractants or repellents. A characteristic feature of this system is that various sensor proteins couple noncovalently (via adaptor proteins) with a separate His kinase, and the response regulators are proteins that control the motility and adaptive behavior of the cell. Adaptation enabling targeted, quasi-intelligent movement is due to sensor desensitization through reversible protein methylation. The phosphorylation and methylation patterns represent a short-term memory store of the cell. In light-sensitive halobacteria the phototactic sensors of two-component systems are coupled with rhodopsin proteins containing *all-trans*-retinal as a light-sensitive pigment.

Upon illumination, the retinal is isomerized to the 13-*cis* form and the chemotaxis system is activated, regulating the movement pattern and the adaptive behavior of the cell. Other microbial rhodopsin types function as ion pumps and use light as an energy source. They are also found in unicellular eukaryotes and fungi. In cyanobacteria, the light-sensitive components coupling with corresponding two-component systems are tetrapyrrole pigments bound by phycochrome proteins, the precursors of plant phytochromes.

3.4 From vagabonds to societies: "bacterial hormones"

If hormones are understood as mediators of *inter*cellular communication in metazoans, a term such as "bacterial hormones" appears to be nonsensical at first glance. However, most prokaryotes are by no means loners but are capable of amazing social performances, thereby acting like multicellular organisms. This process begins with the formation of characteristically shaped colonies and culminates in complex, differentiated structures that consist of millions of cells. Examples are provided by the chainlike colonies of certain cyanobacteria, the predatory spheres of myxococci, the fruiting bodies of myxobacteria, the hunting packs of *Proteus mirabilis*, and the so-called biofilms. Frequently, such structures are aimed at the division of labor through cell differentiation.

3.4.1 Quorum sensing and auto-inducers

For aggregation and differentiation, cells communicate by signals. Among such **social signals**, the so-called quorum sensing has been investigated in detail. Quorum sensing monitors the population density of a collection of cells to find out whether or not it makes sense to start an expensive genetic program—such as sporulation or biofilm formation—which requires a larger group of cells to become effective. The principle of quorum sensing is simple: since cells continuously release the social signal, its concentration is a direct function of cell density, and the target genes become activated above a distinct threshold concentration.

Pioneering experiments were performed on *Vibrio fischeri*. This Gram-negative symbiont inhabits the light-producing organs of certain fishes and cephalopods, causing bioluminescence. In return for this the host protects the settlers from enemies and provides food. For the host, light production makes sense only when the intensity of bioluminescence exceeds a certain value. However, the prokaryote produces light only at a sufficiently high population density (at least 10^{10} cells/mL, as compared with a density of a few hundred cells per milliliter in sea water). Therefore, bioluminescence is triggered by a population-dependent signal that is released by the cells. This auto-inducer stimulates neighboring cells to transcribe the *lux* genes that encode bioluminescence enzymes such as luciferases (more information on such enzymes is found in Section 5.2) and enzymes catalyzing the production of the auto-inducer.

Auto-inducers are not restricted to luminous bacteria but have been found in many other species. As far as Gram-negative bacteria are concerned, auto-inducers are almost exclusively **acyl-homoserine lactones** (Figure 3.26), differing from each other by the acyl residue and exhibiting species-specific effects. This provides quorum sensing with a high degree of privacy. Acyl-homoserine lactones are lipophilic enough to penetrate cellular membranes with increasing cell density, thereby "soaking" the whole population.* The corresponding cellular receptors are transcription factors that by convention are called LuxR in reference to the bioluminescence system, although they are by no means involved

---

*Free diffusion requiring a minimum of water solubility is only one way to distribute social signals within a cell population. Highly hydrophobic molecules have been shown to be packed in vesicles that are pinched off from the outer membrane of certain Gram-negative bacteria to become fused with the membrane of a neighboring cell.

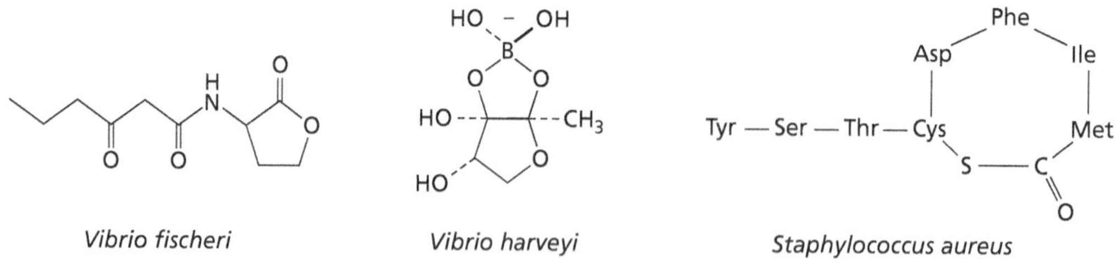

Vibrio fischeri *Vibrio harveyi* *Staphylococcus aureus*

Figure 3.26 Some auto-inducers of quorum sensing Shown are an acyl-homoserine derivative (left), a furanosylborate diester (middle), and a cyclic octapeptide (right) from different microorganisms.

only in light production. They may control many other population-dependent functions including the formation of spores, colonies, and biofilms as well as of virulence factors (these are illness-causing toxins and proteins used by pathogenic microorganisms for the invasion of host cells). Upon binding of a specific auto-inducer, the LuxR factors dimerize and either induce or repress the transcription of individual operons (Figure 3.27). This mechanism of action closely resembles that of steroid hormones, which in higher eukaryotes interact with so-called nuclear receptors (Section 8.3).

Gram-positive bacteria do not produce acyl-homoserine lactones. Instead their auto-inducers are short, partially **modified peptides** released from larger precursor proteins (an example, the cyclic octapeptide of *Staphylococcus aureus*, is shown in Figure 3.26). In some species this signaling system is completed by another one using **furanosylborate diesters** as auto-inducers (Figure 3.26). This provides an interesting and rare example of the utilization of boron in a biological system. Evidently the arsenal of "bacterial hormones" is much more elaborate than hitherto assumed, since novel factors have been found that differ chemically from modified peptides, acyl-homoserine lactones, and furanosylborate diesters. They include derivatives of oligosaccharides, dihydroxypentadienones, butyrolactones, quinolones, and others, with many of them exhibiting quite exotic structures.

In contrast to acyl-homoserine factors the auto-inducers of Gram-positive species are unable to penetrate cellular membranes. Instead, their effects are mediated by highly specific two-component systems controlling the transcription of corresponding genes (Figure 3.27). The analogy to peptide hormones of higher eukaryotes that interact with protein kinase-coupled receptors is striking. Thus, auto-inducers control the activity of transcription factors either directly by binding or indirectly by phosphorylation. These two mechanisms have been preserved in the course of evolution and still represent the major signaling routes leading to gene regulation (Section 8.3.1).

Prokaryotes, but also algae and flowering plants, produce inhibitors of quorum sensing, thus outcompeting and keeping in check populations of competitor, parasitic, and symbiotic bacteria. There is some evidence that human cells also use anti-quorum strategies to fight bacterial infections. For instance, cells from bronchial epithelium have been shown to inactivate the auto-inducer controlling the virulence genes in *Pseudomonas aeruginosa*. It is clear that the investigation of such defense mechanisms and the potential clinical application of synthetic inhibitors offer highly attractive possibilities in medicine.

3.4.2 Biofilms and fruiting bodies: benefits of multicellularity

Population-dependent signals control not only parasitism and symbiosis but also the aggregation of bacteria into multicellular, differentiated structures such as biofilms and fruiting bodies. Biofilms are large bacterial aggregates growing on surfaces. In these complex structures the cells are embedded like tissue cells into an extracellular polysaccharide matrix that is produced by the cell collective. Common examples are tooth plaques, the slimy coat of stream pebbles, mats of bacteria lining water pipes, and the bacterial colonies in the ruminant stomach. The discovery of biofilms has revolutionized the widely held idea of prokaryotes

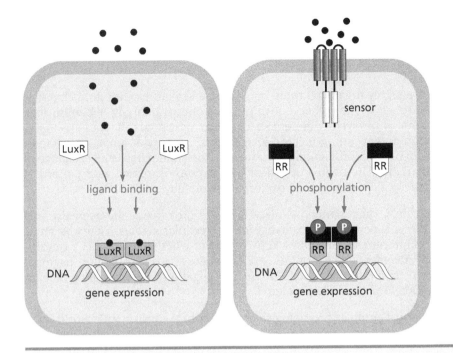

Figure 3.27 Mechanisms of action of auto-inducers: two major pathways along which extracellular signals control gene transcription Depending on the chemical structure, auto-inducers (black dots) activate transcription factors either by direct interaction (left, Gram-negative bacteria) or via a receptor-controlled phosphorylation cascade (right, two-component system of Gram-positive bacteria). LuxR, transcription factors first studied in the bioluminescence system; RR, response regulator; P, phosphate. For other details see text.

Sidebar 3.4 Cyclic di-GMP, a novel second messenger of bacteria Recently an endogenous signaling molecule has been identified that promotes production of the extracellular matrix of biofilms. Cyclic diguanosine monophosphate (c-di-GMP) was originally discovered as an activator of cellulose synthase in certain bacterial species. The number of putative functions has expanded considerably, including, in addition to biofilm formation, intercellular communication in myxobacteria, development of flagella, pathogen–host interactions, phage resistance, and photosynthesis in cyanobacteria.

C-di-GMP is produced from GTP by diguanylate cyclases and inactivated by cognate phosphodiesterases (Figure 3.25). These enzymes are characterized by specific sequence motifs such as GGDEF for the cyclases and EAL for the phosphodiesterases. Such motifs are found in a wide variety of bacterial proteins, including response regulators of two-component systems. The cyclases and phosphodiesterases become activated by external signals. This puts c-di-GMP in the rank of a genuine second messenger, albeit most of its effector molecules are still not known. Bacteria with a "high IQ" may express more than 60 different diguanylate cyclase isoforms, indicating a key role of these enzymes in prokaryotic signal processing. Some of these enzymes are parts of integral transmembrane proteins carrying an extracellular sensor domain, the ligands of which are still unknown.

Considering the high resistance of biofilms to conventional antibacterial drugs and the restriction of c-di-GMP to bacteria, the enzymes controlling the metabolism and effects of c-di-GMP are emerging as attractive targets for novel therapeutic strategies. Cyclic 3′,5′-GMP, a typical second messenger of eukaryotic cells, has not been found in prokaryotes.

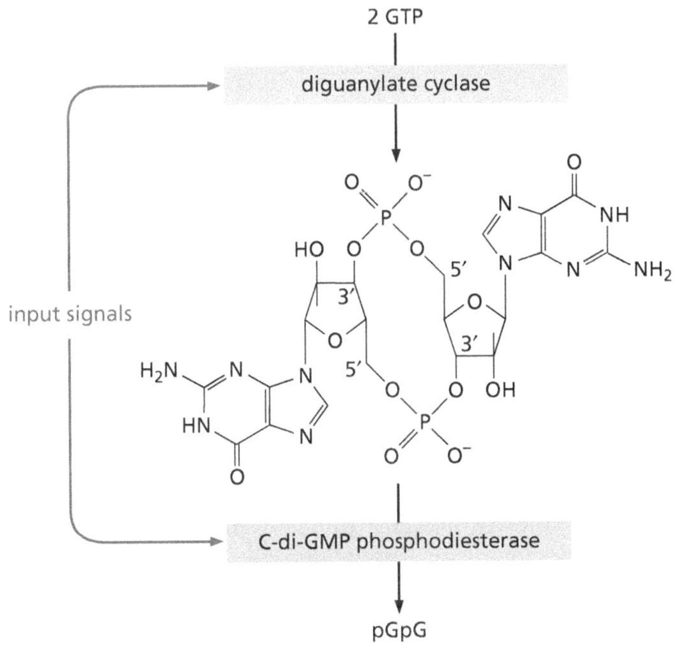

Figure 3.25 Signal-controlled biosynthesis and inactivation of cyclic di-GMP

as "primitive" unicellular organisms. Indeed, rather than single cells living as vagabonds, biofilms are the predominant form of prokaryotic life. As indicated by fossils, the history of such multicellular structures reaches back to the early days of evolution.

Biofilms are highly organized three-dimensional structures with fluid channels running through a spongelike body. The microbes inhabiting a biofilm may belong to one or to several different species. The individual cell types are organized in colonies that communicate with each other and divide labor, for instance, breaking-up of food, protection from stress factors, and parasexual exchange of genetic material. The degree of cell differentiation, expressed as a pattern of genetic activity, depends on the position in the biofilm.

Like metazoans, such highly organized forms of prokaryotic life can exist only when there is a permanent exchange of intercellular signals. Here the auto-inducers of quorum sensing play a key role. Because of their resistance to antibiotics and disinfectants, biofilms are a major cause of acute bacterial intoxications and chronic infections. The investigation of the signaling mechanisms is, therefore, of practical value.

One of the most complex physiological processes in the prokaryotic world is the sophisticated survival strategy of **myxobacteria.** In emergency situations they develop fruiting bodies, consisting of thousands of cells that have become differentiated to spores. This type of cell differentiation is triggered by the ancient prokaryotic stress reaction of the stringent response, resulting in an almost complete stop of cell proliferation (Section 3.1). In addition, stressed myxobacteria release proteases into the extracellular space that hydrolyze extracellular proteins. The amino acids thus produced serve both as food and as quorum signals (called A-signals) that are processed by two-component systems, enabling the cell to monitor whether enough cells are available to generate a fruiting body. When this is the case, the cells emit aggregation signals of still-unknown structure, inducing chemotactic responses. The cell that happens to be the strongest transmitter develops into an aggregation focus, gathering together some 100,000 other cells. The aggregate then differentiates into a fruiting body. The differentiation is controlled by a morphogenetic signal protein (C-signal) exposed at the cell surface, interacting with sensors at the surface of neighboring cells. This juxtacrine signal transduction strongly resembles events regulating the embryonic development of vertebrates (see, for example, Sections 7.1.4, 7.3, and 16.1). By this means, *Myxococcus xanthus* produces fruiting bodies and also differentiates alternatively into another multicellular structure, the so-called predatory sphere, serving the collective catching of prey.

In their mobility mode, myxobacteria differ from less complex species such as *E. coli* in that they have no flagella but execute sliding movements resembling that of amoebae. They possess two genetically fixed programs of mobility, one for single cells and the other for groups of cells sliding like slugs on a slimy secretion. Both programs are probably coordinated by a signaling cascade that formerly had been found only in animals. In this cascade, a transmembrane tyrosine kinase (MasK of *M. xanthus*) interacts with a GTPase (MglA) that, among all prokaryotic G-proteins, is most similar to the small G-proteins of the eukaryotic Ras family. Although a role of MasK as a signal receptor has not been firmly established, these findings strongly indicate a fluent transition between prokaryotic and eukaryotic mechanisms of signal processing.

Myxobacteria developed approximately 2 billion years ago, representing the most ancient multicellular organisms. Independently, multicellularity was also "invented" by eukaryotic myxamoebae or slime molds. The modes of motility, aggregation, and differentiation, most thoroughly studied for *Dictyostelium discoideum*, are quite similar to those of myxobacteria, though they employ other mechanisms. This provides a fascinating example of evolutionary convergence (Figure 3.28).

(A) aggregation phase (B) differentiation phase

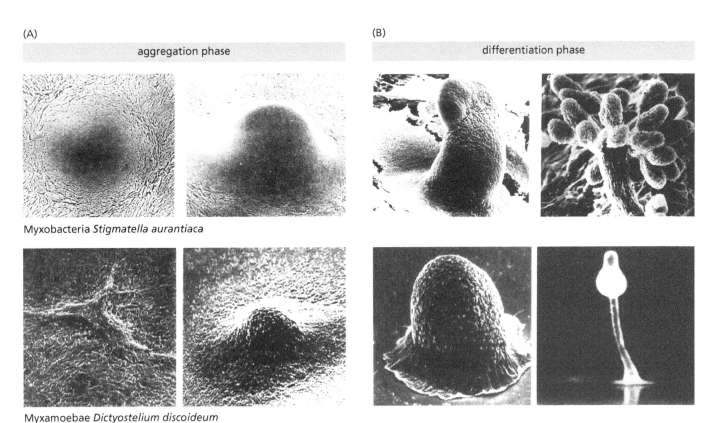

Myxobacteria *Stigmatella aurantiaca*

Myxamoebae *Dictyostelium discoideum*

There is a blurred dividing line between the auto-inducers of quorum sensing and bacterial **pheromones.** These cell-derived attractants are small peptides controlling various aspects of social life. In *Enterococcus faecalis*, for instance, they induce the exposure of surface proteins for "sexual" contacts (serving the exchange of genetic material). This function resembles that of plant gamones or animal sexual hormones. In other species pheromones control the uptake of free DNA or the invasion of tissues by pathogenic bacteria. The effects of the prokaryotic pheromones known thus far are mediated by two-component systems.

Figure 3.28 Aggregation and differentiation of myxobacteria as compared with myxamoebae In both cases, apparently amorphous cellular aggregates (A) develop into fruiting bodies (B). (Myxobacteria, adapted from J.A. Shapiro, *Sci. Am.*, 256, 82–89, June 1988. Myxamoebae, adapted from B. Alberts et al., Molecular Biology of the Cell. New York and London: Garland Publishing, 1983.)

Summary

The formation and function of bacterial colonies and complex multicellular structures (such as biofilms and fruiting bodies) are controlled by intercellular signals including factors that monitor the population density (quorum sensing). In Gram-negative species, such hormone-like auto-inducers are acyl-homoserine lactones able to penetrate the cell membrane and to interact directly with transcription factors. The auto-inducers of Gram-positive species are modified peptides, furanosylborate diesters, and other compounds. Unable to pass the membrane, they instead interact with sensors of corresponding two-component systems that transduce the signal through Asp phosphorylation of transcription factors, or response regulators. Thus, intercellular signaling by "hormones" and the major principles of transcriptional regulation—either ligand binding or phosphorylation of transcription factors—are of prokaryotic origin.

3.5 From bacteria to humans: evolution of signaling mechanisms

Since cells cannot survive without communicating with the environment, the mechanisms of cellular signal processing must have developed in parallel with the mechanisms of food intake, energy metabolism, and reproduction. Indeed, there are fluid transitions between metabolic and signaling reactions, and even the most primitive prokaryotes possess a complex arsenal of signal-processing proteins, while more advanced species communicate via hormone-like factors. A comparison of different species clearly indicates that practically all eukaryotic mechanisms of signal transduction must have emerged from prokaryotic precursors. Examples of these precursors, such as membrane transporters, second messengers, protein kinase-coupled and rhodopsin-like receptors, and the pathways of transcription factor control, have been described above. In the following sections, the discussion of evolutionary aspects is expanded to include standard biochemical devices of signal transduction such as kinases, phosphatases, GTPases, proteases, redox enzymes, and ion channels.

3.5.1 "Prokaryotic" protein phosphorylation in eukaryotes

Phosphorylation is a key reaction of energy metabolism, and at the same time is the most versatile and perhaps most ancient mechanism for post-translational control of protein function. Most probably, protein phosphorylation was applied by the precursors of archaea and bacteria. In the course of evolution, several large families of protein kinases and phosphatases have emerged (see Section 2.6 for an overview).

The His-specific autokinases of prokaryotic two-component systems represent a transition between the phosphotransferases of energy metabolism and signal-processing protein kinases. In the course of prokaryotic evolution, they have become models of success: 1.5% of the genome of *E. coli* includes 30 genes encoding His kinases and 32 genes encoding response regulators. *Nostoc punctiformis*, *Myxococcus xanthus*, and cyanobacteria express more than 150 different two-component systems. Only very simple bacteria such as mycoplasma can survive without two-component systems. They probably have lost the corresponding genes due to their special life conditions (see remarks on the "bacterial IQ" in Section 3.3.4).

Histidine kinases and two-component systems are also expressed by archaebacteria, slime molds, fungi, and plants. It is assumed that they had been taken over from bacteria by horizontal gene transfer. Apart from the mitochondrial genome, no such genes have been found in animals. All eukaryotic two-component systems are constructed as phosphorelays, thus providing more possibilities for interactions and signaling cross talk.

Only assumptions can be made concerning the reason why, on the way to eukaryotes, bacterial His phosphorylation has fallen into oblivion, being replaced by Ser/Thr/Tyr phosphorylation. Because the phosphoric ester bonds are much more stable than the phospho-His and phospho-Asp bonds, they certainly provide a better memory effect, more precise regulation, and a superior signal-to-noise ratio, albeit at the price of speed.

Yeast

The bakers' yeast *Saccharomyces cerevisiae* has only one His kinase gene and three response regulator genes encoding a system for **osmolarity control** (Figure 3.29). A striking feature of this system is its connection with a "modern" Mitogen-Activated Protein (MAP) kinase module (Section 2.6.5), which has not yet been found in prokaryotes. In fact, yeast cells express several MAP kinase modules with different functions (see Section 11.1). By phosphorylating the corresponding transcription factors, one of these modules stimulates genes encoding proteins that protect the cell from osmotic stress, for instance, enzymes of glycerol production. Under normal conditions this module is inhibited by the phosphorylated response regulator SsK1. Hypertonic stress blocks the associated sensor His kinase Sln1, resulting in dephosphorylation of the response regulator and, in turn, activation of the MAP kinase module. Via the interconnected HPt protein Ypd1, the His kinase Sln1 can phosphorylate a second response regulator, Skn7, which does not control a MAP kinase module but functions as a transcription

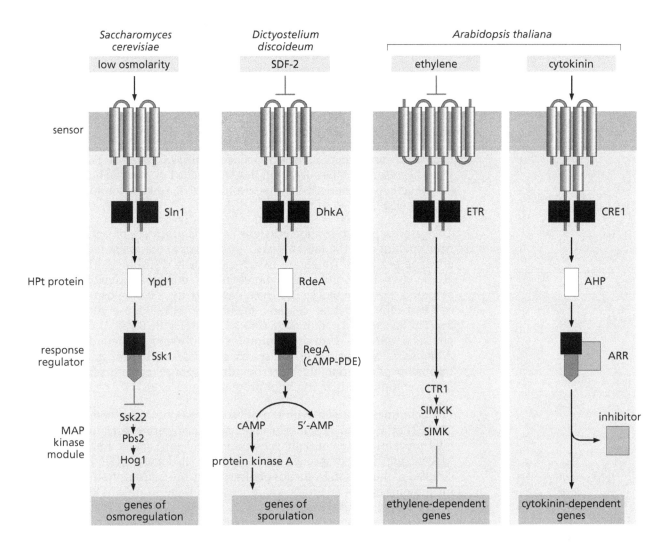

Figure 3.29 "Two-component" phosphorelays of eukaryotes Red cylinder, His autokinase domain; black square, Asp-containing receiver domain; white rectangle, HPt domain; gray wedge-shaped symbol, effector domain of response regulator. For details see text.

factor of additional stress genes. An analogous system has been found in the fission yeast *Schizosaccharomyces pombe*.

Dictyostelium discoideum

Containing 20 His kinase and phosphorelay genes, respectively, this slime mold is presently known to be the eukaryote equipped with the most two-component systems. Among these systems, however, only one is understood in more detail in that it has been found to be involved in the coordination of cell differentiation by controlling the production of the signal molecule cyclic AMP. For *Dictyostelium*, cAMP has two functions: it is a chemotactic attractant for the aggregation of single cells to a multicellular structure and it is a morphogenetic signal inducing the differentiation of spore capsules. These effects of cAMP are mediated by G-protein-coupled receptors (see Chapter 5). The action of cAMP depends on its concentration: at low levels it is an attractant, at high levels a morphogen. The cAMP level is controlled by two enzymes, an adenylate cyclase catalyzing biosynthesis from ATP and a phosphodiesterase catalyzing hydrolysis to inactive 5′-AMP. cAMP phosphodiesterase (cAMP-PDE) is a C-terminal subdomain of a response regulator, RegA, and becomes activated by Asp phosphorylation of RegA. This phosphorylation is stimulated by two sensor His kinases, DhkA and DhkB (*Dictyostelium histidine kinases*), that are inhibited upon binding of their ligands; Slime mold Differentiation Factor SDF-2; and discadenin, a cytokinin derivative. As a result, the interconnected HPt protein RdeA and the response regulator RegA become dephosphorylated, the phosphodiesterase activity is suppressed, and the level of cAMP increases (Figure 3.29). Another sensor His kinase of *Dictyostelium* is probably involved in the adaptation to osmolarity.

115

Plants

Plants express a limited number of two-component systems. For instance, 11 His kinase genes and 16 response regulator genes have been found in the genome of *Arabidopsis thaliana* (as compared with more than a thousand genes encoding "eukaryotic" protein kinases). Some of these systems mediate the response to osmotic stress and to the phytohormones cytokinin and ethylene as well as light adaptation by phytochromes.

In structure and function, the **osmosensor Athk1** (*Arabidopsis thaliana* histidine kinase 1) resembles the corresponding sensor Sln1 of yeast. In both cases the His kinase activity is depressed by hypertonic stress. Athk1 is found predominantly in root cells.

The gaseous plant hormone **ethylene** controls processes of senescence such as fruit ripening and the fall of leaves. *Arabidopsis* expresses at least five different ethylene receptors, which are tissue-specifically distributed. These are sensor His kinases that inhibit the expression of ethylene-inducible genes when activated. Ethylene blocks the His kinase activity. Like the osmosensor system of yeast, the signaling cascade includes a MAP kinase module with the Ser/Thr kinase CTR1 acting as a MAP3 kinase (Figure 3.29). As functional antagonists of ethylene, **cytokinins** promote cell division and retard senescence. In *Arabidopsis*, two sensor His kinases have been identified as cytokinin receptors. Upon phosphorylation, the corresponding response regulators become released from a complex with an inhibitor protein (Figure 3.29).

Phytochromes emerged from the phycochromes of cyanobacteria (see Section 3.3.5). They are used by plants to adapt growth, flowering, germination, and periodic movements to the light situation. The light signals are processed by a "degenerate" two-component system with a His kinase-related Ser/Thr kinase coupled to the sensor. By means of this system, the transcription of genes encoding proteins of light adaptation is controlled. "Degenerate" His kinases have also been found in prokaryotes such as *B. subtilis*, indicating that the change of amino acid specificity is not a result of eukaryotic evolution.

Other light sensors of plants include the phytochrome-related cryptochromes, a group of flavoproteins with unknown functions, and the phototropins. The latter, which are flavoproteins with an additional Ser/Thr kinase activity, are involved in the control of light-directed movements or phototropisms. A response regulator-like protein also plays a role in the regulation of day–night rhythms (photoperiodicity) of *Arabidopsis*.

Animals

The genomes of animals sequenced thus far do not contain His kinase and response regulator genes. An exception is mitochondria, the matrix of which contains two Ser/Thr kinases exhibiting some sequence homology with His kinases. These enzymes are subunits of the large dehydrogenase complexes metabolizing pyruvate and other α-keto acids. The dehydrogenase activity is suppressed by phosphorylation. This reaction is stimulated by protein deficiency, indicating that it provides a protective mechanism against a counterproductive degradation of amino acids.

The occurrence of degenerate His kinases in mitochondria is explained by the evolution of these organelles from endosymbiotic prokaryotes. It should be noted that in contrast to the labile His phosphorylation, the Ser/Thr phosphorylations catalyzed by degenerate His kinases are stable post-translational modifications.

Since conventional two-component systems are not found in animals, they provide interesting targets for novel antibacterial drugs, which are under development.

3.5.2 "Eukaryotic" protein phosphorylation in prokaryotes

At the transition from prokaryotes to eukaryotes, the "prokaryotic" His kinases were widely replaced by Ser/Thr and Tyr kinases. Nevertheless, the adjective "eukaryotic" commonly used for these enzymes is somewhat misleading, since they are by no means restricted to eukaryotes but are found, together with the corresponding protein phosphatases, in the majority of prokaryotes as well. In fact, in 29 out of 35 prokaryotic genomes sequenced, corresponding genes have been identified (for His kinases this ratio is 26:35). Even mycoplasma lacking His kinases express "eukaryotic" Ser/Thr kinases. Obviously these enzymes have a long evolutionary history, probably going back to a primeval form that emerged prior to the separation of bacteria, archaea, and eukaryotes. In contrast, His kinases seem to be genuine bacterial "inventions" taken over by other species through horizontal gene transfer. Such a concept does not contradict the fact that some microbes are devoid of either bacterial His kinases or "eukaryotic" kinases and phosphatases. As a rule, such species are highly adapted pathogenic organisms, such as mycoplasma and rickettsia, as well as inhabitants of isolated biotopes with constant environmental conditions. It is assumed that the corresponding genes devoted to the processing of environmental stimuli turned out to be unnecessary and were eliminated.

The functions of "eukaryotic" protein phosphorylation in prokaryotes are understood only in a few cases. Examples are the kinase Pkn1 of *Myxococcus xanthus*, which is involved in sporulation, and the kinase AfsK, which phosphorylates the transcription factor AfsR and thus regulates the biosynthesis of antibiotics in *Streptomyces coelicolor*. In eukaryotes, with very few exceptions, Ser/Thr phosphorylation and Tyr phosphorylation are catalyzed by different protein kinases. However, in most of the prokaryotic kinases, this specification is not found (thus they are frequently called STY kinases). Nevertheless, Tyr phosphorylation seems to be a rare event in microbes, found only in a very few species such as streptomycetes and myxobacteria. Moreover, *specific* Tyr kinases have been detected as yet only in myxobacteria but in no other prokaryotes. Such enzymes, in particular the receptor-coupled forms, are more or less restricted to animals. Their equivalents in plants (and some prokaryotes) are the "receptorlike" Ser/Thr kinases, of which animals express only a few types (see Sections 2.6.3 and 6.1).

Kinases unrelated to both His kinases and eukaryotic kinases are called atypical. Examples of atypical kinases in prokaryotes are provided by the bifunctional protein kinases/phosphatases (see Section 3.5.3). Atypical kinases of eukaryotes include the myosin heavy-chain kinases of *D. discoideum* and the related elongation factor 2 kinase (eEF2K, see Section 9.3.4) as well as the kinase BCR (known as a fusion partner of the Abl Tyr kinase of chronic myeloid leukemia, see Section 7.1.1) and several protein kinases related to phosphatidylinositol 3-kinase (Section 12.9.1). In Figure 3.30 an attempt is made to schematically derive the evolutionary history of current protein kinase families from their present distribution.

3.5.3 Evolution of protein phosphatases

Among the protein phosphatases, the Ser/Thr-specific enzymes of the PPP family (Section 2.6.4) seem to be phylogenetically most ancient. Like Ser/Thr kinases, they probably were introduced by the common primeval precursor of pro- and eukaryotes. In contrast, the bacterial enzymes of the PPM family (Section 2.6.4) are thought to have been acquired from eukaryotes by horizontal gene transfer since they have not yet been found in archaea. A similar assumption is made concerning Tyr-specific phosphatases, identified in a few, mostly pathogenic, bacterial species where they play a role as virulence factors (see Sidebar 3.5). No genes of the dual-specific Cdc25 phosphatases regulating the cell cycle have yet been found in prokaryotes.

Figure 3.30 Tentative scheme illustrating the evolution of different protein kinase families This schematic sketch (which is not to scale) is based on the distribution of kinase families in today's organisms.

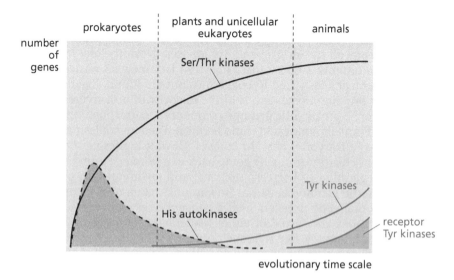

The phospho-His and phospho-Asp residues generated by two-component systems are hydrolyzed either nonenzymatically or by the intrinsic phosphatase activity of the bifunctional His kinases and response regulators. There are, in addition, Asp-specific phosphatases such as CheZ (chemotaxis), SpoOE (sporulation of *B. subtilis*), RapA and B (sporulation), and the His-specific phosphatase SixA (adaptation of *E. coli* to oxygen deficiency). These enzymes are monospecific: they have only one substrate protein each and are considered to be the result of highly specialized development. CheZ dephosphorylates exclusively the response regulator CheY, and SixA affects only the HPt domains of the sensor ArcB, the cytoplasmic part of which is a typical phosphorelay consisting of a His kinase, an Asp kinase, and an HPt domain.

Other extreme specialists are the bifunctional bacterial Ser/Thr kinases/ phosphatases, represented by the isocitrate dehydrogenase kinase/phosphatase (AceK) of *E. coli* and the HPr-kinase/phosphatase of *B. subtilis*. AceK is of historical interest since it was the first protein kinase found in prokaryotes. The enzyme inhibits isocitrate dehydrogenase by phosphorylating a strategic Ser residue in the catalytic center, thus inhibiting a starter reaction of the citrate cycle and shifting the isocitrate metabolism to the glyoxylate pathway:

$$NADP^+ + \text{isocitrate} \rightarrow NADPH + \alpha\text{-ketoglutarate} + H^+ + CO_2$$

In the kinase, the same catalytic center is responsible for both phosphorylation and dephosphorylation. Dephosphorylation occurs through a re-transfer of the protein-bound phosphate residue to ADP. The subsequent hydrolysis of the ATP thus formed renders the reaction exergonic. In other words, the kinase/

Sidebar 3.5 Phosphotyrosine phosphatases as virulence factors: God's scourge of the Middle Ages Some bacteria produce Tyr-specific protein phosphatases of the conventional subfamily (see Section 2.6.4). They are assumed to have acquired this ability from eukaryotes by horizontal gene transfer. The bacterial phosphatases are extremely potent virulence factors. The enzyme YoPH of plague bacteria (*Yersinia*), together with some additional toxins, is thought to be responsible for the depopulation of large areas of Europe and Asia in the Middle Ages. YoPH is injected by the bacteria into the host cell and destroys focal adhesions—the contacts to other cells and to the extracellular matrix— thus inducing cell death by apoptosis. A comparably destructive effect, in this case on the cytoskeleton, is exhibited by the Tyr-specific phosphatases STP of *Salmonella typhimurium* and MPtpB of *Mycobacterium tuberculosis*. Other examples of the fatal consequences of a disturbance by bacteria of cellular data processing are provided by the cholera, pertussis, and diphtheria toxins attacking G-proteins and by the pathogenic adenylate cyclases of anthrax. They will be discussed in more detail in Section 4.3.3.

phosphatase is also an ATPase. Such a mechanism appears to be unique and is not a precedent in evolution. It also holds true for HPr kinase/phosphatase which is, however, devoid of ATPase activity. Obviously both enzymes represent evolutionary impasses.

3.5.4 Evolution of regulatory GTPases

In addition to protein kinases and phosphatases, GTPases constitute a particularly important family of signal-transducing proteins. There are two major types of GTPases: the FtsZ/tubulin family and the G-protein family. Both seem to have evolutionary histories comparable to that of protein kinases, since the corresponding genes have been found in all prokaryotes studied so far.

The prokaryotic **FtsZ proteins** derive their name from a group of *E. coli* proteins involved in cell division. In their three-dimensional architecture and their functions they closely resemble the eukaryotic β-tubulins, though very little sequence homology exists. FtsZ proteins control the directed polymerization of cytoplasmic filaments or protofilaments in prokaryotes, whereas β-tubulins regulate the formation of microtubules in eukaryotic cells. Protofilaments are components of the Z-ring that separates the daughter cells during mitosis, while microtubules constitute the mitotic spindle as well as cilia and flagella. As expected, FtsZ-related proteins have also been found in mitochondria and chloroplasts.

The **G-proteins** are not related to FtsZ proteins and β-tubulins. As biochemical switches they regulate a wide variety of cellular processes, causing spatial and temporal coordination and irreversibility (Section 2.4). First the translation of mRNA has to be mentioned, which is controlled by the GTPase subfamily of initiation and elongation factors. The prokaryotic translation factors are summarized in Table 3.1 (for details see Section 9.1). This mechanism is highly conservative and common to all forms of life. Equally abundant are G-proteins such as the signal recognition particles and their receptors, which manage the transport of newly synthesized polypeptide chains across membranes (see Section 9.2). They have prokaryotic counterparts such as the G-proteins Ffh and FtsY of *E. coli*.

Additional prokaryotic G-proteins seem to be involved in ribosome formation, namely the Era (*E.coli* ras-like) proteins, facilitating the maturation of ribosomal RNA; the EngA proteins and the ribosome small subunit-dependent GTPases RsgA promoting the assembly of large and small ribosomal subunits, respectively; and the ribosome-associated Obg proteins, which may function as stress sensors to measure the fall of the cellular GTP level during a stringent response.

Table 3.1 Prokaryotic translation factors

Factor	Function
IF1	dissociation of the ribosome, binding of initiator tRNA
IF2	GTPase, formation of the 70S initiation complex
IF3	ATPase, binding and scanning of mRNA
EF-Tu	GTPase, binding of aminoacyl tRNA at the A-site
EF-Ts	GDP–GTP exchange factor of EF-Tu
EF-G	GTPase, translocation of peptidyl-tRNA
RF1	release of the completed polypeptide
RF2	release of the completed polypeptide
RF3	GTPase, termination

Homologous GTPases playing a role in ribosome biogenesis are found also in other organisms. Gene mutations have shown that most of the prokaryotic G-proteins are essential.

The search for receptor-coupled heterotrimeric G-proteins in prokaryotes has been in vain. This supports the conclusion that prokaryotic transmembrane signaling is strongly dominated by the two-component mechanism. Counterparts of small G-proteins of the Ras, Rho, Rab, Ran, and Arf types are also widely absent from prokaryotes, perhaps with the exception of the Ras-related GTPase MglA of myxobacteria (mentioned previously) and an Arf-related protein involved in the formation of magnetosomes in magnetic bacteria.

Taken together, the prokaryotic G-protein mechanism appears to be simple when compared with the elaborate G-protein networks in eukaryotic cells. In fact, as explained in Chapter 10, the majority of eukaryotic G-proteins regulate the traffic between intracellular compartments, which are absent from prokaryotes.

3.5.5 Sensors of oxidative stress

Life under an oxygen atmosphere is a risky venture. Organisms may profit from aerobic conditions that enable highly efficient energy production. However, they are also endangered by the aggressive chemical properties of oxygen and its metabolic products (Section 2.2). To deal with oxidative stress, aerobic bacteria have developed simple but efficient protective mechanisms by which they recognize dangerous reactive oxygen species (ROS) and initiate countermeasures. In this situation, the mechanistic principle of signal processing operates, in which certain signaling proteins become post-translationally modified by oxidation of iron-sulfide groups or cysteinyl–SH groups.

To make this clear, let us examine two ROS-controlled transcription factors of *E. coli*: the superoxide anion radical sensor SoxR and the H_2O_2 sensor OxyR. **SoxR** contains as a redox cofactor a $[Fe_2-S_2]$ iron sulfide entity complexed by cysteinyl residues. This so-called iron sulfur cluster (which is found in many redox enzymes) typically undergoes single electron transitions when reacting with free radicals such as superoxide anion radicals. As a result the conformation of SoxR is changed facilitating the binding to DNA. In turn, genes that encode enzymes degrading the highly toxic superoxide anion radical are activated. The prototype of such enzymes is superoxide dismutase catalyzing the reaction $2O_2^{\cdot-} + 2H^+ \rightarrow O_2 + H_2O_2$. The activation of SoxR is reversible because the factor is reduced by NADH/NADPH. The transcription factor **OxyR** takes care of the elimination of hydrogen peroxide that is generated along this and other metabolic pathways. OxyR represents the second family of ROS sensors that contain as a redox center oxidable cysteinyl residues instead of an iron sulfur cluster. Such proteins are less reactive with free radicals but are especially suited to react with peroxides that oxidize the thiol groups of Cys to sulfenic acid and disulfide groups (Section 2.2). As for SoxR this modification is reversible becoming reduced by NADH/NADPH via glutathione as an intermediate reductant. In the oxidized state, OxyR induces the activity of genes controlling peroxide metabolism. Another example of a bacterial redox sensor containing a reactive dithiol configuration is the His kinase ArcB. It will be discussed in more detail in Section 3.5.6.

Saccharomyces cerevisiae possesses a similar protective system; however, the sensor and transcription factors are separated. The yeast peroxide sensor is **Orp1**, a protein that in contrast to OxyR is not a transcription factor but exhibits the properties of a peroxidase, which is an enzyme that catalyzes the oxidation of substrates by hydrogen peroxide. In the course of this reaction, two cysteine residues of Orp1 become oxidized, forming a disulfide bridge or sulfenic acid groups that immediately become reduced again by another SH protein, **Yap1**. Yap1 is a transcription factor controlling several genes taking care of antioxidant production. It is inactivated by thioredoxin-catalyzed reduction. An adaptor protein, Ybp1, brings the sensor Orp1 in close contact with Yap1.

As compared with the bacterial OxyR, the yeast system impressively demonstrates the evolutionary change from a simple peroxide sensor to a H_2O_2-sensitive complex of signal processing consisting of a sensor, an adaptor, and an effector. As discussed in more detail in Section 2.2, in eukaryotes, ROS indeed play a double role as both stress factors that have to be eliminated and intracellular signaling molecules. An example for ROS signaling in higher eukaryotes is found in Section 7.1.

3.5.6 Signal-controlled proteolysis

In eukaryotes, controlled degradation as well as proteolytic activation of signal-transducing proteins is a major regulatory mechanism. As a rule, proteins are marked for degradation by ubiquitylation to become hydrolyzed subsequently in proteasomes. This principle finds its precursor in prokaryotes. An example is the activation of the **alternative sigma factor σ^S** of *E. coli* (see Sidebar 3.6). This factor becomes up-regulated only in stress situations such as starvation, high temperature, or osmotic crisis, and directs the RNA polymerase to genes specialized for stress protection. Under stress-free conditions, σ^S is kept at a low level by permanent proteolytic degradation. To be recognized by the protease ClpXP (a bacterial counterpart of the proteasome), the σ factor needs to be labeled by binding the recognition factor **RssB** (note the striking analogy to ubiquitylation). RssB, a protein, is active only when phosphorylated, thus providing an input terminal for stress signals that are mediated by a two-component system with the His kinase **ArcB** as a sensor and RssB as a response regulator. ArcB is an energy sensor that, under aerobic conditions, exists in an oxidized disulfide form. Upon energy shortage indicated by oxygen deficiency, ArcB is reduced to the dithiol form and its His kinase activity is suppressed. As a consequence, the σ factor becomes resistant to proteolytic degradation. Such a mechanism, which instead of inducing *de novo* synthesis only alters the steady-state concentration of a regulatory protein, has the advantage of responding very rapidly to changes in the environment. This principle is expensive but obviously essential for coping with stress situations.

The release of active signaling proteins from membrane-bound precursors by intramembrane proteolysis (Chapter 13) has also been introduced by prokaryotes. An example is the **alternative sigma factor σ^K** directing RNA polymerase to

---

Sidebar 3.6 Sigma factors The RNA polymerase of *E. coli* consists of a core and a regulatory subunit called σ-factor. The core is a heterotetrameric protein composed of two α-subunits, one β-subunit, and one β'-subunit. It is able to bind to DNA and to catalyze RNA biosynthesis but is unable to recognize the promoter sequences of individual genes. Such selectivity is conferred by the σ-factor. *E. coli* possesses seven different σ-subunits exhibiting different promoter specificities. Everyday needs are satisfied by the σ^{70}-subunit, a protein of 63 amino acids. The additional six σ-factors become activated during stress situations and control the transcription of "emergency genes" such as those encoding heat-shock proteins (Section 2.5.1).

σ^{70} has two binding sites for DNA, with the C-terminal site recognizing double-stranded regions and the N-terminal site recognizing single-stranded regions. The corresponding interaction sequences of DNA are a TTGACA hexamer at −35 bp (35 base pairs upstream of the transcription start) and a TATAAT hexamer, the **TATA-box,** localized in a double-stranded area as well as at −10 bp, where the double strand becomes unwound into single strands. This unwinding, also called melting, is a prerequisite for strong binding between DNA and the RNA polymerase complex. σ^{70} interacts with these two sites only as a complex with the core enzyme due to an auto-inhibitory conformation of the σ-factor where the N-terminal sequence blocks the binding site for the DNA sequence at −35 bp. This inactive form becomes unfolded to the active conformation upon binding to the core enzyme.

sporulation genes. σ^K becomes released from a precursor protein when *B. subtilis* receives environmental signals inducing sporulation. This signaling reaction is a perfect counterpart of the animal notch system (see Section 13.1.1).

3.5.7 Prokaryotic ion channels

Ion channels are absolutely essential for metabolism, equalization of osmotic pressure, and regulation of intracellular acidity. In parallel they were employed for data processing from the very beginning of life. Clearly the evolutionary history of ion channels dates back as far as that of protein kinases and GTPases. Only the simplest prokaryotes, such as some mycoplasma, seem to be devoid of the corresponding genes, leaving open the question whether they had been lost by adaptive processes. Research on prokaryotic ion channels has made an important and significant contribution to our understanding of these signal-propagating devices, in particular since several channel proteins could be crystallized and studied by X-ray analysis. The overall result emerging from these studies is that all basic structures and mechanisms of action of ion channels were introduced by prokaryotes.

A common structural feature of ion channels is a series of transmembrane helices (at least four) forming a central pore that, through its width and electrostatic charge, determines the channel's ion selectivity (Section 2.10.3). Prokaryotic ion channels that have been investigated in great detail include the mechanosensitive channel MscL, several potassium and sodium channels, and the chloride channel ClC. Each of these channels is a prototype of a large and evolutionary extremely successful family of channel proteins. Representatives of prokaryotic cation channels are shown in Figure 3.31.

Mechanosensitive channels

These channels are abundant in both prokaryotes and eukaryotes. Their major role is that of an excessive pressure valve protecting cells from disruption by hypo-osmotic shock. Gating of the channels is triggered by the horizontal tension of the plasma membrane due to a swelling of the cell. This tension causes an irislike twisting of the channel's transmembrane domains, leading to a widening of the central pore (Figure 3.32). Such a response is particularly essential for free-living organisms that are permanently endangered by a hypo-osmotic situation, for instance, rain.

Lacking a ligand-binding site as well as an ion selectivity filter and a voltage sensor, **MscL** (L stands for large conductance) is one of the simplest ion channels. It represents the primeval form of the so-called S5–S6 channels (see Figure 3.31 and Section 2.10.3) and might be a precursor of the ligand-controlled ion channels of higher eukaryotes. In prokaryotes, however, no ligand-controlled channels except the glutamate-dependent potassium channel GluR0 (see below) have been found.

While MscL channels occur predominantly in bacteria, a second family of mechanosensitive channels is distributed more widely. These are the structurally

Figure 3.31 Subunits of prokaryotic ion channels of the S5–S6 family: precursors of eukaryotic channels Mechanosensitive channel MscL is the prototype of the following eukaryotic ion channels: ATP-receptor cation channel P2X (Section 16.6); epithelial sodium channel ENaC (Section 14.3); and several neuropeptide receptor ion channels of invertebrates. Potassium channel KcsA resembles the inwardly rectifying potassium channel K_{ir} of eukaryotes. In other eukaryotic K^+ channels, two KcsA-like subunits (TWIK) or one KcsA-like subunit and one Kch-like subunit (TOK) share one polypeptide chain each (Section 14.4). Potassium channel Kch resembles the cyclic nucleotide-gated ion channels of higher eukaryotes (Section 14.5.7). Voltage-dependent sodium and potassium channels NaChBac and KvAP may be precursors of voltage-dependent Na^+-, K^+-, and Ca^{2+}-channels of eukaryotes (see Sections 14.2, 14.4.1, and 14.5.2). Glutamate-dependent potassium channel GluR0 is probably the precursor of the ionotropic glutamate receptor cation channels of vertebrates (Section 16.4). Gray cylinders, transmembrane domains and P-loops (P); red cylinder, transmembrane helix with voltage sensor; dark gray domains, extracellular Glu-binding site of GluR0. Numbering S1–S6 of the transmembrane domains refers to the voltage-dependent K^+ channel of eukaryotes. The number of subunits constituting a channel is shown beneath the name.

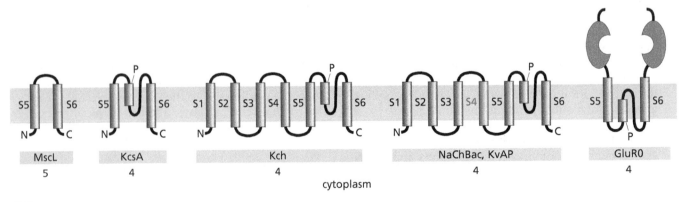

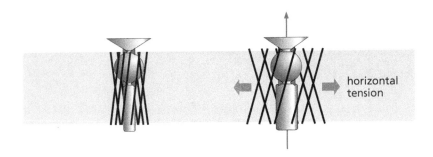

Figure 3.32 Model of the mechanosensitive channel MscL
The channel pore is depicted in a three-dimensional schematic manner (bottleneck structures are shown in red). Transmembrane helices are represented by black lines. The channel is gated by a horizontal tension of the membrane caused by the swelling of the cell in a hypo-osmotic environment.

highly variable **MscS channels** (S stands for small conductance). The corresponding channel of *E. coli* consists of seven symmetrically arranged subunits with three transmembrane helices each and becomes gated by both mechanical stress and membrane depolarization, thus representing a primeval form of voltage-dependent ion channels.

Potassium and sodium channels

A prokaryotic prototype of these channels is **KcsA** ($\underline{K}^+$ channel of streptomyces $\underline{A}$) isolated from *Streptomyces lividans* (homologous channels of *E. coli* are KefB and KefC). X-ray analysis has revealed a subunit structure that is characterized by two transmembrane helices and a P-loop rendering the channel selective for K^+ ions (the problem of channel selectivity is discussed in more detail in Section 2.10.3). Each channel consists of four subunits representing the primeval form of all ion-selective potassium channels of the S5–S6 family (Figure 3.31). The KcsA channel opens at low pH, probably due to an interaction of protons with the extracellular domains. As in the case of the MscL channel, this gating is assumed to be caused by an irislike twisting of the transmembrane helices (Figure 3.33).

The structure of KcsA resembles that of inwardly rectifying potassium channels of vertebrates (Section 14.4.2). A special species of this channel type is the potassium channel **MthK** of *Methanobacterium thermoautotrophicum* (Figure 3.33). It contains an extended cytoplasmic C-terminal domain that binds Ca^{2+} ions resulting in channel gating. X-ray crystallography has provided a deep insight into channel structure and the gating process (Figure 3.34). Functionally MthK resembles the mammalian Ca^{2+}-gated potassium channels BK and SK that, however, have a more complex structure (Section 14.4.1).

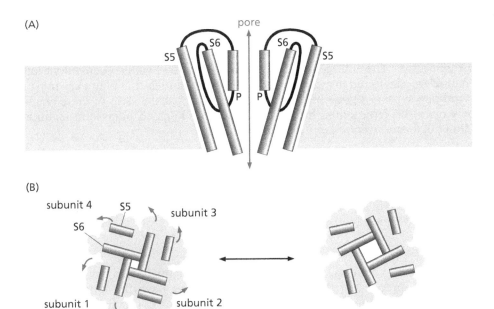

Figure 3.33 Schematic representation of architecture and gating of the KcsA channel (A) Arrangement of transmembrane helices S5 and S6 as well as the P-loop (P) in the plasma membrane (cross section showing two of the four subunits of the channel). (B) Gating of the central pore by an irislike twisting of the transmembrane helices.

Figure 3.34 Gating of the Ca²⁺-dependent K⁺ channel MthK The insert shows the schematic structure of one of the four channel subunits with the transmembrane segments S5 and S6 and the P-loop (P) forming the selectivity filter. The extended C-terminal domain (red) contains Ca²⁺ binding sites. Below a space-filling model of the channel based on X-ray crystallography is shown in a closed (left) and in an open form (right). Channel gating is due to a conformational change of the ring of C-terminal domains (red) that is induced by Ca²⁺ binding. (Modified from M. Schumacher & J.P. Adelman, *Nature* 417, 501–502, 2002.)

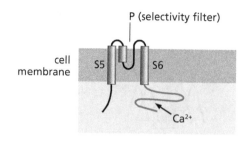

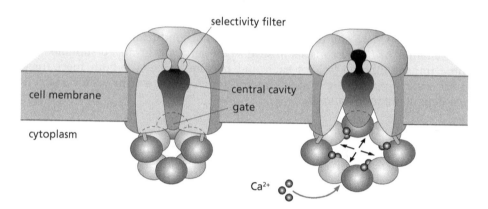

Voltage-dependent cation channels of vertebrates are typically composed of four subunits with six transmembrane domains and a P-loop each whereby transmembrane segment S4 acts as a voltage sensor (Section 14.1). An analogous structure has been found for prokaryotic cation channels, such as the potassium channel **Kch** of *E. coli*. While Kch does not significantly respond to voltage, other bacterial channels do: the **voltage-dependent Na⁺-channel** of *Bacillus halodurans* (Figure 3.31) and the archaeal potassium channel **KvAP** of *Aeropyrum pernix*. Certainly, these channels are evolutionary precursors of vertebrate voltage-dependent ion channels. Their four subunits resemble those of Kch; however, transmembrane segment S4 together with a part of S3 acts as a voltage sensor (see Figure 14.3 in Section 14.1). The elucidation of the KvAP structure has revolutionized our concepts of ion channel gating (see Section 14.1). While the function of the potassium channel is not yet understood, the sodium channel has been proposed to be a part of the flagellar motor, which in this species is supplied by energy derived from a Na⁺ gradient across the cell membrane. It is possible that these channels also participate in regulation of the very negative ion potential (ranging between –100 and –150 mV) of bacterial cells. Cooperating with mechanosensitive channels they may, in addition, protect the cell from osmotic stress. The subunit and domain architecture of voltage-dependent ion channels is also found in cyclic nucleotide-gated channels that play a key role in vertebrate sensory signal processing (Chapter 15). The ancestor of these channels might be represented by the cyclic nucleotide-gated potassium channel **MloK1** of *Mezorhizobium loti*, a plant symbiont.

A **glutamate-dependent potassium channel GluR0**, which has been found in some bacterial species, has evolved from the fusion of a KscA-like channel protein with a periplasmic binding protein for glutamate. GluR0 is considered to represent the evolutionary precursor of the ionotropic Glu receptors of higher eukaryotes (Section 16.4).

Chloride channels

Like the S5–S6 channels, the chloride or anion channels also originate from prokaryotic precursors, though not all prokaryotes have the corresponding genes. X-ray analysis of the channel protein **EcClC** of *E. coli* has revealed an unexpectedly complex structure. This channel is composed of two identical subunits with 18 strongly twisted helices each, 17 of them penetrating the cell membrane partially or completely. In contrast to cation channels, each subunit of EcClC forms a pore; that is, the channel has two pores (Figure 3.35). This structure seems to be prototypical for all chloride channels found as yet (Section 14.7). The function of the prokaryotic chloride channels is unclear. They may participate in regulation of the cell volume. Because of the small volume of a bacterial cell (10^{-12} mL on average), the gating of ion channels must be precisely terminated; otherwise, intolerable changes of the ion concentration and osmotic pressure would occur. How this control is achieved is still an open question.

Water channels

A large and ubiquitous family of membrane channels is specialized for the transport of water and small polyalcohols, which on their own cannot penetrate the lipid bilayer. These channels are called **aquaporins** (Aqp, only for water) and **glyceroaquaporins** (Glp, for water and for polyalcohols such as glycerol). Both types are widely distributed among prokaryotes and participate in osmoregulation and volume control, respectively. Human cells express 10 different aquaporins. The channels AqpZ and GlpF of *E. coli* have been studied in detail. Their structure seems to be representative for all (glycero)aquaporins. Each channel is composed of four subunits, each of which contains four transmembrane helices and two P-loops and forms one pore; that is, the complete channel has four pores (Figure 3.34). Bottlenecks and charged amino acid residues guarantee the selectivity of the pores. Hydrated ions are too large to pass through the pore, and the charge of the pore is insufficient for stripping the ion from water molecules. Therefore, water channels are impassable for ions and other charged molecules.

Summary

While Ser/Thr-specific protein kinases have been found in all species studied thus far, His autokinase-based two-component systems are restricted to prokaryotes and, to a smaller extent, slime molds, fungi, and plants. In animals, only two kinases that structurally resemble prokaryotic His autokinases (but phosphorylating Ser and Thr residues) have been identified in mitochondria. With one exception (myxobacteria), Tyr-specific protein kinases seem to be restricted to animals. Receptor-coupled Tyr kinases are particularly late inventions of evolution. Protein phosphatases are phylogenetically as ancient as protein kinases. His autokinases are bifunctional, exhibiting both kinase and phosphatase activity. In addition, monofunctional His-, Asp-, and Ser/Thr-specific phosphatases have been found in prokaryotes. Some bacterial Tyr phosphatases acquired from animals by horizontal gene transfer play a fatal role as virulence factors. The most ancient G-proteins are translation factors and proteins that catalyze the transmembrane transport of newly synthesized

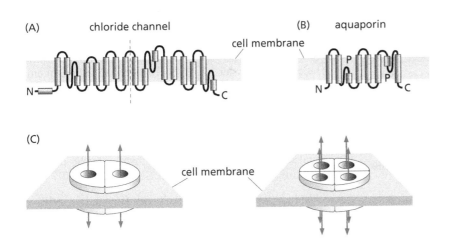

(A) chloride channel

cell membrane

(B) aquaporin

(C) cell membrane

Figure 3.35 Membrane topology of chloride and water channels (A) Schematic representation of a chloride channel subunit showing 18 helices (gray cylinders). Note the internal symmetry caused by an anti-parallel repeat of the first nine helices (see also Section 14.7 for more details). (B) Subunit of an aquaporin water channel showing six transmembrane helices and two P-loops. (C) Channels composed of two and four identical subunits, respectively, exhibiting either two or four pores.

polypeptide chains. Most prokaryotes have been found to express neither small G-proteins of the Ras superfamily nor receptor-coupled trimeric G-proteins. Redox signaling through reversible oxidation of SH groups in proteins is an ancient principle of regulation. In prokaryotes it may have developed from mechanisms that protected the organism from oxidative stress. Mechanisms resembling both the ubiquitin-controlled degradation of signaling proteins and the proteolytic release of active factors from protein precursors are also found in prokaryotes. In bacteria such signaling events play an important role in stress situations. All ion and water channels of eukaryotes can be traced back to prokaryotic precursors. In prokaryotes such channels are involved in cell volume control, regulation of the osmotic pressure, and maintenance of the membrane potential. The role of voltage-sensitive bacterial ion channels is not yet fully understood.

Further reading

Alexandre G & Zhulin IB (2001) More than one way to sense chemicals. *J. Bacteriol.* 183, 4681–4686.

Anantharaman V, Iyer LM & Aravind L (2007) Comparative genomics of protists: new insights into the evolution of eukaryotic signal transduction and gene regulation. *Annu. Rev. Microbiol.* 61, 453–475.

Anishkin A & Kung C (2005) Microbial mechanosensation. *Curr. Opin. Neurobiol.* 15, 397–405.

Baker MD, Wolanin PM & Stock JB (2006) Systems biology of bacterial chemotaxis. *Curr. Opin. Microbiol.* 9, 187–192.

Bart H, Aktories K, Popoff MR et al. (2004) Binary bacterial toxins: biochemistry, biology, and applications of common *Clostridium* and *Bacillus* proteins. *Microbiol. Mol. Biol. Rev.* 68, 373–402.

Bassler BL & Losick R (2006) Bacterially speaking. *Cell* 125, 237–246.

Beier D & Gross R (2006) Regulation of bacterial virulence by two-component systems. *Curr. Opin. Microbiol.* 9, 143–152.

Booth IR, Edwards MD & Miller S (2003) Bacterial ion channels. *Biochemistry* 42, 10045–10053.

Braeken K, Moris M, Daniels R et al. (2006) New horizons for (p)ppGpp in bacterial and plant physiology. *Trends Microbiol.* 14, 45–54.

Bren A & Eisenbach M (2000) How signals are heard during bacterial chemotaxis: protein-protein interactions in sensory signal propagation. *J. Bacteriol.* 182, 6865–6873.

Brückner R & Titgemeyer F (2002) Carbon catabolite repression in bacteria: choice of the carbon source and autoregulatory limitation of sugar utilization. *FEMS Microbiol. Lett.* 209, 141–148.

Caldon CE, Voong P & March PE (2001) Evolution of a molecular switch: universal bacterial GTPases regulate ribosome function. *Mol. Microbiol.* 41, 289–297.

Camilli A & Bassler BL (2006) Bacterial small-molecule signaling pathways. *Science* 311, 1113–1116.

Chang G & Roth CB (2001) Structure of MsbA from *E. coli*: a homolog of the multidrug resistance ATP binding cassette (ABC) transporters. *Science* 293, 1793–1800.

Chow B & McCourt P (2007) Plant hormone receptors: perception is everything. *Genes Dev.* 20, 1998–2008.

Davidson AL & Chen J (2004) ATP-binding cassette transporters in bacteria. *Annu. Rev. Biochem.* 73, 241–268.

D'Autreaux B & Toledano MB (2007) ROS as signalling molecules: mechanisms that generate specificity in ROS homeostasis. *Nat. Rev. Mol. Cell Biol.* 8, 813–824.

Deutscher J, Francke C & Postma W (2006) How phosphotransferase system-related protein phosphorylation regulates carbohydrate metabolism in bacteria. *Microbiol. Mol. Biol. Rev.* 70, 939–1031.

Fuqua C & Greenberg EP (2002) Listening in on bacteria: acyl-homoserine lactone signaling. *Nat. Rev.* 3, 685–692.

Galperin MJ (2005) A census of membrane-bound and intracellular signal transduction proteins in bacteria: Bacterial IQ, extroverts and introverts. *BMC Microbiol.* 5, 35–54.

Galperin MJ (2004) Bacterial signal transduction network in a genomic perspective. *Environ. Microbiol.* 6, 552–567.

Gao R, Mack TR & Stock AM (2007) Bacterial response regulators: versatile regulatory strategies from common domains. *Trends Biochem. Sci.* 32, 225–234.

Gonzalez JE & Keshavan ND (2006) Messing with bacterial quorum sensing. *Microbiol. Mol. Biol. Rev.* 70, 859–875.

Gottesman S (2003) Proteolysis in bacterial regulatory circuits. *Annu. Rev. Cell Dev. Biol.* 19, 565–587.

Heelingwerf KJ (2005) Bacterial observations: a rudimentary form of intelligence? *Trends Microbiol.* 13, 152–158.

Henke JM & Bassler BL (2004) Bacterial social engagements. *Trends Cell Biol.* 14, 648–656.

Hoff WD, Jung KH & Spudich Jl (1997) Molecular mechanism of photosignaling by archaeal sensory rhodopsins. *Annu. Rev. Biomol. Struct.* 26, 223–258.

Jenal U & Malone J (2006) Mechanism of cyclic-di-GMP signaling in bacteria. *Annu. Rev. Genet.* 40, 385–407.

Kaiser D (2004) Signaling in myxobacteria. *Annu. Rev. Microbiol.* 58, 75–98.

Kenelly PJ (2003) Archaeal protein kinases and protein phosphatases: insights from genomics and biochemistry. *Biochem. J.* 370, 373–389.

Kenelly PJ (2002) Protein kinases and protein phosphatases in prokaryotes: a genomic perspective. *FEMS Microbiol. Lett.* 206, 1–8.

Koretke KK, Lupas AN, Warren PV et al. (2000) Evolution of two-component signal transduction. *Mol. Biol. Evol.* 17, 1956–1970.

Laursen BS, Soerensen HP, Mortensen KK et al. (2005) Initiation of protein synthesis in bacteria. *Microbiol. Mol. Biol. Rev.* 69, 101–123.

Lloyd G, Landini P & Busby S (2001) Activation and repression of transcription initiation in bacteria. *Essays Biochem.* 37, 17–31.

Mascher T, Helmann JD & Unden G (2006) Stimulus perception coupling in bacterial signal-transducing histidine kinases. *Microbiol. Mol. Biol. Rev.* 70, 910–938.

Mason P & Kay R (2000) Eukaryotic signal transduction via histidine-aspartate phosphorelay. *J. Cell Sci.* 113, 3141–3150.

Miller MB & Bassler BL (2001) Quorum sensing in bacteria. *Annu. Rev. Microbiol.* 55, 156–199.

Paulsen IT, Nguyen L, Sliwinski MK et al. (2000) Microbial genome analyses: comparative transport capabilities in eighteen prokaryotes. *J. Mol. Biol.* 301, 75–100.

Ryan RR, Fouhy Y, Lucey JF et al. (2006) Cyclic di-GMP signaling in bacteria: recent advances and new puzzles. *J. Bacteriol.* 188, 8327–8334.

Shapiro JA (1988) Bacteria as multicellular organisms. *Scientific Am.* 256, 82–89.

Shenoy AR & Visweswariah SS (2006) New messages from old messengers: cAMP and mycobacteria. *Trends Microbiol.* 14, 543–550.

Sherratt DJ (2003) Bacterial chromosome dynamics. *Science* 301, 780–785.

Shimkets LJ (1999) Intercellular signaling during fruiting-body development of *Myxococcus xanthus*. *Annu. Rev. Microbiol.* 53, 525–549.

Siebold C, Flükiger K, Beutler R et al. (2001) Carbohydrate transporters of the bacterial phosphoenolpyruvate: sugar phosphotransferase system (PTS). *FEBS Lett.* 504, 104–111.

Spudich JL (2006) The multitalented microbial sensory rhodopsins. *Trends Microbiol.* 14, 480–487.

Stock AM, Robinson VL & Goudreau PN (2000) Two-component signal transduction. *Annu. Rev. Biochem.* 68, 183–215.

Stoodley P, Sauer K, Davies DG et al. (2002) Biofilms as complex differentiated communities. *Annu. Rev. Microbiol.* 56, 187–209.

Szurmant H & Ordal GW (2004) Diversity of chemotaxis mechanisms among the bacteria and the archaea. *Microbiol. Mol. Biol. Rev.* 68, 301–319.

Tamayo R, Pratt JT & Camilli A (2007) Roles of cyclic diguanylate in the regulation of bacterial pathogenesis. *Annu. Rev. Microbiol.* 61, 131–148.

Thomasson B, Link J, Stassinopoulos AG et al. (2002) MglA, a small GTPase, interacts with a tyrosine kinase to control type IV pili-mediated mobility and development of *Myxococcus xanthus*. *Mol. Microbiol.* 46, 1399–1413.

Toledano MB, Delaunay A, Monceau L et al. (2004) Microbial H_2O_2 sensors as archetypical redox signaling modules. *Trends Biochem. Sci.* 29, 351–357.

Ulrich LE, Koonin EV & Zhulin IB (2005) One-component systems dominate signal transduction in prokaryotes. *Trends Microbiol.* 13, 52–56.

Ulrich LE & Zhulin IB (2007) MiST: a microbial signal transduction database. *Nucleic Acids Res.* 35, D386–D390.

Urrao T, Yamaguchi-Shinozaki K & Shinozaki K (2000) Two-component systems in plant signal transduction. *Trends Plant Sci.* 5, 67–74.

Warner JB & Lolkema JS (2003) CcpA-dependent carbon catabolite repression in bacteria. *Microbiol. Mol. Biol. Rev.* 67, 475–490.

West AH & Stock AM (2001) Histidine kinases and response regulator proteins in two-component signaling systems. *Trends Biochem. Sci.* 26, 369–376.

Wilson DN & Nierhaus KH (2007) The weird and wonderful world of bacterial ribosome regulation. *Crit. Rev. Biochem. Mol. Biol.* 42, 187–219.

Wolanin PM, Thomason PA & Stock JB (2002) Histidine protein kinases: key signal transducers outside of the animal kingdom. *Genome Biol.* 3, review S3013.

Basic Equipment: G-Proteins, Second Messengers, and Protein Kinases

As explained in Chapter 3, the molecular "nanobrain" of prokaryotes processes signals mainly by means of protein switches operated by reversible protein phosphorylation. The data-processing apparatus of higher cells essentially follows the same principles. In bacteria, one major task of cellular data processing is to cope with environmental stress in order to discriminate between good and bad influences. The same holds true for more advanced organisms, albeit on a much higher level of complexity.

About 75 years ago the term stress was coined by Hans Selye to characterize the various adaptive reactions of the human organism to external and internal stimuli. Today the word stress has a negative overtone since it is understood mostly as a disease-causing alarm reaction. However, when the term is used in a broader sense to characterize *any* type of physiological and psychic mobilization, even the transition from the sleeping to the waking state is considered stress (as probably most readers would confirm).

In our body such a transition from rest to activity leads to stimulation of the three major data-processing networks: the endocrine system, the nervous system, and—in more extreme cases—the nonspecific and specific immune system. These systems produce a wide variety of signaling molecules such as hormones, local mediators including neurotransmitters and antibodies, thus communicating with each other and with their target organs. No matter whether we are concerned with ourselves or with the environment, these factors are always forming interconnected causality chains leading from the macro-world of our environment (or our imaginations) into the nano-world of cellular molecules and back. It is essential to illustrate this principle with the help of a concrete example such as the particular stress situation depicted in Figure 4.1.

4.1 Stress response and the vegetative nervous system

A stress response starts with a sensory impression (leaving aside imaginary stress situations). The environmental signals are transformed by sensory cells into nerve impulses, a process described in detail in Chapter 15. The brain decodes these impulses with the help of stored information and, in the case depicted in Figure 4.1, recognizes the situation as a threat. As a consequence, an appropriate response is selected that, depending on the situation and the psychic constitution, may be either "fight" (consisting of an active change of the environmental condition, either forcibly or verbally) or "flight" (consisting of an adaptation to the situation). In either case motion is required, though it may be of the mouth and vocal cords only, so the skeletal muscles have to be activated by the motor neurons via the neurotransmitter acetylcholine as a stimulatory signal (Figure 4.2; details are found in Sections 14.6.2 and 16.2).

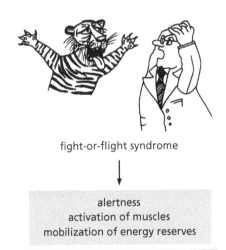

fight-or-flight syndrome

alertness
activation of muscles
mobilization of energy reserves

Figure 4.1 A stress situation The tiger should be taken as a metaphor, since the situation depicted here symbolizes many types of stress including the alarm clock in the morning, political debates, marital crisis, and academic examination.

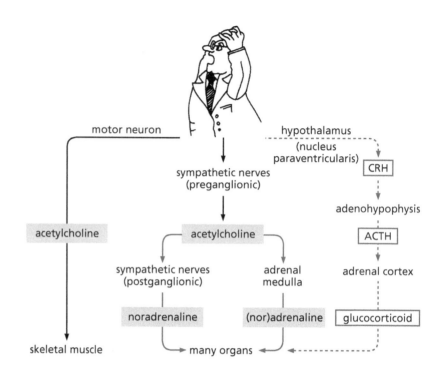

Figure 4.2 Processing of stress signals by the neuronal and endocrine network When the central nervous system recognizes the stress situation, it induces a fight-or-flight reaction (Figure 4.1). Immediately, the skeletal muscles (left) and the sympathetic branch of the vegetative nervous system are activated by means of acetylcholine. By releasing the stress hormones and neurotransmitters adrenaline and noradrenaline, the adrenal medulla and the sympathetic nerves guarantee an overall stimulation of the organism, including mobilization of biochemical energy reserves to supply the energy for muscle contraction. Long-term adaptation is brought about by the endocrine stress axis (right). Here certain neurosecretory cells of the hypothalamus release the peptide hormone CRH (Corticotropin Releasing Hormone), which induces the production of ACTH (AdrenoCorticoTropic Hormone, also known as corticotropin) by the adenohypophysis. In the adrenal cortex ACTH stimulates the biosynthesis of glucocorticoid-type stress hormones.

As a result the energy reserve of the muscle cell, in the form of glycogen, rapidly becomes exhausted and the blood sugar drops, indicating that it is time now to open further energy stores such as liver glycogen and fat reserves. To this end the autonomous nervous system transmits stress signals to the corresponding target organs. These signals include the sympathetic transmitter noradrenaline and the neurosecretions noradrenaline and adrenaline that are released from the adrenal medulla. Via intracellular signaling cascades, these signals induce the metabolic degradation of fat and glycogen to fatty acids and glucose, respectively, which are distributed in the bloodstream and are used by the target organs (for example, muscle cells) for energy production (for details see Section 5.3).

In parallel, dramatic changes in blood circulation occur. To supply muscles and nerves sufficiently, the blood supply is increased for these tissues but decreased for other organs through effects on the vascular smooth muscles. Moreover, to raise the flow rate, the heartbeat becomes faster and stronger. In addition, breathing is deepened for a better oxygen supply and intestinal peristalsis is inhibited, since digestion is a matter of relaxation rather than of activity (see Figure 4.3). For these and many other effects, essentially *one* signal is needed, namely, (nor)adrenaline, the endogenous stress mediator *par excellence*.

Figure 4.3 Effects of sympathetic and parasympathetic stimulation (Adapted from S. Silbernagl and A. Despopoulos, Taschenatlas der Physiologie. Stuttgart: Thieme-Verlag, 1991.)

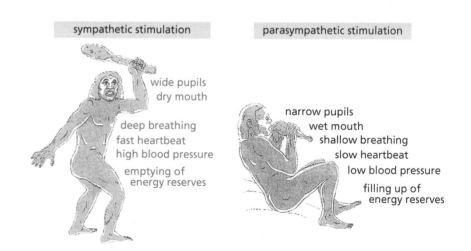

If the stress situation is perpetuated and the energy reserves cannot be refilled fast enough, proteins are "burned" for energy generation. To this end they are fed into the pathways of carbohydrate metabolism. Under normal conditions this process, called gluconeogenesis, is suppressed so as not to waste precious material, but it is activated in emergency situations. The triggering signals are the glucocorticoid hormones of the adrenal cortex. Their biosynthesis is induced along a sequence of orders called the endocrine stress axis, originating in the brain (Figure 4.2). It is easily seen that long-lasting activation of this axis may have deleterious consequences.

The stress response clearly demonstrates how, on the way from the tiger to the molecules, signals are exchanged and transformed until the sensory impression has evoked a response from the whole organism. On the molecular level, three stages of adaptation are evident:

(1) After a moment of shock, the skeletal muscles begin to contract and the peripheral blood flow is changed (visible as turning pale or blushing). Both responses are triggered by the most rapid of all endogenous signals, a change of the membrane potential in nerve and muscle cells mediated by the neurotransmitter acetylcholine.

(2) With some delay, but still within a few minutes, the fat and glycogen reserves are opened. The enzymatic apparatus required is already present, waiting for a stimulation by a stress signal such as (nor)adrenaline.

(3) In contrast, long-term adaptation requires *de novo* synthesis of proteins such as the enzymes of gluconeogenesis, achieved by stimulation of gene expression mainly by glucocorticoids. This may take several hours.

4.1.1 Vegetative nervous system

The short- and medium-term responses are mediated by the **vegetative** or **autonomous nervous system** that in vertebrates controls practically all organ functions and the internal milieu of the body, while at the same time cooperating with the endocrine system and the central nervous system. This cooperation implies that the vegetative nervous system, although frequently reacting via local feedback loops (in a reflectory manner), is by no means "autonomous" but, in fact, is permanently under the control of environmental, hormonal, and psychogenic factors (Figure 4.2).

The contact sites between central and vegetative nervous systems are synapses in ganglions arranged in the spinal marrow. Long axons (nerve fibers) originating from the postsynaptic ganglionic neurons, approach the target organs. These organs include heart muscle; smooth muscles controlling peripheral blood flow, peristalsis, urinary bladder contraction, functions of genital organs (erection,

Sidebar 4.1 Internal milieu The term *"milieu interieur"* was coined by the French physiologist Claude Bernard at the end of the 19th century in order to describe the state of the extracellular milieu in multicellular organisms. The interior milieu is thought to replace the external milieu of unicellular organisms originally living in sea water. Important parameters of the internal milieu are osmolarity (including Ca^{2+} concentration), blood pressure, pH, temperature, and concentration of energy-supplying molecules (such as glucose, fatty acids, and oxygen) as well as of metabolic waste products (such as carbon dioxide). As a rule, the organism tolerates only very small fluctuations of such parameters. Therefore, it needs expensive regulatory mechanisms to guarantee the constancy or homeostasis of the internal milieu. Major control devices are the nervous system, the endocrine–paracrine complex, and the immune system. These systems are not isolated from each other but are consistently cross talking, forming a network that connects the environment with the signal-processing apparatus of the tissue cells, since these are normally blind to environmental stimuli.

ejaculation, uterine contraction) and respiration; metabolically active organs such as liver and fat tissue; and exocrine glands.

Anatomically and functionally there are two branches of the vegetative nervous system, known as **sympathetic** and **parasympathetic**. They operate in an antagonistic way; that is, the vegetative nervous system controls organ function according to a two-rein principle. The sympathetic branch assumes the excitation role and the parasympathetic branch assumes the steadying role (Figure 4.3).

Vegetative neurons do their jobs mainly by use of two transmitters: acetylcholine and noradrenaline. For fine regulation these are joined by a collection of co-transmitters and neuromodulators. Acetylcholine is released by the preganglionic neurons, which transmits the signal from the central to the vegetative nervous system, and by the postganglionic parasympathetic neurons, whereas noradrenaline is the neurotransmitter of the postganglionic sympathetic neurons, which transmits the signal from the nervous system to the target organs (exceptions to this rule are eccrine sweat glands, the periosteum, and blood vessels of skeletal muscles where sympathetic neurons use acetylcholine instead of noradrenaline). Noradrenaline and adrenaline (also known as norepinephrine and epinephrine) are also released into the bloodstream by the adrenal medulla, acting as stress hormones. The adrenal medulla is a sympathetic ganglion that has become anatomically remodeled to a neurosecretory gland (Figure 4.4).

The fact that, for its various functions, the vegetative nervous system gets by with only two major signaling molecules demonstrates very clearly that the medium is not the message. Whether and how a signal is decoded by the receiver cell is dependent solely on the inherited and acquired programming of the data-processing network. This fact had been established early by pharmacological and toxicological experiments. British pharmacologist Sir Henry Hallett Dale (1875–1968; awarded the 1936 Nobel Prize in Physiology or Medicine) is regarded as a pioneer in this field. As early as 1906 he observed that extracts from adrenal medulla (adrenaline and noradrenaline were still unknown at that time) evoked antagonistic sympathetic effects on blood pressure and uterine contraction depending on whether or not the experimental animals had been pretreated with extracts from ergot (*Claviceps purpurea*). Today we know that ergot alkaloids suppress the α-adrenergic effects of (nor)adrenaline, such as uterine contraction, leaving untouched the antagonistic β-adrenergic effects such as uterine dilation. Similar observations were made for acetylcholine: here the poison of the deadly nightshade (*Atropa belladonna*) inhibits selectively the muscarine-like parasympathetic effects but not the nicotine-like effects in the ganglions of the spinal marrow and on skeletal muscles, whereas the South American arrow poison curare exhibits just the reverse effect.

Figure 4.4 Organization of the vegetative nervous system The two branches of the vegetative nervous system, sympathetic and parasympathetic, are controlled by the central nervous system via spinal ganglions. By releasing the neurotransmitter noradrenaline (sympathetic, red) and acetylcholine (parasympathetic, black), the postganglionic fibers regulate the function of almost any organ. The adrenal medulla represents a neurosecretory endocrine gland, evolved from a sympathetic ganglion. It releases adrenaline and noradrenaline into the blood.

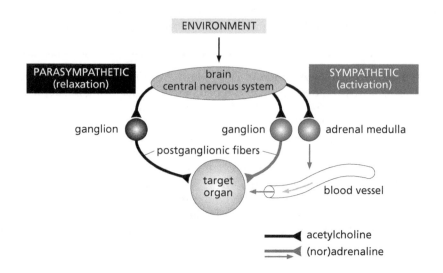

Summary

Environmental or psychogenic stimuli provoke a systemic reaction known as stress response, consisting of a general activation that may culminate in a fight-or-flight syndrome. On the molecular level of endogenous signaling, the response comprises three major stages: a rapid nervous stimulation of muscular activity mediated by acetylcholine, followed by a more delayed mobilization of glycogen and fat reserves due to enzyme activation triggered by (nor)adrenaline, and finally long-term adaptation due to gene transcription induced by glucocorticoids and other hormones. The vegetative nervous system controls the functions of most organs, thus maintaining the constancy of the internal milieu. While its sympathetic branch transmits stimulatory (stress) signals, the parasympathetic branch has a relaxing effect. In general, the vegetative nervous system uses only two neurotransmitters: noradrenaline as a sympathetic signal and acetylcholine as a parasympathetic signal. Acetylcholine is, in addition, the transmitter in the spinal ganglions that connect the vegetative with the central nervous system. The stress hormone adrenaline is released by the adrenal medulla, a sympathetic ganglion anatomically remodeled into a neuroendocrine gland.

4.2 Discovery of intracellular signal transduction

Early investigations with organ extracts and later studies with acetylcholine and (nor)adrenaline had shown that the same signaling substance might have different or even antagonistic effects depending on the tissue and the physiological context. This principle is now firmly established for the majority of hormones, local mediators, and neurotransmitters. There was only one plausible explanation for such a phenomenon: a cell can interpret the meaning of the same signal in different ways. Therefore, different receptor types were postulated, namely, α- and β-adrenergic receptors for (nor)adrenaline and muscarinic and nicotinic receptors for acetylcholine. The existence of these receptors was supported by a wealth of experimental data (these receptors have now been isolated and analyzed in detail; see Chapters 5 and 16). Since acetylcholine and (nor)adrenaline are not lipophilic enough to pass the cell membrane, their receptors were expected to be localized at the cell surface. But how was the signal transduced to the functional apparatus inside the cell? The first and pioneering answer to this question was provided by experiments focusing on an adrenaline-regulated metabolic effect that could be analyzed by the biochemical methods available at that time. This effect was the stress-induced degradation of glycogen. In fact, for the first time these studies gave a hint as to how cellular signal transduction might work.

Glycogen, a polymer of glucose mainly linked via 1,4-α-glycosidic bonds, represents an instantly available energy supply for the body and is found in most tissues, particularly in liver and skeletal muscle. On demand, it becomes degraded via glucose 1-phosphate and glucose 6-phosphate to glucose, which is released by the liver into the bloodstream and oxidatively metabolized for ATP generation by other tissues (Figure 4.5; more details in Section 5.3.1). The course of this fundamental metabolic process was elucidated in the mid-20th century, with Carl Ferdinand Cori and Gerty Theresa Cori at Washington University in St Louis playing a pioneering role (they were awarded the 1947 Nobel Prize in Physiology or Medicine, together with Bernardo A. Houssay). The key reaction of glycogenolysis was found to be a phosphorolytic splitting of the glycosidic bond, catalyzed by the enzyme glycogen phosphorylase and yielding glucose 1-phosphate (Figure 4.5). This reaction became more active in the presence of adrenaline, glucagon, and several other hormones known to induce glycogen degradation.

With regard to the effect of adrenaline, Earl W. Sutherland (1915–1974), at that time a co-worker of Ferdinand and Gerty Cori, made two crucial observations.

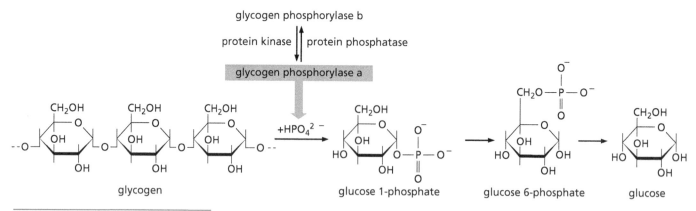

Figure 4.5 Biochemistry of glycogenolysis Glycogen is phosphorolytically degraded to glucose 1-phosphate, which is isomerized to glucose 6-phosphate by phosphoglucomutase. Glucose 6-phosphatase catalyzes the dephosphorylation of glucose 6-phosphate yielding glucose. Glycogen phosphorylase, the key enzyme of signal-induced glycogen degradation, exists in an inactive b-form and an active a-form. The a-form is generated from the b-form by reversible enzymatic phosphorylation of Ser residues.

First, he found that glycogen phosphorylase existed in two forms, an enzymatically active a-form and an inactive b-form. In the 1960s, Edmund H. Fischer and Edwin Krebs demonstrated that, in contrast to the b-form, the a-form was phosphorylated at a strategic serine residue, a reversible reaction catalyzed by an enzyme called glycogen phosphorylase kinase competing with a corresponding protein phosphatase (Figure 4.5). This pioneering discovery of regulatory protein phosphorylation was honored by the 1992 Nobel Prize in Physiology or Medicine.

Sutherland had also shown that sugar-mobilizing hormones such as adrenaline and glucagon stimulated the phosphorylation of glycogen phosphorylase not only in the intact tissue but also in liver homogenate. For the first time a hormonal effect could be studied *in vitro*; the door had been opened for a detailed biochemical analysis carried out in a classical manner. In the course of these studies, the cellular membrane fraction separated by centrifugation was found to contain a Mg^{2+}-dependent enzyme X that was stimulated in the presence of adrenaline, producing from ATP a heat-stable factor of low molecular weight. This factor was able to stimulate the enzymes of glycogenolysis that were enriched in the soluble cell fraction. Referring to adrenaline as the "first messenger," this factor was called a "second messenger" and was considered to be the sought-after signal, which bridged the gap between the hormone receptor at the cell surface and the enzymatic apparatus of glycogen metabolism inside the cell. The chemical identification of the second messenger was facilitated by an individual mixture of intuition and chance. It was determined that the factor had to be a nucleotide consisting of equal portions of adenine, ribose, and phosphoric acid. However, an exact structural analysis was impossible with the methods available at that time because—despite tremendous efforts—only tiny amounts of the factor could be isolated.

An inquiry addressed to the nucleotide expert Leon A. Heppel (National Institutes of Health) about the nature of the second messenger went unanswered at the time. However, according to a rumor, Heppel was in the habit—said to be quite widespread among scientists—of collecting incoming mail in three-month portions before responding. Thus Sutherland's letter was in the same stack of paper as a letter from David Lipkin in St Louis, which reported a novel nucleotide that was obtained upon treatment of ATP with barium hydroxide (Figure 4.6). The nucleotide was cyclic adenosine 3′,5′-monophosphate (cAMP), which ultimately turned out to be identical with Sutherland's second messenger. The adrenaline-activated enzyme X in the liver cell membrane, catalyzing the formation of cAMP from ATP, was then called adenylate cyclase (Section 4.4.1). For his epoch-making discovery, Sutherland was awarded the 1971 Nobel Prize in Physiology or Medicine.

Summary

The first pathway of intracellular signal transduction was discovered during the investigation of stress-induced glycogen degradation in the liver. Upon interaction with the membrane-bound receptor adrenaline, the first messenger, was

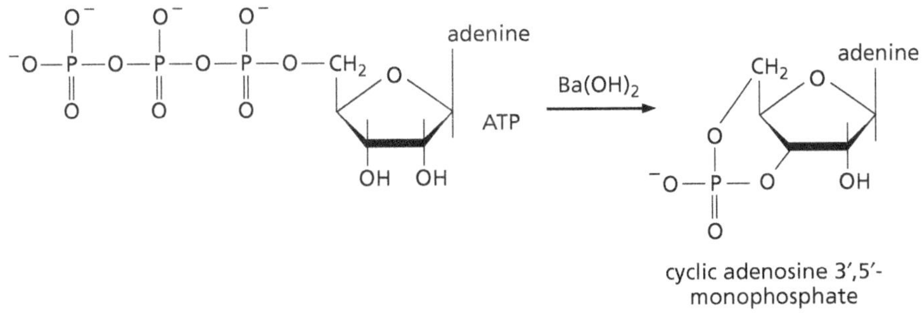

Figure 4.6 Chemical generation of cyclic AMP from ATP

found to evoke the intracellular release of cAMP, the second messenger, produced from ATP by adenylate cyclase. In turn, cAMP stimulated the activation of glycogen phosphorylase by protein phosphorylation, catalyzed by a specific protein kinase and reversed by a protein phosphatase.

4.3 Trimeric G-proteins: coupling of receptors with the protein network

With the discovery of the second messenger mechanism and regulatory protein phosphorylation, Sutherland had identified two central principles of intracellular signal transduction. However, two questions still remained: how the adrenaline receptor activated adenylate cyclase, and how cAMP stimulated the phosphorylation of glycogen phosphorylase. A key observation was that the activation of adenylate cyclase required not only ATP as a substrate but, quite unexpectedly, also GTP, which became hydrolyzed to GDP and inorganic phosphate in the course of cAMP formation. A GTP-hydrolyzing enzyme was found to be interconnected between receptor and adenylate cyclase. This result led to the discovery of the regulatory guanosine triphosphatases (GTPases or G-proteins), which was acknowledged by the 1994 Nobel Prize in Physiology or Medicine, awarded to Martin Rodbell and Alfred G. Gilman. GTP hydrolysis is one of the major switching reactions for transmembrane signaling.

4.3.1 Transmembrane signaling by trimeric G-proteins

The GTPase discovered by Rodbell, Gilman, and co-workers proved to be a protein complex consisting of the three subunits Gα, Gβ, and Gγ. It represents the prototype of the heterotrimeric or "large" G-proteins, the only function of which is to couple a particular class of membrane receptors to the signal-processing network of the cell. The GTPase activity is restricted to the Gα-subunit, being the G-protein proper. As typical for G-proteins, the interaction of Gα with effector proteins such as adenylate cyclase is initiated by the exchange of bound GDP for GTP, catalyzed by a GDP–GTP exchange factor or GEF. The GEFs of the trimeric G-proteins are transmembrane receptors that are characterized by seven transmembrane domains resembling the topography of rhodopsin. These so-called G-protein-coupled receptors (GPCRs), representing the largest of all receptor families in animals, are discussed in detail in Chapter 5.

The binding of GTP catalyzed by the receptor causes a conformational change of the Gα-subunit resulting in its dissociation from the βγ-dimer, which does not dissociate further (Figure 4.7). Upon dissociation, both the Gα- and Gβγ-subunits transmit input signals from the receptors to interacting effector proteins (Figure 4.7). The major effector proteins of Gα subunits are:

- adenylate cyclases
- phosphatidylinositol (PI)-specific phospholipase C (PLC), type β
- cGMP-specific phosphodiesterase in retinal cells
- the GEF of the small G-protein Rho

Typical Gβγ-effector proteins are:

- some isoforms of adenylate cyclase

- PLCβ

- phosphatidylinositol 3-kinase (PI3K), type γ (Section 4.4.3)

- β-adrenergic receptor kinase (Section 5.7)

- $G_{i/o}$-regulated potassium channels (Section 14.4.2)

- voltage-controlled calcium channels in presynaptic membranes of neurons (see Section 14.5.2)

- protein kinase Ste20, involved in pheromone signaling of yeast and probably related kinases of vertebrates (Section 11.4)

Which of these proteins interacts and whether it becomes activated or inactivated depends on the particular isoforms of the Gα- or Gβγ-subunits in the G-protein (see Figure 4.7 and below). G-proteins possess a built-in mechanism of signal termination: within a short time the Gα-subunit undergoes auto-inactivation due to its GTPase activity. As a consequence the trimeric G-protein reassociates and the βγ-dimer becomes inactivated as well.

Membrane anchoring

A characteristic structural feature of large G-proteins is that their subunits are anchored through covalently bound lipid residues at the inner side of the plasma membrane in immediate proximity to the receptor (Figure 4.7). Membrane anchoring of signal-transducing proteins is not restricted to G-proteins but represents a general principle of organization in data-processing protein networks. For this purpose, cells employ several biochemical reactions resulting in post-translational protein modifications (lipidations) that enable membrane associations of different stabilities. The principle of anchoring by covalently bound lipids is briefly summarized in Sidebar 4.2.

Figure 4.7 Trimeric G-proteins: variability and mode of operation
The upper part of the figure shows how G-protein subunits are anchored at the inner side of the cell membrane by tetradecanoyl (or palmitoyl) and farnesyl (or geranylgeranyl) residues (see also Figure 4.8). An activated receptor—of the heptahelical type only—acts as a GDP–GTP exchange factor (GEF) and induces dissociation into α- and βγ-subunits, which both interact with downstream effector proteins. These proteins are listed in the gray boxes, in the case of the Gα-subunits together with the related G-protein types (red, inhibition; black, activation; the lists are not complete but include the major effector types). The GTPase activity of Gα renders the process reversible. The average molecular weights of the subunits are 31-52 kDa for Gα, 37 kDa for Gβ, and 8 kDa for Gγ.

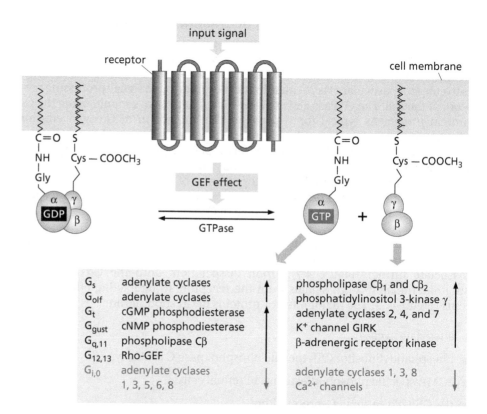

Sidebar 4.2 Lipid anchors Through covalently bound long-chain lipids acting as hydrophobic "anchors," proteins may associate with membranes. This is an important condition for the *ad hoc* assembly of signal-transducing complexes, in particular in the vicinity of transmembrane receptors. The lipid anchor can be hidden inside the protein molecule, becoming reversibly exposed only by a signal-induced conformational change (examples are discussed in Sections 10.4.3 and 14.6.1). By this means, signals may control a reversible membrane anchoring of proteins. The following lipid anchors have been described.

(1) Amides of tetradecanoic acid (myristoyl anchor). Here the acyl residue activated by coenzyme A is bound by an N-terminal amino group, mostly of a glycine (Gly) residue in the sequence Gly-x-x-x-Ser. Frequently, positively charged amino acid residues are found in the immediate vicinity. They promote association with the negatively charged phospholipids of membranes. This modification seems to be irreversible since as yet no enzymes have been found that hydrolyze the amide bond. Examples of proteins anchored by myristoyl residues are the α-subunits of trimeric G-proteins, the small G-protein Arf (Section 10.4.3), and the tyrosine (Tyr) kinases of the Src family (Section 7.2).

(2) Thioesters of palmitoic acid. The acyl residue is bound to the SH group of a (usually C-terminal) cysteine (Cys) residue. In contrast to the amide bond, the thioester bond can be hydrolyzed by the cell, so the modification is reversible. Palmitoylated proteins include α-subunits of trimeric G-proteins, the small G-protein Ras (Figure 4.7 and Section 10.1), and some cytoplasmic protein kinases.

(3) Thioethers of oligoprenols such as farnesol (C_{15}) and geranylgeranol (C_{20}). The prenyl residues are bound to SH groups of C-terminal cysteine residues. The reaction starts from prenyl pyrophosphates and is catalyzed by protein prenyltransferases recognizing the consensus sequence protein-Cys-A-A-X (so-called CAAX box, with A symbolizing an aliphatic amino acid and X a C-terminal serine, threonine, methionine, leucine, alanine, or glutamine residue). After the prenylation of the Cys residue, the AAX tripeptide is removed by a prenylprotein peptidase, and the free carboxyl group of Cys becomes methylated by a highly specific prenylprotein carboxymethyltransferase that uses *S*-adenosylmethionine as a methyl donor (Figure 4.8). Examples of proteins anchored by a prenyl residue are the γ-subunits of trimeric G-proteins, various small G-proteins such as Ras, and kinases phosphorylating GPCRs. Since the oncogenic effect of mutated Ras depends on prenylation, inhibitors of the enzymes involved are considered to be potential anti-cancer drugs.

(4) Amides with glycosylphosphatidylinositol. This rather complex structure called GPI anchor (Figure 4.8) is used by cells to reversibly fix proteins at the *outer* surface of the cell membrane. Examples are acetylcholinesterase and cell-surface antigens.

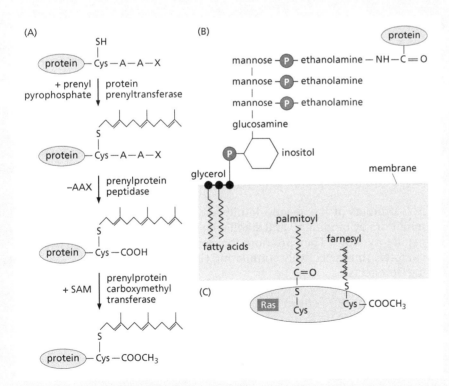

Figure 4.8 Lipid anchors (A) Enzymatic prenylation, including the steps prenylation, removal of AAX, and carboxy-terminal methylation (SAM, *S*-adenosylmethionine). (B) Glycosylphosphatidylinositol anchor at the cell surface. The inositol ring is depicted in a simplified form. (C) Two-fold anchoring of the small G-protein Ras at the inner side of the plasma membrane. P, phosphate group.

Subfamilies

Like most signal-transducing proteins trimeric G-proteins exist in various iso-forms. Theoretically, the 16 Gα, 5 Gβ, and 12 Gγ species expressed by mammalian cells together with numerous splice variants might combine to form more than 1000 individual G-proteins. Even if not all of them are realized, the degree of variability is almost unlimited considering the more than 1000 isoforms of GPCRs (Section 5.2).* On the basis of sequence homologies of the α-subunits and their sensitivity to certain bacterial toxins, the following subfamilies have been defined (Figure 4.7).

(1) **G_s family** (s = stimulatory): the α-subunit *stimulates* adenylate cyclase. This family also includes the olfactory G_{olf} proteins expressed only in the olfactory epithelium. G_s-proteins are selectively stimulated by cholera toxin (Section 4.3.3).

(2) **$G_{i,0}$ family** (i = inhibitory; 0 refers to a subfamily in brain): the α-subunit *inhibits* adenylate cyclase. This family also includes the transducins or G_t-proteins of the retina and the G_{gust}-proteins of the taste cells. G_t- and G_{gust}-proteins activate cGMP-phosphodiesterase and phospholipase Cβ, respectively. With the exception of a G_z-subgroup, these G-proteins are selectively inhibited by pertussis toxin (Section 4.3.3).

(3) **$G_{q,11}$ family**: the α-subunit stimulates phospholipase Cβ. This family includes several subgroups and is insensitive to cholera and pertussis toxin.

(4) **$G_{12,13}$ family**: the α-subunit activates, among others, some GEFs of the Rho family of small G-proteins (Section 10.3.1).

Many G-proteins are expressed ubiquitously, whereas others are restricted to certain cell types and tissues: for instance, G_{t1} are found in retinal rod cells and G_{t2} in retinal cone cells. Most abundant are the $G_{i,0}$-proteins. They provide the major source of Gβγ-subunits, which (as compared with the α-subunits) require considerably higher concentrations to become effective.

AGS proteins: receptor-independent activation of trimeric G-proteins

The only reliable rule of cellular signal transduction is that there is no rule without an exception. Thus, the concept of trimeric G-proteins being stimulated exclusively by receptors has been modified by the discovery of AGS proteins (Activators of G-protein Signaling), which activate trimeric G-proteins independently of receptors. Three families have been identified, each operating through a particular mechanism.

(1) AGS1, representing the first family, is a *bona fide* nucleotide exchange factor that resembles, particularly for $G_{i,0}$-proteins, a GPCR. Interestingly, it is a small G-protein, belonging to the Ras superfamily (see Chapter 10).

(2) AGS proteins of the second family act as guanine nucleotide dissociation inhibitors by binding to and stabilizing the GDP form of Gα via a G-protein regulatory motif. They promote the release of active Gβγ from the trimeric complex, thus selectively stimulating Gβγ-controlled signaling while blocking Gα effects.

(3) AGS proteins of the third family form complexes with Gβγ. Their role in signaling is not yet known.

---

*The nematode *Caenorhabditis elegans* has 20 genes encoding 17 G_i types. In contrast, in plants (at least in *Arabidopsis thaliana*) very few G-proteins have been found. Here signal processing seems to be based primarily on protein phosphorylation.

The wide distribution of these proteins indicates important functions in G-protein-dependent signal transduction that are not yet fully understood. Clearly, they bypass receptor-dependent pathways and possibly control G-protein subunits in cellular compartments that are devoid of receptors. An increasing body of evidence points to a role of AGS proteins in several aspects of neurotransmission. The discovery of AGS proteins indicates that G-protein-dependent signaling is considerably more complex and connected with other signaling pathways to a much higher degree than hitherto assumed.

4.3.2 Regulators of G-protein signaling: GTPase-activating proteins of trimeric G-proteins

As mentioned above, trimeric G-proteins undergo auto-inactivation due to the intrinsic GTPase activity of the α-subunits. In many cases, however, this reaction is not fast enough. For instance, auto-inactivation of the G-protein transducin that mediates light processing in the retina would take around 20 s, whereas an image fades almost immediately when we close our eyes. Therefore, cells strongly stimulate the intrinsic GTPase activity of their G-proteins. This function may be taken over by effector proteins via negative feedback (Figure 4.9). So, phospholipase Cβ is activated by the α-subunit of $G_{q,11}$-proteins, and at the same time stimulates the GTPase activity of that subunit. The same holds true for transducin and its effector protein cyclic GMP-phosphodiesterase.

Even more effective and versatile are special GTPase-activating proteins called RGS (Regulators of G-protein Signaling). They constitute a large protein family found in all eukaryotic cells from yeast to humans. In yeast they counteract the effects of the mating pheromones as well as differentiation of the fermentation phenotype, both of which are mediated by GPCRs (see Section 5.2). Moreover, the deficiency of special RGS proteins leads to defects of reproduction and development in the mold *Aspergillus nidulans* and to malformations of the nervous system in the nematode *C. elegans*. Such observations clearly prove a fundamental role of RGS in eukaryotes.

RGS proteins are quite heterogeneous. The only structural feature they have in common is the RGS domain that, upon interacting with a complementary domain of Gα-subunits, stimulates GTPase activity. This effect may be very specific; for instance, the isoform RGS9 reacts only with transducin. The RGS domain is not related to the GAP (GTPase-Activating Protein) domain interacting with small G-proteins.

Apart from an RGS domain, most RGS proteins have additional interaction domains and, therefore, couple GTPase activation with other effects, thus promoting signaling cross talk. Examples include the RGS protein **AKAP-2** (A-Kinase Anchor Protein 2) and the RGS of $G_{12,13}$ proteins. AKAP-2 anchors protein kinases A (PKA) and C (PKC) and other signal-transducing proteins at their cellular sites of action (see also Section 4.6.1), while the $G_{12,13}$ RGS is not only a GTPase activator but, in addition, a $G_{12,13}$-effector protein that stimulates as a GDP–GTP exchange factor the small G-protein Rho (Figure 4.9).

Figure 4.9 Regulation of trimeric G-protein activity (A) Dual control of Gα activity by a G-protein-coupled receptor (GPCR) acting as GEF and by a GTPase-activating effector (phospholipase Cβ, cGMP phosphodiesterase) and an RGS protein. Some RGS proteins may be also effectors of G-proteins. (B) Sharpening of the Gα signal by an RGS protein. (C) p115 Rho-GEF as an example of a multifunctional RGS protein with various interaction domains (DH, DBL homology domain, see Section 10.3.1; PH, pleckstrin homology domain; PIP_2, phosphatidylinositol 3,4-bisphosphate). More information on Rho and Rho-GEF is found in Section 10.3.

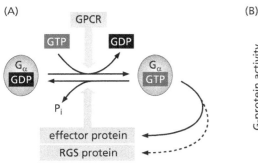

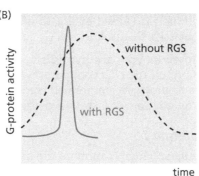

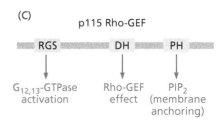

Many RGS proteins owe a special property to this multifunctionality: they facilitate not only the switching-off but also the switching-on mechanism of G-proteins, that is, the effect of the receptor on GDP–GTP exchange. By this means the $G\alpha$-signal is "sharpened" (Figure 4.9B), becoming less sensitive to accidental noise and suited for a more rapid and possibly frequency-modulated transmission.

4.3.3 Diseases caused by disturbances of trimeric G-protein signaling

Considering the central position of trimeric G-proteins in signal transduction across cell membranes, it is conceivable that deregulation of G-protein signaling may have serious consequences. A classical example is cholera, caused by the bacterium *Vibrio cholerae*. The major symptom of this disease is an extreme loss of fluid across the intestinal epithelium: instead of normally 2 L, 20–30 L is excreted per day, resulting in high mortality. The **cholera toxin** is an oligomeric protein interacting with gangliosides at the cell surface. As a result, a subunit expressing ADP-ribosyltransferase activity is released into the cell. There this enzyme specifically catalyzes the ADP-ribosylation of an Arg residue in the active center of $G\alpha_s$-subunits, inhibiting their intrinsic GTPase activity. This effect leads to a continuous activation of the G-protein and its effector adenylate cyclase, and the cell is swamped with cAMP, which stimulates the export of fluid across the cell membrane (a detailed discussion of the mechanism is found in Section 14.7.2). To exert its effect, cholera toxin has to cooperate with a cellular protein, the **ADP-ribosylation factor (Arf).** This is is a small G-protein normally regulating the formation of secretory vesicles (Section 10.4.3). Cholera toxin is highly specific and, therefore, experimentally used to distinguish effects of G_s-proteins from those of other G-proteins. The heat-labile enterotoxin of pathogenic *Escherichia coli* strains has a similar effect as cholera toxin, while the heat-stable enterotoxin of the same organism stimulates an overproduction of cyclic GMP (cGMP) with similar consequences (see Section 14.7.2).

Another bacterial protein disturbing G-protein signaling is **pertussis toxin,** a component of a whole toxin cocktail of *Bordetella pertussis*, the bacterium causing whooping cough. Like cholera toxin, pertussis toxin is an ADP-ribosyltransferase; it selectively attacks a Cys residue in $G\alpha_{i,0}$-subunits thus interrupting the interaction with the receptor. As a result $G_{i/0}$ effects are suppressed. In cells expressing G_s-proteins, this leads to an overactivation of adenylate cyclase (which normally is inhibited along the $G_{i,0}$ pathway) and, as in the case of cholera toxin, a pathogenic accumulation of cAMP. This effect is strongly augmented by a toxic adenylate cyclase of *B. pertussis*. In addition, all signaling reactions mediated by $G\beta\gamma$-subunits, which are predominantly released from $G_{i,0}$-proteins, become blocked.

Examples of other diseases related to disturbances of trimeric G-protein signaling include:

- a form of **night blindness** due to a genetic defect of the α-subunit of transducin

- **McCune–Albright syndrome**, characterized by defects of hormonal regulation. It is caused by a gain-of-function mutation of $G\alpha_s$ that correlates with massive disturbances of G_s-controlled hormone production

- **pseudohypoparathyroidism**, caused by a loss-of-function mutation of $G\alpha_s$, as a result the effects of parathyroid hormone (that interacts with a G_s-coupled receptor) are suppressed leading to premature puberty with obesity, growth retardation, and skeletal deformities

Summary

Trimeric (large) G-proteins are major devices of transmembrane signaling. Upon interaction with an activated receptor, which functions as a guanine nucleotide exchange factor (GEF), G-proteins dissociate into α- and βγ-subunits, which both are signal transmitters that interact with second-messenger-producing enzymes or ion channels. Depending on the downstream coupling, several subfamilies of trimeric G-proteins are distinguished. In addition to transmembrane receptors, certain trimeric G-proteins are also activated by AGS proteins. Trimeric G-proteins may undergo auto-inactivation via their intrinsic GTPase activities. Frequently this process is stimulated in addition by regulators of G-protein signaling (RGS proteins). Due to their interaction domains, RGS proteins enable cross talk of trimeric G-proteins with a wide variety of other signaling reactions. The α-subunits of trimeric G_s- and $G_{i,0}$-proteins are attacked in a highly selective manner by bacterial toxins exhibiting ADP-ribosyltransferase activity. While cholera toxin renders $G\alpha_s$-subunits constitutively active by inactivating the GTPase activity, pertussis toxin suppresses $G_{i,0}$ signaling by interrupting the receptor/G-protein interaction. Both effects are used for a characterization of G-protein functions.

4.4 Downstream of G-proteins: enzymes producing second messengers

As explained above, trimeric G-proteins couple a large number of membrane receptors to the data-processing protein network of the cell. They do this mainly by controlling the activities of enzymes that catalyze the generation of second messengers (Figure 4.10). These effector enzymes of trimeric G-proteins are:

- adenylate cyclases, producing cAMP

- PI3Ks, producing phosphatidylinositol 3,4-bis- and 3,4,5-trisphosphate (PIP_2 and PIP_3)

- PLCs producing diacylglycerol (DAG) and inositol 1,4,5,-trisphosphate ($InsP_3$)

Additional effects of trimeric G-proteins are discussed later in the chapter.

4.4.1 Adenylate cyclases of mammalian cells

Among the effector enzymes of trimeric G-proteins, the adenylate (or adenylyl) cyclases (ACs) were discovered first. They catalyze the hydrolysis of ATP to cAMP and pyrophosphate. The latter is immediately hydrolyzed by a cooperating pyrophosphatase to two molecules of inorganic phosphate, rendering the overall reaction exergonic (see Figure 4.13).

As explained in Chapter 3, adenylate cyclases are ancient enzymes found in almost all cells from the simplest prokaryotes to humans. In the course of evolution they have developed a bewildering diversity of structures (see Sidebar 4.3). The adenylate cyclases of vertebrates are large proteins (up to 140 kDa). Their overall structure, consisting of 2×6 transmembrane domains and a catalytic center with the two ATP-binding sites on the cytoplasmic side (Figure 4.11A), is reminiscent of that of a membrane transporter with an ATP-binding cassette (ABC transporter; see Section 3.2.1), although the two classes of proteins are only distantly related.

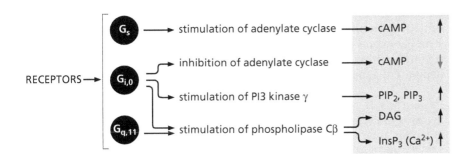

Figure 4.10 Second messenger generation controlled by trimeric G-proteins: an overview Upon activation by membrane-bound receptors, the G-protein subunits change the activity of several enzymes. As a result, cellular levels of the second messengers shown in the red box are up- or down-regulated.

Figure 4.11 Mammalian adenylate cyclases (A) Membrane topology of the two cyclase domains C1 and C2 including the ATP-binding site. In reality, the active enzyme is a dimer or oligomer. Forskolin is a plant-derived adenylate cyclase activator. (B) List of nine adenylate cyclase (AC) isoforms with regulatory factors: (+) stimulation, (−) inhibition, (0) no effect, (?) doubtful or not tested.

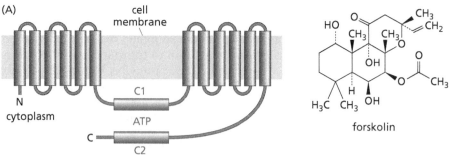

(A) cell membrane, C1, ATP, C2, N cytoplasm, C, forskolin

(B)

	$G\alpha_s$	$G\beta\gamma$	$G\alpha_i$	Ca^{2+}/calmodulin
AC1	+	−	−	+
AC2	+	+	0	0
AC3	+	−	−	+
AC4	+	+	0	0
AC5	+	0	−	0
AC6	+	0	−	?
AC7	+	+	0	0
AC8	+	−	−	+
AC9	+	?	?	0

The mammalian organism expresses at least nine isoforms that are tissue-specifically distributed. The cyclases are integrated as dimers or oligomers in the plasma membrane, where they form signal-transducing complexes with G-proteins, receptors, and ion channels.

As shown in Figure 4.11, all isoforms are activated by the α-subunits of trimeric G_s-proteins; however, they differ in their sensitivity to other signals such as $G\alpha_i$- and $G\beta\gamma$-subunits and Ca^{2+} ions (as calmodulin complexes). In fact, the individual isoforms of adenylate cyclases provide particularly clear examples of logical gates. AC1 constitutes an AND gate or coincidence detector for signals acting via $G\alpha_s$ AND Ca^{2+}/calmodulin and a BUT NOT gate with $G\alpha_{i,0}$-activating signals. On the other hand, AC2 may be activated by G_s- OR $G_{i,0}$-proteins (via $G\beta\gamma$) since it is not inhibited by $G\alpha_i$. These differences in regulation have important physiological consequences, as shown by the following examples.

(1) Isoforms AC5 and AC6, expressed predominantly in heart muscle cells, are activated via the G_s pathway by the sympathetic signal noradrenaline and inhibited via the $G_{i,0}$ pathway by the parasympathetic signal acetylcholine. By this means heart activity is either enhanced or reduced (for details see Section 5.5). In addition, the influx of Ca^{2+} accompanying each heart muscle contraction causes an inhibition of AC5 and perhaps also of AC6, establishing a negative feedback loop.

(2) In the central nervous system, Ca^{2+} controls cellular processes thought to be involved in learning and memory (Section 16.4). The predominant adenylate cyclases in brain neurons are AC1 and AC8, both of which are activated by Ca^{2+}/calmodulin, and AC9, which is inhibited by calcineurin, a Ca^{2+}-dependent protein phosphatase (see Section 14.6.4). These adenylate cyclases are strongly expressed during memory training and, therefore, may play a key role in learning processes, in particular by mediating calcium- and cAMP-dependent gene expression required for strengthening of synaptic contacts.

(3) In neurons, but also in heart muscle cells, the calcium-sensitive adenylate cyclases are found in close proximity to voltage- and ligand-controlled Ca^{2+} channels such as the L-type channel and the NMDA receptor (see Sections 14.5.2 and 16.4, respectively), ensuring particularly rapid interaction.

In addition to the adenylate cyclases of Gram-negative bacteria and the toxic adenylate cyclases of certain pathogenic microbes discussed in Section 3.2.2, the "eukaryotic" adenylate cyclases constitute a separate family. Because "eukaryotic" cyclases have been found also in prokaryotes, the term "adenylate cyclases type III" is now preferred. The common structural element of these enzymes is a catalytic cyclase homology domain. In metazoans, two such domains are combined with 2 × 6 transmembrane domains (Figure 4.12), a membrane topology resembling that of ABC trans-porters, although those are not closely related to cyclases. In contrast, the adenylate cyclases of prokaryotes and unicellular eukaryotes have many additional interaction domains. The bewildering variability of these domains indicates an extended cross talk of adenylate cyclases with other signaling pathways and provides a particularly instructive example of the combinatorial possibilities of protein modules (Figure 4.12). Cyclase homology domains are also found in guanylate cyclases, which are counted as type III cyclases although they differ completely from adenylate cyclases in other structural parameters.

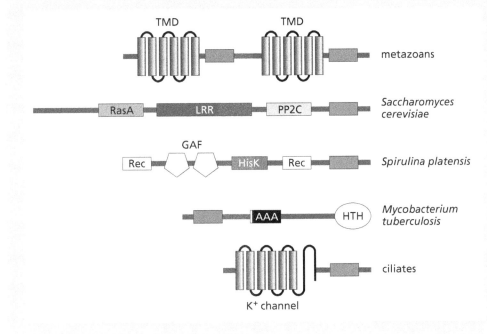

Figure 4.12 Variability of adenylate cyclases: some examples Depending on the species, the cyclase homology domain (red) is combined with various interaction domains. TMD, hexahelical transmembrane domain of metazoan adenylate cyclases (Figure 4.11); RasA, binding domain for the small G-protein Ras of *Saccharomyces cerevisiae*; LRR, leucine-rich repeats; PP2C, protein phosphatase 2C domain; HisK and Rec, His autokinase and Asp receiver domains of a two-component system; GAF, interaction domain for various small signaling molecules (see Sections 3.3.2 and 4.4.2); AAA, triple-A domain with ATPase function (see Section 2.5.2); HTH, helix–turn–helix motif of a DNA-binding site. In ciliates the cyclase domain is linked to the hexahelical transmembrane structure of a voltage-dependent potassium channel.

Contrary to many other signal-transducing proteins, almost no drugs and only a few poisons are known to act on adenylate cyclases. An exception frequently used for experimental purposes is the diterpene **forskolin** (Figure 4.11) from the plant *Coleus forskohlii*, which activates all isoforms except AC9. The cyclases bind forskolin at a pseudosubstrate site that in the absence of forskolin (or a stimulatory G protein) blocks the active center. An endogenous factor fitting this site has been postulated but not yet identified. In Indian folk medicine, extracts from *C. forskohlii* roots have been used against asthma, heart disease, and thyroid insufficiency.

4.4.2 Phosphodiesterases and the importance of signal extinction

The efficacy of a signal-processing apparatus critically depends on signal extinction, since otherwise the system would sink rapidly in a flood of data, becoming paralyzed. Signal extinction may be a signal by itself, since for data processing it is irrelevant whether a binary process proceeds from 0 to 1 or from 1 to 0, or whether a switch is turned to the left or to the right. Therefore, for the cell, signal generation and signal extinction are equivalent events that are both controlled by input signals. This is demonstrated clearly by the cyclic nucleotides: in the liver, the *generation* of cAMP induces glycogenolysis, whereas in the retina, the *degradation* of cGMP causes an impulse in the visual nerve.

The termination of cAMP and cGMP signals is achieved by phosphodiesterases catalyzing their hydrolysis to inactive nucleoside 5'-monophosphates. By controlling (together with the cyclases) the cellular level of cyclic nucleotides, the phosphodiesterases thus adopt a key position in signal processing (Figure 4.13).

In mammalian cells phosphodiesterases are encoded by no less than 20 genes, and by alternative splicing and related mechanisms more than 50 variants may be produced. These isozymes are grouped in 11 subfamilies on the basis of sequence homologies, tissue distribution, and substrate specificity. The fact that there are considerably more phosphodiesterases than cyclases underscores the importance of these enzymes in signal transduction.

As is typical for signal-transducing proteins, phosphodiesterases exhibit a modular structure and mostly exist as homodimers. The catalytic domain common to all isoforms contains two Zn^{2+} ions. The regulatory domains are split, with a large sequence located in the N-terminal portion and a small one in the C-terminal portion of the enzyme molecule. In contrast to the catalytic domain, the regulatory domains are highly variable, reflecting the equally high variability of input signals.

The phosphodiesterases are regulated as strictly as the cyclases. Like protein kinases they are intrasterically inhibited in the absence of an input signal, because an auto-inhibitory sequence blocks the catalytic center. This inhibitory sequence may be a part of the enzyme molecule or may exist as a separate subunit as, for example, in the case of the phosphodiesterase isoform PDE6 degrading cGMP in the retina (see below). Phosphodiesterase-activating input signals

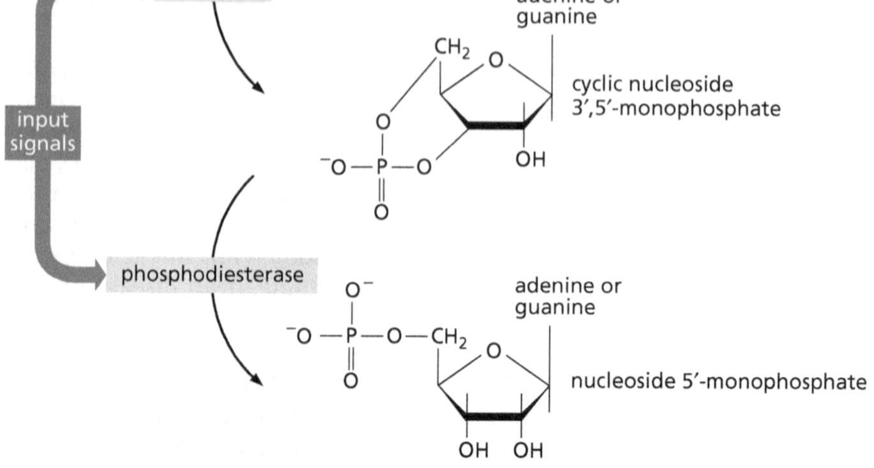

Figure 4.13 Generation and termination of the cAMP/cGMP signal The enzymes involved, cyclases and phosphodiesterases, are expressed in numerous isoforms controlled by various input signals.

overcome the auto-inhibition by inducing an allosteric change of protein conformation. With the exception of the cGMP-specific PDE6 in the retina, which seems to be the only phosphodiesterase controlled by a trimeric G-protein (transducin), such signals are Ca^{2+} ions, cGMP, and protein phosphorylation. The cAMP- and cGMP-degrading enzymes of the **PDE1** subfamily are stimulated, for instance, by Ca^{2+}/calmodulin, thus connecting Ca^{2+}-dependent and cyclic nucleotide-dependent signaling.

PDE2 isoforms also degrade both cyclic nucleotides. They are regulated in another way in that they are allosterically activated by cGMP. The allosteric cGMP-binding site in the regulatory domain is known as the GAF domain, belonging to one of the largest and most ancient domain families for interaction of proteins with smaller molecules such as cGMP (see Sidebar 3.3 in Section 3.3.2). It has a different structure than the cGMP-binding site in the catalytic center of PDE1 and PDE2. PDE2 plays an important role in the hormonal regulation of blood volume and blood pressure. In the adrenal cortex, the pituitary hormone ACTH (adrenocorticotropic hormone) activates adenylate cyclase via a G_s-protein-coupled receptor, and the cAMP generated in turn promotes the production and release of the mineralocorticoid aldosterone. By acting on the kidney, the latter causes an increase of blood volume and blood pressure. These effects are counteracted by the natriuretic peptide ANP released from atrial heart cells. As will be discussed later, the ANP receptor in the adrenal cortex is a receptor guanylate cyclase producing cGMP that activates PDE2. As a result, the cellular cAMP level drops and the release of aldosterone ceases.

The enzymes of the subfamilies **PDE3** and **PDE4** are widely distributed. Being targets of numerous pharmaceuticals with cardiotonic, vasodilatory, and thrombolytic effects, they are of particular clinical interest. PDE3 and PDE4 are activated by phosphorylation catalyzed by the cAMP-dependent protein kinase A. Accordingly, they are negative feedback inhibitors in cAMP-controlled pathways such as glycogenolysis and lipolysis. Another protein kinase that activates PDE3 is PKB/Akt. This enzyme is a component of signaling cascades stimulated by insulin, leptin, and other hormones (see Section 4.6.6). In contrast to PDE2, PDE3 is inhibited rather than stimulated by cGMP.

A phosphodiesterase that has gained considerable attention is **PDE5**. It is absolutely cGMP-specific and, in addition, is allosterically activated by cGMP. The major function of PDE5 is the regulation of vascular tone, being an antagonist of NO. Using cGMP as a second messenger NO induces vasodilatation (see Section 16.5.1). The result is an increased blood flow that causes, among other effects, penile erection. This response is dampened by PDE5 catalyzing degradation of cGMP. PDE5 inhibitors such as Viagra (Sildenafil), which prolong the relaxation phase of vascular smooth muscles, are widely used as erection-promoting drugs.

Another cGMP-specific phosphodiesterase is **PDE6**. This enzyme is more or less restricted to retina cells where it occupies a central position in visual signal processing. The structure of PDE6 is more complex than that of most other phosphodiesterases. The active form is a heterotetramer consisting of two catalytically active subunits, α and β, and two inhibitory γ-subunits. In addition, δ-type subunits seem to bind the complex to disc membranes of photoreceptor cells. PDE6 is allosterically activated by cGMP and by the α-subunit of the trimeric G-protein transducin (more about visual signal processing is found in Section 15.5).

With the exception of PDE8 and PDE9, all PDE isoforms are competitively inhibited by methylxanthines such as caffeine, theophylline, and isobutylmethylxanthine. Many of the effects of these drugs are thus explained. However, additional cellular targets have been identified, in particular for caffeine (see Section 16.9.1).

4.4.3 Phosphatidylinositol kinases

Cyclic nucleotides are not the only second messengers released along G-protein-controlled pathways. In fact, various intracellular signaling molecules are produced from membrane phospholipids, with phosphatidyl inositols (PIs) being a particularly versatile source. Although PIs comprise only about 10% of membrane phospholipids, they are of great importance for signal processing. There are two mechanisms by which membrane-bound signaling molecules are generated from these phospholipids: phosphorylation of the inositol ring and hydrolysis (see Section 4.4.4). Phosphorylation is catalyzed by specific phospholipid kinases that use ATP as a substrate and attack positions 3, 4, and 5 of the inositol ring, thus yielding phosphatidylinositol phosphates (PIPs), also known as phosphoinositides (Figure 4.14). PI phosphorylation is reversed by a series of lipid phosphatases. As a result of these enzymatic reactions in mammalian cells, at least seven types of PIPs have been found that differ in the position and degree of phosphorylation. These relationships are summarized in Figure 4.15.

A characteristic property of these lipids is that they are able to generate specialized membrane domains where they may recruit signal-transducing proteins via specific interaction domains. These proteins thus become concentrated and activated, forming signal-transducing complexes at specific sites. Examples of membrane-targeting protein domains interacting with PIPs are given below.

- **PH domain** (Pleckstrin Homology domain; pleckstrin is a functionally ill-defined PKC substrate protein in blood platelets). This domain is the most abundant PIP-binding motif in human cells. It is absolutely specific for 3-phosphorylated PIPs which are stereospecifically bound . Due to this property proteins with PH domains are major targets of signaling pathways that lead to an activation of PI3K (see below). PH domains exist in various isoforms. In fact, only a minority of them recognizes PIPs, whereas the functions of the rest remain elusive.

- **FYVE domain** (named after the four proteins Fab1, YOTP, Vac1, EEA1 where this domain was originally discovered). This domain binds selectively to PI 3-monophosphate and is particularly enriched in endosomal proteins.

- **PX domain** (PhoX homology domain; the name refers to the enzyme NADPH oxidase) interacting with PI 3-mono-, PI 3,5-bis-, and PI 3,4,5-trisphosphates.

Figure 4.14 Points of attack of phosphatidylinositol kinases
Membrane-bound phosphatidylinositols are a source of various signaling molecules generated by signal-controlled phospholipid kinases that catalyze the ATP-dependent phosphorylation of the inositol ring at the positions shown. Frequently phosphatidylinositols contain an arachidonic acid residue in position 2 of the glycerol moiety.

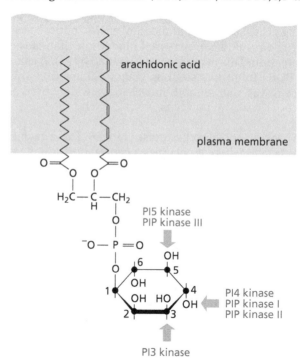

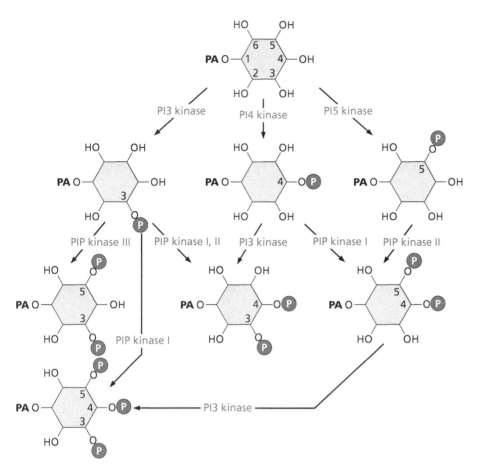

Figure 4.15 Biosynthesis of phosphorylated phosphatidylinositols The six-membered ring of inositol (here shown in a simplified form) is esterified by phosphatidic acid (PA) and phosphoric acid (P, red). The phosphorylation is catalyzed by various lipid kinases and results in phosphatidylinositol mono-, bis-, and trisphosphates (PIP, PIP_2, and PIP_3). Monophosphates are generated by PI3 kinases, PI4 kinases, and PI5 kinases , respectively. The monophosphates are further phosphorylated by type I PIP kinases (positions 4 and 5); type II PIP kinases (position 4); and type III PIP kinases (position 5). Phosphatases that exhibit corresponding specificities render the phosphorylations reversible (not shown).

A common structural feature of such domains is a binding pocket lined by basic amino acids residues that electrostatically interact with the phosphate groups of PIPs. Specificity is provided by additional amino acid residues. Beside PIP-specific domains membrane-targeting protein motifs are known that recognize other lipids. Examples are found in Sections 4.6.3 and 14.6.1. More unspecific membrane targeting is achieved, in addition, by a reversible or irreversible binding of lipid residues to proteins (see Section 4.3.1).

Membrane targeting plays an essential role for the recruitment of signal-processing protein complexes to transmembrane receptors as well as for protein sorting within the different membrane compartments of the cell. Due to strict compartmentalization of lipid kinases and phosphatases, the PIPs are distributed specifically rather than randomly among the cellular membranes. So PI3-monophosphate is found predominantly in endosomes recruiting proteins with FYVE domains, whereas the PI 4-monophosphate prefers the Golgi apparatus, and the bisphosphates are enriched in the plasma membrane. The energy-consuming lipid phosphorylations generate biochemical gradients that together with protein phosphorylation, GTP hydrolysis, and other switching reactions make cellular processes irreversible, pushing them in one direction. This becomes particularly evident in strictly targeted exo- and endocytotic vesicle transport, which is regulated by PIPs (for details see Section 10.4.4).

As shown in Section 4.4.4, PLCs degrade PIPs to diacylglycerols and inositol phosphates, with the 3,4,5-trisphosphate ($InsP_3$) being the most prominent example, whereas almost nothing is known about the putative functions of other inositol phosphates. Recently, however, energy-rich inositol pyrophosphates have gained some attention. These compounds are derivatives of inositol hexakisphosphate ($InsP_6$) that is pyrophosphorylated by an $InsP_6$ kinase. The 5-diphosphoinositol pentakisphosphate may provide a phosphate donor for **nonenzymatic protein phosphorylation** competing with enzymatic phosphorylation. While the reaction has been found to occur in living cells, its physiological significance is not yet clear.

Phosphatidylinositol 3-kinases

Among the phospholipid kinases, the PI3Ks are most prominent and best investigated because they occupy a key position in cellular signal processing. The major reaction catalyzed by these enzymes is the phosphorylation of PI 4-mono- and 4,5,-bisphosphates, produced by position 4- and 5-specific kinases. The products are PIP$_2$ and PIP$_3$. Both stay in membranes, acting as docking sites and activators of protein kinases and many other signal-transducing proteins.

As mentioned above, the binding partners of PIP$_2$ and PIP$_3$ possess a PH domain interacting specifically with the 3-phosphoinositol residue. The enzymes of the PI3K family are, therefore, of great importance for reversible association of proteins with membranes and for the formation of signal-transducing protein complexes in the vicinity of membrane-bound receptors.

On the basis of structural features, the mammalian PI3Ks have been subdivided into three families called PI3K-I, -II, and -III. As yet only the PI3K-I family, including the four isozymes PI3Kα, -β, -γ, and -δ, has been investigated in detail. They have a C-terminal lipid kinase domain in common as well as a C2 domain that facilitates membrane association in a Ca^{2+}-dependent manner (for details see Section 4.6.3). Reversible recruitment to membranes is a critical step in PI3K activation because these cytoplasmic enzymes would not otherwise have access to their phospholipid substrates. Characteristic structural elements of PI3Ks are, in addition, binding sites for the small G-protein Ras, for G$\beta\gamma$-subunits, and for adaptor proteins (Figure 4.16). Since Ras and G$\beta\gamma$-subunits are anchored in the membrane by lipid residues (Section 4.3.1), these interactions promote both membrane recruitment and activation of PI3Ks.

Among the class I PI3Ks, isoform **PI3Kγ** is integrated in a G-protein-controlled pathway. It is recruited to the membrane and activated by G$\beta\gamma$-subunits and therefore is primarily under the control of signals processed by G$_{i,0}$-protein-coupled receptors (remember that G$_{i,0}$-proteins are the major source of G$\beta\gamma$-subunits). The catalytic subunit of PI3Kγ contains two G$\beta\gamma$-binding sites (Figure

Figure 4.16 Domain structure and regulation of phosphatidylinositol 3-kinases Interaction of an α, β, or δ isoenzyme with a p85 adaptor is shown (AdBD, adaptor-binding domain; RasBD, Ras-binding domain; PRO, proline-rich motif for interaction with SH3 domains; IRS, insulin receptor substrate; GAB, Grb2-associated binder). In addition, the domain structure of the γ isoform, including the two sites for interaction with G$\beta\gamma$-subunits, is depicted. Wortmannin is a fungicide from *Penicillium wortmannii*. As a semispecific inhibitor of some PI3K isoforms, it has found broad—and sometimes rather uncritical—application in research.

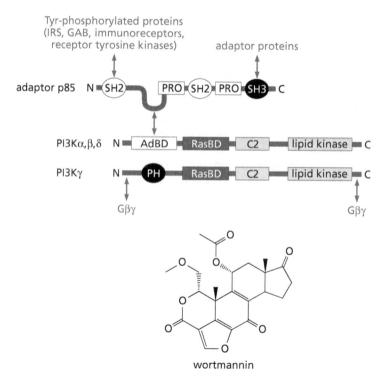

4.16). In some tissues a regulatory subunit p101, lacking known interaction domains, seems to facilitate the interactions of PI3Kγ with Gβγ-subunits.

Isoforms **PI3Kα, -β**, and **-δ** are activated by Tyr kinase-coupled receptors including those of insulin, growth factors, cytokines, antigens, and components of the extracellular matrix (Chapter 7). Activation requires membrane recruitment of PI3K by Ras that is stimulated along such pathways (Figure 4.17).

The interaction with Ras depends on a direct contact of the PI3K with the receptor. How is this achieved? Upon ligand binding, Tyr kinase-coupled receptors undergo autophosphorylation, exposing phospho-Tyr residues as binding sites for proteins with the corresponding interaction domains such as SH2 domains (see Section 7.1.2). Such domains are not found in the PI3Ks proper but in associated regulatory subunits known as adaptor proteins p50, p55, and p85. How these subunits manage enzyme activation has been investigated in detail for the p85 adaptor associating with PI3Kα. This subunit exhibits SH2 and SH3 domains as well as a proline-rich domain as a docking site for proteins with SH3 domains (Figure 4.16). Via these domains the p85/PI3K heterodimer comes in contact with autophosphorylated Tyr kinases, Tyr-phosphorylated docking proteins such as IRS and GAB, and other adaptor proteins like Shc (these proteins are discussed in detail in Section 7.1.2). These interactions induce a conformational change in the PI3K molecule, exposing the binding site for the interaction with Ras.

As mentioned above, little is known about class II and III PI3Ks. Recently, however, the class III enzyme **hVPS34** (human Vacuolar Protein Sorting protein 34) has gained attention since it seems to mediate the effect of nutrients on the protein kinase mTOR, a cellular key sensor of food and energy supply (Section 9.4.1). In contrast to class I PI3Ks, the enzyme prefers PI instead of PI 4-monophosphate and PI 4,5-bisphosphate as a substrate. Moreover, it exhibits a C2 domain like class I enzymes, but lacks a Ras-binding domain, indicating that its activity is regulated along another signaling pathway. Since a homolog of hVPS34 is the only PI3K in yeast, the enzyme probably represents the evolutionary archetype of the PI3K family.

Although showing almost no homology with eukaryotic protein kinases, the catalytic domains of PI3Ks exhibit both lipid kinase and protein kinase activity. Therefore, these enzymes, together with the PI3K-related kinases (Section 12.9), are classified as **atypical protein kinases**. In fact, in cell culture PI3Ks, particularly the γ-form, undergo autophosphorylation and catalyze the phosphorylation of other proteins, thus activating the Ras–RAF–ERK signaling cascade. Experiments on PI3K knockout mice with a deleted catalytic domain of PI3Kγ indicate even more functions, including direct interaction with a cAMP

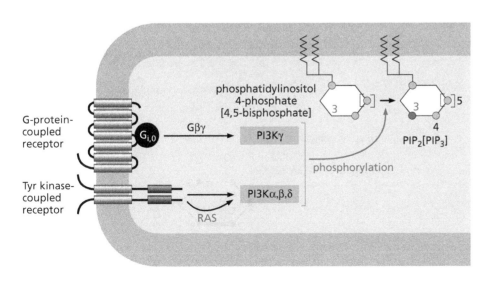

Figure 4.17 Signaling pathways leading to activation of phosphatidylinositol 3-kinases
PI3Ks catalyze the phosphorylation of membrane-bound phosphatidylinositol 4-monophosphate and 4,5-bisphosphate at position 3 of the inositol residue (here depicted in a simplified form as a six-membered ring). The isoform PI3Kγ is activated by Gβγ-subunits derived mainly from $G_{i,o}$-proteins coupled to heptahelical receptors, whereas the isoforms PI3Kα, -β, and -δ are activated by tyrosine kinase-coupled transmembrane receptors in cooperation with active Ras-GTP. Phosphorylated sites are symbolized by coloured dots. For further details, see text.

phosphodiesterase that becomes activated, dampening the stress and sympathetic effect on heart activity by degrading cAMP.

PI3Ks are multifunctional relay stations for a large number of extracellular signals, in particular for those stimulating cell growth and cell proliferation such as growth factors and insulin. In fact, a major function of the PI3K signaling pathway is to coordinate the rates of cell growth and cell proliferation and to adjust them to the energy and food supply in order to ensure the survival of the cell. Cell proliferation critically depends on cell growth, since cell division requires a certain cellular size.

The antagonists of PI3Ks are the **phospholipid phosphatases of the PTEN family** (the name refers to a Phosphatase-deleting mutation of chromosome TEN). These enzymes are tumor suppressor proteins that are deleted in many neoplastic diseases in humans. The tumor-suppressing effect is easily understood: PTENs counteract the proliferative and survival-promoting effects of PI3K signaling by reversing the phosphorylation of PIs in position 3. Accordingly, tumor cells with deleted PTEN respond more sensitively to mitogenic signals and are rescued from apoptosis. Since PTENs belong to the superfamily of protein phosphatases, they will be discussed together with similar enzymes in Section 7.4.

4.4.4 Phosphatidylinositol-specific phospholipases C

In addition to phosphorylation, enzymatic hydrolysis of PIs provides a pathway for second messenger production. The hydrolysis is catalyzed by phospholipases that are grouped in several families according to the preferred splitting site in the phospholipid molecule (Figure 4.18).

PLCs play an important role as producers of the second messengers DAG and InsP3. While DAG stays in the membrane serving as a docking site and activator of PKC and some other enzymes (Section 4.6.3), InsP3 is released into the cytoplasm, where it gates Ca^{2+} channels in the endoplasmic reticulum (Section 14.5.3). Thus, signals stimulating PLC activity increase the cytoplasmic Ca^{2+} level with far-reaching physiological consequences (see Sections 14.5 and 14.6). Particularly in neurons, PLCs have an additional important function: by

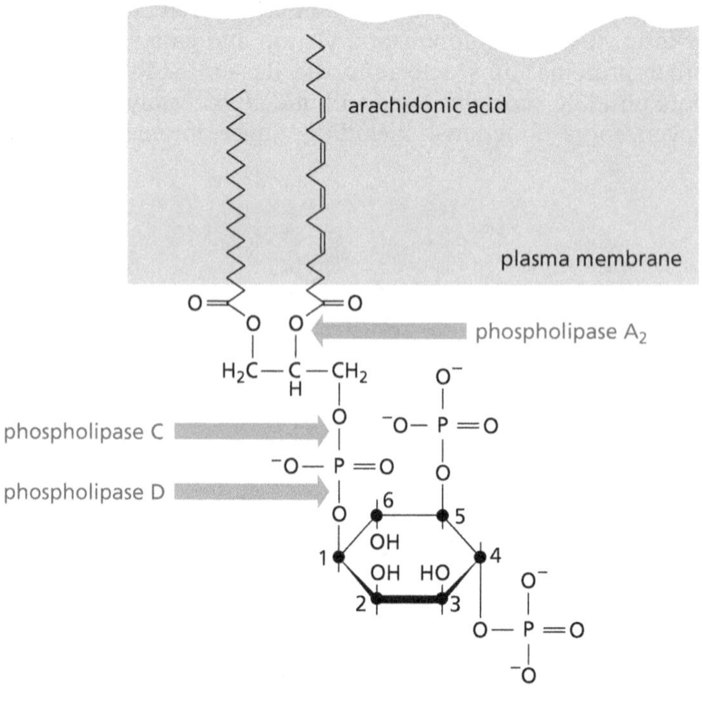

Figure 4.18 Hydrolysis of a phosphatidylinositol phosphate by phospholipases Membrane-bound phosphorylated phosphatidylinositols such as the 4,5-bisphosphate shown are a source of various signaling molecules generated by signal-controlled phospholipases. These include phospholipase A_2, producing arachidonic acid (the precursor of a wide variety of eicosanoids, see Section 4.4.5); phospholipase C, generating the second messengers diacylglycerol (DAG) and inositol 1,4,5-trisphosphate (InsP3); and phospholipase D, catalyzing the formation of phosphatidic acid.

degrading PIP_2, they control the function of a class of nonspecific cation channels known as TRP channels, some of which are inhibited, while others are activated by the phospholipid. As we shall see in Section 14.5.6 and Chapter 15, TRP channels play a key role in sensory signal transduction, fertilization, and calcium homeostasis.

Phospholipase C enzyme family

PI-specific PLCs are Ca^{2+}-dependent cytosolic enzymes. Mammalian cells express at least 13 isoforms as well as a series of splice variants. On the basis of structural similarities, they are divided into the subfamilies PLCβ (four isoforms), PLCγ (two isoforms), PLCδ (three isoforms), PLCε (one isoform), PLCζ (one isoform), and PLCη (two isoforms). PLCα, originally thought to be a separate isoform, is now considered to be a degradation product of PLCδ1. In simple eukaryotes such as yeast, only type δ is found; therefore, it is considered to represent the primeval form of PLC.

PLCs share a catalytic domain separated into two areas X and Y, the active conformation of which is stabilized by Ca^{2+} ions forming saltlike bonds with acidic amino acid residues. All PLC isoforms contain an EF-hand domain for Ca^{2+} binding (see Section 14.6.1) as well as PH and C2 domains for membrane anchoring and activation (Figure 4.19). Through its C-terminal domain containing a motif that interacts with PDZ domains (see Section 7.1.4), PLCβ stimulates the GTPase activity of $G\alpha_{q,11}$, which is the G-protein inducing PLCβ activity; thus it acts as a negative feedback regulator (see also Figure 4.9).

The structural hallmark of PLCγ is a sequence of several hundred amino acids inserted between the catalytic domains X and Y. This insert contains two SH2 and one SH3 domain (Figure 4.19). These domains interact with Tyr-phosphorylated proteins and proline-rich sequences respectively.

While PLCβ and PLCγ are connected with transmembrane signaling, PLCε operates more downstream by cooperating with the small G-protein Ras. To this end it has, in addition to the membrane-targeting and Ca^{2+}-binding domains common to all PLCs, a Ras-GEF domain and, like PI3Ks, Ras-binding domains.

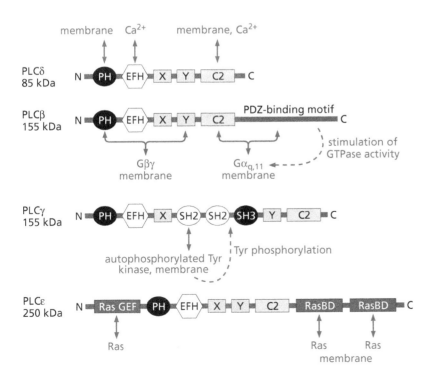

Figure 4.19 Phosphatidylinositol-specific phospholipases C Shown are the domain structures, interacting partners (red), and sites of membrane recruitment of the four PLC subfamilies of mammals. PLCβ interacts with and becomes activated by $G\alpha_{q,11}$- and Gβγ-subunits and, in turn, stimulates the GTPase activity of $G\alpha_{q,11}$.Through its SH2 domains PLCγ binds to autophosphorylated Tyr kinases such as the receptors of insulin and growth factors and becomes activated by phosphorylation. In their domain structures PLCη largely resembles PLCβ, while PLCζ resembles a truncated PLCδ lacking the PH domain (not shown). X, Y (pink), catalytic domains; EFH, EF-hand domain; RasBD, Ras binding domain. Two-headed arrows indicate binding. For details, see text.

Mechanisms of activation

As cytoplasmic enzymes, PLCs are confronted by the same problem as the P13Ks: they must be recruited to the cell membrane to come in contact with their membrane-bound substrates. This recruitment is a key event in enzyme activation. Depending on the PLC isoform, it is brought about by different mechanisms, all based on an interaction with membrane-anchored activators, which reflects the integration of PLC enzymes in a complex network of signaling pathways.

The enzymes of the **PLCβ subfamily** are recruited and activated by the membrane-anchored α-subunits of $G_{q/11}$-proteins. Like the γ-isoforms of PI3Ks, PLCβ1, -β2, and -β3 interact, in addition, with Gβγ-subunits (Figure 4.20). This means that the second messengers DAG and $InsP_3$ are released not only along the $G_{q,11}$-pathway but also through an activation of $G_{i,0}$-proteins, which are the major source of Gβγ-subunits in the cell. PLCβ1 and -3 are widely distributed, whereas PLCβ2 is restricted to the hematopoietic system and PLCβ4 to retina and nerve cells.

The enzymes of the **PLCγ subfamily** are activated by quite another mechanism resembling the activation of PI3Kα, -β, and -δ: by Tyr phosphorylation, catalyzed by either receptor or cytoplasmic Tyr kinases (Figure 4.20). For this purpose they interact via their SH2 domains with the autophosphorylated kinases. In contrast to PI3Ks, they do not depend on the assistance of regulatory adaptor proteins.

The mechanism of activation of **PLCδ** is less clear. Since these enzymes respond quite sensitively to Ca^{2+}, an increase of the cytoplasmic Ca^{2+} concentration as induced by various signals is probably sufficient for activation. Membrane binding is achieved by the C2 domain interacting with phosphatidylserine and Ca^{2+} and by the PH domain binding to PIP_3.

PLCε plays a special role, because it is both an activator (acting as a GEF) and an effector protein of the small G-protein Ras. It becomes activated along the Ras signaling pathways. Since Ras is membrane-anchored, the interaction results in membrane recruitment of PLCε, resembling the effect of Ras on PI3K. The role of PLCε in signal transduction is discussed in more detail in Section 10.1.4.

Only recently an additional isoform, **PLCζ**, has been discovered. This extremely Ca^{2+}-sensitive enzyme seems to play an essential role in reproduction. In the course of fertilization it is injected by the sperm into the egg cell and releases $InsP_3$ and Ca^{2+}, thus stimulating as a "kiss of life" metabolism and mitosis. Research on PLCζ is of considerable medical interest: the enzyme is expected to facilitate artificial fertilization and the production of embryonic stem cells. However, how this phospholipase becomes activated is not known yet.

PLCη, the latest member of the family, is expressed particularly in brain neurons. Since it is the most Ca^{2+}-sensitive PLC, the enzyme is assumed to act as a neuronal calcium sensor that is able to respond to minute changes of the cytoplasmic Ca^{2+} concentration. Its physiological functions are still a matter of conjecture.

4.4.5 Lipid mediators: signaling inside and outside cells

As we have seen, release of signaling molecules from membrane lipids plays a central role in cellular signal transduction. The lipid bilayer of cell membranes is composed mainly of glycerophospholipids and sphingomyelins. Both components have a dual function: as building materials and as a source of signaling molecules. Thus, glycerophospholipids are the starting products for the formation of DAG, inositol phosphates, PIPs, arachidonic acid and its many derivatives, and lysophospholipids, whereas ceramides and sphingosines are generated from sphingomyelins. Since the production of these lipid-derived

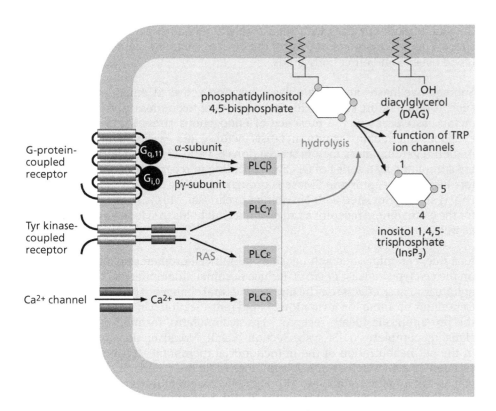

Figure 4.20 Signaling pathways leading to activation of C-type phospholipases Individual PLC isoforms are activated by the membrane receptors and ion channels shown to the left. They all catalyze the hydrolysis of phosphatidylinositol 4,5-bisphosphate, resulting in the formation of DAG and InsP3 as well as in an altered function of TRP cation channels (phosphorylated sites are symbolized by red dots).

mediators is controlled by extracellular signals, they are considered to be second messengers. However, many of them also function as *inter*cellular messengers, being released from cells and interacting with specific cellular receptors located either at the cell surface or inside the cell. This dual function is facilitated by the physicochemical properties of lipids, enabling many of them to pass cellular membranes in both directions. As is typical for signaling molecules, the levels of lipid messengers are extremely low in resting cells but increase rapidly upon stimulation by exogenous and endogenous signals. Effective mechanisms of signal termination, such as enzymatic degradation, guarantee that this increase is transient. Because of the high interconnectivity of lipid metabolism, signaling by lipid mediators occurs in a complex network of interactions that controls a vast number of cellular processes. Moreover, due to their physicochemical properties lipid mediators are enriched in cellular membranes. This provides an additional level of complexity as far as subcellular localization, metabolism and signaling mechanisms are concerned.

Sphingosines and ceramides: cellular stress signals?

Stress as well as a series of endogenous mediators activate **sphingomyelinases** (SMases), thus inducing a hydrolytic degradation of sphingomyelin to ceramide and choline phosphate. SMases are widely distributed from bacteria to humans. There are three enzyme families, the members of which differ in terms of their pH dependency. As far as signal transduction is concerned, the *neutral* SMases are of special interest because they are transmembrane proteins interacting with receptors. *Acid* SMases are required for lipid catabolism in lysosomes and, when expressed at the cell surface, for the extracellular generation of ceramides. *Alkaline* SMases are mainly found in the gastro-intestinal tract and are considered to be digestive enzymes.

Ceramides are hydrolyzed by **ceramidases** to sphingosines and long-chain fatty acids. Like SMases, ceramidases exist in several isoforms and their activities probably are controlled by signals. While *acid* ceramidases are confined to lysosomes, functioning as catabolic enzymes, *neutral* ceramidases are transmembrane enzymes with their catalytic domains facing the extracellular space. They are assumed to control the level of ceramides and sphingosines in blood plasma and other body fluids.

Sphingomyelinases and ceramidases become active in emergency situations such as temperature stress, oxidative stress, genotoxic stress, and lack of growth factors, as well as in the presence of endogenous stress signals such as the cytokines interleukin-1, tumor necrosis factor α, and Fas ligand (for more information regarding these factors see Sections 6.3 and 13.2.3). In addition, SMases are stimulated by a series of GPCRs. According to recent observations, the contact between receptor and SMase is brought about by a special adaptor protein FAN (Factor Associated with N-sphingomyelinase). This is shown in Figure 4.21 for the cannabinoid receptor as an example, but holds true for cytokine receptors as well.

As a rule, ceramides and sphingosines inhibit cell proliferation and, depending on the cell type (at least *in vitro*), induce terminal differentiation, senescence, or apoptosis. These effects may be due to structural changes of cellular membranes. Ceramides promote the formation of lipid rafts (Section 5.8) where, for instance, the pro-apoptotic "death receptor" Fas accumulates, forming death-inducing signaling complexes (DISCs; see Section 13.2.3). Moreover, they may participate in the permeabilization of the mitochondrial membrane, leading to cell death (Section 13.2.4). Whether or not ceramides fulfill the criteria of second messengers is still a matter of debate. Although effector proteins have been identified, no ceramide-specific interaction motif has been found yet in proteins. The ceramide effector proteins include **ceramide-activated protein phosphatases** and **ceramide-activated protein kinases** (CAPKs). The phosphatases belong to the families PP1 and PP2A (see Section 2.6.4 for a classification of phosphatases) and mediate the anti-proliferative and pro-apoptotic effects of ceramides in that they dephosphorylate the transcription factor Rb, thus stopping the cell cycle (Section 12.7), and the anti-apoptotic proteins PKB/Akt and Bcl2, which both become inactivated. The CAPKs include the atypical PKCζ (see below) that, depending on the cell type, is integrated in signaling pathways leading to either

Figure 4.21 Release of intracellular lipid messengers by G-protein-coupled receptor signaling A GPCR (here the cannabinoid receptor CB1, see Section 16.7) activates, via adaptor protein FAN, the enzyme sphingomyelinase, thus generating membrane-bound ceramide and sphingosine, which is phosphorylated to soluble sphingosine 1-phosphate. Alternatively, the receptor may open a conventional G-protein-controlled signaling pathway.

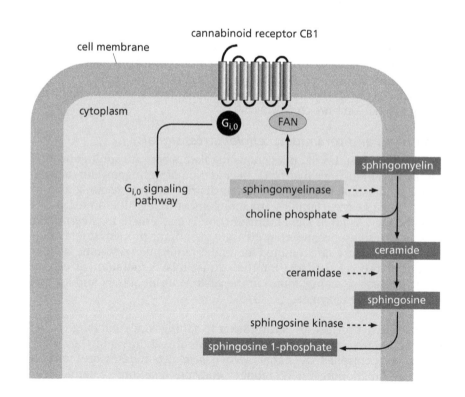

cell proliferation or cell death. Ceramide-activated protein kinase (CAPK proper) is related to the MAP2 kinase RAF and activates the ERK–MAP kinase cascade, which is a survival pathway (Section 11.2). Since CAPK makes a shortcut in the canonical cascade, proceeding from growth factor receptors via the small G-protein Ras to the MAP kinase module, it is also called kinase suppressor of Ras. This protein serves as a scaffold for the components of the RAF–MAP kinase cascade (Section 11.5).

Lysophospholipids and phosphatidic acid: mitogenic survival signals and immune regulators

Sphingosine and ceramide are sources of additional signaling molecules, the **ceramide 1-phosphates (C1Ps)** and **sphingosine 1-phosphates (S1Ps)** that are generated by ceramide- and sphingosine kinases (Figure 4.22). These lipid kinases use ATP as a phosphate donor and are found both inside and outside cells.

C1Ps and S1Ps are functional opponents of ceramides and sphingosines. As typical survival signals, they stimulate cell proliferation and cell motility while inhibiting apoptosis. Sphingosines and S1Ps are partially hydrolyzed phospholipids lacking one fatty acid residue. Such compunds are known as lysophospholipids. Another representative of this family is **lysophosphatidic acid (LPA)** derived from glycerophospholipids. In cellular lipid metabolism, lysophospholipids are biosynthetic intermediates. They are also released by cells and act as auto- and paracrine hormone-like signaling molecules. Many intercellular messengers induce the formation and release of lysophospholipids via mechanisms that still are not known completely. Lysophospholipids are generated both intra- and extracellularly along pathways shown in Figures 4.23 and 4.24 .

For the formation of *extracellular* lysophospholipids, **lysophospholipases D** play a key role. These enzymes are members of a larger family of ectonucleotide

Figure 4.22 Biosynthesis of signaling molecules from sphingolipids Putative functions of the compounds are shown in the gray boxes. Details are found in the text.

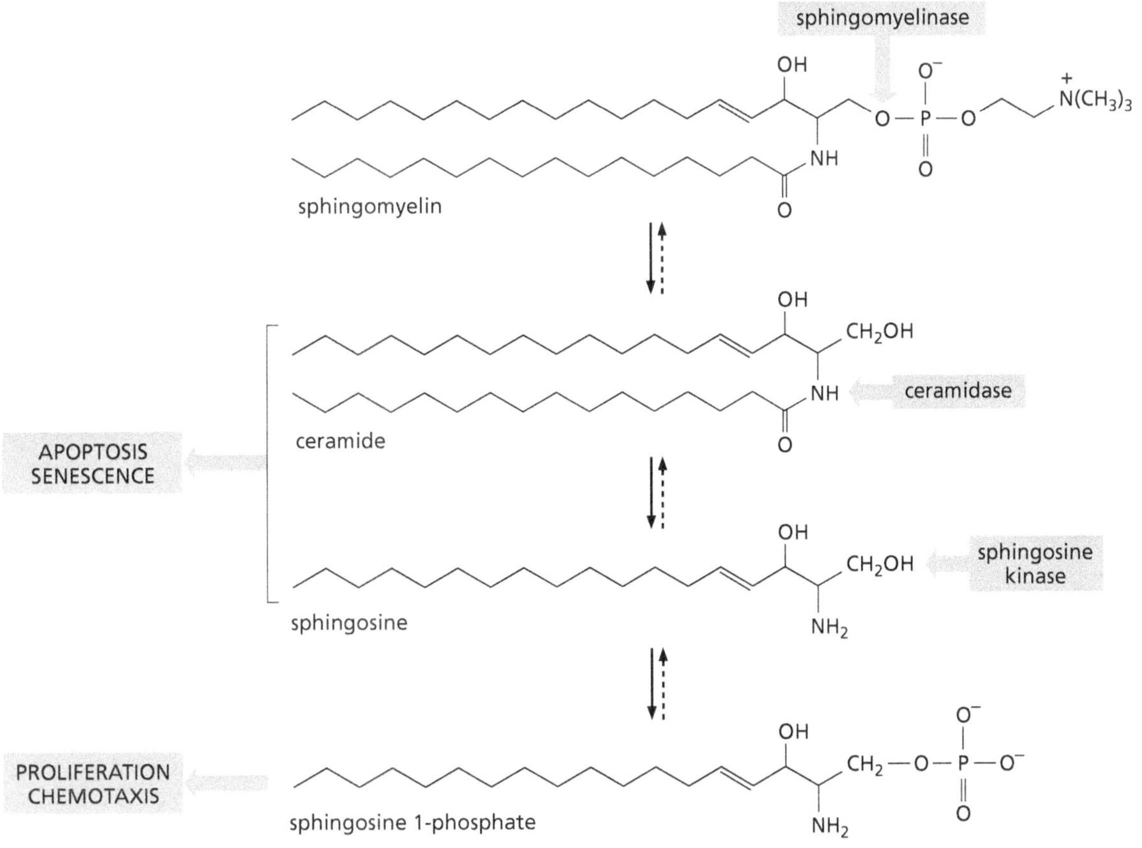

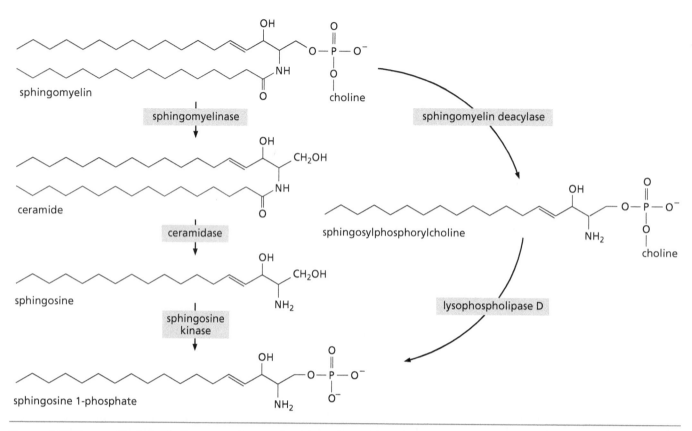

Figure 4.23 Two pathways leading to a release of sphingosine 1-phosphate Sphingosine 1-phosphate (S1P) produced along the sphingomyelinase pathway is probably used mainly as an intracellular messenger. SP1 acting extracellularly is released by lysophospholipases D from sphingosylphosphorylcholine, which is generated from sphingomyelin by enzymatic deacylation.

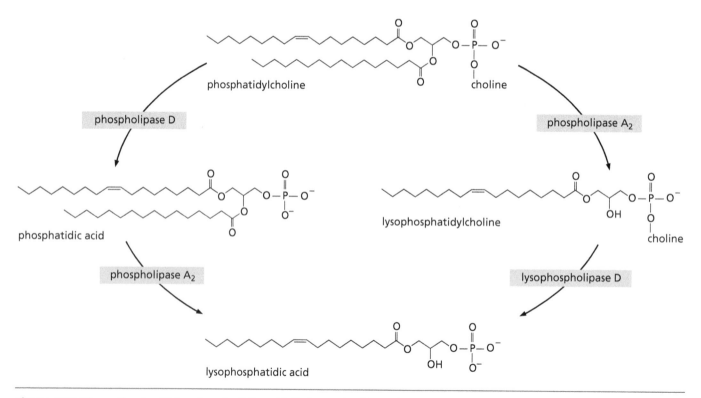

Figure 4.24 Biosynthesis of lysophosphatidic acid Pathways leading to intra- and extracellular formation of lysophosphatidic acid are shown. Phospholipases A₂ and D are signal-activated intracellular enzymes, whereas lysophospholipase D and its substrate lysophosphatidylcholine are found in the blood plasma.

pyrophosphatases/phosphodiesterases found in both pro- and eukaryotes. They catalyze the hydrolysis of both lysophospholipids and nucleotides. A lysophospholipase D isoform called **autotaxin** is of particular interest since it stimulates the proliferation and motility of tumor (melanoma) cells and thus may promote tumor cell invasion and metastasis. Autotaxin is a transmembrane protein exhibiting a single transmembrane domain and an extracellular catalytic domain that may be proteolytically released into the bloodstream as an active enzyme.

Extracellular S1P is also produced from extracellular sphingosine by an extracellular sphingosine kinase and released from cells through an ABC transporter. Extracellular S1P and LPA interact with GPCRs expressed by mammalian cells in three LPA-specific and five S1P-specific isoforms.* Depending on the subtype, $G_{i,0}$, $G_{q,11}$, and $G_{12,13}$ have been identified as effector G-proteins. Along the corresponding signaling cascades, S1P and LPA promote the survival of cells and S1P even stimulates its own biosynthesis. In addition, they evoke smooth muscle contraction along the $G_{q,11}$–InsP$_3$–Ca^{2+} axis.

Lysophospholipids exhibit pronounced effects on the architecture and motility of cells, playing an important role in embryonic development by inducing morphogenetic movements and the formation of individual cell shapes, in particular in the vascular and immune system. Moreover, by augmenting and integrating the effects of pro-inflammatory mediators, they act as chemotactic signals in inflammatory reactions and in the course of wound healing. The significance of lysophospholipids is underlined by the observation that mice with an inactive (knocked-out) S1P$_1$ receptor gene die before birth, since their network of blood capillaries has not developed properly.

In the immune system, S1P and LPA are major regulators of development, chemokine responsiveness, and recirculation of lymphocytes from lymph nodes and thymus to the periphery. In fact, newly developed immunosuppressive drugs (such as FTY720) are S1P$_1$ receptor ligands inducing receptor down-modulation.

What is the mechanistic background of these morphogenetic and immunoregulatory effects? As extracellular signals, LPA and S1P interact with GPCRs and in turn activate Rho proteins along PI3K- and Ras-controlled signaling pathways and the closely related Rac proteins via $G_{12,13}$-proteins. Both Rho and Rac are small G-proteins controlling the dynamics of the actin cytoskeleton (for details see Section 10.3). In parallel, Ca^{2+} ions become released along the $G_{q,11}$–PLCβ–InsP$_3$ cascade, activating the actomyosin complex and, thus, cell motility as well as secretory processes. As an *intracellular* messenger, S1P also directly resembles InsP$_3$ in that it stimulates Ca^{2+} release from the endoplasmic reticulum albeit via an ill-defined mechanism (Figure 4.25 and Section 14.5.3).

LPA exhibits another peculiarity in that it not only interacts with membrane-bound receptors but also, like a steroid hormone, is able to penetrate the cell membrane, binding to ligand-activated transcription factors: the so-called peroxisome proliferator-activated receptors (PPARs) that are involved in the regulation of lipid metabolism (see Section 8.4.1). Such dual receptor interactions have also been reported for other lipid mediators, such as the eicosanoids (see below).

The family of LPA and S1P receptors include the receptor of the **platelet-activating factor** that couples with $G_{i,0}$- and $G_{q,11}$-proteins. Platelet-activating factor is an unusual phospholipid with an ether bond (Figure 4.26). In an auto- and

---

*In the literature these receptors are frequently found under the outdated name EDG (Endothelial Differentiation Gene) because one of the LPA receptor genes was found to become especially active during endothelial differentiation. The modern names of the receptors are respectively LPAR$_{1-3}$ and S1PR$_{1-5}$.

Figure 4.25 Sphingosine 1-phosphate: an intra- and extracellular messenger Upon receptor-mediated activation (here by the autophosphorylated PDGF receptor tyrosine kinase), sphingosine kinase phosphorylates sphingosine. The sphingosine 1-phosphate thus generated may act as an intracellular messenger, gating a special class of Ca^{2+} channels in the endoplasmic reticulum (ER), or may leave the cell as a paracrine or autocrine tissue hormone interacting with a GPCR. Through the activated receptor, various intracellular signaling cascades become stimulated, leading (together with Ca^{2+}) to the physiological responses shown in the box. Arrestin is a scaffold protein relieving the G-protein-coupling of heptahelical receptors and coupling them to other signaling pathways such as the MAP kinase cascades (explained in Section 5.8).

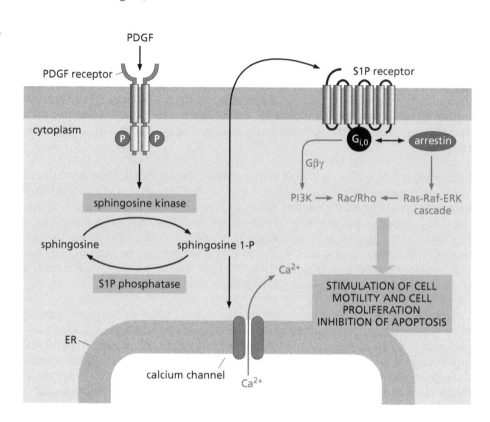

paracrine manner it induces the aggregation of thrombocytes and exhibits other effects as well.

As can be seen in Figure 4.23, **sphingosylphosphorylcholine (SPC)** is a precursor of S1P. This phospholipid, occurring both in blood and in tissues, probably arises from an enzymatic decarboxylation of sphingomyelin. SPC is an intra- and intercellular messenger on its own, since a specific GPCR has been found that stimulates phospholipases A_2, C, and D and the mitogenic Ras–RAF–ERK (MAP kinase) module. In addition, intracellular SPC seems to induce Ca^{2+} release from the endoplasmic reticulum. All in all, the effects of SPC are similar to but not identical with those of S1P.

Another highly efficient membrane-bound messenger molecule with numerous functions is **phosphatidic acid (PA)**, the biosynthetic precursor of LPA. However, in contrast to LPA and S1P, PA is most probably not released by cells but acts exclusively as an intracellular signal. PA is liberated from phospholipids (in particular phosphatidylcholine and phosphatidylethanolamine) by **phospholipase D (PLD)** (Figure 4.27).* The two PLD isoforms found in mammalian cells must not be confused with the above-mentioned lyso-PLDs. PLDs are cytoplasmic proteins of approximately 1000 amino acids containing several domains for interactions with membrane phospholipids (in particular PIP_2), PKC, and small

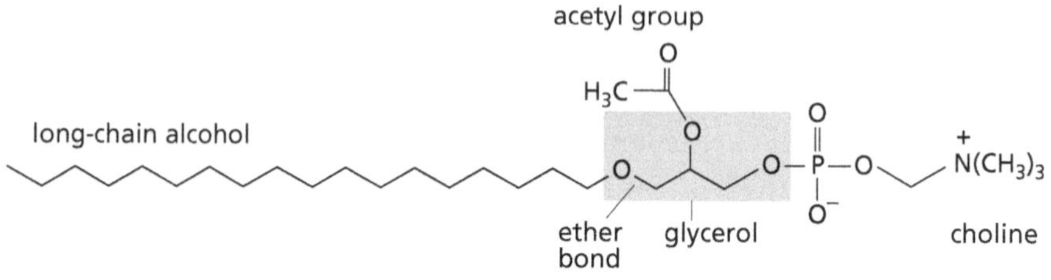

Figure 4.26 Platelet-activating factor PAF

*PA may also be produced by phosphorylation of DAG, catalyzed by DAG kinases (see Sidebar 4.5). However, this metabolic pathway seems to serve primarily signal termination by inactivating DAG.

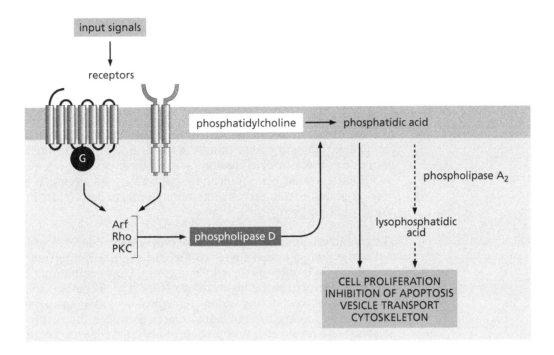

G-proteins of the Arf and Rho families that recruit the enzymes to membranes (Sections 10.4.3 and 10.3). To gain full activity, PLD has to be phosphorylated by either PKCα (see below) or the Rho-activated kinase PRK (see Section 10.3). By this combined action of small G-proteins and protein kinases, PLD is released from a complex with inhibitory proteins involved in the regulation of vesicle transport and cytoskeletal dynamics. Along these pathways PLD is stimulated by a wide variety of signals.

PA is primarily a survival signal that is released under various cellular stress situations. It inhibits apoptosis along the PI3K–PKB signaling cascade, stimulates cell proliferation via the Ras–RAF–MAP kinase pathway, and controls vesicle transport and cytoskeletal dynamics; thus, in its functions it largely resembles LPA (Figure 4.27). Since LPA may be generated from PA, it is sometimes difficult to distinguish between genuine and secondary (LPA-mediated) effects of PA. Among the proteins possibly interacting with PA, only the small G-protein Arf and the protein kinase RAF1 have been identified with certainty.

As another putative signaling molecule released from lysophospholipids by PLD, **cyclic phosphatidic acid (cPA)** has been identified (Figure 4.28). This lipid is widely distributed among eukaryotes. Some of its cellular effects resemble those of PA, LPA, and S1P. cPA induces the release of Ca^{2+} from the endoplasmic reticulum. On the other hand, it inhibits cell proliferation including the invasive growth of tumors. The mechanism of action is still unknown.

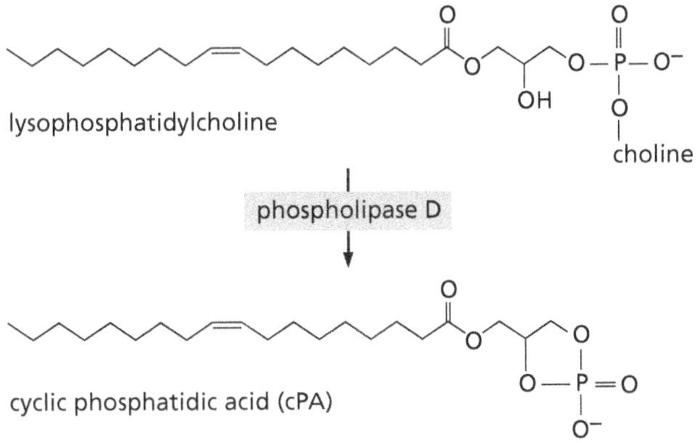

Figure 4.27 Phosphatidic acid as an intracellular messenger Input signals interact with membrane receptors and stimulate (via the small G-proteins Arf and Rho and via protein kinase C) phospholipase D, which hydrolyzes membrane-bound phosphatidylcholine to phosphatidic acid and choline. Phosphatidic acid unfolds its cellular effects either directly or after hydrolysis to lysophosphatidic acid catalyzed by phospholipase A_2. For more details, see text.

Figure 4.28 Cyclic phosphatidic acid

lysophosphatidylcholine

phospholipase D

cyclic phosphatidic acid (cPA)

Lipid mediators derived from unsaturated fatty acids: eicosanoids and their relatives

Polyunsaturated fatty acids such as linoleic acid (C_{18}, two double bonds), α-linolenic acid (C_{18}, three double bonds), and arachidonic acid (C_{20}, four double bonds) are characteristic components of phospholipids. For humans and many animal species, linoleic and α-linolenic acids are essential: they cannot be synthesized but have to be taken up through a vegetable diet. In contrast, our body can synthesize arachidonic acid from linoleic acid by enzymatic lipid-chain elongation. While in earlier times the major function attributed to these compounds was the control of membrane viscosity, they are now known to play a key role in intercellular communication, particularly by serving as precursors of a very large and extremely versatile family of signaling lipids formed primarily along oxidative pathways.

The characteristic structural feature of polyunsaturated fatty acids found in cells is that the double bonds are *cis*-configured and not conjugated: they are separated from each other by a methylene group. Lipids containing such 1,4-*cis,cis*-pentadiene structures are highly oxygen-sensitive, becoming rapidly autoxidized to products of bewildering variety. Biological evolution has benefited from this reactivity by forcing the individual reactions under the control of enzymatic catalysis. As a result, cells now possess a very large collection of lipid messengers called **oxylipins**. Depending on whether C_{18} or C_{20} fatty acids are the precursors, one distinguishes between the subfamilies of **octadecanoids** and **eicosanoids**. While in plants linoleic and linolenic acid-derived octadecanoids predominate, in animals the eicosanoids are the most abundant oxylipins, with arachidonic acid being the major source. The eicosanoid families presently known are summarized in Table 4.1.

To become transformed into oxylipins, the polyunsaturated fatty acids have to be released from membrane phospholipids where they are bound primarily in the *sn*-2 position of glycerol. Mostly (but not exclusively) this reaction is catalyzed by **A_2-type phospholipases (PLA$_2$)** (see Figure 4.18), which in a Ca^{2+}-dependent step are recruited via their C2 domains to membranes. PLA$_2$ are activated by phosphorylation catalyzed by MAP kinases, in particular ERK. In other words, the very large number of signals processed by MAP kinase modules may release polyunsaturated fatty acids and, thus, stimulate the synthesis of octadecanoids and eicosanoids provided Ca^{2+} has been made available by another signal (note that the two signals form a logical AND gate). This means that formation of these lipid mediators is quite a frequent event in cell physiology. In fact, eicosanoids and octadecanoids are released by a wide variety of environmental as well as hormonal, immunological, and neural stimuli.

Free unsaturated fatty acids are extremely short-lived because they are immediately reincorporated into phospholipids or are oxidized to eicosanoids or octadecanoids. Moreover, they undergo rapid autoxidation. Nevertheless, free arachidonic acid has turned out to be a signaling molecule on its own. Two of its effects, the inhibition of Ras–GAP and the activation of Ca^{2+} channels, are discussed in Sections 10.1.3 and 14.5.5, respectively.

Short lifespans hold true also for octadecanoids and eicosanoids, because they also become autoxidized, metabolically inactivated, or incorporated into phospholipids. As a result, their range mostly is limited to the tissue or even the local cell group of origin. They are considered, therefore, to act as local mediators (also called tissue hormones or autocoids), a property they share with other lipid messengers such as S1P and lysophospholipids.

Prostaglandins and thromboxanes

Pharmacological effects of eicosanoids such as stimulation of uterine contraction were discovered in the 1930s and by mistake were attributed to an elusive

Table 4.1 Eicosanoids[a]

Type	Biosynthesis	Major functions
prostaglandins	COX[a], prostaglandin synthases	smooth muscle contraction pain sensation secretory processes luteinization immunosuppression
thromboxanes	COX, thromboxane synthases	platelet aggregation aorta constriction
prostacyclins	COX, prostacyclin synthases	thromboxane antagonists
HPETE and HETE	LOX, cytochrome P450 monooxygenases	inflammation blood pressure control kidney function neurotransmitter?
leukotrienes	LOX, leukotriene synthases	bronchoconstriction leukotaxis
lipoxins	LOX	anti-inflammatory
15-epi-lipoxins	acetylated COX[b]	anti-inflammatory
hepoxilins	LOX	Ca^{2+} release neuromodulation?
trioxilins	LOX	unknown
epoxyeicosatrienoic acids	cytochrome P450 monooxygenases	vasodilatation kidney function
isoprostanes	nonenzymatic	stress signals
anandamide	not clear	neurotransmitter

[a]COX, cyclooxygenase; LOX, lipoxygenase; HPETE, hydroperoxyeicosatetraenoic acid; HETE, hydroxyeicosatetraenoic acid.

[b]Generated by the reaction of COX with acetylsalicylic acid (aspirin).

prostate hormone called "prostaglandin." Although since then every tissue has been found to be able to produce such active agents, the name prostaglandins had become irreversibly established for a group of prototypical eicosanoids. The importance of this field is underlined by the 1982 Nobel Prize in Physiology or Medicine, awarded to Sune K. Bergström, Bengt I. Samuelsson, and John R. Vane for their fundamental research on the biochemistry and physiology of prostaglandins and related compounds.

Characteristic structural elements of prostaglandins are two double bonds and a five-membered ring the structure of which characterizes individual prostaglandin types (Figure 4.28). Together with the related thromboxanes, prostaglandins constitute the prostanoid family of eicosanoids.

Prostaglandins arise by a two-fold addition of molecular oxygen to arachidonic acid, resulting in an unstable cyclic endoperoxide called prostaglandin H_2 that serves further enzymes as a substrate (Figure 4.29). Prostaglandin H_2 synthesis is catalyzed by **cyclooxygenases.** These hemoproteins combine two enzymatic activities, an oxygenase with a peroxidase.

Mammalian cells express two cyclooxygenase isoforms. Cyclooxygenase 1 (COX-1) is a constitutive enzyme satisfying the cell's basic needs for prostanoids. In contrast, cyclooxygenase 2 (COX-2) becomes transiently expressed in most tissues only in situations of increased prostanoid requirement such as infections,

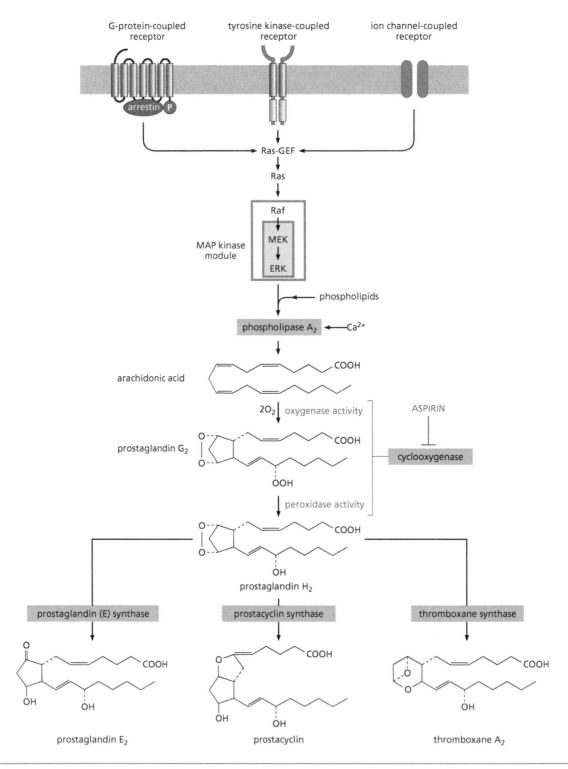

Figure 4.29 Receptor-mediated synthesis of prostanoids Phospholipase A_2, the rate-limiting key enzyme of eicosanoid formation, is activated by Ca^{2+} and phosphorylation (catalyzed by MAP kinases such as ERK) and induces the release of arachidonic acid from membrane phospholipids. By combining the activities of an oxygenase with that of a peroxidase, cyclooxygenase oxidizes arachidonic acid to prostaglandin H_2 which is the common precursor of prostaglandins, prostacyclin, and thromboxanes. Cyclooxygenases are inhibited by aspirin and related drugs. The pathways along which membrane receptors activate Ras are explained elsewhere (see Sections 5.8 and 10.1). The individual prostaglandin types (named prostaglandins A, B, D, E, F, and J) differ from each other by the structure of the oxidized five-member ring system. Types D, E, and F are generated by individual prostaglandin synthases, whereas types A, B, and J are secondary products of types D and E. Prostacyclin synthase and thromboxane synthase convert prostaglandin H_2 into prostacyclin and thromboxane, respectively. The suffix 2 means that a prostanoid carries two double bonds in the side chains; that is, it is derived from arachidonic acid. Prostanoids with one or three double bonds also are found in cells. They are derived from other, less abundant polyunsaturated C_{20} fatty acids.

wounding, and tumor growth. Both COX-1 and COX-2 are inhibited by non-steroidal anti-inflammatory drugs (to be distinguished from anti-inflammatory steroids such as glucocorticoids). The prototype of such drugs is acetylsalicylic acid, better known as **aspirin**. Its well-known effects substantiate the central role eicosanoids play in inflammatory processes and pain (see Section 15.3). COX-2 is overexpressed in most human and animal tumors but also in precancerous lesions where it promotes neoplastic development. Conversely, aspirin and other COX inhibitors have been found to slow down or stop tumor development at a pre-malignant stage, indicating that such drugs may be useful in **cancer prevention**. Consequently, novel synthetic COX-2 inhibitors were greeted with enthusiasm since they lacked the typical side effects of aspirin and its relatives such as gastric and intestinal bleeding. Indeed, these side effects are due to an inhibition of COX-1 rather than of COX-2. Thus, the novel COX-2 inhibitors were expected to represent a breakthrough in the treatment of chronic inflammatory diseases such as rheumatoid arthritis and were heavily promoted to consumers. However, severe and sometimes deadly side effects, in particular on the cardio-vascular system, have brought these "super-aspirins" into disrepute and dramatically dampened the initial enthusiasm.

Another therapeutic effect of aspirin, **prevention of stroke and myocardial infarction**, is due to COX-1 inhibition causing an attenuation of blood clotting. Here the critical step is the aggregation of platelets at sites of blood vessel damage. This reaction, which is induced by a variety of factors, is enormously amplified by an autocrine mechanism involving **thromboxane A$_2$**, a prostanoid produced in platelets from prostaglandin H$_2$ by thromboxane synthase, an enzyme strongly expressed in these cells. The endogenous antagonist of thromboxane A$_2$ is **prostacyclin** (also known as prostaglandin I$_2$), generated from prostaglandin H$_2$ by a prostacyclin synthase that is particularly abundant in endothelial cells. The biosynthesis of both prostanoids starts with COX-1 and is inhibited, therefore, by aspirin and related drugs. However, while endothelial cells overcome this blockade by activating the transcription of the *cox-1* gene, platelets cannot adapt because they lack a nucleus. Thus, repeated intake of aspirin results in a lasting inhibition of platelet aggregation and blood clotting.

The cellular effects of prostanoid signals are primarily mediated by heptahelical transmembrane receptors coupled to G$_s$-, G$_{i/0}$-, and G$_{q,11}$-proteins. The thromboxane receptor, for instance, causes an increase of cytoplasmic Ca^{2+} along the G$_{q,11}$–PLCβ–InsP$_3$ pathway that is the signal for typical thromboxane effects such as platelet aggregation and, in addition, contraction of vascular smooth muscles, in particular those around the aorta. In contrast, the antagonistic effects of prostacyclin are mediated via the G$_s$–adenylate cyclase–cAMP pathway.

15-Deoxy-$\Delta^{12,14}$-prostaglandin J$_2$, an endogenous metabolite of prostaglandin D$_2$, provides an exception to the rule in that it does not interact with a transmembrane receptor but binds directly to, and thus activates, the nuclear transcription factor PPARγ (see Section 8.4.1). This mechanism of action resembles that of a steroid hormone. One may speculate that, besides interacting with membrane receptors, other prostanoids might also be ligands of transcription factors because they are lipophilic enough to penetrate the cell membrane.

Other eicosanoids

Apart from cyclooxygenases, other enzymes catalyze the formation of eicosanoids from arachidonic acid. Several **lipoxygenases** add molecular oxygen stereoselectively to the 1,4-*cis,cis*-pentadiene structures, causing hydroperoxidation together with a shift and a *cis–trans* isomerization of the double bond attacked. The products are hydroperoxyeicosatetraenoic acids (HPETEs), which become rapidly reduced to the corresponding hydroxy compounds (HETEs). HPETEs may also be rearranged into epoxides that, in turn, are either hydrolyzed or added to the tripeptide glutathione. Along this enzymatic pathway 5-HPETE is transformed into **leukotrienes,** a group of physiologically active intercellular messengers. Leukotriene B_4 (Figure 4.30) is one of the most potent chemotactic attractants for leukocytes and plays an important role in inflammatory processes. Via Ca^{2+} release mediated by a $G_{q,11}$-protein-coupled transmembrane receptor and the $PLC\beta$–$InsP_3$ signaling cascade, it facilitates cell motility.

The glutathione adduct leukotriene C_4 evokes, via a $G_{q,11}$-protein-coupled transmembrane receptor and Ca^{2+}, a contraction of airway smooth muscles. Its release—for instance, in the course of an allergic reaction—is considered to be a major cause of bronchial asthma. In this respect leukotriene C_4 cooperates with histamine, the primary mediator of allergy, by positive feedback: each agent promotes the release of the other one. Upon contact with an allergen histamine is released from mast cells and, like leukotriene C_4, causes smooth muscle contraction along the $G_{q,11}$–$PLC\beta$–$InsP_3$–Ca^{2+} pathway.

As compared with the leukotrienes, little is known about the physiological functions of most other lipoxygenase-derived eicosanoids. The same holds true for the receptors aside from 8-HETE, which—similar to deoxyprostaglandin J_2—activates a transcription factor of the PPAR family (in this case isoform $PPAR\alpha$) by direct binding. An analogous effect has been described for leukotriene B_4 and $PPAR\gamma$.

Additional enzymes metabolizing arachidonic acid to eicosanoids are **cytochrome P450-dependent monooxygenases.** They catalyze the formation of a bewildering variety of hydroxylated and epoxidized products with widely unknown functions.

A special class of arachidonic acid metabolites not generated by oxidation are the arachidonyl amides with ethanolamine and Gly. These so-called **anandamides** (referring to the Sanskrit word for "bless") have found particular interest as endogenous ligands of receptors originally characterized as cellular binding sites of tetrahydrocannabinol, the marijuana drug (see Section 16.7). In addition, they interact with a cation channel, the vanilloid receptor, that mediates pain sensations (see Section 15.3).

Finally, the **isoprostanes** should be mentioned. These eicosanoids are endogenous products of a nonenzymatic oxidation of arachidonic acid. Isoprostanes are thought to be messengers signaling emergency situations such as oxidative stress, since they arise upon tissue damage caused by severe intoxication, for instance after misuse of alcohol, tobacco, and drugs.

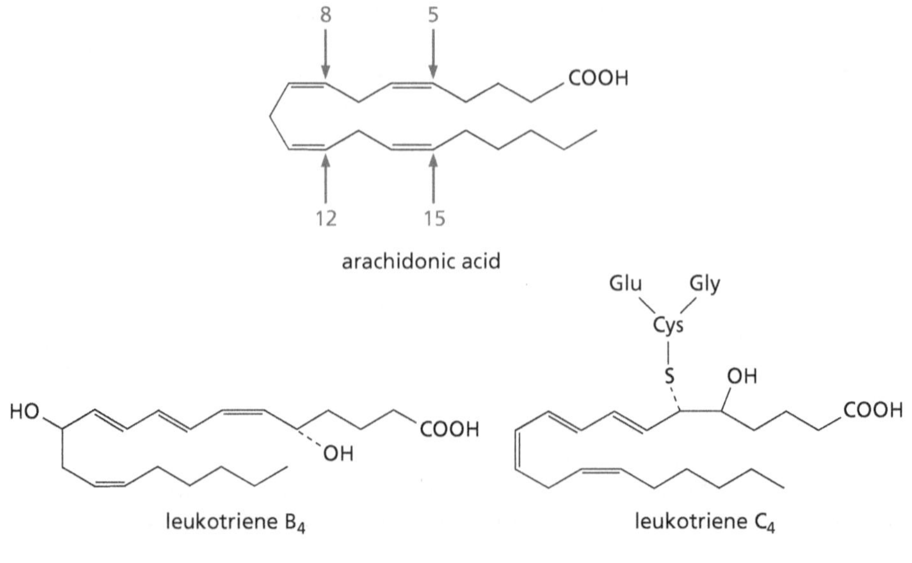

Figure 4.30 Generation of eicosanoids by lipoxygenase-catalyzed oxidation of arachidonic acid Red arrows indicate the points of attack of oxygen catalyzed by individual lipoxygenases of mammals. Leukotrienes B_4 and C_4 are physiologically active eicosanoids generated along the 5-lipoxygenase pathway.

arachidonic acid

leukotriene B_4

leukotriene C_4

Sidebar 4.4 Oxylipins as invertebrate and plant hormones Invertebrates are able to produce and, in contrast to vertebrates, to store eicosanoids. Certain coral species provide probably the richest source of prostaglandins in nature, using them to deter predatory fishes. The structural variability of invertebrate eicosanoids seems to be almost unlimited, whereas our knowledge of their functions is rather fragmentary. Such fragments include the neuromodulatory effect of hepoxilins in the sea snail *Aplysia californica* and a hatching effect of the same compounds on barnacles, as well as the induction by 8-HETE of larval metamorphosis of freshwater and sea water polyps and of egg cell maturation in sea stars. In plants, arachidonic acid is quite rare. Instead, lipid messengers are produced from linoleic and linolenic acid along lipoxygenase-catalyzed pathways. The octadecanoids thus obtained are metabolized to a wide variety of biologically active compounds acting, among others, as stress signals and wound hormones (Figure 4.31). Sea algae use the lipoxygenase pathways to produce small hydrocarbons serving the gametes as sexual pheromones.

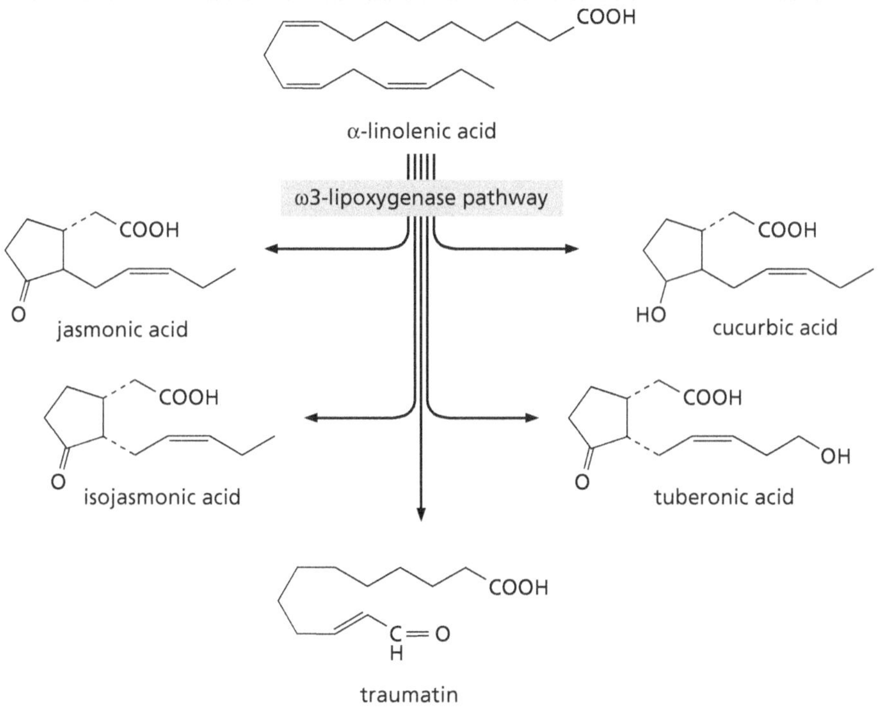

Figure 4.31 Linolenic acid-derived stress signals of plants By oxidation of α-linolenic acid along the ω3-lipoxygenase pathway, signaling lipids are produced that trigger defense reactions and tissue repair. Jasmonic, isojasmonic, tuberonic, and cucurbic acids represent the large family of jasmonate-type phytohormones. They induce the formation of chemical compounds counteracting attacks by fungi, insects, and microbes. Volatile derivatives, such as jasmonic acid methyl ester, seem to act as long-distance signals between remote parts of a plant. In perfumery, such compounds are used as aromatic essences. Traumatin is released upon wounding, stimulating wound repair.

Conclusion: lipid mediators are higher-class second messengers

At first glance the various types and functions of lipid mediators present quite a confusing picture. However, a common denominator is beginning to emerge. In fact, most lipid mediators represent a new type of second messengers that, upon stimulation of a cell by a primary signal (environmental stimuli, hormones), are released into both the cytoplasm and the extracellular space. While in the cell the lipids function as conventional second messengers, their task outside the cell is to coordinate and modulate (amplify, integrate, or dampen) the tissue's overall response to the primary stimulus. Thus they play a key physiological role.

An example is provided by functional blood circulation. When, for instance, a hormone stimulates the function of a gland, an increase in blood flow is essential to rapidly distribute the glandular products released, while muscle activity triggered by a neurotransmitter requires an increased supply of oxygen and glucose via blood circulation. Another example is the inflammatory response, which requires the cooperation of various cell types including white blood cells, platelets, endothelial cells, vascular smooth muscle cells, pain-sensing neurons, fibroblasts, epithelial cells, and others. For the coordination and modulation of this complex interplay, lipid mediators such as the prostanoids play a critical role. Since lipid mediators assist rather than replace primary stimuli, many of them are not essential, although deficiencies may result in pronounced functional disorders.

Summary

Mammalian adenylate cyclases catalyze the formation of cAMP from ATP. Numerous isoforms are known that are either activated or inhibited by a variety of input signals, such as subunits of trimeric G-proteins and calcium ions. Type III cyclases, characterized by a cyclase domain, include mammalian adenylate cyclases and also guanylate cyclases. Phosphodiesterases terminate cyclic nucleotide signaling by catalyzing the hydrolysis of cGMP and cAMP. They are expressed in numerous isoforms that exhibit different specificities and are controlled by a wide variety of input signals. Phosphodiesterases are important targets of pharmaceutical drugs. Mammalian cells reversibly produce a variety of phosphatidylinositol (PI) derivatives that are nonrandomly distributed in cellular membrane systems, generating distinct lipid domains that serve as docking sites for proteins with corresponding interaction domains. The phosphorylated lipids are involved in targeted vesicle transport and signal transduction, with at least some of them playing the role of second messengers. Phosphatidylinositol 3-kinases (PI3Ks) catalyze the generation of membrane-bound second messengers phosphatidylinositol 3,4-bis- and 3,4,5-trisphosphate, which are membrane docking sites and activators of proteins with pleckstrin homology (PH) domains. Individual PI3K isoforms differ in their mechanisms of activation. Phospholipid phosphatases such as PTEN, which antagonize PI3K effects, are tumor suppressor proteins. PI-specific phospholipases C (PLCs) are Ca^{2+}-dependent signal-transducing enzymes that catalyze the release of second messengers diacylglycerol (DAG, a stimulator of protein kinase C) and inositol 1,4,5-trisphosphate (which increases cytoplasmic calcium). Thus, signaling pathways leading to PLC activation are of critical importance for induction of Ca^{2+}-dependent cellular processes. Individual PLC isoforms differ in their mechanisms of activation and tissue distribution. Enzymes of sphingomyelin metabolism are signal-controlled, and their products may mediate stress signals affecting both cell survival and programmed cell death. The physiological significance of these interactions is not entirely clear. Lysophospholipids function both as intracellular second messengers and as auto- and paracrine tissue hormones. They stimulate cell proliferation, cell movements, and changes of the cell shape, particularly during embryonic development, inflammation, and wound repair. Intracellular sphingosine 1-phosphate (S1P) is a Ca^{2+}-mobilizing signal, while phosphatidic acid (PA) released by signal-controlled phospholipase D is a membrane-bound second messenger promoting cell survival in stress situations. Polyunsaturated fatty acids are the source of inter- and intracellular messengers known as oxylipins. The key event in their biosynthesis, release of the precursor fatty acid from membrane phospholipids, is catalyzed by phospholipase A_2, a signal-controlled enzyme activated by Ca^{2+} ions and MAP kinase-catalyzed phosphorylation. Prostanoids, oxylipins that are produced from arachidonic acid via cyclooxygenases, play an important role as endogenous modulators of inflammatory and blood-coagulating effects. Cyclooxygenases are inhibited by nonsteroidal anti-inflammatory drugs and pain relievers such as aspirin and are targets of preventive measures against stroke, myocardial infarction, and cancer. In animal cells, eicosanoids are also generated along lipoxygenase pathways. While leukotrienes have been identified as powerful leukotactic and bronchoconstrictive agents, the physiological effects of most other lipoxygenase products are unknown. In plants, lipoxygenases catalyze the production of a series of α-linolenic acid derivatives involved in microbial and fungal defense and wound repair. Eicosanoids and related lipid mediators seem to be *inter*cellular second messengers modulating the effects of primary signals.

4.5 Guanylate cyclases: not controlled by trimeric G-proteins

The second cyclic nucleotide to which the function of an intracellular messenger has been attributed is cyclic guanosine 3′,5′-monophosphate (cGMP). It is as widely distributed as cAMP, but the tissue concentrations are much lower with the exception of brain and retina, where very high levels are found. cGMP formation from GTP is catalyzed by guanylate (or guanylyl) cyclases. Apart from a similar catalytic domain, guanylate cyclases have little in common with adenylate cyclases. Although guanylate cyclases are not controlled by trimeric G-proteins, they are discussed here in more detail.

There are two families of guanylate cyclases: soluble and receptor types (Figure 4.32). The **soluble guanylate cyclases** (sGCs) are cytoplasmic enzymes found in practically all vertebrate tissues. Their active form is an α,β-heterodimer linked by disulfide bridges. In the N-terminal regulatory domain a heme is bound as a prosthetic group, interacting with nitrogen monoxide (NO) in nanomolar concentrations and, with much lower affinity, with carbon monoxide (CO) but unlike most other hemoproteins not with oxygen. Both NO and CO are gaseous intercellular mediators with a wide spectrum of physiological activities (for details see Section 16.5). Binding of NO/CO causes a conformational change resulting in cyclase activation. Physiologically significant functions of sGCs are relaxation of vascular smooth muscles (resulting in increased blood supply), retrograde regulation of synaptic signal transmission, inhibition of platelet aggregation, and immunomodulatory effects.

Receptor guanylate cyclases exhibit quite a different structure. They do not contain heme (and therefore are not activated by NO and CO) but resemble receptor protein kinases in their dimeric structure and the way enzyme activity is regulated. In fact, the cytoplasmic part of these enzymes harbors both a cyclase and a protein kinase homology domain, the latter being inactive, as it lacks an essential aspartate residue in subdomain VI (see Section 2.6.1). Nevertheless, this kinase homology domain plays a key role in regulating cyclase activity, because upon phosphorylation by an unknown protein kinase of several Ser and Thr residues, it blocks the access of GTP. This auto-inhibition is relieved by an allosteric interaction with ATP that occurs when the ligand is bound to the extracellular receptor domain. The active state of a receptor guanylate cyclase is short-lived because the enzyme becomes rapidly desensitized by dephosphorylation rendering it insensitive to further stimulation. Resensitization requires new phosphorylation. This built-in switch resembling an automatic fuse prevents a prolonged overstimulation of the system. Related safety measures albeit

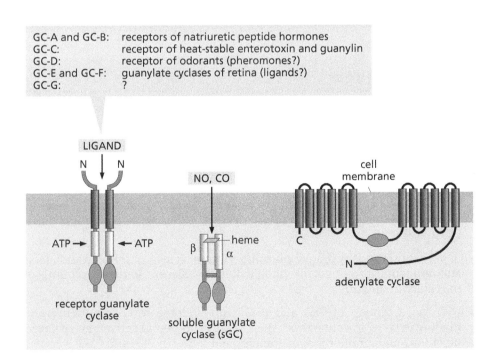

GC-A and GC-B:	receptors of natriuretic peptide hormones
GC-C:	receptor of heat-stable enterotoxin and guanylin
GC-D:	receptor of odorants (pheromones?)
GC-E and GC-F:	guanylate cyclases of retina (ligands?)
GC-G:	?

LIGAND

NO, CO

cell membrane

ATP → ← ATP

β —heme
α

C

N

adenylate cyclase

receptor guanylate cyclase

soluble guanylate cyclase (sGC)

Figure 4.32 Guanylate cyclases
(Left) Receptor guanylate cyclase. The isoforms and corresponding ligands are shown in the box. (Middle) Soluble guanylate cyclase. (Right) Mammalian adenylate cyclase for comparison. Cyclase domains are symbolized by red ellipses, inactive ATP-binding protein kinase domains of the receptor guanylate cyclase by light gray cylinders, and transmembrane domains by dark gray cylinders.

following other rules, are known for a variety of other receptors (see, for instance, Section 16.2).

In mammals, seven isoforms of receptor guanylate cyclases have been found. Isoforms A and B are widely distributed, serving as receptors for the **natriuretic peptides ANP, BNP, and CNP.** These hormones, produced by atrial cells of the heart, control salt and water homeostasis (osmolarity) and thus blood pressure. A major effect is relaxation of vascular smooth muscles resulting in vascular dilation, increased blood flow, and decreased blood pressure. Receptor isoform C is concentrated in epithelial cells of the small intestine. Its ligands are bacterial **enterotoxin** and the tissue hormone **guanylin**, which both activate the chloride ion channel CFTR along a signaling pathway including a cGMP-dependent protein kinase (details in Section 14.7.2). Expression of isoform D is restricted to a special subgroup of olfactory neurons; it probably serves as a receptor of pheromone-like odorants (Section 15.4). Isoforms E and F are found only in the retina (and pineal gland) and play a key role in **visual signal processing** (Section 15.5). Their endogenous ligands, if they exist at all, are not known. The same holds true for the more widely distributed isoform G.

Invertebrates seem to have more guanylate cyclase genes than vertebrates *(C. elegans*, for instance, has 27). They may have quite another structure than the vertebrate enzymes (a form with 2×6 transmembrane domains, resembling the structure of adenylate cyclase, has been found). The functions of the invertebrate enzymes are more or less obscure with the exception of a receptor guanylate cyclase in the membranes of **sea urchin spermatozoa**. Its ligands are peptides (called speract and resact) that are released from the egg cell, stimulating the motility and acrosomal reaction (rendering the sperm cell able to penetrate the egg cell) of sperm cells. The question as to whether such proteins are also expressed by mammalian egg cells is still open. It is easily conceivable that a more in-depth investigation of this subject may have far-reaching consequences for contraception.

Summary

cGMP is a second messenger resembling cAMP. Its bioformation is catalyzed by guanylate cyclases that are expressed either as soluble cytoplasmic hemoproteins or as transmembrane receptors. Soluble guanylate cyclases are specifically activated by NO and CO, whereas receptor guanylate cyclases interact with endo- and paracrine peptide mediators.

4.6 The next level: protein kinases as sensors of second messengers

The second messengers produced along receptor- and G-protein-controlled pathways are input signals of various Ser/Thr-specific protein kinases, namely PKA (cAMP), PKG (cGMP), PKB/Akt and phospholipid-dependent kinase 1 (PIP_2 and PIP_3), and PKC and PKD (DAG; the Ca^{2+} ions released by $InsP_3$ are also activators of several protein kinases). These kinases belong to the large family of AGC kinases (Section 2.6.3) and exhibit some common features:

- they are relatively nonspecific but gain substrate specificity by **anchor proteins** directing the kinases to distinct intracellular sites and to individual substrates

- to become active they require **phosphorylation**, at least of the activation loop and frequently also of a hydrophobic domain adjacent to the C-terminus of the catalytic domain

- with the exception of PKA and PKG, they undergo reversible **membrane binding** to come in contact with their membrane-bound activators (achieved by various membrane-binding domains, not found in PKA and PKG)

Because of their low but regulated substrate specificity, AGC kinases are multi-purpose enzymes and universal relay stations in the data-processing protein network. As such they distribute, transform, and modulate most of the signals transmitted by membrane-bound receptors.

4.6.1 Protein kinase A: a cAMP sensor

Several receptor proteins of cAMP have been identified in eukaryotes, including, first and foremost, PKA (also found as cAMP-dependent protein kinase, cAPK, in the literature), cation channels of sensory cells (Section 14.5.7), and the GDP–GTP exchange factor EPAC of the small G-protein Rap (Section 10.2). To this list the G-protein-coupled cAMP receptor found in the cell membrane of *Dictyostelium discoideum* (for the slime mold, *extracellular* cAMP is a chemotactic aggregation signal) should be added. These proteins have in common a conservative cAMP-binding domain that can be traced back to the predecessor of all cAMP-binding proteins, the transcriptional co-activator CRP of bacteria (Section 3.2.2). As shown by crystallographic studies, binding of cAMP to those domains induces far-reaching conformational changes of the cAMP effector proteins.

The ubiquitously distributed PKAs, expressed in 11 mammalian isoforms, differ from other AGC kinases by two special features: they undergo autophosphorylation of the activation loop instead of being phosphorylated by a separate kinase and the regulatory and catalytic domains are located on separate polypeptides. In fact, PKA is a tetrameric protein consisting of two cAMP-binding (regulatory) and two kinase (catalytic) subunits each. In the absence of cAMP, a pseudosubstrate sequence in the regulatory subunits blocks the catalytic center. The high-affinity binding of two molecules of cAMP per regulatory subunit induces a conformational change resulting in dissociation of the tetramer and release of the active kinase subunits (Figure 4.33).

Protein kinase A modules

Though it is a soluble enzyme, PKA interacts via its regulatory subunits with anchor proteins and thus becomes concentrated at distinct sites and in cellular organelles. These proteins are called **AKAPs** (A-Kinase Anchor Proteins). Mammalian cells express more than 30 isoforms with quite variable interaction domains. The term AKAP is somewhat misleading, because these proteins also anchor other protein kinases and other signal-transducing proteins in addition to PKA. In fact, AKAPs are typical multivalent scaffold proteins connecting signal-transducing proteins to complexes or modules that are found, for instance, in the neighborhood of receptors. Such an arrangement allows individual signal transduction in spite of identical G-protein coupling. Although two

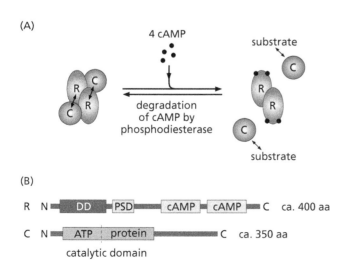

Figure 4.33 Domain structure and activation of protein kinase A
(A) In the absence of cAMP, catalytic subunits (C) of PKA form a stable complex with a homodimer of the regulatory subunits (R). In this complex the substrate-binding site of C is blocked by a pseudosubstrate sequence of R (symbolized by two-headed arrows). Upon binding of four molecules of cAMP, the complex dissociates, exposing the catalytic domains of C. cAMP molecules leaving R are immediately hydrolyzed by phosphodiesterases. (B) Domain structure of R and C. The catalytic domain contains binding sites for ATP and substrate protein. DD, dimerization domain; PSD, pseudosubstrate domain; cAMP, cAMP-binding domain.

Figure 4.34 "Individualization" of transmembrane signaling by A-kinase anchor proteins Different AKAPs couple protein kinase A (PKA) with different G$_S$-protein-coupled receptors (R1, R2) and different substrate proteins (S1, S2). Thus, different signals may evoke individual effects despite the fact that they activate the same G$_S$–adenylate cyclase–cAMP pathway.

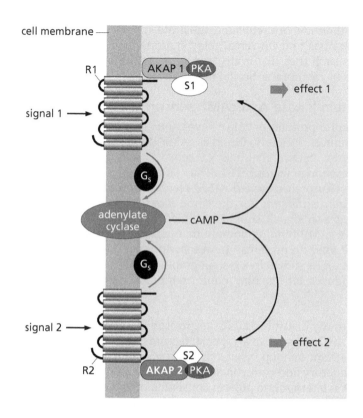

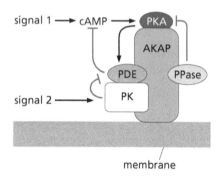

Figure 4.35 Signal-transducing protein module organized by an A-kinase anchor protein In this model, which is based on experimental evidence, the AKAP associated with a membrane brings together protein kinase A (PKA) with another protein kinase (PK; an example is provided by the MAP kinase ERK5), a cAMP-phosphodiesterase (PDE) inhibiting PKA activation, and a protein phosphatase (PPase) inhibiting protein phosphorylation. PDE is assumed to be activated by PKA and inhibited by PK. The module processes at least two different input signals acting on PKA and PK, respectively, and might function as an oscillator, generating a frequency-modulated output signal.

different signals, A and B, both stimulate the G$_S$–adenylate cyclase–PKA pathway, they may evoke different cellular responses (Figure 4.34). Thus, AKAPs provide signaling with a high degree of privacy (see Section 1.6).

Many AKAPs bring together PKA and other kinases with enzymes of signal extinction such as protein phosphatases and cAMP phosphodiesterases. Through such a construction, the PKA signal is dramatically sharpened and may become subject to frequency modulation (for more details see Section 14.5.4). Figure 4.35 shows the prototype of an AKAP module exhibiting a complex pattern of positive and negative feedback loops. The resemblance to a neural or electronic switching element can hardly be ignored.

Selective anchoring in the vicinity of individual substrate proteins is a crucial condition of PKA specificity, since the phosphorylation consensus sequence Arg-Arg-x-Ser/Thr-x is rather abundant and, in the test tube PKA is all but choosy. The number of PKA substrate proteins is large, including structural proteins, enzymes, and transcription factors. Because of this versatility, PKA is involved in a wide variety of physiological processes extending from simple metabolic reactions up to complex brain functions such as learning and memory fixation. Examples are found throughout this book.

The ability of the cell to equip, by specific localization, a relatively nonselective signaling protein with specificity, efficacy, and privacy is not restricted to PKA but represents a general motif of cellular signal transduction. It is one of the combinatorial tricks enabling the cell to achieve a large number of effects with a rather limited arsenal of tools.

4.6.2 cGMP-activated protein kinases

Three types of intracellular cGMP receptors have been identified: cGMP-activated protein kinases, phosphodiesterases, and cGMP-controlled ion channels. In this section only the kinases are treated in detail; information about phosphodiesterases is found in Section 4.4.2, and ion channels are discussed in Section 14.5.2. cGMP-activated protein kinases or PKG (another abbreviation found in the literature is cGK) catalyze Ser/Thr phosphorylation. In contrast to PKA, the catalytic and regulatory domains of PKG share a single polypeptide chain. The enzymes form dimers, ensuring activation by trans-autophosphorylation of the activation loop upon interaction with cGMP. Mammalian cells express two isoforms, PKGI and PKGII (with PKGI existing in two splice variants, α and β). PKGI is widely distributed and has a large number of substrate proteins including several transcription factors (Figure 4.36). In smooth muscle cells, the enzyme transduces the relaxing effect of the NO signal to the contractile apparatus by counteracting myosin light-chain phosphorylation and $InsP_3$-induced Ca^{2+} release (Section 16.5). To this end PKGI phosporylates and activates myosin light chain phosphatase and inactivates $InsP_3$ receptor Ca^{2+} channels in the endoplasmic reticulum. Along this pathway the kinase controls blood pressure and penile erection. For PKGII only one physiological effect is known: phosphorylation and activation of the CFTR chloride channel (Section 14.7.2). The role of PKG in NO-dependent long-term potentiation of synaptic signaling in brain neurons as implied by the high cGMP level in nerve cells is a matter of debate.

An illustrative example of the role of cGMP in general and of PKGI in particular is provided by the hormonal regulation of blood pressure and kidney function. As shown in Figure 4.37, situations leading to an increase of blood pressure stimulate atrial heart cells to release hormones—atrial natriuretic peptides (ANPs)—which in the kidneys induce vasodilatation and water and salt excretion. This effect is mediated by cGMP, produced by an ANP receptor guanylate cyclase, and by PKGI. The major ANP antagonists are the steroid hormone aldosterone and the peptide hormone angiotensin II. The latter is derived from a precursor protein (angiotensinogen) that is cleaved by the proteases renin and angiotensinogen-converting enzyme. Angiotensin II strongly elevates blood pressure by causing vasoconstriction along the $G_{q,11}$–$InsP_3$–Ca^{2+} axis. Moreover, it stimulates the release of aldosterone from the adrenal cortex, where aldosterone production is controlled by the pituitary hormone ACTH. ANP counteracts angiotensin II by inducing vasodilatation and switches off the aldosterone effects by inhibiting, along the cGMP–PKG pathway, the release of renin from the kidneys and by suppressing aldosterone synthesis in the adrenal cortex. Here the major effect of cGMP is to activate phosphodiesterase type 2 that hydrolyzes cAMP, the second messenger of ACTH.

4.6.3 Protein kinases C: a family of multipurpose enzymes

The second messenger DAG is an activator of PKCs. These enzymes are found in all eukaryotes from yeast to humans. Mammalian cells contain nine *pkc* genes and express 10 isotypes including two splice variants (PKCβI and -βII). Some of them, such as PKCα, -δ, and -ζ, are found in almost every tissue, whereas others are expressed by certain cell types only; for instance, PKCγ is produced exclusively by neurons.

In contrast to PKA, regulatory and catalytic domains in PKC share one polypeptide chain. Differences in the regulatory domains lead to a definition of three subfamilies (Figure 4.38).

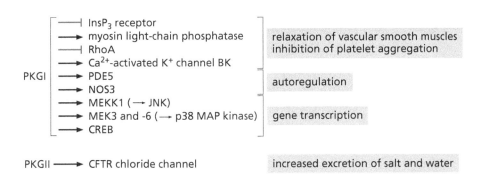

Figure 4.36 Substrates and effects of cGMP-activated protein kinases
Black arrows indicate activation; red bars show inhibition. NOS, NO synthase; CREB, cAMP response element-binding protein; MEK and MEKK, protein kinase components of MAP kinase modules (see Chapter 11).

Figure 4.37 Blood pressure control through cGMP signaling (A) Atrial cells respond to an increase of blood volume, blood pressure, and blood osmolarity by releasing natriuretic peptide (ANP), thus enhancing the excretion of water and salt in the kidney. ANP also inhibits the synthesis of aldosterone in the adrenal cortex. (B) In the kidney ANP activates a trimeric receptor guanylate cyclase (GC). The cGMP formed stimulates cGMP-dependent protein kinase (PKG), which phosphorylates a hyperpolarizing potassium channel (resulting in activation) and the InsP$_3$ receptor Ca^{2+} channel of the endoplasmic reticulum (resulting in inhibition). These effects culminate in increased blood flow due to vasodilatation and inhibition of renin secretion and NaCl reabsorption. Renin is a protease that, together with angiotensin-converting enzyme, releases angiotensin II from a precursor protein. Angiotensin II induces vasoconstriction and promotes the release of aldosterone from the adrenal cortex. In the adrenal cortex, ACTH triggers aldosterone synthesis via the G$_s$–cAMP–protein kinase A pathway. cGMP activates PDE2, thus lowering the cAMP level. This results in a diminished release of aldosterone.

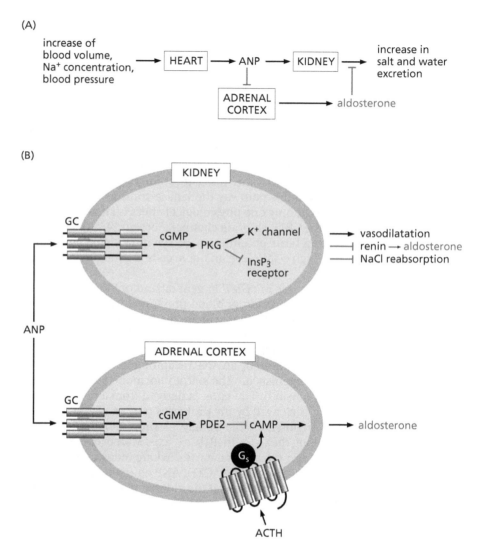

(1) **Conventional PKCs (cPKCs)**. These enzymes require both DAG and Ca^{2+} ions for activation. Association with membrane-bound DAG is accomplished by two **C1 domains** arranged in a tandem structure. C1 domains are named after "<u>C</u>onserved region <u>1</u> of PKC." They are Cys-rich sequences of about 50 amino acids with a DAG-binding pocket stabilized by two zinc ions. The role of DAG is to shield the hydrophilic region of the domain, rendering it hydrophobic enough to interact with membrane phospholipids, in particular with phosphatidylserine. **C2 domains** provide a second binding site for acidic phospholipids, enabling a Ca^{2+}-dependent association of cPKC with membranes in addition to the C1-mediated binding. The Ca^{2+} ions are bound by negatively charged amino acid residues connecting these with the anionic headgroup of phosphatidylserine. As already discussed, C2 domains play an important role in the reversible docking of proteins to membranes. The name protein kinase <u>C</u> refers to the <u>C</u>a^{2+} dependency of these domains, which were originally discovered in PKC. However in contrast to C1 domains, C2 domains are not restricted to DAG-interacting proteins such as PKC but are found in many other proteins (see Section 14.6.1).

(2) **New PKCs (nPKCs).** These enzymes are activated by DAG but do not require Ca^{2+}. The reason is an altered C2 domain that lacks acidic amino acid residues needed for Ca^{2+} binding, whereas the C1 domain is intact.

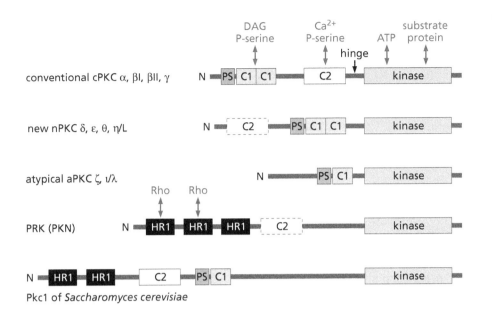

conventional cPKC α, βI, βII, γ

new nPKC δ, ε, θ, η/L

atypical aPKC ζ, ι/λ

PRK (PKN)

Pkc1 of *Saccharomyces cerevisiae*

Figure 4.38 Protein kinase C families The 10 mammalian isoforms of protein kinase C are marked by Greek letters. For activation, conventional cPKCs require diacylglycerol (DAG), phosphatidylserine (P-serine), and Ca^{2+}; new nPKCs require only DAG and P-serine (but not Ca^{2+}); and atypical aPKCs require only P-serine. Like nPKCs, PKC-related kinases (PRK or PKN) have an atypical C2 domain exhibiting a low affinity for Ca^{2+}. In addition, PRKs possess HR1 domains that interact with the small G-protein Rho. In the considerably larger Pkc1 of yeast, the interaction domains of PKC and PRK are combined, indicating that this enzyme is an evolutionary precursor of the PKC families. In PKC the C-terminal kinase domain is separated from the N-terminal regulatory domains by a hinge sequence; proteases may split the molecule here, releasing a constitutively active kinase fragment. In fact, protein kinase C was discovered as a proteolytically activated "protein kinase M" (refers to Mg^{2+} dependency) in brain preparations in 1979. Whether the proteolytic activation has a physiological function is not clear. PS (red box), pseudosubstrate motif.

(3) **Atypical PKCs (aPKCs).** These enzymes require neither DAG nor Ca^{2+} for activation, since they lack a C2 domain and the C1 domain exists only as a single copy that cannot bind to DAG but still interacts with phosphatidylserine.

Mechanism of activation

PKC becomes activated in a rather complicated multi-step process. First the enzyme must associate with the inner side of the cell membrane. To this end, membrane-binding domains such as C1 and C2 are indispensable because all PKC isoforms are intrasterically blocked by a pseudosubstrate sequence that is displaced from the catalytic center only when the kinase comes in contact with acidic membrane phospholipids such as phosphatidylserine, causing a conformational change (Figure 4.39). This enables another protein kinase to gain access and phosphorylate the activation loop of PKC. The "activation loop kinase" has turned out to be identical with the phospholipid-dependent kinase **PDK1** (see below). To gain full activity, PKC needs to be phosphorylated, in addition, at two sites in the substrate-binding domain. This is achieved by autophosphorylation.

The simple phosphorylation consensus sequence Arg-x-x-Ser/Thr-x-Arg-x of PKC is found in many proteins, making them potential PKC substrates. Thus, in

---

Sidebar 4.5 Diacylglycerol effector proteins: more than just protein kinase C High-affinity C1-binding sites for phosphatidylserine and DAG are found not only in PKC but also in other proteins. These include:

- PKD (see below)

- DAG kinases, which catalyze phosphorylation of DAG to PA, and thus extinction of the DAG signal (but at the same time generation of PA signals); mammals express nine isoforms, but only DAGKα, -β, and -γ are activated by DAG

- GRP, a GDP–GTP exchange factor of the small G-protein Ras (see Section 10.1.3)

- chimaerins, GTPase-activating proteins of the small G-protein Rac (Section 10.3.1)

- Munc-13, Mammalian homolog of the protein uncoordinated 13 of *C. elegans*, restricted to neurons; DAG promotes association with the presynaptic membrane, where Munc-13 acts as a chaperone for syntaxin, thus facilitating neurotransmitter release

A characteristic feature of these proteins is that they interact via their C1-domains with plant-derived phorbol esters that mimic the effects of DAG (for details see Sidebar 4.6).

Figure 4.39 Activation of conventional protein kinase C
Activation is a multi-step process. (A) In the inactive enzyme, a pseudosubstrate motif occupies the binding sites for protein kinase PDK1 (which phosphorylates the activation loop of PKC) and for the substrates. (B) The blockade is lifted by binding to membrane-bound DAG and phosphatidylserine (PS). PKC can be transferred into the substrate-binding open conformation by autophosphorylation (auto-P) and phosphorylation by PDK1. (C) In the open conformation, PKC interacts with the substrates ATP and effector protein. P, phosphate group. For the sake of clarity only one of the two C1 domains of cPKC are depicted.

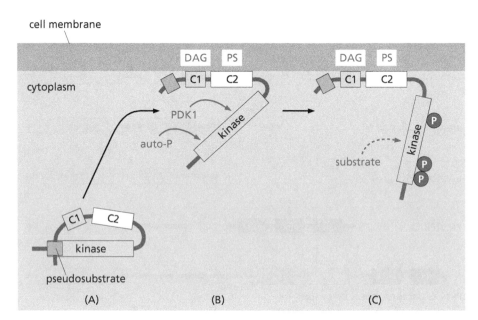

principle PKC is as nonselective as PKA. But again specificity is gained by various anchor proteins. These include the AKAPs, already mentioned above, as well as receptors of active C-kinases (RACKs) and substrates of inactive C-kinases (STICKs). In addition, anchor proteins have been found that bind individual PKC isotypes and modulate their activities. Such anchor proteins direct PKC not only to substrates but also to regulatory proteins, thus acting as scaffolds of signal-transducing protein complexes. A striking example is provided by **InaD.** In the *Drosophila* eye, this scaffold protein assembles a complex of light-processing proteins including PKC, which here has a negative control function (see Section 15.5).

Protein kinase C-related kinases

More distant relatives of PKC are the kinases of the **PRK family** (Protein kinase C-Related Kinases, also known as protein kinases N; Figure 4.38) occurring in several isoforms in all eukaryotes. Like PKC, these versatile enzymes must associate with membranes in order to be phosphorylated by PDK1. However, the mechanism of membrane anchoring differs from that of PKC in that it depends on several N-terminal **HR1 domains** rather than on C1 and C2 domains. HR1 domains specifically interact with the small G-protein Rho, which in its GTP-loaded form reversibly binds to membranes, thus providing a membrane docking site for PRK. In other words, PRKs are Rho effectors: their activation depends on an active Rho signal (for more details on Rho signaling, see Section 10.3). In yeast a primeval Pkc1 has been found. This enzyme combines the properties of PKC and PRK and is considered to represent an evolutionary precursor of both kinase families (Figure 4.38).

4.6.4 Activation loop kinase PDK1: a master key for activation of AGC kinases

By catalyzing activation loop phosphorylation, the kinase PDK1 (Phospholipid-Dependent Kinase 1) is a key regulator of most kinases of the AGC superfamily. Gene-knockout experiments have shown the kinase to be essential for survival. PDK1 differs from most other protein kinases in that there is only one corresponding gene in the human genome and that no PDK1 isoforms (such as a "PDK2") have been found. The name of the enzyme indicates that it interacts with membrane phospholipids in order to phosphorylate its substrates recruited to membranes. However, PDK1 activity does not depend on phospholipids since the enzyme is constitutively active due to autophosphorylation of its activation loop. Thus, input signals only control the intracellular distribution of PDK1 by

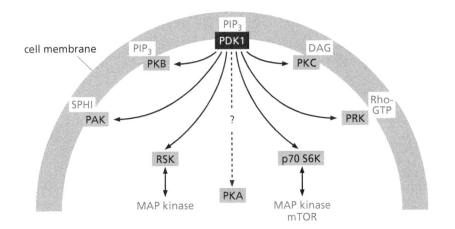

Figure 4.40 Activation loop kinase PDK1 and its substrates PDK1 is reversibly bound at the inner side of the plasma membrane by an interaction of its PH domain with phosphatidylinositol 3,4,5-trisphosphate (PIP_3) and catalyzes the activation loop phosphorylation of various kinases of the AGC family shown in the red boxes. These kinases are recruited to the membrane and prestimulated by membrane-bound activators (red letters; SPHI, sphingosine) or primed by phosphorylation (catalyzed by MAP kinases and mTOR kinases; see Section 9.4.1). The question as to whether PKA is activated by PDK1 or by autophosphorylation only is still open.

generating docking sites at the membrane. These docking sites are 3-phosphorylated PIs such as PIP_3 produced along the PI3K signaling pathways. In fact for PIP3 binding, PDK1 contains a PH domain.

PDK1 substrates, in addition to PKC and PRK, include PKB/Akt, ribosomal S6 kinases p90 RSK and p70 S6K (see Section 9.4.2), p21-activated kinase PAK (see Section 11.4), and others (summarized in Figure 4.40). To be phosphorylated by PDK1, these substrate kinases have to undergo preactivation to displace the pseudosubstrate sequences from the activation loops. This is achieved by an interaction with membrane-bound activators such as DAG and phosphatidylserine (for PKC), PIP_3 (for PKB), Rho-GTP (for PRK), and sphingosine (for PAK) or by phosphorylation (for ribosomal S6 kinases). In other words, the effects of PDK1 depend on both a special conformation and a specific cellular localization of the substrate protein.

4.6.5 Protein kinases D: integrators of phosphatidylinositol 3-kinase, protein kinase C, and G-protein signaling

In the course of PKC research, a novel kinase was encountered that, because of its DAG-binding C1 domains, was thought to be a new PKC isoform and therefore was called PKCμ. However, the enzyme differed so much from PKC that now it is considered to be the representative of a separate protein kinase family. This family is called protein kinase D (PKD) and includes several isoforms.

Like c- and nPKC, PKD has two C1 domains, whereas the auto-inhibitory role of a pseudosubstrate domain is taken over by a **PH domain** interacting with the catalytic center. This PH domain becomes displaced by signals activating PKD, such as the membrane-bound mediators PIP_2/PIP_3 and Gβγ (Figure 4.41). The function of the C1 domains, as in the case of PKC, is to reversibly anchor PKD to membrane-bound DAG, thus fulfilling the condition for activation loop phosphorylation, catalyzed here by DAG-bound a- and nPKC rather than by PDK1. Thus the enzymes of the PKD family are downstream effectors of PKC as well as of PI3K (providing PIP_2/PIP_3), G-proteins (providing DAG and Gβγ), and Tyr kinase-coupled receptors (providing DAG and activating PI3K). This multiplicity of input signaling predestines the PKDs to function as versatile logical gates. Nevertheless, the physiological function of PKD is still not entirely clear. Some data suggest a role in mitogenic signaling and in the control of the Golgi apparatus.

4.6.6 Protein kinase B/Akt: processing of survival and insulin signals

A major performer in the signaling cascades controlled by PI3K is protein kinase B (PKB). Since this Ser/Thr-specific enzyme may mutate to the oncoprotein Akt (referring to a spontaneous thymoma of the AKR mouse strain where this mutation

continued on page 178

Figure 4.41 Protein kinase D
(A) Domain structure of PKD with interacting factors. (B) Activation by DAG and PKC. Via its second C1 domain, PKD binds to DAG in the membrane. This neutralizes the intrasteric inhibition by the PH (pleckstrin homology) domain (which subsequently may interact with Gβγ or PIP₃ localized in the membrane), thus opening the catalytic center for activation loop phosphorylation by PKC and substrate binding.

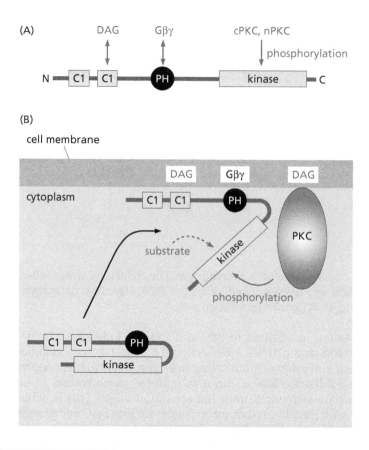

Sidebar 4.6 Tumor-promoting phorbol esters mimic diacylglycerol effects Among second messengers, DAG is unique in that natural products have been found that specifically imitate its intracellular effects. The most prominent of these DAG mimics are the phorbol esters. Phorbol is a diterpene alcohol found as a diester in certain plants of the spurge family (Euphorbiaceae). The prototype of these esters is 12-*O*-tetradecanoylphorbol 13-acetate or **TPA** (frequently found in the literature under the synonym phorbol myristate acetate, PMA; Figure 4.42). TPA is the toxic ingredient of the so-called croton oil from *Croton tiglium*, which was used as a drastic purgative in folk medicine (in alpine countries also known as *Malefizöl* and said to be employed for the elimination of unpleasant relatives). As a highly efficient tumor promoter, TPA had gained great significance in experimental cancer research long before its role as a DAG imitator was discovered.

Tumor promoters are substances that accelerate the development of oncogenically mutated cells into tumors without causing oncogenic mutations by themselves. This effect is explained by the theory of **multi-stage carcinogenesis.** According to this concept several (up to about six) genetic changes such as mutations of proto-oncogenes, deletions of tumor-suppressor genes, and chromosomal defects are required to generate a malignant tumor. The accumulation of such molecular defects correlates with clinical symptoms of cancer development starting from a single mutated cell and proceeding via various forms of tissue dysplasia, benign tumors, and carcinomas *in situ*. The probability is extremely low that in a single cell such subsequent mutations become induced by accidental environmental stimuli such as chemicals, radioactivity, UV light, X-rays, and viruses, to mention the most frequent carcinogenic agents. The role of tumor promoters is to dramatically increase this probability. They do this by inhibiting cell death (apoptosis) and stimulating proliferation, thus enlarging the clone of genetically pre-damaged cells. This **clonal expansion** is thought to find its explanation in the genetic defect of the pre-malignant cells, rendering them less sensitive for apoptotic and more sensitive for mitogenic signals. When continuously stimulated by a tumor promoter, the pre-malignant cells therefore gain a selective advantage over their nonmutated neighbors, although the latter respond as well to the treatment.

In addition to clonal expansion, tumor promoters also increase **genetic instability**, or the probability of genetic damage occurring faster than expected from the burden of environmental genotoxic agents. To this end, tumor promoters stimulate metabolic processes that generate potentially genotoxic agents such as reactive oxygen species and free radicals, and, in addition, may disturb mechanisms of DNA repair.

It is easily conceivable that tumor promoters play a critical role in cancer development, since in their absence the disease most probably would not proceed beyond a harmless stage. A large number of so-called **nongenotoxic carcinogens** in the environment as well as endogenous factors such as sexual hormones, bile acid metabolites, and growth factors or "wound hormones" exhibit tumor-promoting efficacy. All these stimuli have in common up-regulation of signaling cascades that promote cell proliferation at the expense of physiological cell death, as well as metabolic processes yielding genotoxic products. There is a close relationship between tumor promotion and wound healing. In fact, repeated wounding exhibits a strong tumor-promoting effect, while tumor promoters such as TPA provoke a wound response (Figure 4.42). TPA mimics the effects of DAG, thus overactivating, in particular, PKC-controlled pathways. This result obtained by a Japanese group in the 1980s has put PKC research into the center of interest. Indeed, the TPA molecule not only fits the DAG-binding C1 domains but even has a higher binding affinity and is metabolically much more stable than DAG. For these reasons TPA has become the epitome of an experimental tumor promoter and, in addition, a preferred tool of experimental cell biology. Its effect is indeed unique, since no other substance is known to imitate so specifically the effects of a second messenger.

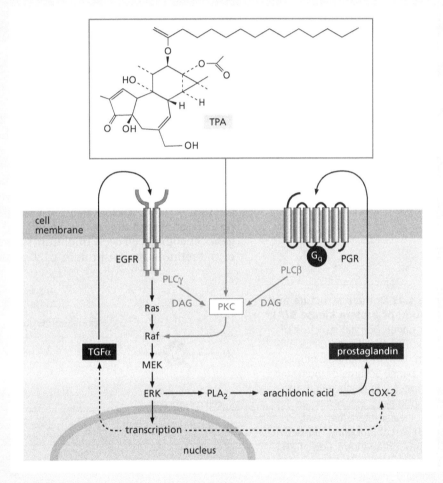

Figure 4.42 Signaling pathways of tumor promotion Repeated wounding is a tumor-promoting stimulus, while the phorbol ester TPA promotes tumor development by interfering with signaling cascades along which the effects of wound hormones are processed. The figure shows two auto-amplifying signaling cascades that, upon wounding, become activated by transforming growth factor α (TGFα), a mitogenic wound hormone, and prostaglandin, a pro-inflammatory mediator. TGFα interacts with the epidermal growth factor receptor (EGFR), a receptor Tyr kinase, and prostaglandin with a G_q-protein-coupled prostaglandin receptor (PGR). In both cases, different isoforms of phospholipase C (PLC) become activated, releasing diacylglycerol (DAG) that activates PKC (red arrows). Among the various downstream reactions of PKC activation, the stimulation of the mitogenic MAP kinase cascade, depending on phosphorylation of the kinase RAF, is shown (for details see Section 10.1.4). This reaction enhances the direct effect of TGFα on the MAP kinase cascade, which is due to an interaction between EGFR and the small G-protein Ras. The cascade leads to an activation of numerous genes (see Section 11.7), including that of TGFα, resulting in positive feedback (left). At the same time, the biosynthesis of prostaglandin is stimulated by phosphorylation and activation of phospholipase A_2 (PLA$_2$), catalyzed by the MAP kinase ERK and transcriptional up-regulation of cyclooxygenase 2 protein (COX-2; for details see Section 4.4.5). As a result, a second positive feedback loop becomes established (right). Both signaling cascades exhibit a pro-mitogenic and anti-apoptotic effect. The metabolism of arachidonic acid is also a potential supply of genotoxic metabolites. In premalignant cells these effects on signal transduction are intensified by an oncogenically mutated Ras protein. Therefore, the conditions of tumor promotion are fulfilled. Both feedback loops may be stimulated by the phorbol ester TPA, rather than by repeated wounding, which specifically mimics the activating effect of diacylglycerol on PKC. The effects of both TPA and DAG critically depend on long-chain fatty acid residues for membrane anchoring and PKC recruitment. DAG and phorbol esters with short acyl groups are inactive, as are nonesterified glycerol and phorbol, respectively.

was found for the first time), it is frequently called PKB/Akt or simply Akt. Mammalian cells express three isoforms. Like other AGC kinases, PKB/Akt is activated in several steps, including at least two phosphorylations and a specific interaction with cell membranes. PKB has no pseudosubstrate sequence. Instead, as in PKD, the catalytic center is blocked by an N-terminal PH domain (Figure 4.43). This auto-inhibition is relieved when the PH domain associates with the second messengers PIP_2/PIP_3 at the inner side of the plasma membrane. So, the activation of PKB always requires cooperation with PI3Ks supplying PIP_2/PIP_3 as input signals (Figure 4.43). In membrane-bound PKB, the active center is accessible for PDK1-catalyzed activation loop phosphorylation. To open the kinase domain for substrate proteins, a still-elusive protein kinase has to phosphorylate, in addition, a Ser residue in a C-terminal hydrophobic motif of PKB (Figure 4.43). Recently some evidence has accumulated to indicate that this elusive kinase might be mTOR, which is itself a PKB substrate (Section 9.4.1).

Like the other AGC kinases, PKB is a multipurpose kinase that occupies a key position in vertebrate signal processing. Among the various effects, two functions stand out (Figure 4.44): transduction of survival signals by suppression of apoptosis in favor of cell proliferation and transduction of insulin signals.

Protein kinase B, cell survival, and cancer

The survival effect is based on the phosphorylation by PKB of diverse pro-apoptotic proteins (see Section 13.2.4) that are thus inactivated (whereas ubiquitin ligase MDM2 inducing the degradation of the pro-apoptotic transcription factor p53 is stabilized; see Section 12.9.3). These factors include the mitochondrial protein Bad, the enzyme caspase 8, and the **FOXO transcription factors** (Forkhead box, type O). The latter, a subgroup of the forkhead factors, are widely distributed in eukaryotes. Their major function is to promote anti-proliferative and anti-survival strategies as an adaptive response to stress, for instance due to starvation. They stimulate the transcription of genes encoding pro-apoptotic proteins such as Bim (Section 13.2.4) and Fas ligand (Section 13.2.3) as well as of genes encoding cell cycle inhibitors such as p21^{Kip} and p27^{Kip} (Section 12.5), and of the retinoblastoma protein p130 (blocking the cell cycle in the absence of

Figure 4.43 Domain structure and activation of protein kinase B/Akt (Upper panel) Domain structure of PKB/Akt (PH, pleckstrin homology domain; HM, hydrophobic motif). (Lower panel) Scheme of multi-step activation. (A) In inactive PKB, the kinase domain is blocked by an intramolecular interaction with PH and HM. (B) Partial unfolding of inactive PKB, through an interaction of PH with membrane-bound phosphatidylinositol 3,4,5-trisphosphate (PIP₃), enables activation loop phosphorylation by PDK1 and phosphorylation of HM by an unknown kinase. (C) PKB with fully unfolded substrate-binding site. P, phosphate.

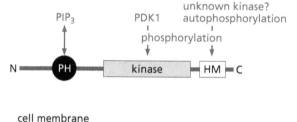

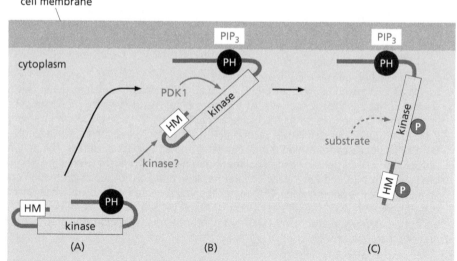

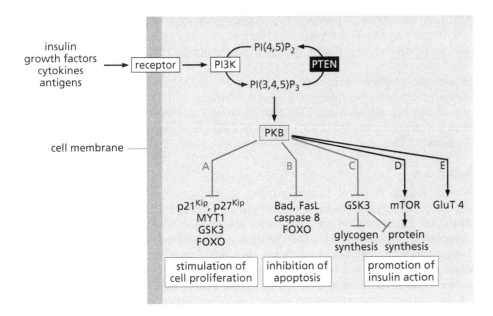

Figure 4.44 Protein kinase B-controlled signaling pathways In a shortened way, the figure shows the activation of PKB by PIP$_3$ generated along the receptor–phosphatidylinositol 3-kinase (PI3K) pathway. The phospholipid phosphatase PTEN antagonizes PKB activation and acts as a tumor suppressor protein. (A) Pro-mitogenic signaling: PKB phosphorylates and inactivates several anti-proliferative factors such as cell cycle inhibitors p21, p27, and MYT1; protein kinase GSK3, suppressing the activity and synthesis of cyclins and mRNA translation; and FOXO transcription factors, inhibiting cyclin transcription and promoting synthesis of the cell cycle inhibitor retinoblastoma protein (see also Section 12.7). (B) Anti-apoptotic signaling: PKB phosphorylates and inactivates pro-apoptotic proteins and FOXO transcription factors that stimulate pro-apoptotic genes (see Section 13.2.4). FasL, Fas ligand. (C–E) Insulin signaling: PKB promotes glycogen synthesis by inactivating GSK3 (C) and recruiting the glucose transporter GluT4 to the membrane (E) and stimulates protein synthesis by inactivating GSK3 and indirectly activating the protein kinase mTOR (D; see Section 9.4.1). For details, see text.

mitogenic signals; see Section 12.7). FOXO factors *inhibit* the transcription of the *cyclin D* gene, a major input terminal of mitogenic signals (Section 12.6). Through PKB-catalyzed phosphorylation, FOXO is hindered from entering the nucleus since it is sequestered in the cytoplasm by **14-3-3 proteins** (see Section 10.1.4 for more details on this sequestering mechanism). This inhibitory effect is seen under nutrient-rich conditions, resulting in an increased release of anabolic hormones such as insulin that activate the PI3K–PKB cascade.

PKB suppresses not only *de novo* synthesis but also activity of the above-mentioned cell cycle inhibitors. At the same time, the kinase turns off glycogen synthase kinase 3 (GSK3), an anti-proliferative enzyme that inhibits mRNA translation (Section 9.4.4) and interrupts the cell cycle by phosphorylation of cyclins D and E thus priming them for ubiquitylation and proteasomal degradation. MYT1, another anti-proliferative protein kinase, is also inactivated by PKB. By phosphorylating the cyclin B–CDK1 complex, MYT1 prevents the transition from G2- into M-phase (see Chapter 12 for details on cell cycle regulation). These observations indicate a key role of PKB in the regulation of tissue homeostasis, the steady state between the birth and decay rates of cells. Thus, it is easily conceivable that a permanent overactivation of PKB due to oncogenic mutation promotes tumor development. The complex activation mode of the enzyme, including at least two "safety locks" (two-fold phosphorylation), may provide a measure against accidental stimulation. Moreover, a protein preventing PKB activation is expected to function as a tumor suppressor. Such a protein is the phospholipid phosphatase **PTEN**, the antagonist of PI3K. In fact, the *pten* gene has been found to be inactive in many human cancers.

Protein kinase B and insulin action

By promoting major anabolic effects such as glycogen and protein synthesis, PKB occupies a central position in the mechanism of action of insulin (Figure 4.44). The insulin receptor is an oligomeric transmembrane protein with cytoplasmic Tyr kinase domains (Section 7.1.3) activating PI3Kα and -β and, in turn, PKB. A major PKB substrate related to insulin action is GSK3, which becomes inactive upon phosphorylation. Since GSK3 inhibits glycogen synthase (Section 5.3.1), PKB indirectly stimulates the deposition of glycogen, an insulin effect *par excellence*. To supply glycogen synthase with its substrate glucose, PKB promotes, in addition, the exposure at the cell surface of the glucose transporter GluT4, that facilitates the uptake of glucose from blood into muscle cells, another characteristic insulin effect (for details see Section 7.1.3). To this end PKB probably phosphorylates and inhibits AS160, a GTPase-activating protein

(GAP) of the small G-protein Rab. As a result Rab is prevented from auto-inactivation and can now stimulate the translocation of GluT4-loaded secretory vesicles to the cell membrane (for details see Section 10.4.4).

Finally, the anabolic actions of insulin on protein synthesis and cell growth are mediated by PKB: the kinase activates a signaling reaction leading to a stimulation of the protein kinase mTOR that increases mRNA translation, at the same time suppressing the inhibitory effect of GSK3 on ribosomal protein synthesis (Section 9.4.1).

Summary

Protein kinase A (PKA) is one among several cellular targets of cAMP. The tetrameric protein dissociates upon binding of cAMP. Because of a simple and abundant phosphorylation consensus sequence, PKA has a wide variety of potential substrate proteins. Specificity is achieved by anchor proteins concentrating the kinase at distinct cellular sites and in the vicinity of selected substrates and other signal-transducing proteins. cGMP-dependent protein kinases are widely distributed and exhibit various effects including vasodilatation, diuresis, and water transport in mucous epithelia. Protein kinase C (PKC) subfamilies differ in their mechanisms of activation. While each requires activation loop phosphorylation by phospholipid-dependent kinase 1 (PDK1) and attachment to acidic membrane lipids via a C1 domain, nPKC needs, in addition, an interaction with membrane-bound diacylglycerol (DAG), and cPKC requires both DAG and Ca^{2+} ions interacting with a C2 domain. Thus, all PKC isoforms operate downstream of PI3K (activating PDK1) and cPKC and nPKC, in addition, operate downstream of PLC (providing DAG and Ca^{2+}). Like PKA, the rather nonselective PKCs gain substrate specificity through anchor proteins. PKC-related kinases operate downstream of signaling pathways leading to activation of the small G-protein Rho. PDK1 operates downstream of PI3Ks since it requires membrane attachment via 3-phosphorylated phosphatidylinositols (PIs). Its main function is to activate protein kinases of the AGC group by catalyzing activation loop phosphorylation. Protein kinases D (PKDs), like PKCs, need DAG for activation. Membrane attachment is facilitated by 3-phosphorylated PIs and Gβγ-subunits, while activation loop phosphorylation is catalyzed by PKC. Thus PKDs are a common downstream target of several signaling pathways, but their cellular functions of the kinases are not clear. Protein kinases of the B/Akt family (PKB/Akt) require 3-phosphorylated PIs for membrane attachment and PDK1 for activation loop phosphorylation; they operate downstream of PI3Ks. They significantly contribute to cell survival (but also to tumorigenesis). A second major function is transduction of anabolic insulin signals.

Further reading

Ahn N (Guest Ed.) (2001) Protein phosphorylation and signaling. *Chem. Rev.* 101, issue 8.

Anliker B. & Chun J (2004) Lysophospholipid G protein coupled receptors. *J. Biol. Chem.* 279, 20555–20558.

Beene DL & Scott JD (2007) A-kinase anchoring proteins take shape. *Curr. Opin. Cell Biol.* 19, 192–198.

Biondi RM & Nebreda AR (2003) Signaling specificity of Ser/Thr protein kinases through docking-site-mediated interactions. *Biochem. J.* 372, 1–13.

Biondi RM (2004) Phosphoinositide-dependent protein kinase 1, a sensor of protein conformation. *Trends Biochem. Sci.* 29, 136–142.

Brazil DP, Yang ZZ & Hemmings B (2004) Advances in protein kinase B signaling: AKTion on multiple fronts. *Trends Biochem. Sci.* 29, 233–242.

Brazil DP & Hemmings BA (2001) Ten years of protein kinase B signaling: a hard Akt to follow. *Trends Biochem. Sci.* 26, 657–671.

Brose N, Betz A & Wegmeyer H (2004) Divergent and convergent signaling by the diacylglycerol second messenger pathway in mammals. *Curr. Opin. Neurobiol.* 14, 328–340.

Cabrera-Vera TM, Vanhauwe J, Thomas TO et al. (2003) Insights into G protein structure, function, and regulation. *Endocr. Rev.* 24, 765–781.

Cantley LC (2002) The phosphoinositide 3-kinase pathway. *Science* 296, 1655–1657.

Carlton JG & Cullen PJ (2005) Coincidence detection in phosphoinositide signalling. *Trends Cell Biol.* 15, 540–547.

Cary SPL, Winger JA, Derbyshire ER et al. (2006) Nitric oxide signaling: no longer simply on or off. *Trends Biochem. Sci.* 31, 231–239.

Cockcroft S (2006) The latest phospholipase C, PLCη, is implicated in neuronal function. *Trends Biochem. Sci.* 31, 4–7.

Cooper DMF & Crossthwaite AJ (2006) Higher-order organization and regulation of adenylyl cyclases. *Trends Pharmacol. Sci.* 27, 426–432.

Cooper DF (2003) Regulation and organization of adenylyl cyclases and cAMP. *Biochem. J.* 375, 517–529.

De Matteis MA & Godi A (2004) PI-loting membrane traffic. *Nat. Cell Biol.* 6, 487–492.

Dohlman HG & Thorner JW (2001) Regulation of G protein-initiated signal transduction in yeast. *Annu. Rev. Biochem.* 70, 703–754.

Francis SH, Turko IV & Corbin JD (2001) Cyclic nucleotide phosphodiesterases: relating structure and function. *Prog. Nucleic Acid Res.* 65, 1–53.

Fukami K (2002) Structure, regulation, and function of phospholipase C isozymes. J. *Biochem. (Tokyo, Jpn)* 131, 293–299.

Goetzl EJ & Rosen H (2004) Regulation of immunity by lysosphingolipids and their G protein-coupled receptors. *J. Clin. Invest.* 114, 1531–1537.

Hall A (ed) (2000) GTPases. Oxford University Press.

Hannun YA & Obeid LM (2008) Principles of bioactive lipid signalling: lessons learned from sphingolipids. *Nature Rev. Mol. Cell Biol.* 9, 139–150.

Hollinger S & Hepler JR (2003) Cellular regulation of RGS proteins: modulators and integrators of G protein signaling. *Pharmacol. Rev.* 54, 527–559.

Kopperud R, Karkstad C, Selheim F et al. (2003) cAMP effector mechanisms. Novel twists for an old signaling system. *FEBS Lett.* 546, 121–126.

Kuhn M (2003) Structure, regulation, and function of mammalian membrane guanylyl cyclase receptors, with a focus on guanylyl cyclase A. *Circ. Res.* 93, 700–709.

Lemmon MA (2008) Membrane recognition by phospholipid-binding domains. *Nat. Rev. Mol. Cell Biol.* 9, 99–11.

Linder JU & Schultz JE (2003) The class III adenylyl cyclases: multi-purpose signaling modules. *Cell. Signalling* 15, 1081–1089.

Lipid Signaling (2002) JB minireviews. *J. Biochem. (Tokyo, Jpn)* 131, 283–306.

Luquain C, Sciorra VA & Morris AJ (2003) Lysophosphatidic acid signaling: how a small lipid does big things. *Trends Biochem. Sci.* 28, 377–383.

Malbon CC, Tao J & Wang H (2004) AKAPs (A-kinase anchoring proteins) and molecules that compose their G-protein-coupled receptor signaling complexes. *Biochem. J.* 379, 1–9.

Manning BD & Cantley LC (2007) AKT/PKB signaling: navigating downstream. *Cell* 129, 1261–1274.

Marks F (2000) Der Stoffwechsel der Arachidonsäure: ein riskantes Bravourstück der Evolution. *Biol. Unserer Zeit* 30, 342–353.

Marks F & Fürstenberger G (eds) (1999) Prostaglandins, Leukotrienes and Other Eicosanoids. Wiley–VCH.

McCudden CR, Hains MD, Kimple RJ et al. (2005) G-protein signaling: back to the future. *Cell. Mol. Life Sci.* 62, 551–577.

Michell RH (2008) Inoositol derivatives: evolution and functions. *Nature Rev. Mol. Cell Biol.* 9, 151–161.

Mukai H (2003) The structure and function of PKN, a protein kinase having a catalytic domain homologous to that of PKC. *J. Biochem. (Tokyo, Jpn)* 133, 17–27.

Neves SR, Ram PT & Iyengar R (2002) G protein pathways. *Science* 296, 1636–1639.

Newton AC (2001) Protein kinase C: structural and spatial regulation by phosphorylation, cofactors, and macromolecular interactions. *Chem. Rev.* 101, 2353–2364.

Newton AC (2003) Regulation of the ABC kinases by phosphorylation: protein kinase C as a paradigm. *Biochem. J.* 370, 361–371.

Rebecchi MJ & Pentyala SN (2000) Structure, function, and control of phosphoinositide-specific phospholipase C. *Physiol. Rev.* 80, 1291–1335.

Riobo NA & Manning DR (2005) Receptors coupled to heterotrimeric G proteins of the G_{12} family. *Trends Pharmacol. Sci.* 26, 146–154.

Smith FD, Langeberg LK & Scott JD (2006) The where's and who's of kinase anchoring. *Trends Biochem. Sci.* 31, 316–320.

Spiegel S & Milstien S (2003) Sphingosine-1-phosphate: an enigmatic signaling lipid. *Nat. Rev. Mol. Cell Biol.* 4, 397–408.

Spiegel S, English D & Milstien S (2002) Sphingosine-1-phosphate: providing cells with a sense of direction. *Trends Cell Biol.* 12, 236–242.

Takuwa Y, Takuwa N & Sugimoto N (2002) The Edg family G protein-coupled receptors for lysophospholipids: their signaling properties and biological activities. *J. Biochem. (Tokyo, Jpn)* 131, 767–771.

Tani M, Ito M & Igarashi Y (2007) Ceramide/sphingosine/sphingosine 1-phosphate metabolism on the cell surface and in the extracellular space. *Cell. Signalling* 19, 229–237.

Tigyi G & Goetzl EJ (eds) (2002) Lysolipid mediators in cell signalling and disease. *Biochim. Biophys. Acta* 1582.

Toker A & Newton AC (2000) Cellular signaling: pivoting around PDK-1. *Cell* 103, 185–188.

Van Blitterswijk WJ, van der Luit AH, Veldman J et al. (2003) Ceramide: second messenger or modulator of membrane structure and dynamics? *Biochem. J.* 369, 199–211.

Van Lint J, Rykx A, Maed J et al. (2002) Protein kinase D: an intracellular traffic regulator on the move. *Trends Cell Biol.* 12, 193–200.

Wettschureck N & Offermanns S (2005) Mammalian G proteins and their cell type specific functions. *Physiol.Rev.* 85, 1159–1204

Woodgett J (2005) Recent advances in the protein kinase B pathway. *Curr. Opin. Cell Biol.* 17, 150–157.

Wymann MP, Zvelebil M & Laffargue M (2003) Phosphoinositide 3-kinase signaling—which way to target? *Trends Pharmacol. Sci.* 24, 366–375.

Yang CF & Kazanietz MG (2003) Divergence and complexities in DAG signaling: looking beyond PKC. *Trends Pharmacol. Sci.* 24, 602–608.

Signal Transduction by Receptors with Seven Transmembrane Domains

In the last chapter we saw that trimeric G-proteins generally receive their input signals from a distinct class of receptors localized in the cell membrane. These are characterized by seven transmembrane domains and are called heptahelical, seven-pass, 7TM, or serpentine receptors. In animal cells they constitute the largest receptor family. They interact with a huge number of intercellular and environmental ligands and in medicine are important targets of therapeutic measures.

In this chapter, the basic mode of action of this receptor type will be explained with the help of a few examples. Moreover, it will be shown that such receptors not only activate trimeric G-proteins but transduce exogenous signals to the signal-processing protein network of the cell also along other pathways.

5.1 An evolutionary model of success

Heptahelical receptors represent one of the most successful models of evolution. Since its emergence as archaeal rhodopsin, this particular protein structure has evolved in animals into the largest receptor family and one of the largest gene and protein families overall. While yeast cells possess only three, the nematode *Caenorhabditis elegans* has more than 1000 such genes, representing the largest gene family of this animal. Encoding mainly chemoreceptors for environmental odorants and taste stimuli, these genes guarantee the special way of life of the worm.

In the human genome, around 800 active and several hundred pseudogenes of heptahelical receptors have been found. The active genes constitute the fourth largest gene family and encode 460 olfactory and 342 non-olfactory receptors. Due to post-transcriptional (such as alternative splicing of pre-mRNA) and post-translational modifications, the actual number of receptors is considerably higher.

It appears that the structural principle of seven interconnected transmembrane domains (resembling the magic number seven!) was "invented" several times in the course of evolution, since no significant sequence homologies exist between the prokaryotic, invertebrate, and mammalian proteins. Such a pronounced evolutionary convergence indicates that this structure is uniquely suited for the interaction with signaling molecules and the transduction of signals across the cell membrane. The circular arrangement of the seven transmembrane domains linked by polypeptide loops starting in an N-terminal extracellular polypeptide tail of variable length obviously offers nearly unlimited variability to adjust to any possible shape of signaling molecules (Figure 5.1). Technology has realized this principle in the combination lock to be opened by a discrete series of numerals. It must be remembered, however, that this story of success holds true for animals only. In plants and prokaryotes the heptahelical receptor type seems

Figure 5.1 Arrangement of transmembrane helices in a heptahelical receptor (mammalian rhodopsin) In this schematic representation of a model based on X-ray analysis, the circular arrangement of the transmembrane helices (cylinders) within the protein (gray cloud) is shown. Note that the ligand (red), a small molecule, is bound in the cavity formed by the transmembrane helices (whereas more bulky ligands such as peptides and proteins interact with the extracellular domains of the receptor protein).

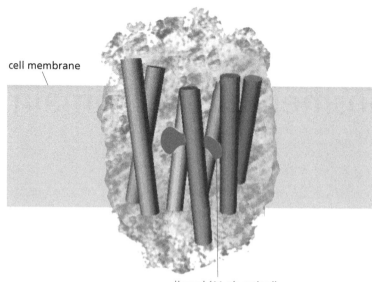

cell membrane

ligand (11-*cis*-retinal)

to be quite rare or absent and signal transduction is dominated by protein kinase-coupled receptors.

Summary

In animal cells, receptor proteins anchored in the cell membrane by seven transmembrane domains represent the most abundant receptor type encoded by one of the largest gene families. Phylogenetic predecessors are the rhodopsins of archaebacteria.

5.2 G-protein-coupled receptors: structure and mode of operation

The best-known function of most receptors with seven transmembrane domains is to catalyze (as GDP–GTP exchange factors) the loading of trimeric G-proteins with GTP upon binding of the ligand. Thus they are called G-protein-coupled receptors (GPCRs). The major signaling cascades originating from these receptor/G-protein couplings have been described in Chapter 4. A schematic overview is seen in Figure 5.2.

It should be rememberd that there is practically no aspect of cellular life – whether development, survival, motility, metabolism or genetic readout – that is not addressed by these signaling cascades.

Table 5.1, which presents a selection of GPCR ligands acting as input signals and their corresponding receptors, illustrates the variability and versatility of this

Figure 5.2 Major signaling pathways downstream of G-protein-coupled receptors: a recapitulation Upon ligand binding, G-protein-coupled receptors (GPCRs) activate the G_s-, $G_{i,0}$-, and $G_{q,11}$-dependent signaling cascades regulating second messenger release and the membrane potential. An individual GPCR may interact either selectively with one or promiscuously with more than one G-protein (see Figure 5.3). The second messengers generated and the downstream actions of the effector proteins (in the red boxes) are shown to the right. As can be seen, the $G_{i,0}$ family comprises the most versatile G-proteins. Black arrows indicate activation; red pistils symbolize inhibition. GIRK, $G_{i,0}$-protein-gated inwardly rectifying K^+ channel.

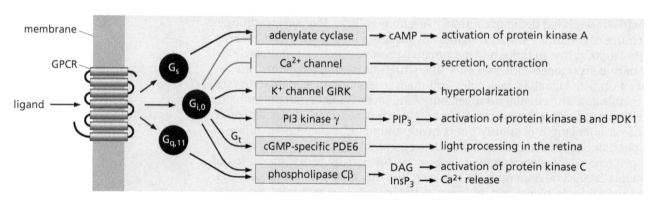

Table 5.1 G-protein-coupled receptors of mammals: a selection[a,b]

Ligand	$G_{q,11}$	$G_{i,0}$	G_s	Alternative receptor coupling
acetylcholine (5)	+	+		cation channel
adenosine (4)		+	+	
adrenaline (9)	+	+	+	
angiotensin (2)	+	+		
bradykinin (2)	+			
calcitonin and relatives (4)	+		+	
calcium ions (1)	+	+		
cannabinoid (2)		+		
chemokine (17)		+		
cholecystokinin/gastrin (2)	+		+	
CRH (2)			+	
dopamine (5)		+	+	
endothelin (2)	+	+	+	
fatty acids (5)	+	+	+	
FSH			+	
GABA (2)		+		anion channel
galanin (3)	+	+		
glucagon			+	
glutamate (8)	+	+		cation channel
histamine (4)	+	+	+	
leukotriene	+	+		
LH	+	+	+	
lysophospholipid	+	+		
melanocortin (5)			+	
melatonin (3)		+		
neuropeptide Y (6)		+		
neurotensin (2)	+			
opioid (4)		+		
oxytocin	+	+		
platelet-activating factor	+	+		
prostanoid (15)	+	+	+	
purine (4)	+			cation channel
rhodopsin (retinal)		+		
serotonin (13)	+	+	+	cation channel
somatostatin (5)		+		
sphingosine 1-phosphate	+	+		
tachykinin (3)	+			
thrombin (4)	+	+		
TRH (2)	+			
TSH	+	+	+	
vasopressin (3)	+		+	
VIP (3)			+	

[a]The number of isoforms unequivocally identified is given in parentheses. For a more comprehensive list the reader is referred to Wettschureck & Offermanns (2005).

[b]CRH, corticotropin-releasing hormone; FSH, follicle-stimulating hormone; GABA, γ-aminobutyric acid; LH, luteinizing hormone; TRH, thyrotropin-releasing hormone; TSH, thyroid-stimulating hormone; VIP, vasoactive intestinal peptide.

receptor type. The number of ligands is striking indeed, ranging from ions, nucleotides, amino acids, amines, and lipids to peptides and proteins. It must not be overlooked, however, that many of the GPCRs are still "orphans": their natural ligands are unknown. Elucidation of their physiological functions is a major task in the field of molecular pharmacology.

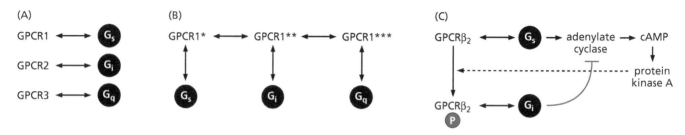

Figure 5.3 Mechanisms of variable G-protein-coupling of GPCRs (A) Different isoforms of a receptor interact selectively with individual G-proteins. (B) A receptor exists in several conformations, each of which interacts with an individual G-protein. (C) Change of G-protein-coupling by post-translational modification shown for the β$_2$-adrenergic receptor as an example. Upon feedback phosphorylation (P) by protein kinase A and exchange of the G-protein, the adenylate cyclase-activating becomes an adenylate cyclase-inactivating receptor.

As can be seen in Table 5.1, almost every ligand interacts with more than one receptor isoform. This **principle of multiple receptors** is reinforced by the principle of **variable G-protein-coupling**. Accordingly, G-protein-coupling is not necessarily restricted to a specific receptor isoform, but the same isoform may interact with different G-proteins.

Many GPCRs exist in several G-protein-specific conformations in equilibrium with each other (Figure 5.3). A change of G-protein specificity may be due to post-translational modifications of the receptor protein. So the β$_2$-adrenergic receptor turns from G$_s$- into G$_i$-coupling upon phosphorylation by protein kinase A (PKA). Such modifications provide almost unlimited possibilities for tissue-specific fine-tuning of signaling. In view of this variability it is not surprising that the pharmaceutical industry is developing countless synthetic ligands of GPCR. They comprise almost half of all medical drugs.

In the postgenomic era it has become possible to construct phylogenetic trees for individual gene families. By this means, five GPCR families have been identified for human cells, which may have evolved from a single primeval gene by gene doubling, exon shuffling, and other molecular mechanisms of evolution. Only 24 out of almost 1000 GPCR genes cannot be assigned to these families. The GPCR families of man are summarized in Table 5.2 (among the ligands, only the most frequent ones are quoted). The largest family by far is the rhodopsin group, covering about 700 receptors arranged in four subgroups.

Table 5.2 Human GPCR families[a]

Rhodopsin family (702)
α-group (89): amine receptors (40) of serotonin, histamine, (nor)adrenaline, dopamine, acetylcholine; prostanoid receptors (15), opsins of retina, receptors of adenosine, melatonin, melanocortin, cannabinoids, lysophosphatidic acid
β- and γ-group (95): receptors of peptide hormones and peptide neurotransmitters
δ-group (518): receptors of pituitary gland hormones, purine/pyrimidine receptors, thrombin receptor, odorant receptors (460)

Secretin receptor family (15)
receptors of secretin, calcitonin, glucagon, parathyroid hormone, corticotropin releasing hormone

Glutamate receptor family (15)
metabotropic glutamate receptors, metabotropic GABA$_B$ receptors, calcium receptors, T1 taste receptors

Adhesion receptor family[b] (24)

Frizzled/TAS2 family (24)
Wnt receptors (10), T2 taste receptors (13), Smo proteins of Hedgehog signaling[c]

[a]The number of different receptor types is given in parentheses.

[b]These receptors are also known as LNB-7TM proteins, with LN referring to an extraordinarily long N-terminal domain and B to a former classification of GPCR families. The extracellular N-terminal domain contains interaction motifs that are found also in cell adhesion molecules and receptors acting in a juxtacrine manner. It has been proposed, therefore, that LNB-7TM proteins mediate interactions between adjacent cells by binding homotypically to each other or heterotypically to other cell surface proteins.

[c]For these receptors G-protein-coupling has not been shown yet.

Sidebar 5.1 G-protein-coupled receptors in yeast and plants and as viral pathogens In the yeast *Saccharomyces cerevisiae,* only three GPCRs have been found that are related neither to bacterial rhodopsins nor to GPCRs of vertebrates. They are sensors of sexual pheromones and of glucose. The action of **yeast pheromones** ranks among the best-investigated cellular events controlled by GPCRs. The pheromones are peptides produced as a- and α-factors by the haploid gametes (a- and α-cells) in the course of sexual reproduction. In the corresponding partner cell, each pheromone induces readiness for mating, cumulating in the fission to a diploid wild-type cell. To this end a-cells express an α-receptor Ste2 and α-cells express an a-receptor Ste3 (the term Ste refers to sterile mutants). Both are heptahelical proteins coupling with the trimeric G-protein Gpa1. The βγ-subunit of Gpa1 activates the transcription of mating-specific genes via a protein kinase Ste20 and a mitogen-activated protein (MAP) kinase module (Figure 5.4). Investigation of this signaling pathway has led to the discovery of the regulators of G-protein signaling (RGS proteins).

The **glucose receptor Gpr1 of yeast** is also G-protein-coupled. It activates a signaling cascade resembling the β-adrenergic signaling pathway in vertebrate cells, though Gpr1 is not related to the β-adrenergic receptor. So, the α-subunit of the Gpr1-coupled G-protein Gpa2 activates adenylate cyclase and the cAMP thus produced induces a fermentation phenotype via a yeast homolog of PKA and a series of protein phosphorylations (Figure 5.4).

As compared with animals, the family of heptahelical receptors is entirely underdeveloped in **plants**. Although about 50 GPCR-like genes have been identified in the *Arabidopsis* genome, their sequences are widely unrelated to those of animal GPCR genes and only two of the corresponding proteins (GCR1 and GCR2) have been found to interact with a Gα subunit. As far as G-proteins are concerned there is little choice, since *Arabidopsis* possesses only one Gα-, one Gβ-, and two Gγ genes, and practically the same holds true for all other plants the genomes of which have been sequenced. Moreover, plants lack the typical downstream effectors such as adenylate cyclase and phospholipase Cβ as well as regulatory proteins such as arrestins and GPCR kinases (see Sections 5.7 and 5.8).

The Gα-subunit (called GPA1 for *Arabidopsis*) seems to fulfill several purposes. One of its functions is to control a mitogenic signaling cascade which probably is induced by plant hormones of the brassinosteroid family. However, the brassinosteroid receptor is coupled to a protein kinase rather than to a G-protein (Section 6.1) and the mechanism of signal transduction remains a contentious issue.

GPA1 is also involved in the control of stomata opening. Stomata are pores at the surface of leaves that serve as the exchange mediums of water and gases. They open and close depending on environmental conditions. These movements are controlled by ion channels regulating the turgor pressure in the stomata cells. **Abscisic acid,** a multifunctional plant hormone, inhibits one of these channels and thus prevents the gating of stomata. This reaction is mediated by GPA1. In fact, abscisic acid interacts with one of the above-mentioned GPCR-like receptors (GCR2).

continued

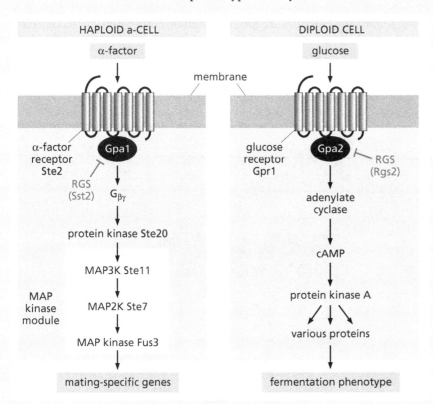

Figure 5.4 GPCR-regulated signaling pathways in yeast (Left) Pheromone signaling pathway of a haploid a-cell. (Right) Fermentation pathway of a diploid cell. Inhibition of G-protein signaling by regulator of G-protein signaling (RGS) proteins, activating intrinsic GTPase activity, is shown in red. For more details, see text.

Several types of human **herpesviruses**, including the cytomegalovirus as well as the tumorigenic forms of Kaposi sarcoma virus HHV8 and Epstein-Barr virus, captured vertebrate GPCR genes (especially those of the chemokine receptor family) in the course of evolution and misuse them as pathogens. As compared with their mammalian counterparts, the viral GPCRs are mutated in such a way that they do not require a ligand; they are constitutively active. Depending on the virus type, the receptors have been found to activate $G_{i,0}$-, $G_{q,11}$-, $G_{12,13}$-, and G_s-controlled pathways, thus triggering cell proliferation, inhibiting apoptosis, promoting cell migration and angiogenesis, and escaping immune surveillance. Considering the role herpesviruses play in a wide variety of potentially life-threatening diseases, the receptors may turn out to be highly important targets of therapeutic measures.

5.2.1 A closer look at the architecture and mode of operation of G-protein-coupled receptors

Figure 5.5 Constructive principle of G-protein-coupled receptors
(A) Three-dimensional scheme. Note the tilted (nonparallel) transmembrane helices (as shown also in Figure 5.1) and the lipid anchor. (B) Schematic "serpent model" of the β-adrenergic receptor including the lipid anchor and the phosphorylation sites of PKA (red) and β-adrenergic receptor kinase (β-ARK, black), as well as the domains of G-protein interaction in the second and third cytoplasmic loops and in the C-terminal part of the molecule. Phosphorylation mostly inhibits the interaction with the G-protein (however, see Figure 5.3 for an exception). Receptors of this type bind small ligands such as adrenaline or acetylcholine in the central channel formed by the helices, whereas more bulky ligands such as peptides and proteins interact with the extracellular loops and the amino-terminal extension.

GPCRs are type 1 transmembrane proteins (N-terminus outside, C-terminus inside the cell). Each transmembrane helix of a GPCR is about 25–35 amino acids long and penetrates the plasma membrane. This membrane binding is frequently reinforced by a long-chain lipid anchor (Figure 5.5). Structural investigations have shown that the interaction of the receptor with the ligand causes a conformational change that induces the GDP–GTP exchange and the dissociation of the G-protein into subunits. As a result, the binding site for the effector protein becomes uncovered in the Gα- or Gβγ-subunit.

How this conformational change proceeds has been investigated in detail by use of rhodopsin as a model. It was found that the ligand – here retinal – is bound by a lysine residue of transmembrane helix 7 as a protonated Schiff base, resembling the situation in bacteriorhodopsin (Section 3.3.5). The positively charged amide group interacts through a "salt bridge" (an electrostatic bond) with a negatively charged glutamate (Glu) residue of transmembrane helix 3. Light induces a *cis–trans* isomerization of retinal, causing the hydrolysis of the Schiff base bond. As a result the salt bridge breaks down and helices 6 and 7 move outward opening a cleft in the cytoplasmic surface of the receptor. This cleft harbours the binding site for the Gα-subunit and the catalytic center for GDP–GTP exchange (Figure 5.6). Although the activation of a receptor by *hydrolysis* of a covalent bond between ligand and protein represents a rare exception, it is assumed that other GPCRs activated by noncovalent ligand *association* undergo similar conformational changes. For a switching process, the direction of switching does not matter.

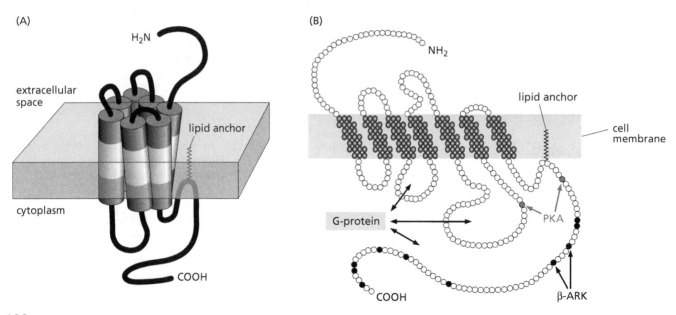

(A)

(B)

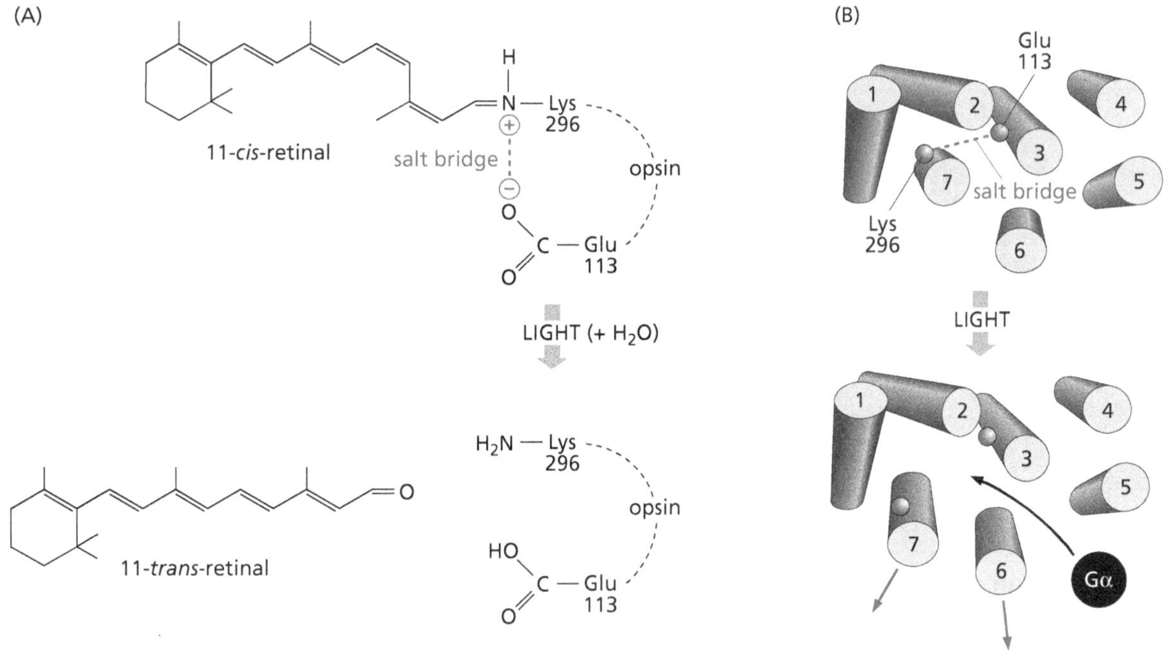

(A)

(B)

5.2.2 Benefits of receptor oligomerization

An increasing body of evidence indicates that, like most receptors, GPCRs exist as homo- and heterodimers (Figure 5.7). This dimerization is mostly constitutive but in some cases may be induced also by ligands. Whether the dimers or the monomers or both are the active configurations remains a matter of debate. Nevertheless, heterodimerization, in particular, has far-reaching consequences for receptor function, altering ligand specificity or modulating G-protein coupling of one monomer by the other one. Below are some examples.

(1) **Change of ligand specificity.** The G_s-coupled T1 taste receptors respond as T1R1–T1R3 heterodimers to certain amino acids mediating the taste sensation "umami," whereas in the combination T1R2–T1R3 they are sweet sensors (Section 15.1.2).

(2) **Change of sensitivity.** Certain $G_{i,0}$-coupled receptors in brain neurons such as dopamine D2 receptor, somatostatin SSTR5 receptor, and opioid receptors DOP and KOP are much more effective as heterodimers of different isoforms than as homodimers.

(3) **Function.** The function of the metabotropic B-type receptors of the neurotransmitter γ-aminobutyric acid (GABA) critically depends on the heterodimerization of the isoforms GABA$_B$-R1 and GABA$_B$-R2, which show only 35% sequence homology. The corresponding homodimers are inactive (Section 16.3).

A special case is the hetero-oligomerization of certain GPCRs with **RAMPs** (Receptor Activity Modifying Proteins). These are single-pass transmembrane proteins with a large extracellular domain. The three RAMP isoforms known have been shown to facilitate the transport of some GPCRs, particularly those of the secretin receptor family (such as receptors for the thyroid gland hormone calcitonin and related peptides), to the cell surface. In addition, they modulate the functions of those GPCRs in such a way that the same receptor acquires different ligand specificity depending on the interacting RAMP isoform. These findings indicate a much greater diversity of GPCR functions than previously imagined.

In the context of receptor oligomerization, the reader should remember that oligomeric protein molecules provide a noise filter, which is an important condition of efficient and specific signal processing (see Section 1.3).

Figure 5.6 Mechanism of rhodopsin activation (A) Breakdown of the intramolecular salt bridge between helices 3 and 7 of the opsin protein caused by hydrolysis of the covalent Schiff base bond between opsin and retinal. (B) View of the seven transmembrane helices (cylinders) from below. The conformational change due to breakdown of the salt bridge opens the interaction site for the α-subunit of the G-protein. Not shown are additional interactions between amino acid residues in helices 2 and 7 as well as in helices 3 and 6, which further stabilize the conformation of the inactive receptor and become disrupted upon activation. For other details, see text.

189

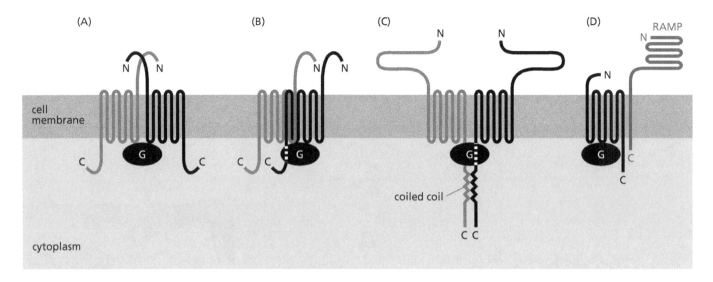

Figure 5.7 Dimerization of G-protein-coupled receptors Dimers are formed by an interaction (A) between extracellular domains (for instance, homodimerization of the β_2-adrenergic receptor); (B) between transmembrane domains (found in various GPCRs); or (C) between intracellular domains (for instance heterodimerization of GABA$_B$ receptor isoforms through a coiled-coil structure). (D) Heterodimerization of a class 2 GPCR (secretin receptor family) with a receptor activity modifying protein (RAMP) is shown. Depending on the RAMP type, a receptor interacts with different ligands. Each dimer probably activates only one G-protein molecule.

Summary

Most heptahelical receptors interact with trimeric G-proteins (G-protein-coupled receptors or GPCRs) catalyzing GDP–GTP exchange and subunit dissociation. By interacting with different receptor isotypes or conformations, most ligands activate more than one G-protein type. While in animals GPCRs are the most abundant receptors, plants express very few GPCRs. Yeast possesses two GPCRs that interact with pheromones and glucose, respectively. Mutated GPCRs are pathogens of herpesviruses. GPCRs are type 1 transmembrane proteins incorporated in membranes through a circularly arranged group of seven transmembrane domains that may be completed by a long-chain lipid residue. Upon ligand binding, this structure is rearranged in such a way that the trimeric G-protein gains access to the catalytic center of the receptor's guanine nucleotide exchange factor domain. Through dimerization of GPCRs, with each other or with receptor activity modifying proteins (RAMPs), a noise filter is established. Moreover, ligand specificity and sensitivity may be changed and in some cases a heterodimer is the active conformation.

Sidebar 5.2 Analysis of protein–protein interactions To demonstrate noncovalent interactions between proteins, such as the heterodimerization of receptors, several methods have been developed. They are of utmost importance for research because cellular signal processing is primarily due to interactions of proteins aiming at network formation.

Preliminary indications of a noncovalent binding of two proteins are obtained in the course of protein purification, when, for example, different proteins cannot be separated by chromatography (**co-chromatography**) or when an antiserum against a protein 1 also precipitates a protein 2 (**co-immunoprecipitation**). The molecular-biological method of the **yeast two-hybrid system** is more reliable (Figure 5.8).

Here one makes use of the fact that the activity of certain transcription factors depends only on the correct distance between DNA-binding and transactivating domains; it is insignificant what type of polypeptide sequence bridges the gap between the domains. An example is provided by the transcription factor GAL4 of yeast, inducing transcription of the β-galactosidase gene. Here the sequence between the N-terminal DNA-binding domain and the C-terminal transactivating domain can be cut out enzymatically and replaced by a foreign protein. To prove an interaction between two foreign proteins 1 and 2, protein 1 is fused with the DNA-binding domain and protein 2 with the transactivating domain and both fusion products are expressed in yeast cells. Upon interaction, the so-called reporter gene *gal4* becomes activated and β-galactosidase is synthesized. For a quantitative evaluation of enzyme activity, a simple test is at hand: upon hydrolysis of the synthetic substrate *o*-nitrophenyl β-D-galactoside, the product *o*-nitrophenol is generated, which is easily assayed due to its yellow color. When the transactivat-

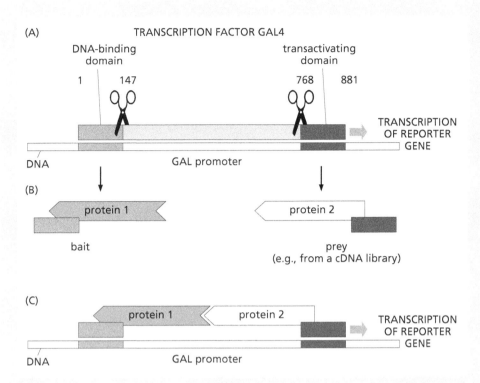

(A) TRANSCRIPTION FACTOR GAL4

DNA-binding domain

transactivating domain

1   147   768   881

TRANSCRIPTION OF REPORTER GENE

DNA   GAL promoter

(B) protein 1

bait

protein 2

prey (e.g., from a cDNA library)

(C) protein 1   protein 2

TRANSCRIPTION OF REPORTER GENE

DNA   GAL promoter

Figure 5.8 Yeast two-hybrid system for analysis of protein–protein interactions The DNA-binding domain (light gray) and the transactivating domain (dark gray) are (A) separated enzymatically from transcription factor GAL4 of yeast at the sites indicated and then (B) fused with proteins to be tested. If the bait and the prey thus obtained interact with each other (C), linking the functional domains of the transcription factor in a correct distance, the reporter gene of the yeast cell transformed with the fusion products becomes activated. For further details, see text.

ing domain of GAL4 is fused with all proteins obtained from a cDNA library, the hybrid molecule of protein 1 with the DNA-binding domain can be used as "bait" to fish out the "prey"; that is, all interacting proteins. Although the method is very sensitive, it frequently produces false-positive results due to nonspecific interactions. Moreover, the results obtained do not necessarily reflect the situation in the living cell. Thus, they have to be verified by microscopic techniques such as **immunofluorescence microscopy** or **multicolor fluorescent protein imaging** (which, on the other hand, both suffer from rather poor spatial resolution).

A method to prove protein–protein interactions more precisely is based on a quantum-physical phenomenon discovered by the German physical chemist Theodor Förster in 1946. It is known as **fluorescence resonance energy transfer (FRET**; a variant is called bioluminescence resonance energy transfer or BRET). The method is based on the following principle. Fluorescent protein 1 transfers a certain portion of its fluorescence energy to protein 2 fluorescing at another wavelength, provided the emission spectrum of the donor overlaps with the excitation spectrum of the acceptor and the distance between the proteins is less than 100Å. If this is the case, the fluorescence intensity of protein 1 is diminished exactly as much as the fluorescence intensity of protein 2 increases (Figure 5.9). This energy transfer can be measured. Since most proteins do not fluoresce on their own, they have to be labeled with

suitable chromophores that either are synthetic compounds or are derived from biological sources.

In fact, certain organisms make use of resonance energy transfer to produce bioluminescence. The light energy is obtained by the oxidation of a luciferin, catalyzed by an oxygen- and ATP-consuming luciferase. Luciferins are prosthetic groups of photoproteins (Figure 5.10). A well-studied example is provided by the pacific jellyfish *Aequorea victoria*. This animal uses the luciferase-catalyzed oxidation of the luciferin coelenterazine embedded in the photoprotein aequorin (emitting blue light) to stimulate, by resonance energy transfer, a green fluorescent protein (GFP) to emit green-yellow light (Figure 5.9). The fluorophore of GFP is a cyclic tripeptide integrated in the protein (Figure 5.10). This system finds many and diverse applications in research. The luciferase reaction may be employed as an extremely sensitive ATP assay (the lower limit is less than 0.1 pmol of ATP). Moreover, since light emission by aequorin is also stimulated by Ca^{2+} ions, it can be used to monitor local changes of cellular Ca^{2+} (see Section 14.5 for an example). Finally, GFP is a highly suitable probe to localize proteins in cells. For this purpose, fusion proteins with GFP are prepared. These hybrid molecules are expressed in the cell and their fluorescence is stimulated by UV light. It has turned out that the tagging with GFP does not significantly change the intracellular distribution of most proteins.

continued

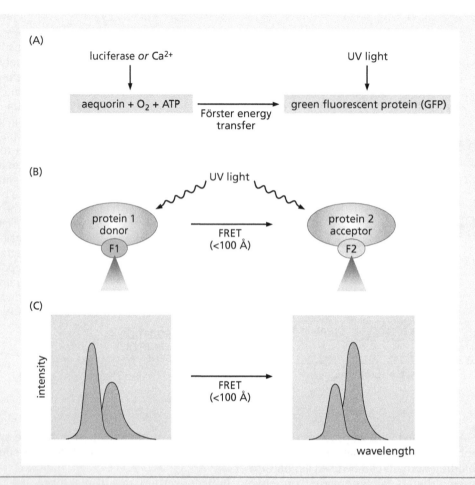

Figure 5.9 Experimental proof of protein–protein interactions by fluorescence resonance energy transfer
(A) Biological resonance energy transfer between the protein aequorin (containing the fluorophore coelenterazine) and the green fluorescent protein GFP in the luminous pacific jellyfish *Aequorea victoria*. The fluorescence of aequorin is excited by the ATPase luciferase (or by Ca^{2+} ions) under oxygen consumption. By Förster resonance energy transfer, the excited aequorin induces fluorescence of GFP. Independently, GFP fluorescence is also induced by UV light. (B, C) Proof of protein–protein interaction by fluorescence resonance energy transfer (FRET). The two proteins expected to interact have been labeled with different fluorophores (F1 and F2), sending out light of different colors, and fluorescence is activated by UV light (B). If the conditions of FRET are fulfilled, the fluorescence intensity of the donor protein decreases to the same degree as the fluorescence intensity of the acceptor protein increases (C).

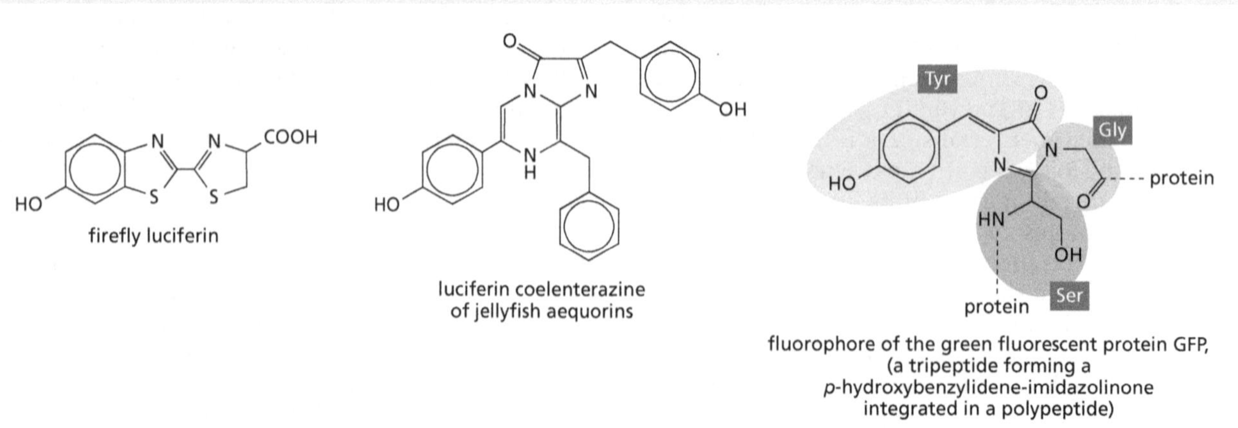

firefly luciferin

luciferin coelenterazine of jellyfish aequorins

fluorophore of the green fluorescent protein GFP, (a tripeptide forming a *p*-hydroxybenzylidene-imidazolinone integrated in a polypeptide)

Figure 5.10 Biological fluorophores

5.3 Adrenergic receptors: sensors of stress and sympathetic signals

To treat the GPCR family in all details would break up the frame of this book and make little sense, since the basic functions common to all receptors of this type may be explained with the help of a few examples. Particularly good examples are the stress effects and the interplay between sympathetic and parasympathetic systems explained in Chapter 4. In fact, the adrenergic receptors of adrenaline and noradrenaline that process stress and sympathetic effects are GPCRs. Vertebrate cells express these receptors in nine isoforms. In the first half of the 20th century, pharmacological studies led to the definition of two receptor families, called α- and β-adrenergic. Since then it has become clear that this classification is based on different G-protein couplings: the receptors of the α_1-family (three isotypes) activate the $G_{q,11}$–PLCβ pathway, α_2-receptors (three isotypes) activate $G_{i/0}$-proteins, and β-receptors (three isotypes) are coupled to G_s-proteins stimulating cAMP formation (Figure 5.11). The receptor isoforms differ in their sensitivities for adrenaline versus noradrenaline as well as in tissue distribution. β_1-Receptors prefer noradrenaline and β_2-receptors prefer adrenaline, whereas β_3-receptors do not distinguish between the two. β_1-Receptors are found predominantly in heart muscle cells; β_2-receptors are particularly abundant in liver, smooth muscle, skeletal muscle, and glandular cells; and β_3-receptors are characteristic of adipose tissue. While α-adrenergic receptors principally mediate sympathetic and stress effects on blood pressure, major functions of β-adrenergic receptors are the stimulation of heart activity and the mobilization of the body's energy reserves.

5.3.1 Biosynthesis and degradation of glycogen: how signals control metabolic processes, part I

To open the endogenous energy reserves in stress situations, β-adrenergic receptors transduce the (nor)adrenaline signal to the metabolic apparatus of the cell, stimulating glycogen and fat metabolism. The investigation of this process represents a milestone in science history. In fact, in the mid-20th century it was shown for the first time how a metabolic process, glycogen metabolism, is controlled by a signal-processing protein network. As pointed out in Section 4.2, these investigations culminated in the discovery of regulatory protein phosphorylation and of cAMP as a second messenger of hormonal signals. Moreover, they led to the first verifiable theory of how a signal is transduced from the cellular periphery to the metabolic machinery inside the cell.

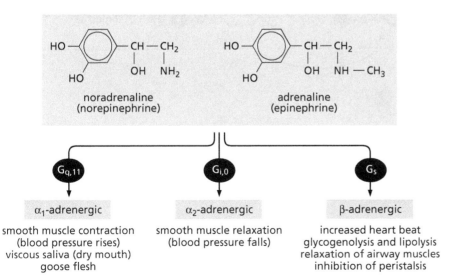

Figure 5.11 Stress and sympathetic effects mediated by G-protein-coupled adrenergic receptors

Glycogen metabolism

Glycogen, a polymer of glucose molecules mainly linked by 1,4-α-glycosidic bonds, is the primary energy depot of the body. While most cells can accumulate small amounts of glycogen, the major stores are in the liver and the skeletal muscles (only when these stores are filled will the energy-richest food component, fat, be deposited). After a meal, the liver and muscles take up glucose from the blood, the liver by a constitutively expressed membrane transporter (the protein GluT2) and the muscle by a related transporter (GluT4), the expression of which has to be induced by insulin. Inside the cell, glucose is phosphorylated at position 6 by hexokinase. The glucose 6-phosphate thus generated either may be degraded along the glycolytic pathways, destined for ATP formation or it becomes transformed into glucose 1-phosphate and UDP-glucose, destined for incorporation into glycogen. The last step is catalyzed by the enzyme glycogen synthase (Figure 5.12).

The synthase has an opponent in glycogen phosphorylase, which catalyzes the phosphorolytic breakdown of glycogen to glucose 1-phosphate. The latter is rearranged to the 6-phosphate, which may be either utilized for glycolysis or dephosphorylated to glucose to be released into the bloodstream (Figure 5.13).

Both glycogen synthase and glycogen phosphorylase are under the strict control of the signal-processing protein network. They are "switched" by signal protein phosphorylation, albeit with an important difference: phosphorylation *activates* the phosphorylase but *inactivates* the synthase. While the activation of glycogen phosphorylase is specifically catalyzed by a phosphorylase kinase, glycogen synthase may be inactivated by various kinases. In stress situations; however, first and foremost, it is controlled by cAMP-dependent PKA activated along the β-adrenergic route.

Figure 5.12 Breakdown and biosynthesis of glycogen For details, see text. P, phosphate residue.

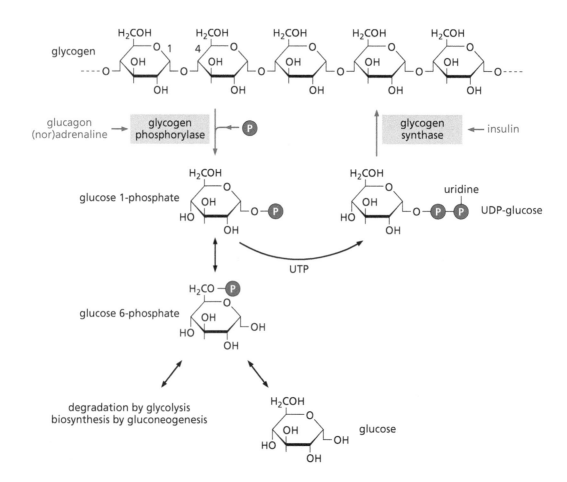

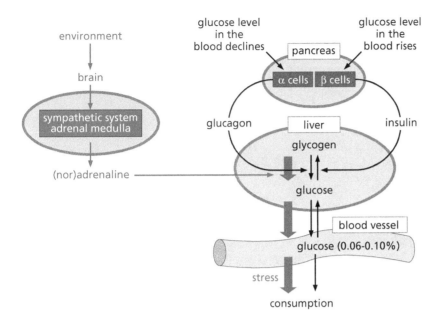

Figure 5.13 Regulation of blood sugar level The glucose level in the blood is an essential parameter of the internal milieu. Normally, it is precisely adjusted to 0.06–0.10% due to a steady state between glucose release (mainly from the liver) and glucose consumption by skeletal muscle and other tissues. A deviation from this mean value may result in hyper- or hypoglycemic shock. Therefore, the steady state is controlled continuously by the two peptide hormones glucagon (promoting glycogen breakdown and glucose release) and insulin (promoting glucose uptake and glycogen formation). Both hormones are synthesized by individual pancreatic cell populations that monitor the rise or fall of the glucose level by means of appropriate sensor proteins. To match the strongly increased glucose consumption in stress situations, glycogen breakdown and glucose release are reinforced by (nor)adrenaline (red). While glucagon and insulin constitute an autonomous regulatory system (being, of course, under the supervision of the vegetative nervous system), the stress axis is dominated by the central nervous system monitoring the environmental (or emotional) situation.

The cellular signal-processing network controlling the activity of the glycogen-metabolizing enzymes is under the supervision of numerous hormones and neurotransmitters and, in addition, is influenced by physical activity. The major players in this game are **glucagon**, promoting glycogen degradation, and **insulin**, promoting glycogen formation. Both are peptide hormones released by individual cell types of the pancreatic Langerhans islets. These cells are equipped with glucose-sensing proteins monitoring any rise or fall of the blood sugar level (see also Section 14.4.2). In a competitive manner both hormones adjust the glucose concentration in blood to a value of 0.06–0.10%. When the energy demand increases, for example, under stress, this regulatory feedback loop becomes overtaxed and has to be reinforced by additional hormones and neurotransmitters, first of all (nor)adrenaline (Figure 5.13).

Although glucagon and (nor)adrenaline interact with different GPCRs, both use the same G_s–cAMP–PKA signaling pathway to stimulate glycogen degradation. In fact, PKA thus activated has two major substrates: glycogen synthase, which is turned off, and the above-mentioned phosphorylase kinase, which is turned on by phosphorylation.

Phosphorylase kinase

Phosphorylase kinase has become famous as the first protein kinase identified. In contrast to PKA, it is a very specific enzyme with only one major substrate, glycogen phosphorylase, which by phosphorylation of a Ser residue is transformed from an inactive b-form into an active a-form (see Chapter 4, Figure 4.5). With four different subunits, phosphorylase kinase has a rather complex quaternary structure. One of these subunits is the Ca^{2+}-sensor protein calmodulin (CaM) that plays an important role especially in skeletal muscle, where it couples the Ca^{2+}-dependent muscle contraction, an energy-consuming process, with glycogen degradation, an energy-supplying process (Figure 5.14; more details on muscle contraction are found in Section 14.6.2).

Glycogen synthase kinase 3

As mentioned above, PKA is only one among several kinases inhibiting glycogen synthase. Other Ser/Thr-directed protein kinases turning down glycogen synthesis are glycogen synthase kinase 3 (GSK3), the "casein kinases" CK1 and CK2 and the 5′-AMP-dependent kinase AMPK. As discussed in Section 9.4, these kinases are regulated along signaling pathways other than those for PKA.

Figure 5.14 Coupling of contraction and glycogen degradation in skeletal muscle Phosphorylase kinase is activated by protein kinase A as well as Ca^{2+}, the signal of muscle contraction. Ca^{2+} enters the cytoplasm through acetylcholine-controlled cation channels (nicotinic receptors) as well as Ca^{2+} channels in the endoplasmic reticulum (not shown; see Section 14.5.3). The calcium sensor is the calmodulin subunit of phosphorylase kinase. The effect of phosphorylation is to increase the Ca^{2+} sensitivity of the enzyme.

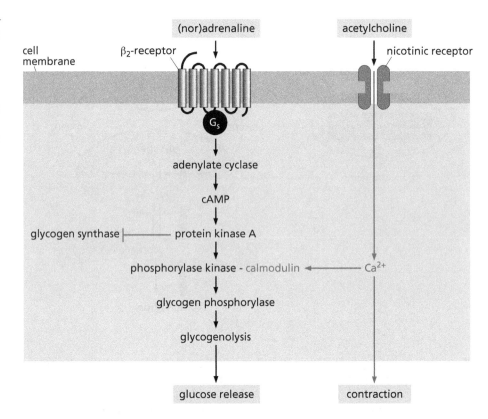

For an understanding of insulin action, GSK3 is of special interest. Until recently, insulin was thought to directly stimulate protein phosphatases, thus reversing the phosphorylation of glycogen synthase and glycogen phosphorylase. Today another mechanism is favored where insulin via PI 3-kinase activates protein kinase B (PKB)/Akt, which in turn inhibits GSK3 (see Section 4.6.6 for more details on PKB/Akt). As a result, constitutively active protein phosphatases gain the upper hand and dephosphorylate the enzymes of glycogen metabolism. This holds true, in particular, for **protein phosphatase PP1**, a (in principle) nonspecific enzyme that nevertheless is selectively bound to glycogen by regulatory glycogen-targeting subunits (see also Section 2.6.4 for the control of protein phosphatases).

As far as phosphorylation of glycogen synthase is concerned, the situation depicted in Figure 5.14 is somewhat oversimplified. In fact, for the inhibition of glycogen synthase, several protein kinases have to cooperate. In a first step the enzyme becomes phosphorylated by CK1 at a specific serine (Ser) residue. This priming provides the condition for additional phosphorylations by PKA. The whole sequence resembles a logical AND gate and has been called **hierarchical phosphorylation**. In a similar manner, GSK3 and AMPK cooperate with protein kinase CK2 (Figure 5.15). CK1 and CK2 are constitutively active regulator kinases, whereas AMPK represents an energy sensor of the cell, becoming active upon a fall of the ATP level and interrupting energy-consuming metabolic reactions such as glycogen synthesis. These three kinases as well as GSK3 are treated in detail in Section 9.4.

Figure 5.15 "Hierarchical phosphorylation" of glycogen synthase Shown are the phosphorylation sites for protein kinase A (PKA), AMP-dependent protein kinase (AMPK), casein kinases CK1 and CK2, and glycogen synthase kinase 3β (GSK3).

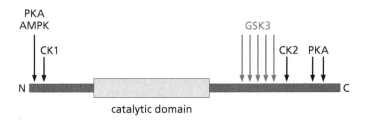

5.3.2 Hormone-stimulated mobilization of fat deposits: how signals control metabolic processes, part II

Besides glucose, long-chain fatty acids are major energy sources of the body. They are stored as glycerol esters in fat cells. The initial step of fat mobilization consists of a hydrolytic degradation of these triglycerides, yielding glycerol and free fatty acids. Like glycogen degradation, this reaction is stimulated by various hormones and catalyzed by a **hormone-sensitive lipase**, an enzyme found predominantly in adipose tissue but also in skeletal muscle and steroid hormone-producing glands. While in adipose and muscle tissue the enzyme is mainly stimulated by (nor)adrenaline or glucagon, in the glands it becomes activated by pituitary hormones such as adrenocorticotropin and releases cholesterol, the precursor of steroids, by hydrolyzing cholesteryl esters. Together these hormones induce lipolysis along the G_s–cAMP–PKA pathway. In fact, as far as the principles of signal transduction are concerned, the hormone-controlled lipolysis represents a simplified version of glycogenolysis. Like glycogen phosphorylase, the hormone-sensitive lipase is activated by protein phosphorylation catalyzed by PKA. In other words, every extracellular signal stimulating cAMP formation along the G_s–adenylate cyclase pathway triggers lipolysis, provided the fat cell expresses the corresponding receptor. Under stress, the mobilization of fat is provoked by (nor)adrenaline interacting with a β_3-adrenergic receptor (Figure 5.16).

However, the phosphorylation of lipase by PKA is not sufficient: to gain access to its substrates, the lipase has to be docked to the fat droplets of adipocytes, where its binding sites normally are blocked by the regulatory protein **perilipin**. To make room for the lipase, perilipin also has to be phosphorylated by PKA. As soon as the hormonal stimulus has ceased, both lipase and perilipin are dephosphorylated again by constitutively active protein phosphatases. This dephosphorylation is augmented by **insulin**, probably through a direct stimulation of protein phosphatases or indirectly through an activation of PKB/Akt that, in turn, phosphorylates and activates a cAMP phosphodiesterase, thus suppressing the cellular cAMP level.

5.3.3 α-Adrenergic receptors: blood pressure control by antagonistic signaling

Under stress, the blood pressure rises: (nor)adrenaline promotes the contraction of vascular smooth muscles, thus reducing the vascular volume. This effect is mediated by α_1-adrenergic receptors and the $G_{q,11}$–PLCβ signaling cascade generating InsP$_3$, which leads to a rise of the cytoplasmic Ca^{2+} level by gating Ca^{2+} channels in the endoplasmic reticulum. Ca^{2+} stimulates the protein kinase **MLCK (Myosin Light-Chain Kinase)** that, like phosphorylase kinase, contains the Ca^{2+} sensor protein CaM as a subunit. In turn, the kinase phosphorylates the regulatory myosin light chains, thus lifting their inhibitory effect on the contractile actomyosin complex (for details see Section 14.6.2).

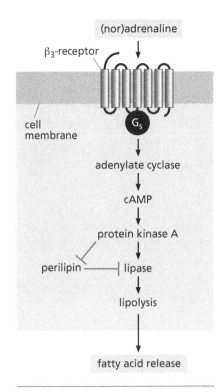

Figure 5.16 Stress hormone-stimulated lipolysis Protein kinase A, activated along the β-receptor–G$_S$ signaling pathway, phosphorylates lipase, which becomes stimulated, and the lipase blocker perilipin, which becomes inactivated. The fatty acids produced thereupon are released into the bloodstream.

Sidebar 5.3 Fatty acid receptors: new targets of antidiabetic drugs? Like sugars, fatty acids also stimulate the release of insulin, which promotes their incorporation into triglycerides and inhibits lipolysis. As has become clear only recently, this effect seems to be controlled by a special class of GPCRs that are activated specifically by saturated and nonsaturated fatty acids ranging from acetic acid to long-chain species. Some of these receptors are strongly expressed in pancreatic β-cells, where they stimulate insulin secretion in response to high-fat food. Another fatty acid receptor known as GPR120 is found, in particular, in intestinal cells. Upon activation it induces, via the PLCβ–$G_{q,11}$–inositol 1,4,5-trisphosphate (InsP$_3$)–Ca^{2+} cascade, the release of a glucagon-like peptide, GLP1. GLP1 acts as a hormone, augmenting (again along a $G_{q,11}$- and Ca^{2+}-controlled signaling pathway) glucose-dependent insulin secretion and anti-apoptotic signaling in the pancreas. As potential targets of novel antidiabetic drugs, such fatty acid receptors presently are attracting much attention.

Even during permanent stress, the blood pressure does not increase indefinitely, thus preventing fatal consequences. This self-control is managed by a negative feedback mechanism by which noradrenaline turns down its own release. To this end noradrenaline interacts with an α_2-adrenergic receptor incorporated in the presynaptic membrane, that is, in the membrane of the same sympathetic neuron that releases noradrenaline (Figure 5.17). The presynaptic receptor is coupled to a $G_{i/0}$-protein and inhibits noradrenaline release along three pathways:

- a $G_{i,0}$-protein-gated inwardly rectifying K^+ channel (Section 14.4.2) becomes gated, hindering the development of an action potential by hyperpolarization of the neuron

- a voltage-dependent Ca^{2+} channel in the presynaptic membrane, which transports the Ca^{2+} required for exocytosis of noradrenaline-filled vesicles, is inhibited

- the same Ca^{2+} channel is *activated* by PKA-catalyzed phosphorylation; this is the target of the third inhibitory pathway, $G_{i/0}$-dependent suppression of adenylate cyclase activity (for details on Ca^{2+} channels see Section 14.5).

Because of its dual function, noradrenaline is called a mixed $\alpha_1\alpha_2$-agonist. Its structure can be altered synthetically in such a way that pure α_1- and α_2-agonists become available. This process had led to a broad field of clinical applications, particularly, of course, in blood pressure therapy (Figure 5.18).

Summary

The constancy of the blood sugar level is based on a steady state between glycogen degradation and glycogen formation. Glycogenolysis is stimulated along a cAMP-dependent pathway involving protein kinase A (PKA) and phosphorylase kinase. Inducers are all signals interacting with G_s-coupled receptors. These effects are opposed by insulin. Due to its multidomain structure, phosphorylase kinase, the key enzyme of glycogenolysis, is controlled by both PKA-dependent phosphorylation and Ca^{2+} ions. By this means, contraction and glycogen degradation are coupled in skeletal muscle. Glycogen synthase is inactivated by

Figure 5.17 α-Adrenergic signal transduction As an example, blood pressure control by regulation of the volume of peripheral blood vessels is shown. In the smooth muscle cell equipped with α_1-adrenergic receptors, contraction is induced along the $G_{q,11}$–PLCβ–InsP$_3$–Ca^{2+} pathway. In the sympathetic neuron innervating the smooth muscle cell, noradrenaline inhibits its own release via a $G_{i,0}$-coupled α_2-adrenergic receptor incorporated in the presynaptic membrane (see Figure 5.18). MLCK, myosin light-chain kinase; DAG, diacylglycerol; ER, endoplasmic reticulum.

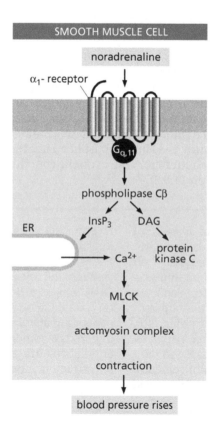
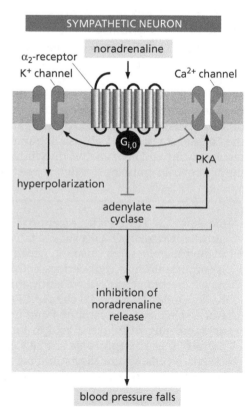

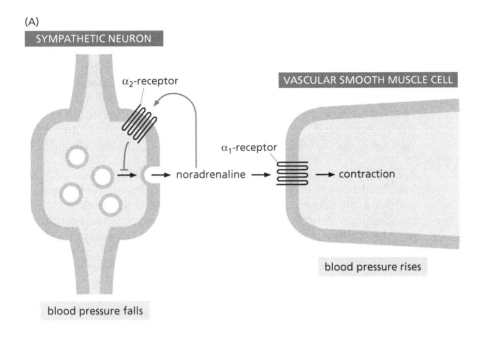

Figure 5.18 Blood pressure control by antagonistic α-adrenergic effects

phosphorylation catalyzed by various protein kinases. Among these, glycogen synthase kinase 3 (GSK3), a multifunctional enzyme, becomes switched off by insulin along the phosphatidylinositol 3-kinase (PI3K)–protein kinase B (PKB)/Akt pathway. This provides a major condition of insulin-dependent glycogen synthesis. Lipolysis, the release of free fatty acids from triglycerides, is catalyzed by hormone-controlled lipases. β-Adrenergic and other hormonal stimuli induce lipolysis through PKA-catalyzed phosphorylation of both the lipase that becomes activated and the inhibitor perilipin that becomes inactivated. In the presence of insulin, both proteins are dephosphorylated and inactivated. α-Adrenergic receptors play an important role in blood pressure control by the sympathetic nervous system. While postsynaptic α_1-receptors coupled to $G_{q,11}$-proteins provoke an increase of the blood pressure through a Ca^{2+}-dependent contraction of vascular smooth muscles, presynaptic α_2-receptors are components of a negative feedback loop in that they inhibit noradrenaline release from the sympathetic neuron along $G_{i,0}$-controlled pathways.

Figure 5.18 Blood pressure control by antagonistic α-adrenergic effects (A) α-Adrenergic synapse between a sympathetic neuron and a vascular smooth muscle cell. (B) α_1- and α_2-adrenergic agonists as compared with noradrenaline.

5.4 Muscarinic acetylcholine receptors: sensors of parasympathetic signals

Acetylcholine is the standard neurotransmitter of the parasympathetic nervous system and, therefore, the major endogenous opponent of noradrenaline. In addition, it transmits contraction-inducing signals at neuromuscular synapses and interneuronal signals in spinal ganglions and in various parts of the central nervous system. In contrast to noradrenaline, acetylcholine interacts with two quite different types of "cholinergic" receptors: while the parasympathetic effects are mediated by a G-protein-coupled type (also called metabotropic type) found in virtually all tissues, the other effects are transmitted by an ion channel-coupled type (also known as ionotropic type, Figure 5.19). For historical reasons the GPCRs are called "muscarinic" and the receptor ion channels are known as "nicotinic" acetylcholine receptors. The latter are treated in detail in Section 16.2.

In the beginning of the 20th century, when acetylcholine was still unknown, cellular binding sites of muscarine, one of the poisons of the fly agaric mushroom, had been postulated in order to explain the characteristic toxic effects of this compound. Although muscarine was later found to be an acetylcholine agonist and the binding sites turned out to be acetylcholine receptors, the name **muscarinic (or muscarine) receptors (M-receptors)** had become firmly established (and the same holds true for the nicotinic receptors, originally postulated to explain the pharmacological effects of nicotine).

Mammals possess five genes encoding M-receptors that are tissue-specifically expressed. The M-receptors couple with different G-proteins: types M_1, M_3, and M_5 interact with $G_{q,11}$, and types M_2 and M_4 activate $G_{i/0}$. Some typical parasympathetic effects triggered by M-receptors are shown in Figure 5.20. Like adrenergic receptors M-receptors are targets of many pharmaceutical drugs. Of historic interest is the contraction of the eye pupils, since it provides an early example of how a signaling pathway was manipulated. In the past, women used an M-receptor blocker, the plant poison atropine of the deadly nightshade *Atropa belladonna*, to counteract pupil contraction for a radiant and tempting look.

Summary

Acetylcholine interacts with both ion channel-coupled (nicotinic) and G-protein-coupled (muscarinic) receptors. Parasympathetic effects are transmitted by M-receptors expressed in five isoforms. They are either $G_{i,0}$- or $G_{q,11}$-coupled and are selectively inhibited by atropine.

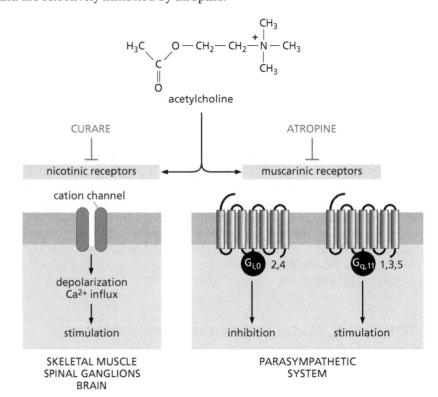

Figure 5.19 Acetylcholine (or cholinergic) receptors Acetylcholine opens cation channels (via nicotinic receptors) and activates $G_{i,0}$- and $G_{q,11}$-controlled signaling cascades (via muscarinic receptors). Nicotinic receptors are specifically blocked by the arrow poison curare, and muscarinic receptors are blocked by atropine, the poison of the deadly nightshade (*Atropa belladonna*). The numbers beneath the muscarinic receptors indicate the five different receptor isotypes M_1–M_5.

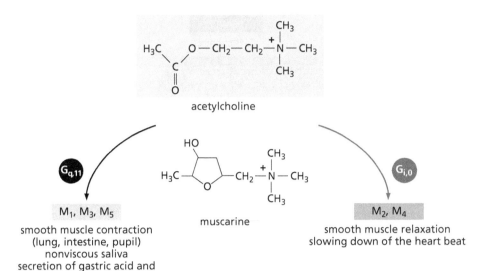

smooth muscle contraction
(lung, intestine, pupil)
nonviscous saliva
secretion of gastric acid and
digestive enzymes

smooth muscle relaxation
slowing down of the heart beat

Figure 5.20 Parasympathetic effects mediated by muscarinic acetylcholine receptors Depending on the receptor type, acetylcholine may exhibit antagonistic effects, for instance, on smooth muscles.

5.5 Stress and the heart: the competition between sympathetic and parasympathetic signals

Our treatment of stress and signaling in the vegetative nervous system will be concluded by an instructive example demonstrating the interplay between cholinergic and adrenergic signals, that is, between parasympathetic and sympathetic stimuli, in the regulation of heart activity. In a particularly clear manner it shows how organ function is controlled by a network of antagonistic signaling pathways.

As for most target organs of the vegetative nervous system, the heart is subject to control "by two reins:" upon parasympathetic activation it is slowed down, whereas upon sympathetic activation it beats faster and stronger. Under stress the heart "is pounding up to the neck." The discovery that, upon parasympathetic stimulation, an isolated frog heart released a substance into the culture medium that dampened the frequency of a second heart was a scientific milestone. Using this experimental approach, Otto Loewi (Nobel Prize in Physiology or Medicine for 1936) in Graz demonstrated that neurons transmit chemical messengers instead of communicating exclusively through electrical signals as had been assumed before. Loewi called the signaling substance (today classified as a neurotransmitter) "Vagusstoff," referring to the synonym nervus vagus for parasympathetic nerves. Not until years later was "Vagusstoff" identified as acetylcholine.

Each heartbeat is triggered by an action potential that is generated by the pacemaker cells in the sinoatrial knot and induces in the heart muscle cells an influx of Ca^{2+} through voltage-controlled membrane channels (for details see Section 14.6.2). The *power of contraction* depends on the strength of the Ca^{2+} influx, that is, the length of time in which the Ca^{2+} channels remain open. The opening period is extended by a PKA-catalyzed phosphorylation of the channel protein, which is activated along a β-adrenergic signaling pathway. This provides the mechanism by which the sympathetic nerves and stress, or (nor)adrenaline, increase the *power* of heart contraction (Figure 5.21).

The *heart frequency* is controlled by HCN (Hyperpolarization- and Cyclic Nucleotide-activated) ion channels in the pacemaker cells (see Section 14.6.2 for a more in-depth discussion of this subject). They become opened at the end of an action potential (at maximal hyperpolarization) and, being Na^+ channels, have a depolarizing effect. The more active the pacemaker channels, the shorter the intervals between action potentials and the faster the heartbeat. Pacemaker

201

Figure 5.21 Control of heart muscle activity Schematic representation of the heart muscle equipped with a β_1-adrenergic receptor for the sympathetic signal (noradrenaline, left) and an M_2-muscarinic receptor for the parasympathetic signal (acetylcholine, right, in red) as well as with three different ion channels: a voltage-controlled Ca^{2+} channel regulating heart contraction or power (above); a hyperpolarizing K^+ channel (right); and a depolarizing pacemaker channel HCN controlling heart frequency (below). cAMP released by the sympathetic signal activates the Ca^{2+} channel (via PKA-catalyzed phosphorylation) and the HCN channel (directly), thus increasing the contraction power and frequency of heart beats. This effect is antagonized by the parasympathetic signal that induces hyperpolarization through K^+ channel gating (pathway 1), hinders Ca^{2+} influx by hyperpolarization and a direct effect of $G_{i,0}$ on the Ca^{2+} channel (pathways 2), and inhibits adenylate cyclase (pathway 3). The figure illustrates the regulatory principle without laying claim to anatomical correctness. In fact, HCN channels are restricted to the pacemaker cells in the sinoatrial knot that, through gap junctions, are coupled electrically and metabolically with the rest of the heart.

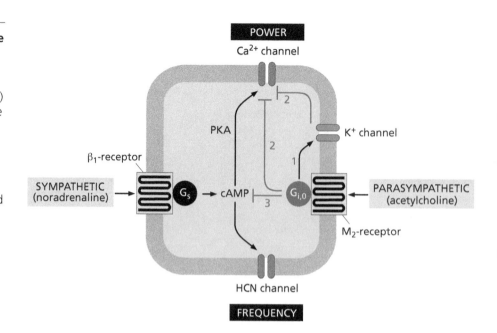

channels belong to the family of cyclic nucleotide-gated cation channels and are sensitized by cAMP binding (see Section 14.5.7). This is the signaling pathway along which sympathetic and stress signals increase the heart frequency (Figure 5.21).

Acetylcholine activates the inhibitory $G_{i/0}$-cascade via an M_2 receptor, resulting in hyperpolarization, blockade of voltage-dependent Ca^{2+} channels, and inhibition of cAMP production. As a consequence, the heart activity is turned down.

Summary

The heart muscle provides a classical example of an organ controlled by antagonistic signals of the vegetative nervous system. Via a β-adrenergic receptor, the sympathetic transmitter noradrenaline stimulates both contraction power and frequency by activating voltage-dependent Ca^{2+} channels and cAMP-controlled hyperpolarization- and cyclic nucleotide-activated (HCN) pacemaker channels. These effects are antagonized by the parasympathetic transmitter acetylcholine that, along a $G_{i,0}$-controlled pathway, inhibits cAMP formation, blocks the Ca^{2+} channels, and causes hyperpolarization (by gating K^+ channels).

5.6 G-protein-coupled receptors activated by proteolysis: major players in blood clotting and inflammation

For most GPCRs, the ligands are separate molecules that interact spontaneously. There are two exceptions: rhodopsins, with their covalently bound ligand retinal, and a special class of receptors that are activated enzymatically by extracellular proteases. The latter play an important role in inflammatory processes, blood clotting, and wound healing. In mammals four isoforms of **protease-activated receptors (PARs)** have been found. They have become known as "thrombin receptors" although they react with other proteases as well.

Proteases are not the ligands proper of these receptors. Instead, PARs contain a tethered ligand, a hexapeptide incorporated in the N-terminal extracellular domain. In the resting state this domain is prevented from interacting with the ligand-binding site (localized in the second extracellular peptide loop) by the N-terminal peptide residue of the receptor protein. This intramolecular blockade is abolished by proteases splitting off the auto-inhibitory domain (Figure 5.22).

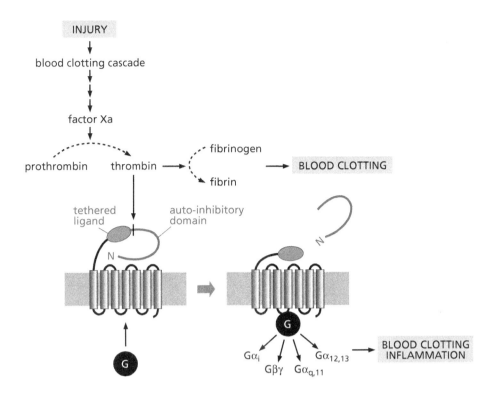

Figure 5.22 Activation of thrombin receptor PAR1 As an example, the proteolytic release of thrombin in the course of blood clotting is shown. Thrombin has two effects: it catalyzes the formation of fibrin from fibrinogen, and it activates PAR1 in various cell types by removing an N-terminal auto-inhibitory domain, thus enabling an interaction of the tethered ligand with the receptor. PAR1 may couple with various G-proteins. For more details, see text.

Protease-activated receptors exhibit "G-protein promiscuity": they interact with G_q-, $G_{i/0}$-, and $G_{12,13}$-proteins, activating numerous signaling pathways.

For the cell the activation of PARs is an expensive venture because it is irreversible: the regeneration of the system requires the synthesis of new receptor protein. Therefore, it makes sense that PARs are employed predominantly in emergency situations where the costs are unimportant. Such an emergency situation is wounding, with **blood clotting** being an immediate response. In fact, thrombin becomes activated by the blood clotting cascade and has two functions: it catalyzes the proteolysis of fibrinogen to fibrin, thus leading to clot formation, and it activates PARs in cells and tissues that support clotting and perform defense functions.

A key event of blood clotting, including the pathological forms of myocardial infarction and stroke, is the aggregation of platelets, promoted by an emptying of secretory vesicles containing aggregation-inducing substances such as ATP. The Ca^{2+} ions needed for this exocytotic process are provided by the $G_{q,11}$–PLCβ–$InsP_3$ signaling cascade triggered by a PAR. In fact, thrombin ranks among the most potent inducers of platelet aggregation.

Cells involved in defense reactions are endothelial and epithelial cells, vascular smooth muscle cells, leukocytes, and sensory neurons. Their stimulation evokes the typical symptoms of **inflammation** such as redness and heat (due to dilation of blood vessels), swelling (due to increased vascular permeability), and pain. In these responses PARs expressed by these cell types play an important role, although it must be emphasized that inflammation is an extraordinarily complex process that is by no means regulated by proteases alone but by many additional factors (see, for instance, Section 7.3.2).

The widening of blood vessels due to a dilation of vascular smooth muscles is induced indirectly by thrombin, since the protease activates, via the $G_{q,11}$-coupled receptor PAR1, the Ca^{2+}-dependent biosynthesis of nitric oxide and some other mediators evoking smooth muscle relaxation (for details see Section 16.5.1). In addition, thrombin stimulates, again through activation of PAR1, the proliferation of endothelial and smooth muscle cells, thus promoting the formation of new blood vessels (angiogenesis). To this end the protease turns up several signaling cascades controlled by PKB/Akt, protein kinase C (PKC), and arrestin (more about arrestin is found in Section 5.8). In parallel the *de novo* synthesis of the vascular endothelial growth factor (VEGF) is induced, which additionally stimulates these signaling cascades.

Inflammatory pain is caused by an activation of temperature-sensitive heat and pressure receptors in sensory neurons (Section 15.3). Such neurons are stimulated by several pro-inflammatory mediators including thrombin and related proteases. In fact, neurons as well as neuron-associated astrocytes express several PAR isoforms. In sensory neurons the receptor PAR2 activates along the $G_{q,11}$–PLCβ–diacylglycerol pathway PKC that sensitizes the vanilloid receptor VR1 by phosphorylation. VR1 is a receptor ion channel evoking pain sensations (Sections 15.3 and 16.7). Certain sensory neurons respond to thrombin and other proteases by a release of pro-inflammatory neuropeptides that cause a "neurogenic inflammation" in adjacent tissues. The cellular events of inflammation are, in addition, dramatically amplified by the Ca^{2+}-dependent release of prostaglandins and related lipid mediators. This response provides the target of most widely used anti-inflammatory drugs such as aspirin and its relatives (see Section 4.4.5).

PARs are found not only in peripheral nerves but also in the **brain**. Brain cells produce small amounts of thrombin that may activate survival, (anti-apoptotic) pathways such as the PI3K–PKB/Akt and RAF–MAP kinase cascades. This effect helps to protect neurons from death due to glucose and oxygen deficiency or oxidative stress, as occurring, for instance, in the course of stroke. In higher concentrations, however, thrombin may promote cell death by inducing Ca^{2+} release as mentioned above. To suppress this pro-apoptotic function of thrombin, neurons produce several antidotes including the protein nexin-1, the strongest thrombin inhibitor known.

Summary

Protease-activated receptors (PARs) constitute a subfamily of G-protein-coupled receptors predominantly involved in the handling of emergency situations such as wounding and inflammation. They are stimulated by a tethered ligand that is incorporated in the extracellular domain and are activated when the auto-inhibitory N-terminus is removed by proteolysis.

5.7 Adaptation of G-protein-controlled signaling: a matter of feedback

Rapid signal extinction and adaptation protect the signal-processing apparatus from overloading and collapsing. Moreover, these mechanisms help the apparatus to respond to oscillating signals that are typical within the nervous system. As we have learned from bacterial chemotaxis (Section 3.3.4), adaptation also ensures maximal sensitivity over a broad range of signal intensity and may function as a molecular short-term memory store.

Signaling pathways controlled by G-proteins are highly adaptive. Let us look at a prototypical example: adaptation of β-adrenergic signal transduction. Principally, adaptation occurs at two levels: extinction of the input signal, and adjustment of the signal-processing apparatus. The standard mechanisms for the extinction of chemical messengers are dilution and metabolic inactivation. At adrenergic synapses, dilution of the noradrenaline released into the synaptic cleft is mainly due to a recapture mechanism. For this purpose the presynaptic membrane of the neuron is equipped with specific transporter proteins that catalyze the re-uptake of the neurotransmitter (Section 16.9). In addition, neurons and other cells express enzymes catalyzing the inactivation of noradrenaline. Two of them, the catechol O-methyltransferase COMT and the monoamine oxidase MAO (Figure 5.23), are targets of drugs used to treat high blood pressure (due to α-adrenergic effects) and depression (as is discussed in more detail in Section 16.9, noradrenaline is critically involved in the processing of emotions in the brain).

Adaptation of the signal-processing apparatus takes place primarily at the receptor according to a mechanistic scheme that probably holds true for most GPCRs. A strong input signal leads to a **desensitization** of the receptor within seconds. In the minutes thereafter, the receptor becomes internalized via uptake into endocytotic vesicles, removing it from the membrane. The internalized receptors may be either re-exposed through exocytosis or degraded in lysosomes. The whole sequence of events, called **down-modulation,** is a major reason for tolerance and tachyphylaxis, or addiction to pharmaceutical drugs and narcotics.

The desensitization starts with a phosphorylation of several Ser and threonine (Thr) residues in the cytoplasmic domains of the receptor. It is controlled by two negative feedback loops. In one of these loops, phosphorylation is catalyzed by downstream effector kinases of G-protein-controlled signaling such as PKA (for G_s-signaling) or PKC (for $G_{q,11}$-signaling). Since such kinases may also become activated along other signaling pathways, this type of desensitization is said to be heterologous.

Somewhat more selective is another type of desensitization. Here the receptor phosphorylation is catalyzed by a **G-protein-coupled Receptor Kinase (GRK)** that in mammals is found in seven isoforms belonging to the AGC kinase family. Some GRKs are restricted to individual tissues and are relatively specific: for

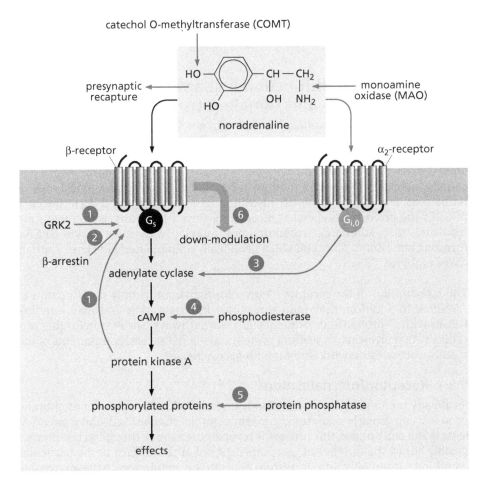

Figure 5.23 Adaptation of β-adrenergic signal transduction The signaling molecule noradrenaline may be inactivated enzymatically by O-methylation, catalyzed by catechol O-methyltransferase, or by the oxidative deamination of the monoamine to an aldehyde group, catalyzed by monoamine oxidase. However, at both α- and β-adrenergic synapses the predominant mechanism of signal extinction consists of the recapture of noradrenaline by the presynaptic neuron. At the level of intracellular signal processing, various mechanisms of adaptation exist. The most important ones are shown here: (1) inactivation of the receptor by phosphorylation, catalyzed by both PKA and a specific GPCR kinase GRK2; (2) interruption of G-protein interaction by β-arrestin; (3) inhibition of adenylate cyclase along an antagonistic signaling pathway; (4) extinction of the cAMP signal by phosphodiesterase; (5) dephosphorylation of PKA substrate proteins; and (6) receptor down-modulation. These pathways are interlinked and connected with additional pathways controlled by other signals, such as those controlling the activities of cAMP phosphodiesterases and protein phosphatases. For the sake of clarity, this signaling cross talk is not shown.

example, rhodopsin and iodopsin kinases GRK1 and GRK7 are found only in retina cells. In contrast, other GRKs such as the β-adrenergic receptor kinases GRK2 and GRK3 (also known as βARK1 and βARK2) are widely distributed, interacting not only with β-adrenergic receptors but with with most other GPCRs.

To phosphorylate the membrane-integrated receptor, a GRK has to associate with the cell membrane. For this purpose, GRK2 and 3 possess pleckstrin homology (PH) domains that interact with the membrane-bound second messenger phosphatidylinositol 3,4-bisphosphate (PIP_2) and with membrane-anchored Gβγ-subunits that both serve as GRK activators. In contrast, other GRKs such as the rhodopsin kinases are constitutively active because they are anchored in the membrane by their own covalently bound lipid residues and do not need the assistance of PIP_2 or Gβγ. Such permanent GRK activity is essential for the most rapid light adaptation of the retina (see Section 15.6).

It must be emphasized that GRKs attack exclusively GPCRs activated by the corresponding ligand. In the case of GRK2 and 3 this is easily conceivable, since only an active receptor is able to release the Gβγ-subunits required for GRK activation. Moreover, Gβγ stimulates PI3K type γ, which catalyzes the formation of PIP_2 (Section 4.4.3). Since, however, both Gβγ-subunits and PIP_2 are generated along a wide variety of other pathways, GPCR desensitization is expected to be integrated in a network of signaling cross talk.

Summary

Via a series of feedback loops, G-protein-dependent signaling is under strict negative control. Rapid desensitization is caused by receptor phosphorylation catalyzed by specific G-protein-coupled receptor kinases (GRKs) or other downstream kinases. As a result, the contact with the G-protein becomes interrupted and the receptor undergoes internalization through endocytosis. This reversible down-modulation may become irreversible when the receptor protein is degraded in lysosomes.

5.8 Arrestins: multifunctional adaptors for receptor down-modulation and signaling cross talk

To desensitize a receptor, phosphorylation is not sufficient but provides a prologue for the next step, which involves a displacement of the G-protein by an arrestin. Arrestins are eukaryotic proteins that bind specifically to *phosphorylated* GPCRs, thus cooperating with GRKs and other protein kinases. Again one can distinguish between two rhodopsin-specific visual arrestins, found only in retina cells, and two ubiquitously distributed β-arrestins 1 and 2. As in the case of GRKs the prefix β is somewhat misleading since β-arrestins by no means are restricted to β-adrenergic receptors but interact with most other GPCRs. Arrestins are essential proteins since genetically manipulated knockout animals die as embryos.

The interruption of the receptor/G-protein interaction is only one function of arrestins. In addition, they induce receptor internalization and may interlink GPCRs with G-protein-independent signaling pathways. The reason for this versatility is that arrestins, as scaffold proteins, assemble several components of the signal-processing network into functional complexes.

5.8.1 Receptor internalization

As already mentioned, internalization is understood as an uptake of membrane proteins, particularly receptors, into endocytotic vesicles and endosomes. At least in the early phase, this process is reversible, enabling the cell to rapidly and flexibly adjust the number of receptors exposed at the surface to the particular conditions. Internalization starts from dents in the membrane. In these **caveolae** and **coated pits**, receptors accumulate and are internalized according to the mechanism of receptor-induced endocytosis (see Sidebar 5.4). The role of arrestins is to connect as scaffolds the phosphorylated receptor protein with a complex of endocytotic control proteins (Figure 5.24). The arrestin-binding site of the receptor is called, therefore, the internalization motif. In addition, arrestins promote endocytosis directly: the separation of the vesicle from the membrane is controlled by the small G-protein Arf6 (Section 10.4.3) that together with its GDP–GTP exchange factor ARNO (ARf6 Nucleotide binding site Opener) is bound by arrestin and thus concentrated in the coated pit. The ability of arrestins to promote endocytosis is modulated by phosphorylation and ubiquitylation, indicating that these proteins are knots in a more complex signaling network with various feedback and feedforward loops.

The final step of adaptation is the proteolytic degradation of the receptor protein in lysosomes. It represents the cell's ultimate answer to overstimulation. In the subsequent recovery phase, receptors have to be newly synthesized. The individual stages of a membrane receptor's life cycle are summarized in Figure 5.25.

Sidebar 5.4 Microdomains and endocytosis On membranes, receptors and other membrane proteins are concentrated in microdomains rather than distributed randomly. Such microdomains are due to a demixing of membrane lipids: because of their higher melting temperature, sphingomyelins and cholesterol have the tendency to aggregate to **lipid rafts** swimming in the phospholipid bilayer, which has an oily consistency at body temperature. In these semi-solid structures the lateral diffusion of membrane-bound proteins is strongly reduced and the assembly of stable signal-processing protein complexes is facilitated. Such complex formation seems to be supported by intrinsic raft proteins such as caveolins, annexins, and flotillins. How the cell controls the formation and disappearance of lipid rafts is still widely unknown.

A type of lipid raft visible under the microscope is the **caveolae**. These are small dents in the cell surface where the lipid rafts are pulled together and stabilized at the inner side of the membrane by the membrane protein caveolin. Caveolae may be precursors of endocytotic vesicles. Similar membrane invaginations are the **coated pits**. At the inner side of the membrane they are coated by the structural protein **clathrin**, which due to its triskelion structure forms basket-like aggregates, stabilizing the invagination. Like caveolae, coated pits are precursors of endocytotic vesicles. In both cases the separation of the vesicle from the membrane is controlled by **dynamin.** This monomeric G-protein aggregates at the neck of the invagination, forming a collar that contracts under GTP hydrolysis (Figure 5.24). Endocytosis of clathrin-coated pits is the major mechanism by which cells take up macromolecules and distribute them between individual membrane compartments such as plasma membrane, endosomes, lysosomes, Golgi apparatus, and endoplasmic reticulum.

Separation of the vesicles and their transport between cellular compartments is regulated by a special class of adaptor proteins (APs). These proteins form complexes with clathrin and control vesicle trafficking between plasma membrane and endosomes (protein AP2) or endosomes and Golgi apparatus (proteins AP1 and AP3). Before the vesicle fuses with its target membrane, the clathrin coat has to be removed. This **uncoating** is under the control of several proteins, with the chaperone Hsc70 and its co-chaperone auxilin playing key roles. Auxilin is the ATPase-activating protein of Hsc70 (itself an ATPase) that is active only in the ADP-loaded form. Hsc70 changes the conformation of clathrin in such a way that the vesicle coat disintegrates. During this uncoating, the ADP bound to Hsc70 is replaced by ATP (for details on the mechanism of action of chaperone ATPases, see Section 2.5.1).

Extracellular macromolecules to be taken up by the cell are bound by transmembrane receptors that are enriched in lipid rafts and coated pits undergoing endocytosis. This **receptor-mediated endocytosis** (Section 10.4.1) supplies cells with essential nutrients such as iron and cholesterol. In blood, iron is bound to the transport protein transferrin and taken up via a specific transferrin receptor, while cholesterol is concentrated in low-density lipoprotein (LDL) particles that interact with an LDL receptor. Both receptors are homodimeric transmembrane proteins with one transmembrane domain each per subunit. Receptor-mediated endocytosis is also used by cells to clear active compounds such as antigen–antibody complexes and hormones from the blood. Finally, it represents the standard pathway for the internalization of signal receptors.

continued

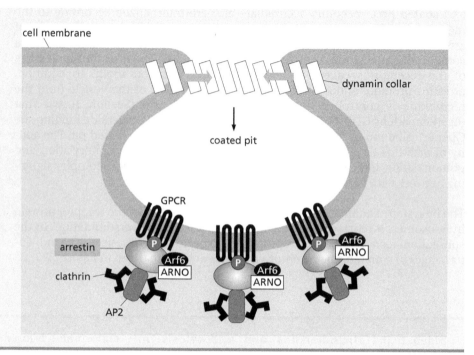

Figure 5.24 Internalization of G-protein-coupled receptors by clathrin-dependent endocytosis Activated receptors assembled in coated pits become phosphorylated (P) and bound to the proteins of the endocytotic apparatus (clathrin, Arf6, ARNO) by use of arrestin and the protein AP2 as adaptors. For endocytosis, the invaginated area of the cell membrane is released as a vesicle into the cytoplasm by means of the GTPase dynamin. Afterward the vesicle is stripped of the clathrin coat, which is reutilized. Arf6 is a small G-protein required for vesicle formation (see Section 10.4.3), and ARNO is the corresponding GDP–GTP exchange factor.

5.8.2 Reprogramming of receptors: from G-protein coupling to arrestin coupling

As we have seen, GRKs and arrestins block G-protein-dependent signaling pathways. However, arrestins can do more: upon displacement of the G-protein, they interlink *phosphorylated* GPCRs, which are unable to interact with G-proteins,

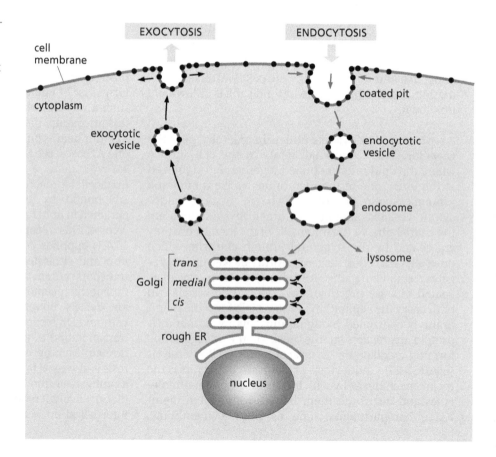

Figure 5.25 Life cycle of transmembrane receptors Receptor proteins (black dots) are synthesized at the rough endoplasmic reticulum (ER), packed into secretory vesicles in the Golgi apparatus, and exposed at the cell surface by exocytosis (left). During adaptation by down-modulation (red arrows), the membrane receptors assemble in membranous microdomains and membrane invaginations such as clathrin-coated pits, which as endocytotic vesicles become released into the cytoplasm. Several vesicles fuse to endosomes that deliver their content either to the Golgi apparatus for re-utilization or to lysosomes for final degradation.

with signaling cascades, which until recently have been thought to be stimulated by tyrosine (Tyr) kinase-coupled receptors only (Section 7.1). A typical example is the mitogenic MAP kinase module consisting of the protein kinases RAF1, MEK1, and ERK. The conventional pathway leading to a stimulation of this module starts from Tyr kinase-coupled growth factor receptors and proceeds via the small G-protein Ras that assembles the three kinases of the module to a functional complex at the inner side of the cell membrane (details in Section 10.1.4 and Chapter 11). The organizing role of Ras can be taken over by an arrestin that acts as a scaffold protein for the module. This works only when the arrestin is bound to a phosphorylated GPCR and means that the input signal stimulating the MAP kinase module is identical with the GPCR ligand that induces adaptive receptor phosphorylation and arrestin binding. In a similar manner, the stress-activated MAP kinase modules (Section 11.1) become stimulated by arrestin-coupled receptors.

Such signaling cascades are activated along still another route. In this case the arrestins recruit cytoplasmic Tyr kinases of the Src family, which phosphorylate Tyr kinase-coupled receptors, such as those of cytokines, antigens, and growth factors, as well as the substrate proteins thereof. This phosphorylation has the same activating effect on the receptor as autophosphorylation, as it is normally induced by the receptor ligand. In other words, Tyr-kinase-coupled receptors (and MAP kinase modules) are not only activated in the conventional way by their genuine ligands, but also by ligands interacting with GPCRs and transforming them into arrestin-coupled receptors. Such cross talk between different signaling cascades has been called **receptor transactivation** (Figure 5.26). It enables GPCR ligands to secure a direct grasp on the genome, because a major effect of MAP kinase modules is modulation of gene transcription.

Latest observations have added additional pieces to the arrestin puzzle. Not only interact β-arrestins with GPCRs but also with several cytokine receptors, the Toll-like receptors of bacterial components (Section 6.3.2), the protease-coupled receptor Notch (Section 13.1.1) and others. Moreover, arrestins even seem to translocate into the nucleus, augmenting gene transcription by supporting the

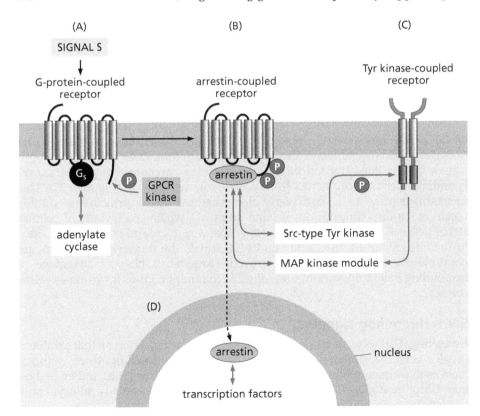

Figure 5.26 Reprogramming of heptahelical receptor by phosphorylation and arrestin (A) Signal S activates, via a GPCR, a G-protein-controlled signaling cascade such as the G_S–cAMP pathway. In the course of adaptation, the receptor becomes phosphorylated by a GPCR kinase. (B) Arrestin has displaced the G-protein from the phosphorylated (P) receptor, thus interrupting G-protein-controlled signaling. Instead, arrestin-controlled signaling becomes active. Due to the scaffold function of arrestin, the receptor is coupled to the components of MAP kinase modules or Tyr kinases of the Src type that both become activated. (C) Src-type kinase activates a dimeric Tyr kinase-coupled receptor by phosphorylating in the cytoplasmic domains those sites that, upon ligand binding, would undergo trans-autophosphorylation. In turn, the receptor stimulates the MAP kinase modules and activates other signaling pathways. (D) Arrestin translocates into the nucleus, acting as transcriptional co-activator. In summary, one signal S has cleared a variety of signaling routes.

interaction between transcription factors and transcriptional co-activators such as histone acetyltransferases, thus acting as transcriptional co-activators on their own. These observations indicate signaling functions of arrestins that go far beyond those currently appreciated.

5.8.3 Receptor transactivation by ectodomain shedding

GPCRs may also transactivate receptor Tyr kinases, independently from arrestin coupling, by triggering the release of the corresponding ligand. Such a mechanism has been described for the epidermal growth factor (EGF) receptor family. Here an ADAM-type protease (Section 13.1.2), activated along a still ill-defined GPCR-controlled pathway, detaches EGF or related growth factors from membrane-bound precursor proteins through a mechanism known as ectodomain shedding (see Section 13.1 for more information). The process seems to play a role in hyperproliferative diseases such as tumors and cardiac hypertrophy.

Summary

Arrestins are proteins playing a key role in receptor desensitization by displacing G-proteins from phosphorylated, desensitized G-protein-coupled receptors (GPCRs). They may induce receptor down-modulation through endocytosis, or they may cause receptor reprogramming by acting as scaffold proteins to couple the phosphorylated GPCRs with Tyr kinase-coupled receptors and cytoplasmic Tyr kinases. Upon translocation into the nucleus, arrestins may act also as transcriptional co-activators.

5.9 Heptahelical receptors without G-protein coupling: hedgehog and Wnt signaling pathways

While practically all trimeric G-proteins are activated by heptahelical receptors, not all receptors of this type are coupled to G-proteins. The most prominent examples of such a case are provided by **Smo** and **Fz**, two receptors originally discovered in the *Drosophila* mutants "smoothened" and "frizzled" but since then found in homologous forms in all animals hitherto studied. The activity of Smo and Fz is controlled by the peptide factors "hedgehog" and "wingless." Together with ephrins (Section 7.1.4), several growth factors such as fibroblast growth factor and platelet-derived growth factor (Section 7.1), and factors of the transforming growth factor β-family (Section 6.2), these peptides rank among the most important and most abundant morphogens controlling embryonic tissue development. In addition, they regulate the structure and function of adult tissues.

As far as the mechanistic principle of signal transduction is concerned, the hedgehog and Wnt pathways are closely related: in both cases the primary input signals induce gene transcription along signaling cascades that protect corresponding transcription factors from proteolytic degradation (Figure 5.27). This mechanism provides a particularly clear example of the principle of double negation ("minus times minus is equal to plus"), which is a favorite of cellular signal processing. Certainly, it is an expensive way to control gene activity, since in the absence of the hedgehog and Wnt signals the transcription factors are continuously synthesized and destroyed. However, it has the advantage of responding in a rapid, variable, and flexible manner and thus is obviously worth the cost.

5.9.1 Hedgehog signaling

Hedgehog (Hh) received its name from mutated *Drosophila* larvae that exhibited bristles similar to those on hedgehogs. Due to gene doubling, three isoforms have evolved in vertebrates. Their names—sonic hedgehog (Shh), indian hedgehog (Ihh), and desert hedgehog (Dhh)—obscure the already idiosyncratic nomenclature used by the *Drosophila* community.

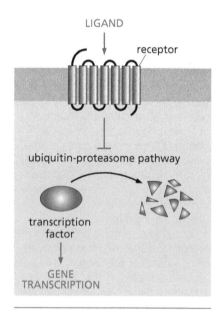

Figure 5.27 Principle of hedgehog and Wnt signaling The ligand activates a heptahelical receptor that, in turn, produces a G-protein-independent signal interrupting the continuous degradation of transcription factors along the ubiquitin–proteasome pathway. As a consequence, the transcription of specific genes is up-regulated.

Hedgehog controls, among other processes, the formation of body segments in insects and, in vertebrates, the development of the neural tube, limbs, and right–left asymmetry. Disturbances of hedgehog signaling result, therefore, in massive malformations. In adult tissues, hedgehog is involved in the control of homeostasis, the equilibrium between cell gain and cell loss that ensures a constant tissue mass and function. Accordingly, an overactivation of the hedgehog system may result in tumor development.

Hedgehog is a 19 kDa protein carrying a cholesteryl residue at the C-terminus and a palmitoyl residue at the N-terminus. Due to these modifications, it is sufficiently lipophilic to accumulate in lipid rafts at the surface of its target cells, where it also finds its receptors. On the other hand, hedgehog acts as a paracrine messenger, diffusing from cell to cell. This is made possible through an interaction with strongly hydrophilic proteoglycans that shield the lipid residue, rendering hedgehog water-soluble.

The hedgehog receptor called **"patched" (Ptc)** is a protein with 12 transmembrane domains structurally resembling bacterial membrane transporters that derive their energy from proton gradients. It is highly conserved among species. The role of Ptc is somewhat unusual since it binds hedgehog specifically but does not directly transmit the hedgehog signal into the cell. In fact, the signal transducer proper is Smo, a heptahelical transmembrane protein. Ptc inhibits the signaling activity of Smo, and this blockade is relieved by the interaction with hedgehog.

Since Ptc does not bind directly to Smo, the mechanism by which Ptc suppresses Smo activity is not entirely clear. A currently favored hypothesis proposes that Ptc, being a putative membrane transporter, releases a cholesterol-like lipophilic signaling molecule such as vitamin D3 that interacts with a corresponding sensor domain of Smo, thus inhibiting signal transduction. When hedgehog binds to the large extracellular domains of Ptc, the transmission of this inhibitory signal would be interrupted (Figure 5.28).

Figure 5.28 Hedgehog signaling (in *Drosophila*) The receptor of the lipoprotein hedgehog is the protein Ptc (patched), with 12 transmembrane domains resembling a membrane transporter. (A) In the absence of hedgehog, Ptc inhibits the heptahelical transmembrane protein Smo (smoothened), perhaps by releasing an inhibitory lipid signal S. Under these conditions, the cellular level of the transcription factor Ci is very low because Ci is hierachically phosphorylated by the protein kinases depicted in the box and partially degraded along the ubiquitin (UB) pathway, yielding fragment CiR, which acts as a transcriptional repressor. (B) Hedgehog neutralizes the inhibitory effect of Ptc. As a result, Smo aggregates to active complexes, leading to stabilization of Ci along a still ill-defined pathway. (C) Schematic domain structure of Ci and CiR with the DNA-binding site (zinc finger, light gray box, DBD), the transactivating domain (TD) that interacts with the transcriptional co-activator CBP/p300, and the repressor domain (RD). The red arrow shows the site of proteolytic cleavage; black arrows indicate phosphorylation sites.

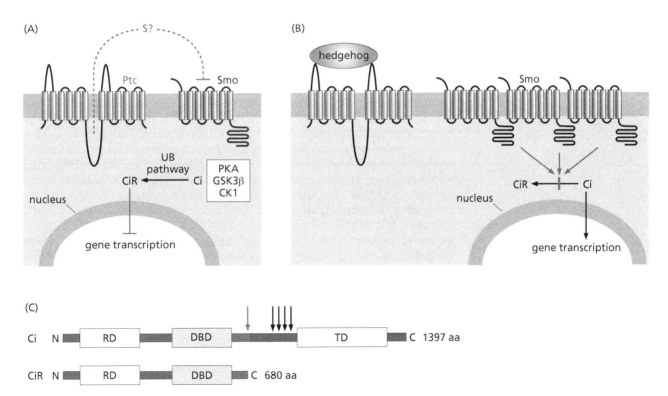

In an adaptive response, internalized Smo becomes exposed and the concentration of Smo in the membrane rises considerably, together with an increased transcription of hedgehog-responsive genes. The corresponding transcription factors are called **Ci** in the case of *Drosophila* (referring to a mutant <u>Cubitus</u> <u>interruptus</u>) and **Gli** in the case of vertebrates (referring to <u>Gli</u>oblastoma cells, where these factors were found for the first time). Due to the structure of their DNA-binding sites, these factors belong to the large family of zinc-finger proteins that includes many transcription factors.

How does Smo transduce the hedgehog signal to the transcriptional apparatus? This question has not been settled. Apparently, G-proteins are not involved despite the heptahelical structure of Smo. Instead, Smo stabilizes the metabolically labile transcription factors along a still rather ill-defined pathway.

In the absence of a hedgehog signal, the cellular levels of Ci and Gli are very low since both factors become degraded rapidly along the ubiquitin–proteasome pathway. In the course of degradation an N-terminal fragment is generated that still carries the DNA-binding domain as well as a repressor domain, whereas the transactivating domain has been destroyed together with the C-terminal fragment. As a result the N-terminal fragment, called **CiR** and **GliR**, respectively, acts as a *repressor* of hedgehog-responsive genes (Figure 5.28).

The formation of CiR/GliR is strictly regulated. It occurs only in a complex with other regulatory proteins and depends on a hierarchical phosphorylation of Ci/Gli by the kinases PKA, GSK3β, and CK1 (see Section 5.3.1). As characteristic for many ubiquitylation reactions phosphorylation generates a docking motif that is recognized by a specific S3 ubiquitin ligase of the SCF family (for more details see Section 12.16). This complex is joined by the Ser/Thr-specific protein kinase "fused" (FU) and an inhibitor thereof, called SuFU (<u>Su</u>ppressor of <u>FU</u>) and assembled at least in *Drosophila*, by the scaffold protein "coastal-2" (COS-2), a kinesin-like motor protein interacting with microtubules. In the presence of hedgehog, Smo transmits a signal that induces the dissociation of this complex. The nature of this signal is not known. It may consist of several protein phosphorylations. As a consequence Ci/Gli becomes stabilized (because its degradation to CiR/GliR is inhibited) and activated. The stabilized Ci/Gli is still prevented from translocation into the nucleus, probably by binding to SuFu. This blockade is overcome by SuFu phosphorylation, probably catalyzed by the kinase FU that becomes activated upon dissociation of the multiprotein complex. Thus, two safety locks prevent an accidental activation of Ci/Gli.

Hedgehog's function as a morphogen depends on a strictly controlled distribution in the developing organism and on steep extracellular concentration gradients. Such conditions may be fulfilled by a distinct expression pattern of proteins that sequester hedgehog, delivering it to the receptor complex on demand. Very recently such proteins have been found in *Drosophila*, known as **iHog** (interference hedgehog) and **Boi** (brother of iHog), while their mammalian counterparts are known as **Cdo** and **Boc**. With a single transmembrane domain and several extracellular immunoglobin- and fibronectin-like domains, they structurally resemble cell adhesion proteins (see Section 7.3.1). The proteins seem to be expressed preferentially in hedgehog target cells, forming a specific pattern. As hedgehog co-receptors they may selectively enrich the morphogen at the cell surface in the immediate neighborhood of Ptc and Smo. Experiments with genetically manipulated animals have revealed an essential role of Cdo and Boc in the development of the central nervous system.

As already mentioned, defects of hedgehog signaling may lead to tumorigenesis. Thus a permanent overactivation of sonic hedgehog signaling correlates with **basal cell carcinoma**, the most frequent type of skin cancer. The reason is a gain-of-function mutation of the receptor Smo. On the contrary, Ptc and SuFU have been identified as tumor suppressor proteins. Loss-of-function mutations of their genes are associated with medulloblastoma (a brain tumor), with prostate carcinoma, and with basal cell carcinoma.

For their work on *Drosophila* development including the discovery of the hedgehog system Edward B. Lewis, Christiane Nüsslein-Volhard, and Eric F. Wieschaus were awarded the Nobel Prize in Physiology and Medicine in 1995.

5.9.2 Wnt signaling

For embryonic development and the maintenance of tissues, the paracrine **Wnt factors** of vertebrates are as important as hedgehog. Their *Drosophila* prototype, **wingless**, is a palmitoylated glycoprotein that, among others, controls the correct development of cell patterns in wings and cuticle. For vertebrates, 19 isoforms of homologous Wnt proteins have been described that are tissue-specifically expressed and exhibit different functions. The term Wnt is a combination of <u>W</u>ingless and <u>Int</u>, the chromosomal locus of integration of a murine mammary tumor virus the proliferative effect of which led to the discovery of the Wnt/Int factors being homologs of *Drosophila* wingless. The Wnt signaling pathway is highly conserved and apparently goes back into evolutionary history to the first appearance of multicellular animals.

Wnt factors interact with various receptor types including receptor Tyr kinases of the Ryk and probably of the Ror family (Section 7.1). However, the canonical Wnt receptors are the ubiquitously distributed heptahelical **Fz proteins.** Depending on the Wnt isotype, these receptors either stimulate the $G_{q,11}$–PLCβ signaling cascade, acting as genuine GPCRs, or, similar to Smo, stabilize and activate transcription factors in a G-protein-independent way. In the latter case they require the assistance of a co-receptor **LRP** (<u>L</u>DL receptor-<u>R</u>elated lipo<u>P</u>rotein), a single-pass transmembrane protein. Interestingly, some other developmental factors (norrin and R-spondin) have been found to interact with Fz, confirming the notion of multipurpose usability of signaling proteins.

The transcription factors targeted by Fz are the **β-catenins** (alias "armadillo" in *Drosophila*). The corresponding genes include those of proteins controlling cell proliferation such as cyclin D (Section 12.3), transcription factors cJun and cMyc, and growth factors. In the absence of β-catenin, these genes are repressed by transcription factors of the **TCF/LEF** group (<u>T</u>-<u>C</u>ell <u>F</u>actor/<u>L</u>ymphoid <u>E</u>nhancer binding <u>F</u>actor). When these factors heterodimerize with β-catenin, they become transcriptional activators, recruiting transcriptional co-activators such as the histone acetyltransferase CBP/p300 and the nucleosome remodeling complex SWI/SNF (Figure 5.29). As found for many other transcription factors, the activity of **TCF/LEF** is fine-tuned by additional phosphorylation.

The activity of β-catenins is as strictly controlled as that of Ci, involving a dual safety lock due to two different β-catenin pools in the cell. In the first pool β-catenin exists as a complex with cell adhesion molecules of the cadherin family (see Section 7.3 for details). In this complex it is sequestered and thus not available as a transcription factor. The second pool consists of free cytoplasmic β-catenin. In nonstimulated cells this pool is kept extremely small by proteolytic degradation, promoted by a β-catenin destruction complex containing, in addition to β–catenin, the scaffold proteins **axin** (the name is reminiscent of a characteristic Wnt effect in vertebrate development, duplication of the body axis) and **APC** (referring to the disease <u>A</u>denomatous <u>P</u>olyposis of the <u>C</u>olon; see below), as well as the protein kinases GSK3 and CK1 and the protein phosphatase PP2A (Figure 5.29). Axin and APC organize and stabilize the complex in such a way that β-catenin undergoes phosphorylation by GSK3 and CK1, which is the primer kinase of GSK3 (see Section 9.4.4). Phosphorylation labels the factor for degradation along the ubiquitin–proteasome pathway. Interestingly, the SCF-type E3 ubiquitin ligase β-TrCP triggering β-catenin degradation seems to be identical

Figure 5.29 Wnt-β-catenin signaling (A) In the absence of Wnt, β-catenin is bound, together with GSK3 and CK1, to the scaffold proteins APC and axin. In this destruction complex, β-catenin is phosphorylated by GSK3 and CK1 and thus labeled for ubiquitylation and subsequent proteolytic degradation. Under this condition the expression of Wnt-responsive genes is repressed by TCF/LEF in cooperation with co-repressors such as histone deacetylases (see Section 8.2.3). (B) Wnt binds to its receptor Fz and to the co-receptor LRP (LDL receptor-related lipoprotein), thus activating the protein DSH (which is phosphorylated by CK2). DSH inhibits GSK3/CK1, while upon phosphorylation by GSK3 together with a membrane-bound CK1 isoform (not shown), LRP recruits axin, inducing its proteolytic degradation. As a result, the destruction complex dissociates and β-catenin becomes dephosphorylated and stabilized; it is now able to neutralize the repressor effect of TCF/LEF by complex formation. In parallel, transcriptional co-repressors are exchanged for transcriptional co-activators such as the histone acetyltransferase CBP/p300 and nucleosome remodeling complexes so that the WNT-responsive genes can be transcribed.

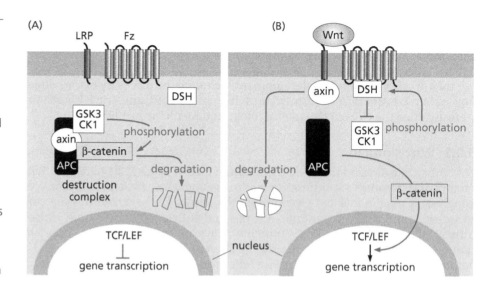

with the enzyme inducing partial proteolysis of the transcription factor Ci of the hedgehog system.

Wnt signals induce the dissociation of the destruction complex. As a result, the phosphorylation is stopped and dephosphorylation by PP2A gains the upper hand, leading to an increased level of free β-catenin and, subsequently, to a transcriptional activation of the corresponding genes. The pathway connecting the Fz receptor with the destruction complex is known only in fragments. It includes a protein **"dishevelled" (DSH, also known as Dvl)** rather than a trimeric G-protein. Upon interaction with the receptor (probably with subsequent phosphorylation by the protein kinase CK2), DSH recruits the destruction complex to Fz, where axin comes in contact with the co-receptor LRP. This interaction triggers the proteolytic degradation of axin, resulting in complete disassembly of the destruction complex.

The sophisticated and expensive regulation of Wnt signaling reflects the central role of this pathway in the development and maintenance of tissues. While defects of the signaling cascade result in severe embryonic malformations, Wnt signaling is required throughout life for the homeostatic balance of rapidly self-renewing tissues such as intestinal crypts and hair follicles as well as for the regulation of bone mass, and it has been suggested as a factor in the maintenance and control of the stem cell pool of a tissue. As would be expected, disturbances of Wnt signaling may cause uncontrolled cell proliferation and cancer. Indeed, in various human tumors β-catenin has been found to be mutated to an oncoprotein. The mutation renders β-catenin widely resistant to proteolytic degradation. The reason is that it cannot be phosphorylated by GSK3 since the target Ser and Thr residues have become replaced by other amino acids. This gain-of-function mutation is accompanied by loss-of-function mutations of axin and APC, which both are tumor suppressor proteins. The deletion of APC, in particular, ranks among the most frequent causes of benign and malignant tumor growth.

Adenomatous polyposis of the colon is a relatively rare hereditary pre-malignant condition characterized by benign tumors in colon epithelium. Due to the very high number of these polyps, the tendency for malignant development is practically 100%, being only a matter of time. While the polyps are due to a germline deletion of the *APC* gene, their malignant transformation results from an accumulation of additional genetic defects occurring accidentally (see also Section 4.6.5, Sidebar 4.6). A loss-of-function mutation of APC is the triggering event not only of polyposis but also of the great majority of spontaneous colorectal carcinomas that also develop from polyps and rank among the most frequent and dangerous forms of cancer. Since such polyps are found in about half of the older population, the deletion of *APC* seems to be by far the most abundant tumorigenic gene defect in man. Therefore, it is not surprising that the Wnt signaling system is in the forefront of experimental and clinical cancer research.

Summary

While almost all trimeric G-proteins are coupled to heptahelical receptors, a few members of this receptor family are linked to G-protein-independent pathways. Examples are provided by the receptors of the morphogenetic and mitogenic hedgehog and Wnt proteins. These hormone-like factors are widely distributed in eukaryotes. Both stabilize and activate certain transcription factors of the Ci/Gli and β-catenin type, respectively. Deregulation of the signaling cascades may result in tumor development such as basal cell carcinoma (hedgehog) and large bowel cancer (Wnt).

5.10 Other G-protein-independent effects of heptahelical receptors: lessons from slime molds and Homer

Apart from the hedgehog and WNT systems, other signaling pathways have been described that start with heptahelical receptors but do not (or not exclusively) proceed via trimeric G-proteins. An interesting example is provided by the slime mold ***Dictyostelium discoideum***, a social amoeba switching between uni- and multicellularity. Under stress, tens of thousands of cells aggregate to form a multicellular structure that develops into a spore capsule (see Section 2.14 and Figure 3.28). The signal for both aggregation and differentiation is cAMP, released from a few pacemaker cells and received by adjacent cells. The latter respond in three ways: they move towards the pacemaker cells (chemotactic effect); they start to produce cAMP on their own, thereby attracting further cells (relay effect); and they activate the genes controlling aggregation and differentiation (morphogenetic effect). Thus, for the slime mold, cAMP is both a pheromone and a hormone.

Depending on the phase of differentiation, *Dictyostelium* cells express four different cAMP receptors (CAR1–4) of the GPCR type and possess 11 genes encoding Gα-subunits and one gene each for Gβ- and Gγ-subunits. The rearrangement and activation of the contractile actomyosin complex required for the chemotactic response is induced by CAR1 and the Gβγ-subunits, whereas the other cAMP receptors evoke the obligatory Ca^{2+} influx and most of the gene activations along G-protein-independent pathways, which are still not known in their details.

In contrast to the process occurring in *Dictyostelium*, the mechanism of a G-protein-independent Ca^{2+} mobilization by another type of heptahelical receptors is known in more detail. These receptors are the **metabotropic glutamate receptors** of neurons. In addition to coupling in the conventional manner with $G_{i/0}$- and $G_{q,11}$-proteins, they have been found also to interact directly with those Ca^{2+} channels of the endoplasmic reticulum that are normally controlled by $InsP_3$. The gap between receptor and channel is bridged by an adaptor protein "**Homer**" interacting with proline-rich binding sites (called "Homer ligands") localized both in the cytoplasmic C-terminus of the receptor protein and in the $InsP_3$ receptor and ryanodine receptor Ca^{2+} channels (Figure 5.30).

This mechanism offers the neuron two possibilities to gate the Ca^{2+} channels in the endoplasmic reticulum: via $InsP_3$ (or the still-elusive ryanodine receptor ligand; see Section 14.5.3) or via Homer. Upon excitation, neurons produce an isoform (a-type Homer) that interrupts Homer signaling and Ca^{2+} release, thus protecting the cells from an overstimulation by an excessive Ca^{2+} signal that otherwise would result in cell death, for instance as a consequence of stroke and other neurodegenerative diseases (Section 16.4.2). Moreover, this feedback

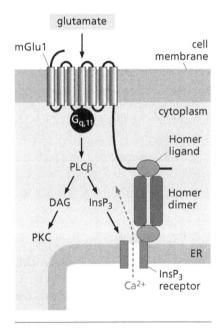

Figure 5.30 Two pathways leading to Ca^{2+} release by metabotropic glutamate receptor mGlu1 in neurons The protein Homer bridges the gap between a G-protein-coupled receptor and the $InsP_3$-activated Ca^{2+} channel in the endoplasmic reticulum, thus inducing channel gating in a G-protein-independent way. Homer signaling is interrupted by a protein, Homer type a, produced upon overstimulation (not shown). ER, endoplasmic reticulum.

control has been proposed to play an important role in memory fixation. A steadily increasing body of evidence indicates, on the other hand, that hereditary or acquired defects of Homer signaling are associated with psychoses, epilepsy, drug addiction (particularly alcoholism), and other neuropsychiatric disorders.

Some GPCRs induce the release of the intracellular messenger ceramide along a G-protein-independent signaling pathway. To this end they activate the membrane-bound enzyme **sphingomyelinase** by means of a special adaptor protein, a mechanism that has been discussed in Section 4.4.5.

Summary

Heptahelical receptors interacting with the chemotactic and morphogenetic signal cAMP in *Dictyostelium discoideum* stimulate gene transcription and mobilize intracellular Ca^{2+} along G-protein-independent pathways. Heptahelical metabotropic glutamate receptors in brain neurons activate both G-protein-dependent and -independent signaling. In the independent pathway, the protein Homer interconnects the receptor with Ca^{2+} channels in the endoplasmic reticulum.

Further reading

Alexander SPH, Mathie A & Peters JA (2001) 2001 nomenclature (receptor and ion channel) supplement. *Trends Pharmacol. Sci.*

Baillie GS & Houslay MD (2005) Arrestin times for compartmentalised cAMP signalling and phosphodiesterase-4 enzymes. *Curr. Opin. Cell Biol.* 17, 129–134.

Beaulieu JM & Caron MG (2005) β-arrestin goes nuclear. *Cell* 123, 755–757.

Bockaert J, Claeysen S, Becamel C et al. (2002) G protein-coupled receptors: dominant players in cell–cell communication. *Int. Rev. Cytol.* 212, 63–132.

Brzostowski JA & Kimmel AR (2001) Signaling at zero G: G-protein-independent functions for 7-TM receptors. *Trends Biochem. Sci.* 26, 291–297.

Bulenger S, Marullo S & Bouvier M (2005) Emerging role of homo- and heterodimerization in G-protein-coupled receptor biosynthesis and maturation. *Trends Pharmacol. Sci.* 26, 131–136.

Clevers H (2006) Wnt/β-catenin signaling in development and disease. *Cell* 127, 469–480.

Couty JP & Gershengorn MC (2006) G-protein-coupled receptors encoded by human herpesviruses. *Trends Pharmacol. Sci.* 26, 405–411.

DeWire SM, Ahn S, Lefkowitz J et al. (2007) β-Arrestins and cell signaling. *Annu. Rev. Physiol.* 69, 483–510.

Ferguson SSG (2001) Evolving concepts in G protein-coupled receptor endocytosis: the role in receptor desensitization and signaling. *Pharmacol. Rev.* 53, 1–24.

Ferrer JC, Favre C, Gomis RR et al. (2003) Control of glycogen deposition. *FEBS Lett.* 546, 127–132.

Franco R, Canals M, Marcellino D et al. (2003) Regulation of heptaspanning-membrane-receptor function by dimerization and clustering. *Trends Biochem. Sci.* 28, 238–243.

Gilchrist A (2007) Modulating G-protein-coupled receptors: from traditional pharmacology to allosterics. *Trends Pharmacol. Sci.* 28, 431–437.

Hooper JE & Scott MP (2005) Communicating with hedgehogs. *Nat. Rev. Mol. Cell Biol.* 6, 306–316.

Hupfeld CJ & Olefsky JM (2007) Regulation of receptor T tyrosine kinase signaling by GRKs and β-arrestins. *Annu. Rev. Physiol.* 69, 561–577.

Kalderon D (2002) Similarities between the hedgehog and Wnt signaling pathways. *Trends Cell Biol.* 12, 523–531.

Kallal L & Benovic JL (2000) Using green fluorescent proteins to study G-protein-coupled receptor localization and trafficking. *Trends Pharmacol. Sci.* 21, 175–180.

Kirstein SL & Insel PA (2004) Autonomic nervous system pharmacogenomics: a progress report. *Pharmacol. Rev.* 65, 31–52.

Kukkonen JP, Näsman J & Akerman KEO (2001) Modelling of promiscuous receptor–G_i/G_s-protein coupling and effector response. *Trends Pharmacol. Sci.* 22, 616–622.

Lefkowitz RJ (2004) Historical review: a brief history and personal retrospective of seven-transmembrane receptors. *Trends Pharmacol. Sci.* 25, 413–422.

LeRoy C & Wrana JL (2005) Clathrin- and non-clathrin-mediated endocytotic regulation of cell signalling. *Nat. Rev. Mol. Cell Biol.* 6, 112–126.

Lu ZL, Sakdanha JW & Hulme EC (2002) Seven-transmembrane receptors: crystals clarify. *Trends Pharmacol. Sci.* 23, 140–146.

Lum L & Beachy PA (2004) The hedgehog response network: sensors, switches, and routers. *Science* 304, 1755–1759.

Marchese A, Chen C, Kim YM et al. (2003) The ins and outs of G protein-coupled receptor trafficking. *Trends Biochem. Sci.* 28, 369–376.

Meyer T & Truel MN (2003) Fluorescence imaging of signaling networks. *Trends Cell Biol.* 13, 101–106.

Milligan G, Stoddart LA & Brown AJ (2006) G protein-coupled receptors for free fatty acids. *Cell Signalling* 18, 1360–1365.

Moon RT, Kohn AD, De Ferrari GV et al. (2004) Wnt and β-catenin signaling: diseases and therapies. *Nat. Rev. Genet.* 5, 691–701.

Moore CAC, Milano SK & Benovic JL (2007) Regulation of receptor trafficking by GRKs and arrestins. *Annu. Rev. Physiol.* 69, 451–482.

Ogden SK, Ascano M Jr, Stegman MA et al. (2004) Regulation of hedgehog signaling: a complex story. *Biochem. Pharmacol.* 67, 805–814.

Oldham WM & Hamm HH (2008) Heterotrimeric G protein activation by G-protein-coupled receptors. *Nat. Rev. Mol. Cell Biol.* 9, 60–71.

Parton RG & Simons K (2007) The multiple faces of caveolae. *Nat. Rev. Mol. Cell Biol.* 8, 185–194.

Perez DM (2003) The evolutionary triumphant G-protein-coupled receptor. *Mol. Pharmacol.* 63, 1202–1205.

Pierce KL, Premont RT & Lefkowitz RJ (2002) Seven-transmembrane receptors. *Nat. Rev. Cell Biol.* 3, 639–651.

Piston DW & Kremers GJ (2007) Fluorescent protein FRET: the good, the bad and the ugly. *Trends Biochem. Sci.* 32, 407–414.

Premont RT & Gainetdinov RR (2007) Physiological roles of G protein-coupled receptor kinases and arrestins. *Annu. Rev. Physiol.* 69, 511–534.

Sakmar TP (2002) Structure of rhodopsin and the superfamily of seven-helical receptors: the same and not the same. *Curr. Opin. Cell Biol.* 14, 189–195.

Schöneberg T, Schulz A & Gudermann T (2002) The structural basis of G-protein-coupled receptor function and dysfunction in human diseases. *Rev. Physiol. Biochem. Pharmacol.* 144, 143–227.

Sexton PM, Morfis M, Tilakaratne N et al. (2006) Complexing receptor pharmacology. Modulation of family B G protein-coupled receptors by RAMPs. *Ann. N.Y. Acad. Sci.* 1070, 90–104.

Stagljar I & Fields S (2002) Analysis of membrane protein interactions using yeast-based technologies. *Trends Biochem. Sci.* 27, 559–563.

Tsao P, Cao T & von Zastrow M (2001) Role of endocytosis in mediating downregulation of G-protein-coupled receptors. *Trends Pharmacol. Sci.* 22, 91–96.

Vergnolle N, Ferazzini M, D'Andrea MR et al. (2003) Proteinase-activated receptors: novel signals for peripheral nerves. *Trends Neurosci.* 26, 496–501.

Versele M, Lemaire K & Thevelein JM (2001) Sex and sugar in yeast: two distinct GPCR systems. *EMBO Rep.* 2, 574–579.

Wettschureck N & Offermanns S (2005) Mammalian G proteins and their cell type specific functions. *Physiol. Rev.* 85, 1159–1204.

Willets JM, Challiss RAJ & Nahorski SR (2003) Non-visual GRKs: are we seeing the whole picture? *Trends Pharmacol. Sci.* 24, 626–633.

Wilson CW & Chuang PT (2006) New "hogs" on hedgehog transport and signal reception. *Cell* 125, 435–438.

Wojcikiewicz RJH (2004) Regulated ubiquitination of proteins in GPCR-initiated signaling pathways. *Trends Pharmacol. Sci.* 25, 35–41.

Yeaman SJ (2004) Hormone-sensitive lipase—new roles for an old enzyme. *Biochem. J.* 379, 11–22.

Signal Transduction by Serine/Threonine Kinase-Coupled Receptors

In Chapter 5, heptahelical receptors coupling with trimeric G-proteins have been hailed as an evolutionary model of success originating from archaeal rhodopsins and dominating signal transduction in animals. The same holds true for protein kinase-coupled receptors occupying the first rank in prokaryotes and plants and the second rank in animals. In fact, signal transduction across membranes by means of protein phosphorylation is the most abundant method of communication between cells and environment, and the corresponding receptors have a venerable evolutionary history as the most ancient devices of transmembrane signaling. Constituting a large and heterogeneous protein family, they are found in all living beings. This is in clear contrast to G-protein-coupled receptors, which seem to be restricted to eukaryotes, particularly to animals (leaving aside the special case of archaebacterial rhodopsins).

6.1 Protein kinase-coupled receptors: the principle of dimerization-driven signal transduction

We have already encountered ancient prototypes of protein kinase-coupled receptors in prokaryotes: the histidine (His) kinase-coupled receptors of two-component systems. Here the kinase activity is mostly intrinsic, harbored in the cytoplasmic domain of the protein, while only a few receptors (particularly the chemotaxis receptors) associate noncovalently with a separate His kinase. These two principles—intrinsic and associated protein kinase activity—have been preserved throughout evolution, even though during eukaryotic development the His-specific kinases were replaced by serine/threonine- (Ser/Thr-) and tyrosine- (Tyr-) specific kinase activities (Figure 6.1).

All protein kinase-coupled receptors are transmembrane proteins possessing one (or, as in the case of His kinase-coupled receptors, more than one) transmembrane domain. The protein kinase-coupled receptors of eukaryotes share a common mechanism of activation. In the absence of a ligand, they are blocked in an inactive conformation. To transduce signals they have to assemble into **dimers** (or higher oligomers). Depending on the receptor type, the role of the ligand is to induce (or stabilize) a conformational change that promotes dimerization or stabilizes preformed dimeric receptor complexes. As a result the intrinsic or associated kinase domains come into a juxtaposition to activate each other. In most cases (but not always) this activation requires activation loop phosphorylation that is generally achieved by **trans-autophosphorylation**. Depending on the receptor type, this occurs according to two different principles: either the monomers phosphorylate each other (principle of equal partnership) or – rather an exception – one monomer phosphorylates the other one, which transmits the signal downstream (master–slave principle). By phosphorylating effector proteins, the activated kinases then induce intracellular signaling cascades

Figure 6.1 Protein kinase-coupled receptors: an overview The receptors exist as oligomers (at least as dimers) of subunits with at least one transmembrane domain each. The colored rectangles symbolize the protein kinase domains (left) and subunits (right).

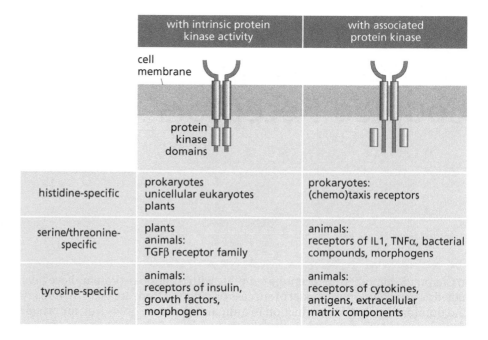

	with intrinsic protein kinase activity	with associated protein kinase
histidine-specific	prokaryotes unicellular eukaryotes plants	prokaryotes: (chemo)taxis receptors
serine/threonine-specific	plants animals: TGFβ receptor family	animals: receptors of IL1, TNFα, bacterial compounds, morphogens
tyrosine-specific	animals: receptors of insulin, growth factors, morphogens	animals: receptors of cytokines, antigens, extracellular matrix components

connecting the periphery with the metabolic apparatus and the genome. Frequently the cytoplasmic domain of the receptor protein itself becomes phosphorylated in addition to activation loop phosphorylation (Figure 6.2).

This generates docking sites for interacting proteins, an effect that is particularly pronounced for Tyr kinase-coupled receptors (Chapter 7). Moreover, phosphorylation may prolong signal transduction across the cell membrane: even when the ligand has dissociated, the receptor or the associated protein kinase, respectively, stays active until it is dephosphorylated by a cognate protein phosphatase or otherwise down-modulated. This is in clear contrast to the prokaryotic receptor His kinases, where the phosphorylated form is a short-lived transition state of a phosphotransfer reaction between the receptor and the downstream substrate protein, the response regulator (Section 3.3). The evolutionarily more ancient Ser/Thr kinase-coupled receptors are discussed below, while the more modern Tyr-kinase-coupled species are described in Chapter 7.

Summary

Protein kinase-coupled receptors of eukaryotes are proteins with one transmembrane domain and either an intrinsic cytoplasmic kinase domain or a docking domain for interaction with a separate kinase. Mutual activation of the

Figure 6.2 Mode of action of protein kinase-coupled receptors The receptor protein acts as an adaptor linking an extracellular signal with intracellular protein kinases. (A) The inactive receptor dimer (or oligomer) with intrinsic or associated kinase domains (red) encounters an extracellular signal molecule S. (B) Binding of S causes a conformational change inducing kinase activity by trans-autophosphorylation (or bringing together constitutively active kinase domains). (C) The activated kinase domains may catalyze the phosphorylation of the receptor protein, thus generating docking sites for the recruitment of cellular signaling proteins, which may become phosphorylated by the receptor-associated kinase or otherwise activated. Until signal extinction by dephosphorylation, the receptor–kinase complex remains active even upon dissociation of S.

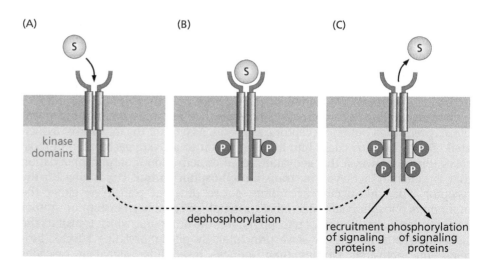

kinase domains, mostly by trans-autophosphorylation, requires oligomerization (mostly dimerization).

6.2 Receptors with intrinsic serine/threonine kinase activity: transforming growth factor β receptor family

Transmembrane proteins with intrinsic kinase activity represent the simplest molecular architecture of kinase-coupled receptors. The Ser/Thr-specific forms are by far the most abundant receptor type in plants (see Sidebar 6.1), whereas in animals they are restricted to the rather small family of TGFβ receptors. TGFβ stands for **Transforming Growth Factor** β, a historically based term since the factor was discovered by the observation that, together with the unrelated TGFα (which interacts with a receptor with intrinsic Tyr kinase activity; see Section 7.1), it induced the neoplastic transformation of cell cultures. However, this is by no means the main function of TGFβ. Instead, the protein has turned out to be a multifunctional tissue hormone regulating many physiological processes from embryogenesis to wound healing. TGFβ ranks among the most potent inducers of extracellular matrix proteins and seems to be the strongest endogenous inhibitor of epithelial, endothelial, and lymphatic cell proliferation. However, depending on tissue and physiological context, TGFβ also promotes cell proliferation and cell differentiation.

Beside TGFβ proper, the TGFβ family comprises at least 30 additional factors, mainly exhibiting morphogenetic functions such as the **activins**, the **inhibins,** and the **bone morphogenetic proteins (BMP)**. The latter are expressed in more than 20 isoforms and, contradicting their name, are involved in more than just bone formation. In fact, gene knockout experiments demonstrate an essential role of TGFβ and related factors in various stages of embryonic development where they cooperate with other morphogenetic signals such as Wnt and hedgehog (Section 5.9). Particularly exciting is their function as key regulators of stem cell renewal and proliferation.

Functionally and structurally closely related are the morphogenetic factors DPP (referring to the gene locus _decapentaplegic_ that regulates the development of 15 imaginal discs in _Drosophila_ larvae) and corresponding proteins of

---

Sidebar 6.1 Receptorlike protein kinases in plants
Arabidopsis thaliana was the first plant with a completely sequenced genome. It possesses no less than 417 genes that encode proteins with a single transmembrane domain and a cytoplasmic Ser/Thr kinase domain, a topography resembling that of the TGFβ receptors, whereas Tyr kinase-coupled receptors have not been found (although some of the _Arabidopsis_ kinases seem to phosphorylate both Ser/Thr and Tyr residues). As compared with plants, animals are rather low in receptor kinase genes: _C. elegans_ contains 3 and 40 (receptor Ser/Thr kinases and receptor Tyr kinases), _Drosophila_ 5 and 20, and humans 12 and 58 genes of this type. However, a ligand has been identified for only one of the plant kinases: the steroid hormone brassinolide, which controls longitudinal plant growth. Since no ligands have been found yet for the other plant enzymes, it is preferable to refer to them as receptorlike kinases. Nevertheless, structural features of the extracellular domains allow us to draw some conclusions on the mechanisms of activation. Most abundant are motifs characteristic of cell adhesion molecules (CAMs) such as leucine-rich repeats, lectin sequences, and epidermal growth factor-like repeats, but there are also motifs normally found in extracellular matrix proteins. Thus, potential ligands might be cell surface proteins and carbohydrates (interacting with lectin sequences) as well as components of the extracellular matrix. Physiological processes that have been tentatively connected with receptorlike kinases include pollen recognition, embryogenesis, organ differentiation, development of gametophytes, control of cell volume, and defense of pathogens. Little is known about the mechanisms of cellular signal processing. Nevertheless, one may assume that, as far as complexity is concerned, such mechanisms certainly rival those found in animals.

Caenorhabditis elegans, indicating the evolutionary ancientness of this receptor type. The pathways of signal transduction are identical in principle for all these factors and are explained in the following section.

6.2.1 Receptor activation and downstream signaling

Receptors of the TGFβ family are grouped into three types: I, II, and III. All types have a single transmembrane motif, and types I and II also have cytoplasmic Ser/Thr kinase domains. Upon binding of the dimeric ligands (the active forms of TGFβ and its relatives are homodimers, in some cases also heterodimers), each receptor type dimerizes and cooperates with the other types. Signal transduction occurs according to the master–slave principle. The cytoplasmic kinase domain of receptor type II, the master, is constitutively active. It has only one substrate: receptor type I, the slave. The job of the ligand is to join both receptor types, thus enabling phosphorylation of type I by type II. To this end, both receptor types have ligand-binding sites. Type I receptors are expressed in several isoforms that are subsumed under the term **activin-like kinases** (ALK1–6). These isoforms are specifically phosphorylated by individual type II receptors.

Phosphorylation results in a stimulation of the type I kinase, which is the actual signal transducer. Deviating from the rule, this phosphorylation does not occur in the activation loop but at multiple Ser and Thr residues in another segment of the kinase molecule. The effect is the same: the active center of the kinase is made accessible for the substrates ATP and protein. One function attributed to the kinase-inactive receptor type III (also known as β-glycan, since chemically it is a membrane-anchored proteoglycan) is to concentrate ligands near the type II receptor. Through this effect, type III receptors are thought to potentiate, for instance, the effects of inhibins. These hormone-like proteins are distantly related to TGFβ. They have become known as endogenous inhibitors of follicle hormone release by the anterior pituitary gland. By blocking the corresponding type II receptors, inhibins may antagonize the effects of another group of TGFβ-related hormones, the activins. The inhibin–activin system plays an important role in the regulation of the menstrual cycle and thus is of interest in reproductive medicine.

In its simple straightforward strategy, the TGFβ system is reminiscent of a bacterial two-component system: the major substrates of the type I receptor kinase are transcription factors. These are known as **SMADs** (the term is composed of Sma, referring to the small body size of *Dauer* larvae of *C. elegans* induced by stress-activated TGFβ signaling, and MAD, referring to Mothers-Against-DPP mutants of *Drosophila*). SMAD proteins are linked with the receptor by the membrane-associated adaptor protein SARA (SMAD Anchor of Receptor Activation; see Figure 6.3) but they dissociate upon phosphorylation.

Eight human SMAD isoforms are known. Their different functions gave rise to a distinction between regulatory R-SMADs (isoforms 1–3, 5, and 8), co-regulatory Co-SMADs (isoform 4), and inhibitory I-SMADs (isoforms 6 and 7). Only R-SMADs are phosphorylated by type I receptor. As a result, they form dimers with the Co-SMAD that enter the nucleus, whereas I-SMADs are negative regulators that prevent phosphorylation of R-SMADs and the interaction with Co-SMAD. Thus, the receptor–SMAD complex represents a sophisticated system of logical gates. SMAD signals are terminated by dephosphorylation and degradation along the ubiquitin–proteasome pathway (the SMAD-specific ubiquitin ligase is called SMAD-Ubiquitylation Regulating Factor, or SMURF).

6.2.2 Transcriptional control by SMADs

R-SMAD/Co-SMAD complexes help to control the transcription of numerous genes and thereby interact with other signaling pathways. For this purpose they bind to specific DNA sequences that are found in many transcriptional enhancers and promoters, which also contain binding sites for other transcription factors

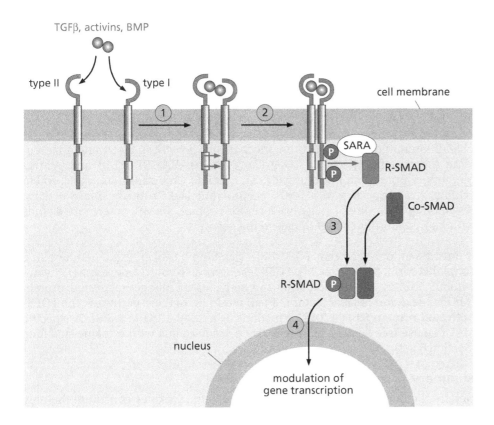

Figure 6.3 Signal transduction by transforming growth factor β receptors (Step 1) Upon binding of the dimeric ligand, receptor types I and II associate; type II activates type I by phosphorylation (red arrows). (Step 2) Assisted by the membrane-attached adaptor protein SARA, phosphorylated (P) receptor I binds and phosphorylates an R-SMAD protein. (Step 3) The phosphorylated R-SMAD dimerizes with a Co-SMAD protein. (Step 4) The R-SMAD/Co-SMAD complex accumulates in the nucleus and modulates gene transcription. Type I and II receptors exist as homodimers. For the sake of clarity, only the monomeric forms are shown.

such as AP1, ATF2 (see Section 11.7), the nuclear receptors of vitamin D and glucocorticoid (see Section 8.3), and Forkhead activin signal transducers. The task of the SMAD proteins is to assemble transcriptional co-activators (such as the histone acetyltransferase CBP/p300 and the SWI-SNF chromatin remodelling complexes) or co-repressors (such as histone deacetylases) at the initiation sites marked by those transcription factors (Figure 6.4). Since the latter are activated along additional signaling pathways and, moreover, SMAD activity is fine-tuned by phosphorylation catalyzed by various protein kinases, TGFβ signaling is integrated in a complex network of signaling cross talk enabling all kinds of logical decisions.

Interaction of SMAD with co-repressors requires the assistance of adaptor proteins that displace co-activators from the complex. These adaptors are called **Ski**, referring to their discovery as oncoproteins of the Sloan-Kettering retrovirus that mainly infects birds. However, overexpression of Ski has also been described for

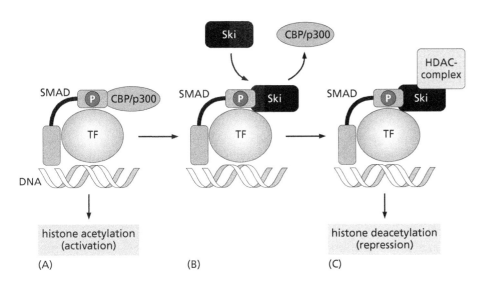

Figure 6.4 Modulatory effect of SMAD and Ski on gene transcription Most R-SMAD proteins have two conserved domains (light gray rectangles) that interact, as a complex with Co-SMAD (not shown), with DNA (albeit some R-SMADs do not bind to DNA) and another transcription factor (TF). (A) Phosphorylated SMAD recruits the co-activator histone acetyltransferase CBP/p300, thus activating the gene marked by TF. (B) Ski displaces the co-activator from the complex. (C) The SMAD–Ski complex recruits a co-repressor, histone deacetylase (HDAC), which slows down gene expression. P symbolizes the phosphorylation of SMAD by the TGFβ receptor.

223

human tumors. Ski probably facilitates tumor growth by suppressing the anti-proliferative effect of TGFβ in epithelia, which is due to the activation of genes encoding cell cycle inhibitors such as p15^{INK4} (see Section 12.5). Under certain circumstances, Ski may also slow down tumor development, probably by eliminating stimulatory effects of TGFβ on invasive growth and metastasis. Normally Ski seems to be essential for development of the nervous system and skeletal muscles, as indicated by experiments with Ski knockout animals.

In addition to the Ski system, other mechanisms have been discovered that regulate the various effects of TGFβ-like proteins. **BAMBI** (<u>B</u>MP and <u>A</u>ctivin <u>M</u>embrane-<u>B</u>ound <u>I</u>nhibitor) provides an example. This transmembrane protein exhibits the properties of a BMP receptor but lacks protein kinase activity. Resembling a dominant negative receptor mutant, BAMBI intercepts ligands without transmitting SMAD signals to the genome.

It must be emphasized that TGFβ/BMP signaling has been explained here in a simplified form. Most probably, TGFβ receptors also control signaling pathways without SMADs, thus modulating the activities of mitogen-activated protein (MAP) kinase and nuclear factor (NF) κB modules. In these pathways, the **TGFβ-Activated protein Kinase TAK1** functions as a relay. TAK1 is a Ser/Thr-specific MAP3 kinase (see Section 11.3) that also provides a link with cytokine signaling (Section 6.3.2).

Summary

Hetero-oligomeric receptors of TGFβ and related cytokines constitute the only family of Ser/Thr-coupled transmembrane receptors in animals. Ligand binding induces both receptor phosphorylation and, in turn, substrate phosphorylation. Major substrate proteins are the SMAD transcription factors, which, in complexes with other transcription factors, stimulate or repress the activity of a wide variety of genes, the majority of which regulate morphogenetic processes. In plants, receptorlike Ser/Thr kinases represent one of the largest protein families. Their functions are still widely obscure.

6.3 Cytokine receptors associated with serine/threonine kinases: key players in defense reactions

Protection from microbial and parasitic attacks is an essential and basic condition of survival. For this purpose, organisms make use of an arsenal of defense reactions such as the production of toxic compounds: for instance, reactive oxygen species and antibacterial or antifungal metabolites. This rather unspecific "chemical warfare," known as **innate immunity**, is employed even by simple animals and by plants. In fact, innate immunity provides the first and evolutionarily most ancient line of defense found throughout the metazoan kingdom. In higher animals the major effector cells of the innate immune system are phagocytic white blood cells, but most other cell types are able to respond as well. This holds true, in particular, for epithelia of skin, eyes, lung, cervix, vagina, intestine, and other tissues, which function as barriers to microbial invasion.

In vertebrates, the innate defense system is complemented by the highly specific **adaptive** (or **acquired**) **immune system**. This late achievement of evolution is based on the generation of tailor-made antibodies and antigen receptors by lymphocytes. In contrast to the immediately responding innate immune system, adaptive immunity reacts with some delay and is also characterized by a long-term memory effect. Adaptive and innate immunity are interlinked in that they depend on each other. Thus, phagocytes provide antigenic peptide fragments, and a major task of antibodies is to mark invaders as targets of killer cells, including those of the innate immune system.

Both the innate and the adaptive immune systems communicate through and are under the control of an intercellular signaling network that makes use of a wide variety of peptide factors, most of which interact with protein kinase-coupled receptors. The evolutionary progress from innate to adaptive immunity is mirrored by the mechanisms of signal transduction: while the factors stimulating and controlling innate immunity use "old-fashioned" Ser/Thr phosphorylation, adaptive immunity is mainly regulated by "modern" Tyr kinase-coupled receptors, in particular those interacting with antigens (Section 7.2.2). A practical consequence is that the signaling mechanisms controlling innate immunity can be studied in nonvertebrate models such as worms and insects, an approach that is not feasible for adaptive immunity.

Most of the factors that control innate immunity and (in vertebrates) the communication between both innate and adaptive immunity are **cytokines**. Cytokine is a collective term for a large and heterogeneous group of protein and peptide mediators that exhibit autocrine, paracrine, and endocrine effects. While some of them are circulating hormones, most are local mediators or tissue hormones (also called autocoids). Cytokines transmit differentiation, proliferation, and death signals, in particular in the hematopoietic and lymphatic systems.

There are distinctions among chemokines, interleukins (IL), interferons, colony-stimulating growth factors (CSF), monokines, and cytokines of the tumor necrosis factor α (TNFα) family. Chemokines are cytokines with chemotactic activity that recruit defense cells such as granulocytes, monocytes, macrophages, and lymphocytes to the sites of action. Both the major sources and major targets of interleukins are white blood cells (one subgroup secreted mainly by lymphocytes is also known as lymphokines). Interferons are cytokines specialized for viral defense, while monokines are produced by monocytes and macrophages. Finally, CSFs are multipurpose growth factors secreted by and acting on a variety of cell types, in particular on white blood cells. A particular subgroup called neurokines provides intercellular communication in the nervous rather than in the hematopoietic system.

There are four major families of protein kinase-coupled cytokine receptors characterized by common structural features:

- family I: Tyr kinase-coupled receptors of most interleukins, growth factors G-CSF and GM-CSF, and circulating hormones such as prolactin, leptin, erythropoietin, and thrombopoietin
- family II: Tyr kinase-coupled receptors of interferons and IL10, -19, -20, -22, and -24
- family III: Ser/Thr kinase-coupled receptors of TNF and related proteins and low-affinity receptor of nerve growth factor*
- family IV: Ser/Thr kinase-coupled receptors of IL1 and microbial surface components

If one defines TGFβ and its relatives as cytokines (as is frequently done), then the TGFβ receptors would constitute a fifth family. However, in contrast to TGFβ receptors, the receptors of families I–IV do not harbor intrinsic kinase activity but associate with separate cytoplasmic protein kinases. Still other cytokines, in particular chemokines, interact with G-protein-coupled receptors. Finally, some authors also define growth factors that interact with receptor Tyr kinases as

---

*Nerve growth factor also interacts with the high-affinity receptor Trk, which is a receptor Tyr kinase (see Section 7.1.1)

cytokines. We do not follow this classification but discuss this topic separately in Chapter 7.

While family I and II receptors directly bind their cognate Tyr kinases (Section 7.2), family III and IV receptors use various adaptor proteins that serve as specific links to Ser/Thr kinases (Figure 6.5). This situation is reminiscent of the bacterial chemotaxis receptors, where the adaptor protein CheW brings together the His kinase CheA with the sensor proteins. As discussed in Section 3.3.4, such a mechanism enables the cell to couple a single signaling pathway with a variety of receptors and extracellular signals. In the case of Ser/Thr kinase-associated cytokine receptors, this common signaling pathway is the NFκB cascade described in Section 11.8. Most cytokines stimulate, in addition, MAP kinase cascades, particularly those with the stress-activated MAP kinases p38 and JNK (cJun N-terminal Kinase, see Chapter 11). A particular subgroup is also specialized for transmission of pro-apoptotic signals. Upon stimulation by cytokines a major task of these pathways is to process signals that control the activity of white blood cells, thus triggering the above-mentioned defense reactions. The perceptible outcomes of these signaling events are the various symptoms of inflammation and illness. Thus, many cytokines belong to the large and heterogeneous family of pro-inflammatory mediators. It must be emphasized, however, that, in higher animals, Tyr kinase-coupled receptors also participate in the regulation of inflammatory responses (see Sections 7.2.3 and 7.3.2).

In Figure 6.5 the principle of cytokine signal processing is depicted. In the following discussion we shall present selected examples, each illustrating an individual subtype of signaling. As we shall see, cytokine signaling is a complicated issue. One reason for this is the complexity of the inflammatory response, involving numerous cell types and a wide variety of cellular responses. This situation requires extended signaling cross talk at both intercellular and intracellular levels. Another reason is that inflammation in the absence of a pathogen would be a dangerous event for the body to be avoided at all costs, given that the hallmarks of inflammation (the activation of aggressive cell types and the production of highly cytotoxic metabolites) are processes that result in tissue destruction. An even more dangerous effect of an acute overproduction of many pro-inflammatory cytokines is shock, the potentially fatal breakdown of blood

Figure 6.5 Basic theme of type III and IV cytokine signal transduction (A) Upon binding of a ligand, the oligomeric cytokine transmembrane receptor undergoes a conformational change. Consequently, interaction domains of the cytoplasmic extension become exposed and bind corresponding adaptor proteins. These recruit cytoplasmic Ser/Thr-specific protein kinases that come in close contact, enabling activation through mutual (auto)phosphorylation. (B) The kinases, in turn, stimulate pro-apoptotic and/or immunomodulatory and pro-inflammatory signaling devices such as the NFκB modules (left) and the MAP kinase modules (right) primarily aiming at defense reactions (P, phosphorylation).

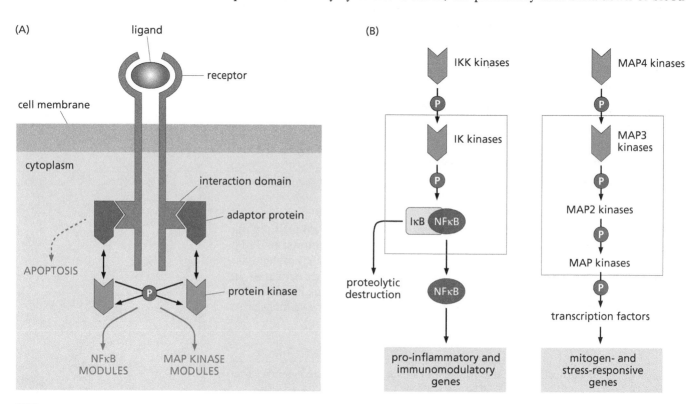

pressure. In fact, cytokine signaling provides a particularly illustrative example of how signaling cross talk and fine-tuning collaborate to safeguard a risky venture. Understanding these signaling interactions is crucial to any progress in treating some of the most widespread and dangerous diseases.

6.3.1 Tumor necrosis factor receptors: mediators of inflammation and apoptosis

On the basis of structural similarities, around 20 different cytokines with immunomodulatory, pro-inflammatory, and pro-apoptotic properties have been brought together in the so-called TNF superfamily. The name comes from the prototypical representative of this family, **tumor necrosis factor α (TNFα)**, and is somewhat misleading, since TNFα, in addition to its predominant immunomodulatory effects, causes tumor cell death by apoptosis rather than by necrosis. The cytokines of the TNF superfamily are expressed as homotrimeric transmembrane proteins of type 2 (N-terminus inside and C-terminus outside the cell), particularly by leukocytes and lymphocytes. They exert juxtacrine effects on adjacent cells. However, many of them also may be shed by membrane-bound proteases, in particular those of the ADAM family (see Section 13.1.2), to diffuse freely and act as paracrine signaling molecules. An overproduction of TNFα contributes to chronic inflammatory diseases such as rheumatoid arthritis, asthma, and Crohn's disease (bowel inflammation) and may have fatal consequences such as emaciation (cachexia) of tumor patients.

Receptors

In vertebrates, more than 40 different TNF receptors have been found. Most of them are transmembrane proteins of type 1 (C-terminus inside, N-terminus outside the cell) penetrating the membrane with a single transmembrane domain. The extracellular domains of the receptors are quite similar, reflecting the close relationship of the ligands. Their characteristic structural elements are cysteine-rich repeats. The mostly homotrimeric ligands assemble three receptor molecules into a homotrimeric complex, thus gating various signaling pathways.

The intracellular extensions of the receptors contain domains that interact with adaptor proteins and, thus, establish contact with downstream protein kinases. In some of the receptors a special type of contact domain is found: the so-called **death domain**. It enables the receptor to trigger programmed cell death or apoptosis. In fact, the strongest endogenous inducers of apoptosis are found within the TNF family of cytokines (see Section 13.2.3).

Signal transduction

The mechanisms of TNF signal transduction are rather complicated and not yet completely understood. A typical feature of the TNF receptors is their ability to establish large signal-transducing protein complexes by means of several adaptor proteins that are recruited upon ligand binding. Depending on the receptor type and the adaptors, TNF receptors control

- signaling devices such as stress-activated MAP kinase and NFκB modules that control the transcription of genes required for inflammation and immune modulation

- signaling cascades leading to apoptosis

How these processes occur will be explained for the **TNFα receptors 1** and **2**, the most thoroughly investigated members of the family. While TNFR1 is expressed by most tissues, TNFR2 is restricted to the cells of the immune system. The major structural difference between the receptors is that TNFR1 contains a cytoplasmic death domain, which is not found in TNFR2. By means of this death domain, TNFR1 is able to induce apoptosis. It shares this ability with a group of other death receptors that all belong to the TNF receptor family (a prototype, the death receptor Fas that interacts with the cytokine Fas ligand, is discussed in Section 13.2.3). To induce apoptosis, TNFR1 binds the adaptor **TRADD** (TNF Receptor-Associated Death Domain protein) by a homotypic interaction of death domains. When, in turn, TRADD interacts with another adaptor, **FADD** (Fas-Associated Death Domain protein), the caspase cascade is triggered. This standard route leading to cell death is described in detail in Section 13.2.3.

Alternatively, depending on the cell type, TRADD may bind the adaptor **TRAF2** (TNF Receptor-Associated Factor 2), thus activating the pro-inflammatory JNK/p38 MAP kinase and NFκB modules instead of pro-apoptotic signaling (Figure 6.6). Signal transduction depends on auto-ubiquitylation of TRAF2, a ubiquitin ligase (see Sidebar 6.2). To activate NFκB signaling, an additional adaptor **RIP1** (Receptor-Interacting Protein 1) is required. RIP1 is essential, since corresponding gene-knockout animals die prior to birth due to severe retardation of lymphatic system development. RIP1 contains a death domain and exhibits protein kinase activity, properties which do not appear to be required for stimulation of NFκB signaling. Instead, RIP1 functions as a scaffold for TRAF2 and NFκB module proteins (Section 11.8), recruiting them to the TNFR1–TRADD complex by means of a homotypic death domain interaction (Figure 6.6).

In humans, seven different RIP kinases have been found. They are closely related to another family of protein kinases, IRAKs, which are treated in more detail below. While RIP1 is engaged in TNF signal transduction, isoform RIP2 transmits signals from the proteins NOD1 and NOD2, two cytoplasmic receptors of bacterial peptidoglycans, to the NFκB module. This pathway described in Section 6.3.2 plays an important role in nonspecific defense of pathogenic microorganisms.

Since it lacks death domains, the receptor **TNFR2** (and various other receptors of the family) cannot interact with TRADD and FADD and therefore is unable to induce cell death. Instead, TNFR2 directly binds the adaptor ubiquitin ligase TRAF2. As in the case of TNFR1 signaling, TRAF2 undergoes auto-ubiquitylation and brings the receptor in contact with protein kinases that exhibit MAP3 kinase activity. As a result the signal is transduced to MAP kinase and NFκB modules. Possible candidates for such kinases are ASK1, NIK, MEKK1, and several MAP3 kinases of the GCK family (see Section 11.3).

Figure 6.6 Signal transduction by tumor necrosis factor receptors: the latest state of the art The trimeric ligand TNFα binds to the trimeric receptor TNFR1 (left) or TNFR2 (right). By its cytoplasmic death domains (DD, dark gray arrows), TNFR1 recruits the adaptor protein TRADD, which has a choice between two different downstream adaptors: FADD or TRAF2. Via FADD, the pro-apoptotic caspase cascade is activated; via TRAF2, the MAP kinase and NFκB cascades are stimulated. The latter effect requires an auto-polyubiquitylation of TRAF2 and, for NFκB activation in particular, the support of the adaptor kinase RIP1, a death domain protein. Both TRAF2 and RIP1 function as scaffolds, recruiting the proteins of the MAP kinase and NFκB modules to the receptor/TRADD complex, where they become activated. Receptor TNFR2 directly interacts with TRAF2, which again has to polyubiquitylate itself. TNF receptors may also bind the adaptor protein Grb2, thus activating the Ras–RAF–MAP kinase cascade (not shown). The protein kinases associated with the receptor/adaptor complexes are shown in the colored box (P, phosphorylation).

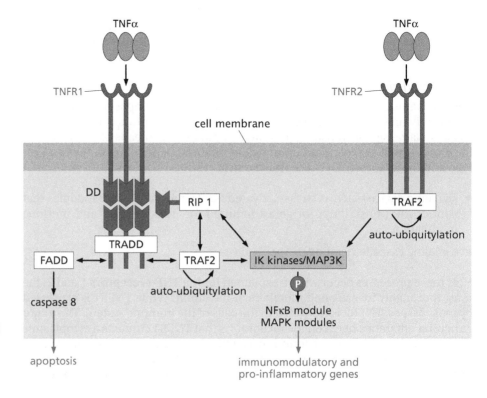

Sidebar 6.2 TRAF adaptors and the dual role of ubiquitylation Mammals express six different TRAF proteins that interact with receptors of the TNF receptor family. Moreover, TRAF3 and TRAF6 also link IL1- and Toll-like receptors to downstream protein kinases (see Section 6.3.2). TRAFs have a C-terminal TRAF domain for receptor binding and homo- or heterotrimerization as well as (with the exception of TRAF1) a **RING domain** (Figure 6.7). The latter is important for the binding and activation of protein kinases leading, in particular, to an activation of MAP kinase and NFκB modules. RING domains are, furthermore, characteristic structural elements of many E3 ubiquitin ligases (Section 12.16). In fact, TRAF2 and TRAF6 (at least) can fulfill their adaptor functions only when they have catalyzed their own polyubiquitylation (note the analogy to protein kinases, most of which have to undergo autophosphorylation to become active). For this modification, TRAF must be released from the complex with the receptor that inhibits the ubiquitin ligase activity. This dissociation is induced by ligand binding. TRAF1 lacking the RING domain seems to function as a competitive inhibitor of TRAF-controlled signaling reactions. In addition to receptors and protein kinases, other signaling proteins are bound by TRAFs, for instance, the adaptor TRADD and the caspase inhibitors IAP1 and IAP2 (Section 13.2.2).

The signaling pathway connecting TNF receptors with the NFκB module provides an instructive example of the dual effect of ubiquitylation, either being a stimulatory covalent signaling reaction or labeling the protein for degradation in proteasomes. As explained in Section 2.8, for stimulatory ubiquitylation the polyubiquitin residue transferred to the protein is linked via Lys residue 63. In this case, the reaction is catalyzed by the Lys63-specific E3 ubiquitin ligase activity of TRAF2. As a result, the structure of TRAF2 becomes modified in such a way that it can interact with MAP3 kinases or the IK kinases of the NFκB module, recruiting them to the receptor complex. The link is probably provided by the MAP3 kinase TAK1 together with its regulator proteins TAB1 and TAB2, which both have polyubiquitin-binding domains (see Section 6.3.2). In a Lys 63-specific reaction TRAF2 also polyubiquitylates RIP and the scaffold protein NEMO of the NFκB module.

To mark a protein for proteolysis, the polyubiquitin residue has to be linked via Lys48. This reaction plays an important role in NFκB activation since it eliminates the IKK inhibitors. On the other hand, it terminates signal transduction by removing NFκB. Both reactions are catalyzed by two different Lys48-specific E3 ubiquitin ligases that are not identical with TRAFs (Section 11.8).

Polyubiquitin chains are degraded by de-ubiquitylating proteases. One of these enzymes, called CYLD, has been identified as a tumor suppressor protein. A loss-of-function mutation of CYLD is the cause of cylindromatosis, a benign neoplastic disease of hair follicles and sebaceous glands. CYLD selectively degrades Lys63-linked polyubiquitin residues, thus opposing the effects of TRAF2.

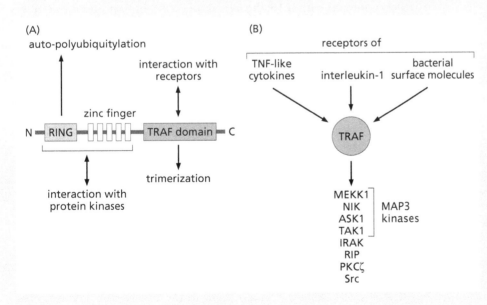

Figure 6.7 Coupling of receptors with cytoplasmic protein kinases by TRAF adaptors (A) Schematic domain structure of TRAFs 2–6. (B) Receptors upstream and protein kinases downstream of TRAF. For details, see text.

6.3.2 Interleukin 1 receptors and Toll-like receptors: mediators of defense reactions

The second largest family of mammalian receptors associated with Ser/Thr kinases is divided into two subfamilies: interleukin 1 (IL1) receptors and receptors of microbial structures. The latter are known as Toll-like receptors or TLRs, referring to the prototype *Toll* of a *Drosophila* mutant (see Sidebar 6.3). The predominant function of IL1 and TLRs is to regulate innate defense reactions, with IL1 receptors transducing endogenous signals and TLRs transducing exogenous signals. Like other cytokine receptors, they primarily stimulate the pro-inflammatory NFκB and MAP kinase cascades and, in addition, aid in the production of the antiviral interferons α and β.

IL1 receptors and TLRs are type 1 transmembrane proteins. They are expressed in various isoforms that form homo- and heterodimers upon ligand binding. The receptors are characterized by a cytoplasmic **TIR domain** (Toll-like–Interleukin 1 Receptor domain; see Figure 6.8) of about 200 amino acids that allows for the binding of adaptor proteins and is essential for signal transduction. The main differences between the two receptor types are found in the extracellular domains, reflecting the different chemical structures of the ligands: three immunoglobulin (Ig)- like domains in IL receptors as compared with leucine-rich repeats in TLRs.

Interleukin 1 receptors: master regulators of inflammation

The cytokine **IL1**, a 17 kDa protein expressed in two isoforms, α and β, is a key signal for the coordination of endogenous defense reactions. IL1 is released by leukocytes in response to injury, infections, and inflammatory reactions. It prompts a wide variety of tissues to produce pro-inflammatory tissue hormones (such as prostaglandins and cytokines), leukotactic chemokines, cell adhesion molecules, and hydrolytic enzymes catalyzing tissue degradation; in other words, the whole molecular arsenal of inflammation. Therefore, IL1 is considered to be the major trigger of the **classical symptoms of inflammation**: heat (including fever), redness (due to increased blood flow), swelling (due to increased vascular permeability), pain, and tissue destruction. In fact, IL1 was initially described as an endogenous pyrogen, a factor that, via effects on the thermoregulatory center in the hypothalamus, induced fever.

Mammalian cells express one IL1 receptor and at least eight related receptors. One of these interacts with **IL18**; the ligands of the others—probably members of

---

Sidebar 6.3 Toll-like receptors in *Drosophila*: dual role in embryogenesis and defense Among the TLRs of invertebrates, the prototype *Toll* (means "terrific") of *Drosophila* has been studied in the most detail. Its ligand is an endogenous peptide called *Spätzle* (referring to the form of the larvae that is reminiscent to a Swabian noodle speciality) that, upon demand, is released by controlled proteolysis from a precursor protein. This occurs either during larval development or as a response to fungal infections. Triggering signals, either morphogens or fungal molecules, activate different proteases such as the enzyme Persephone.

By this means the *Spätzle–Toll* system controls two entirely different physiological processes: dorso-ventral differentiation of the larvae and biosynthesis of fungicidic peptides such as drosomycin. In both cases, gene transcription is stimulated along the NFκB pathways including the same IκB-homologous inhibitor Cactus but differing from each other by the isoforms of the Rel transcription factors (for details see Section

11.8). The corresponding signaling cascades resemble those of vertebrate cells (Figure 6.9). By means of the adaptor protein dMyD88 (and an additional adaptor Tube found only in insects) the receptor recruits the kinase *Pelle* (homologous to IRAK), dTRAF (homologous to TRAF6), and the adaptor ECSIT, whereas the kinases resembling the vertebrate IκB kinases, MEKK1, and TAK1 have yet to be identified. Other TLRs of insects seem to participate in defense reactions as well.

TLRs are also present in other nonvertebrates lacking an adaptive immune system. In plants (at least in *A. thaliana*) a subgroup of the large family of transmembrane receptors containing a cytoplasmic Ser/Thr kinase domain seem to be involved in defense reactions. Approximately 80 corresponding "receptors" have been characterized as cytoplasmic rather than transmembrane proteins. Their ligands are unknown, but there is some evidence indicating a role in protection from pathogenic organisms such as fungi.

the IL1 family as is IL18—have yet to be identified. IL18, a potent pro-inflammatory mediator induces T-helper cell differentiation and activates natural killer cells. Both IL1 and IL18 are released from precursor proteins by the protease caspase 1, which has therefore been called IL1-Converting Enzyme (ICE; for a detailed discussion of caspases see Section 13.2.1). The mechanism by which caspase 1 becomes activated is explained in Sidebar 6.4.

The active form of the IL1 receptor is a heterodimer of two quite similar transmembrane proteins, IL1 R1 and IL1 R-associated protein. The conformational change induced by the ligand IL1 enables the receptor to bind the **adaptor protein MyD88** (the name refers to Myeloid cell Differentiation induced by IL1), which for IL1 signaling plays a similar role as the TRAF adaptors for TNF signaling. MyD88 contains an N-terminal death domain and a C-terminal TIR domain. The latter interacts with the corresponding TIR domains of the receptor, thus transmitting the IL1 signal to the MAP kinase and NFκB modules that, in turn, activate the transcription factors of the IL1-responsive genes.

As with TNFR, the mechanism of signal transduction is rather sophisticated (Figure 6.8). The major task of MyD88 is to recruit two Ser/Thr-specific protein kinases at the receptor. These kinases are called **IRAK1** and **IRAK4** (Interleukin 1 Receptor-Activated Kinases). They are close relatives of the RIP kinases involved in TNF signaling. IRAK4 links IRAK1 with MyD88 and catalyzes the phosphorylation of IRAK1. This enables IRAK1 to dissociate from the receptor and form a complex with the adaptor protein **TRAF6**. Like TRAF2, TRAF6 is a Lys 63-specific ubiquitin ligase (see Sidebar 6.2) that, upon stimulation, activates itself by auto-ubiquitylation. In the cytoplasm, the ubiquitylated TRAF6 then interacts with the MAP3 kinase **TAK1** bound to the regulator proteins TAB1 and TAB2 (see Section 11.3). As a result, TAK1 becomes activated by (auto?)phosphorylation and transmits the signal to the pro-inflammatory p38 MAP kinase and NFκB modules. At the same time, IRAK1 is degraded along the ubiquitin–proteasome pathway.

A parallel pathway leads to activation of another MAP3 kinase, MEKK1, resulting in signal transduction to the JNK and ERK MAP kinase modules. This seems to require an additional adaptor protein, **ECSIT** (Evolutionarily Conserved Signaling Intermediate of Toll pathways), linking TRAF6 with MEKK1 and promoting the autophosphorylation of MEKK1 (Figure 6.8).

As in the case of RIP1, the role of the kinase activity of IRAK1 is not essential for signal transduction but remains to be elucidated. IL1 signaling is protected from accidental stimulation at different stages of signal transduction and, in addition, by **TOLLIP** (Toll-Interacting Protein), which sequesters inactive IRAK1. Upon activation, simultaneously with the phosphorylation of IRAK1, TOLLIP is also phosphorylated, resulting in its degradation along the ubiquitin–proteasome pathway.

Toll-like receptors: defense against microorganisms and viruses

In the past, innate immune surveillance was assumed to function in an entirely nonspecific way. More recently, however, the system was found to be equipped with a sophisticated mechanism of signal transduction that ensures a very selective response to a wide variety of stimuli. In this context, TLRs have turned out to function as cellular sensors for unique chemical compounds that are restricted to and essential for parasitic and pathogenic viruses, bacteria, protozoans, and fungi. These **pathogen-associated molecular patterns** comprise cell surface components as well as bacterial and viral nucleic acids.

In mammals, TLRs are expressed in at least 12 isotypes that, like the related IL1 receptors, are active only as homo- or heterodimers. Some of these isoforms are ubiquitously expressed, whereas others are restricted to distinct tissues. The

Figure 6.8 Signal transduction by interleukin 1 receptor By its TIR domains (cylinders), the activated receptor heterodimer assembles a complex consisting of adaptor protein MyD88 and protein kinases IRAK1 and IRAK4. Through phosphorylation, IRAK4 induces the degradation of the IRAK1 inhibitor TOLLIP and activates IRAK1. Upon dissociation from the receptor complex, IRAK1 associates with TRAF6, which becomes activated by auto-ubiquitylation and transmits the signal via the adaptor ECSIT to the MEKK1-JNK and MEKK1-ERK modules and via the TAK1/TAB1,2 complex to the p38 MAP kinase and NFκB modules. As a result, pro-inflammatory mediators such as cytokines and arachidonic acid metabolites are synthesized (P, phosphorylation). In the box (red dotted line) the core of the signal-transducing complex assembing at the receptor is shown.

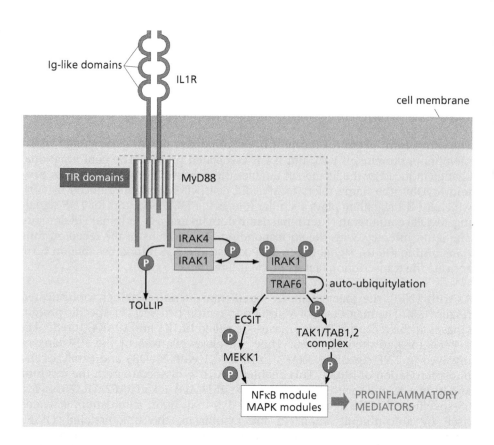

individual TL receptor types exhibit different ligand specificities. As a rule, TLR2, -4, and -5 recognize lipids, normal and modified peptides, and sugar derivatives, while TLR3, -7, -8, and -9 are nucleic acid receptors.

TLR4 has attracted particular interest because it is the cellular receptor of **lipopolysaccharides (LPS)**. These macromolecules are constituents of the cell wall of Gram-negative bacteria including the highly pathogenic species *Salmonella*, *Legionella*, *Bordetella pertussis*, and certain *Escherichia coli* strains. LPS, also known as **endotoxins**, cause severe symptoms of illness ranging from fever to **septic shock** (which still is one of the most frequent causes of death). The reason for this toxicity is an overactivation of MAP kinase and NFκB signaling. This results in a massive release of pro-inflammatory cytokines (such as TNFα and IL1, with the effect enhanced by positive feedback), arachidonic acid metabolites (such as leukotriene C_4 and thromboxane A_2; see Section 4.4.5), and nitric oxide. These factors, in particular NO, cause a dramatic and frequently fatal drop in blood pressure (see Section 16.5.1).

Gram-positive bacteria such as *Bacillus*, *Streptococcus*, and *Staphylococcus* species do not produce LPS. Instead, the innate immune system recognizes other cell wall components such as **lipoteichoic acids**, **lipoproteins**, and **peptidoglycans** (see Sidebar 6.5). The corresponding cellular receptors are TLR1 and TLR2. TLR2 (and TLR4) also interacts with special cell wall and cell membrane components (lipoglycans and proteins) of mycobacteria and infectious fungi. The flagellar protein **flagellin**, another pathogen-associated molecular pattern of bacteria, binds to TLR5. TLR9 is a specific receptor for **CpG-rich DNA** found in a nonmethylated form only in prokaryotes, whereas in eukaryotes it is always methylated (Section 8.2.8). In contrast to TLR2 and TLR4, TLR9 is concentrated in the endosomal rather than in the plasma membrane. In endosomes, bacterial DNA is degraded, yielding single-stranded fragments with CpG-rich DNA, which are the TLR9 ligands proper. TLR9 as well as TLR2 and TLR4 in addition recognize surface molecules of infectious protozoans such as *Trypanosoma* and *Plasmodium*. TLRs are also the body's outpost in the battle against **viruses**. In

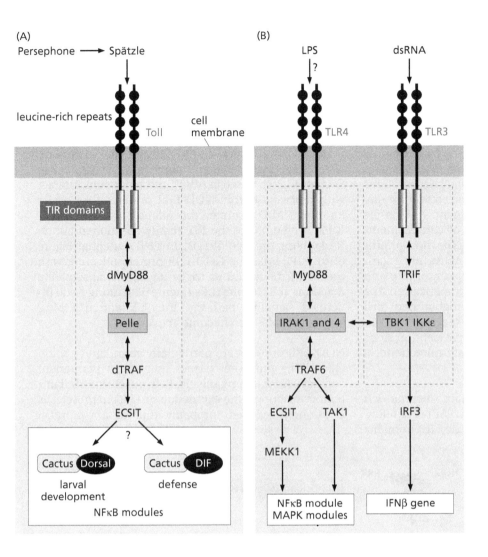

(A)

Persephone → Spätzle

leucine-rich repeats — Toll

cell membrane

TIR domains

dMyD88

Pelle

dTRAF

ECSIT

?

Cactus Dorsal — Cactus DIF

larval development — defense

NFκB modules

(B)

LPS — dsRNA

?

TLR4 — TLR3

MyD88 — TRIF

IRAK1 and 4 ↔ TBK1 IKKε

TRAF6

ECSIT — TAK1 — IRF3

MEKK1

NFκB module MAPK modules — IFNβ gene

Figure 6.9 Signal transduction by Toll and Toll-like receptors (A) Toll-activated signaling cascades of *Drosophila*. Persephone is one of the proteases releasing the ligand Spätzle from a precursor protein. Dorsal and DIF are two Rel isoforms of the NFκB modules addressing different genes, while Cactus is the IκB-homologous inhibitor becoming degraded upon activation of the modules (see Section 11.8). (B) Two TLR-activated signaling cascades of vertebrates. One of the cascades that is typically induced by TLR4 completely resembles IL1 signaling. It may be stimulated by lipopolysaccharide (LPS) and leads to a release of pro-inflammatory mediators. Rather than interacting directly with the receptor, LPS might provoke the release of a Spätzle-like peptide. The other cascade, shown here for the receptor TLR3 stimulated by double-stranded viral RNA (dsRNA), leads to the synthesis of the antiviral interferon β (IFNβ). The protein kinases associated with the receptor/adaptor complexes are shown in the colored boxes.

fact, together with adaptive immunity and inhibition of viral protein synthesis by the eIF2 kinase PKR (Section 9.3.1) and by RNA interference (Section 8.8), these receptors and the corresponding signaling pathways represent the body's arsenal against viral infections.

While in some cases TLR4 recognizes glycoproteins of viral envelopes, other TLRs respond specifically to viral nucleic acids. Thus, double-stranded RNA, a characteristic product of most viral infections, selectively activates TLR3, while TLR7 and -8 are sensors of single-stranded RNA of RNA viruses (for example, HIV and influenza virus) and TLR9 recognizes nonmethylated CpG sequences found in the genomes of DNA viruses (for example, herpes and papilloma virus). Like other TLRs, these receptors stimulate nonspecific defense reactions via the TRAF6-NFκB cascade and the stress-activated MAP kinase modules. In addition, they gate a specific antiviral pathway leading to the transcription of the genes of the type 1 interferons α and β. These cytokines are antiviral key regulators (Sidebar 6.6). The interferon genes are specifically upregulated by interferon response element-binding factors (IRFs). These transcription factors are expressed in various isoforms and require protein phosphorylation to become active. TLR7, -8, and –9 seem to use the conventional signaling pathway via MyD88, TRAF6 and IRAK1 and –4. As a result not only NFκB and MAP kinase modules are activated but also the transcription factor IRF7 that controls interferon α and -β genes. The TLR3 signaling pathway is a variation of this theme (Figure 6.9). The receptor interacts via its TIR domains with the adaptor protein TRIF (TIR domain inducer of interferon β) rather than with MyD88 and organizes

Sidebar 6.4 NOD-LRR proteins: an intracellular line of defense against microorganisms TLRs represent an antimicrobial defense line at the cell surface. But what happens to microbes that escape this surveillance and invade the cell? For quite some time plants have been known to possess various cytoplasmic proteins (called resistance or R-proteins) for the recognition and elimination of such intruders. Only recently have similar proteins been found in animals and humans. They belong to a family called NOD proteins because of a characteristic <u>N</u>ucleotide-binding <u>O</u>ligomerization <u>D</u>omain. This domain is closely related to the ATP-binding cassette of certain membrane transporters (see Section 3.2.1). The most prominent NOD protein is the <u>AP</u>optosis-<u>A</u>ctivating <u>F</u>actor **APAF1** that, upon stimulation by cytochrome c and ATP/dATP (binding to NOD) and oligomerization to a so-called apoptosome, induces programmed cell death by activating procaspase 9 (Section 13.2.4). Among the NOD proteins, APAF1 is an exception since it is a sensor of stress effects that damage mitochondria rather than of microbes.

Antimicrobial NOD proteins contain leucine-rich repeats (LRRs) that are not found in APAF1. Therefore they are called **NOD-LRR proteins**. The LRR domains resemble the extracellular domains of TLRs and, in fact, enable the NOD-LRR proteins to recognize bacterial components such as peptidoglycans and possibly also LPS. Interaction with these ligands results in a conformational change that exposes the NOD domain for oligomerization and binding of ATP, which are conditions of the subsequent reactions. Like APAF1, the major NOD-LRR proteins (NOD1 and NOD2) contain N-terminal <u>CA</u>spase <u>R</u>ecruitment <u>D</u>omains (CARD, see Section 13.2.3). While in APAF1 these domains interact with procaspase 9, in NOD-LRR proteins they prefer another CARD protein, the adaptor/protein kinase RIP2, a member of the RIP family mentioned above. Resembling the role of RIP1 in TNF signaling (Figure 6.10), RIP2 links the NOD-LRR protein oligomer with the NFκB module, thus triggering the same signaling reactions as TNF and TLRs (Figure 6.9). Along a still ill-defined signaling pathway, the JNK and p38 MAP kinase modules also become activated.

NOD-LRR proteins are particularly abundant in colon epithelium and other tissues that are in permanent contact with microorganisms. To avoid chronic inflammatory reactions, the expression of TLRs in these cells is strongly reduced, probably due to a reciprocal *continued*

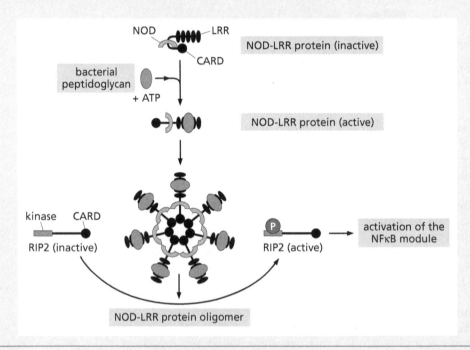

Figure 6.10 Activation and function of a cytoplasmic pathogen receptor protein of the NOD-LRR family Binding of bacterial peptidoglycan to the regulatory domain (leucine-rich repeats, LRR, shown as black ellipses) of NOD-LRR induces, via a conformational change, oligomerization to a wheel-like structure (the nucleotide-binding oligomerization domains are shown in light gray). Working like a machine, the oligomer recruits several molecules of protein kinase RIP2 by homotypic interaction of CAR domains (CARD, black circles) and enables their activation, probably through proximity-induced trans-autophosphorylation (P) or by another mechanism. The active RIP2 stimulates NFκB signaling. Activation of the NOD-LRR protein requires, in addition, binding (and hydrolysis?) of ATP or GTP at the NO domains (NOD). The mechanism depicted here applies to the CARD-containing NOD-LRR isoforms such as NOD1 and NOD2.

interaction between the NOD-LRR protein and the TLR systems. In fact, an inactivation of the prototype NOD2 resulting from a loss-of-function mutation has been found to promote the development of chronic inflammatory bowel diseases such as **Crohn's disease**.

An additional line of defense consists of the release of pro-inflammatory cytokines such as IL1 and IL18, which prime white blood cells and lymphocytes for antimicrobial defense. This is achieved by a partial proteolytic degradation of precursor proteins (called pro-interleukins) catalyzed by the protease ICE, which is identical with **caspase 1**, a member of the family of inflammatory caspases (Section 13.2.1). Like RIP2, caspase 1 contains a CARD motif for homotypical interactions.

As is typical for caspases (and most proteases), caspase 1 is synthesized as an inactive zymogen (procaspase) that, upon activation, releases the active enzyme by partial proteolysis, here by autoproteolysis. Resembling RIP2 activation, caspase 1 activation is promoted by a series of protein complexes consisting of particular NOD-LRR isoforms (known as NALPs, NAIPs, and IPAFs) that respond to microbial and cytotoxic factors. These complexes, known as **inflammosomes,** are analogous to the apoptosome, and bring the procaspase 1 molecules in close contact, thus enabling activation by partial trans-autoproteolysis. Like RIP2, procaspase 1 binds to the complexes through homotypical CARD–CARD interactions assisted by special adaptor proteins.

a signaling complex together with the factor IRF3 and a protein kinase resulting in IRF3 phosphorylation and activation of the interferon-β gene. Two IRF3 kinases are known: TBK1 (TANK-binding kinase 1; TANK, the TRAF associated NFκB activator, is a scaffold protein) and IKKε. They also activate the NFκB pathway. Since the expression of TLR3 is induced by interferon β, this defense mechanism amplifies itself by positive feedback.

In contrast to TLR4 that is exposed at the cell surface TLR3, -7, -8 and -9 are found primarily in endosomal membranes with their ligand binding sites oriented towards the endosomal lumen where they interact with viral components taken up by endocytosis. This limits their antiviral efficacy because they are unable to detect viral nucleic acids in the cytoplasm. To cope with this problem, cells have developed an additional defense strategy based on cytoplasmic proteins that recognize double-stranded RNA. For RNA binding, these proteins contain an RNA helicase domain. Two of these RNA sensors are known: RIG1 (<u>R</u>etinoic acid-<u>I</u>nducible <u>G</u>ene 1) and MDA5 (<u>M</u>elanoma <u>D</u>ifferentiation-<u>A</u>ssociated gene 5). Upon interaction with a mitochondria-derived adaptor these proteins activate NFκB and MAP kinase modules as well as IRF3 along a pathway that resembles that of TLR3 signaling. A deregulation of cytoplasmic virus defense is thought to be responsible for the catastrophic influenza epidemic in 1918.

Summary

Receptors of the pro-inflammatory, immunomodulatory, and pro-apoptotic cytokines couple via adaptor proteins with Ser/Thr-specific protein kinases, thus inducing signal transduction along NFκB and MAP kinase pathways. The receptors of the tumor necrosis factor (TNF) family are oligomeric transmembrane proteins that use TNF receptor-associated factor (TRAF) adaptor proteins to interact with kinases incorporated in MAP kinase and NFκB modules. A subgroup of TNF receptors exhibits cytoplasmic death domains that couple with other adaptor proteins and establish contact with pro-apoptotic signaling pathways; these receptors are the major input terminals of endogenous death signals. Interleukin 1 (IL1) is a master regulator of the inflammatory response. Its receptor is a heterodimer that, upon ligand binding, activates protein kinases of the IL1 receptor-activated kinase (IRAK) family via several adaptor proteins. Toll-like receptors (TLRs) are specialized to recognize characteristic components of infectious bacteria, viruses, and parasites in order to induce a nonspecific defense reaction (innate immunity). They are homodimeric transmembrane proteins that stimulate, in addition to NFκB and MAP kinase cascades, the antiviral interferon β system. Overactivation may result in endotoxin shock. In insects, TLRs are also activated by endogenous ligands (such as *Spätzle*) that control morphogenetic processes.

Sidebar 6.5 Bacterial surface molecules: unique ligands of Toll-like receptors Macromolecules activating TLRs are present in the cell wall of Gram-negative and Gram-positive bacteria as well as in the peptidoglycan layer covering the membranes of both types. Peptidoglycan consists of an oligosaccharide backbone that is composed of disaccharides containing *N*-acetylgalactosamine and *N*-acetylmuramic acid with interlinked oligopeptide chains bound to the muramic acid residues (see textbooks of biochemistry and microbiology). The peptidoglycan layer of the cell is anchored in the (inner and outer) membrane by lipoproteins (in Gram-negative species) or lipoteichoic acid (in Gram-positive species). The outer membrane of Gram-negative bacteria is covered to about 75% by LPS, which exhibits a uniquely complex structure not found elsewhere in nature (Figure 6.11). For this reason, the interaction with TLR4 is absolutely specific.

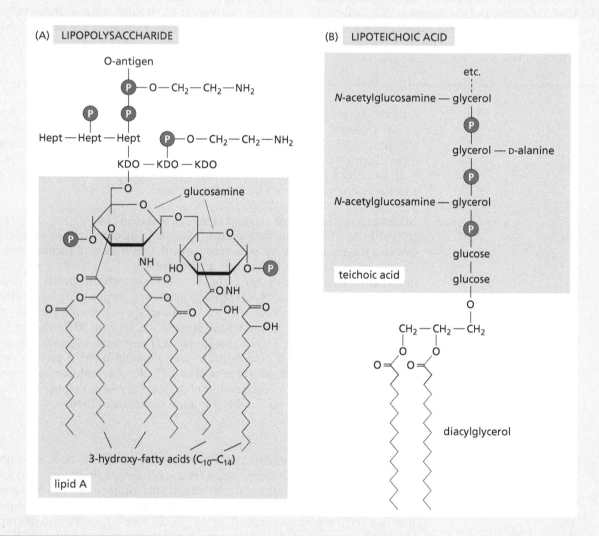

Figure 6.11 Two bacterial surface molecules acting as ligands of Toll-like receptors (A) **Lipopolysaccharide (LPS)**. By its hydroxy-fatty acid residues, the molecule is anchored in the outer membrane of Gram-negative bacteria. The lipid A moiety, mediating the interaction with receptor TLR4, is responsible for the endotoxin effect. The O-antigen is a branched oligosaccharide composed of glucose, galactose, and *N*-acetylglucosamine. KDO (2-keto-3-deoxyoctulosonic acid), Hept (L-glycero-D-manno-heptose), and 3-hydroxy-fatty acids are characteristic and unique components of LPS. A large number of LPS types are known that differ from species to species in their chemical composition and their endotoxic potency. (B) **Lipoteichoic acid**. This molecule is anchored by a diacylglycerol residue in the cell membrane and by the teichoic acid residue in the peptidoglycan layer of Gram-positive bacteria. Teichoic acid may also be found covalently bound at the outside of the peptidoglycan layer. The chemical composition of lipoteichoic acid varies from species to species (the glycerol may be replaced, for instance, by ribitol). The structures recognized by the TLRs are shown in the colored boxes (P, phosphate residues).

Sidebar 6.6. Antiviral strategies For all organisms the defense against pathogenic viruses is an essential condition of survival. In vertebrates this strategy is based on the functions of the **interferons**. These cytokines exist in two types. Type I comprises the interferons α (23 isoforms) and β that are produced by all cell types in response to viral (and bacterial) infections. Type II is identical with interferon γ, also called immune interferon, because it is exclusively produced by T lymphocytes and natural killer cells. Interferon γ activates macrophages to generate cytotoxic metabolites such as nitric oxide and reactive oxygen species (Section 2.2) and exhibits immunoregulatory and anti-tumor activities. The antiviral interferons proper are those of type I. They enable the innate and adaptive immune systems to fight viruses. Interferon receptors couple with cytoplasmic Tyr kinases of the JAK family that phosphorylate and activate the STAT transcription factors (see Section 7.2.2). They control the transcription of more than 100 different genes along this pathway, thus generating an antiviral state that extends to the tissue surrounding virus-infected cells. Key events are those that prompt the infected cells to commit suicide and the surrounding cells to activate signaling reactions that intervene with virus invasion and replication. A powerful strategy is to sequester double-stranded viral RNA or to prevent its translation into protein. While sequestering is achieved by a dynamin-related GTPase Mx, RNA translation is inhibited by the double-stranded RNA-activated protein kinase PKR that phosphorylates and inactivates the translation factor eIF2 (see Section 9.3.1). The genes encoding Mx and PKR are activated by interferons. In addition, interferons stimulate cytotoxic T lymphocytes and natural killer cells and control the immune system at many other points aiming at the production of virus-specific antigens.

Viruses (and bacteria) have developed various strategies to subvert the innate immune system. These include reduction of the CpG sequences in viral DNA, sequestering or proteolytic degradation of host signaling proteins, and production of modified ligands altering or blocking TLR activation. A major target of viral subversion is the protein kinase PKR, which is prevented from activation or deceived by false activators and substrates. In addition, viruses may protect double-stranded RNA from sequestering and destruction. Moreover, by binding to certain proteins, they may render the RNA "invisible" for the cellular RNA sensors TLR3 and RIG1/MDA5, which may also be blocked separately by inhibitory proteins. Viruses also secrete soluble truncated forms of interferon receptors that intercept circulating interferons or they produce inactive forms of IRFs that compete – as dominant negative mutants – with the cellular transcription factors. Since cells always try to cope with these viral tricks, the situation ends up in an arms race. An understanding of this battle is an essential condition for the clinical control of viral diseases.

Further reading

Akira S, Uematsu S & Takeuchi O (2006) Pathogen recognition and innate immunity. *Cell* 124, 783–801.

Alonso A, Sasin J, Bottini N et al. (2004) Protein tyrosine phosphatases in the human genome. *Cell* 117, 699–711.

Beutler B & Rietschel EZ (2003) Innate immune sensing at its roots: the story of endotoxin. *Nat. Rev. Immunol.* 3, 169–176.

Bodmer JL, Schneider P & Tschopp J (2002) The molecular architecture of the TNF superfamily. *Trends Biochem. Sci.* 27, 18–26.

Caestecker M (2004) The transforming growth factor-β superfamily of receptors. *Cytokine Growth Factor Rev.* 15, 1–11.

Chen G & Goeddel DV (2002) TNF-R1 signaling: a beautiful pathway. *Science* 296, 1634–1635.

Dunne A & O'Neill LAJ (2003) The interleukin-1 receptor/Toll-like receptor superfamily: signal transduction during inflammation and host defense. *Sci. STKE* 2003/171/re333333.

Garcia-Sastre A & Biron CA (2006) Type 1 interferons and the virus-host relationship: a lesson in détente. *Science* 312, 879–882.

Inohara N, Chamaillard M, McDonald C, Nuñez G (2005) NOD-LRR proteins: role in host–microbial interactions and inflammatory disease. *Annu. Rev. Biochem.* 74, 355–383.

Janssens S & Beyaert R (2002) A universal role for MyD88 in TLR/IL-1R-mediated signaling. *Trends Biochem. Sci.* 27, 474–482.

Kawai T & Akira S (2006) Innate immune recognition of viral infection. *Nat. Immunol.* 7, 131–137

Langer JA, Cutrone EC, & Kotenko S (2004) The class II cytokine receptor (CRF2) family: overview and patterns of receptor–ligand interactions. *Cytokine Growth Factor Rev.* 15, 33–48.

Luo K (2004) Ski and SnoN: negative regulators of TGFβ signaling. *Curr. Opin. Genet. Dev.* 14, 65–70.

MacEwan DJ (2002) TNF receptor subtype signaling: differences and cellular consequences. *Cell. Signalling* 14, 477–492.

Moustakas A & Heldin CH (2003) Ecsit-ement on the crossroads of Toll and BMP signal transduction. *Genes Dev.* 17, 285–289.

Nadiri A, Wolinski MK & Saleh M (2006) The inflammatory caspases: key players in the host response to pathogenic invasion and sepsis. *J. Immunol. 177*, 4239–4245.

Nohe A, Keating E, Knaus P et al. (2004) Signal transduction of bone morphogenetic protein receptors. *Cell. Signalling* 16, 291–299.

O'Shea JJ, Gadina M & Schreiber RD (2002) Cytokine signaling in 2002: new surprises in the Jak/Stat pathway. *Cell* 109, S121–S131.

Pitha PM (2004) Unexpected similarities in cellular responses to bacterial and viral invasion. *Proc. Natl Acad. Sci. U.S.A.* 101, 695–696.

Raetz RHC & Whitfield C (2002) Lipopolysaccharide endotoxins. *Annu. Rev. Biochem.* 71, 635–700.

Schlessinger J (2000) Cell signaling by receptor tyrosine kinases. *Cell* 103, 211–225.

Schlessinger J (2002) Ligand-induced, receptor-mediated dimerization and activation of EGF receptor. *Cell* 110, 669–672.

Schmierer B & Hill CS (2007) TGFßβ-SMAD signal transduction: molecular specificity and functional flexibility. *Nat. Rev. Mol. Cell Biol.* 8, 970–982.

Schmidt AM, Yan SD, Yan SF et al. (2001) The multiligand receptor RAGE as a progression factor amplifying immune and inflammatory responses. *J. Clin. Invest.* 108, 949–955.

Shi Y & Massague J (2003) Mechanisms of TGF-β signaling from cell membrane to the nucleus. *Cell* 113, 685–700.

Silverman N & Maniatis T (2001) NFκB signaling pathways in mammalian and insect innate immunity. *Genes Dev.* 15, 2321–2341.

Stroud RM & Wells JA (2004) Mechanistic diversity of cytokine receptor signaling across cell membranes. *Sci. STKE* 2004/231/re7.

Takeda K, Kaisho T & Akira S (2003) Toll-like receptors. *Annu. Rev. Immunol.* 21, 335–376.

ten Dijke P & Kill CS (2004) New insights into TGF-β-Smad signalling. *Trends Biochem. Sci.* 29, 265–273.

Vossheinrich CAJ & DiSanto JP (2002) Interleukin signaling. *Curr. Biol.* 12, R760–R763.

Wajant H, Henkler F & Scheurich P (2001) The TNF-receptor-associated factor family: scaffold molecules for cytokine receptors, kinases and their regulators. *Cell. Signalling* 13, 389–400.

Zapata JM & Reed JC (2002) TRAF1: lord without a ring. *Sci. STKE* 2002/133, pe27.

Signal Transduction by Tyrosine Kinase- and Protein Phosphatase-Coupled Receptors: A Late Invention of Evolution

While serine/threonine (Ser/Thr) kinase-coupled receptors are evolutionarily ancient devices of transmembrane signaling found also in invertebrates and plants, tyrosine-coupled receptors—and tyrosine (Tyr) kinases in general—are later developments of evolution. This receptor type has been found only in animals to date, where tyrosine-specific protein phosphorylation plays a key role in the development and maintenance of tissues and in the control of the specific immune system. In fact, there are fluid transitions between Tyr kinase-coupled receptors and cell adhesion proteins that are primarily responsible for the formation and cohesion of tissues. Conversely, such receptors are major targets of defects that lead to chronic metabolic diseases, immunological disorders, embryonic malformations, and cancer.

7.1 Receptor tyrosine kinases: the most modern cellular sensors

The family of mammalian Tyr kinases is composed of 85 members, 58 of which are receptors. Their ligands are hormones such as **insulin** and a large and heterogeneous family of **growth factors** (Figure 7.1). The latter regulate embryonic organ development and, in the mature organism, wound repair and other proliferative processes. This physiological function is mirrored by the downstream targets of receptor Tyr kinase signaling: these are first of all mitogen-activated protein (MAP) kinase cascades and the anti-apoptotic phosphatidylinositol 3-kinase (PI3K)–protein kinase B (PKB)/Akt complex (which is also a key mediator of insulin effects). Both signaling pathways lead to an activation of a wide variety of genes that control cell survival and proliferation (see Chapter 11). Most of the receptor Tyr kinases have been found to undergo oncogenic mutations.

All receptor Tyr kinases are oligomeric type 1 transmembrane proteins with one transmembrane domain per subunit. Their C-terminal cytoplasmic regions are quite similar, containing the kinase domains, whereas the N-terminal extracellular domains with the ligand-binding sites are highly diverse (Figure 7.2). This diversification is mirrored by functional variability, which has led to a classification into 20 subfamilies. As a rule each receptor type is expressed in several isoforms that differ in properties and tissue distribution.

All receptor Tyr kinases undergo dimerization-induced activation as shown in Figure 7.3. To this end the individual isoforms of each receptor type associate into homodimers or heterodimers that are stabilized either by two ligands or by one ligand with two binding sites, one for each receptor subunit. Dimerization accompanies a conformational change that activates the receptor by allosteric interactions of the cytoplasmic domains. Presently, this mechanism is best

Figure 7.1 Receptors with intrinsic Tyr kinase activity Shown are those receptor families for which the ligands are known, together with the corresponding ligands. The numbers of receptor isoforms identified to date are given in parentheses.

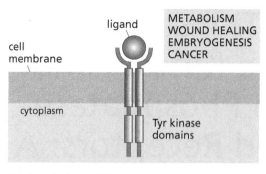

Ephrin-R (14)	ephrins
PDGF-R (6)	platelet-derived growth factor
	keratinocyte growth factor (KGF)
	colony-stimulating factor (CSF)
	stem cell factor (SCF)
EGF-R (4)	epidermal growth factor family
	transforming growth factor α (TGFα)
FGF-R (4)	fibroblast growth factor family
VEGF-R (3)	vasculo-endothelial growth factors
NGFR/Trk (3)	nerve growth factor (NGF), neurotrophins
Ins-R (3)	insulin, insulin-like growth factors (IGF)
HGFR/Met (3)	hepatocyte growth factor (HGF)
Ret	neurotrophic factors from glia cells
Tie/Tek (2)	angiopoietin 1
Musk	agrin
Axl/Tyro (3)	Gas-6, protein S
Ryk	Wnt-type morphogens

Figure 7.2 Diversification of receptor Tyr kinases Note the variability of the extracellular sensor domains as compared with the monotony of the cytoplasmic kinase domains, resembling the situation found for prokaryotic sensor His kinases (Chapter 3). The ligands of receptors Klg, Ros, Ltk/Alk, and Ror are still unknown. Many of the receptors are named after corresponding oncogenes.

understood for the epidermal growth factor receptors (see Section 7.1.1). For most receptor Tyr kinases, the active state is then stabilized by trans-autophosphorylation of the activation loops. Additional phosphorylations of the cytoplasmic domains generate docking sites for the recruitment of signal-transducing proteins, many of which are Tyr kinase substrates themselves. In other words, receptor Tyr kinases are both enzymes and scaffold proteins that, upon phosphorylation, organize signal-processing protein complexes at the inner side of the cell membrane.

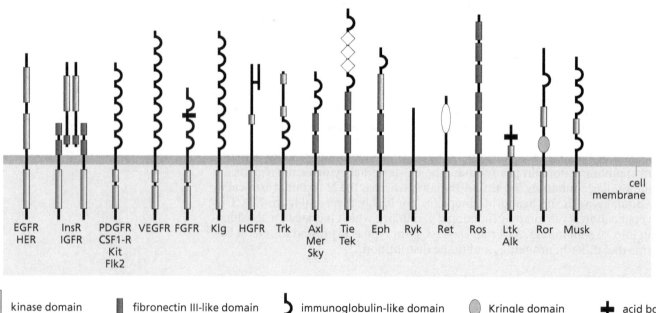

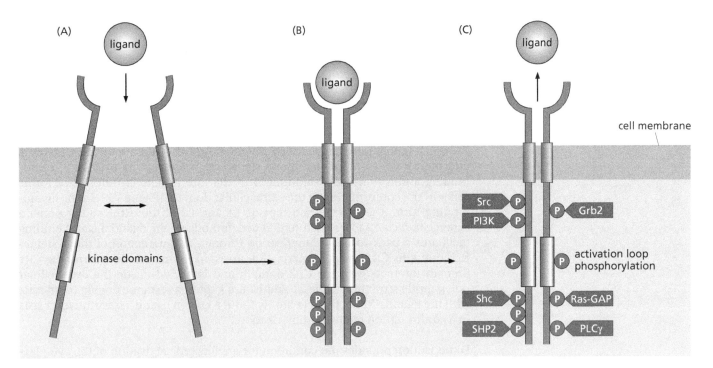

The Tyr-phosphorylated docking sites of the receptors specifically bind proteins with domains that recognize phospho-Tyr residues. Even though an active receptor is phosphorylated at multiple Tyr residues, this binding does not occur randomly but in a highly selective manner (Figure 7.3). As explained in Section 7.1.2, the reason for this selectivity is found in the variability of phospho-Tyr-binding domains.

7.1.1 Signaling by growth factors
Epidermal growth factor receptors

For a more in-depth understanding of receptor Tyr kinases, we shall have a closer look at the most thoroughly investigated prototype of this family, the epidermal growth factor (EGF) receptor. The EGF receptor was one of the first receptor Tyr kinases characterized and purified. Moreover, its mutated form ranks among the first oncoproteins discovered. In fact, oncogenic v-ErbB protein of avian erythroblastosis virus (which provokes a neoplastic blood disease in chickens) was found to be identical with a truncated EGF receptor. This defective protein is constitutively active because it lacks the extracellular ligand-binding domain that, in the absence of the ligand, exerts an auto-inhibitory effect on the receptor's Tyr kinase activity (Figure 7.4).

The EGF receptor represents a family of four related receptors that are collectively called ErbB1–4 (also known as HER1–4, Human EGF Receptor 1–4). Their ligands are EGF and EGF-like growth factors, a family of 13 hormone-like proteins that evoke a wide variety of cellular responses, particularly cell proliferation but also cell migration, cell differentiation, and cell death. These proteins perform these functions not only in epidermis but in most other epithelia. Which of these effects is realized depends on the cell type, the physiological context, and the ligand and receptor isoforms. Gene knockout experiments show that each ErbB isoform is important for embryonic development, particularly of epithelia. A constitutive overexpression of ErbB proteins (for instance, resulting from gene amplification) or hyperactivity due to oncogenic point mutations, particularly in the ATP-binding domain, are associated with several human malignancies such as cancer of the lung, pancreas, head and neck, and brain. Frequently such defects of the EGF receptor system correlate with a poor prognosis. Therefore, the ErbB system is a preferred target of novel cytotoxic anti-cancer drugs, some of which have been approved for clinical use (see below).

Figure 7.3 Three steps of signal transduction by a receptor Tyr kinase (A) Interaction of the receptor with a bivalent ligand (or two ligand molecules) leads to a conformational change that promotes dimerization, bringing the cytoplasmic Tyr kinase domains in close contact. (B) Trans-autophosphorylation of the activation loops in the kinase domains and of other sites. The phosphate residues are symbolized by P. (C) Site-specific interactions of individual phospho-Tyr residues with some selected proteins carrying Src homology 2 (SH2) or phospho-Tyr-binding (PTB) domains. Src is a cytoplasmic Tyr kinase, Grb2 and Shc are adaptor proteins linking the receptor to MAP kinase pathways, SHP2 is a phospho-Tyr-specific protein phosphatase, PI3K is phosphatidylinositol 3-kinase, Ras-GAP is a GTPase-activating protein, and PLCγ is phospholipase C type γ. Additional proteins undergoing phosphorylation by the receptor are not shown. The autophosphorylated receptor remains active even in the absence of the ligand until it becomes dephosphorylated.

Although closely related in structure, the four ErbBs differ functionally. ErbB1/HER1, the classical EGF receptor, represents a prototypical receptor Tyr kinase that, upon ligand binding, homo- or heterodimerizes. The same holds true for ErbB4. For ErbB2, no ligand is known. Instead, this receptor forms an active heterodimer with ErbB1 or ErbB4, thus acting as a signal amplifier. ErbB3, finally, has an inactive kinase domain and for signaling requires ErbB1, -2, or -4 as a partner.

X-ray crystallography has revealed much about the structure of EGF receptors and their mode of transmembrane signaling. The receptors exist in a ligand-binding and a non-ligand-binding form that are in equilibrium. These forms differ in the conformation of the extracellular domain (Figure 7.4). In the ligand-binding form, the leucine-rich repeats L1 and L2 of the extracellular domain arrange to a clamplike structure that accommodates the ligand. Ligand binding facilitates a back-to-back dimerization through an interaction of the cysteine-rich domains CR1, called dimerization arms. In the nonbinding form, the CR1 domain interacts with the CR2 domain and is buried inside the extracellular polypeptide structure, while L1 and L2 are kept at a distance unable to interact with the ligand. ErbB2, the receptor without a known ligand, constitutively exists in a conformation ready for dimerization.

Dimerization provides the condition for an allosteric activation of the cytoplasmic kinase domains. In the resting state, hydrophobic leucine residues block the active center. These keep the activation loop and a strategic α-helix in the N-lobe (see Section 2.6.1) in a position that does not allow ATP binding (the "mouth" of the substrate binding site is closed). With the conformational change induced by dimerization, these structural elements turn into the correct position required for kinase activation (the "mouth" opens). To this end the C-lobe of one monomer interacts with the N-lobe of the other monomer, provided the kinase domains are juxtaposed asymmetrically (Figure 7.4). The activation mechanism involves a special feature: the EGF receptors belong to those few kinases that do not require activation loop phosphorylation for activation. This is in contrast to other receptor Tyr kinases (such as the insulin receptor) that do need activation loop phosphorylation. Nevertheless, in principle the activation mechanism of ErbB is assumed to be representative for receptor Tyr kinases in general.

A key event of ErbB signaling is the phosphorylation of several Tyr residues in the cytoplasmic tail regions of the receptor protein. By this means, contact sites for proteins with corresponding phospho-Tyr recognition sites are generated. As far

Figure 7.4 Activation of the EGF receptor ErbB1 (A) Two monomeric ErbB1 proteins in the inactive conformation: the kinase domains (KD) cannot interact with each other. L1 and L2, leucine-rich repeats; CR1 and CR2, cysteine-rich domains. (B) Active dimer stabilized by ligand binding. "Back-to-back" dimerization occurs through an interaction of the CR1 domains: the kinase domains are in close contact, enabling one domain to activate the other one. (C) Active autophosphorylated dimer (P, phospho-Tyr residues). ErbB1 does not need activation loop phosphorylation.

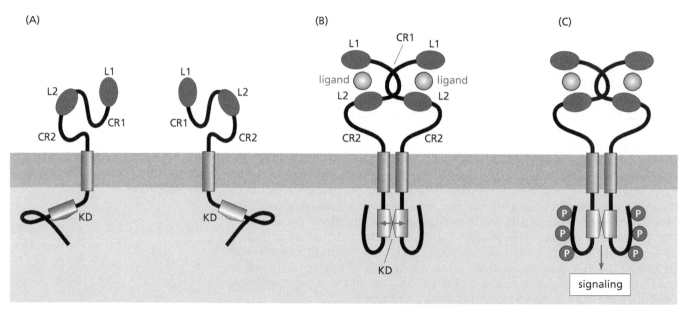

as the interaction partners are concerned, the individual ErbB isoforms exhibit some selectivity. Thus, each ErbB type contributes specifically to signaling. Together, the four ErbBs exhibit about 90 potential Tyr phosphorylation sites that have been estimated to interact with no less than 80 different partner proteins. This network of interactions is controlled by 13 different input signals (EGF and its relatives) and regulates the activity of a wide variety of genes. The organization of the ErbB network clearly demonstrates what has been discussed in Chapter 1: signal processing means an excitation of a complex "neural" network rather than a process involving a few individual sequences of biochemical reactions. An in-depth understanding of such a situation demands a systems-based approach because conventional methods rapidly reach their limits.

An additional signaling pathway has been characterized for ErbB4 consisting of a proteolytic release of the cytoplasmic domain. The fragment then migrates into the nucleus where, together with other transcriptional regulators such as STAT5 (Section 7.2.1), it modulates gene expression (for more details see Section 13.1.4). Whether this indicates a more general mechanism of receptor Tyr kinase action or is restricted to this particular receptor type is a matter of debate.

An interesting variation of this mechanism is found in nerve cells that are confronted with the problem of signaling across long distances. The growth of their axons is stimulated by nerve growth factor interacting with the receptor Tyr kinase Trk at the tip of the nerve fiber. This turns up both local signaling and gene transcription, even though the nucleus may be centimeters or even meters away. To bridge this distance, the ligand–receptor complex is taken up in endocytotic vesicles (endosomes) that, with the help of motor proteins, migrate to the nucleus by a process called **retrograde transport**.

Reactive oxygen species facilitate signaling by receptor tyrosine kinases

Since Tyr phosphorylation occupies a central position in the regulation of tissue formation, tissue maintenance, and tissue regeneration, the cell must keep it under strict control. In this context protein Tyr phosphatases play a particularly important role as the most efficient opponents of Tyr kinases (see Section 7.4). In fact, in most cases, activation of a receptor Tyr kinase is inadequate for efficient signal transduction, unless the phosphatases are switched off. This is achieved by the redox signal H_2O_2, which inactivates Tyr phosphatases by oxidizing catalytically relevant sulfhydryl groups to disulfide and sulfenic acid groups. The thioredoxin cycle renders the reaction reversible (Figure 7.5 and Section 2.2). How does the cell switch on H_2O_2 production simultaneously with the activation of Tyr kinase-coupled receptors? A major source of hydrogen peroxide is the

Figure 7.5 Enhancement by redox signaling of signal transduction by a receptor Tyr kinase The Tyr kinase-coupled receptor activated by a signal molecule S and subsequent autophosphorylation (P) interacts with and activates phosphatidylinositol 3-kinase (PI3K). The phosphatidylinositol 3,4,5-trisphosphate (PIP$_3$) thus generated recruits the small G-protein Rac to the cell membrane, where it stimulates nicotinamide adenine dinucleotide phosphate (NADPH) oxidase activity. The reactive oxygen species (ROS) produced by this enzyme oxidize and inactivate a protein Tyr phosphatase (PTP) and the phospholipid phosphatase PTEN, which both would inhibit receptor-dependent signaling (red pistils). By means of thioredoxin (Trx) and NADPH, both enzymes become reactivated. Thus, signaling by receptor Tyr kinases depends on the redox status of the cell.

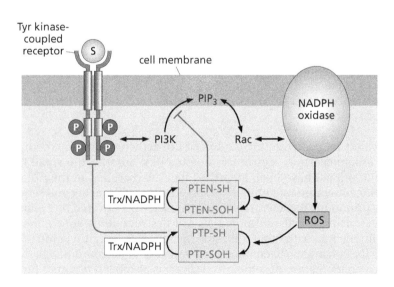

243

membrane-bound enzyme NADPH oxidase, which is activated by the small G-protein Rac (Section 10.3.4). Rac is a target of various signaling cascades including the PI3K pathway controlled by Tyr kinase-coupled receptors. Thus, the release of hydrogen peroxide is a general response to receptor activation. To augment the effect, H_2O_2 simultaneously inhibits the phospholipid phosphatase PTEN, an opponent of PI3Ks.

Receptor inactivation and down-regulation: an extravagant venture

In its autophosphorylated form a Tyr kinase-coupled receptor stays active even after the ligand has dissociated. A standard mechanism of rapid signal termination is, therefore, receptor dephosphorylation by Tyr-specific protein phosphatases. As we have seen, the activity of these is limited, however, by redox signaling activated by the receptor. This situation appears to be rather tricky, indicating a continuous oscillation between phosphorylation and dephosphorylation.

Upon prolonged stimulation, Tyr kinase-coupled receptors become internalized by endocytosis and destroyed inside the cell—similar to the process occurring with G-protein-coupled receptors (GPCRs, see Section 5.7). At least for some Tyr kinase-coupled receptors, this down-regulation is achieved by monoubiquitylation and subsequent degradation in lysosomes or by polyubiquitylation followed by proteolysis in proteasomes. The E3 ubiquitin ligase that has been found to play a key role in this process is **Cbl** (the name refers to the Casitas B-cell lymphoma of mice, where Cbl has undergone oncogenic mutation). Its mechanism of action has been investigated in some detail. We shall look more closely at it because Cbl provides a particularly clear example of the function of a ubiquitin ligase in signal transduction.

Cbl-type proteins have been found in multicellular animals; humans express three isoforms. In experimental animals the complete loss of Cbl by gene knockout results in prenatal death, indicating a key role of these enzymes in cellular physiology. A hint to their function is given by the Cbl structure: it includes a Tyr kinase-binding motif containing an SH2 domain as a docking site for Tyr-phosphorylated sequences, a proline-rich domain that is a binding site for SH3 domains, a ubiquitin-associating domain probably interacting with regulatory proteins, and a RING domain that is essential for ubiquitylation (Figure 7.6 and Section 12.16). By means of these domains, Cbl interacts with both Tyr-phosphorylated (activated) receptors and autophosphorylated cytoplasmic Tyr kinases. Due to this interaction Cbl is Tyr-phosphorylated, and its conformation becomes unfolded in such a way that an E2 ubiquitin-conjugating enzyme can bind to the RING domain and ubiquitylate the Tyr kinase-coupled receptor. In other words, Cbl exhibits the properties of an E3 ubiquitin ligase (Section 12.16). The viral oncoprotein vCbl of mice is a truncated form lacking the RING domain (Figure 7.6). Through its Tyr kinase-interacting domain it still binds substrate proteins but is unable to establish contact with the E2 ubiquitin-conjugating enzyme; thus it acts as a dominant negative mutant. The oncogenic effect results from a continuous overactivation of mitogenic and anti-apoptotic signaling pathways stimulated by Tyr kinases and Tyr kinase-coupled receptors that are rescued from downmodulation.

Recent results indicate that the control of Cbl activity is far more complex than hitherto assumed. Thus, regulatory proteins (Sts, Sprouty, the small G-protein Cdc42, and others) have been identified that are recruited to the Cbl substrate and specifically counteract ubiquitylation, either by displacing and sequestering Cbl or by promoting its down-regulation. In fact, Cbl activated by a substrate Tyr kinase is short-lived since it undergoes self-destruction by auto-ubiquitylation or by an interaction with other ubiquitin ligases. Auto-ubiquitylation means that Cbl and its substrates become downregulated in a coordinated, reciprocal manner, a phenomenon of self-control that has also been reported for other E3

(A)

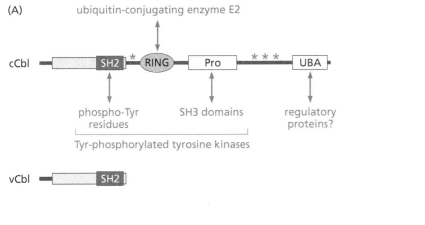

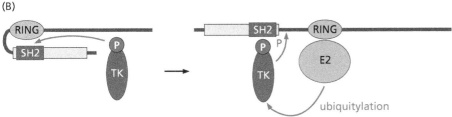

Figure 7.6 Domain structure and activation of the E3 ubiquitin ligase Cbl (A) cCbl is one of several human Cbl isoforms. Interaction partners are depicted in red. The gray box is the Tyr kinase-interacting domain, Pro is a proline-rich domain, and UBA is a ubiquitin-associating domain. Locations that become Tyr-phosphorylated are marked by asterisks. vCbl, the viral oncogene product, is a truncated Cbl acting as a dominant negative mutant. (B) Mechanism of Cbl activation by a Tyr-phosphorylated (P) Tyr kinase (TK). In the inactive conformation of Cbl (left, depicted in a simplified form), the RING domain is blocked. This blockade is lifted by the attachment of TK to the Tyr kinase-interacting domain, in particular to its SH2 domain, and Tyr phosphorylation of Cbl in a "linker domain" (red P). As a result, an E2 ubiquitin-conjugating enzyme can bind to the RING domain and ubiquitylate TK (right). Tyr kinases such as Src containing an SH3 domain interact, in addition, with the proline-rich domain of Cbl (not shown).

ubiquitin ligases. Quite an opposite effect is observed when Cbl is downregulated by another ubiquitin ligase. Because of the high substrate specificity of E3 ligases, such a reaction is selective, meaning it does not include the Cbl substrate that, instead, is protected from ubiquitylation and destruction through Cbl.

Clearly, protein down-regulation is quite an expensive venture, since phosphorylation, ubiquitylation, and in particular the resynthesis of downregulated proteins require a large amount of ATP energy. Obviously, the precise regulation of cellular signaling is anything but simple and is worth any extravagance!

Signal transduction therapy of cancer

Cancer, a characteristic disease of complex multicellular organisms, is caused by somatic or hereditary gene mutations and other persisting defects that disturb cellular signal processing. As stated in Section 12.1, a tumor cell misinterprets the communicative signals of the body, thus to some extent resembling a psychotic person. In the last 10 years, efforts have been made to correct such oncogenic defects of signal transduction. For this approach, Tyr kinases rank among the most attractive targets for two reasons.

(1) Tyrosine phosphorylation is inseparably linked with multicellularity; it is not found in unicellular organisms but has entered the stage of evolution with the appearance of multicellular animals.

(2) Most Tyr kinases have been found to be targets of oncogenic mutations.

Synthetic Tyr kinase inhibitors are called **tyrphostins** (<u>tyr</u>osine <u>ph</u>osphorylation <u>in</u>hibitors). Several hundred such compounds have been synthesized, but as yet only three drugs have progressed to clinical trials. The first of these drugs is Imatinib, also known as Gleevec, which was developed by Novartis and received FDA approval in May 2001. It has been used successfully for the treatment of chronic myelogenous leukemia. Imatinib/Gleevec inhibits the BCR-Abl Tyr kinase (see Sidebar 7.1) by blocking the ATP-binding site. This kinase's activity is essential for the survival of leukemia cells. Related drugs are Gefitinib and Erlotinib, two inhibitors of the EGF receptor Tyr kinases. They are used to treat non-small-cell lung cancer, but have proved to be of benefit only for a small subgroup of patients.

Even though the anti-tumor effects of these tyrphostins are quite impressive, their long-term usage is hampered by the development of resistance due to additional mutations in the kinases' ATP-binding sites. Presently, efforts are being made to overcome this obstacle, both by developing novel drugs and by combining the administration of tyrphostins with other anti-cancer treatments (for instance, by application of receptor or kinase antibodies). Admittedly, the signal transduction therapy of cancer is still in its infancy. But when the rapidly progressing elucidation of the cell's signal-processing network is considered, the approach promises to have great potential for the future.

Sidebar 7.1 The Philadelphia translocation: cause of leukemia The oncogene of the Abelson leukemia virus of mice encodes a Tyr kinase **vAbl**. The corresponding proto-oncogene *c-abl* is distributed over the entire animal kingdom. It has gained much attention because of the so-called **Philadelphia translocation,** a chromosomal rearrangement considered to be the primary cause of chronic myelogenous leukemia. In this rearrangement the *c-abl* gene is translocated from chromosome 9 to chromosome 22, where it fuses with

a truncated form of a gene called *bcr.* The hybrid gene thus generated gives rise to a cAbl/BCR fusion protein exhibiting deregulated cAbl Tyr kinase activity (Figure 7.7). **BCR** (**B**reakpoint **C**luster **R**egion) is an atypical Ser/Thr-directed protein kinase containing a Rho guanine nucleotide exchange factor (GEF) domain and a Rac GTPase-activating protein (GAP) domain. Since the small G-proteins Rho and Rac are key regulators of the cytoskeleton (Section 10.3), BCR may have a similar function.

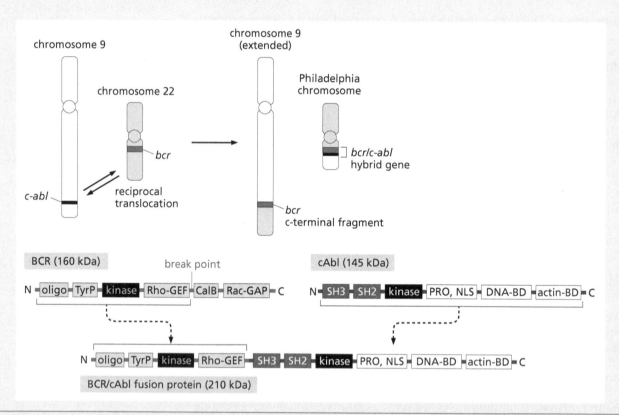

Figure 7.7 Oncogenic activation of Tyr kinase cAbl by the Philadelphia translocation Due to a reciprocal translocation of fragments of chromosomes 9 and 22, a so-called Philadelphia chromosome is generated in which the *c-abl* gene is fused with the N-terminal fragment of the *bcr* gene. In the constitutively expressed BCR/cAbl fusion protein encoded by the hybrid gene, cAbl has oncogenic potential and is a major cause of chronic myelogenous leukemia. The anti-leukemic drug Gleevec blocks the ATP-binding site of the oncogenic cAbl. Abbreviations: oligo, oligomerization domain; TyrP, domain with phosphorylated Tyr interacting with SH2 domains; CalB, Ca^{2+}-dependent lipid-binding domain; Rac-GAP, domain activating the GTPase activity of Rac; Rho-GEF, domain catalyzing the GDP–GTP exchange of Rho; PRO, proline-rich domain interacting with SH3 domains; NLS, nuclear localization sequence; DNA-BD, DNA-binding domain; actin-BD, actin-binding domain; SH2 and SH3 = Src-homology domains 2 and 3.

cAbl is not a receptor Tyr kinase. Instead it exhibits binding domains for DNA and the actin cytoskeleton as well as SH2, SH3, and proline-rich (SH3-interacting) domains and is found in the nucleus and in the cytoplasm, as well as in a lipid-anchored form at the membrane. The kinase activity is strictly autoregulated, mainly by an intramolecular interaction between the SH3 and the Pro-rich domains. As with most Tyr kinases, cAbl must be phosphorylated at a Tyr residue in the activation loop to gain full activity. This occurs either by autophosphorylation, by a Tyr kinase of the Src family (Section 7.2), or by the DNA damage checkpoint kinase ATR (Ataxia telangiectasia mutated Rad3-related; see Section 12.9). The active cAbl is a short-lived enzyme that is rapidly inactivated by dephosphorylation and proteolytic degradation. Both autoregulation and degradation are impaired in the oncogenic cAbl-BCR fusion protein. A variety of functions has been attributed to cAbl. First of all, the enzyme transmits proliferative and survival signals, which explains its oncogenic potential. On the other hand, cAbl also seems to participate in apoptosis. In fact, in response to DNA damage, its nuclear form becomes stimulated by ATR-catalyzed phosphorylation and subsequently phosphorylates ATR and other stress proteins such as the p53-homologous factor p73 (Section 12.9). However, since ATR is also activated independently from cAbl, the role of the latter in the cell's DNA damage response is not clear yet.

cAbl is an essential enzyme because corresponding knockout animals are not viable. Indeed, the kinase (particularly isoform Abl2) is thought to play a critical role in the development of the nervous system, controlling the formation of synaptic contacts brought about by targeted growth of nerve fibers or axons. As explained in Section 16.1, this axonal navigation is due to interactions between receptors at the axonal growth cone with stimulatory or inhibitory signaling molecules transmitted by adjacent cells. A major intracellular target of those interactions is the actin cytoskeleton, producing membrane extrusions at the top of the growing nerve fiber. These events critically depend on cAbl activity, as has been shown in particular for the *Drosophila* model. By means of its actin-binding domains, cAbl directly interacts with the cytoskeleton, phosphorylating a series of actin-regulatory proteins and navigation receptors.

7.1.2 Binding domains for phosphotyrosine residues: the art of complex formation

Whether or not a protein becomes bound to an autophosphorylated Tyr kinase-coupled receptor depends on interaction domains that specifically recognize phosphorylated Tyr residues in proteins. These binding sites are called SH2 and PTB domains.

Src homology domains

The acronym SH2 means Src Homology 2. It refers to Src, a cytoplasmic Tyr kinase that in a mutated form was discovered as an oncoprotein of a chicken sarcoma virus (see Section 7.2). Src contains sequences that are also found in homologous forms in many other proteins. These Src homology domains are:

- SH1, a domain resembling the kinase domain of Src

- SH2, a domain interacting with sequences containing phospho-Tyr residues

- SH3, a domain interacting with proline-rich sequences with the recognition site -Pro-X-X-Pro-

SH3 domains have been identified in about 200 and SH2 domains in about 100 human proteins (whereas *Saccharomyces cerevisiae* expresses only one protein with an SH2 domain, reflecting the almost complete lack of Tyr phosphorylation in unicellular eukaryotes). Proteins with SH2 and SH3 domains may be enzymes, transcription factors, or adaptors. Some examples are shown in Figure 7.8.

SH2 domains recognize sequences with phosphotyrosine residues

An SH2 domain comprises about 100 amino acids. The key structural element is a bundled four-stranded β-sheet flanked by a three-stranded β-sheet and two α-helices (Figure 7.9). The four-stranded sheet contains conserved amino acids binding the phospho-Tyr residue, while the loops linking the sheet to the other

Figure 7.8 A selection of proteins with Src homology domains (A) Enzymes Src, Ras-GAP, PLCγ, Tyr phosphotase SHP2, and transcription factor STAT. SH3, SH2, and SH1, Src homology domains 3, 2, and 1; PTK, protein Tyr kinase domain; PH, pleckstrin homology domain; C2, Ca²⁺/phospholipid-binding domain; GAP, GTPase-activating protein domain; PLC, phospholipase C domain; PTPase, protein Tyr phosphatase domain; DNA-BD, DNA-binding domain. The proteins are not shown to scale. (B) Adaptor proteins without enzymatic activity. Domains that recognize phospho-Tyr residues are shown in color. PTB, phosphotyrosine-binding domain.

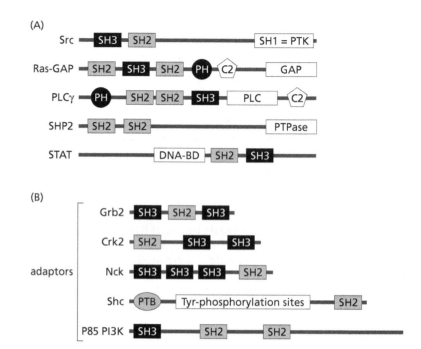

structures interact with 2–6 amino acids located C-terminally of the phospho-Tyr residue. These structural elements may differ among SH2 domains and determine the specificity of the interaction, which goes far beyond recognition of the phospho-Tyr residue. On this basis, three major families of SH2 domains have been identified and are named for prototypical proteins:

- Src-type SH2 domains recognize charged residues in positions +1 and +2 as well as a hydrophobic residue in position +3 C-terminally from phospho-Tyr

- phospholipase Cγ-type SH2 domains interact with aliphatic residues in positions +1 to +5

- Grb2-type SH2 domains recognize an Asp residue in position +2 (as explained below, Grb2 is an adaptor protein connecting receptor Tyr kinases with downstream effectors)

Figure 7.9 Structure of a Src-typical SH2 domain The four β-sheets with the central phospho-Tyr (P-Tyr, red) binding site are seen.

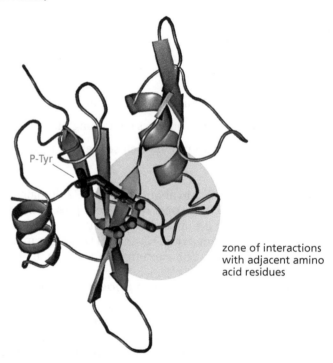

In addition, other SH2 domains recognize nonphosphorylated Tyr residues, for instance, in the intrasteric inhibition of protein phosphatase SHP1 (Figure 7.10).

SH2 domains have two major functions: formation of signal-processing complexes, in particular at autophosphorylated receptors, and intrasteric control of the activity of signal-processing proteins. Frequently both effects are combined. An instructive example is provided by the Tyr-specific protein phosphatases of the type SHP1 (SH2-containing protein Tyr Phosphatase1, pronounced "ship") that, among other functions, take care of the dephosphorylation of autophosphorylated receptors, that is signal termination. In a negative feedback loop, the same receptors activate SHP1 by lifting an intrasteric blockade (Figure 7.10). In a similar but more sophisticated way, Src-type Tyr kinases are activated. As explained in Section 7.2, these cytoplasmic enzymes are integrated in signaling cascades controlled by Tyr kinase-coupled receptors (Figure 7.11). Src activity is controlled by both inhibitory and stimulatory Tyr phosphorylation providing a particularly clear example of the ambiguousness of protein phosphorylation. While a C-terminal phospho-Tyr residue keeps the enzyme in an inactive conformation by interacting with the SH2 domain, phosphorylation of a Tyr residue in the activation loop is required for full activity. Therefore, Src activation proceeds in two steps. For activation the interaction between the SH2 domain and the C-terminal phospho-Tyr residue is first interrupted by a competing Tyr-phosphorylated protein or by enzymatic dephosphorylation. In the second step, activation loop phosphorylation occurs by trans-autophosphorylation. The enzyme is switched off again by dephosphorylation of the activation loop and re-phosphorylation of the C-terminal Tyr residue. The latter reaction is catalyzed by a specific C-terminal Src kinase (Csk). Csk is a cytoplasmic enzyme that, in order to establish contact with Src, interacts with a Csk-binding single-pass transmembrane protein. The binding protein also associates with the SH3 domain of Src and, in turn, is phosphorylated by Src at several Tyr residues. The phospho-Tyr groups thus generated function as anchoring sites for additional proteins with SH2 domains. Thus, the Csk-binding protein is a multifunctional scaffold that organizes via SH2-specific interactions large signal processing complexes at Src-type kinases.

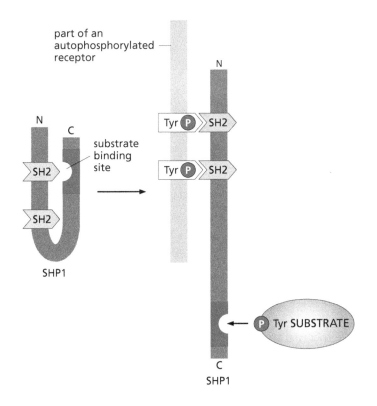

Figure 7.10 Role of SH2 domains in activation of the protein Tyr phosphatase SHP1 In inactive SHP1, one of the two tandem SH2 domains shields the substrate binding site (left; no P-Tyr interaction). This blockade is abolished by the conformational change induced by the interaction with two tandem P-Tyr residues of the autophosphorylated receptor (right).

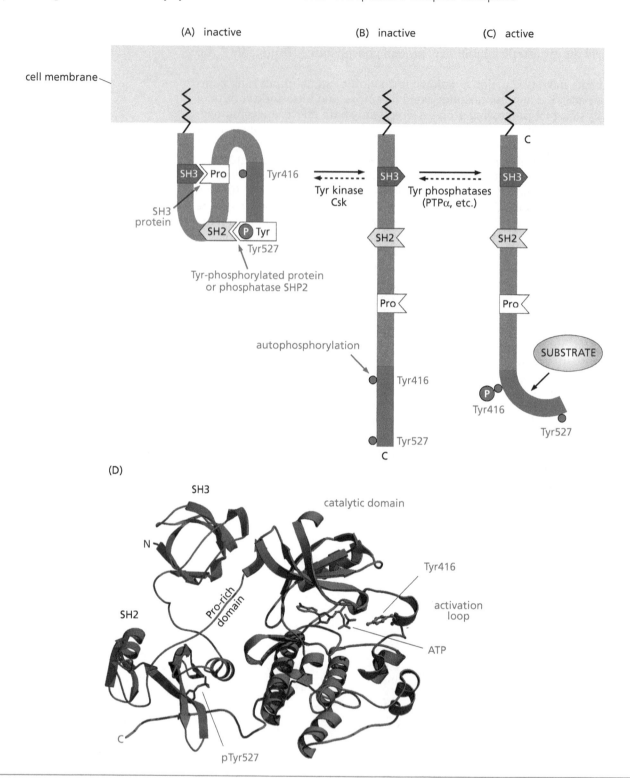

Figure 7.11 Role of SH2 and SH3 domains in activation of the cytoplasmic protein Tyr kinase Src (A) In the absence of an input signal, Src (membrane-anchored by a tetradecanoyl residue symbolized by the zigzag line) is intrasterically inhibited by two interactions, one between an SH2 domain and a C-terminal phospho-Tyr residue (Tyr527 in humans) and the other between an N-terminal SH3 domain and a proline-rich domain (Pro). (B) Src has become unfolded due to competitive interactions of an SH3 protein with the Pro domain and of a Tyr-phosphorylated protein (such as a receptor) with the SH2 domain (for the sake of clarity, the interaction partners are not shown). Alternatively or in addition, phosphatase SHP2 may have dephosphorylated Tyr527 (for an example of this mechanism, see Section 7.2.2). (C) Src has become activated by autophosphorylation of Tyr416 in the activation loop. The activation process is rendered reversible by dephosphorylation of Tyr416 and rephosphorylation of Tyr527 catalyzed by the C-terminal Src kinase (Csk). To become recruited to the membrane-bound Src, the cytoplasmic Csk interacts with a Csk-binding protein, a single-pass transmembrane protein that functions as a scaffold (not shown). The Src oncoprotein lacks Tyr527; therefore, it is partially activated. (D) Molecular model of Src in the inactive conformation (based on X-ray crystallography).

An example of the role of SH2 and SH3 domains in adaptor proteins is provided by receptor-induced activation of the Ras–MAP kinase signaling cascade. For mitogenic stimuli such as growth factors, this pathway represents a signal-transducing main street from the cell periphery to the genome; therefore, it is one of the major downstream effector systems of receptor Tyr kinases (Figure 7.12). Here the adaptor protein **Grb2** (Growth factor receptor-binding protein 2) connects the autophosphorylated receptor, bound via an SH2 domain, with the Ras-GEF mSOS, bound via an SH3 domain, thus bringing mSOS into the immediate vicinity of the receptor and of membrane-anchored Ras. Through this interaction mSOS, and in turn Ras, becomes activated. By means of a second SH3 domain (see Figure 7.8), Grb2 may interact with additional proteins such as the docking protein GAB (Grb2-Associated Binding protein, see Section 7.1.3). As a result, large signal-processing protein complexes may be assembled at the receptor. Under certain circumstances (probably at low intensities of the input signal), the receptor does not interact with Grb2 directly but via the scaffold protein **Shc** (SH2 domain-containing protein, pronounced "shic"). Shc binds to the receptor by a PTB domain and to Grb2 by a proline-rich sequence interacting with an SH3 domain. By means of its SH2 domain, Shc may recruit further proteins. Moreover, it becomes phosphorylated at multiple Tyr residues acting as binding sites for other proteins with SH2 domains (see Figure 7.8).

In contrast to SH2 domains, **PTB domains** recognize phospho-Tyr residues together with amino acids located *amino-terminally,* but in some cases they also recognize sequences with nonphosphorylated tyrosine or even without tyrosine. Their structure is characterized by a central seven-stranded β-sandwich and differs considerably from that of SH2 domains. The human genome encodes about 30 proteins with PTB domains. Most of them are adaptor and docking proteins that help to assemble large protein complexes at Tyr-phosphorylated proteins, in particular at autophosphorylated receptor Tyr kinases.

7.1.3 Signal transduction by the insulin receptor: a matter of docking

For some receptor Tyr kinases, the main purpose of autophosphorylation is activation of the enzymatic domain and association with a docking protein. By multiple phosphorylations, the docking protein is then rendered into a recruitment center for a wide variety of proteins with SH2 and PTB domains, thus taking over

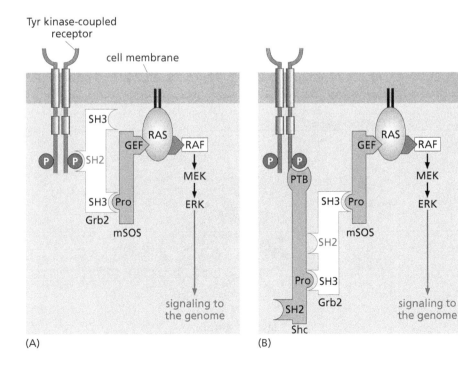

Figure 7.12 Role of adaptor proteins in signal transduction by receptor Tyr kinases (A) Membrane-associated signal-transducing complex between autophosphorylated (P) receptor, adaptor Grb2, Ras-GEF mSOS (colored; Pro, proline-rich sequence), membrane-anchored Ras, and the MAP kinase module RAF–MEK–ERK. (B) The same complex but coupled to the receptor by the scaffold protein Shc.

251

the role of the autophosphorylated receptor protein. Most known docking proteins have a PTB domain by which they become fixed at the autophosphorylated receptor. In addition, many of them are anchored in the membrane by lipid residues, noncovalent interactions with membrane phospholipids (see Figure 7.13), or transmembrane domains.

The best-known examples of docking proteins are the **insulin receptor substrates (IRS)**. These are large proteins (130 kDa) that can accommodate many effector proteins. An IRS protein binds through a PTB domain to the autophosphorylated receptors of insulin, insulin-like growth factors, and interleukin 4 (IL4). It is phosphorylated at multiple Tyr residues by the receptor itself or, as in the case of the IL4 receptor lacking an intrinsic Tyr kinase domain, by receptor-associated Tyr kinases of the JAK family (see Section 7.2.1). IRS is attached to the membrane through a pleckstrin homology (PH) domain (Figure 7.13). In addition to phospho-acceptor Tyr residues, IRS has many Ser and Thr residues that are phosphorylated by a wide variety of protein kinases, resulting mostly in inhibition of signal transduction. Thus, graded Ser/Thr phosphorylations may fine-tune signaling by IRS and provide a mechanism of feedback control (an example is discussed in Section 9.4.3).

The insulin receptor has a more complex structure than other receptor Tyr kinases. Each monomer consists of two polypeptide chains held together by disulfide bonds (see Figures 7.2 and 7.13). The monomers are linked to each other through disulfide bonds. Thus, in contrast to EGF receptors, the insulin receptor has a preformed dimeric structure that interacts with one ligand molecule. The conformational change resulting from ligand binding triggers a mutual activation loop phosphorylation of the cytoplasmic kinase domains. To evoke the anabolic effects of insulin, the receptor activates a particular set of signaling cascades by using, in addition to IRS, other adaptor and docking proteins such as Shc to stimulate the Ras–RAF–ERK MAP kinase module. Along this mitogenic pathway, insulin-like growth factors (and in cell culture insulin by itself) can induce cell proliferation.

Mechanism of insulin-dependent glucose absorption

Following a meal, insulin is released from pancreatic β-cells and stimulates tissues, first and foremost skeletal muscle and adipose tissue, to absorb glucose from the blood. To this end the hormone activates the **glucose transporter GluT4** and promotes **glycogen synthesis** for glucose storage. GluT4 is a protein with 12 transmembrane domains forming an aqueous membrane channel that regulates glucose influx into muscle and fat cells (another glucose transporter, GluT1, is constitutively, that is, insulin-independently, expressed, particularly in the liver). The transporter shuttles between the cell membrane and intracellular

Figure 7.13 Docking protein IRS, insulin receptor substrate The insulin receptor consists of four polypeptide subunits held together by disulfide bonds (red). Upon autophosphorylation, the cytoplasmic domain of the receptor binds to the phospho-Tyr-binding (PTB) domain of the docking protein IRS (interacting P-Tyr, colored circle; P-Tyr in the activation loop of the receptor, black circle). In addition, IRS is attached to the membrane by a pleckstrin homology (PH) domain and becomes phosphorylated by the receptor at multiple Tyr residues (red circles). Thus it is transformed into a docking center for proteins with SH2 and PTB domains.

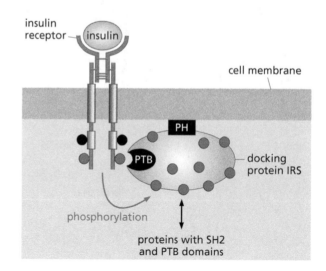

compartments. Insulin induces its translocation to the cell surface, whereas in the absence of the hormone it becomes internalized by clathrin-dependent endocytosis.

A major pathway activated by insulin involves PI3K and PKB/Akt (Section 4.6.6). Along this route the translocation of GluT4 is induced, albeit in a manner still not fully understood. Potential substrates of PKB/Akt include a GAP (called AS160) for a small G-protein of the Rab family and a protein (called Synip) controlling the SNARE interaction of vesicle fusion. As discussed in Sections 10.4.2 and 10.4.4, Rab proteins are master regulators of intracellular vesicle transport and exocytosis. AS160 seems to be inactivated by PKB/Akt-catalyzed phosphorylation, a reaction that would make sense since GluT4 is exposed at the cell surface along a Rab-controlled exocytotic pathway, which would be blocked by AS160. Another signaling pathway leading to GluT4 activation starts from the small G-protein **TC10**, a member of the Rho family that interacts with the insulin receptor through a special adaptor protein. The physiological significance of this pathway is not entirely clear.

PKB/Akt also phosphorylates and inactivates glycogen synthase kinase 3β (GSK3β), thus leading to glycogen synthesis, a major anabolic effect of insulin (which otherwise would be inhibited by GSK3β; see Section 5.3.1). To stimulate the PI3K–PKB cascade, the small G-protein **Ras** is needed (Section 10.1.4). As discussed in Section 7.1.2, Ras is activated by receptor Tyr kinases such as the insulin receptor by means of the adaptor proteins Grb2 and Shc and the Ras-GEF mSOS.

When the insulin level drops, the signaling system is immediately shut off. This is achieved by the protein Tyr phosphatases PTP-1B and SHP1, which catalyze the dephosphorylation of the receptor and docking proteins and are in control when the kinases become inactive. Simultaneously, phosphatidylinositol 3,4,5-trisphosphate (PIP$_3$) is inactivated by lipid phosphatases such as PTEN (which dephosphorylates PIP$_3$ at the 3-OH group of inositol) and SHIP2 (SH2-containing Inositol Phosphatase 2, which dephosphorylates PIP$_3$ at the 5-OH group of inositol; not to be confused with SHP2). As a result, PKB/Akt activity is terminated (Section 4.6.6) and IRS is released from the membrane. In addition, the insulin receptor and IRS are inactivated through phosphorylation of Ser and Thr residues catalyzed by the protein kinases GSK3β and mTor (Section 9.4.1). This context is schematically summarized in Figure 7.14. In diabetes, insulin signaling is strongly impaired, a subject that is treated in more detail in Section 9.4.3.

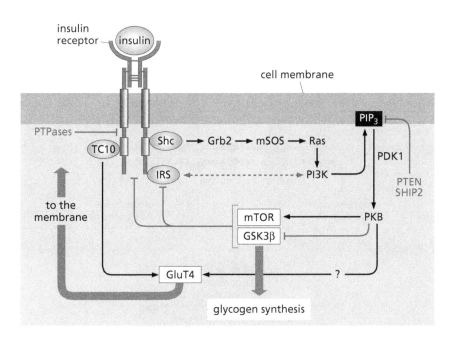

Figure 7.14 Signaling pathways mediating insulin effects Shown to the left is the tetrameric insulin receptor that, upon ligand binding and autophosphorylation, binds (and phosphorylates) the scaffold/docking proteins Shc and IRS as well as the GTPase TC10. As a result, signaling cascades become activated, leading to exposure of GluT4 at the cell surface, followed by glycogen synthesis and feedback inhibition of signal transduction. Further details are found in the text.

Other docking proteins

In addition to IRS, other docking proteins have been identified.

(1) **FRS**, <u>F</u>ibroblast growth factor (FGF) <u>R</u>eceptor <u>S</u>ubstrates. Like the closely related IRS, they bind to receptors (mainly those of fibroblast and nerve growth factor) by a PTB domain, whereas membrane attachment is due to a fatty acid anchor.

(2) **GAB**, Grb2-<u>A</u>ssociated <u>B</u>inders. They interact with various receptor and receptor-associated Tyr kinases. The membrane-binding site is a PH domain.

(3) **DOK** (<u>D</u>ownstream <u>O</u>f Tyr <u>K</u>inases) proteins. They preferentially bind Ras-GAP and are therefore negative regulators of Ras signaling. Membrane binding is due to a PH domain.

(4) **LAT**, <u>L</u>inker for <u>A</u>ctivation of <u>T</u> cells. This docking protein with a transmembrane domain is associated with the T-cell receptor and plays an important role in T-cell activation (Section 7.2.2).

(5) **Cas**, <u>C</u>rk-<u>a</u>ssociated kinase <u>s</u>ubstrate. This docking protein lacks a membrane-binding site but has proline-rich sequences and SH3 domains for interaction with *cytoplasmic* Tyr kinases (Section 7.2.3).

(6) **Cbl**. This adaptor and docking protein lacks a membrane binding site but has E3 ubiquitin ligase activity that induces the endocytosis and degradation of Tyr kinase-coupled receptors, Tyr kinases of the Src family, and SH2 proteins bound by phospho-Tyr residues (for more information see Section 7.2.2).

7.1.4 Ephrin receptors: signaling in two directions controls embryonic development

The essential role of receptor Tyr kinases in animal development is strikingly exemplified by the ephrin receptors. With 16 isoforms, they constitute the largest subfamily of Tyr kinase-coupled receptors. Their ligands, the ephrins, are membrane-bound proteins that are exposed at the cell surface and interact with the receptors in a juxtacrine manner, limiting their effects to direct cell–cell communication (Figure 7.15). However, like the factors of the tumor necrosis factor (TNF) family, they may be released by ADAM proteases from the membrane by ectodomain shedding (see Section 13.1.2). The name ephrin comes from an <u>e</u>rythropoietin-<u>p</u>roducing <u>h</u>epatoma cell line where the proteins were discovered in 1987.

Figure 7.15 Signal transduction by the ephrin B system Cell 1 (left) expresses ephrin B; cell 2 (right) expresses the corresponding receptor Tyr kinase. Through a juxtacrine interaction, both partners activate each other. Signal transduction is induced by autophosphorylation of the receptor and by phosphorylation of ephrin B by associated Tyr kinases. As a result, proteins with SH2 and PTB domains become assembled (and possibly phosphorylated) at the intracellular regions of both ephrin B and its receptor (not shown). Through their intracellular carboxy ends, ephrins and ephrin receptors recruit, in addition, proteins with PDZ domains. The dimeric receptor structure shown here is a borderline case because normally ephrin receptors exist as higher-order oligomers.

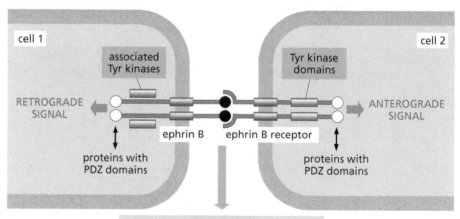

A-type and B-type ephrins have several distinguishing features. A-type ephrins are bound at the cell surface by a glycosylphosphatidylinositol anchor (Section 4.3.1) but are easily detached enzymatically by ADAM proteases, whereas B-type ephrins are genuine transmembrane proteins. A characteristic feature of the ephrin B system is that it transmits signals in both directions: the B-ephrins and the ephrin B receptors are both ligands and receptors. While the receptors, like other receptor Tyr kinases, are activated by ligand-induced autophosphorylation, the ephrins, which lack intrinsic kinase domains, are phosphorylated by associated Tyr kinases, probably by those of the Src family (Figure 7.15). The triggering signal for these phosphorylations is the interaction with the ephrin B receptor. Thus, in the ephrin B system both types of protein kinase-coupled receptors, those with intrinsic and those with associated kinases, are combined. Like other Tyr kinase-coupled receptors, the phosphorylated ephrins and their receptors interact with proteins carrying SH2 and PTB domains. In addition, they are able to bind proteins with PDZ domains (see Sidebar 7.2) via C-terminal motifs.

Ephrins control first of all **targeted growth processes** in the course of embryogenesis rather than cell proliferation. These include the directed outgrowth of nerve fibers (a process called axon guidance) together with the formation and stabilization of synapses and the sprouting of blood vessels. Thus, ephrins are essential for cell differentiation and interlinkage formation in both the nervous and vascular systems. The growth of nerve fibers is a strictly targeted process that involves both attractive and repulsive forces. Attraction is supported by juxtacrine binding of ephrins (and other morphogenetic factors; see Section 16.1) to their receptors. Consequently, repulsion seems to depend on the proteolytic release of ephrins from the membrane, as well as on endocytotic down-modulation of ephrins and their receptors.

The cellular effects of ephrins are mirrored by the signaling cascades they activate, which target predominantly the cytoskeleton and influence cell size, adhesion, and motility. In this respect they differ from the pathways activated by other receptor Tyr kinases, which primarily stimulate the mitogenic MAP kinase cascades and the PI3K–PKB/Akt system. The protein **PAR3**, for instance, interacting with ephrins through PDZ domains, activates the small G-proteins Cdc42 and

---

Sidebar 7.2 PDZ domains PDZ domains rank among the most abundant interaction domains, found in animals, plants, and prokaryotes (the human genome contains about 230 genes encoding PDZ domains). The acronym PDZ refers to three proteins—P̲ostsynaptic density protein 95 and D̲iscs large protein of *Drosophila* and Z̲ona occludens protein of vertebrates—where these domains were found for the first time. Proteins may contain more than 10 PDZ domains and function as perfect scaffolds for the formation of multiprotein complexes.

PDZ domains, consisting of 80–90 amino acids, exhibit quite variable sequences. They possess a characteristic three-dimensional structure that enables them to interact heterotypically with both carboxy-terminal tetrapeptide motifs and internal sequences, as well as homotypically with PDZ domains of other proteins, thus participating in the formation of multiprotein complexes. One distinguishes between pure PDZ proteins, acting as adaptors and scaffolds, and PDZ proteins with additional interaction or functional domains, for instance the widely distributed **membrane-associated guanylate kinase homologs (MAGUKs)**. These are scaffold proteins that assemble multiprotein signaling complexes rather than catalyzing the phosphorylation of GMP by ATP as genuine guanylate kinases do.

PDZ proteins play an important role in the development of cell polarity, that is, the nonequal distribution of cellular components, particularly in epithelial cells and neurons. The postsynaptic density complex is an example. This assembly of proteins is found beneath the postsynaptic membrane of frequently used neurons and stabilizes the synaptic contact (Sections 16.1 and 16.4). It contains a large number of proteins with PDZ domains including the prototypical protein PSD95, a MAGUK. Another function of PDZ domains is to organize receptor complexes at cell membranes. This holds true not only for ephrins but equally well for other receptors. The signal-transducing protein Dsh of the Wnt cascade (Section 5.9), for instance, is a PDZ protein.

Rac, which are key regulators of cytoskeletal dynamics (see Section 10.3.4). As a result, asymmetrical cell divisions are induced, leading to cell polarity. Another protein with PDZ domains, syntenin, is a scaffold for proteins controlling the formation of intercellular contacts, in particular those of synapses.

Proteins interacting with Tyr-phosphorylated ephrins through SH2 domains also include important regulators of the cytoskeleton such as Grb4, an adaptor protein involved in the control of the GTPase dynamin, the Rac/Cdc42-activated protein kinase PAK1 (Section 10.3.4), and the scaffold proteins Dock180 and axin (Section 5.9), as well as Vav, a GEF of small G-proteins of the Rho family. One negative characteristic of ephrins is that they may promote invasive cancer growth and metastasis as well as (by stimulating angiogenesis) the blood supply of tumors. In fact, ephrins have been found to be overexpressed in neoplastic lesions.

7.1.5 Receptor tyrosine kinases as cell adhesion molecules: how to kill two birds with one stone

The juxtacrine interactions of the ephrin system provide an example of a role for receptor Tyr kinases in cell adhesion. In fact, some receptor Tyr kinases contain extracellular domains that are characteristic for bona fide cell adhesion molecules (CAMs; see Section 7.3 for more information). Such motifs are, for instance, the Ig- and fibronectin-like domains found in platelet-derived growth factor and FGF receptors and the receptor Axl as well as the cadherin-like domains of the receptor Ret (these receptors are depicted schematically in Figure 7.2). Vice versa, proteins classified as CAMs may function as signal receptors, indicating that there are fluid transitions between the two types of proteins. The following three examples illustrate the dual function of specific receptor Tyr kinases and their role in embryogenesis.

Ret

The receptor Tyr kinase Ret plays an important role in embryonic development of the nervous system. In cooperation with ephrins, it controls the outgrowth and branching of neurites and the formation and function of synapses. In addition, Ret participates in the differentiation of kidney and sperm cells. In nervous tissue, Ret is activated by **Glial cell-Derived Neurotrophic Factors** (GDNF) that are expressed in the four isoforms GDNF, neuroturin, persephin, and artemin. They interact with Ret only when they have been bound first by **GDNF Family Receptors, GFRα,** a class of glycosylphosphatidylinositol-anchored membrane proteins (Figure 7.16). The active receptor is a Ret homodimer flanked by two GFRα co-receptors and requires, in addition, heparan sulfate proteoglycans to facilitate the interaction with the GDNF ligand. The primary downstream effectors of this complex are docking proteins such as FRS2 as well as scaffold and adaptor proteins such as Shc and Grb2, gating the usual signal-transducing pathways. A unique feature of Ret is the extracellular cadherin-like domains, which indicate a role in cellular adhesion. In fact, another receptor of GDNFs is the bona fide neuronal cell adhesion molecule N-CAM; like Ret, N-CAM is activated by the GDNF–GFR complex. In contrast to Ret, however, N-CAM is devoid of intrinsic Tyr kinase activity. Instead, it is an example of a kinase-associated receptor that transduces the signal by interacting with the cytoplasmic Tyr kinase Fyn, a Src family member (see Section 7.2). Due to chromosomal rearrangements, Ret (Rearranged during transfection) may mutate into a constitutively active oncoprotein found in thyroid tumors and multiple endocrine neoplasias. Conversely, given the effects of Ret on nervous tissue, drugs that treat Parkinson's disease and drug addiction may be obtained by developing synthetic Ret agonists.

Fibroblast growth factor receptors

The four isoforms of the FGF receptor interact with 21 different isoforms of FGF that, in addition to their effects as wound hormones, play a key role in early phases of embryogenesis and organ development, particularly of the skeleton. This is drastically demonstrated by malformations caused by an overactivation

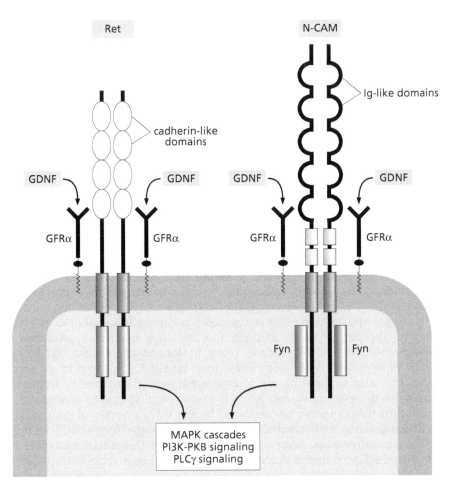

Figure 7.16 Transmission of glial cell-derived neurotrophic factor signals by receptor Tyr kinase Ret (left) and cell adhesion molecule N-CAM (right) Tyr kinase domains are symbolized by colored rectangles. Details are found in the text.

of FGF receptors due to point mutations, especially of the extracellular Ig-like domains (Figure 7.17). These domains are essential for ligand binding, dimerization, and *cis* interactions (interactions within the same membrane plane) with Ig-CAMs (Section 7.3.1). Whether or not they also contribute to cell adhesion is still not clear. Like Ret, FGF receptors also need heparan sulfate, probably to open the ligand-binding site. This restricts FGF effects to tissue areas exhibiting a high local concentration of this type of proteoglycan.

Figure 7.17 Malformations of the skeleton due to point mutations of the fibroblast growth factor receptor (A) The receptor is depicted schematically to the right (TK, colored rectangles, Tyr kinase domains; I, II, and III, IgG-like domains). Colored circles indicate the positions of point mutations in the extracellular and transmembrane domain leading to deregulation and overactivation of the receptor. (B, C) The syndromes shown here are hereditary diseases. They are characterized by a disproportionate growth of extremities and body (frequently dwarf-like) and severe malformations of limbs, skull, and face (craniosynostosis). Achondroplasia obtained by breeding is also the reason for the short legs of dachshunds and Basset hounds. (B and C, from A.O.M. Wilkie, G.M. Moriss-Kay, E.Y. Jones et al., *Curr. Biol.* 5:500–507, 1995.)

Axl

The receptor Tyr kinase Axl and its relatives Tyro3 and Mer are strongly expressed by monocytes, macrophages, and cells of the vascular, nervous, and reproductive

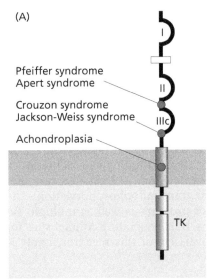

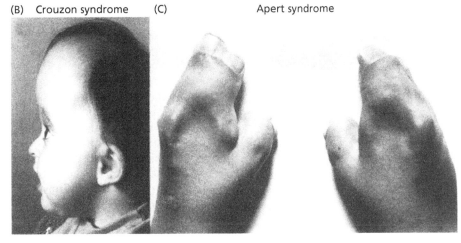

systems. The extracellular extensions of these receptors contain two Ig- and fibronectin-like domains, each possibly interacting with CAMs. As ligands, the proteins Gas6 and S have been described. Gas6 is encoded by a Growth arrest-specific gene, and protein S inhibits the blood-clotting cascade. Like Ret and FGF receptors, receptors of the Axl type are also engaged in embryonic development. In addition, they inhibit the function of antigen-presenting cells. Indeed, upon gene knockout, an extreme hyperplasia of lymphoid organs, accompanied by autoimmune diseases and massive disturbances of the nervous system, has been observed. Oncogenic mutations of Axl have been found in several human neoplasias.

Summary

Receptors of insulin and a series of growth and morphogenetic factors contain an intrinsic cytoplasmic Tyr kinase domain. Receptor activation requires ligand-induced dimerization or oligomerization of receptor subunits. As a result the cytoplasmic kinase domains activate each other, mostly by trans-autophosphorylation of the activation loops, and phosphorylate the cytoplasmic extension of the receptor. Thus, numerous phospho-Tyr residues are generated that serve as docking points for further signaling and substrate proteins. Tyr dephosphorylation is suppressed and the effect of receptor Tyr kinases is augmented by inactivation of Tyr phosphatases through reactive oxygen species produced by receptor-activated NADPH oxidase. Upon prolonged stimulation, Tyr kinase-coupled receptors become internalized; they are either destroyed in lysosomes or, after cessation of the stimulus, re-exposed at the cell surface. The signal that induces this receptor down-regulation is monoubiquitylation of several sites in the receptor molecule, mainly catalyzed by the RING domain ubiquitin ligase Cbl, which also triggers proteolytic degradation. Truncated forms of Cbl acting as dominant negative mutants are oncogenic in animals. Domains such as SH2 and PTB allow docking proteins to recognize and bind to sequences with phospho-Tyr residues. Such domains serve two major purposes: formation of signal-transducing multiprotein complexes and intrasteric control of protein function. Apart from autophosphorylation, the insulin receptor Tyr kinase binds a membrane-associated docking protein IRS and phosphorylates it at numerous Tyr residues. A multiprotein complex is formed that is essential for insulin effects such as activation of glycogen synthesis and expression of the glucose transporter GluT4. Ephrins are membrane-bound (or soluble) proteins that interact with receptor Tyr kinases and evoke morphogenetic responses. The membrane-bound ephrins are Tyr kinase-*associated* receptors, themselves activated by the ephrin receptors. Thus, the ephrin system transduces signals in both directions. Its pathological overactivation promotes invasive tumor growth. Receptor Tyr kinases with extracellular domains containing cell–cell interaction motifs (fibroblast growth factor receptor, Ret, Axl) combine the effects of signal transducers with the properties of cell adhesion molecules (CAMs). Expressed on the surface of adjacent cells, they may interact with each other or with other CAMs. Vice versa, CAMs may transduce signals by associating with cytoplasmic Tyr kinases.

7.2 Receptors associated with tyrosine kinases

The advantage of separating a receptor's ligand-binding domain from its effector domain has been discussed in Section 3.3.4: by this "trick," a wide variety of input signals can be processed by the same intracellular apparatus for signal transduction. This occurs, of course, at the expense of a selective targeting of signals. Receptors associated with Tyr kinases function according to the same principles as receptors with intrinsic Tyr kinase activity. Ligand binding causes homo- or hetero-oligomerization (or activates preformed oligomers) in such a way that associated cytoplasmic Tyr kinases undergo trans-autophosphorylation. The kinases thus activated phosphorylate effector proteins as well as multiple Tyr residues in the cytoplasmic receptor domains, thus generating contact sites for proteins with SH2 and PTB domains. An overview of this receptor family is

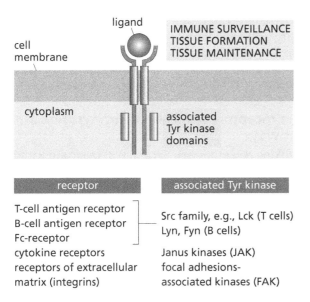

Figure 7.18 Receptors associated with Tyr kinases and their overall functions

receptor | associated Tyr kinase

T-cell antigen receptor
B-cell antigen receptor
Fc-receptor
cytokine receptors
receptors of extracellular
matrix (integrins)

Src family, e.g., Lck (T cells)
Lyn, Fyn (B cells)
Janus kinases (JAK)
focal adhesions-
associated kinases (FAK)

Figure 7.19 Cytoplasmic Tyr kinases associated with transmembrane receptors To the left are the names of the kinases; to the right are the receptors that predominantly interact with them. Kinases of the Frk/Brk, Fes/Fer, and ACK subfamilies are found mostly downstream of kinases that directly associate with receptors. The Tyr kinase domain is symbolized by a gray rectangle. C_{14}, tetradecanoyl residue anchoring the kinases of the Src family at the inner side of the cell membrane; BD, interacting domain; FAT, focal adhesion targeting domain; CRIB, Cdc42/Rac-interactive binding domain; NLS, nuclear localization sequence; PH, pleckstrin homology domain; PRO, proline-rich motif that interacts with SH3 domains; RTK, receptor Tyr kinase.

presented in Figure 7.18. The cytoplasmic Tyr kinases interacting with these receptors or positioned downstream of such interactions are summarized in Figure 7.19. Many of them have been found to undergo oncogenic mutations.

The prototype of a non-receptor Tyr kinase is **Src** (pronounced "sark"). Its mutated form encoded by the viral v-*src* gene has become famous as the first oncoprotein discovered in the Rous sarcoma virus in 1970. This chicken retrovirus was characterized by Peyton Rous (Nobel Prize in Physiology or Medicine for 1966) as the first transmissible agent inducing cancer. Further studies led to a theory on the genetic roots of cancer that states that oncogenes in general are gain-of-function mutations of proto-oncogenes that encode signaling proteins involved in normal cell physiology. For this discovery, the Nobel Prize in Physiology or Medicine for 1989 was awarded to J. Michael Bishop and Harold E. Varmus. Src represents a major subfamily of non-receptor Tyr kinases that include the enzymes Src, Yes, Fyn, Fgr, Lyn, Hck, Lck, and Blk, as well as some

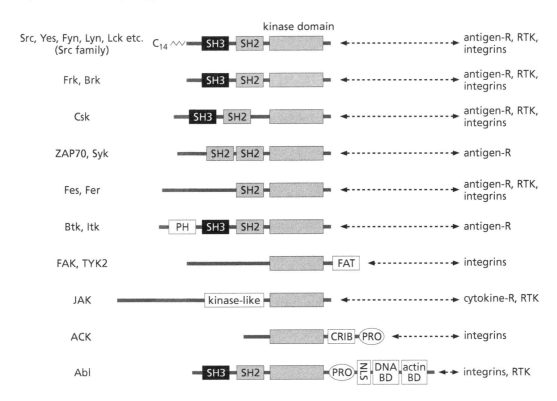

closely related enzymes. Common features of these kinases are membrane anchoring by fatty acid residues and a characteristic arrangement of SH2, SH3 and Pro-rich domains as well as a C-terminal phospho-Tyr residue. As explained for Src (Section 7.1.2), the intramolecular interactions between these domains keep the enzymes in an inactive state, which is unlatched by input signals.

7.2.1 Cytokine receptors associated with tyrosine kinases: the JAK–STAT pathway

Type I and II cytokine receptors (see Section 6.3) associate with cytoplasmic Tyr kinases. In their active form, these receptors are homo- and heterodimers or oligomers assembled according to a unit construction system. Four receptor subunits called α_C, β_C, γ, and GP130 (glycoprotein of 130 kDa) are known. They are expressed in various isotypes combining to form a large number of receptors with different ligand specificities. There are, however, deviations from this scheme. These are the receptors of growth hormone, prolactin, erythro- and thrombopoietin, and granulocyte growth factor G-CSF (colony-stimulating factor; this cytokine was discovered by its ability to stimulate the growth of granulocyte colonies in tissue culture) that are homo-oligomers of a special set of subunits. While some receptors (for example, erythropoietin and growth hormone receptor) are dimerized by a single cytokine molecule "bridging" the monomers, others interact with one ligand per subunit (for instance, the G-CSF receptor).

Each receptor subunit is a type 1 transmembrane protein with one transmembrane domain. Other structural features are several conserved Cys residues and, only in type I receptors, one or two Trp-Ser-X-Trp-Ser (WSxWS) motifs in the extracellular domains as well as signal-transducing "boxes" of conserved amino acid sequences in the cytoplasmic part. Through these boxes, cytokine receptors associate with cytoplasmic Tyr kinases of the **JAK** family (the term stands for JAnus Kinase, although another interpretation is said to be "just another kinase"). They exist in four isoforms, JAK1–3 and TYK2 (TYrosine Kinase 2). Through specific N-terminal interaction domains called JAK homology domains, they bind to the receptor subunits and become allosterically activated in such a way that they trans-autophosphorylate their activation loops. Thereupon the activated kinases catalyze the phosphorylation of the receptor (rendering it a docking site for SH2 and PTB proteins) and other substrate proteins, thus opening a series of signaling pathways including MAP kinase cascades and the PI3K–PKB axis. Principally, these signaling responses resemble those evoked by receptor Tyr kinases.

In addition, JAKs gate a route that directly leads from the cell's periphery to the genome and thus is strongly reminiscent of the straightforward strategy of prokaryotic signal transduction. This "two-component system" is called the **JAK–STAT pathway**. STATs (Signal Transducers and Activators of Transcription) are transcription factors that become activated by tyrosine phosphorylation. STAT-like proteins are found throughout the animal kingdom but not in unicellular eukaryotes, fungi, or plants, which are mostly devoid of Tyr kinase-mediated signaling. The mammalian genome encodes seven isoforms. STATs contain both SH2 and SH3 domains. Through the SH2 domains they bind to the phosphorylated receptor, where they become phosphorylated by the associated JAKs. As a result, they form homo- or heterodimers by mutual interactions between SH2 domains, and phospho-Tyr groups and become enriched in the nucleus. Their target genes are characterized by a response element called γ-**interferon activation sequence**, although by no means does it respond solely to γ-interferon but in addition to a wide variety of other cytokines. This versatility is due to numerous sequence variations that are specifically recognized by individual STAT isoforms. Thus, a STAT1 homodimer controls another set of genes than a STAT3 homodimer or a STAT1,2 heterodimer.

The principle of STAT activation is shown in Figure 7.20 for the interleukin 2 (IL2) receptor as an example of a type I cytokine receptor. Other cytokine receptors

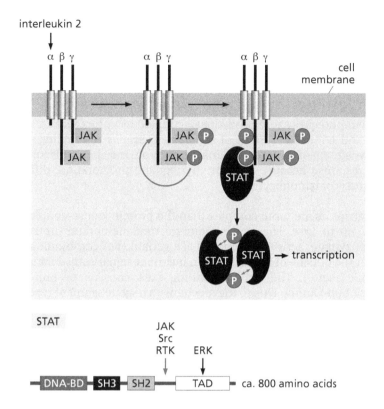

Figure 7.20 Signal transduction by the interleukin 2 receptor The receptor is a heterotrimer. Binding of interleukin 2 induces an activation by trans-autophosphorylation of the associated Tyr kinase JAK that, in turn, phosphorylates the receptor (P, phosphate). It generates numerous binding sites (for the sake of clarity only two are shown) for proteins with SH2 and PTB domains and triggering corresponding signaling cascades (not shown). These proteins include the transcription factors STAT. Upon phosphorylation by JAK, STAT dimerizes by a mutual interaction between SH2 domains and P-Tyr groups. The dimer accumulates in the nucleus, inducing the transcription of genes. The domain structure of STAT is depicted below. The red arrow symbolizes Tyr phosphorylation; the black arrow indicates Ser/Thr phosphorylation; red double-headed arrows mean interactions. DNA-BD, DNA-binding domain; TAD, transactivating domain; RTK, receptor Tyr kinase.

transmit signals in an analogous way. STATs may also be phosphorylated and activated by other Tyr kinases such as receptor Tyr kinases. In addition, certain GPCRs have been found to activate STATs, possibly along the arrestin–Src signaling pathway (Section 5.8). Moreover, STAT effects are augmented by Ser phosphorylation of the transactivating domain catalyzed, for instance, by the MAP kinase ERK. Therefore, STATs seem to resemble logical AND gates that gain full activity only when they have received two different signals.

The JAK–STAT system is controlled by sophisticated mechanisms of signal termination. In general, the degree of phosphorylation of the components is kept at a low level by Tyr-specific phosphatases. Moreover, STATs induce the transcription of genes encoding **suppressors of cytokine signaling (SOCS)**. These proteins block the catalytic domain of JAKs, at the same time triggering ubiquitylation and proteolytic degradation of the receptor and the associated JAK. That is, they play a similar role as arrestins in GPCR signaling. SOCS knockout animals are not viable, giving some idea of how essential this negative feedback control is. The characteristic structural element, called a SOCS box, comprises about 40 amino acids. It is found not only in SOCSs but also in many other proteins, most of which exhibit E3 ubiquitin ligase activity (for more details see Section 12.16). Recently, SOCS have gained particular attention since they seem to be involved in obesity. As shown in Section 9.4.3, excessive accumulation of body fat is prevented by adipokinetic hormones such as **leptin**, a cytokine-like protein that interacts with a class I cytokine receptor and activates the JAK–STAT pathway. Chronic stimulation resulting from overnutrition may lead to leptin resistance, due to an accumulation of SOCS and receptor down-modulation. The consequence would be obesity.

Another family of inhibitors hinders the binding of STATs to DNA. These **protein inhibitors of activated STAT (PIAS)** have functions going far beyond regulation of STAT activity. Some of them control, for instance, the activity of ion channels, while others are E3 ubiquitin ligases that catalyze the sumoylation of transcription factors such as cJun and p53, which become inhibited.

Finally, STAT forms that lack a transactivating domain are significant as well. They arise from alternative RNA splicing and function as competitive inhibitors resembling dominant negative mutants. In Section 7.2.3, we shall encounter another example of this regulatory principle.

7.2.2 Antigen receptors: a highlight of sophistication

B- and T-lymphocytes express Tyr kinase-coupled receptors that are activated by antigens and control a system of signaling cascades resembling in principle those downstream of receptor Tyr kinases. They are indispensable for the specific immune response because they strictly regulate proliferation, differentiation, and apoptosis of lymphocytes.

Antigen receptors are more complex than the protein kinase-coupled receptors discussed up to now. They are oligomeric transmembrane proteins without intrinsic enzymatic activity composed of a variable antigen-binding part that is exposed at the outside of the cell and an invariant signal-transducing part in the cytoplasmic region. The antigen-binding part consists of proteins of the immunoglobulin family. Due to the mechanism of gene segment recombination, its variability is almost unlimited (the number of variants amounts to 10^{15}; for details the reader should consult immunology textbooks).

All antigens evoke more or less identical cellular responses. From the beginning, this situation excludes any fixed "wiring" between the sensor and the effector domains of the antigen receptors. In fact, upon ligand binding, all antigen receptors couple with the same signal-transducing complex of various cytoplasmic Tyr kinases and their substrates, thus activating a rather complicated network of signal-processing reactions. This network—which is by no means understood in all details—will be presented here in a summary fashion by use of T-cell activation as an example.

T-cell receptor and its co-receptors

As shown in Figure 7.21, the T-cell receptor is composed of six different transmembrane proteins, subunits α, β, γ, δ, ε, and ζ, that are linked noncovalently. Two of them form the α,β-heterodimer, which is the variable part of the receptor. In a juxtacrine manner this heterodimer interacts with the antigen exposed at the surface of an antigen-presenting cell together with the major histocompatibility complex (MHC). This results in a conformational change of the invariant part of the receptor (called CD3 complex), which consists of a γ,ε heterodimer and a δ,ε heterodimer as well as a ζζ homodimer. As a result these subunits become accessible for an associated Tyr kinase that phosphorylates their cytoplasmic domains at so-called **ITAM sequences** (Immune receptor Tyrosine Activation Motifs) with two Tyr residues each. Since ITAM sequences resemble the autophosphorylated domains of receptor Tyr kinases serving as docking sites for proteins with SH2 and PTB domains, their phosphorylation is absolutely essential for signal transduction.

The Tyr kinase responsible for ITAM phosphorylation is **Lck**, a member of the Src subfamily of cytoplasmic Tyr kinases. As a primary target of input signals, Lck plays a key role in antigen-dependent signal transduction. The kinase is – like Src – anchored in the cell membrane by a fatty acid residue and concentrated in the immediate vicinity of the T-cell receptor by a noncovalent binding to **CD4** (in T-helper cells) or **CD8** (in cytotoxic T cells). These two transmembrane proteins belong to the large family of Ig-CAMs and interact with MHC (see Section 7.3), thus functioning as both co-receptors and adaptors. Another co-receptor required for the contact between the T-cell and antigen-presenting cell is the transmembrane protein **CD28**. Its ligands are the proteins CD80 or CD86 exposed at the surface of antigen-presenting cells. Like CD4/CD8, CD28 also binds Lck that catalyzes Tyr phosphorylation, thus rendering CD28 an additional docking site for proteins with SH2 and PTB domains. This signaling pathway becomes

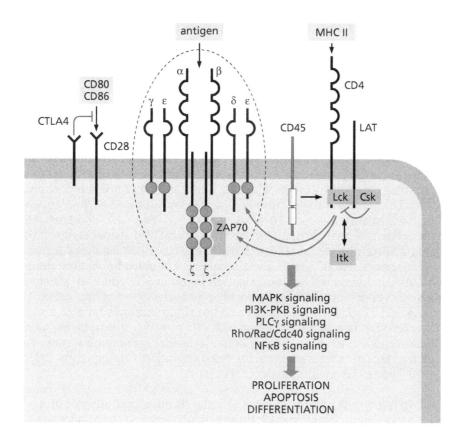

<antigen   MHC II

Figure 7.21 Signal transduction by the antigen receptor complex of T cells: the basic events The core of the receptor complex, shown in the oval, is flanked on both sides by co-receptors CTLA4, CD28, and CD4. Colored circles symbolize the ITAM sequences, which upon phosphorylation by the Tyr kinase Lck primarily bind the Tyr kinase ZAP70 via a tandem of SH2 domains. Upon activation by Lck and autophosphorylation, ZAP70 serves as a docking site for proteins with SH2 and PTB domains. CD45 is a protein Tyr phosphatase with two catalytic domains (light gray rectangles) that activates Lck by hydrolyzing a C-terminal phospho-Tyr residue. This reaction is reversed by the Tyr kinase Csk. Lck and its opponent Csk are bound by the transmembrane scaffold protein LAT. Upon Tyr phosphorylation by Lck and ZAP70, LAT binds additional proteins with SH2 and PTB domains including the Csk-related Tyr kinase Itk that further amplifies T-cell receptor signaling. For more details, see the text.

activated mainly during long-lasting immune reactions. CD28 activity is inhibited by the transmembrane protein **CTLA4** that probably sequesters the ligands.

Regulation of tyrosine kinase Lck

Like most other Tyr kinases Lck is not constitutively active but has to be switched on in the course of T-cell receptor activation. This occurs according to the scheme shown for Src in Figure 7.11. A key event is the release from auto-inhibition by the hydrolysis of a C-terminal phospho-Tyr residue. The dephosphorylation is catalyzed by the **Tyr-specific phosphatase CD45**, a transmembrane protein and further component of the T-cell receptor complex (Figure 7.21). In nonstimulated T cells CD45 undergoes auto-inhibition, probably by dimerization. This blockade is overcome by an allosteric interaction due to receptor and co-receptor activation (more details about this mechanism are found in Section 7.4).

By re-phosphorylating and inactivating Lck, the cytoplasmic Tyr kinase Csk opposes the effect of CD45. Csk is identical with the kinase that inactivates Src by catalyzing C-terminal phosphorylaton (mentioned in Section 7.1.2). As in the case of Src the interaction of Lck with Csk occurs at the surface of membrane-bound adaptor and scaffold proteins. These are **LAT** (Linker of Activated T-cells) and **LIME** (Lck-Interacting MEmbrane protein), two T-cell-specific Csk-binding proteins. Like other Csk-binding proteins, LAT and LIME are multifunctional scaffold proteins. By means of SH2 domains and several phosphorylatable Tyr residues, they recruit large signal-processing protein complexes in the neighborhood of the T-cell receptor. These complexes contain, among others, phospholipase Cγ and the adaptor Grb2 that connects the mitogenic Ras-RAF-MAP kinase cascade with the receptor (Section 7.1.2).

Csk is a target of immunosuppressive signals. One of these, prostaglandin E_2, induces the release of cAMP in T cells by interacting with a G_S-protein-coupled receptor. cAMP activates protein kinase A that, in turn, inactivates Csk by

Ser/Thr phosphorylation. The recruitment and activation of Csk ensures a negative feedback control of T-cell receptor signaling. Moreover, upon prolonged stimulation, Lck is – like other Tyr kinases of the Src family – degraded in proteasomes. As for receptor tyrosine kinases, this down-regulation is triggered by polyubiquitylation catalyzed by E3 ubiquitin ligase Cbl (Section 7.1.1).

Role of tyrosine kinases ZAP70 and Itk

The signal generated by the T-cell receptor/Lck complex is amplified and further processed by at least two additional Tyr kinases. With their ITAM sequences, the invariable subunits of the T-cell receptor functionally resemble docking proteins. As such, they assemble proteins with SH2 and PTB domains when phosphorylated by Lck. One of these proteins is the Tyr kinase ZAP70 (Zeta-chain Associated Protein kinase of 70 kDa). ZAP70 has two SH2 domains. These domains exactly fit two phospho-Tyr residues of an ITAM sequence, specifically bringing ZAP70 into the immediate vicinity of Lck, which catalyzes activation loop phosphorylation of ZAP70. By subsequent autophosphorylation, the activated ZAP70 renders itself a docking station for other SH2 and PTB proteins. To augment this effect further, ZAP70 phosphorylates the Csk-binding scaffold proteins LAT and LIME, thus generating additional binding sites for SH2 and PTB proteins. In other words, the activation of ZAP70 provides an amplifying mechanism that results in the formation of signal-transducing complexes at the T-cell receptor, a response that is essential for T-cell activation. Consequently, deletion of ZAP70 leads to severe immune deficiencies.

LAT/LIME not only bind and activate Csk but – via additional adaptor proteins – another cytoplasmic Tyr kinase, **Itk** (Interleukin 2-activated tyrosine kinase). As shown by inhibitor experiments, this Csk-related enzyme plays an essential role in T-cell signal processing.

Summarizing these effects, one easily recognizes that the activation of the T-cell receptor triggers an avalanche of Tyr phosphorylations in the signal-processing protein network, which in the cell leads to an extended excitation pattern. As a result, cell proliferation (clonal expansion) and defense competence are stimulated dramatically. Termination of T-cell receptor signaling follows the route that is typical for Tyr kinase-coupled receptors: Tyr-specific protein phosphatases (such as SHP1 and others; see Section 7.4.2) cooperate with the E3 ubiquitin ligase Cbl in inactivating and downregulating the tyrosine-phosphorylated components of the receptor complex.

Immunological synapses

When an immune response is prolonged, the T cell develops a special contact zone with the antigen-presenting cell, thus ensuring and reinforcing a long-lasting signal transduction. This "immunological synapse" contains a large aggregate of T-cell receptor complexes and is stabilized by an interaction between **ICAM1** (Immune Cell Adhesion Molecule 1), expressed at the surface of the antigen-presenting cell, and the integrin **LFA** (Lymphocyte Function-associated Antigen), exposed by the T cell. The situation somewhat resembles the formation of strong synaptic contacts in frequently used neurons, where aggregates of neurotransmitter receptors assemble to form postsynaptic densities (Section 16.4.1).

Other immune receptors

Receptors of the Ig type that resemble the T-cell receptor to some degree are the B-cell receptors, Fc receptors, and myeloid cell receptors. Like the T-cell receptor, these receptors are transmembrane complexes consisting of ligand-binding and signal-transducing subunits. The latter bear cytoplasmic ITAM sequences that are phosphorylated by cytoplasmic Tyr kinases recruited upon receptor activation. Again this is the trigger of a signaling avalanche, which resembles that described for T-cell activation.

Like the T-cell receptor, the **B-cell receptor** consists of variable and invariant parts and processes signals in an analogous manner. The variable part is the membrane-bound IgM, and the invariant part is an Ig α,β heterodimer with ITAM sequences. The role played in T cells by the Tyr kinase Lck is taken over here by the closely related Tyr kinases Lyn and Fyn, while ZAP70 is replaced by the related kinase Syk, Itk by the kinase Btk, and the co-stimulator CD28 by the transmembrane protein CD19 (the activation mechanism of which is still unknown).

Fc receptors are expressed primarily by natural killer cells, granulocytes, macrophages, and mast cells. Their major effect is to stimulate the killer activities of those cells. To this end they interact with antibodies that are bound via their variable Fab regions to pathogens. The receptors recognize these antibodies by their invariant Fc or tail region. As a result, the pathogen is bound at the surface of the receptor-bearing cell, which finally destroys the invader. Fcγ receptors, which are the most abundant type, recognize IgG-type antibodies; Fcα receptors are specific for IgA; and Fcε receptors interact with IgE. Fcα and Fcγ receptors are expressed by almost all types of phagocytic white blood cells, where they induce phagocytosis and the release of antimicrobial factors such as reactive oxygen species. In contrast, Fcε receptors are more or less restricted to mast cells and baso- and eosinophilic granulocytes. The mast cell receptors respond to complexes of IgE antibodies with allergens. By recruiting and activating Tyr kinases such as Lyn and Syk, they induce the degranulation of exocytotic vesicles that release histamine and other factors, thus evoking an allergic reaction. The receptors on eosinophilic granulocytes recognize large parasites such as worms coated with IgE antibodies, and they trigger the release of enzymes such as peroxidases and other factors toxic for the invader. Since both phagocytosis and vesicle degranulation require an increase in the cytoplasmic Ca^{2+} concentration, the phospholipase $C\gamma$–$InsP_3$–Ca^{2+} pathway activated by Tyr phosphorylation occupies a central position in Fc receptor signaling.

A few years ago a novel type of immune receptors was discovered and called **TREM** (<u>T</u>riggering <u>R</u>eceptors <u>E</u>xpressed on <u>M</u>yeloid cells). They are structurally related to Fc receptors. TREMs are expressed in several isoforms and are particularly abundant in monocytes and neutrophils, bone-dissolving osteoclasts, and brain microglia. As their ligands are unknown, a conclusive assessment of their physiological function is lacking. Their major role is thought to be that of amplifiers of inflammatory reactions, for instance, by pushing up the release of cytokines and other pro-inflammatory mediators. To this end they seem to cooperate with TL and NOD-LRR-type receptors, thus effectively lowering the threshold of detection of bacterial and viral pathogens. Their putatively central role in infectious diseases renders TREM receptors a highly interesting subject of clinical research.

7.2.3 Integrins: sensors of signals from the extracellular matrix

In tissues, cells are embedded in a matrix of proteins and mucopolysaccharides. The contact between cells and matrix is not passive. In fact, the function of a multicellular structure critically depends on a continuous communication between both partners: cells permanently remodel the extracellular medium that, in turn, continuously sends signals to the cells. The ancient form of such interactions is represented by the bacterial biofilms, the cells of which produce an extracellular matrix that controls their behavior (Section 3.4.2).

Animal cells communicate with the extracellular matrix via the Tyr kinase-coupled receptors of the integrin family (unicellular eukaryotes, fungi, and plants that lack Tyr kinases use other mechanisms). Integrins are heterodimers of two proteins, α and β, exhibiting one transmembrane domain each. Mammalian cells express 18 α- and 8 β-subunits, combining these into at least 24 receptors with different ligand specificities.

Integrin activation

Integrins are receptors. Therefore, the frequently heard term "integrin receptors" is somewhat misleading because it insinuates that integrins are ligands of such receptors. In fact, the ligands of integrins are proteins of the extracellular matrix such as collagen, fibronectin, vitronectin, and laminin as well as proteins of the blood-clotting cascade such as fibrinogen and von Willebrand factor. In addition, CAMs of the Ig-CAM and ADAM families exposed by neighboring cells interact with integrins in a juxtacrine mode. A particularly instructive example of such an effect is the egg–sperm interaction, where sperm proteins known as **fertilins** activate integrins of egg cells. The three mammalian fertilin isoforms (fertilin α, fertilin β, and cyritestin) are transmembrane proteins. They are derived from transmembrane proteases of the ADAM family by partial proteolytic degradation of the extracellular domains uncovering integrin binding sites (see Section 13.1.2). There is convincing experimental evidence that fertilization depends on the fertilin–integrin interaction, which in animals is achieved by both a fertilin α,β-heterodimer and cyritestin. In humans, however, the genes of fertilin α and cyritestin have degenerated to inactive pseudogenes and it is assumed that the contact between egg cell and sperm is brought about either by a fertilin β,β-homodimer or by another member of the ADAM family.

Most integrins establish contacts with their ligands only upon demand, staying in a resting state unless they receive an additional input signal. Such a signal may be sent by the ligand on its own but can also be transmitted by the signal-processing network of the cell, through what is known as "**inside-out signaling**." The key to understanding this regulation is the structure of the cytoplasmic domains of the α-subunits: by strongly interacting with the β-subunits, they keep the integrin in an inactive conformation characterized by a bending of the extracellular domains that hinders the access of an exogenous ligand (Figure 7.22). The inhibition is overcome by proteins that interact with the cytoplasmic integrin domains. These integrin-activating proteins are targets of intracellular signaling cascades, becoming activated by post-translational modifications such as phosphorylation (or dephosphorylation) and partial proteolysis or by binding of second messengers such as phosphoinositides. In most cases these signaling reactions result in the exposure of a PTB domain by which the regulatory protein binds to the integrin β-subunit, displacing the inhibitory α-subunit. Proteins such as talin and α-actinin fulfill the function of intracellular integrin activators

Figure 7.22 Model of integrin activation by "inside-out signaling" Inactive integrin differs from active integrin by a "knee-bent" conformation of the ectodomains of the α,β-heterodimer. This structure is due to an inhibitory interaction between the cytoplasmic domains that hinders ligand binding. This inhibition is relieved by an intracellular activator A that is activated by extracellular signals along signal-processing pathways (symbolized by the red arrow). The nature of A is still somewhat obscure. For some integrin types it is identical with the cytoskeletal protein talin.

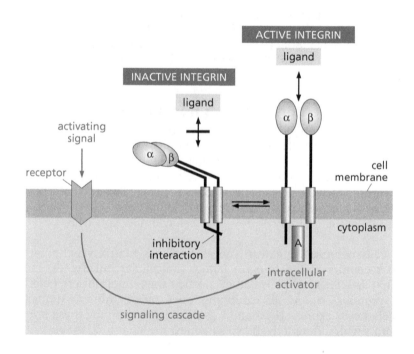

by connecting integrins with the actin cytoskeleton. The activation of these proteins is still poorly understood.

Inside-out signaling prevents accidental integrin activation

Thrombocytes, or blood platelets, provide an example of strict control of integrin activity and the importance of inside-out signaling. Their integrin, termed αIIbβ3, is responsible for the interaction of the cell with proteins of the blood-clotting cascade that induce thrombocyte aggregation at the site of clotting. An uncontrolled integrin effect would have fatal consequences such as thrombosis. Therefore, platelet integrins cannot be allowed to bind their ligands unless they are activated by thrombogenic stimuli such as thrombin, adrenaline, or collagen that are released in sufficient amounts or come in contact with platelets only upon injury. Adrenaline and thrombin interact with GPCRs and activate integrin along intracellular pathways of inside-out signaling, whereas collagen interacts with a special integrin subtype to stimulate the integrin αIIbβ3 in an elusive way. Another process requiring a strictly controlled inside-out activation of integrins is **leukocyte invasion** during an inflammatory response; this will be explained in more detail in Section 7.3.2. Finally, the interaction between the adhesion molecule ICAM1 and the **T-cell integrin LFA** should be recalled. As explained above, this interaction is required for the formation of immunological synapses between T cells and antigen-presenting cells. It strictly depends on T-cell activation.

The cytoplasmic domains of integrins are themselves substrates of protein kinases. For instance, the phosphorylation of distinct Ser and Thr residues by the MAP kinase ERK stabilizes the inactive conformation. Since ERK is activated by integrin signaling, this provides a means of negative feedback control (Figure 7.23).

Integrin signaling

The activation of integrins has far-reaching consequences for cell proliferation and differentiation. Particularly striking are the effects on the cytoskeleton, becoming visible as changes of cell shape, motility, and adhesion. As explained in Section 10.3, such changes critically depend on the formation of focal adhesions, not to mention special cases such as immunological synapses. At these

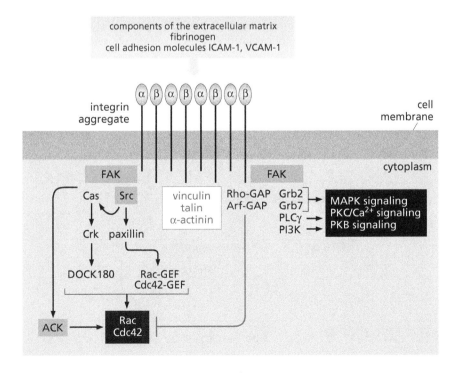

Figure 7.23 Signal transduction by integrins The input signals cause integrin aggregation at focal adhesion sites. At the cytoplasmic side, the integrin aggregates assemble large complexes of cytoskeletal proteins (red, shown in the white box) and the Tyr kinase FAK. Upon autophosphorylation, FAK functions as a docking protein for proteins with SH2 and PTB domains. These are summarized beneath the FAK symbol to the right. They lead to an activation of the signaling cascades shown in the black box. Moreover, FAK cooperates with downstream Tyr kinases such as Src and ACK. In addition, FAK controls signaling pathways leading either to activation or to inhibition of the small G-proteins Rac and Cdc42. Other details are found in the text.

cellular contact sites, active integrins assemble into large aggregates that anchor the cytoskeleton to structures of the extracellular matrix and to neighboring cells, thus contributing significantly to the integrity and function of a tissue. At their cytoplasmic domains, the integrins accumulate large complexes of signal-processing proteins, providing the link with the cytoskeleton. Such complexes are estimated to comprise more than 30 different proteins.

For the transduction of integrin signals, cytoplasmic Tyr kinases play a key role. Since these enzymes are concentrated at the contact sites, they are called **focal adhesion kinases (FAK)**. Two isoforms are known: the ubiquitously distributed FAK1 and the tissue-specifically expressed FAK2. The domain structure of these kinases is shown in Figure 7.24. When the FAKs bind to activated integrins by their N-terminal regions, their conformation is changed, enabling a trans-autophosphorylation of Tyr residues in the activation loop and at other sites of the molecule. As in the case of other receptor-coupled Tyr kinases, contact sites for proteins with SH2 and PTB domains are generated and large signal-transducing complexes are formed around the receptor. Among the proteins recruited, Src and related Tyr kinases are key effectors, phosphorylating further proteins (together with FAK). Thus, as in the case of immunological receptors, the activation of integrins triggers an avalanche of Tyr phosphorylation in the signal-transducing protein network.

Activation of G-proteins Rac and Cdc42: a key event in integrin signaling

Three substrates of Src, namely ACK, Cas, and paxillin, are the origins of signaling pathways leading to an activation of Rac and Cdc42. As explained in Section 10.3, these small G-proteins are master regulators of actin dynamics and focal adhesions.

ACK (Activated Cdc42-associated Kinase) is a cytoplasmic Tyr kinase that is phosphorylated and activated by Src. Being a component of the integrin signaling complex, it binds via a CRIB (Cdc42/Rac-Interactive Binding) domain to the GTP-loaded form of Cdc42, inhibiting its GAP-promoted GTPase activity, thus preventing signal termination (see Figure 7.19). As a result, the signal-transducing efficacy of Cdc42 becomes augmented.

Cas (Crk-associated kinase substrate) is a multifunctional adaptor/docking protein binding through an SH3 domain to autophosphorylated FAK in the immediate vicinity of Src. Upon phosphorylation by Src, it interacts with **Crk**. Crk was

Figure 7.24 Focal adhesion kinase and its inhibitor Upon interacting with integrins via its N-terminal region, focal adhesion kinase (FAK) autophosphorylates five Tyr residues (red arrows), generating binding sites for proteins with SH2 domains including the Tyr kinase Src. A C-terminal focal adhesion targeting domain (FAT) binds the adaptor/docking protein paxillin, linking FAK with focal adhesions. Two proline-rich domains (PRO) provide interaction with SH3 proteins such as Cas. FAK contains, in addition, four Ser/Thr phosphorylation sites (not shown). The catalytic domain is symbolized by the gray rectangle. The box in the lower part of the figure shows a truncated FAK lacking the catalytic domain. This FAK-related nonkinase (FRNK) resembles a dominant negative mutant acting as a competitive inhibitor of FAK.

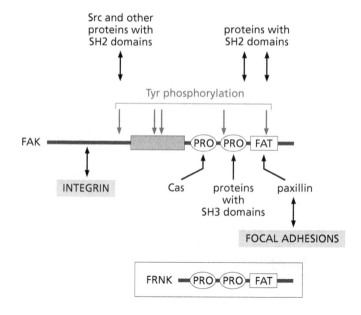

originally discovered as a viral oncoprotein and erroneously assumed to be a kinase (therefore the name "Chicken tumor virus regulator kinase"). Crk represents a family of widely distributed docking proteins with SH2 and SH3 domains as well as proline-rich sequences for an SH3 interaction. In addition, it is phosphorylated at around 15 Tyr residues serving as binding sites for a wide variety of proteins with SH2 and PTB domains including Src (providing positive feedback), PI3K, GAB, and Grb2. By this means, Crk may activate the PI3K/PKB as well as the GAB/Grb2–Ras–MAP kinase cascades, that is, the "survival program" of the cell. This explains the oncogenic potency of mutated Crk. In the context of integrin signaling, the protein **DOCK180** (Dedicator Of CytoKinesis) is an important interaction partner of Crk binding to the SH3 domain. Probably by recruiting a Rac-GEF, it specifically activates the small G-protein Rac.

Paxillin, another starting point of a Rac-activating pathway, is a scaffold protein that links integrins with proteins of the cytoskeleton. Upon multiple Tyr phosphorylations by FAK and Src, it binds and activates proteins with SH2 domains including Src, Crk, DOCK180, the Rac-GEF C3G, and a GEF of Cdc42. Paxillin-associated FAK activates GAPs and attenuates Rac and Cdc42 activities by negative feedback. Thus, the FAK–paxillin complex is an activator of Rac and Cdc42 with a built-in autocontrol mechanism.

Cross talk with other signaling cascades

Like other receptor-coupled Tyr kinases, FAK may stimulate signaling cascades including the Ras–RAF–MAP kinase cascade (Figure 7.23). Moreover, there is direct cross talk between integrins and receptor Tyr kinases because these are activated by integrins even in the absence of the corresponding ligands. In this case, Src takes over the role of a kinase phosphorylating the activation loops of the receptor kinases. Conversely, growth factor receptors are able to activate signaling pathways controlled by integrins.

Summary

Most cytoplasmic Tyr kinases exhibit motifs that enable interaction with transmembrane receptors lacking an intrinsic Tyr kinase domain. Receptor–kinase complexes transduce signals in a similar way as receptor Tyr kinases. Family I and II cytokine receptors comprise a large family of homo- and heterodimers or oligomers assembled according to a unit construction system. For signal transduction, they associate with cytoplasmic Tyr kinases of the JAK family. Upon autophosphorylation, these complexes may activate the same signaling pathways as receptor Tyr kinases. Major JAK substrates that transmit cytokine signals to the genome are transcription factors of the STAT family that bind to genes with γ-interferon activation sequences. STAT signaling is inhibited by suppressors of cytokine signaling (SOCS) and by protein inhibitors of activated STAT (PIAS). Antigen receptors of lymphocytes associate for signal transduction with cytoplasmic Tyr kinases that generate interaction sites for SH2 proteins. The signal is amplified and distributed by downstream Tyr kinases. In addition, lymphocyte activation requires co-receptors to be stimulated by proteins exposed at the surface of antigen-presenting cells. Structurally related receptors are Fc receptors of phagocytes, which interact with pathogen- and allergen-bound antibodies, and TREM receptors, which amplify inflammatory reactions evoked by bacteria and viruses. Signal transduction by all these immune receptor types critically depends on phosphorylation of ITAM sequences catalyzed by cytoplasmic Tyr kinases. Integrins are heterodimeric transmembrane proteins that establish contact to other cells and, in particular, to the extracellular matrix. They oscillate between inactive and active conformations. Activation is thought to be induced by extracellular signals along receptor-dependent signaling cascades (known as inside-out signaling). Integrins combine the properties of cell adhesion molecules (CAMs) with those of signal transducers. Upon activation by their ligands, they aggregate (in particular at focal adhesion sites) and recruit cytoskeletal proteins and cytoplasmic Tyr kinases that are essential for the activation of various signaling cascades. Integrin activation has widespread effects on cell proliferation and cell differentiation and, in particular, on the actin cytoskeleton. The latter are mediated by small G-proteins of the Cdc42/Rac family that become activated by integrin signaling.

7.3 Signal transduction by cell adhesion molecules

As we have seen, integrins and some receptor Tyr kinases (such as ephrin receptors, Axl, Ret, and FGF receptors) are both sensors of extracellular signals and mediators of cellular adhesion, thus supporting the concept of co-evolution of Tyr kinase-coupled receptors and multicellular animals since a selective cohesion of tissue cells is the primary condition of multicellularity. In addition to these receptors, other transmembrane proteins have been defined as CAMs because of their ability to interlink cells (Figure 7.25). At first glance, such CAMs appeared to be nothing but passive "glue." However, more in-depth studies have revealed that they are able to transduce signals, resembling both receptors and juxtacrine ligands. To discuss CAMs in this chapter is justified because many of them exhibit properties of Tyr kinase-coupled receptors or are ligands of such receptors.

7.3.1 Cellular adhesion proteins: intercellular glue and signaling receptor combined in one molecule

The most abundant CAMs, besides integrins, are cadherins, selectins, and Ig-CAMs. In addition, there are a series of more specialized, tissue-specific adhesion proteins.

Cadherins

Cadherins are responsible for the cohesion and tensile strength of tissues and are involved in embryogenesis, for instance, in gastrulation and morula formation They are concentrated, in particular, at stable intercellular contact sites such as adherens junctions and desmosomes.

Figure 7.25 Basic types of the most abundant cell adhesion molecules Parts of two cells are shown. The membrane of the lower cell is equipped with individual cell adhesion molecules; the membrane of the upper one has the corresponding ligands. While Ig-CAMs and cadherins interact with their own kind, the ligands of selectins and integrins are membrane-bound carbohydrates and extracellular matrix compounds (ECM), respectively. Integrins and some other Tyr kinase-coupled receptors also interact with individual Ig-CAMs (not shown). Further details are found in the text. CRP complement regulatory protein.

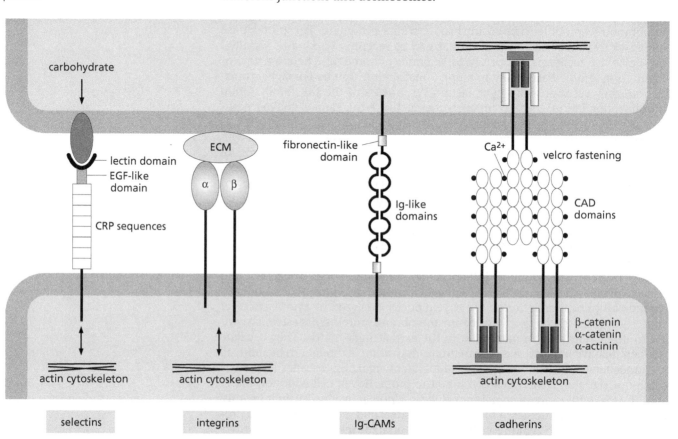

Mammalian cells express more than 80 different cadherins, which are tissue-specifically distributed. Cadherins are their own receptors; that is, they interact homotypically. This interaction is based on typical structural features consisting of five repeats of an extracellular cadherin domain (CAD domain) that comprises 115 amino acids. Binding one Ca^{2+} ion each, the domains of interacting cadherins stick together like a Velcro fastening (*Klettverschluss*). The cytoplasmic domains of most cadherins bind **α- and β-catenins**. These are adaptor proteins that provide contact with the cytoskeleton. In fact, numerous interactions between catenins and the small G-proteins of the Rho family, the major regulators of cytoskeletal dynamics, have been described. As explained in Section 5.9, β-catenin is, in addition, a transcription factor that becomes activated along the Wnt signaling pathway.

Cadherins also activate receptor Tyr kinases independently from the ligands proper and may also become phosphorylated by such receptors. The (neuronal) N-cadherin, for instance, cooperates with one of the FGF receptors to induce the growth of neurites, and the (vasculo-endothelial) VE-cadherin triggers the outgrowth of blood capillaries together with vasculo-endothelial growth factor, a ligand of a receptor Tyr kinase.

A particularly clear example of the close relationship between signal receptors and CAMs is provided by **Ret,** which is both a receptor Tyr kinase and a special cadherin characterized by an extracellular CAD motif (see Figure 7.2). As described in Section 7.1.5, its ligands are neurotrophic factors of glial cells.

Close relatives of the cadherins are the **protocadherins**. These CAMs are mainly expressed by nerve cells. In contrast to cadherins, they do not seem to bind catenins but instead interact with cytoplasmic Tyr kinases such as Fyn.

Selectins

These adhesion proteins use an ancient mechanism of cell–cell contact, the **lectin interaction**, and therefore are also called Lec-CAMs. Lectins are sugar-binding proteins of plants and have become known as phytohemagglutinins because in experiments they cause an aggregation of erythrocytes. However, the lectin interaction, which provides a means of specific intercellular communication, is not restricted to plants but is widely distributed in the animal kingdom as well, although for vertebrates a physiological role has been shown unequivocally only for endothelial cells, leukocytes/lymphocytes, and platelets. In these cases, selectins bring about the adhesion of blood cells to the inner layer of arterial blood vessels in the course of an inflammatory and immune response such as that evoked by injury or infection. This adhesion is a first step in blood coagulation and, as explained below, in leukocyte invasion and lymphocyte homing. Like other CAMs, selectins may be considered to be receptors with signaling properties. They interact with and are phosphorylated by cytoplasmic Tyr kinases. However, this is still widely unexplored terrain.

Cell adhesion proteins with immunoglobulin-like domains

Proteins with Ig-like domains are derived from one of the largest gene families of vertebrates (comprising 765 genes in humans as compared with 140 genes in *Drosophila* and 64 genes in *C. elegans*). Most of them contain various additional interaction domains characterizing them as components of the signal-processing apparatus. Typical examples are the antigen receptors and the MHC receptors CD4 and CD8 of lymphocytes. Moreover, receptor functions have also been attributed to several hundred **Ig-CAMs**. The canonical feature of these adhesion receptors is approximately seven Ig-like domains in the extracellular region, which in many types are flanked by fibronectin III-like domains (resembling a structure found in the extracellular matrix protein fibronectin). An Ig-like domain consists of about 100 amino acids arranged in several β-sheets and held together by a disulfide bridge.

A subfamily of Ig-CAMs is represented by **junctional adhesion molecules (JAMs),** concentrated mostly but not exclusively in tight junctions. These cellular contact sites are permeability barriers of endothelial and epithelial cell layers sealing off individual areas or the inner area of the body against each other and the environment, such as in the blood–brain barrier and in kidney and intestinal epithelia. Tight junctions prevent a mingling of apical and basolateral membrane proteins, thus maintaining cell polarization (see Figures 10.17 and 14.9). The junctions are composed of many different proteins. Major components are the claudins, exhibiting four transmembrane domains and interacting with each other in a homotypical juxtacrine mode. Claudins do not belong to the Ig-CAM family.

JAMs seem to control the stability and permeability of the junctions. To this end they dimerize, like other Ig-CAMs, at the cell membrane and bind homotypically to corresponding dimers as well as heterotypically to integrins exposed by adjacent cells. The prototypical tight junction JAM (JAM-A) exhibits all properties of a receptor. Like ephrin receptors, it binds various adaptor proteins with PDZ domains at its intracellular C-terminus, thus recruiting signal-transducing protein complexes. By this means JAM-A modulates both the dynamics of the actin cytoskeleton and the transcription of certain genes.

As discussed in Section 7.1.5, adhesion molecules with intrinsic Tyr kinase activity such as Ret, Axl, and FGF receptors are critically involved in embryonic tissue development. The same holds true for Ig-CAMs, cadherins, and related proteins. Their role in embryogenesis is vital for the nervous system and the formation of synaptic contacts. This is described in Section 16.1.

7.3.2 The inflammatory response: how cell adhesion molecules cooperate

The inflammatory reaction clearly shows how individual CAMs and receptors work together to bring about a complex physiological response. A major event in inflammation is the release of leukocytes from the blood vessels through the endothelium into the surrounding tissue. This trans-endothelial cell migration or diapedesis is managed by a wide variety of soluble factors (pro-inflammatory mediators) and adhesion molecules. It is tightly regulated because an accidental or incorrectly regulated leukocyte infiltration of healthy tissue would be deleterious, as shown by severe diseases such as rheumatoid arthritis, psoriasis, and asthma.

Initially, the cells of the damaged or infected tissue produce an "inflammatory soup" of mediators including chemotactic factors, such as leukotriene B (Section 4.4.5) and chemokines, which attract white blood cells. At the same time, endothelial cells become activated by other ingredients of the inflammatory soup, such as histamine, TNFα, thrombin, and complement factor C5a. In addition, bacterial components by stimulating TL receptors trigger the exposure of selectins that bind to glycoproteins on the surface of white blood cells. The weak interaction thus established slows down the velocity of the cells in the bloodstream to what is called "leukocyte rolling" over the inner surface of the blood vessel. To firmly fix the cells, an Ig-CAM called PE-CAM1 (Platelet-Endothelial Cell Adhesion Molecule 1), expressed by both leukocytes and endothelial cells, undergoes a homotypic interaction. In addition, leukocyte integrins bind to vasculo-endothelial VE-CAMs and immunological ICAMs, which are enriched in docking structures that develop on the surface of endothelial cells upon contact with leukocytes. For this purpose, the integrins must undergo inside-out activation (as shown in Figure 7.22) by endothelial chemokines, the effects of which are mediated by GPCRs (Section 5.2). In leukocytes, the CAM–integrin interaction seems to trigger the whole cascade of integrin signaling explained above, including, in particular, the pronounced effects on the cytoskeleton. This strengthens the contact with the endothelial cells and stimulates leukocyte mobility to such an extent that more and more cells assemble at the site of injury and prepare to penetrate the vessel wall. The tight junctions between the endothelial cells are loosened transiently, a response that seems to be brought about by JAMs activated by an interaction with leukocyte integrins. Finally the leukocytes can slip through the endothelial cell layer and leave the blood vessels, invading the injured or infected tissue and becoming visible, at the end, as pus. A closely related process is **lymphocyte homing**, where lymphocytes escape from blood vessels to enter the lymphatic system and to become enriched in lymph nodes.

7.3.3 AGE and RAGE in diabetes and Alzheimer's dementia

Among the receptorlike CAMs, a special type of Ig-CAM is gaining increasing attention in biomedical research. These are the **Receptors of Advanced Glycation End products (RAGEs).** Their ligands are thought to be involved in the development of severe diseases including atherosclerosis, Alzheimer's dementia, diabetes, and probably cancer.

Feared long-term effects of diabetes mellitus include disturbances of the blood supply, neuropathies, and a dimming of the eye lens. These symptoms are traced back to tissue defects caused by irreversible protein modifications. The glycation of proteins, a nonenzymatic reaction resembling the Maillard reaction (which is involved in the caramelization of sugar), is a key step in facilitating these modifications. For glycation, glucose primarily interacts with proteins, establishing Schiff-base linkages with free amino groups. In the course of an Amadori rearrangement, such compounds then become transformed into fructosamine derivatives, which are deaminated and oxidized, yielding a large number of secondary products, known as the **Advanced Glycation End products (AGEs)**. Being chemically very reactive, they may covalently cross-link polypeptide chains. Moreover, they induce inflammatory reactions and stimulate the proliferation of connective tissue and vascular smooth muscle cells, resulting in vascular constriction and other tissue damage. Such mechanisms may also be involved in nondiabetic atherosclerosis and other diseases of advanced age. To induce these cellular effects, the AGEs interact with RAGEs.

RAGEs have been found in various tissues and are expressed in response to AGE production. Upon ligand binding, RAGEs activate intracellular signaling cascades such as the MAP kinase and NFκB cascades, probably via associated protein kinases. As a result, many genes are stimulated, including those of pro-inflammatory mediators such as TNFα, IL1, and IL6.

In addition to AGEs, the leukocyte integrin Mac1 (reacting also with ICAM1 and fibrinogen), certain pro-inflammatory tissue hormones, and the non-histone protein amphoterin have been identified as RAGE ligands. **Amphoterin**, a member of the high-mobility-group proteins (see Section 8.2.1), is concentrated in the nucleus and probably plays a role in the control of chromatin function, but it has also been found in the extracellular matrix, in particular during embryonic development. By interacting with RAGEs, the extracellular amphoterin activates the small G-proteins Rac and Cdc42, thus promoting the outgrowth of neurites. In some tumors, RAGEs and amphoterin are overexpressed and are assumed to facilitate invasive growth and the formation of metastases. Since RAGEs are also activated by β**-amyloid** (Section 13.1.3) they may be involved in Alzheimer's dementia as well. In fact, high concentrations of AGEs and RAGEs have been measured in the damaged brain areas.

The *physiological* role of RAGEs remains an open question. Possibly, like many other Ig-CAMs, they are important for developmental processes during embryogenesis (such as neurite outgrowth as mentioned above) and adolescence, whereas at advanced age the pathological effects become more evident. In the face of the frequent and severe diseases that involve RAGEs, strong efforts are being made in clinical research to bring this receptor type and its cellular effects under control.

Summary

The most abundant cell adhesion molecules (CAMs), besides integrins, are cadherins, selectins, and immunoglobulin-like (Ig-)CAMs. Most of these transmembrane proteins have dual functions, namely, as a glue to establish intercellular contact sites and as receptors transducing signals. By interacting with both cytoplasmic and transmembrane Tyr kinases, many CAMs exhibit properties of Tyr kinase-associated receptors. Other CAMs activate ligands of receptor Tyr kinases or activate intracellular signaling cascades along other pathways. There are fluid transitions between the structures of CAMs and transmembrane receptors. CAMs play a critical role in inflammation. In particular, the migration of leukocytes through the blood vessel wall into the injured or infected tissueis regulated by an interaction between leukocyte and endothelial CAMs cooperating with soluble pro-inflammatory mediators. Receptors for advanced glycation end products (RAGEs), a special type of Ig-CAMs, interact with toxic metabolites (advanced glycation end products, AGEs), certain integrins, pro-inflammatory mediators, amphoterin, and β-amyloid. As a result, cellular signaling cascades become activated, which control the formation of pro-inflammatory and pro-apoptotic factors. AGEs and RAGEs are thought to play a significant role in diseases such as diabetes, atherosclerosis, Alzheimer's dementia, and cancer.

7.4 Protein tyrosine phosphatases and phosphatase-coupled receptors

In a mammalian cell, more than a third of all proteins may become reversibly phosphorylated. This permanently fluctuating level of protein phosphorylation resembles a type of short-term memory or, if one prefers the computer metaphor, a random access memory. Biochemically, it represents a steady state resulting from the competing effects of protein kinases and protein phosphatases.

Generally, the steady-state level of Ser/Thr phosphorylation is much higher than that of Tyr phosphorylation. The reason for this is the high number of Ser/Thr kinases, comprising 428 types in humans as compared with only 85 Tyr kinases. In addition, however, Tyr phosphorylation is controlled by phosphatases in a particularly strict manner. In fact, Tyr phosphorylation occurs in an extended "inhibitory field" of phosphatase activity. Moreover, the phosphatases are frequently activated by the kinases, while the latter stimulate their own activities by autophosphorylation. In other words, a positive feedback of short range is coupled with a negative feedback of long range. Such a situation is described by a set of nonlinear differential equations that give rise to different solutions depending on the initial conditions. Correspondingly, the system may either oscillate, self-organize into a defined pattern, or respond chaotically (see Sections 1.6 and 14.5.4). Oscillation offers the fascinating possibility of frequency-modulated signaling that would enormously increase the efficacy and versatility of cellular data processing. In fact, signal transmission between neurons is based on this principle, and intracellular cAMP and Ca^{2+} signals are also frequency-modulated (Section 14.5.4). Since the necessary conditions are fulfilled, the same may hold true for Tyr phosphorylation, although convincing experimental proof is still lacking.

7.4.1 Cytoplasmic and receptorlike protein tyrosine phosphatases

The human genome contains 107 genes and some inactive pseudogenes with sequences characteristic for protein Tyr phosphatases. However, only 81 of them encode bona fide protein Tyr phosphatases, whereas the others are genes of phospholipid and RNA phosphatases as well as of some enzymatically inactive proteins. Thus, the number of genuine protein Tyr phosphatases corresponds well with the number of Tyr kinases (85). In reality, however, the number of kinase and phosphatase isoforms is much higher than the number of genes due to alternative splicing of pre-mRNA, alternative gene promoters, and other mechanisms of diversification. Thus, both enzyme families show a high degree of variability, with their individual members exhibiting specific and frequently tissue-restricted functions, which do not really overlap.

Protein Tyr phosphatases are regulated differently than Ser/Thr-specific phosphatases. The latter are mostly small proteins consisting only of a catalytic domain. As explained in Section 2.6.4, they gain specificity by combining *ad libitum* with various regulatory subunits. In contrast, in most protein Tyr phosphatases, the catalytic and regulatory domains are firmly wired, sharing a single polypeptide chain. In other words, they are multidomain proteins. Similar to Tyr kinases, the regulatory domains of Tyr phosphatases seem to have a dual function: on the one hand they control enzymatic activity, and on the other hand they regulate interaction with the substrate and intracellular localization of the Tyr phosphatase. As an example, the regulation of the Tyr phosphatase SHP2 has been discussed above (see Figure 7.9).

On the basis of amino acid sequences of the catalytic domains, protein Tyr phosphatases (PTPs) are categorized in several families, which are summarized in Table 7.1. Only the so-called classical PTPs are strictly Tyr-specific, while the dual-specific PTPs also dephosphorylate Thr and Ser residues, as well as RNA and phospholipids.

The classical protein Tyr phosphatases are subdivided into cytoplasmic and membrane-bound enzymes. Cytoplasmic Tyr phosphatases are responsible for the strong inhibitory field of Tyr phosphorylation in the cell. Their efficacy is so pronounced that frequently they have to be inactivated transiently in order to ensure signal transduction from a Tyr kinase-coupled receptor into the cytoplasm or nucleus. As explained above, this inactivation is brought about by a reversible oxidation by the redox signal H_2O_2, at least as far as PTP-1B and SHP (but also the lipid phosphatase PTEN) are concerned.

With each exhibiting a single transmembrane domain and one or two catalytic domains in the cytoplasmic part, the 21 membrane-bound members of the classical protein Tyr phosphatase family have a topography resembling that of receptor Tyr kinases. The corresponding ligands can be identified in only a few cases.

Table 7.1 Protein tyrosine phosphatases (PTPs) and related enzymes

Enzyme	Number of human genes	Specificity
Classical PTPs		
receptorlike PTP	21	P-Tyr
cytoplasmic PTP	17	P-Tyr
Dual-specific PTPs		
MAP kinase phosphatases	11	P-Tyr, P-Thr
atypical dual-specific PTPs	19	P-Tyr, P-Thr, RNA
PRL	3	P-Tyr
slingshots	3	P-Ser
Cdc14	4	P-Ser, P-Thr
PTEN and myotubularins	21	PI 3-P
Others		
Cdc25[a]	3	P-Tyr, P-Thr
LMPTP	1	P-Tyr
EyA[b]	4	P-Tyr, P-Ser

[a]Cdc25 enzymes are not related to other dual-specific PTPs.

[b]EyA enzymes catalyze dephosphorylation through an aspartate rather than a cysteine residue (see also Section 2.6.4).

Thus, pleiotrophin, a neurite growth-promoting cytokine, interacts with PTPζ, inhibiting the phosphatase activity. As a consequence signaling through Tyr phosphorylation is facilitated and leads to effects on cell adhesion and cytoskeletal dynamics. However, in most cases the ligands are not known and the transmembrane Tyr phosphatases are therefore referred to as **receptorlike**. In many of them the extracellular extensions show structural features typical of CAMs, such as Ig- and fibronectin III-like domains (Figure 7.26), supporting the notion that, like certain receptor Tyr kinases, these phosphatases participate in cell–cell adhesion and interactions of cells with the extracellular matrix. In fact, several receptorlike Tyr phosphatases show high binding affinities for typical matrix components such as proteoglycans and collagen, which are known to play a role in tissue development.

A characteristic structural element of many receptorlike protein Tyr phosphatases is a doubled catalytic domain. Only one partner of this tandem, the membrane-proximal domain, exhibits phosphatase activity, whereas the other one is called a pseudophosphatase domain. It has an auto-inhibitory control function blocking the enzymatically active domain. For steric reasons this is possible only in a dimeric molecule, by a *trans*-interaction. It is assumed, therefore, that activating signals trigger the dissociation of the inactive dimer into two active monomers (Figure 7.27) whereas inhibitory signals stabilize the dimeric form. This is in contrast to receptor Tyr kinases, which have to dimerize to become activated by *trans*-autophosphorylation.

7.4.2 Tyrosine phosphatases in health and disease: a short overview

Although protein Tyr phosphatases are extremely important for signal processing, our knowledge of their physiological and pathological functions resembles a patchwork rather than a clear picture. Indeed, as yet only a few enzymes can be associated unequivocally with physiological processes such as cell proliferation,

Figure 7.26 Variability of protein Tyr phosphatases Receptorlike protein Tyr phosphatases (PTPs) and a selection of cytoplasmic PTPs are shown. PRO, proline-rich interaction domains for proteins with SH3 domains; PBM, phospholipid-binding motif; C2, Ca^{2+}/phospholipid-binding domain.

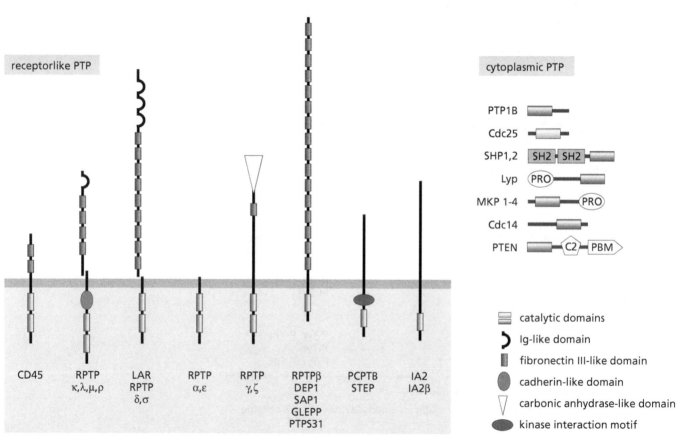

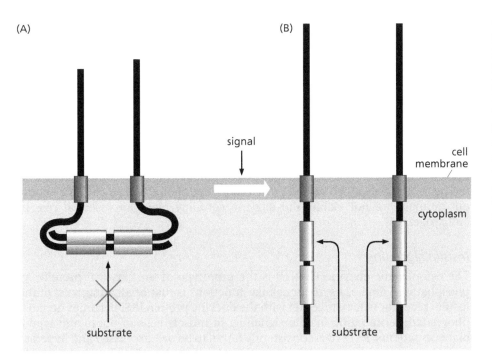

Figure 7.27 Model of the mechanism of activation of receptorlike protein Tyr phosphatases (A) In the dimeric molecule a trans-interaction between the regulatory (light gray cylinders) and catalytically active (colored cylinders) phosphatase domains is assumed to block access of the substrate to the catalytic centers. (B) An input signal has triggered dissociation into monomers, thus exposing the enzymatically active domains.

cell adhesion, immune response, and insulin action, all of which are under the control of Tyr kinases.

Cell cycle control

In Section 12.4, we shall examine the dual-specific protein phosphatases of the **Cdc25** subfamily as key regulators of the cell cycle. They catalyze the dephosphorylation of a Thr-Tyr pair in cyclin-dependent kinases, which thus become activated and promote cell cycle progression. In contrast, the phosphatases **Cdc14** of the dual-specific Tyr phosphatase family *deactivate* cyclin-dependent kinases by dephosphorylating the activation loops. This reaction is important for the termination of cell cycle phases. The Cdc25 phosphatases have ancient precursors in the bacterial rhodanases.

Immune response

Another phosphatase for which our knowledge of its physiological role is supported by experimental observations is the receptorlike enzyme **CD45**. Its expression is restricted to hematopoietic cells. As an activator of the receptor-coupled Tyr kinase Lck, CD45 is a master switch in T-cell activation (Section 7.2.2). Consistently, deletion of the enzyme results in severe disturbances of immune surveillance. There is also evidence indicating a connection between CD45 and multiple sclerosis, an autoimmune disease leading to destruction of nerve cell myelin. While CD45 is involved in lymphocyte *activation*, other Tyr phosphatases expressed in lymphocytes exhibit *inhibitory* effects. Such enzymes include PEP and the related PEST (which inactivate Csk and Lsk by dephosphorylating the activation loops) as well as LMPTP and PTPH1. Another opponent of lymphocyte activation is the classical Tyr phosphatase **SHP1**, which inactivates the Tyr kinases ZAP70 (in T cells) and Syk (in B cells) by activation loop dephosphorylation. As expected, SHP1 knockout mice suffer from immune defects but also from other disorders such as hair follicle development (because of their holey coat they are called "motheaten").

Another cytoplasmic Tyr phosphatase that inhibits T-cell activation by catalyzing the dephosphorylation of ZAP70, T-cell receptor, and the ITAM-sequence kinase Lck (Section 7.2.2) is **Lyp** (lymphoid phosphatase). This enzyme has received attention because a gain-of-function mutation is a risk factor for **autoimmune diseases** such as rheumatoid arthritis, type 1 diabetes, and Graves–Basedow disease, the most common cause of hyperthyroidism and goiter. The Lyp mutation may prevent a sufficient activation of regulatory T cells.

Mitogenic signaling

In contrast to SHP1, the sister enzyme **SHP2** does not inhibit lymphocyte activation. Moreover, it does not seem to antagonize the effects of other Tyr kinases. Instead, it stimulates Ras–RAF–ERK signaling by a still-elusive mechanism, possibly involving a stimulatory dephosphorylation of a Tyr kinase of the Src family. SHP2 may even mutate to an oncoprotein, causing the so-called **Noonan syndrome**. This disorder is characterized by pronounced developmental defects and has been connected with juvenile myelomonocytic leukemia. Since SHP2 knockout animals are not viable, the enzyme certainly plays an essential role in embryogenesis.

Insulin signaling

The cytoplasmic enzyme **PTP-1B** is the prototype of the classical protein Tyr phosphatases. One of its major cellular functions is that of an antagonist of the insulin receptor which, together with the docking protein IRS, becomes dephosphorylated, terminating insulin signaling. In muscle cells of obese and type 2 diabetic patients, PTP-1B is consistently found to be overexpressed and hyperactive. Moreover, in persons with congenital obesity or type 2 diabetes, the *ptp-1b* gene is frequently mutated. Conversely, PTP-1B knockout mice are particularly insulin-sensitive, even at a high intake of carbohydrate and fat. This indicates that, in these animals, the insulin receptor is not desensitized by excessive insulin as it normally occurs under carbohydrate/fat overload. Therefore, great hopes are placed on clinically applicable inhibitors of PTP-1B; these are not yet on the market, however. Another antagonist of the insulin receptor and, in particular, of the integrin–FAK complex is the receptorlike Tyr phosphatase **LAR** (Leukocyte Antigen Related). The enzyme is expressed in various isoforms and is especially abundant in neurons where it participates in the control of synapse formation and neurite outgrowth.

Cell adhesion, cancer, and other realms of protein tyrosine phosphatases

The receptorlike **RPTPμ** is assumed to stabilize cadherin linkages. In fact, the phosphatase antagonizes Tyr phosphorylation of the cadherin–catenin complexes, which is catalyzed by Src-type and receptor Tyr kinases and destabilizes adherens junctions. Such destabilization of intercellular contact sites has been observed in tumors exhibiting excessive Tyr kinase activity, where it may contribute to metastasis.

Since many Tyr kinases have been found to mutate to oncoproteins, one would expect their opponents, the Tyr phosphatases, to be tumor suppressors. In fact, several classical Tyr phosphatases are candidates for tumor suppressors and are thought to be involved in a variety of malignancies. However, this role has been proven unequivocally for only one of these phosphatases, specifically the receptorlike **DEP1** that is deleted in colorectal carcinomas and some other tumors. DEP1 is thought to antagonize the receptor Tyr kinase Met that is hyperactive in many human neoplastic diseases. Protein Tyr phosphatases may even mutate to oncoproteins. This unexpected finding is understandable when the role of Tyr phosphatases as indirect activators of Tyr kinases, particularly those of the Src type, is considered (see Figure 7.6). One example is the overexpression of **PTPα,** which in colon carcinomas correlates with a hyperactive Src. Another example is the oncogenic mutation of **SHP2** in patients suffering from Noonan syndrome as mentioned above.

A tumor suppressor *par excellence* is the phospholipid phosphatase **PTEN** (Phosphatase- and TENsin-homologous enzyme; tensin is a catalytically inactive

Tyr phosphatase). This enzyme specifically removes phosphate groups from the 3-hydroxyl group of inositol residues, thus inactivating the second messengers phosphatidylinositol 3,4-bisphosphate (PIP_2) and 3,4,5-trisphosphate (PIP_3). As a result, the anti-apoptotic PKB/Akt pathway is closed (Section 4.6.6). This explains why loss-of-function mutations of PTEN protect cells from apoptosis. As a result, the mutated cells gain an advantage over their normal neighbors. Such mutations correlate with the development of certain brain tumors (gliomas), endometrial carcinomas, and some benign neoplasias. The molecular event underlying PTEN inactivation is due to a defect of the C2 domain, which results in the mutated PTEN losing its ability to associate with its membrane-bound substrates.

The largest family of phospholipid phosphatases is the **myotubularins**. As indicated by the name, myotubularins of vertebrates are involved in muscle cell differentiation. In fact, mutations of the corresponding genes have been found to correlate with severe neuromuscular diseases such as myotubular myopathy. The substrates of the myotubularins are phosphatidylinositol 3-phosphate and 3,5-bisphosphate rather than PIP_2 and PIP_3. These phospholipids serve as membrane-bound anchoring sites for proteins with lipid interaction domains such as PH, PX, and FYVE (Sections 2.6.2 and 4.4.3).

Tyr phosphatases with still-unknown functions are the slingshots, PRL, LMPTP, and EyA enzymes. LMPTP, the Low Molecular weight Protein Tyrosine Phosphatase, is a representative of an ancient protein family that is closely related to prokaryotic arsenate reductases.

As mentioned in Section 3.5.3, certain bacteria produce Tyr-directed protein phosphatases as extremely powerful **virulence factors**. The disastrous effects of such enzymes, for instance in the great plague epidemics of the Middle Ages, impressively demonstrate the essential role tyrosine phosphorylation plays in human health and survival.

Summary

Protein Tyr phosphatases, the opponents of Tyr kinase-coupled receptors, constitute a large family of rather diverse cytoplasmic and transmembrane proteins including both Tyr-specific and dual-specific phosphatases but also RNA and phospholipid phosphatases. They are multidomain proteins with catalytic and regulatory domains combined in one polypeptide chain. Transmembrane phosphatases structurally resemble receptors and CAMs, although only very few ligands are known. Protein Tyr phosphatases are critically involved in all signaling processes mediated by Tyr phosphorylation. The dual-specific phosphatases of the Cdc25 family are, in addition, key regulators of the cell cycle. Deregulation of tyrosine (and phospholipid) dephosphorylation due to gene mutation, bacterial infection, or other reasons has been linked with numerous diseases.

Further reading

Ahmed N & Thornalley PJ (2003) Quantitative screening of protein biomarkers of early glycation, advanced glycation, oxidation and nitrosation in cellular and extracellular proteins by tandem mass spectrometry multiple reaction monitoring. *Biochem. Soc. Trans.* 31, 1417–1422.

Alberti L, Carniti C, Miranda C et al. (2003) RET and NTRK1 proto-oncogenes in human diseases. *J. Cell. Physiol.* 195, 168–186.

Alonso A, Sasin J, Bottini N et al. (2004) Protein tyrosine phosphatases in the human genome. *Cell* 117, 699–711.

Andersen JN, Jansen PG, Echwald SM et al. (2004) A genomic perspective on protein tyrosine phosphatases: gene structure, pseudogenes, and genetic disease linkage. *FASEB J.* 18, 8–30.

Bazzoni G (2003) The JAM family of junctional adhesion molecules. *Curr. Opin. Cell Biol.* 15, 525–530.

Bollen M & Beullens M (2002) Signaling by protein phosphatases in the nucleus. *Trends Cell Biol.* 12, 138–145.

Bourdeau A, Dube N & Tremblay Ml (2005) Cytoplasmic protein tyrosine phosphatases, regulation and function: the roles of PTP1B and TC-PTP. *Curr. Opin. Cell Biol.* 17, 203–209.

Brümmendorf T & Lemmin V (2001) Immunoglobulin super-family receptors: cis-interactions, intracellular adapters and alternative splicing regulate adhesion. *Curr. Opin. Cell Biol.* 13, 611–618.

Bubil EM & Yarden Y (2007) The EFG receptor family: spear-heading a merger of signaling and therapeutics. *Curr. Opin. Cell Biol.* 19, 124–134.

Carpenter G (2003) Nuclear localization and possible functions of receptor tyrosine kinases. *Curr. Opin. Cell Biol.* 15, 143–148.

Chen G & Goeddel DV (2002) TNF-R1 signaling: a beautiful pathway. *Science* 296, 1634–1635.

Cheng A, Dube N, Gu F et al. (2002) Coordinated action of protein tyrosine phosphatases in insulin signal transduction. *Eur. J. Biochem.* 269, 1050–1059.

Citri A & Yarden Y (2006) EGF–ERBB signalling: towards the systems level. *Nat. Rev. Mol. Cell Biol.* 7, 505–516.

Comoglio PM, Boccaccio C & Trusolino L (2003) Interactions between growth factor receptors and adhesion molecules: breaking the rules. *Curr. Opin. Cell Biol.* 15, 565–571.

Cowan CA & Henkemeyer M (2002) Ephrins in reverse, park and drive. *Trends Cell Biol.* 12, 339–346.

Cross MJ, Dixelius J, Matsumoto TL et al. (2003) VEGF-receptor signal transduction. *Trends Biochem. Sci.* 28, 488–494.

Egea J & Klein R (2007) Bidirectional Eph-ephrin signaling during axon guidance. *Trends Cell Biol.* 17, 230–238.

Evans JP (2002) The molecular basis of sperm-oocyte membrane interactions during mammalian fertilization. *Hum. Reprod. Update* 8, 297–311.

Gold MR (2002) To make antibodies or not: signaling by the B-cell antigen receptor. *Trends Pharmacol. Sci.* 23, 316–324.

Greer P (2002) Closing in on the biological functions of Fps/Fes and Fer. *Nat. Rev. Mol. Cell Biol.* 3, 278–288.

Hermiston ML, Xu Z & Weiss A (2003) CD45: a critical regulator of signaling thresholds in immune cells. *Annu. Rev. Immunol.* 21, 107–137.

Hordijk PL (2006) Endothelial signalling events during leukocyte transmigration. *FEBS J.* 273, 4408–4415.

Huang EJ & Reichardt LF (2003) Trk receptors: roles in neuronal signal transduction. *Annu. Rev. Biochem.* 72, 609–642.

Humphries MJ, McEwan PA, Barton SJ et al. (2003) Integrin structure: heady advances in ligand binding, but activation still makes the knees wobble. *Trends Biochem. Sci.* 28, 313–320.

Hynes RO (2002) Integrins: bidirectional, allosteric signaling machines. *Cell* 110, 673–687.

Ingley E (2008) Src family kinases: regulation of their activities, levels and identification of new pathways. *Biochim. Biophys. Acta* 1784, 56-65.

Juliano RL (2002) Signal transduction by cell adhesion receptors and the cytoskeleton: functions of integrins, cadherins, selectins, and immunoglobulin-superfamily members. *Annu. Rev. Pharmacol. Toxicol.* 42, 283–323.

Kerr IM, Costa-Pereira AP, Lillemeier BF et al. (2003) Of JAKs, STATs, blind watchmakers, jeeps and trains. *FEBS Lett.* 546, 1–5.

Kile BT, Schulman BA, Alexander WS et al. (2002) The SOCS box: a tale of destruction and degradation. *Trends Biochem. Sci.* 27, 235–241.

Kinbara K, Goldfinger LE, Hansen M et al. (2003) Ras GTPases: integrins' friends or foes? *Nat. Rev. Mol. Cell Biol.* 4, 767–776.

Klesney-Tait J, Turnbull IR & Colonna M (2006) The TREM receptor family and signal integration. *Nat. Immunol.* 7, 1266–1273.

Kullander K & Klein R (2002) Mechanisms and functions of Eph and Ephrin signaling. *Nat. Rev. Mol. Cell Biol.* 3, 475–486.

Kung HJ (Ed.) (2002) Tyrosine kinases. *Oncogene Reviews. Oncogene* 19 (49).

Lemke G & Lu Q (2003) Macrophage regulation by Tyro 2 receptors. *Curr. Opin. Immunol.* 15, 31–36.

Levitzki A & Mishani E (2006) Tyrphostins and other tyrosine kinase inhibitors. *Annu. Rev. Biochem.* 75, 93–109.

Levy DE & Darnell JE Jr (2002) STATs: transcriptional control and biological impact. *Nat. Rev. Mol. Cell Biol.* 3, 651–662.

Marie PJ (2003) Fibroblast growth factor signaling controlling osteoblast differentiation. *Gene* 316, 23–32.

Monteiro RC & van de Winkel JGJ (2003) IgA Fc receptors. *Annu. Rev. Immunol.* 31, 177–204.

Mustelin T & Tasken K (2003) Positive and negative regulation of T-cell activation through kinases and phosphatases. *Biochem. J.* 371, 15–27.

Nadiri A, Wolinski MK & Saleh M (2006) The inflammatory caspases: key players in the host response to pathogenic invasion and sepsis. *J. Immunol.* 177, 4239–4245.

Neel BG, Gu H & Pao L (2003) The Shp-ping news: SH2 domain-containing tyrosine phosphatases in cell signaling. *Trends Biochem. Sci.* 28, 284–293.

Nourry C, Grant SGN & Borg JP (2003) PDZ domain proteins: plug and play! *Sci. STKE* 2003/179/re7.

Pawson T, Gish GD & Nash P (2001) SH2 domains, interaction modules and cellular wiring. *Trends Cell Biol.* 11, 512–518.

Pendergast A (2002) The Abl family kinases: mechanisms of regulation and signaling. *Adv. Cancer Res.* 85, 51–100.

Petri B & Bixel MG (2006) Molecular events during leukocyte diapedesis. *FEBS J.* 273, 4399–4407.

Ponta H, Sherman L & Herrlich PA (2003) CD44: from adhesion molecules to signaling regulators. *Nat. Rev. Mol. Cell Biol.* 4, 33–45.

Rhee SG, Kang SW, Jeong W et al. (2005) Intracellular messenger function of hydrogen peroxide and its regulation by peroxiredoxins. *Curr. Opin. Cell Biol.* 17, 183–189.

Ryan PW, Davies GC, Nau MN et al. (2006) Regulating the regulator: negative regulation of Cbl ubiquitin ligases. *Trends Biochem. Sci.* 31, 79–88.

Saltiel AR & Pessin JE (2002) Insulin signaling in time and space. *Trends Cell Biol.* 12, 65–71.

Samelson LE (2002) Signal transduction by the T cell antigen receptor: the role of adapter proteins. *Annu. Rev. Immunol.* 20, 371–394.

Sariola H & Saarma M (2003) Novel functions and signaling pathways for GDNF. *J. Cell Sci.* 116, 3855–3862.

Scheiffele P (2003) Cell-cell signaling during synapse formation in the CNS. *Annu. Rev. Neurosci.* 26, 485–508.

Schlessinger J (2000) Cell signaling by receptor tyrosine kinases. *Cell* 103, 211–225.

Schlessinger J (2002) Ligand-induced, receptor-mediated dimerization and activation of EGF receptor. *Cell* 110, 669–672.

Schmidt AM, Yan SD, Yan SF et al. (2001) The multiligand receptor RAGE as a progression factor amplifying immune and inflammatory responses. *J. Clin. Invest.* 108, 949–955.

Silverman N & Maniatis T (2001) NFκB signaling pathways in mammalian and insect innate immunity. *Genes Dev.* 15, 2321–2341.

Simm A, Bartling B & Silber RE (2004) RAGE: a new pleiotropic antagonistic gene? *Ann. N.Y. Acad. Sci.* 1019, 228–231.

Tiganis T & Bennett AM (2007) Protein tyrosine phosphatase function: the substrate perspective. *Biochem.J.* 402, 1–15.

Tonks NK (2003) PTP1B: from the sidelines to the front lines! *FEBS Lett.* 546, 140–148.

Tonks NK (2006) Protein tyrosine phosphatases: from genes, to function, to disease. *Nat. Rev. Mol. Cell Biol.* 7, 833–846.

van der Merwe PA & Davis SJ (2002) The immunological synapse—a multitasking system. *Science* 295, 1479–1480.

van der Merwe PA & Davis SJ (2003) Molecular interactions mediating T cell antigen recognition. *Annu. Rev. Immunol.* 21, 659–684.

Ward CW, Lawrence MC, Streltsov VA et al. (2007) The insulin and EGF receptor structures: new insights into ligand-induced receptor activation. *Trends Biochem. Sci.* 32, 129–137.

Watson RT & Pessin JE (2006) Bridging the GAP between insulin signaling and GLUT4 translocation. *Trends Biochem. Sci.* 31, 215–222.

Wheelock MJ & Johnson KR (2003) Cadherin-mediated cellular signaling. *Curr. Opin. Cell Biol.* 15, 509–514.

Wishart MJ & Dixon JE (2002) PTEN and myotubularin phosphatases: from 3-phosphoinositide dephosphorylation to disease. *Trends Cell Biol.* 12, 579–585.

Yaffe MB (2002) Phosphotyrosine-binding domains in signal transduction. *Nat. Rev. Mol. Cell Biol.* 3, 177–186.

Yamagata M, Sanes JR & Weiner JA (2003) Synaptic adhesion molecules. *Curr. Opin. Cell Biol.* 15, 621–632.

Yu TW & Bargmann CI (2001) Dynamic regulation of axon guidance. *Nat. Neurosci. Suppl.* 4, 1169–1176.

Zhang X, Gureasko J, Shen K et al. (2006) An allosteric mechanism for activation of the kinase domain of epidermal growth factor receptor. *Cell* 125, 1137–1149.

Zhou FQ, Zhong J & Snider WD (2003) When GDNF meets N-CAM. *Cell* 113, 814–815.

Zinn K (2007) Dscam and neuronal uniqueness. *Cell* 129, 455–456.

Gene Transcription: The Ultimate Target of Signal Transduction

For all living beings, the ability to adjust the genetic readout to the environmental conditions is an essential condition of survival. Therefore, most signaling pathways end in the nucleus, and all types of exogenous signals have an impact on the genome. For humans this holds true even for a misplaced word or a panicked thought.

In both pro- and eukaryotes, gene transcription is regulated according to the same principle: transcription-controlling proteins rather than genomic DNA are the targets of input signals. These proteins, the transcription factors, interact with regulatory DNA motifs known as promoter, enhancer, and silencer sequences. By this interaction, genes are specifically marked for the enzyme RNA polymerase, the central player of the game, which catalyzes the formation of RNA from nucleoside triphosphates.

Two general pathways exist for signal transduction from the cell's periphery into the nucleus (Figure 8.1): signals that are unable to penetrate the cell membrane are intercepted by transmembrane receptors and give rise to the formation of second messengers or activate other secondary reactions. As a result, intracellular signaling cascades are stimulated, leading to modulation of transcription factor activity by post-translational modification such as protein phosphorylation. Only when the exogenous signaling molecules are lipophilic enough to pass the membrane can they interact directly with transcription factors.

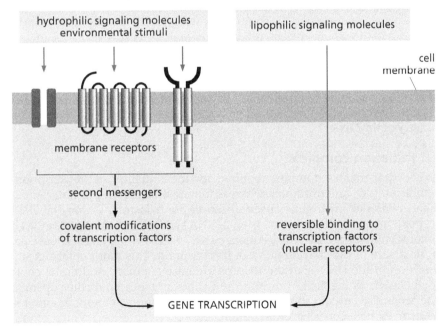

Figure 8.1 Two standard pathways of transcriptional regulation by exogenous signals
The effects of hydrophilic signaling molecules that are unable to penetrate the cell membrane are mediated by second messengers and signaling cascades controlled by covalent modifications (frequently phosphorylations) of transcription factor activity. Most lipophilic signaling molecules penetrating the cell membrane are noncovalent ligands of certain transcription factors called nuclear receptors.

Transcription consists of three phases.

(1) **Initiation.** During this phase a multiprotein complex is formed, composed of RNA polymerase associated with several regulatory proteins. This complex is recruited and noncovalently bound to a specific site of an individual gene called a control or promoter sequence, which has been marked by gene-specific transcription factors. As a consequence, the double-stranded DNA becomes unwound at the beginning of the coding sequence to enable the polymerase to begin transcription.

(2) **Elongation.** Upon completion of initiation, the regulatory proteins interacting with RNA polymerase in the initiation complex are replaced by corresponding proteins to form an elongation complex. Bound in this complex, RNA polymerase moves along the coding sequence, generating the complementary RNA strand.

(3) **Termination.** The end of the coding DNA sequence is marked by a specific nucleotide sequence. When it arrives at this point of termination, the complex dissociates, releasing the completed RNA strand.

8.1 Initiation of transcription

Most signals that control gene activity act at the stage of initiation. To initiate transcription, the enzymatic machinery of RNA biosynthesis is recruited to a regulatory gene sequence, the promoter. For the synthesis of ribosomal, transfer, and messenger RNA (rRNA, tRNA, and mRNA), eukaryotic genes have different types of promoters recognized by three different RNA polymerases, types I, II, and III. Here the interaction of **RNA polymerase II** with genes encoding mRNA is discussed in more detail since it represents the primary target of signaling cascades. RNA polymerase II consists of 10 subunits and has a molecular mass of about 500 kDa. In its active form, the enzyme is associated with a large number of additional proteins. It is estimated that, for the start of transcription, up to 50 different protein molecules are assembled at the gene promoter.

Eukaryotic **gene promoters** are divided into different functional regions: the core promoter, adjoining the start point of transcription to the 5′-end ("upstream"), and the proximal/distal promoter, located further upstream. In addition to a conventional promoter, certain genes contain alternative promoters located in the coding region. These regulatory sites control the production of truncated protein variants that may act as antagonists of the complete protein (for an example see Section 8.6). Such inhibitory fragments (which resemble dominant-negative mutants) may also be generated by alternative splicing of pre-mRNA (Section 8.10).

An **enhancer** is a regulatory sequence separated from the corresponding promoter. Enhancers may be found at far distances either upstream or downstream of a gene; nevertheless, they can interact with the gene due to a special folding geometry of the DNA.

8.1.1 Initiation complex

The RNA polymerase complex required for the initiation of transcription is assembled at the core promoter, whereas the proximal/distal promoter and the enhancer interact with gene-specific transcription factors, which "label" the start point and control the progression of RNA synthesis. In most genes the core promoter contains a TATA box [sequence TATA(T)AAG(A)] at 30 to 50 base pairs (bp) upstream of the starting point of transcription. This motif determines the interaction of the DNA with the RNA polymerase complex. Additional control elements such as the CAAT box and the GC box are located further upstream. Some genes also contain so-called downstream promoter elements at about +30 bp. Normally such genes lack a TATA box.

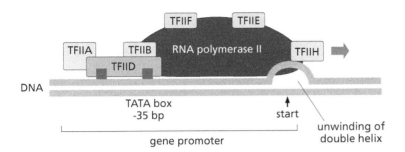

Figure 8.2 Schematic representation of the eukaryotic RNA polymerase II complex of transcriptional initiation The core enzyme (black) is shown with the six general transcription factors of the TFII family. The DNA-binding TFIID complex is shown in color. For further details, see text.

In eukaryotes, the selective interaction of RNA polymerase with a promoter is not controlled by a σ factor as in bacteria (Section 3.5.6). Instead, in a defined sequence, the enzyme forms a multiprotein complex with **general transcription factors** that, in contrast to other transcription factors, are not gene-specific. Their role is to assemble the initiation complex at the transcription start site of a promoter, which is marked by a gene-specific transcription factor.

General transcription factors are grouped in the TFII family. Among the six members complexed with RNA polymerase, TFIID binds directly to promoter DNA at the TATA box and upstream sequences. This factor is not a simple protein but a complex of at least 12 subunits including a TATA box-binding protein and associated proteins. Some of the associated proteins are co-activators of transcription, such as enzymes catalyzing the covalent modification of histones (see below). TFIIA and -B stabilize the interaction of TFIID with DNA, while TFIIB and -F enforce the binding of the polymerase to the complex (Figure 8.2). To this end TFIIA and -B interact with both DNA and the TATA box-binding protein.

TFIIF can be compared with a bacterial σ factor since it hinders the polymerase from binding to DNA sequences outside of gene promoters. TFIIF facilitates the binding of TFIIH, a large enzyme complex consisting of nine subunits and containing a protein kinase that catalyzes a C-terminal phosphorylation of RNA polymerase II (see below), as well as helicases, which under ATP hydrolysis wind up the DNA double strands at the start point of transcription.

8.1.2 RNA polymerase phosphorylation: triggering the elongation phase

The activity of RNA polymerase II is controlled by signaling cascades. In the C-terminal domain of the large subunit, multiple copies of a heptapeptide with three serine (Ser) residues are found. Two of these residues are phosphorylated at the end of the initiation and during the elongation phase. As a result, the polymerase dissociates from the gene promoter, a process known as promoter clearance or promoter escape. The initiation complex is replaced by an elongation complex, with the phosphorylated C-terminal region of RNA polymerase serving as a scaffold for the recruitment of elongation factors and proteins that promote the maturation of mRNA by splicing, capping of the 5′-end by methylguanosine, and polyadenylation of the 3′-end (see textbooks of molecular biology for details).

Phosphorylation of the polymerase is catalyzed by several protein kinases, which are under the control of exogenous signals (Figure 8.3). One of these is a subunit of the general transcription factor TFIIH and is identical with the **CDK-Activating Kinase (CAK)**. In addition to catalyzing the phosphorylation of RNA polymerase, CAK, a complex of the Cyclin-Dependent Kinase CDK7 and the regulatory protein cyclin H is also involved in cell cycle control. Two other kinases that phosphorylate RNA polymerase are **CDK8** (as a complex with cyclin C) and **CDK9** (as a complex with cyclin K or T; for more details about cyclins and CDKs, see Sections 12.3 and 12.4). CDK9 and cyclin T (K) are subunits of the positive transcription elongation factor b. Besides the polymerase, this complex also phosphorylates elongation factors, thus controlling their activities.

285

Figure 8.3 Effects of signal-induced protein phosphorylation on different components of the transcription complex (Left) Kinases of the CDK family existing as complexes with regulatory cyclins catalyze the phosphorylation of RNA polymerase II in the C-terminal domain and of elongation factors. This occurs at the transition from the initiation to the elongation stage and results in a clearance of the promoter from the initiation complex, allowing the formation of the elongation complex. (Right) Transmodulating activities of many gene-specific transcription factors are under the control of various protein kinases.

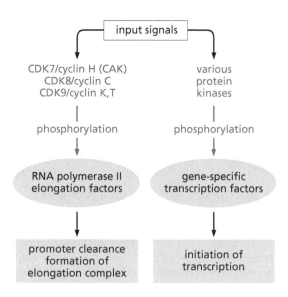

8.1.3 Gene-specific transcription factors

The gene- or sequence-specific transcription factors are predominant target proteins of signals regulating gene activity. By interacting with regulatory gene sequences (and in some cases with the polymerase complex), transcription factors either promote or hinder gene transcription.

The DNA sequences of proximal/distal promoters and enhancers specifically binding transcription factors are called **response elements**. A gene is generally controlled by several response elements, each of which interacts with an individual transcription factor; that is, genes are regulated by a combinatorial strategy (see also Section 3.2.2). This enables the cell to modulate the activity of thousands of genes through a relatively small number of transcription factors (approximately 6% of the human genome). The factors are organized on the regulatory gene sequences into complex patterns rather than binding side by side. Additional proteins such as co-activators and co-repressors (see below) are integrated into this puzzle. These multiprotein complexes are called **enhanceosomes**.

The role of transcription factors is either to assemble the initiation complex at a distinct gene locus or to prevent such an assembly. They possess a modular structure typical for signal-transducing proteins that consists of at least three domains: a regulatory domain functioning as a receptor of input signals, a DNA-binding domain for specific interaction with a response element, and a trans(cription)-modulating domain that either activates or deactivates gene transcription (Figure 8.4). Upon activation by the input signal, the transcription factor binds noncovalently to the corresponding response element through an interaction of amino acid residues with complementary nucleotide residues in the major groove of the DNA double helix. By means of their transmodulating (here transactivating) domains, the transcription factors that act as inducers now recruit the RNA polymerase complex as well as transcriptional co-activators. These co-activators are proteins that change the chromatin structure in such a way that transcription is facilitated because the RNA polymerase complex can bind to the promoter region and read the gene. Inhibitory transcription factors known as repressors have an opposite effect: they recruit proteins called co-repressors that catalyze the formation of an inactive chromatin structure or block the access of the RNA polymerase complex to DNA in another way. In a particularly clear manner, transcription factors demonstrate that data processing in the protein network is not simply the result of an interaction between biochemical reactions but critically depends on the assembly of the components at a specific site at the right time.

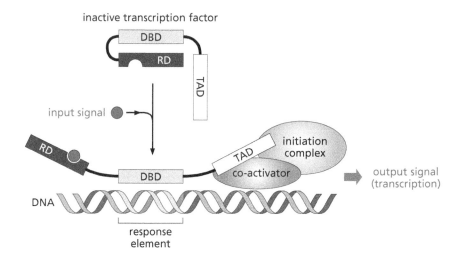

inactive transcription factor

DBD

RD

TAD

input signal

RD

DBD

TAD

initiation complex

co-activator

output signal (transcription)

DNA

response element

Figure 8.4 Mode of action of a gene-specific transcription factor (activating type) In the inactive form, the transcription factor is folded in such a way that the regulatory domain (RD) blocks the DNA-binding domain (DBD) and the transactivating domain (TAD). An input signal (red circle) induces a conformational change that relieves the blockade. The activated transcription factor binds by its DBD to a response element in a regulatory gene sequence and at the same time recruits by its TAD the initiation complex together with a transcriptional co-activator such as a histone acetyltransferase (see Figure 8.7).

The active form of a transcription factor is a homo- or heterodimer. Such a structure strengthens the binding to DNA and allows subtle allosteric regulation and efficient suppression of signaling noise. Frequently the DNA-binding domains are involved in dimerization. This dual function is due to specific structural elements such as zinc finger, leucine zipper, and helix–turn–helix motifs (see Section 8.3.2 for examples of such domains; more details may be found in textbooks of molecular and cell biology).

The input signals controlling transcription factor activity consist of interactions with other proteins or low-molecular weight ligands, or of covalent modifications such as phosphorylation (or a combination of both). The controlled translocation between cytoplasm and nucleus also plays an important regulatory role. The lifespan of transcription factors is strictly regulated by degradation along the ubiquitin–proteasome pathway competing with *de novo* synthesis. Examples are discussed below.

Summary

mRNA transcription is initiated by a complex of RNA polymerase II with six types of TFII general transcription factors binding at a gene promoter. This complex is a target of signaling reactions, particularly phosphorylations catalyzed by several cyclin/cyclin-dependent kinase (CDK) pairs that induce the replacement of the initiation complex by the elongation complex. Gene-specific transcription factors specifically bind to response elements in gene promoters and enhancers, thus labeling individual genes as targets of the RNA polymerase complex and associated proteins (co-activators) required for transcription. In contrast, inhibitory transcription factors (repressors) recruit proteins (co-repressors) that hinder RNA polymerase binding and transcription. Most gene promoters exhibit binding sites for several different transcription factors. Gene-specific transcription factors are the ultimate targets of cellular signal transduction.

8.2 Histones, nucleosomes, and chromatin

Transcription factors are unable to regulate gene transcription on their own. Instead, they rely on the assistance of either co-activators or co-repressors. The reason for this is the special physicochemical structure of the eukaryotic genome. Electrostatic repulsion would make it impossible to package the huge DNA molecules (amounting to a total length of 2 m/cell in man) in the tiny volume of a nucleus (10 μm diameter). Thus, in eukaryotic cells, strongly basic proteins, the histones, neutralize the negative charge of DNA by forming a highly organized nucleoprotein complex known as chromatin. [Histone-like packing proteins have evolved with archaea; bacteria use other proteins for chromosome

Figure 8.5 Inactive and active chromatin Inactive chromatin has a highly condensed structure due to strong interactions between DNA and the histone core of nucleosomes as well as the binding of histone 1 (H1). At sites that have become labeled by gene-specific stimulatory transcription factors, this structure is unfolded by H1 phosphorylation to such a degree that transcriptional co-activators inducing covalent modifications of the core histones gain access to the core histones of the nucleosomes. Chromatin activation is reversed by H1 phosphatases and enzymes eliminating the covalent modifications of core histones.

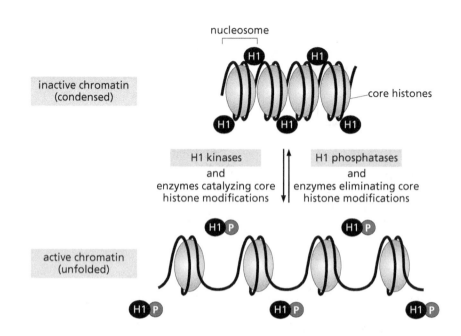

formation.] This DNA packaging creates a problem for gene transcription because the strong interactions with histones hinder access of the RNA polymerase complex to the genes, even when these have been marked by transcription factors. Transcription requires, therefore, a temporally and spatially limited weakening of these interactions. Such a mechanism provides an ideal target for regulatory signals. In fact, in a most ingenious way, nature has killed two birds with one stone by combining the process of DNA packaging with transcriptional control. To this end, the physicochemical properties of chromatin are changed in such a way that transcription is either facilitated or repressed. Therefore, all input signals that modulate gene transcription induce changes of chromatin structure aiming at an opening or closing of genes.

There are five major types of histones: "linker histones" such as histone 1 (H1) and "core histones" such as the histones H2a, H2b, H3, and H4. Core histones are phylogenetically highly conserved proteins characterized by a C-terminal histone fold, a structure that enables interactions with each other and with DNA. These interactions give rise to particular complexes consisting of two copies of each core histone species. In chromatin, this complex is wrapped by about 1.8 windings (approximately 146 bp) of the DNA double helix. This structure, called a **nucleosome**, is the basic building block of chromatin. Individual nucleosomes are connected by a linker DNA sequence comprising about 50 bp; thus, chromatin resembles a pearl necklace (Figure 8.5). The linker histone H1 is not a component of nucleosomes but binds to the linker DNA, causing an additional condensation of the chromatin structure by contracting the pearl chain. H1 exists in several variants and exhibits greater phylogenetic variability than the core histones.

8.2.1 Transcriptional co-activators and co-repressors control DNA–histone interactions

The strong interactions between DNA and histones are broken up by covalent modifications that diminish the basic character of the histones and "soften" the compact nucleosomal structure. Such modifications include acetylation, methylation, ADP-ribosylation, ubiquitylation, and phosphorylation. The enzymes catalyzing these reactions function as transcriptional co-activators. Their opponents, the co-repressors, stabilize the histone–DNA complex by reversing these modifications. The work of the histone-modifying enzymes is further supported by translocases that interrupt the histone–DNA bonds by mechanical force (as represented in Figure 8.6). Since transcriptional regulation would be impossible

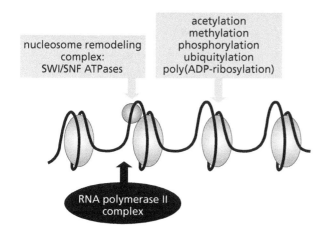

nucleosome remodeling complex: SWI/SNF ATPases

acetylation
methylation
phosphorylation
ubiquitylation
poly(ADP-ribosylation)

RNA polymerase II complex

Figure 8.6 Mechanisms of chromatin activation: unfolding of the nucleosomal structure The condensed structure of the nucleosome is weakened by mechanical force exerted by SWI/SNF ATPases (left) and covalent modifications (right). These alterations enable the interaction of DNA with the RNA polymerase complex.

without co-activators and co-repressors, gene-specific transcription factors need to recruit histone-modifying enzymes to individual gene promoters in order to clear the path for the RNA polymerase complex. Thus, both transcription factors and cofactors are targets of gene-regulatory signaling cascades.

To induce transcription, the compact structure of inactive chromatin is destabilized in a stepwise manner. In a first step, the linker histone H1 is removed from DNA by phosphorylation (Figure 8.5). Thus, in transcriptionally active chromatin (called euchromatin), H1 is found to be phosphorylated at many serine and threonine residues. These patches of negative charge weaken the interaction with DNA to such an extent that H1 may become displaced by **HMG proteins** (High-Mobility Group; the name refers to electrophoretic mobility), which belong to the large collection of non-histone proteins that modify the architecture of chromatin and its interactions with other regulatory proteins.

The goal of H1 phosphorylation and HMG protein interactions is to make the nucleosomal core accessible for enzymes catalyzing those covalent modifications that are required to break up the nucleosomal structure itself.

8.2.2 Histone acetyltransferases: co-activators of transcription

The major covalent modification neutralizing the positive charge of core histones is acetylation. For the initiation of transcription, histone acetylation is indispensable, whereas gene repression is always accompanied by histone deacetylation (Figure 8.8). The targets of acetylation are the ε-amino groups of conserved lysine (Lys) residues in the N-terminal regions of the four core histones (Figure 8.9). These regions are exposed at the surface of the nucleosomes and as carrieres of positive charge are essential for the interaction with DNA.

The acetyl groups supplied by acetyl-coenzyme A are introduced by **histone acetyltransferases (HAT).** This enzyme family comprises a large number of members exhibiting different specifications. Quite abundant are **CBP** [CREB-Binding Protein; CREB is a cAMP-activated transcription factor (see Section 8.6)] and the closely related **p300** (protein with a molecular mass of 300 kDa). Both are multipurpose histone acetyltransferases exhibiting numerous regulatory domains (Figure 8.7) by which they interact not only with CREB but also with many other transcription factors and control proteins. In fact, CBP/p300 combines the properties of an enzyme with those of a multifunctional scaffold protein, thus recruiting large protein complexes at individual gene promoters and softening the nucleosomal structure by histone acetylation. CBP/p300 and the proteins assembled by these enzymes are targets of numerous signaling cascades.

Apart from being essential partners of stimulatory transcription factors, histone acetyltransferases are also involved in other processes requiring a change of

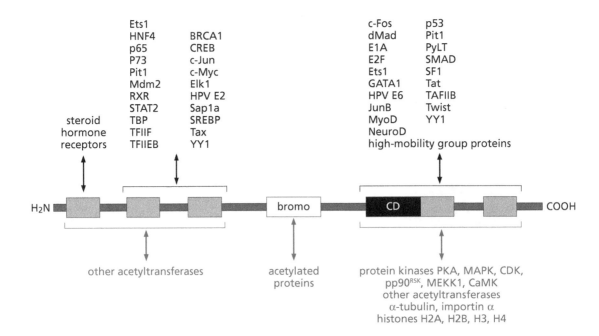

Figure 8.7 Histone acetyltransferase CBP/p300: a versatile transcriptional co-activator and scaffold protein
Schematic multidomain structure of the 2441 aa protein is shown with the interaction partners identified so far (black, transcription factors and related non-histone proteins of chromatin; red, other interaction partners). Note that phosphorylation by PKA and binding of CREB are only two among numerous interactions of CBP/p300. CD, catalytic domain; bromo, bromo-domain; gray rectangles, other interaction domains

chromatin structure, including DNA replication, DNA repair, and genetic recombination. The acetyltransferases acetylate Lys residues of histones selectively rather than randomly, thus generating a characteristic acetylation pattern. Together with other post-translational modifications, this pattern resembles an epigenetic, DNA-independent code (see Section 8.2.8), which can be deciphered by proteins that possess interaction domains for sequences with acetylated Lys residues. A typical recognition motif for acetyl-Lys is the **bromodomain**, a tetrahelical peptide sequence of about 110 amino acids found from yeast to humans in many transcription factors and other chromatin-associated proteins including CBP/p300 (Figure 8.7). Via bromodomains, complexes of histone acetyltransferases with other proteins are bound to acetylated histones.

8.2.3 Histone deacetylases: co-repressors of transcription

The opponents of histone acetyltransferases are the histone deacetylases (HDAC). As co-repressors, they inhibit or terminate gene transcription (Figure 8.8), forming complexes with various other proteins such as inhibitory transcription factors. Histone deacetylases also play a key role in gene silencing and the formation of heterochromatin, highly condensed and transcriptionally inactive areas of chromatin arising in the course of tissue differentiation. Such zones are characterized by DNA sequences with methylated CpG residues (see Section 8.2.8), and histone deacetylases are directed to those sites by methyl-CpG-binding adaptor proteins, thus guaranteeing continuous gene silencing.

As inhibitors of gene transcription, histone deacetylases may act as tumor suppressors. An interesting example is **BRCA1** (BReast CAncer gene 1). This enzyme associates with a specific repressor of estrogen-sensitive genes. A loss-of-function mutation of BRCA1 may cause breast and ovarian cancer and is thought to be one reason for the increased incidence of these diseases in certain families. Complexes of repressors with histone deacetylases also attenuate gene transcription activated by other steroid hormone receptors (see Section 8.3).

Another tumor suppressor protein, the **retinoblastoma protein Rb,** forms a complex with HDAC1 and inhibits the transcription of genes required for a proliferating cell to enter the DNA replication phase. Upon phosphorylation of Rb, catalyzed by cell cycle-specific cyclin-dependent protein kinases, the complex dissociates and the repressor activity is lost. A deleted *rb* gene is a risk factor for retinoblastoma, a tumor of the eye (for details see Section 12.7).

Exogenous signals regulate the function of histone deacetylases predominantly by phosphorylation. In muscle cells, for instance, calmodulin-dependent protein kinases provoke a translocation of histone deacetylases from the nucleus into the cytoplasm, where they become sequestered by proteins of the 14-3-3 family (Section 10.1.4). The enzymes return to the nucleus only upon dephosphorylation, catalyzed by signal-controlled phosphatases. Phosphorylation also modulates the enzymatic activity of histone deacetylases and their interactions with other chromatin-associated proteins.

Sirtuins

Conventional histone deacetylases are zinc proteins that catalyze the hydrolysis of the amide bond between Lys and the acetyl residue. Another family of deacetylating enzymes called **Sir2-like histone deacetylases** or **sirtuins** (referring to the homologous yeast enzyme Sir2, $\underline{S}$ilent mating type $\underline{i}$nformation $\underline{r}$egulator $\underline{2}$) operate by quite a different mechanism: they are ADP-ribosyltransferases that cleave the amide bond by use of nicotinamide adenine dinucleotide (NAD^+) instead of water, yielding nicotinamide and 2'-O-acetyl-ADP-ribose (Figure 8.9). Sirtuins have been found from bacteria to humans and may play a key role in gene silencing and cell senescence. In addition to histones, other proteins such as transcription factor p53, α-tubulin, and TATA box binding proteins are deacetylated by sirtuins. Since sirtuins need NAD^+ as a co-substrate, they are sensors of the metabolic redox state, that is, the energy status of the cell, and they probably coordinate it with repression of transcriptional activity. This may be one of the reasons for the evolution of this costly, energy-consuming mechanism of deacetylation.

Competitors for NAD^+ are the poly(ADP-ribosyl)polymerases (PARPs) (see Section 8.2.6). Since these enzymes are involved in gene activation while sirtuins are involved mostly in gene inactivation, a reverse regulation has been proposed. The consumption of NAD^+ by PARPs would turn down sirtuin activity and vice versa. Moreover, the reaction product 2'-O-acetyl-ADP-ribose might be an

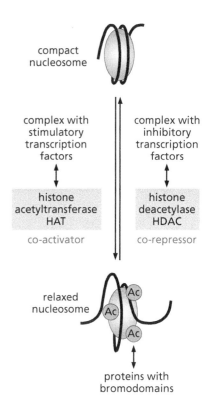

Figure 8.8 Modulation of transcriptional activity by histone acetylation By acetylation (Ac) of Lys residues of core histones, the compact structure of the inactive nucleosome is weakened, enabling transcription. Therefore, histone acetyltransferases function as co-activators and histone deacetylases as co-repressors. Both enzyme families are recruited to individual gene promoters by stimulatory and inhibitory transcription factors, respectively. Via bromo domains, additional regulatory proteins may become bound to acetylated histones.

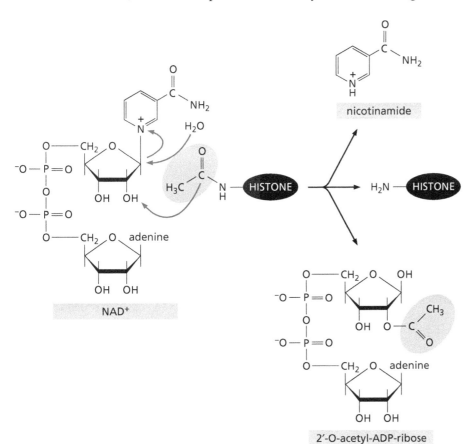

Figure 8.9 Histone deacetylation by Sir2-like ADP-ribosyltransferases (sirtuins) The enzymes transfer the acetyl residue (highlighted in color) from histone to NAD^+ simultaneously with the separation of the nicotinamide residue.

intracellular signaling molecule resembling cyclic ADP-ribose, which is generated in an analogous manner (Section 14.5.3). However, in contrast to cyclic ADP-ribose, 2′-O-acetyl-ADP-ribose has not been found to induce a release of Ca^{2+} ions from the endoplasmic reticulum.

Only recently, the activity of sirtuins was found to correlate with an increase of the lifespan in *Saccharomyces cerevisiae*, *Caenorhabditis elegans*, and *Drosophila melanogaster*, at least under situations of caloric restriction that are monitored by sirtuins. Moreover, genetic studies have prompted some authors to refer to sirtuins as products of **longevity genes**. Whether or not this also holds true for vertebrates, including humans, remains to be shown. Notwithstanding this reservation, pharmaceutical companies have begun to develop drugs that potentiate sirtuin effects. Their prototype is resveratrol, a phenolic compound (3,4′,5-trihydroxystilbene) produced by certain plants to fight parasitic organisms and found in higher doses in grapes and red wine.* In cell cultures as well as in *C. elegans* and *Drosophila*, resveratrol increases sirtuin activity and prolongs the lifespan. Whether or not there is a causal relationship between both effects is not entirely clear, because resveratrol also acts as a potent scavenger of reactive oxygen species and free radicals, which are thought to play a key role in aging and cell death.

8.2.4 Two faces of histone methylation

The core histones, in particular H3 and H4, are not only acetylated but also methylated in their exposed N-terminal regions. This reaction is catalyzed by **histone methyltransferases** with *S*-adenosylmethionine as a methyl donor.

---

*The high concentration of resveratrol in red wine may explain, at least partially, the so-called French paradox: that in France the considerably lower incidence of coronary heart diseases despite a high-fat and high-cholesterol diet correlates with a higher consumption of red wine.

Figure 8.10 Biochemistry of histone demethylation (A) Oxidative demethylation of an *N*-methyllysine residue catalyzed by a mixed-function oxygenase; the methyl residue is released as formaldehyde. This type of reaction resembles the oxidative demethylation of drugs and other xenobiotics in the liver. (B) Demethylation of a methylarginine residue catalyzed by a peptidylarginine deiminase; arginine is transformed into citrulline and the methyl residue is released as methylamine.

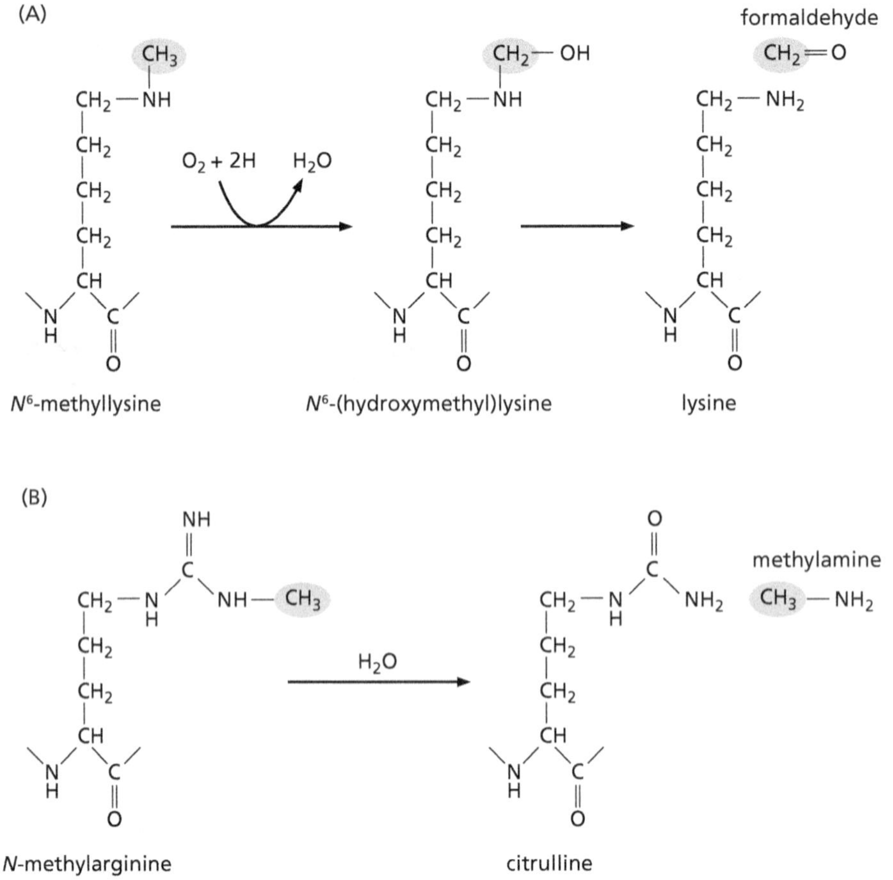

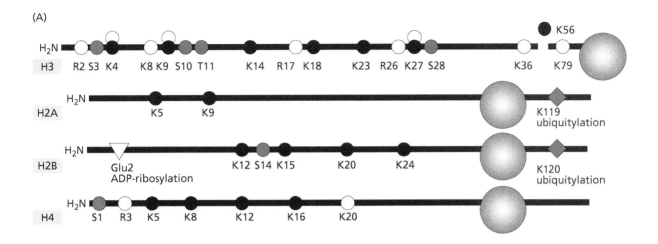

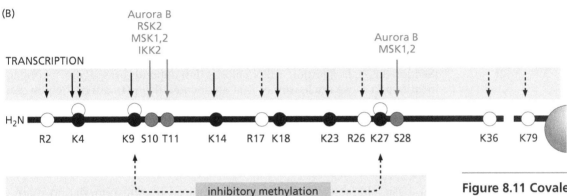

Figure 8.11 Covalent histone modifications (A) Proven modifications of core histones. The black lines stand for the N-terminal regions directly contacting DNA; the larger balls (right) indicate globular domains. Acetylation is symbolized by small black circles, methylation by small white circles, and phosphorylation by small colored circles; ubiquitylation (of H2A and H2B) is shown by a colored rhombus; and poly(ADP-ribosylation) (of H2B) is indicated by a triangle. (B) Association of individual modifications of histone 3 with transcription and gene silencing. Beside methylation of K9 and K27 silencing requires complete dephosphorylation and deacetylation (not shown). Phosphorylation is symbolized by red arrows (with the corresponding protein kinases), acetylation by black arrows, and methylation by black broken arrows. K, lysine; R, arginine; S, serine; T, threonine. An additional modification found in H3 is a *cis–trans* isomerization of peptidyl–prolyl bonds (not shown). A comprehensive list of histone modifications and the corresponding enzymes is found in T. Kouzarides (2007) *Cell* 128:802–803.

A large variety of histone methyltransferases exhibiting different specificities have been identified. Depending on the enzyme type, either ε-amino groups of Lys residues or guanidino groups of arginine (Arg) residues are methylated one-, two-, and in the case of Lys, three-fold. Until recently, histone methylation was considered to be irreversible, since enzymes catalyzing a demethylation had been found only for drug metabolism in the liver but not for methylated nuclear proteins. However, two types of histone demethylases have now been identified that are expressed in various isoforms: peptidyl-arginine deiminases, which deiminize methylated Arg residues to citrulline residues, and ε-*N*-methyllysine demethylases, which remove methyl groups by oxidation to formaldehyde (Figure 8.10). Therefore, histone methylation is as reversible as any other post-translational modification.

Histone methylation has either a stimulatory or an inhibitory effect on transcription. Thus, methylation cooperates with acetylation in reversible gene activation but also plays a key role in gene silencing and the formation of heterochromatin—that is, in the permanent inactivation of certain chromatin areas in the course of tissue differentiation—thereby cooperating with DNA methylation. Which of the two effects is expressed depends on the site of methylation (Figure 8.11). As a rule, methylation of Arg residues in H3 is required for *transcriptional activation*. The same holds true for methylation of H3 Lys residues 4, 36, and 79, catalyzed by multiprotein complexes consisting of strictly site-specific methyltransferases. These modifications may hinder access of repressor histone deacetylase complexes and inhibit competitive methylation of Lys9 and Lys27.

In fact, methylation of Lys9 and Lys27 in H3 leads to *gene silencing*. The reaction is catalyzed again by highly specific methyltransferases that generate interaction sites in H3 for a protein HP1, which has been shown to bind to methylated H3 Lys 9 via a **chromodomain** (<u>chrom</u>atin <u>o</u>rganization <u>modifier</u> domain), thus triggering heterochromatin condensation. Chromodomains are 30–70 amino acids long and are found in various, mostly transcription-repressing chromatin proteins. Methylation of Lys9 occurs only when Lys14 has *not* been acetylated first, a situation resembling a logical BUT NOT gate. To make the situation even more complex, the acetylation of Lys14 depends on phosphorylation of Ser residue 10, resembling a logical AND gate. A similar pattern of spatially and temporally coordinated acetylations and methylations is seen for H4.

The different covalent modifications of the core histones are mutually dependent. This spectrum of covalent signaling reactions has been investigated in detail for steroid hormone-dependent gene promoters. According to these studies, a steroid hormone receptor acting as a transcription factor recruits the initiation complex including RNA polymerase II to a gene promoter that is associated with H3 acetylated at Lys14. This complex contains acetyltransferases that catalyze the acetylation of additional Lys residues in H3 and other core histones, thus weakening their interactions with DNA. These changes create the conditions for an activating methylation of H3 catalyzed by a methyltransferase also bound in the transcription complex. Consequently, transcription is accelerated by making room for translocases that catalyze a mechanical rearrangement of nucleosomes (Section 8.2.7). In addition, mRNA elongation depends on specific histone modifications in those nucleosomes that are located downstream of the polymerase complex. These modifications include a trimethylation of Lys4 and a methylation of additional Lys residues (Lys36 and -79) in H3.

8.2.5 Histone kinases: teammates of acetyl- and methyltransferases

The basic character of histones is neutralized by phosphorylation more efficiently than by acetylation and methylation. Moreover, selective phosphorylation of individual Ser residues promotes the interaction of histones with acetyl- and methyltransferases. Of particular importance for transcriptional activation is the phosphorylation of H3 at Ser10 since it is, as mentioned, the condition of acetylation at Lys14 and methylation at Lys4. The phosphorylation generates a signal for the recruitment of histone acetyl- and methyltransferases. Ser10 phosphorylation is catalyzed by the mitotic kinase Aurora B (Section 12.12), which directly associates with chromatin. Corresponding phosphorylation sites are found in the N-terminal regions of H2b and H4 (H2A is not phosphorylated).

Frequently, the phosphorylation of core histones is catalyzed by protein kinases that also activate a variety of transcription factors. In other words, for gene activation, the decondensation of chromatin through histone phosphorylation/acetylation and the marking of individual genes by transcription factors are strictly coupled events. Protein kinases falling into this category are the mitogen/stress-activated kinases MSK1 and MSK2 and the ribosomal S6 kinase RSK2. These enzymes are under the control of MAP kinase cascades (details in Section 11.6) and as complexes with the cognate MAP kinase isoforms bind to chromatin. Other histone kinases are IKK2, a component of the NFκB cascades where it phosphorylates the inhibitor IκB (Section 11.8), and the "casein kinase" CK2 (Section 9.4.5). Both MAP kinase and NFκB cascades are major signalling pathways transducing exogenous signals to the genome.

Extensive histone phosphorylation occurs in the course of cell division. Here the removal of the linker histone H1 is of particular importance since DNA replication and sister chromatid separation are blocked as long as chromatin exists in a condensed form. Therefore, H1 is phosphorylated across the whole genome in two steps in strict correlation with DNA replication and mitosis. The reaction is catalyzed by cyclin-dependent protein kinases such as CDK1 and CDK2 that are key regulators of the cell cycle (see Section 12.4).

8.2.6 Poly(ADP-ribosylation) and ubiquitylation of histones

In addition to acetylation, methylation, and phosphorylation, histones are covalently modified by poly(ADP-ribosylation) and ubiquitylation. **Poly(ADP-ribosylation)** is a reversible signaling reaction by which an ADP-ribosyl residue is transferred from NAD^+ to the carboxy groups of aspartate and glutamate residues and enlarged by polymerization (Section 2.9.2). Targets are, in particular, H1 and H2B (Figure 8.11). The key enzyme is poly(ADP-ribosyl) polymerase1 (PARP1), found mainly in the nucleus. It contains a DNA-binding zinc finger domain and an automodification domain, the poly(ADP-ribosylation) of which results in auto-inhibition of PARP1, demonstrating control of enzyme activity by negative feedback.

The poly(ADP-ribosyl) residue carries a strong negative charge. Poly(ADP-ribosylation), therefore, weakens the histone–DNA interaction in a particularly effective manner. In fact, PARP1 activity is stimulated by certain gene-regulatory proteins and correlates with high transcriptional activity, for instance, in the puffs of insect chromosomes (Section 8.3.1). Moreover, PARP1 is connected with DNA repair since it is strongly activated by binding to damaged DNA. The enzyme functions mainly as a transcriptional co-activator, although in some cases it is also involved in transcriptional inhibition. The reverse holds true for poly(ADP ribose) glycohydrolase, the opponent of PARP, which renders the modification reversible.

A post-translational modification of Lys residues by **mono-ubiquitylation** is seen for core histones H2A and H2B and linker histone H1. In contrast to poly-ubiquitylation, this reaction does not trigger proteolytic degradation but correlates with histone methylation and gene silencing, particularly in the course of chromosome condensation during mitosis. The individual modifications clearly depend on each other. In yeast cells, for instance, ubiquitylation of Lys123 in H2B is a prerequisite for methylation of Lys4 in H4. Histone mono-ubiquitylation is a reversible modification catalyzed by highly specific E3 ubiquitin ligases and reversed by de-ubiquitylating isopeptidases. As potential transcriptional co-activators and co-repressors, these enzymes resemble acetyltransferases and deacetylases.

8.2.7 Nucleosome-remodeling complexes: molecular machines of gene regulation

The covalent modifications of histones by acetylation, methylation, phosphorylation, poly(ADP-ribosylation), and ubiquitylation are not sufficient for a complete disruption of the interactions between DNA and histones. They need to be supported by a mechanical separation of DNA from core nucleosomes. This energy-consuming process of nucleosome or chromatin remodeling is catalyzed by at least four families of ATPases, with the **SWI/SNF** family providing the best known example. The name refers to the yeast mutants mating type SWItching and Sucrose Non-Fermenting, where the function of such enzymes was found for the first time. [A comprehensive list of chromatin remodeling complexes is found in Bao & Shen (2007).] Recognizing acetylated histones, these enzymes are molecular machines that migrate along the nucleosome, pushing the zone of the double-stranded DNA that is detached from the histone core like a "bow wave" (see Figure 8.6), thus giving access for the RNA polymerase complex to interact with the DNA. The mechanism resembles that of motor proteins: upon ATP hydrolysis, the ATPase anchored to both the nucleosomal core and the surrounding DNA undergoes a transformational change resulting in a translocation of the DNA strand (Figure 8.12). The abundance of remodeling ATPases (about one machine per 10 nucleosomes) underlines their indispensable role. In fact, nucleosome-remodeling machines are involved not only in gene expression but also in other processes such as DNA repair, gene replication, gene recombination, and chromosome assembly. Moreover, in tumor tissue, overexpression of these enzymes is frequently observed.

Figure 8.12 Model of ATP-dependent nucleosome remodeling The remodeling ATPase thought to be anchored at the nucleosomal core particle interacts with the surrounding DNA strand at two sites. Due to the conformational change caused by ATP hydrolysis, the DNA strand is shifted, allowing interaction with the proteins of the transcriptional machinery.

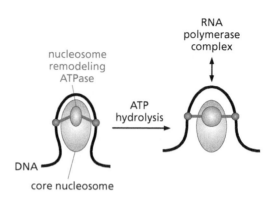

SWI/SNF ATPases have a DNA-binding domain and a bromodomain that recognizes acetylated histones. By this means the remodeling complex cooperates with the histone acetyltransferase complex, becoming active only at genes primed for transcription. On the other hand, some SWI/SNF isoforms also interact with histone deacetylases involved in transcriptional repression. In this case, they are clearing the gene for an interaction with DNA methyltransferases (see Section 8.2.8).

During the elongation phase of transcription, additional measures are required to clear a path for the RNA polymerase complex along the genomic DNA. **FACT** (**FA**cilitator of **C**hromatin **T**ranscription), a component of the elongation complex, is a heterodimeric molecular machine or "histone chaperone" that, by means of strongly acidic groups, releases H2A and H2B from the nucleosome, reintegrating them when the RNA polymerase complex has passed. FACT is also involved in DNA repair and replication. **Elongator**, a multiprotein complex with histone acetyltransferase activity, plays a similar role.

8.2.8 Epigenetic long-term memory stabilizes tissue functions

Almost all somatic cells possess the complete genome. Nevertheless, they perform quite different functions depending on the tissue. These tissue-specific patterns of differentiation are maintained over the whole lifespan of the organism. They are passed during cell division in an unchanged form from cell generation to cell generation, thus maintaining tissue individuality and continuity. The differentiation pattern is guaranteed by a set of active genes, whereas the rest of the genome is silenced by stable modifications of chromatin and DNA. This occurs in the course of embryonic tissue development. Tissue-specific gene expression is mirrored by patterns of transcriptionally active euchromatin and transcriptionally inactive (silenced) heterochromatin. Since such patterns are transferred from mother to daughter cells, although they are not encoded in genomic sequences, they are called epigenetic.

Epigenetic heredity depends on covalent modifications of both DNA and histones. A reaction that plays a key role is **DNA methylation**. In mammals the 5′-position of cytidine in the dinucleotide residue CpG is affected almost exclusively. Such CpG sequences are concentrated at distinct sites of the genome, such as gene promoters, forming **CpG islets**. CpG methylation facilitates the formation of highly condensed heterochromatin. In the genome of mature mammalian tissue cells, 70% of all CpG elements are methylated on average, with the methylation pattern changing from tissue to tissue and in the course of embryonic development. As expected, an exceptionally high degree of DNA methylation is found in inactive regions of the genome, whereas nonmethylated CpG elements are concentrated in the regulatory and coding domains of active genes.

DNA methylation hinders the access of stimulatory transcription factors. Moreover, methylated CpG islets are binding sites of histone deacetylase complexes, nucleosome remodeling ATPases, and other transcriptional co-

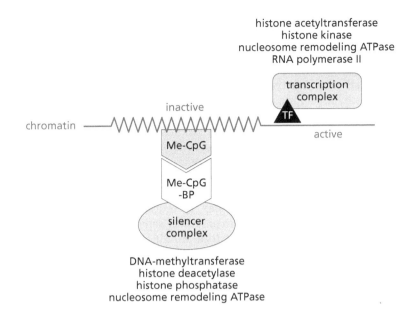

histone acetyltransferase
histone kinase
nucleosome remodeling ATPase
RNA polymerase II

transcription complex

chromatin

inactive

active

Me-CpG

Me-CpG -BP

silencer complex

DNA-methyltransferase
histone deacetylase
histone phosphatase
nucleosome remodeling ATPase

Figure 8.13 DNA methylation as a signal of ontogenetic gene silencing The chromatin region (red) includes transcriptionally active (straight line) and transcriptionally inactive (zigzag line) zones. The inactive zone contains an embryonically determined pattern of methylated CpG islets (symbolized as Me-CpG) that, via Me-CpG-binding proteins (symbolized as Me-CpG-BP), interacts with a silencer complex. This complex, containing DNA methyltransferases, histone deacetylases, nucleosomal remodeling complexes, and other inhibitory proteins, reestablishes the pattern of DNA methylation and inhibitory histone modifications after each cell division in a site-specific manner. In an active area of chromatin, DNA is not methylated, enabling the binding of a transcription factor (TF) that recruits the transcription complex.

repressors. For complex formation, **methyl-CpG-binding proteins** are required, which act as adaptors binding to DNA via a methyl-CpG-binding domain and to the co-repressors via a transcription-repressing domain (Figure 8.13). Thus, in the vicinity of a methylated CpG element the histones become modified in such a way that transcription is rendered impossible. Therefore, the pattern of DNA methylation together with the pattern of histone modifications represents a code of tissue-specific gene expression.

DNA methylation is catalyzed by several DNA methyltransferases controlling *de novo* methylation in the course of tissue formation and transferring the methylation pattern from the mother DNA to the daughter DNA during cell division. Nucleosome-remodeling ATPases facilitate the access of these enzymes to DNA. Modified histones and methyl-CpG-binding proteins, together with repressors and histone deacetylases with which the DNA methyltransferases form complexes, direct the whole process. This guarantees that the enzymes required for gene silencing stay concentrated near methylated DNA sequences so that the pattern of differentiation is maintained over many cell generations. Disturbances of this epigenetic information transfer leads to decreased DNA methylation, resulting in serious consequences such as cancer.

Summary

In eukaryotic chromatin, DNA is densely packed by histones. H2a, H2b, H3, and H4 combine to form complexes around which DNA is wound (nucleosomes); H1 occupies the linker DNA between the nucleosomes. Eukaryotic histones fulfill two purposes: they pack DNA and they hinder the access of RNA polymerase. For gene transcription, covalent modifications that weaken the basic character of histones and mechanical forces temporarily lift this blockade. Enzymes that catalyze such modifications are transcriptional co-activators, whereas proteins that stabilize histone–DNA interaction are co-repressors. Acetylation of Lys residues is a major histone modification that weakens the interaction with DNA, enabling transcription. Histone acetyltransferases rank among the most important transcriptional co-activators. They are expressed in a large number of isoforms that, as scaffold proteins, combine transferase activity with a wide variety of other interactions. As opponents of histone acetyltransferases, histone deacetylases are important transcriptional co-repressors. Like acetyltransferases, they combine enzyme activity with various other interactions. Sirtuins are nonconventional histone deacetylases that transfer the acetyl group to the ADP-ribose moiety derived from NAD^+. They are thought to coordinate gene transcription with the metabolic redox status, and their activity seems to correlate with longevity. Methylation of H3 and H4 at Lys and Arg residues weakens histone–DNA interaction but, on the other hand, generates binding sites for proteins that cause gene silencing by promoting chromatin condensation. Thus, histone methyltransferases may act either as co-activators or as co-repressors depending on the enzyme type and the methylation site in the histone molecule. Histone demethylases oppose these effects. Phosphorylation by histone kinases, in cooperation with acetyl- and methyltransferases, also weakens histone–DNA interaction. Phosphorylation by cyclin-dependent kinases is indispensable for DNA replication and cell division. By poly(ADP-ribosylation), a strong negative charge is transferred to the histone molecule, abolishing the interaction with DNA and facilitating gene transcription. In contrast, ubiquitylation of histones generates conditions for chromatin condensation and gene silencing. Complete disruption of the DNA–histone complex, as required for transcription, occurs through nucleosome remodeling: a mechanical separation of histones and DNA supporting the effects of covalent histone modifications. The process requires molecular machines driven by ATP hydrolysis. Permanent gene silencing results from specific DNA methylation. The methylated sites are recruitment centers for transcriptional co-repressors, such as histone-modifying enzymes and DNA methyltransferases. The particular pattern of histone modifications and DNA methylation represents a self-renewing epigenetic code of gene activity that guarantees the constancy of tissue differentiation.

8.3 Transcription factors as hormone receptors

Transcription factors are major targets of signaling cascades activated by extracellular signals. As pointed out in Section 2.12, there are the two basic mechanisms of signal transduction connecting the cell periphery with the genome, which have been preserved from prokaryotes (Section 3.4) to humans.

(1) Extracellular messengers (such as peptide hormones) that are not lipophilic enough to penetrate the cell membrane interact with **membrane receptors**

Figure 8.14 Signaling cascades controlling gene transcription
The majority of extracellular signals interact with receptors at the cell surface. The most abundant types are shown here. Directly or via second messengers, these receptors activate signaling pathways that lead to phosphorylation of transcription factors. More lipophilic signal molecules may penetrate the cell membrane and activate transcription factors by direct binding (right). Signaling pathways not fitting this scheme will be discussed elsewhere. The figure represents an oversimplification, since in reality complex cross talk exists between the individual pathways, ensuring that the activity of a transcription factor is controlled by a distinct pattern of signals transmitted by different receptor types. ANP, atrial natriuretic peptide; ECM, extracellular matrix; TGFβ, transforming growth factor β; G, G-protein; GPCR, G-protein-coupled receptor; GC, guanylate cyclase; InsP₃, inositol 1,4,5-trisphosphate; DAG, diacylglycerol; P, phosphate residue. Note that protein kinase-coupled receptors exist in two forms: receptors with intrinsic and receptors with associated protein kinase activities.

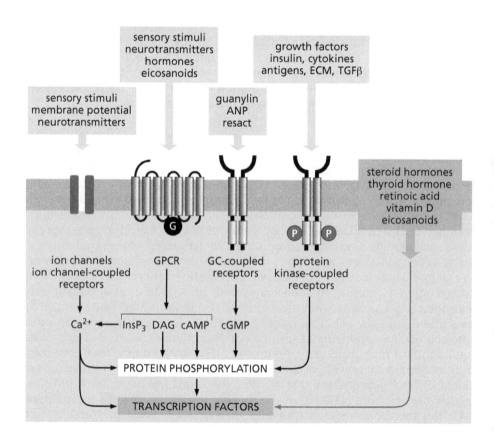

at the cell surface, activating downstream signaling cascades that, in most cases, lead to phosphorylation of transcription factors.

(2) Lipophilic messengers (such as steroid hormones) able to pass the cell membrane are bound by a class of transcription factors known as **nuclear receptors** (Figure 8.14). These receptors demonstrate in an exemplary manner how extracellular stimuli influence genetic activity.

8.3.1 The nuclear receptor family

Prototypes of nuclear receptors are the **steroid hormone receptors**. Steroid hormones are of historical significance because they were the first hormones seen to have an effect on genes. This was a breakthrough in our understanding of hormone action comparable with the discovery of transmembrane signal transduction (Section 4.2). In this context, the pioneering studies on insect molting and pupation should be mentioned. In the course of those developmental events, puffed regions arise at specific sites of the giant or polytene chromosomes* that are characteristic for such species. Quite early, these "puffs" observed under the light microscope had been interpreted as sites of increased genetic activity.

Puffs may be induced experimentally by the steroid ecdysone, the molting and pupation hormone of insects (Figure 8.15). Almost 50 years ago, this observation gave rise to the hypothesis that hormones were transcriptional activators. Radioactively labeled steroid hormones were then found to bind to cytoplasmic proteins in their target organs, subsequently migrating into the nucleus, where their accumulation correlated with increased mRNA synthesis.

Among the transcription factors, nuclear receptors constitute a major family. *Drosophila* has 21 corresponding genes, including those for the receptors of ecdysone and of juvenile hormone, a lipophilic sesquiterpene derivative that as an ecdysone antagonist prevents molting and premature metamorphosis.

---

*Giant chromosomes are generated by DNA replication without subsequent cell division. The chromatids do not become separated but stay together, forming a "cable bundle." The number of such "endomitoses" depends on the insect species and on the tissue. For instance, in the salivary glands of *Drosophila*, a preferred experimental model, nine endomitoses yield the 1024-fold enlargement of the haploid chromosome set.

Figure 8.15 Puffing of giant chromosomes (A) Section of a giant chromosome from the salivary gland of *Acricotopus lucidus* (fourth larval stage) in condensed and puffed state (5000-fold magnification). (B) Simplified scheme of a partially puffed giant chromosome; note the four chromatid strands twisted to chromomeres (resembling the strongly dyed zones in panel A) and unwound in the puff. (C) Development of a puff (shown in color) in the first salivary gland of *Chironomus* larvae 0, 30, and 60 min after injection of the steroid hormone ecdysone. (A, from F. Mechelke, in P. Karlson, ed., *Funktionelle und morphologische Organisation der Zelle*. Berlin: Springer-Verlag, 1963; B, from E.J. DuPraw, *Cell and Molecular Biology*. New York: Academic Press, 1968; C, from U. Clever, *Verh. Dtsch. Zool. Ges.* 62:75–92, 1961.)

(A)

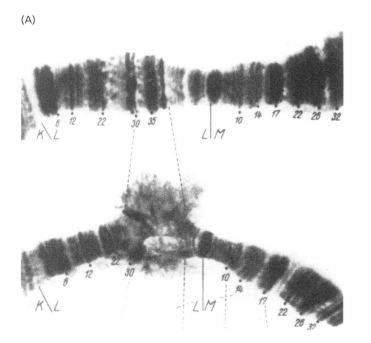

(B)

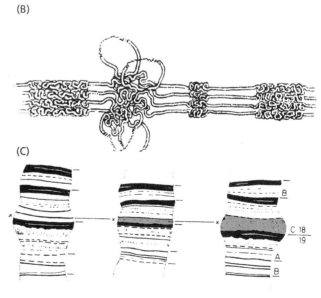

(C)

Figure 8.16 Mode of action of nuclear receptors The ligands (colored ellipses) penetrate the cell membrane and bind to their receptors in the cytoplasm. The ligand–receptor complexes oligomerize and accumulate in the nucleus, where they bind to hormone-responsive elements on DNA and act as transcription factors.

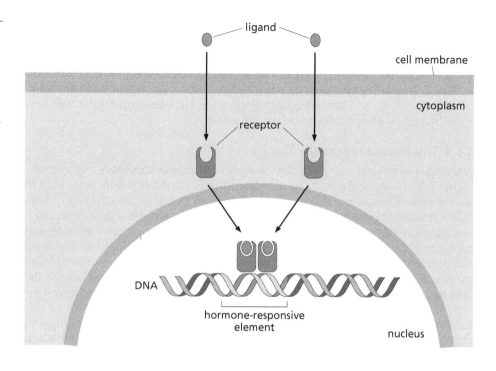

Mammals have about 50 genes encoding nuclear receptors. For many of them, the functions and ligands are still unknown. Some of these "orphan receptors" may be activated by covalent modifications or "tethered ligands" (a mechanism resembling that described for protease-activated receptors; see Section 5.6) rather than by exogenous ligands. A comprehensive list of nuclear receptors is found in Germain et al. (2006).

Evaluation of the genome of *C. elegans* yielded an unexpected result: no less than 270 genes encoding nuclear receptors were found. Their functions are widely unknown, however. In contrast, the genomes of plants as well as of molds and unicellular eukaryotes studied so far do not contain any such genes. It is hypothesized, therefore, that the primeval gene of nuclear receptors entered the stage of evolution with the first multicellular animals. On the other hand, steroid hormones are also found in algae and plants. However, here they do not interact with nuclear receptors but activate other signaling pathways. Thus, in the weed *Arabidopsis thaliana* the multifunctional phytohormone brassinolide, a steroid, binds to a membrane-bound receptor Ser/Thr kinase that indirectly induces gene transcription by activating protein phosphatases. This promotes a dephosphorylation of certain transcription factors that are inactive in the phosphorylated state.

The general mode of action of nuclear receptors is shown in Figure 8.16. Receptors, which are found both in the cytoplasm and in the nucleus, dimerize upon ligand binding and interact with gene regulatory DNA sequences.

8.3.2 Receptor structure

Nuclear receptors have the characteristic overall structure of transcription factors (as shown in Figure 8.4) with a central DNA-binding domain flanked by a C-terminal regulatory domain interacting with the ligand and a transactivating domain called AF1 (transcriptional Activation Function 1; see Figures 8.17 and 8.18) positioned N-terminally from the DNA-binding domain. In many receptors, a second transactivating domain AF2 is found, overlapping with the ligand-binding domain (Figure 8.18). A major function of AF1 and AF2 is to recruit transcriptional co-activators such as histone acetyltransferases. The DNA-binding domain, a two-fold zinc finger sequence of 66 amino acids, interacts specifically with hormone-responsive elements in enhancer and promoter regions of genes. It may be

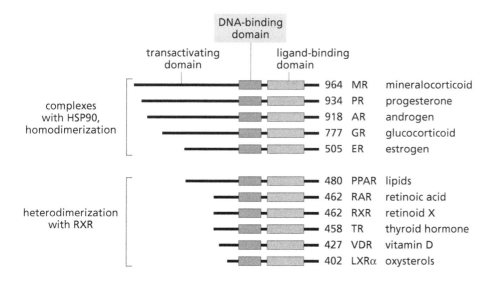

Figure 8.17 Ligand-activated nuclear receptors of vertebrates Shown are the domain structures of the receptors (amino terminus to the left), the numbers of amino acids (for human receptors), and the major ligands. For an explanation of abbreviations, see Table 8.1. Other details are explained in the text.

assisted in recognizing its cognate response element by the AF1 domain. This is necessary because the response elements of most steroid hormones are identical (see below). AF1 domains fulfill the requirements of *specific* interaction sites because they are quite heterogeneous, differing from receptor to receptor. To a

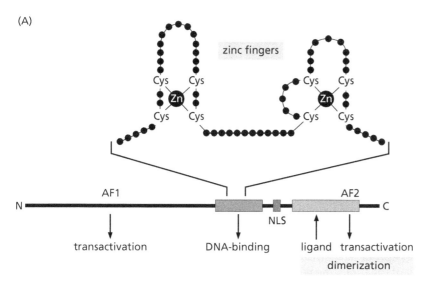

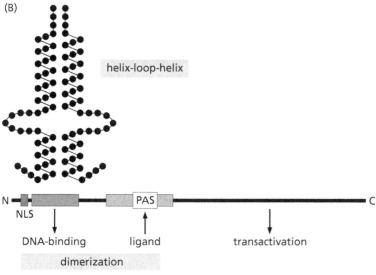

Figure 8.18 Two major types of ligand-controlled transcription factors Shown are the domain structures of (A) nuclear receptors and (B) arylhydrocarbon receptor. The characteristic structural elements of the DNA-binding domains, zinc finger and helix–loop–helix, are depicted in a simplified manner. Cys, cysteine residue; AF1 and AF2, activation functions 1 and 2; NLS, nuclear localization sequence.

Table 8.1 Ligand-dependent transcription factors of vertebrates: a selection

Transcription factor	Dimer	Preferred ligand
Family of nuclear receptors		
Subfamily of steroid hormone receptors		
estrogen receptor ER	ER-ER	17β-estradiol
progesterone receptor PR	PR-PR	progesterone
androgen receptor AR	AR-AR	testosterone
glucocorticoid receptor GR	GR-GR	cortisol
mineralocorticoid receptor MR	MR-MR	aldosterone
Subfamily of non-steroid receptors		
retinoic acid receptor RAR	RAR-RXR	*all-trans*-retinoic acid
retinoid X receptor RXR	RXR-RXR	9-*cis*-retinoic acid
thyroid hormone receptor TR	TR-RXR	3,5,3'-triiodothyronine
vitamin D receptor VDR	VDR-RXR	1,25-dihydroxy-vitamin D_3
peroxisome proliferator-activated receptor		
PPARα	PPAR-RXR	fatty acids, eicosanoids
PPARβ	PPAR-RXR	fatty acids, eicosanoids
PPARγ	PPAR-RXR	fatty acids, eicosanoids
pregnane-X receptor PXR	PXR-RXR	xenobiotics
constitutively activated receptor CAR	CAR-RXR	phenobarbital
farnesoid receptor FXR	FXR-RXR	bile acids
liver-X receptor LXR	LXR-RXR	oxysterols
Family of arylhydrocarbon receptors		
arylhydrocarbon receptor ArH	ArH-ARNT	dioxin

certain extent they resemble docking domains of protein kinases (Sections 1.8 and 11.2).

The receptors have, in addition, a nuclear localization sequence located in the center of the molecule within a binding-site of the chaperone heat-shock protein 90 (HSP90). As shown below, HSP90 plays an important role in the activation of steroid hormone receptors (see Section 8.5).

Like other receptor types, most nuclear receptors exist in various isoforms that frequently are generated after transcription by alternative mRNA splicing or alternative start codon usage for translation. These isoforms are tissue-specifically expressed and differ in their properties, thus contributing considerably to the functional versatility of the receptor ligands.

Generally, steroid hormone receptors are active only as homodimers, whereas other nuclear receptors form heterodimers with the 9-*cis*-retinoic acid receptor RXR (Table 8.1). The dimerization results in a pronounced cooperative effect. With its response element, a dimer interacts 100 times more strongly and more specifically than a monomer. Some "orphan receptors," however, bind to DNA as monomers.

8.3.3 Hormone response elements

All hormone response elements exhibit a symmetrical structure that fits the receptor dimer and comprises two identically or palindromically arranged hexa-

(A)

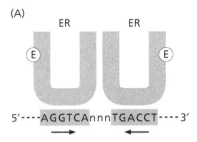

5'----AGGTCAnnnTGACCT----3'

(B)

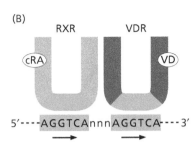

5'----AGGTCAnnnAGGTCA----3'

Figure 8.19 Dimeric nuclear receptors and their response elements (A) Estrogen receptor (ER; E, estradiol), an example of steroid hormone receptors, binds as a homodimer to a palindromic response element that is arranged as an inverted repeat separated by three arbitrary nucleotide residues n. (B) Non-steroid hormone receptors, here the vitamin D receptor (VDR) as an example, form heterodimers with the 9-*cis*-retinoic acid receptor (RXR) as a partner (cRA, 9-*cis*-retinoic acid; VD, 1,25-dihydroxy-vitamin D). The response elements are two parallel oriented identical repeats separated by 1–5 arbitrary nucleotide residues n (n = 3 for VDR-RXR).

nucleotide sequences separated by 1–5 nucleotide residues. A palindromic head-to-head arrangement as inverted repeats is typical for steroid hormone response elements. Thus, the estrogen response element has the structure 5'-AGGTCA-n-n-n-TGACCT-3' (n stands for any nucleotide). The response elements of other steroid hormone receptors share the common sequence 5'-AGAACA-n-n-n-TGTTCT-3'. To confer specificity, the AF1 domains probably assist the receptors in recognizing their cognate response elements.

In contrast to steroid hormone response elements, the binding sites for nuclear receptors forming heterodimers with RXR are characterized by two parallel or head-to-tail oriented (direct) repeats of the sequence AGGTCA, which differ from receptor to receptor only by the number of connecting nucleotides n (ranging from 1 to 5 according to the so-called 1–5 rule). The structure of the vitamin D response element, for instance, is 5'-AGGTCA-n-n-n-AGGTCA-3': one of the hexanucleotide sequences binds the complex of RXR and 9-*cis*-retinoic acid and the other one interacts with the liganded vitamin D receptor (Figure 8.19).

X-ray analysis has provided deep insight into the receptor–DNA interaction. The DNA-binding site of nuclear receptors contains two α-helices that are oriented by the zinc finger in a rectangular manner. One of the two α-helices dips into the major groove of the DNA double helix while the other α-helix binds at the outside, thus contributing to the specificity of the interaction between receptor and DNA (Figure 8.20).

8.3.4 Nuclear translocation and activation of gene transcription

Steroid hormone receptors are found in both the cytoplasm and the nucleus. Their active conformation is quite unstable. Therefore, for ligand binding and nuclear translocation they need the assistance of the chaperones HSP70 and HSP90 (see Section 8.5). Upon arrival in the nucleus, the chaperones are replaced by transcriptional co-activators and the nuclear receptor binds to the corresponding response element, marking the gene for the RNA polymerase II complex. Co-activators are first of all histone acetyltransferases such as the enzyme CBP/p300 and more specialized steroid receptor co-activator proteins. As mentioned before, these proteins are both enzymes and scaffolds that assemble additional factors required for transcriptional activation.

The non-steroid nuclear receptors have not been found to interact with chaperones but are converted into the active conformation by other means, probably phosphorylation. Even in the absence of their ligands, most of these receptors are located at the nuclear membrane or in the nucleus where they are bound to chromatin. In this form, they are frequently complexed with transcriptional co-repressors that inhibit the transcription of their target genes but also of other genes. Upon ligand binding, the complex disintegrates and co-activators replace the co-repressors. Most co-repressors are histone deacetylases or at least recruit such enzymes. Examples are the silencing mediator of retinoic acid and the thyroid hormone receptor (SMRT) and the nuclear receptor co-repressor (NCOR).

For steroid hormone receptors, co-repressors have also been described. The most prominent example is the above-mentioned estrogen receptor co-repressor

Figure 8.20 Nuclear receptors: interactions between DNA-binding domains and response elements Shown in a strongly simplified mode are the palindromically arranged DNA-binding domains of (A) the glucocorticoid receptor (GR) homodimer and (B) the parallel binding domains of the 9-*cis*-retinoic acid receptor (RXR)–thyroid hormone receptor (TR) heterodimer in front of the DNA double helix. Note that the α-helices (cylinders) are placed at right angles by the zinc fingers with one helix each dipping into the major groove of the DNA double helix. (C) Molecular model of the glucocorticoid receptor-DNA complex based on X-ray crystallography.
The red balls symbolize the zinc ions.

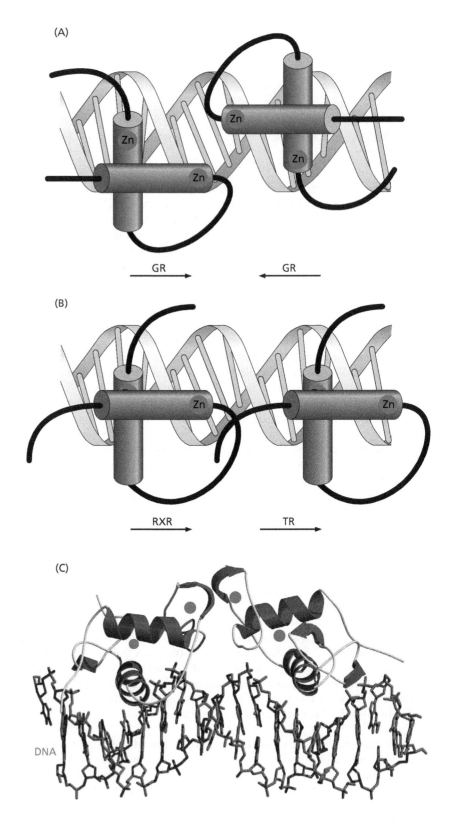

BRCA1, a tumor suppressor protein the deletion of which may cause breast cancer.

Nuclear receptors also interact with other transcription factors such as NFκB and activator protein 1, AP1 (Chapter 11) forming hetero-oligomers that either promote or attenuate the transcription of thousands of target genes. Such interactions indicate cross talk between various signaling pathways.

The concept of the receptors interacting only with co-repressors or co-activators is an oversimplification. In reality, large multiprotein complexes are assembled at the responsive promoter elements. Apart from the receptors, these complexes contain enzymes that regulate acetylation, methylation, and other forms of covalent modifications of histones, such as nucleosome-remodeling proteins; factors involved in RNA processing and protein degradation; adaptors for binding of the complexes to RNA polymerase; additional transcription factors; and others. Moreover, the transcriptional activity induced by a nuclear receptor is not static; instead it is a process oscillating in periods of seconds to minutes due to a permanent interaction between stimulatory and inhibitory influences causing cycles of receptor binding and release. These may be superimposed with phases of receptor degradation and *de novo* synthesis controlled by a system of positive and negative feedback loops. An increasing body of evidence indicates that **oscillating transcription** is not restricted to nuclear receptors but might be a common feature of inducible genes. Oscillating activity of transcription factors (and of signal-transducing proteins in general) may provide a way to escape desensitization, thus insuring long-lasting signaling. In other words, the effect induced by a gene-regulatory signal may resemble a stroke of a gong rather than a shot. The idea that, under such conditions, gene transcription represents a frequency-modulated process is easily seen (see also Section 11.7).

8.3.5 Covalent modifications of nuclear receptors

To consider nuclear receptors as transcription factors that are exclusively controlled by ligand binding is another unjustified oversimplification, because in reality their activities are fine-tuned by the data-processing protein network. In this context, post-translational modifications such as phosphorylation, ubiquitylation, and sumoylation play an important role. Each receptor has several phosphorylation sites, particularly in the transactivating domains AF1 and AF2 as well as in the DNA-binding domain. These sites are targets of MAP kinases, cyclin-dependent kinases, protein kinases A, B, and C, and tyrosine kinases that all are components of other signaling pathways. The particular effect depends on which domain of the receptor is phosphorylated. A phosphorylation of the AF domains mostly activates the receptor, for example by reinforcing the affinity for a co-activator, whereas a phosphorylation of the DNA-binding site hinders the binding to the response element by electrostatic repulsion.

Phosphorylation is also important for signal extinction because it labels the receptor for ubiquitylation and subsequent degradation by proteasomes. Under experimental conditions, *activated* nuclear receptors may have a very short lifespan, as they are degraded after one round of transcription (the so-called suicide mechanism; see Section 2.8.2). However, in the living cell, receptor-dependent signaling, once triggered, may escape desensitization by oscillating for a longer period between phases of activity and inactivity.

As co-activators and co-repressors are substrates of protein kinases (see, for instance, Figure 8.7) and other protein-modifying enzymes, highly sophisticated cross talk occurs between different signaling pathways. Therefore, genes controlled by nuclear receptors are, in addition and in parallel, targets of those receptors inducing second messenger formation and protein phosphorylation. Or in other words: nuclear receptors are embedded in extended signaling networks controlling a wide variety of cellular functions.

Summary

Nuclear receptors are a family of closely related transcription factors, activated by interaction with ligands (such as steroid hormones), that are able to penetrate the cell membrane. They seem to be restricted to animals. A common structural element of nuclear receptors is a DNA-binding zinc finger domain flanked by one or two transactivating domains and a C-terminal ligand binding and dimerization site. Ligands include steroid hormones, thyroid hormones, dihydroxy-vitamin D_3,

Sidebar 8.1 Non-genomic effects of steroid hormones
In textbooks the theory that steroid hormones are direct modulators of gene transcription is frequently treated like a dogma. However, this is a misconception, because steroid hormones can do more. Half a century ago it was observed that certain steroid hormone effects occurred within seconds or minutes, whereas for gene transcription and subsequent protein biosynthesis the cell needed hours. Since then a whole series of these rapid effects have been identified, which are not suppressed by inhibitors of transcription (such as actinomycin D) or translation (such as cycloheximide) and are called, therefore, non-genomic effects. Three of these non-genomic effects are worth examining in more detail (Figure 8.21):

- meiosis of amphibian oocytes
- acrosomal reaction of sperm cells
- anesthetizing effect

Maturing **amphibian oocytes** are arrested in the G2 phase of the cell cycle. This blockade is relieved by progesterone, which enables the cells to enter meiosis. The critical step is activation of the cyclinB–CDK1 complex, which induces the transition of the cell from G2 into the mitosis phase. As discussed in Chapter 12, this reaction can occur only when an inhibitory phosphorylation of CDK1, caused by the protein kinase MYT1, is reversed by the protein phosphatase Cdc25. Progesterone seems to affect this regulatory mechanism at two sites: it stimulates a protein kinase (called polo-like kinase) that activates Cdc25, and at the same time, it inactivates MYT1 along the MAP kinase–p90^{RSK} pathway (see Section 11.6). The second effect is mediated by the conventional nuclear progesterone receptor, which in its N-terminal region has a proline-rich sequence. Via this sequence, the receptor binds to an SH3 domain of the cytoplasmic Tyr kinase Src thus inducing enzymatic activity (for the mechanism of Src activation, see Section 7.1.2). By means of the scaffold protein Shc, the activated Src stimulates the Ras-GEF mSOS and, thus, the MAP kinase module RAF–MEK–ERK, which transduces mitogenic signals (see Figure 7.12 in Section 7.1.2). In an analogous way, the nuclear estrogen receptor activates the MAP kinase cascade.

The **acrosomal reaction** of sperm is a Ca^{2+}-dependent exocytotic process by which the cell releases lytic enzymes to penetrate the zona pellucida (jelly coat) of the egg cell. The response is triggered by oocyte factors. One of these factors, progesterone, induces a Ca^{2+} influx into the sperm cell within seconds. Since the acrosomal reaction is resistant to contraceptive drugs that inhibit the nuclear progesterone receptor, it has been postulated to be mediated by a nonclassical receptor, which has not yet been identified but might be a G-protein-coupled receptor. Such an assumption is supported by the discovery of a GPCR called GPR30 that specifically interacts with estradiol. This receptor is found in the endoplasmic reticulum of a variety of tissue cells and stimulates – via G-proteins – signaling reactions such as cAMP formation and others.

The **anesthetizing effect of steroid hormones** has been used in human and especially in veterinary medicine for more than 50 years. The rapid effect resembles that of barbiturates. Indeed, like barbiturates, steroids interact specifically with a subunit of the γ-aminobutyric acid (GABA$_A$) receptor anion channel, stimulating the influx of chloride ions and, as a consequence, hyperpolarization. By this, the excitation of nerve cells is dampened (for more details on GABA receptors see Section 16.3).

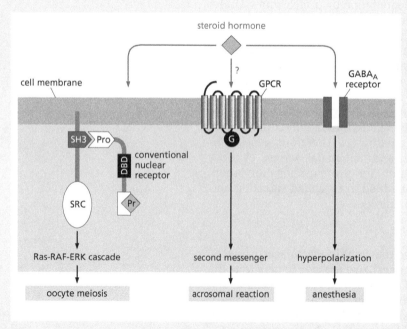

Figure 8.21 Non-genomic effects of steroid hormones (Left) Upon progesterone binding the conventional progesterone nuclear receptor activates, by means of an N-terminal proline-rich domain (Pro), the membrane-anchored tyrosine kinase Src and the Ras–RAF–ERK (MAP kinase) cascade. This signaling pathway plays a role in meiosis of amphibian oocytes. Pr, progesterone; DBD, DNA-binding domain; SH3, Src homology domain 3. (Middle) Steroid hormone controls, via a membrane receptor, the generation of second messengers and induces through Ca^{2+} influx the acrosomal reaction of sperm cells. Whether or not the receptor is G-protein-coupled, as depicted here (GPCR), remains a matter of debate. (Right) Via a γ-aminobutyric acid (GABA$_A$) receptor anion channel, the steroid induces membrane hyperpolarization and thus inhibits neuronal excitation. This effect may result in anesthesia.

retinoids, certain eicosanoids, oxysterols, bile acids, and xenobiotics as well as juvenile hormone (in insects). While steroid hormone receptors are homodimers, the other receptors are active only as heterodimers with the retinoid receptor RXR. Binding sites of nuclear receptors on gene promoters consist of two identically or palindromically arranged hexanucleotide sequences separated by 1–5 nucleotide residues. Steroid hormone response elements differ in the hexanucleotide sequences, whereas the other response elements share an AGGTCA hexanucleotide sequence but differ in the number of nucleotide residues linking the hexanucleotides. Steroid hormone receptors are localized in the cytoplasm but translocate into the nucleus upon activation. For ligand binding and activation, they require the assistance of chaperones. Most non-steroid receptors are concentrated in or around the nucleus. Their activation is supported by other means such as phosphorylation. Nuclear receptors, as well as the transcriptional co-modulators they recruit to their effector genes, are targets of a wide variety of signaling reactions that fine-tune the activity and control the lifespan.

8.4 Ligand-controlled transcription factors: xenosensors of the toxic stress response

Some nuclear receptors are strongly activated by foreign compounds (xenobiotics) such as pharmaceutical drugs and environmental poisons and are, therefore, termed xenosensors. Upon heterodimerization with RXR, these receptors evoke the toxic stress response of cells that consists of *de novo* synthesis of proteins regulating the metabolism, transport, and excretion of xenobiotics but also of endogenous factors such as steroids and bile acids. For the metabolism of these compounds, the large enzyme family of **cytochrome P450-dependent oxygenases** is of central importance. Using molecular oxygen, they oxidize their (mostly lipophilic) substrates to products that, as a rule, are less toxic or biologically inactive and can be rendered water-soluble by conjugation with glucuronic or sulfuric acid and excreted by the kidneys. The nuclear receptors that control genes encoding cytochrome P450 enzymes and other proteins of drug metabolism include PPARβα, PXR, CAR, FXR, and LXR.

8.4.1 Peroxisome proliferator-activated receptors

The induction of adipose tissue differentiation is the most prominent among the numerous functions of this receptor type. Peroxisome proliferator-activated receptors (PPARs; frequently pronounced "peepars") exist in three isoforms. The isoform **PPARα** is expressed by tissues with active fat metabolism and regulates the transcription of the corresponding genes. The receptor was discovered as a cellular binding protein of blood fat-lowering drugs such as clofibrate, of the industrial solvent trichloroethylene, and of plasticizers. In animal experiments, these compounds stimulate an increase of peroxisomes in the liver and may promote tumor development. Peroxisomes are organelles that are derived from the endoplasmic reticulum and are found in all eukaryotic cells. They contain enzymes of oxidative lipid and amino acid metabolism together with catalase, which destroys the toxic hydrogen peroxide generated along such pathways.

In contrast to PPARα, the isoforms **PPARβ** and **PPARγ** are not activated by peroxisome proliferators. The ubiquitously expressed PPARβ induces early steps while the more restricted PPARγ activates late steps of fat cell differentiation. In addition, PPARβ promotes the differentiation of skin keratinocytes in the course of wound healing, myelinization of nerve fibers, and placenta development during pregnancy, whereas PPARγ is involved in inflammatory processes and insulin action (this receptor is activated by thiazolinediones, drugs that are employed for the treatment of insulin-resistant type 2 diabetes). Defects of PPAR signaling also play important roles in other disorders such as atherosclerosis and cancer. Endogenous ligands of PPARs are unsaturated fatty acids and their derivatives, particularly certain prostaglandins and other arachidonic acid-derived metabolites (Section 4.4.5).

8.4.2 Receptors PXR, CAR, FXR, and LXR

PXR (the human homolog is called SXR) was originally described as a receptor of pregnane derivatives, but it must not be confused with the progesterone receptor PR. PXR is, in addition, activated by glucocorticoids, drugs, and poisons. It occupies a central position in detoxification and metabolism of both endogenous compounds and xenobiotics. The cytochrome P450 type 3A induced by PXR is of great medical interest since it controls more than 60% of drug metabolism and, in addition, plays a significant role in bile excretion. An adaptive overproduction of cytochrome P450 3A along the PXR pathway is a major reason for drug tolerance or tachyphylaxis.

Another typical xenobiotic receptor is **CAR** (<u>C</u>onstitutive <u>A</u>ndrostane <u>R</u>eceptor). The promoters of its target genes are characterized by a phenobarbital response element (phenobarbital is a pharmaceutical drug of the barbiturate family). CAR also binds androstane derivatives, indicating a role in the metabolism of endogenous steroids. It is constitutively active (in the absence of ligands) but ligand binding facilitates its accumulation in the nucleus.

The closely related farnesoid receptor **FXR** is activated by derivatives of farnesol and vitamin A as well as by bile acids. FXR interacts with genes carrying a bile acid response element and encoding proteins that control the biosynthesis, metabolism, and transport of bile acids. Another receptor involved in the regulation of bile acid and cholesterol metabolism is the liver X-receptor **LXR**. Its ligands are oxysterols, which are products of oxidative cholesterol metabolism. The role of this receptor in the metabolism of xenobiotics is disputed.

8.4.3 "Dioxin receptor" ArH

A ligand-controlled transcription factor that does not belong to the family of nuclear receptors proper is the **arylhydrocarbon receptor ArH.** It differs from the nuclear receptors by the arrangement and structure of its functional domains

Figure 8.22 Mode of action of the arylhydrocarbon receptor
Lipophilic xenobiotics (such as dioxin or benzopyrene, symbolized as a colored rhombus) penetrating the cell membrane are bound by the receptor in the cytosol. Upon translocation into the nucleus and heterodimerization with ArNT, the complex activates the transcription of genes with dioxin-responsive elements, thus inducing the synthesis of detoxifying enzymes. As indicated by the broken arrow, these enzymes also metabolize the xenobiotics by which they are induced.

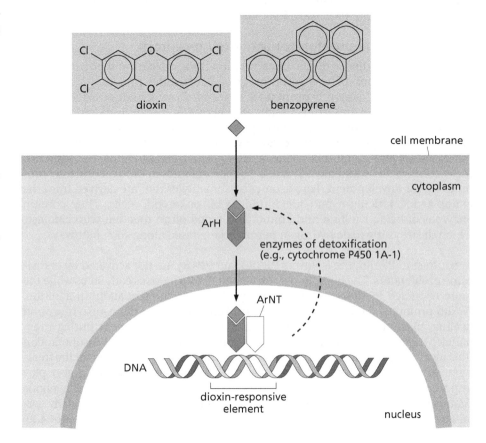

(Figure 8.18). Here, the N-terminal DNA-binding domain has a helix–loop–helix rather than a zinc finger structure, and the ligand-binding site is a PAS domain (see Section 3.3.2). Nevertheless, ArH shares many features with nuclear receptors. For instance, in the cytoplasm, ArH forms complexes with the chaperones HSP70 and HSP90. This results in a more stable conformation and enables ligand binding. As a typical xenosensor, ArH interacts with a wide variety of drugs, toxic products including aromatic arylhydrocarbons such as benzopyrene, and food ingredients. No endogenous ligands have yet been defined. By far the most potent ArH activator is 2,3,7,8-tetrachlorodibenzo-*p*-dioxin, briefly called **dioxin** (Figure 8.22). This extremely toxic substance, also known as Seveso poison, has gained an unfortunate notoriety: its accidental release by a chemical plant near the Italian village Seveso in 1976 was one of the worst environmental disasters in history.

Upon binding of the ligand, ArH translocates into the nucleus, where it heterodimerizes with the closely related protein ArNT (<u>Ar</u>H <u>N</u>uclear <u>T</u>ranslocator). [ArNT factors are also involved in the stress response to oxygen deficiency; see Section 8.7.] This complex activates genes carrying a dioxin or xenobiotics response element. Altogether more than 100 genes are directly controlled and an additional 200 genes are indirectly influenced by dioxin, including those for certain cytochrome P450 isoforms. As mentioned above, cytochrome P450 normally catalyzes the detoxification of xenobiotics. Under certain circumstances, however, the reaction products may turn out to be even more toxic: for example, in the case of dioxin or polycyclic aromatic hydrocarbons such as benzopyrene. These products are activated in such a way that they covalently bind to DNA, thus causing gene mutations that can result in cancer (benzopyrene is the carcinogenic component of coal tar).

Summary

Some types of nuclear "non-steroid" receptors interact with lipophilic xenobiotics and, in turn, stimulate the transcription of genes that encode detoxifying enzymes. Their endogenous ligands are lipids such as eicosanoids, certain steroids, farnesol and vitamin A derivatives, bile acids, and cholesterol metabolites. The arylhydrocarbon receptor ArH is a ligand-binding transcription factor that differs from nuclear receptors by the nature of its DNA-binding domain and its overall domain structure. It is activated by xenobiotics such as carcinogenic hydrocarbons and dioxin and induces the synthesis of enzymes that normally produce less toxic, (however, sometimes also more toxic) metabolites. Endogenous ligands are not known.

8.5 Chaperones and peptidyl-prolyl isomerases: how signaling proteins are prepared for work

As mentioned above, the chaperones HSP70 and HSP90 are essential for the activation of steroid hormone receptors. This function indicates that HSPs can do more than just protect essential proteins from thermal denaturation. Indeed,

Sidebar 8.2 Antioxidant response element With the signaling pathways discussed above, the cell's arsenal against foreign compounds is by no means exhausted. Many xenobiotics activate genes containing an antioxidant response element (ARE). The name originates with investigations on the detoxification of synthetic antioxidative compounds used in the cosmetics and food industries. It is somewhat misleading, because genes containing an ARE respond to oxidative stress by inducing the synthesis of protective factors such as proteins with sulfhydryl groups and enzymes of glutathione metabolism. Stimulators are, therefore, xenobiotics evoking oxidative stress or sequestering antioxidative proteins. The endogenous ligand of ARE is the transcription factor Nrf2. In contrast to ArH, Nrf2 is activated by MAP kinase-mediated phosphorylation rather than by ligand binding. How this signaling cascade is stimulated by xenobiotics is unknown.

Figure 8.23 Stepwise activation of a steroid hormone receptor by the chaperones HSP70 and HSP90 The principle of the effect is shown in (A). In the cytoplasm most of the receptor molecules exist in an inactive conformation that cannot bind the ligand. Activation is an ATP-consuming process that is catalyzed by a chaperone complex. Ligand binding shifts the equilibrium between inactive and active receptors into the direction of the active form. The details of the reaction are shown in (B). Inactive receptor R with a closed ligand-binding pocket interacts with the ATP-loaded HSP40/HSP70 complex (step 1). This results in a partial opening of the ligand-binding pocket, powered by HSP70-catalyzed ATP hydrolysis, and binding of HSP90 dimer, Hop, and p23 (step 2). Next HSP90 is activated by ADP–ATP exchange and the ligand-binding pocket of the receptor becomes fully opened (step 3). Upon ATP hydrolysis catalyzed by HSP90, the complex dissociates (step 4), enabling the primed receptor to interact with the ligand and with chromatin (step 5). During the disintegration of the complex, HSP70 is assumed to become reloaded with ATP. Upon dissociation or degradation of the ligand, the receptor automatically relapses into the inactive conformation (see panel A); that is, the role of the ligand is to stabilize the structure of the active receptor.

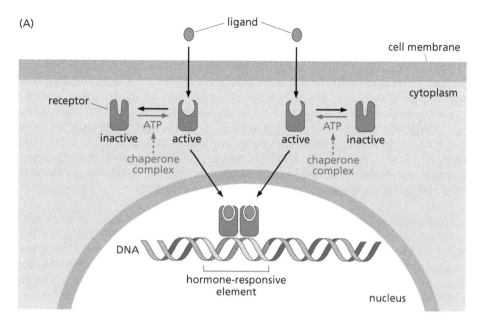

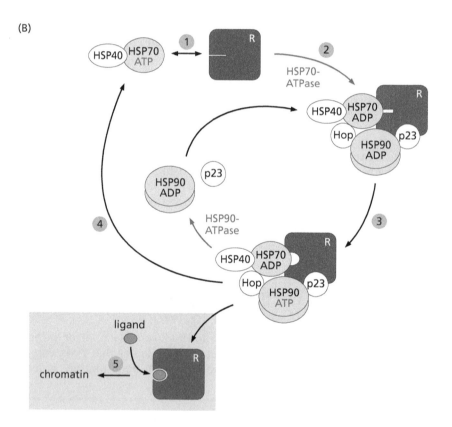

more and more client proteins have been found, the functions of which are controlled by chaperones even in the absence of stress situations. These proteins are almost exclusively components of the signal-processing network. They include transcription factors, membrane receptors, protein kinases, G-proteins and the corresponding GDP–GTP exchange factors (GEFs) and GTPase-activating proteins (GAPs), signal-generating enzymes, and ion channels. Thus, stress protection is only one aspect in a much larger functional repertoire of HSPs.

The mode of function of the two chaperones has been studied most thoroughly with steroid hormone receptors as models. These studies have revealed an unexpectedly complex mechanism by which HSP70 and HSP90 change the receptor conformation in such a way that a hydrophobic binding pocket is opened for the ligand (Figure 8.23). Since such a conformation is unstable in aqueous solution, it does not appear spontaneously but is induced in an energy-dependent process consuming two molecules of ATP, the hydrolysis of which is catalyzed by the intrinsic ATPase activities of HSP70 and HSP90.

The first ATP hydrolysis promotes the formation of a multiprotein complex and is catalyzed by HSP70 (Section 2.5.1). The components of the complex are the steroid receptor, HSP70, a HSP90 dimer, an adaptor protein (HSP-complex organizing protein HOP), and the co-chaperone HSP40, which stimulates the ATPase activity of HSP70 and thus functionally resembles a GAP of a G-protein. In this complex HSP90 is activated by ADP–ATP exchange and stabilized by an interaction with another co-chaperone, p23, in such a way that it can bind the steroid hormone receptor. HSP90 then transmits the conformational tension produced by the nucleotide exchange to the receptor, a reaction requiring another ATP hydrolysis catalyzed by HSP90 itself. As a result, HSP90 returns into the ADP-loaded state and the chaperone complex disintegrates.

As indicated by the large number of client proteins, it seems to be a general function of the HSP70/HSP90 pair to generate active but energetically unfavorable conformations of signal-transducing proteins. Thus, the HSP70/HSP90 complex represents an important switching element in the signal-processing protein network.

8.5.1 Peptidyl-prolyl *cis–trans* isomerases

Frequently the chaperones form complexes with other proteins that support their effects on protein conformation. These assistants, known as **immunophilins**, cyclophilins, or FK-binding proteins, were discovered as cellular binding proteins of immunosuppressants such as cyclosporine, FK 506, and rapamycin. They are widely distributed: the human genome encodes about 30 isoforms.

All immunophilins exhibit the activity of peptidyl-prolyl *cis–trans* isomerases (PPIases). These enzymes, found in pro- and eukaryotes, catalyze the isomerization of peptide bonds at the imide nitrogen of prolyl residues (Figure 8.24). *Cis* and *trans* conformations of proteins are in equilibrium. However, under cellular conditions the setting of this steady state is very slow, requiring catalysis. Whether the equilibrium is *cis*- or *trans*-directed depends on the type of substrate protein. Therefore, the result of a PPIase reaction resembles either a *cis* to *trans* or a *trans* to *cis* rearrangement. Since the isomerization causes a pronounced conformational change, PPIases are counted among the (co)chaperones.

Although immunosuppressants inhibit the enzymatic activity of immunophilins, their clinical effects seem to be due primarily to another mechanism: inhibition

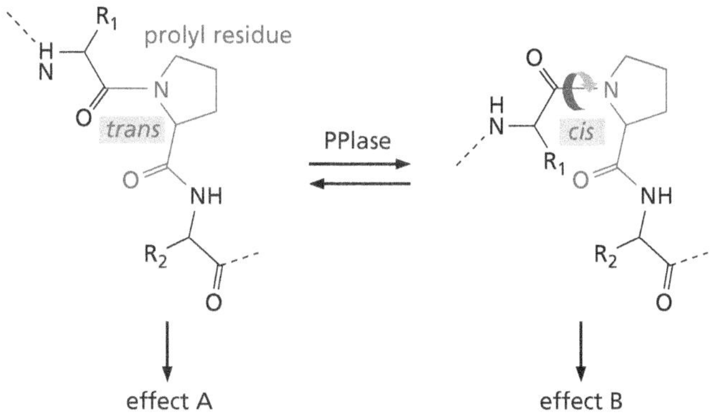

Figure 8.24 Peptidyl-prolyl *cis–trans* isomerase reaction

of the Ca^{2+}-dependent protein phosphatase calcineurin, which activates the NFAT transcription factors that are essential mediators of the immune response (Section 14.6.4).

In the steroid receptor–HSP90 complex, specific immunophilins contribute to a stabilization of the active receptor conformation. This function is not restricted to nuclear receptors but is described for other transcription factors as well, which become either inhibited or activated. Thus, like the chaperones, immunophilins play an important role in cellular signal processing.

In contrast to cyclophilins and FK-binding proteins, another subfamily of PPIases, the parvulin-like enzymes (the name comes from a bacterial protein), do not react with immunosuppressants. Certain members of this family called **PIN1** (meaning Protein Interacting with NIM-A, a mitotic kinase discussed in Section 12.12) have gained considerable attention as proteins forming logical AND gates with protein kinases: they differ from other PPIases in that they interact exclusively with prolyl residues adjacent to phosphorylated Ser and Thr residues. In other words, PIN1 PPIases operate downstream of those protein kinases that catalyze Pro-directed phosphorylation. Such kinases include cyclin-dependent kinases of cell cycle and transcription control (Sections 8.1.2 and 12.4), MAP kinases (Section 11.2), and glycogen synthase kinase 3 (Section 9.4.4). These enzymes phosphorylate only proteins with a *trans*-prolyl–peptidyl bond, and the same conformation-specificity has also been found for the protein phosphatases PP2A, calcineurin, and the dephosphorylating enzyme of RNA polymerase II (see Section 8.1). Subsequent to phosphorylation or dephosphorylation, PIN1 catalyzes a *trans–cis* rearrangement of the substrate protein by which the stability, activity, and subcellular distribution are changed.

Considering the enormous versatility of kinases and phosphatases, it is not surprising that additional PIN1 substrates are being identified. These include (just to mention a few examples) the cell cycle regulators cyclin D1, Cdc25, WEE1, and MYT1, (Chapter 12); the transcription factors cJun (Section 11.7.2), NFκB (Section 11.8), p53 (Section 12.9.4), NFAT (Section 14.6.4), and β-catenin (Section 5.9.2); and RNA polymerase II and some histone deacetylases as well as the protein kinase CK2 (Section 9.4.5). Phosphorylated NFκB, p53, and β-catenin are protected from proteolytic degradation, and cJun and cyclin D are activated by PIN1-catalyzed isomerization. As discussed in more detail in the following chapters, these stabilizing and activating effects are controlled by additional input signals, rendering PIN1 a component of extended signal-processing networks. Diminished PIN1 activity correlates with neurodegenerative diseases such as Alzheimer's dementia where the *cis–trans* isomerization of certain phosphorylated proteins plays a crucial role in pathogenesis (Section 13.1.3), whereas overexpression of PIN1 is a common feature of cancer. PIN1 probably stimulates oncogenic signaling by rendering phosphorylated key proteins of mitogenic and anti-apoptotic pathways resistant to degradation along the ubiquitin–proteasome pathway. Examples of such key proteins are β-catenin and NFκB.

Summary

Activation of nuclear receptors includes the opening of a lipophilic ligand-binding pocket. In the case of steroid hormone receptors, this is achieved by the chaperone ATPases HSP70 and HSP90. Activation of steroid hormone receptors demonstrates a more general effect of these chaperones: control of signaling protein conformation by catalyzing the generation of an active but energetically unfavorable form. Enzymatic isomerization of the peptidyl–prolyl bond is a means of conformational change that supports and stabilizes the effects of chaperones. Peptidyl-prolyl *cis–trans* isomerases (PPIases) are found in pro- and eukaryotes and, as immunophilins, bind to immunosuppressive drugs. The PIN1 subfamily of PPIases isomerizes only prolyl residues adjacent to phosphorylation sites; that is, it cooperates with Pro-directed protein kinases. PIN1 enzymes are involved in cancer, diabetes, and Alzheimer's disease.

8.6 Transcription factors as substrates of protein kinases

As already mentioned, most transcription factors are controlled by phosphorylation. This is a theme with many variations. Phosphorylation may modulate such diverse features as nuclear translocation, oligomerization, interaction with cofactors, DNA binding, transactivating activity, and lifespan. Phosphorylation may occur in the cytoplasm or, upon translocation of the protein kinase, in the nucleus. In some cases, preformed complexes of transcription factors with corresponding protein kinases and phosphatases as well as other proteins required for transcriptional regulation are associated with chromatin and await an input signal. Their particular construction renders such complexes potential oscillators, indicating a frequency-modulated effect. The principle of this signaling reaction for the transcription factor CREB (cAMP Response Element Binding protein) is discussed below. For other factors, the reader is referred to later chapters.

CREB and its isoform CREM (cAMP Response Element Modulator) are both abundant transcription factors, whereas the third protein of this family, ATF1 (Activating Transcription Factor 1), is mainly restricted to nerve and glandular cells. Each of the three factors has a leucine zipper domain for dimerization, a DNA-binding domain, and a kinase-inducible domain, the phosphorylation of which results in activation. The DNA-binding domain contains several basic amino acids for the specific recognition of cAMP response elements (CREs) consisting of the palindromic sequence TGACGTCA or half of it (CGTCA). Such sites have been found in the promoters of more than 100 different genes. They interact with the dimeric transcription factors, provided these have been phosphorylated at a Ser residue in the kinase-inducible domain.

CREB, CREM, and ATF1 exist in many variants generated by alternative splicing of the corresponding pre-mRNAs and via intronic gene promoters. The latter are sequences located in an intron of a gene and induce a truncated form of mRNA. These variants lack the DNA-binding and dimerization domains and may act as competitive inhibitors resembling dominant negative mutants. An example is the truncated CREM variant **ICER** (Inducible cAMP response Element Repressor) that suppresses the transcription of certain CRE-containing genes during cell maturation.

The names CREB and CREM indicate that these factors are regulated by cAMP and protein kinase A (PKA), respectively (Figure 8.25). Along this standard pathway of signal transduction, all signals activating adenylate cyclase via G_S-protein-coupled receptors (Section 5.2) or Ca^{2+} (Section 4.4.1) may influence gene expression. However, CREB and CREM are phosphorylated and activated not only by PKA, but also by various other protein kinases (summarized in Figure 8.26). Although these phosphorylations all occur at the same strategic Ser residue, cells are able to distinguish between the different input signals, probably because additional factors direct CREB and CREM to individual genes or, vice versa, protect certain genes from the interaction with CREB and CREM (as a rule, genes are controlled by a collection of different regulatory factors rather than by a single transcription factor). Upon phosphorylation CREB/CREM recruit the initiation complex of transcription to the genes with CREs. To this end, they cooperate with the co-activator CBP/p300 (CBP stands for CREB-binding protein), which like CREB/CREM is the target of various protein kinases, or signaling pathways (see Figure 8.7). The phosphorylation signals are extinguished by the phosphatases PP1 and PP2A.

Figure 8.25 Activation of transcription factor CREB by protein phosphorylation Via a G_S-protein-coupled receptor, an extracellular signal induces the formation of cAMP (colored rhombus), which activates protein kinase A (PKA), resulting in a dissociation of the catalytic (C) and regulatory (R) subunits. The catalytic subunits of PKA accumulate in the nucleus, where they catalyze the phosphorylation and activation of CREB (P, phosphate residue), thus enabling the transcription factor to interact, as a dimer, with cAMP-responsive elements found in more than 100 genes.

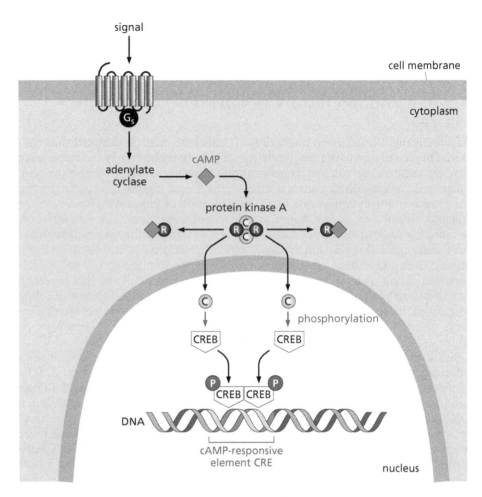

More recently, CREB has found particular attention since there is evidence that it plays a central role in **memory fixation and learning** processes. The factor stimulates the *de novo* synthesis of proteins that reinforce and stabilize synaptic contacts (synaptic plasticity; see Section 16.4.1). Moreover, CREB promotes the transcription of anti-apoptotic genes such as *bcl-2* and thus prolongs the life-span of neurons. In such cases, neuronal survival factors such as nerve growth factor and brain-derived neurotrophic factor are major input signals leading to CREB activation.

As shown in later chapters, the phosphorylation of transcription factors is a theme with many variations. By no means is the target always the transactivating domain, with activation being the outcome. In fact, there are many examples of phosphorylation of the DNA-binding domain, mostly but not always resulting in

Figure 8.26 Signaling pathways leading to CREB phosphorylation and activation Along the signaling pathways shown, protein kinases (box) are activated, phosphorylating CREB at a Ser residue in the transactivating domain. Ca^{2+} may stimulate certain isoforms of adenylate cyclase as well as Ras-controlled signaling cascades (Section 10.1). More details are found in Section 4.6.1 (protein kinase A), Section 4.6.6 (protein kinase B), Section 11.6 (protein kinases MSK1 and p90^{RSK}), and Section 14.6.3 (CaM kinases).

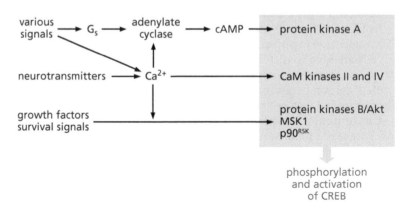

inactivation (see steroid hormone receptors). This means that transcription factors may also be stimulated by dephosphorylation (NFAT provides an example; see Section 14.6.4). Another variant is that both the transcription factor and an accessory protein, for instance an inhibitor, are controlled by phosphorylation. NFκB is a typical example (see Section 11.8).

Summary

The function of most transcription factors is controlled by phosphorylation. An example is provided by cAMP response element binding protein (CREB), which is activated by phosphorylation catalyzed by a variety of protein kinases. It is a target of several signaling cascades. Upon activation, CREB recruits the transcriptional co-activator CBP/p300 to genes containing the cAMP response element.

8.7 Multiple post-translational modifications of transcription factors: the hypoxic stress response

Due to their multidomain structure, transcription factors may simultaneously transform several input signals into one output signal, working as molecular "nanoneurons." An example is steroid hormone receptors: as discussed previously, they are not only activated by their ligands but their functions are modified by various additional signals such as phosphorylation and other covalent alterations. The hypoxic stress response provides another impressive and relatively well-studied example. This emergency reaction enables the cell to cope with a diminished oxygen supply.

The critical event of the hypoxic stress response is the rescue from inactivation of the **Hypoxia-Inducible transcription Factors (HIFs)**. In fact, when sufficient oxygen is available, the cellular concentration of HIFs is kept at a very low level due to continuous degradation along the ubiquitin–proteasome pathway. This situation is reminiscent of Wnt and hedgehog signaling, where input signals also protect the corresponding transcription factors from destruction (Section 5.9).

The input signal leading to a stabilization of HIFs is a decline of the cellular oxygen concentration. As a result, the HIF isoforms associate into α/β-heterodimers, which become enriched in the nucleus and stimulate the transcription of more than 70 genes with hypoxia-responsive elements. Such genes give rise to the formation of glucose transporters and of enzymes of anaerobic energy production, which is due to the formation of lactic acid from glucose yielding 2 ATP (as compared with 38 ATP molecules obtained by oxidative glucose metabolism to CO_2 and water). At the same time, HIFs induce the synthesis of membrane transporters and ion exchangers required to protect the cell from acidification due to lactic acid production.

Other HIF-sensitive genes encode transferrin and its receptor, facilitating the intake of iron ions (required for enzymes of oxidative metabolism), as well as of factors that improve the tissue's blood and oxygen supply. The latter include vasodilatory enzymes such as NO synthase (Section 16.5) together with factors inducing the formation of blood capillaries (vasculo-endothelial growth factor and its receptor) and of erythrocytes (erythropoietin).

A number of covalent signaling reactions control the stability and activity of the HIFs: hydroxylation, acetylation, phosphorylation, ubiquitylation, sumoylation, and S-nitrosylation (see Figure 8.27). As expected, HIFs respond very sensitively to changes of the cellular oxygen concentration. The oxygen sensor of the hypoxic stress response is an oxygenase, **prolyl 4-hydroxylase.** This type of enzyme is well known for its important role in collagen biosynthesis, where it stabilizes the triple helix by Pro hydroxylation. In the hypoxic stress response, a

Figure 8.27 Domain structure, interactions, and post-translational modifications of the hypoxia-inducible transcription factor HIF1α Inhibitory modifications are depicted in red; stimulatory modifications are shown in black. The interaction partners are given in gray boxes. HLH, basic helix–loop–helix motif; TD, transactivating domain; CBP, CREB-binding protein.

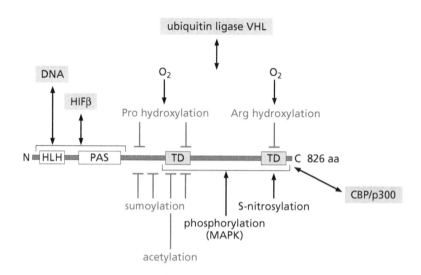

particular isoform (known as prolyl hydroxylase domain protein, PHD) catalyzes the hydroxylation of two Pro residues in the transactivating domain of HIFα. This unusual covalent signaling reaction occurs under normal oxygen supply and signals the ubiquitin ligase VHL (see below) to mark the transcription factor for degradation. Under oxygen deficiency, Pro hydroxylation does not take place and HIFα (which is continuously produced by the cell) is not recognized by VHL. In contrast to the α subunit, HIFβ (which is identical with the factor ArNT mentioned above) is neither hydroxylated nor degraded by the VHL–proteasome system.

A second hydroxylation occurs at an Arg residue in the C-terminal transactivating domain of HIFα. This causes the transcription factor to lose its activity, since it cannot interact any longer with the co-activator CBP/p300. The hydroxylation is catalyzed by the oxygenase **FIH** (Factor Inhibiting HIF) and provides an additional on–off switch. Under hypoxic conditions, both the prolyl 4-hydroxylase PHD and, to a lesser extent, the Arg hydroxylase FIH become less active because of a lack of substrate oxygen.

Like many other transcription factors, HIF becomes **acetylated**. This covalent modification, catalyzed by a specific acetyltransferase, takes place at a Lys residue in the transactivating domain. Its effect resembles that of Pro hydroxylation in that HIF is attacked by VHL and is degraded in proteasomes. The signaling pathways linked to HIF acetylation are not yet known.

Another frequent modification of transcription factors is **sumoylation**. Here, an oligomer of the ubiquitin-related peptide SUMO (Small Ubiquitin-like MOdifier) is bound to a Lys residue (Section 2.8), resulting in inactivation, but not degradation, of HIF. In contrast to acetylation, ubiquitylation, and sumoylation, **phosphorylation** activates HIF. Only MAP kinases and CK2 have been identified as corresponding kinases thus far. An activating signal is also provided by **S-nitrosylation,** which is due to a nonenzymatic reaction of a cysteine residue in the C-terminal transactivating domain with nitric oxide (for more details on this reaction, see Section 16.5).

8.7.1 Hypoxic stress response and cancer

The ubiquitin ligase VHL is a tumor suppressor protein. Its deletion causes **Von Hippel–Lindau disease** and contributes, in addition, to the development of primary kidney and pancreatic carcinomas. The tumorigenic effect is caused by an accumulation of HIF in the affected cells, that results in uncontrolled proliferation due to an oversupply of metabolic energy. Von Hippel–Lindau syndrome is characterized by tumors in the kidneys, adrenal glands, spinal bone marrow, pancreas, and eyes. However, the inglorious role played by a defective hypoxic stress response is not restricted to this rare hereditary disease. **Angiogenesis**, the supply of a tumor with new blood capillaries, is essential for tumor development since otherwise the cells would die from starvation. This condition is prevented by hypoxia prevailing in the fast-growing tumor mass. As a response, the HIF system is activated, which in turn triggers angiogenesis. Most cancer cells exhibit, in addition, a particular metabolic defect discovered in 1924 by Otto Heinrich Warburg (who was awarded the 1931 Nobel Prize in Physiology or Medicine) and known as the **Warburg effect**: even when the oxygen supply is sufficient, they prefer anaerobic to oxidative glucose metabolism, thus maintaining the angiogenetic stimulus. Moreover, the lactic acid produced from glucose is one reason for the characteristic acidification (acidosis) of tumor cells, which may promote invasive growth by activating acidic hydrolases that degrade the extracellular matrix. These observations prompted Warburg to postulate that cancer is a disease of mitochondrial respiration, a hypothesis that recently has regained attention. Given the central position the HIF system occupies in low-oxygen adaptation and pH control, the Warburg effect may be due to defective regulation of the hypoxic stress response, although the underlying mechanism is unknown. When its role in tumorigenesis is considered, the HIF system provides an important target of future anti-cancer strategies.

Summary

The hypoxia-inducible transcription factor HIF controls genes that rescue the cell from hypoxic stress. HIF provides a typical example of a "molecular neuron," processing a variety of stimulatory and inhibitory input signals. Oxygen inactivates the factor through a rather unusual post-translational modification: protein hydroxylation at Pro and Arg residues. Stabilization of HIF caused by an inactivation of the ubiquitin ligase VHL may lead to cancer.

8.8 Annulment of gene transcription by RNA interference

From yeast to humans, cells possess a very efficient mechanism for inactivating or destroying newly synthesized mRNA, thus annulling gene transcription. This process is called RNA interference. The corresponding signaling molecules are short (21–25 nucleotides long) double-stranded RNA species termed small interfering RNA (siRNA) and micro-RNA (miRNA). For the formation of siRNA, double-stranded RNA—as generated, for instance, in the course of viral infections—is partially degraded by a ribonuclease type III, also known as "**dicer**." In contrast, miRNA is derived from cellular precursor RNA (pri-RNA), which is encoded by about 1% of all protein-encoding genes (there are about 250 pri-mRNA-encoding genes in humans). However, miRNA is not translated into protein. Pri-RNA has several hairpin loops generated by partially complementary sequences. These loops are cut off by another type II ribonuclease called "**drosha**." The remaining pre-miRNA is then cut open by dicer, yielding miRNA.

siRNA and miRNA direct a multiprotein RNA-induced silencing complex to mRNA molecules with complementary sequences, thus hindering their translation or promoting their degradation. Typical components of this complex are proteins of the **argonaut family**, required for interaction of the complex with RNA (Figure 8.28). "Slicer" endonucleases, which are one type of argonaut protein, catalyze degradation of the complementary mRNA. Dicer and argonaut proteins have an additional function, which seems to be independent of mi- and siRNA formation: at least in yeast, they are required for cell cycle arrest upon genotoxic stress (see Section 12.9).

A special function of RNA interference is the removal of foreign RNA, which frequently is present in a double-stranded form—for example, after viral infection.

Figure 8.28 RNA interference
Small interfering RNA (siRNA) and micro-RNA (miRNA) are released from high molecular weight double-stranded RNA or hairpin loop RNA (pre-miRNA) by the RNase III "dicer" (pre-miRNA is obtained from high molecular weight pri-miRNA by the RNase "drosha"). Upon dissociation into single strands, si- and miRNA form RNA-Induced Silencing Complexes (RISC) with argonaut proteins; these complexes are directed to mRNA species with complementary sequences. As a result, those mRNAs are degraded by the endonuclease "slicer" (left) or are hindered from becoming translated (right).

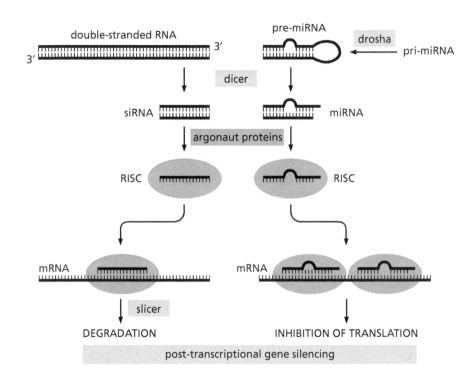

This "nucleic acid immune system" is part of the body's **antiviral strategy** that also includes formation of antibodies, production of antiviral cytokines (interferons; see Section 6.3), and inhibition of viral protein synthesis by protein kinase PKR phosphorylating translation factor eIF2 (Section 9.3.1).

In addition, miRNA and siRNA are involved in the formation of transcriptionally inactive heterochromatin. Here their effects correlate with DNA methylation, and the idea that they may form complexes with chromodomain proteins and DNA methyltransferases is as attractive as it is speculative. In addition, the mechanism by which the small RNA species inhibit mRNA translation is not yet clear. What is known is that they have to bind to 3'-untranslated regions of mRNA, thus probably hindering both the initiation and elongation phases of polypeptide synthesis.

It is easy to see that RNA interference provides an efficient tool for the investigator to experimentally annul the effects of gene transcription in a specific way. For the discovery of RNA interference, Andrew Z. Fire and Craig C. Mello were honored by the Nobel Prize in Physiology or Medicine for 2006.

Summary

Cells produce short RNA molecules that recruit proteins to degrade mRNA or inhibit translation at complementary sequences in mRNA molecules. This regulatory process, known as RNA interference, enables post-transcriptional annulment of gene transcription.

8.9 Post-transcriptional splicing of pre-mRNA: a powerful mechanism to generate variability

Covalent modification of proteins is only one way in which the degree of proteomic complexity is increased in the face of a limited number of genes. Another possibility is the post-transcriptional processing of mRNA. While bacterial genes are composed of noninterrupted coding sequences, the genes of archaea and eukaryotes are assembled in such a way that the coding sequences, called **exons,** are interrupted by noncoding sequences, called **introns**. The number of introns

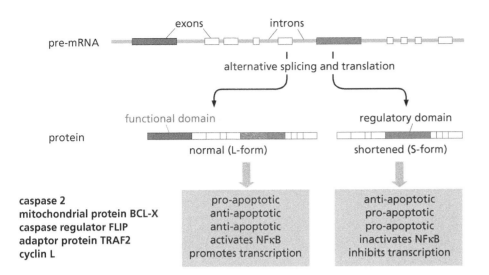

Figure 8.29 Generation of antagonistically acting proteins by alternative splicing of pre-mRNA By translation of the mRNA species generated by alternative splicing, a normal protein (left) and a truncated protein lacking the functional domain (right) are generated. Via its regulatory domain, the truncated protein scavenges input signals, acting as a competitive inhibitor of the normal protein. In the boxes, some examples are shown.

is variable; some genes may contain up to 50 introns. The primary transcript, the pre-mRNA, mirrors the intron–exon pattern of the gene. During maturation to functional mRNA, the introns are cut out enzymatically, resulting in splicing of the exons. In the course of this process, the exons may be either combined in new patterns, provided with a new reading frame, or partially lost. Due to these rearrangements, known as **alternative splicing,** a gene may give rise to a wide variety of protein isoforms exhibiting modified functions. It has been estimated that at least half of all pre-mRNAs in humans undergo alternative splicing. The muscle proteins troponin, tropomyosin, and titin as well as potassium channels are examples of alternative splicing. All exist in many splice variants, which are tissue-specifically distributed and differ in their functional properties (see, for instance, Section 14.4). Splice variants may even exhibit antagonistic effects, thus playing important roles in regulation (Figure 8.29).

In extreme cases, one gene may supply thousands of protein isoforms, even outdoing the number of genes (for an example see Section 7.4). Therefore, alternative splicing is one of the most efficient mechanisms by which the complexity of the proteome is increased, as compared with the genome. Alternative splicing and exon shuffling between different genes seem to be major reasons for the evolution of interrupted genes.

Splicing of pre-mRNA occurs in the nucleus. The biochemical course is shown in Figure 8.30. Similar to gene transcription and mRNA translation, splicing is under the control of a large number of regulatory proteins (estimated at more than 100) forming **spliceosomes** with small nuclear RNA. The basic constituents of these organelle-like complexes are five **small nuclear ribonucleoprotein** particles (snRNPs). The corresponding RNA components (U1, U2, U4, U5, and U6) occupy the pre-mRNA via complementary sequences at critical splice sites in a fixed order and catalyze splicing due to their ribozyme activities (for additional information, the reader is referred to textbooks of cell and molecular biology). Small nuclear ribonucleoproteins cooperate with many regulatory proteins that, like transcription factors, stimulate or inhibit splicing depending on whether they bind to enhancer or silencer sequences in the exons. In their function, such sequences resemble the response elements of gene promoters.

Abundant splice factors are the so-called **SR proteins** [Ser (S)- and Arg (R)-rich proteins], which possess RNA-binding domains and protein interaction motifs. They are mostly activators of splicing. Typical splicing repressors are, in contrast, proteins of the heterogeneous nuclear ribonucleoprotein family. The number of individual splice factors is relatively small, raising the question as to how gene- and sequence-specific splicing is achieved. The explanation may be found in a

Figure 8.30 Splicing of pre-mRNA
(A) Pre-mRNA consists of coding sequences (exons, gray rectangles) and noncoding sequences (introns, red lines). During the maturation process, the introns are cut out enzymatically and the exons are spliced, forming the mature mRNA. (B) Splicing includes two nucleophilic attacks. At first the 2'-OH group of a conserved adenine residue (A) in the so-called branching point of an intron attacks the phosphate linkage between the intron and exon 1, leading to the release of exon 1. At the remaining intron, a lasso structure is formed by a 2'–5' linkage between the adenosine residue and a terminal guanine residue (G) called the splice donor. In the second step, the free 3'-OH-group of exon 1 attacks the phosphate linkage (splice acceptor) between the intron and exon 2. This results in a 3'–5' bond between exon 1 and exon 2 and the release of the intron sequence, which undergoes degradation. The entire splicing process depends on ATP hydrolysis. It is catalyzed by the ribozyme activities of special RNA species and controlled by at least 100 different proteins (for more details, see text).

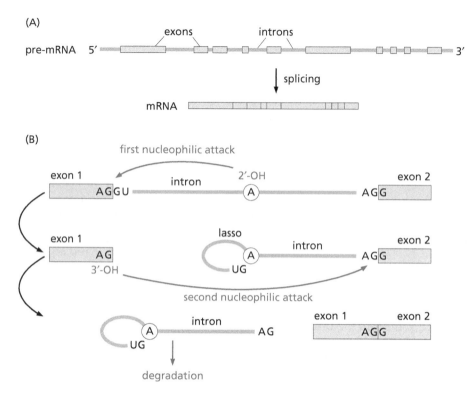

combinatorial mechanism, which, similar to transcriptional control, is based on cooperation of sets of factors rather than on the effect of a single regulatory protein. In addition, post-translational modifications may change the efficacy and specificity of splice factors. In other words, like transcription factors, splice factors are also targets of signaling cascades. However, the investigation of these regulatory pathways is still in its infancy. The few cases studied in detail indicate that phosphorylation and dephosphorylation are central regulatory events, involving MAP kinases, calcium/calmodulin-dependent kinases, protein kinase C, cyclin-dependent kinases, and Tyr kinases of the Src type.

It should be mentioned that alternative splicing of pre-mRNA is only one of the powerful mechanisms for the generation of protein variants from a single gene. Another means is based on **alternative gene promoters**, located at internal intron–exon transitions rather than at the start point of a gene like conventional promoters. The proteins thus produced are amino-terminally truncated forms that functionally may resemble dominant-negative mutants as they are widely used in the laboratory. An example is the generation of the CREB antagonist ICER mentioned previously. Corresponding sites are also found in certain mRNA species, enabling the generation of truncated protein isoforms by alternative translation initiation. mRNAs allowing such **alternative start codon usage** encode transcription factors (including nuclear receptors) and growth factors.

Summary

Pre-mRNA derived from the transcription of eukaryotic split genes is processed by splicing: removal of noncoding intron sequences and combination of the remaining exon sequences into new patterns. As a result, one gene may give rise to a wide variety of proteins differing in their functions. Like transcription, splicing is regulated by numerous proteins (splice factors), which are potential targets of signaling cascades. Splicing is one of the most efficient mechanisms by which the complexity of the proteome is increased in spite of a limited set of genes. Other mechanisms that generate (truncated) protein isoforms are the usage of alternative gene promoters and alternative start codons.

Further reading

Akiyama TE & Gonzalez FJ (2003) Regulation of P450 genes by liver-enriched transcription factors and nuclear receptors. *Biochim. Biophys. Acta, Gen. Subj.* 1619, 223–234.

Bain DL, Henghan AF, Connaghan-Jones KD et al. (2007) Nuclear receptor structure: implications for function. *Annu. Rev. Physiol.* 69, 201–230.

Bao Y & Shen X (2007) Snapshot: chromatin remodeling complexes. *Cell* 129, 632–633.

Belandia B & Parker MG (2003) Nuclear receptors: a rendezvous for chromatin remodeling. *Cell* 114, 277–280.

Bernstein BE, Meissner A & Lander ES (2007) The mammalian epigenome. *Cell* 128, 669–681.

Brahimi-Horn C, Mazure N & Pouyssegur J (2005) Signaling via the hypoxia-inducible factor-1α requires multiple post-translational modifications. *Cell. Signal.* 17, 1–9.

Brahimi-Horn MC, Chiche J & Pouyssegur J (2007) Hypoxia signaling controls metabolic demand. *Curr. Opin. Cell Biol.* 19, 223–229.

Brivanlou AH & Darnell JE Jr (2002) Signal transduction and the control of gene expression. *Science* 295, 813–818.

Bukrinsky MI (2002) Cyclophilins: unexpected messengers in intercellular communications. *Trends Immunol.* 23, 323–325.

Carrozza MJ, Utley RT, Workman JL et al. (2003) The diverse functions of histone acetyltransferase complexes. *Trends Genet.* 19, 321–329.

Clayton AL & Mahadevan LC (2003) MAP kinase-mediated phosphoacetylation of histone H3 and inducible gene regulation. *FEBS Lett.* 546, 51–58.

Contreras A, Hale TK, Stemoien DL et al. (2003) The dynamic mobility of histone H1 is regulated by cyclin/CDK phosphorylation. *Mol. Cell. Biol.* 23, 8626–8636.

Corton JC, Anderson SP & Stauber A (2000) Central role of peroxisome proliferator-activated receptors in the actions of peroxisome proliferators. *Annu. Rev. Pharmacol. Toxicol.* 40, 491–518.

Denison MS & Nagy SR (2003) Activation of the aryl hydrocarbon receptor by structurally diverse exogenous and endogenous chemicals. *Annu. Rev. Pharmacol. Toxicol.* 43, 309–334.

Denly AM & Hannon GJ (2004) RNAi: an ever-growing puzzle. *Trends Biochem. Sci.* 28, 196–201.

DeRuijter AJM, van Gennip AH, Caron HN et al. (2003) Histone deacetylases (HDACs): characterization of the classical HDAC family. *Biochem. J.* 370, 737–749.

Dornan DJ, Taylor P & Walkinshaw MD (2003) Structure of immunophilins and their target ligand complexes. *Curr. Top. Med. Chem.* 3, 1392–1409.

Dou Y & Gorovsky MA (2000) Phosphorylation of linker histone H1 regulates gene expression in vivo by creating a charge patch. *Mol. Cell* 6, 225–231.

Dykxhoorn DM, Novina CD & Sharp PA (2003) Killing the messenger: short RNAs that silence gene expression. *Nat. Rev. Mol. Cell Biol.* 4, 457–467.

Eberharter A & Becker PB (2002) Histone acetylation: a switch between repressive and permissive chromatin. *EMBO Rep.* 3, 224–229.

Fernandes I & White JH (2003) Agonist-bound nuclear receptors: not just targets of coactivators. *J. Mol. Endocrinol.* 31, 1–7.

Fu M, Wang C, Wang J et al. (2002) Acetylation in hormone signaling and the cell cycle. *Cytokine Growth Factor Rev.* 13, 259–276.

Galat A (2003) Peptidylprolyl cis/trans isomerases (immunophilins): biologic diversity – targets – functions. *Curr. Top. Med. Chem.* 3, 1315–1347.

Geiman TM & Robertson KD (2003) Chromatin remodeling, histone modifications, and DNA methylation – how does it all fit together? *J. Cell. Biochem.* 87, 117–125.

Germain P, Staels B, Dacquet C et al. (2006) Overview of nomenclature of nuclear receptors. *Pharmacol. Rev.* 58, 685–704.

Gill G (2001) Regulation of the initiation of eukaryotic transcription. *Essays Biochem.* 37, 33–44.

Goodwin B & Moore JT (2004) CAR: detailing new models. *Trends Pharmacol. Sci.* 25, 437–441.

Grewal SIS & Moazed D (2003) Heterochromatin and epigenetic control of gene expression. *Science* 301, 798–802.

Haushalter KA & Kadonaga T (2003) Chromatin assembly by DNA-translocating motors. *Nat. Rev. Mol. Cell Biol.* 4, 613–620.

Hobert O (2004) Common logic of transcription factor and microRNA action. *Trends Biochem. Sci.* 29, 462–468.

Holmberg CI, Tran SEF, Eriksson JE et al. (2002) Multisite phosphorylation provides sophisticated regulation of transcription factors. *Trends Biochem. Sci.* 27, 619–632.

Jaronczyk K, Carmichael JB & Hobman TC (2005) Exploring the functions of RNA interference pathway proteins: some functions are more RISCy than others? *Biochem. J.* 387, 561–571.

Jenuwein T & Allis CD (2001) Translating the histone code. *Science* 293, 1074–1080.

Khorasanizadeh S & Rastinejad F (2001) Nuclear-receptor interactions on DNA-response elements. *Trends Biochem. Sci.* 26, 384–390.

Khorasanizadeh S (2004) The nucleosome: from genomic organization to genomic regulation. *Cell* 116, 259–272.

Kraus WL & Lis JT (2003) PARP goes transcription. *Cell* 113, 677–683.

Lee M & Goodbourn S (2001) Signaling from the cell surface to the nucleus. *Essays Biochem.* 37, 71–85.

Li B, Carey M & Workman JL (2007) The role of chromatin during transcription. *Cell* 128, 707–719.

Lonard DM & O'Malley BW (2005) Expanding functional diversity of the coactivators. *Trends Biochem. Sci.* 30, 126–132.

Lones PA, Baylin SB (2002) The fundamental role of epigenetic events in cancer. *Nat. Rev. Genet.* 3, 415–428.

Longo VD & Kennedy BK (2006) Sirtuins in aging and age-related diseases. *Cell* 126, 257– 268.

Losel R & Wehling M (2003) Nongenomic actions of steroid hormones. *Nat. Rev. Mol. Cell Biol.* 4, 46–56.

Lu NZ & Cidlowski JA (2006) Glucocorticoid receptor isoforms generate transcription specificity. *Trends Cell Biol.* 16, 301–307.

Lu KP & Zhou XZ (2007) The prolyl isomerase PIN1: a pivotal new twist in phosphorylation signalling and disease. *Nat. Rev. Mol. Cell Biol.* 8, 904–916.

Martens JA & Winston F (2003) Recent advances in understanding chromatin remodeling by Swi/Snf complexes. *Curr. Opin. Genet. Dev.* 13, 136–142.

Mayr B & Montminy M (2001) Transcriptional regulation by the phosphorylation-dependent factor CREB. *Nat. Rev. Mol. Cell Biol.* 2, 599–608.

McDonnell DP & Norris JD (2002) Connections and regulation of the human estrogen receptor. *Science* 296, 1642–1644.

McEwan LJ (ed) (2004) The nuclear receptor superfamily. *Essays Biochem.* No 40.

Meehan RR & Stancheva I (2001) DNA methylation and control of gene expression in vertebrate development. *Essays Biochem.* 37, 59–70.

Mimura J & Fujii-Kuriyama Y (2003) Functional role of AhR in the expression of toxic effects by TCDD. *Biochim. Biophys. Acta, Gen. Subj.* 1619, 263–268.

Moore JT, Moore LB, Maglich JM et al. (2003) Functional and structural comparison of PXR and CAR. *Biochim. Biophys. Acta, Gen. Subj.* 1619, 235–238.

Nguyen T, Sherratt PJ & Pickett CB (2003) Regulatory mechanisms controlling gene expression mediated by the antioxidant response element. *Annu. Rev. Pharmacol. Toxicol.* 43, 233–260.

Nilsen TW (2007) Mechanisms of microRNA-mediated gene regulation in animal cells. *Trends Genet.* 23, 234-249.

Nowak SJ & Corces VG (2004) Phosphorylation of histone H3: a balancing act between chromosome condensation and transcriptional activation. *Trends Genet.* 20, 214–220.

Pearl LH & Prodromou C (2006) Structure and mechanism of the Hsp90 molecular chaperone machinery. *Annu. Rev. Biochem.* 75, 271–294.

Perissi V & Rosenfeld MG (2005) Controlling nuclear receptors: the circular logic of cofactor cycles. *Nat. Rev. Mol. Cell Biol.* 6, 542–553.

Porcu M & Chiarugi A (2005) The emerging therapeutic potential of sirtuin-interacting drugs: from cell death to lifespan extension. *Trends Pharmacol. Sci.* 26, 94–102.

Pratt WB & Toft DO (2003) Regulation of signaling protein function and trafficking by the hsp90/hsp70-based chaperone machinery. *Exp. Biol. Med.* 228, 111–133.

Prossnitz ER, Arterburn JB, Smith HO et al. (2008) Estrogen signaling through the transmembrane G protein-coupled receptor GPR30. *Annu. Rev. Physiol.* 70, 165-190.

Ptashne M & Gann A (2001) Transcription initiation: imposing specificity by localization. *Essays Biochem.* 37, 1–15.

Rangwala SM & Lazar MA (2004) Peroxisome proliferator-activated receptor γ in diabetes and metabolism. *Trends Pharmacol. Sci.* 25, 331–336.

Ratajczak T, Ward BK & Minchin RF (2003) Immunophilin chaperones in steroid receptor signaling. *Curr. Top. Med. Chem.* 3, 1348–1357.

Reese JC (2003) Basal transcription factors. *Curr. Opin. Genet. Dev.* 13, 114–118.

Rochette-Egly C (2003) Nuclear receptors: integration of multiple signaling pathways through phosphorylation. *Cell. Signal.* 15, 355–366.

Russell DW (2003) The enzymes, regulation, and genetics of bile acid synthesis. *Annu. Rev. Biochem.* 72, 138–171.

Saunders A, Core LJ & Lis JT (2006) Breaking barriers to transcription elongation. *Nat. Rev. Mol. Cell Biol.* 7, 557–567.

Sauve AA, Wolberger C, Schramm VL et al. (2006) The biochemistry of sirtuins. *Annu. Rev. Biochem.* 75, 435–465.

Schofield CJ & Ratcliffe PJ (2004) Oxygen sensing by HIF hydroxylases. *Nat. Rev. Mol. Cell Biol.* 5, 343–354.

Shabazian MD & Grunstein M (2007) Functions of site-specific histone acetylation and deacetylation. *Annu. Rev. Biochem.* 76, 75–100.

Shaw PE (2002) Peptidyl-prolyl-isomerases: a new twist to transcription. *EMBO Rep.* 3, 521–526.

Shilatifard A (2006) Chromatin modifications by methylation and ubiquitination: implications in the regulation of gene expression. *Annu. Rev. Biochem.* 75, 243–269.

Shilatifard A, Conaway RC & Conaway JW (2003) The RNA polymerase II elongation complex. *Annu. Rev. Biochem.* 72, 693–715.

Shin C & Manley JL (2004) Cell signaling and the control of pre-mRNA splicing. *Nat. Rev. Mol. Cell Biol.* 5, 727–738.

Smale ST & Kadonaga JT (2003) The RNA polymerase II core promoter. *Annu. Rev. Biochem.* 72, 449–479.

Smith S (2001) The world according to PARP. *Trends Biochem. Sci.* 26, 174–179.

Sontheimer EJ (2005) Assembly and function of RNA silencing complexes. *Nat. Rev. Mol. Cell Biol.* 6, 127–138.

Spector DL (2003) The dynamics of chromosome organization and gene regulation. *Annu. Rev. Biochem.* 72, 573–608.

Stetefeld J & Ruegg MA (2005) Structural and functional diversity generated by alternative mRNA splicing. *Trends Biochem. Sci.* 30, 515–521.

Swigut T & Wysocka J (2007) H3K27 demethylase, at long last. *Cell* 131, 29–32.

Tang G (2005) siRNA and miRNA: an insight into RISCs. *Trends Biochem. Sci.* 30, 106–113.

Tat JR (2002) Signaling through nuclear receptors. *Nat. Rev. Mol. Cell Biol.* 3, 702–710.

Thomas MC & Chiang CM (2006) The general transcription machinery and general cofactors. *Crit. Rev. Biochem. Mol. Biol.* 41, 105–178.

Tsukiyama T (2002) The in vivo functions of ATP-dependent chromatin remodelling factors. *Nat. Rev. Mol. Cell Biol.* 3, 422–429.

Turner BM (2002) Cellular memory and the histone code. *Cell* 111, 285–291.

Wade PA (2001) Methyl CpG-binding proteins and transcriptional repression. *Bioessays* 23, 1131–1137.

Wärnmark A, Treuter E, Wright APH et al. (2003) Activation functions 1 and 2 of nuclear receptors: molecular strategies for transcriptional activation. *Mol. Endocrinol.* 17, 1901–1909.

Wolffe AP (2001) Transcriptional regulation in the context of chromatin structure. *Essays Biochem.* 37, 45–57.

Xu L & Massague J (2004) Nucleocytoplasmatic shuttling of signal transducers. *Nat. Rev. Mol. Cell Biol.* 5, 209–219.

Young JC, Moarefi I & Hartl FU (2001) Hsp90: a specialized but essential protein folding tool. *J. Cell Biol.* 154, 267–274.

Zoete V, Grosdidier A & Michielin O (2007) Peroxisome proliferator-activated receptor structures: ligand specificity, molecular switch and interactions with regulators. *Biochim. Biophys. Acta* 1771, 915–925.

Signals Controlling mRNA Translation

Gene expression culminates in the translation of a nucleotide sequence into the amino acid sequence of the complementary polypeptide. This translation occurs at the ribosomes and is controlled by various regulatory proteins or translation factors, which are targets of cellular signaling cascades. In other words, exogenous input signals are controlling protein biosynthesis on both the level of gene transcription and the level of mRNA translation. In fact, the regulation of translation is clearly superior as far as speed is concerned, as is required, for example, in acute stress situations. Moreover, protein biosynthesis is dependent on controlled translation in tissues whose genetic activity is diminished or switched off. A classical example that will be discussed in more detail below is hemoglobin synthesis in the mammalian reticulocyte. Finally, strict control of translation is essential since protein synthesis is by far the most expensive cellular process, consuming – depending on the conditions, up to 50% of the ATP and GTP supply. In emergency situations, protein biosynthesis, therefore, becomes drastically turned down; we have seen this already for the stringent response of bacteria (Section 3.1).

9.1 Eukaryotic mRNA translation: the essentials*

Ribosomes are biochemical machines specialized for protein synthesis. The eukaryotic ribosome (80S) is composed of a small (40S) and a large (60S) subunit. These subunits are huge multiprotein complexes: the small subunit is made up of an 18S ribosomal RNA and 33 different proteins, and the large subunit contains 28S, 4.8S, and 2S ribosomal RNA and 49 different proteins.

To start translation, the small subunit recruits an **initiation complex** consisting of at least 11 initiation factors (IF), with some of them being multiprotein complexes themselves (the functions of the individual factors are summarized in Table 9.1). This assembly occurs in a stepwise manner. In the beginning, the initiation factors eIF1A (e, eukaryotic) and eIF3 induce the dissociation of the ribosomal subunits so that mRNA can bind to the ribosomal translation apparatus. The initiation complex recognizes mRNA by a 5′-terminal 7-methylguanosine nucleotide residue called the **m⁷GpppX cap**, which binds the heterotrimeric initiation factor eIF4F. The latter is composed of the cap-binding protein eIF4E, the RNA helicase eIF4A (a DEAD-box ATPase breaking up secondary structures in the mRNA molecule that otherwise would hinder translation; see Sidebar 9.1), and the scaffold protein eIF4G. In cooperation with eIF3, a putative ATPase

---

*Note that the principles of translation are summarized here only in a condensed form sufficient for an understanding of signal-controlled regulatory processes. More details can be found in textbooks of biochemistry and molecular biology.

Table 9.1 Eukaryotic translation factors

Factor	Function
eIF1	recognition of the start codon AUG
eIF1A	Met-tRNA binding to 40S ribosomal subunit
eIF2	GTPase, methionyl-tRNA binding
eIF2B	GEF of eIF2
eIF3	ATPase(?), association of 40S ribosomal subunit with mRNA
eIF4A	ATPase, helicase
eIF4E ⎫ eIF4F	binding to m⁷GpppX cap of mRNA
eIF4G ⎬	scaffold protein
eIF4B	activator of helicase eIF4A
eIF4H	activator of helicase eIF4A
eIF5	GAP of eIF2
eIF5B	GTPase, association of ribosomal subunits
eEF1A	GTPase, binding of aminoacyl-tRNA
eEF2	GTPase, translocation of peptidyl-tRNA
eEF1B	GEF of eEF1A
eRF1	recognition of stop codon
eRF3	GTPase, disintegration of translation complex

(made up of 12 subunits), this complex recruits the 40S ribosomal subunit, which has bound additional initiation factors such as eIF1, eIF1A, and eIF5 as well as a complex of initiator tRNA (in eukaryotes, methionyl-tRNA) and the GTP-loaded GTPase eIF2, a typical G-protein.

The 40S ribosomal subunit thus prepared now scans the mRNA under ATP consumption until it has reached the start codon AUG, where it associates with the large subunit. This process and the subsequent disintegration of the initiation complex are targeted (energy-consuming) events that require GTP hydrolysis catalyzed by the two G-proteins eIF2 and eIF5B. In this context, another factor, eIF5, plays the role of an eIF5 GTPase-activating protein (GAP), while eIF2B, as a GDP–GTP exchange factor (GEF), reloads eIF2 with GTP for the next initiation cycle. Therefore, eukaryotic cells are consuming two molecules of GTP per initiation step. In addition, for mRNA scanning, two molecules of ATP are hydrolyzed, probably by the ATPases eIF4A and eIF3.

Initiation does not always depend on an interaction between the 5′-cap of mRNA and eIF4F. In fact, in about 5% of all mRNA species, so-called internal ribosomal

Sidebar 9.1 DEAD-box proteins These highly conserved ATPases found in pro- and eukaryotes got their name from a characteristic D-E-A-D (Asp-Glu-Ala-Asp) sequence in their catalytic domain. Most of them are **RNA helicases** involved in pre-mRNA splicing, maturation of ribosomal RNA, removal of loops and helices that are hindering mRNA translation, etc. Moreover, the enzymes of a subfamily called SNF2p catalyze the **restructuring of nucleosomes** in the course of gene transcription (for details see Section 8.2.7). The energy needed for such structural changes is derived from ATP hydrolysis. One should not confuse the DEAD box with the *death domain* of apoptosis-inducing proteins!

entry sites (IRES) have been found (mostly in the 5′-untranslated region), which enable cap-independent translation. These regulatory sites are controlled by proteins called IRES-transacting factors, which probably are chaperones stabilizing a functional 3D structure of mRNA. The input signals controlling these factors are not yet known. IRES-mediated translation enables the cell to survive and adapt under conditions where normal translation is suppressed, particularly under stress. A similar emergency reaction enabling a preferred translation of certain mRNA species in spite of an overall inhibition of protein biosynthesis will be discussed in Section 9.3.2.

Like initiation complex formation, peptide chain elongation also requires GTP hydrolysis: elongation factor eEF1A, a G-protein, transfers the aminoacyl-tRNA to the A-site of the ribosome, while another GTPase (eEF2) promotes the translocation of the peptidyl-tRNA. A third elongation factor (eEF1B) plays the role of an eEF1A-GEF. Finally, the termination factor eRF3 (cooperating with another factor eRF1) is a GTPase, too. Thermodynamically the energy obtained from GTP hydrolysis is needed because peptide chain elongation generates a state of higher order or decreased entropy.

Summary

Eukaryotic mRNA translation at ribosomes occurs in initiation, elongation, and termination phases. These phases are controlled by a series of translation factors, some of which are regulatory GTPases. Like transcription factors, the translation factors are targets of signaling pathways.

9.2 Release of newly synthesized protein by the endoplasmic reticulum: an ancient playground of G-proteins

Ribosomes are associated with the membrane system of the endoplasmic reticulum (ER). This characteristic structure of eukaryotic cells is connected with the extracellular space; thus its lumen represents external conditions inside the cell. Therefore, the ER membranes are the site where the release of secretory and transmembrane proteins starts. Such proteins have been estimated to amount to about a quarter of all proteins produced by a cell. They are directly synthesized into the ER lumen by the ribosomes attached, and they subsequently are transported to the Golgi apparatus where they become packed into secretory vesicles (for details see Section 10.4.1).

To penetrate the ER membrane, the growing polypeptide chain is equipped with an N-terminal lipophilic amino acid sequence called signal peptide. This sequence binds a **signal recognition particle (SRP),** which interacts with an **SRP receptor** at the ER membrane (Figure 9.1). The receptor forms a complex with a membrane pore, the **translocon**, through which the growing polypeptide chain becomes threaded. At the same time a signal peptidase associated with the translocon is removing the signal peptide and further post-translational modifications such as N-glycosylation occur, preventing a re-transport of the protein.

Transportation of the translating ribosome to the translocon and the linearization and subsequent threading through the pore of the polypeptide chain are targeted movements and forced conformational changes requiring energy. This energy is supplied by the hydrolysis of two molecules of GTP. For this purpose, SRP and its receptor have G-protein domains; that is, they are regulatory GTPases. The translating ribosome functions as a SRP-GEF, and the free translocon as a GEF of the SRP receptor: both activate their effector G-proteins by catalyzing GDP–GTP exchange. Therefore, the system becomes active only in the presence of an mRNA-loaded ribosome and a free translocon. The receptor–GTP complex facilitates GTP hydrolysis at the SRP molecule: it plays the role of an

Figure 9.1 Protein biosynthesis at the endoplasmic reticulum (1) The ribosome carrying the growing polypeptide chain (dotted line; the broken arrow symbolizes the signal peptide) activates the G-protein <u>S</u>ignal <u>R</u>ecognition <u>P</u>article (SRP) by GDP–GTP exchange (GEF effect). (2) GTP-loaded SRP binds to the signal peptide and recruits the translating ribosome at the GTP-loaded <u>S</u>ignal <u>R</u>ecognition <u>P</u>article <u>R</u>eceptor (SRPR) and the adjacent translocon in the endoplasmic reticulum (ER) membrane. (3) By means of the signal peptide, the growing polypeptide chain is threaded through the pore of the translocon. At the same time, SRP and SRPR stimulate their GTPase activities in a reciprocal manner (GAP effect). (4) While polypeptide synthesis is progressing, the signal peptidase of the translocon removes the signal peptide and the SRP–SRPR complex disintegrates. Upon termination of translation, the naked translocon reactivates SRPR by GDP–GTP exchange (GEF effect; not shown).

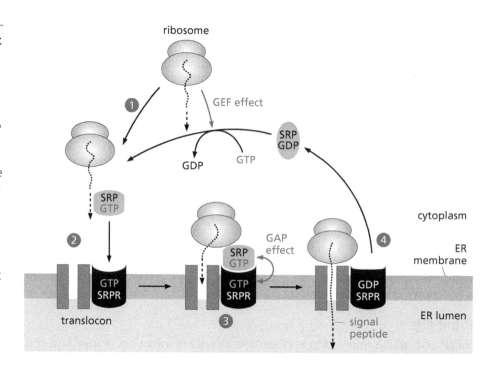

SRP-GAP that induces the dissociation of the ribosome–SRP complex when the polypeptide chain has entered the translocon pore. Vice versa, SRP-GTP is a GAP of the SRP receptor. Thus SRP and its receptor are interconnected in a reciprocal regulatory circuit (see Figure 9.1).

Sidebar 9.2 Signal recognition particles and their receptors Eukaryotic signal recognition particles are ribonucleoprotein complexes consisting of six different protein subunits (according to their molecular weights termed SRP9, -14, -19, -54, -68, and -72) and a 7S RNA forming some sort of a backbone. The subunit SRP54 harbors the GTPase domain of the SRP and contains an unusually high number of methionine residues (so-called M-domain) for the hydrophobic interaction with the signal peptide, as well as a nucleic acid-binding helix–turn–helix motif for the interaction with 7S RNA. The eukaryotic SRP receptor is made up of two subunits, SRPRα and SRPRβ, both exhibiting GTPase domains. SRPRβ also exhibits a transmembrane domain. The basic structure of the translocon is a heterotrimer of the transmembrane proteins Sec61p (with 10 transmembrane domains) and Sec61α as well as Sec61β (with one transmembrane domain each). These proteins assemble forming a membrane pore of 5 nm diameter which is opened by the SRP–receptor complex. The translocon is a component of a large complex with numerous other proteins such as a signal peptidase, an oligosaccharyltransferase, and the chaperone BiP (see Section 9.3.2).

Proteins homologous to SRP, SRP receptor, and the Sec61 complex are found in all pro- and eukaryotes examined thus far. However, the translational machinery of bacteria is considerably simpler than that of eukaryotes. The SRP of *Escherichia coli* consists only of the SRP54-homologous GTPase Ffh (<u>F</u>ifty-<u>f</u>our <u>h</u>omologue) and a 4.5S RNA, and the SRP receptor of the SRPα-homologous GTPase FtsY. The latter is associated with the inner side of the plasma membrane (prokaryotes lack an endoplasmic reticulum). The basic structure of the prokaryotic translocon widely resembles that of its eukaryotic counterpart.

Summary

The release of newly synthesized protein into the lumen of the ER is a targeted process controlled and promoted by two regulatory GTPases: signal recognition particle (SRP) and SRP receptor, a transmembrane protein associated with a

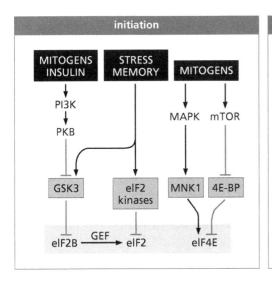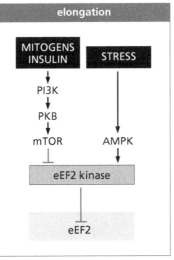

Figure 9.2 Major signaling pathways controlling translation: an overview Both peptide chain initiation and elongation are stimulated by mitogens (and insulin) and partially inhibited by stress factors and during memory fixation. Shown are the translation factors known to be targeted by such signals and the signaling pathways involved. In addition, other translation factors have been found to be substrates of protein kinases, albeit with less conclusive evidence as far as the physiological role of phosphorylation is concerned. More details are found in the text. PI3K, phosphatidylinositol 3-kinase; PKB, protein kinase B; GSK3, glycogen synthase kinase 3; eIF, eukaryotic initiation factor; GEF, GDP–GTP exchange factor; MAPK, mitogen-activated protein kinase; MNK, an effector kinase of the mitogenic MAP kinase module; mTOR, protein kinase mammalian target of rapamycin; 4E-BP, eIF4E binding protein; eEF, eukaryotic elongation factor; AMPK, 5'-AMP-dependent protein kinase.

peptide channel (translocon) in the ER membrane. At the SRP, GDP–GTP exchange is catalyzed by the ribosome; at the SRP receptor, it is catalyzed by the translocon. Homologous GTPases and translocons are also found in prokaryotes where they control peptide transport across the cell membrane. Prokaryotic GTPases are the functional predecessors of eukaryotic G-proteins.

9.3 Signaling cascades controlling translation

Like gene transcription, mRNA translation critically depends on the environmental conditions. As a general rule, translation is up-regulated by survival signals such as growth factors and hormones stimulating cell proliferation and cell function and turned down by stress factors. These signals primarily affect the initiation phase but also the phase of polypeptide-chain elongation. Major target proteins are the cap-binding complex and the GTPases eIF2 and eEF2. An overview is provided by Figure 9.2.

9.3.1 eIF2 kinases: stress-sensitive repressors of protein biosynthesis

To save energy, protein biosynthesis is suppressed in stress situations. This takes place at an early stage of translation, when the initiator tRNA loads the initiation complex, a step catalyzed by the initiation factor eIF2. Under stress (and in neurons also during learning), this GTPase is rendered inactive through phosphorylation of a strategic serine (Ser) residue catalyzed by an eIF2 kinase. As a result, the interaction of eIF2 with its GEF eIF2B is interrupted and translation is blocked at the initiation stage. This effect, occurring in a few minutes, is by far the most rapid cellular stress response.

In mammals, eIF2 kinase is expressed in four isoforms that differ in cellular functions and mechanisms of regulation (Figure 9.3). The classical isoform is the **heme-regulated inhibitor HRI**. Its major task is to adjust protein synthesis, particularly of hemoglobin, to requirements such as the availability of heme, iron, and energy. Hemoglobin synthesis is regulated at the level of mRNA translation – because, at least in mammals, the genome of the mature reticulocyte is silent. HRI, expressed primarily in reticulocytes, binds heme via a specific domain and thus becomes inhibited. As a consequence, eIF2 phosphorylation is suppressed and the rate of translation remains high as long as heme is available. Vice versa, heme deficiency leads to a deblockade of HRI. This holds true also for other stress situations because HRI adjusts translation not only to the availability of heme and iron but also to the physiological situation in general. Such an adaptive response provides an economical advantage and, in addition, protects

Figure 9.3 eIF2 kinases as stress-activated translational repressors
The four kinases Heme-Regulated Inhibitor (HRI), Protein Kinase R (PKR), homolog of *Drosophila* protein General Control Non-derepressible 2 (GCN2), and Protein kinase R-related Endoplasmic Reticulum Kinase (PERK) respond to different stress factors and inhibit protein biosynthesis by phosphorylating a specific Ser residue of eukaryotic initiation factor 2 (eIF2).

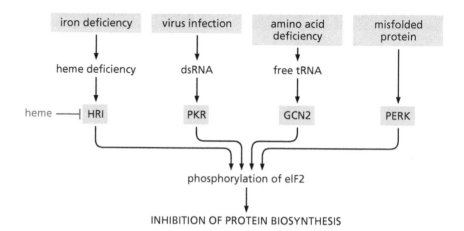

the chaperone system from an overload that would result in intoxication of the cell by incorrectly folded proteins (see Section 9.3.2).

Another eIF2 kinase isoform is **protein kinase R, PKR** (do not confuse with PKC-related kinase, PRK). It is closely related to HRI, but instead of a heme-binding motif it has an interaction domain for double-stranded RNA (therefore PKR) which activates the enzyme. Double-stranded RNA arises primarily in the course of viral infections. Therefore, PKR is a component of cellular defense: by turning down translation, it prevents the synthesis of viral proteins and spares the energy supply of the infected cell. Appropriately, the *de novo* synthesis of PKR is induced by antiviral interferons. Viruses have developed various mechanisms to outmaneuver PKR: for instance, by inhibitor proteins, by double-stranded RNA that is not recognized by the kinase, or by phosphatases that counteract the phosphorylation of eIF2.

The third eIF2 kinase isoform is a dimeric transmembrane protein of the ER called **GCN2** (referring to the homologous *Drosophila* protein General Control Non-derepressible 2). GCN2 is activated by unloaded tRNA and phosphorylates eIF2 at the same Ser residue as HRI and PKR. Since unloaded tRNA signals amino acid shortage, GCN2 helps the cell to adjust the rate of translation to the availability of amino acids (another mechanism aiming at this purpose is discussed in Section 9.4.1). At first glance, this adaptive reaction to starvation seems to resemble the stringent response of bacteria as explained in Section 3.1. It follows, however, another mechanistic route. Amazingly, GCN2 inhibits overall translation but, nevertheless, stimulates the translation of mRNA species encoding those transcription factors that activate the genes of amino acid biosynthesis and of the unfolded-protein response. The mechanism of this apparently paradoxical but physiologically meaningful effect is explained below. It seems to play an important role in memory fixation in the brain, where the phosphorylation of eIF2α serves as a molecular switch between short- and long-term memory (see Section 16.4.1).

The fourth eIF2 kinase isoform is the PKR-related ER Kinase **PERK**. Like GCN2, PERK is a transmembrane protein of the ER. The enzyme plays an important role in the so-called unfolded-protein response to be discussed in Section 9.3.2.

The stress-induced inactivation of eIF2 is augmented by an inhibition of eIF2B (the eIF2-GEF) that also is due to phosphorylation. The corresponding kinase is glycogen synthase kinase 3 (GSK3), which becomes activated under stress but inhibited by insulin (see Section 9.4.4). In fact, eIF2B phosphorylation is involved in a typical anabolic insulin response, the increase of protein biosynthesis along the phosphatidylinositol 3-kinase–protein kinase B (PI3K–PKB) signaling pathway. Insulin abolishes the effect of GSK3, thus facilitating the dephosphorylation (by omnipresent phosphatases) and reactivation of eIF2B.

9.3.2 Unfolded-protein response: how the endoplasmic reticulum copes with stress

In the ER, a group of chaperones takes care of the correct folding of proteins, since un- or misfolded proteins are not only useless but are toxic and possibly may kill the cell by evoking apoptosis (so-called proteotoxicity). The accumulation of such defective proteins in the ER may result in what is known as **ER stress**: an overload of the chaperone system. Such a situation may have several causes: overshooting protein synthesis, production of mutated proteins, viral infections, disturbed redox state (for instance, due to a defective blood supply or ischemia), disturbed Ca^{2+} homeostasis (the ER is a major Ca^{2+} store of the cell; see Section 14.5), and toxic substances inhibiting chaperone activity or blocking glycosylation of newly synthesized proteins. When confronted by ER stress, the cell is not helpless but makes use of a set of efficient repair and rescue mechanisms known as unfolded-protein response (UPR), which together with DNA repair represents the hard core of the cellular stress defense. The ultimate goal of the UPR is to increase the protein folding capacity of the ER by promoting the synthesis of chaperones.

The UPR involves the following sequence of events:

- recognition of the defective protein by a chaperone

- stop of counterproductive *de novo* synthesis of housekeeping proteins

- transcriptional activation of stress-responsive genes (for instance, those encoding chaperones)

- elimination of defective proteins along the ubiquitin–proteasome pathway or, if this turns out to be impossible, apoptosis

ER stress sensor BiP

The major stress sensor of the ER is the **chaperone BiP** (<u>B</u>inding <u>P</u>rotein). BiP is an ATPase of the HSP70 type, which, like other members of this family, interacts with an HSP90-type chaperone and various co-chaperones in order to bring its misfolded client proteins into the native conformation (for a more detailed treatment of chaperones see Section 2.5.1). The energy required for this process comes from ATP hydrolysis.

Under ER stress due to an overloading by client proteins, the stock of BiP is used up, resulting in a shortage of free BiP. This situation calls upon the cell to activate signaling cascades, which turn down translation and thus stop a continuous supply of new proteins to the chaperone system. Simultaneously, a large number of anti-stress genes become activated, which encode (among others) chaperones. In their regulatory sequences, these genes carry UPR or ER stress response elements that are recognized by three transcription factors called ATF4, ATF6, and XBP1.

The signaling cascades controlling these responses start from three transmembrane proteins of the ER. These proteins are inactivated when they bind BiP but become reactivated when BiP is intercepted by a surplus of misfolded proteins. The three BiP receptors are the eIF2 kinase **PERK** mentioned above, a serine/threonine (Ser/Thr)-specific **protein kinase IRE1** that is also an endoribonuclease, and the precursor protein of the **transcription factor ATF6** (Figure 9.4).

Effects on gene transcription and mRNA translation

In the case of ER stress, the kinases PERK and IRE1 are cleared from BiP and undergo oligomerization, facilitating a trans-autophosphorylation of their activation loops. While the phosphorylation and inactivation by PERK of eIF2 is the most rapid response to ER stress, IRE evokes a more delayed effect. Here

Figure 9.4 Signal transduction of the unfolded-protein response
Incorrectly folded proteins accumulating in the lumen of the endoplasmic reticulum sequester the chaperone BiP, thus rescuing the transmembrane protein kinases PERK and IRE1, and the ATF6 precursor from inhibition by BiP. This results in an activation of the transcription factors ATF4, XBP1 and ATF6 controlling the transcription of unfolded-protein response (UPR) genes, as well as in an inhibition of eIF2; that is, in an interruption of protein biosynthesis. The ATF6 precursor translocates from the endoplasmic reticulum to the Golgi apparatus, where it undergoes maturation by limited proteolysis catalyzed by proteases S1P and S2P. Dark gray cylinders, transmembrane domains; light gray cylinders, kinase domains; red cylinders, endoribonuclease domains; ellipses, active ATF6.

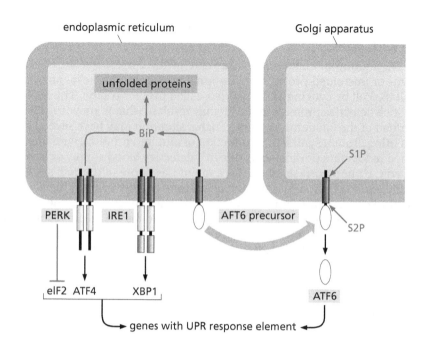

trans-autophosphorylation results in a stimulation of IRE's intrinsic endoribonuclease activity. This activity is used to release the mRNA of the **transcription factor XBP1** from a precursor mRNA. For this purpose, IRE1 cuts out an intron sequence of 26 nucleotides. After ligation of the remaining fragments, a reading frame for XBP1 synthesis arises. This signaling pathway is conserved in all eukaryotes and has been investigated in detail in yeast. Mammals express two IRE isoforms. The important role played by IRE1 is mirrored by gene knockout experiments: mice lacking an active *IRE1α* gene die before birth.

In its function, the **transcription factor ATF6** resembles XBP1, though it is activated more rapidly along quite a different pathway: by release from a precursor protein. This precursor is a transmembrane protein of the ER. Under stress-free conditions it forms a complex with BiP; upon ER stress, the complex dissociates because BiP becomes monopolized by misfolded proteins. As a result the ATF6 precursor is released from the ER and, guided by a specific Golgi localization motif, migrates to the Golgi apparatus, where it is cut by two proteases in the lumeinal domain and the transmembrane domain. This leads to a release of active ATF6 into the cytoplasm, where it migrates into the nucleus and activates a series of genes, including that of XBP1. Thus the activation of ATF6 provides an example of a signal-transducing reaction involving partial proteolysis (more in Section 13.1.4).

While IRE1 and ATF6 are responsible exclusively for the regulation of UPR-inducible gene transcription (genes containing an ER stress response element), the protein kinase PERK inhibits translation by selective eIF2 phosphorylation, thus in a most rapid way relieving the stressed chaperone system from the supply of new client proteins. Notwithstanding this inhibition of overall translation, PERK signaling leads to a *de novo* synthesis of the **transcription factor ATF4**, which controls the activity of a third of all UPR-inducible genes (including those of XBP1 and ATF6). This selective translation is due to a special structure of ATF4 mRNA consisting of several upstream open reading frames in the 5′-region that are not in frame with the ATF4 sequence, the translation of which is thus prevented. This inhibitory effect is relieved by phosphorylation of eIF2, which retards the formation of a ribosomal initiation complex. Consequently, the scanning ribosome skips the open reading frames, becoming active only when it has arrived at the ATF4 sequence. In addition to its function in the unfolded-protein response, the eIF2–ATF4 system plays an important role in brain neurons, where it regulates memory fixation (see Section 16.4.1).

The eIF2 kinase GCN2 activated upon amino acid deficiency resembles PERK in activating ATF4 translation though inhibiting overall translation. Moreover, at least in insects, GCN2 also stimulates the translation of mRNAs encoding transcription factors that regulate genes of amino acid synthesis. This combined action of the two eIF2 kinases, PERK and GCN2, is known as the **integrated stress response**.

PERK not only is an emergency enzyme but also controls the equilibrium between the rate of translation and the efficacy of chaperones under normal conditions. As an example, the β-cells of the Langerhans islets of the pancreas may be mentioned, where PERK signaling harmonizes the glucose-induced synthesis of pro-insulin with protein folding in the ER. In fact, PERK mutations are one of the various causes of diabetes.

Adaptation and apoptosis

The unfolded-protein response is limited in time because the cell adapts to the stress situation. This may occur, for instance, by providing more chaperones or by the destruction of the misfolded proteins along the ubiquitin–proteasome pathway. For this waste disposal, known as **ER-associated degradation**, the proteins have to be re-translocated through the translocons from the ER into the cytoplasm where the proteasomes are located. Whether or not a protein is recognized by the system seems to depend on its glycosylation status.

Another possibility to adapt is by signal extinction. In fact, one of the UPR genes encodes a protein, GADD34, which is a regulatory subunit of the protein phosphatase PP1C directing the enzyme to the phosphorylated eIF2. The degradation of the extremely protease-sensitive transcription factors of the UPR also aims at signal extinction.

Like genotoxic stress, ER stress may finally culminate in programmed cell death (apoptosis). To this end, several signaling pathways connect the unfolded-protein response with the apoptotic machinery, which in fact is partially localized in the ER. The ER-bound caspase 12 is activated by IRE1, triggering the caspase cascade (see Section 13.2.1). Moreover, one of the UPR-inducible genes produces a protein that strongly stimulates apoptosis. Finally, by interacting with the adaptor protein TRAF2, IRE1 may activate the pro-apoptotic protein kinase ASK1. Being a MAP3 kinase, ASK1 is a component of the JNK–MAP kinase module, which typically becomes activated in stress situations and induces apoptosis (see Section 11.1). Apoptosis is also caused by an overload of the ER with *correctly* folded proteins; for instance, upon viral infection. In this case, however, the NFκB signaling cascade rather than the UPR is activated (Section 11.8).

Cytosolic protein quality control

Protein quality systems such as those evoking the UPR are not restricted to the ER but also are found in the cytosol, nucleus, and mitochondria. They are hard-working devices, since it is estimated that about a third of all newly synthesized proteins are defective and, thus, have to be eliminated. Such defects are due to unavoidable errors of translation as well as all sorts of stress, first and foremost reactive oxygen species, which are a major risk factor of aerobic life. The cytosolic sensors of damaged or incompletely folded proteins are again BiP-related chaperones of the HSP70 and HSP90 families. Under stress-free conditions they sequester the transcription factor **HSF1** (Heat-Shock transcription Factor 1), keeping it inactive. Defective proteins intercept the chaperones, and HSF1 now can enter the nucleus to activate (similar to ATF4, ATF6, and XBP1) the transcription of stress-responsive genes.

How does such a system distinguish defective from nascent polypeptide chains? The answer to this question is found probably in a set of ubiquitin ligases (such as the Carboxyterminus of HSP70 Interacting Protein, CHIP), which selectively interact with chaperones bound to misfolded proteins. When the stress has ceased, the same ubiquitin ligases also trigger the destruction of the chaperones, thus reestablishing the normal situation.

9.3.3 Translation factors eIF4E and eIF2B: targets of mitogens and insulin

In addition to eIF2, several other factors controlling the initiation of polypeptide synthesis are targets of regulatory signals transduced by protein phosphorylation. As mentioned above, **eIF2B**, the GEF of the GTPase eIF2, is inactivated by GSK3. This inhibitory effect augments the suppression of translation by eIF2 phosphorylation. It is overcome by mitogens and insulin, which inhibit GSK3 along the PI3K–PKB pathway (see Section 4.6.6).

eIF4E is the subunit of the cap-binding complex that directly interacts with the m^7-GpppX sequence of mRNA. Stimulatory hormones and growth factors induce a phosphorylation of the factor, causing the detachment of the cap-binding complex from mRNA and an increase of translational efficiency (Figure 9.2). While the mechanism of this effect is still not clear, the protein kinase catalyzing it has been identified as MNK1, an effector kinase of the mitogenic MAP kinase module (Section 11.6.2). Genes encoding key proteins of cell cycle progression, such as cyclin D, are activated along this pathway. Through eIF4E phosphorylation, mitogenic signals turn up this effect by stimulating the translation of the corresponding mRNA species (Section 11.6.2). Because of these pro-mitogenic effects, eIF4E may promote tumor development. In fact, human tumors frequently show high levels of eIF4E, and in cell culture and transgenic animals, overexpression of this factor facilitates neoplastic transformation and tumorigenesis.

An opponent of eIF4E is the **translational repressor 4E-BP** (eIF4E-Binding Protein), which exists in several tissue-specific isoforms. 4E-BP sequesters eIF4E, thus preventing formation of the initiation complex. Exogenous signals that induce eIF4E phosphorylation also inactivate 4E-BP through phosphorylation, which in this case is catalyzed by the protein kinase mTOR. As shown in Section 9.4.1, this mechanism plays a particularly important role in the control of protein synthesis.

9.3.4 eEF2 kinase: a repressor of peptide chain elongation

Apart from the initiation phase, peptide chain elongation is a target of signaling reactions. In this context, the eEF2 kinase plays a key role. This Ser/Thr-specific enzyme, found in all eukaryotic cells, is activated by Ca^{2+} ions interacting with the multifunctional Ca^{2+}-binding protein calmodulin (CaM) that is a subunit of eEF2 kinase. In the literature eEF2 kinase may be found, therefore, also under the name CaM kinase III, although it does not relate to the CaM kinases proper but belongs to the atypical protein kinases (see Section 2.6.3 for a classification of protein kinases). eEF2 kinase is as specific as eIF2 kinase: the only substrate known is the elongation factor eEF2 that, upon phosphorylation of a Thr residue, loses contact with the ribosome, resulting in premature termination of peptide chain elongation.

The activity of eEF2 kinase is controlled by phosphorylation according to the principle of two reins: depending on the kinase type and the site of phosphorylation, it is either inhibited or activated. Signals driving up the rate of translation *suppress* eEF2 kinase activity along a kinase cascade. Such signals include growth factors as well as insulin and other hormones. The kinase cascade proceeds mainly along the axis PI3K–PKB/Akt–mTOR (Figure 9.2) and is discussed below in more detail. Other kinases inhibiting eEF2 kinase are the ribosomal S6 kinases (see below) as well as some MAP kinases.

An *activating* phosphorylation of eEF2 kinase is catalyzed by the 5′-AMP-dependent protein kinase AMPK (Figure 9.2 and Section 9.4.3). By this mechanism, protein biosynthesis, as a major ATP-consuming process, is suppressed in situations of ATP deficiency such as starvation and hypoxia.

An inhibition of peptide chain elongation also occurs upon phosphorylation of another elongation factor, the GEF eEF1B, catalyzed by the cyclin-dependent kinase CDK1. This reaction has been connected with the suppression of protein synthesis observed at the beginning of mitosis, a phase of the cell cycle that is controlled by CDK1 (see Chapter 12).

Summary

Protein biosynthesis, one of the most energy-consuming cellular processes, is under tight control by signaling cascades that adjust mRNA translation to cell growth and proliferation and prevent wasted energy in emergency situations. Ser/Thr-specific protein kinases HRI, PKR, GCN2, and PERK repress mRNA translation by phosphorylating and inactivating eukaryotic initiation factor 2 (eIF2). These highly specific enzymes are activated in stress situations, such as iron and amino acid deficiency, virus infection, and accumulation of incorrectly folded proteins. To protect the cell from damage by proteotoxicity, an ER-based signaling mechanism monitors any accumulation of incorrectly folded proteins. The sensing mechanism consists of sequestering the chaperone BiP, resulting in activation of the protein kinases PERK and IRE1 as well as of the transcription factor ATF6 precursor protein in the ER membrane. Consequently, mRNA translation becomes inhibited by the eIF2 kinase PERK and a series of genes are activated that encode (among others) chaperones. In the cytosol, defective proteins activate heat-shock transcription factor 1 (HSF1), which also stimulates stress-responsive genes. If these countermeasures turn out to be in vain, apoptosis is induced. Mitogens and insulin stimulate elongation factors eIF2B and eIF4E, thus adjusting the rate of protein synthesis to the rate of cell proliferation and cell growth. While eIF2B is activated by protein kinase B-catalyzed inactivation of the eIF2B inhibitor glycogen synthase kinase 3 (GSK3), eIF4E becomes phosphorylated directly through MAP kinase signaling, in parallel with inactivation of the eIF4E repressor 4E-binding protein (4E-BP). The highly specific eukaryotic elongation factor 2 (eEF2) kinase is a repressor of peptide elongation. The enzyme is itself a target of protein phosphorylation, adjusting its activity to the physiological context. Depending on the phosphorylation site, it is either inhibited (by mitogenic signals and insulin) or activated (under stress situations such as starvation).

9.4 Network for adjustment of cell growth to the supply situation

Signaling systems controlling ribosomal protein synthesis fulfill an important task: they adjust the rate of translation and thus the state of vitality and the growth rate of cells to the food and energy supply. As discussed above, this system responds with economic measures to stress factors overstraining those parameters. For this purpose, it makes use primarily of protein phosphorylation, which is catalyzed by kinases serving as sensors of energy and nutrients. Disturbances of this network rank among major causes of metabolic diseases, particularly obesity and diabetes.

9.4.1 TOR kinases: sensors of food supply control protein biosynthesis

Mammalian target of rapamycin

A key signaling mechanism controlling protein synthesis in response to the state of nutrition has been discovered during studies both on yeast cells and on a particular neoplastic disease.

Hamartomas are benign tumors, which may arise in various tissues. The tumor cells are unusually large and obviously overfed. Hamartomas are induced by the deletion of two tumor suppressor genes encoding the so-called tuberous sclerosis proteins TSC1 (alias hamartin) and TSC2 (alias tuberin). Both together inhibit a signaling cascade, which adjusts the rates of cell growth and cell proliferation to food and energy supply. In this signaling pathway, a central role is played by the Ser/Thr-specific protein kinase mTOR, an enzyme conserved from yeast to mammals in all eukaryotes examined. mTOR stands for **m**ammalian **T**arget **O**f **Rapamycin**. Rapamycin is an immunosuppressive antibiotic from *Streptomyces hygroscopicus* that in cells interacts with the immunophilin FKBP12, a peptidyl-prolyl *cis–trans* isomerase (see Section 8.5). The complex of rapamycin and FKBP12 is a highly specific inhibitor of mTOR. It reduces both cell size and cell proliferation, mainly by inhibiting ribosomal protein synthesis. One of the reasons for the immunosuppressive effect of rapamycin is seen here: the drug suppresses the activation and proliferation of T-lymphocytes.

mTOR signal transduction

The protein kinase mTOR is a key regulator of cell growth. The enzyme (higher vertebrates contain only one *mTor* gene) belongs to the atypical PI3K-like protein kinases. In contrast to PI3K, it does not exhibit lipid kinase activity but is a pure protein kinase. Nevertheless, mTOR connects PI3K signaling with the regulation of protein synthesis (Figure 9.5).

mTOR stimulates mRNA translation. Major substrates of the kinase include the ribosomal S6 kinase type p70, 4E-BP1, and (indirectly) eEF2 kinase (Figure 9.2). As shown in Section 9.4.2, S6 kinase promotes mRNA translation and cell survival and is activated by mTOR, whereas the translational repressors 4E-BP1 and eEF2 kinase become inactivated.

mTOR also promotes the formation of ribosomes. An important condition of cell growth and proliferation is to adjust the number of ribosomes to the actual physiological requirements. In rapidly growing and proliferating cells, for instance, up to 8000 new ribosomes are produced per minute, whereas in resting or stressed cells, ribosome formation almost comes to a standstill. Ribosome synthesis requires the translation of a special family of mRNAs encoding all ribosomal proteins and several translation factors. The peculiarity of these mRNA species is that they carry a 5′-TOP (**T**ract **O**f **P**yrimidines), an oligopyrimidine sequence adjacent to the 5′-terminal m^7-GpppX-cap. When cells are deprived of energy and amino acids, as well as in the absence of mitogenic and growth-promoting signals (or when treated with inhibitory drugs), polysomes (mRNA occupied by actively translating ribosomes) with 5′-TOP-mRNA disintegrate. As a result, formation of ribosomes is blocked and overall protein synthesis comes to a halt, a reaction strongly reminiscent of the stringent response of bacteria (see Section 3.1). Certainly, the mTOR signaling pathway is critically involved in the control of 5′-TOP-mRNA translation. However, the precise mechanism is still a matter of conjecture.

Through its effects on the ribosomal machinery, mTOR considerably increases the translational capacity of the cell, provided that the conditions are favorable. To this end, the mTOR system not only is under the control of signaling factors but, in addition, monitors the availability of nutrients and metabolic energy. mTOR is, in particular, a sensor for glucose and amino acids, with leucine being a stimulatory key signal, and it measures the energy status of the cell in two ways. First, since its K_m value for ATP is 5–10 times higher than that of other protein kinases, mTOR becomes active only at a particularly high concentration of intracellular ATP. Second, mTOR activity is suppressed by the kinase AMPK, a cellular energy sensor that becomes active at a low ATP:AMP ratio indicating energy shortage (see below). Of course, monitoring both nutrient availability and energy status makes sense, since without a sufficient energy supply any activation of

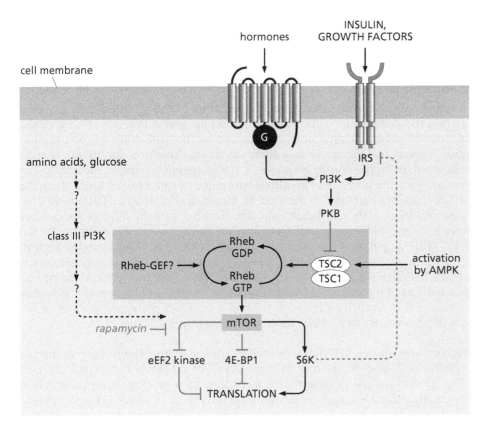

Figure 9.5 Activation and effects of the mTOR protein kinase By inactivating the GAP TSC2 of the small G-protein Rheb, extracellular signals stimulating the PI3K–PKB signaling cascade prompt Rheb to activate mTOR. mTOR enhances the activity of the protein kinase S6K and represses 4E-BP1 and eEF2 activities, resulting in an increased rate of translation (whether 4E-BP1 and eEF2 kinase are phosphorylated directly by mTOR, as shown here, or by S6K or by both kinases is not entirely clear). mTOR may also be directly phosphorylated and activated by PKB (not shown). A stimulatory effect resembling that of PKB has the MAP kinase ERK connecting mTOR signaling with mitogenesis (not shown). mTOR is also activated by nutrients such as amino acids and sugars along an ill-defined pathway that seems to include a class III PI3K. The red dotted line shows the negative feedback of insulin signaling: S6K phosphorylates and inactivates the insulin-specific docking protein IRS. This effect is augmented by overnutrition (leading to increased insulin release) and provides one of the causes of diabetes. Also shown is the *activation* of the Rheb-GAP TSC2 by 5'-AMP-dependent protein kinase (AMPK) that results in an inhibition of mTOR signaling and protein synthesis and protects the cell in situations of energy deficiency (see Section 9.4.3).

protein biosynthesis would be idling (the reader may recall that about 80% of cellular energy is used for cell growth rather than cell proliferation, being consumed primarily by protein synthesis).

The pathway along which amino acids and glucose stimulate mTOR is still not fully known. It seems to include a class III PI3K (known as hVPS34) that differs from the conventional class I PI3Ks by a lack of a Ras-binding domain and an altered substrate specificity (preferring phosphatidylinositol instead of phosphatidylinositol 4,5-bisphosphate).

Better known is the pathway along which hormones and growth factors activate mTOR. Here the Ras-related small G-protein **Rheb** (Ras homology enriched in brain) plays a key role as an immediate mTOR activator, provided it is loaded with GTP. Therefore, stimulatory signals are expected either to promote the GDP–GTP exchange by activating a still ill-defined Rheb-GEF or to stabilize the Rheb–GTP complex by inhibiting a Rheb-GAP, whereas inhibitory signals have an opposing effect. At this stage, one of the TSC tumor suppressor proteins, **TSC2**, enters the scene: it acts as a Rheb-GAP and has turned out to be a key switch in mTOR signaling (Figure 9.5).

The function of TSC2 is regulated by protein phosphorylation, which may be either stimulatory or inhibitory depending on the kinase and the site of phosphorylation. Thus, TSC2 is *inhibited* (and mTOR activity and protein synthesis are stimulated) by multiple phosphorylation catalyzed by **PKB/Akt** or **ERK**, the executor kinase of the mitogenic Ras–RAF–MAP kinase module. Both kinases are activated simultaneously by insulin, growth factors, and many other hormones and mediators that transduce anabolic, growth-promoting, and mitogenic signals. However, since mTOR signaling resembles a logical AND gate, these signals become active only when amino acids and ATP are available in sufficient concentrations. In fact, TSC2 becomes *stimulated* (and protein synthesis inhibited) by a shortage of ATP energy, turning down Rheb activity and mTOR signaling.

This is where **AMPK** comes into play: it activates the GAP activity of TSC2 by phosphorylating Ser and Thr residues that differ from those phosphorylated by PKB/Akt or ERK.

mTOR and disease

In summary, this positive and negative regulation of the mTOR signaling system enables the cell to adjust the effects of anabolic and mitogenic stimuli to the capacity of protein biosynthesis determined by the availability of food (amino acids, sugars) and energy. It is easily understood, therefore, that mTOR is critically involved in the control of anabolic hormone effects and in metabolic diseases. In fact, the function of **insulin** is controlled by negative feedback along the mTOR signaling pathway. To this end, S6 kinase activated by mTOR phosphorylates and inactivates the insulin-specific docking protein IRS1 at several Ser residues, thus blocking signal transduction from the insulin receptor to the PI3K–PKB/Akt pathway (Figure 9.5; more details about IRS1 are found in Section 7.1.3). With an increasing supply of fat and carbohydrates and, as a consequence, a rising insulin level, this inhibitory effect becomes more pronounced, being further reinforced by an adaptive down-modulation of insulin receptors. This may explain why overnutrition is one of the major causes of **diabetes.** Another cause, lack of physical activity, will be discussed in Section 9.4.3.

As mentioned above, mTOR is inhibited by clinical immunosuppressants such as rapamycin. Moreover, its role as a promoter of diabetes, tumorigenesis, and other growth-related diseases (such as cardiac hypertrophy) makes the mTOR system an attractive target for further inhibitory drugs that are in ongoing clinical trials.

Outlook

The story of mTOR is not at an end yet. The active mTOR kinase forms heterotrimeric complexes with regulatory proteins. It has been found that the substrate specificity of mTOR depends on the individual composition of such a complex: translation (and thus cell size) is selectively controlled by the complex **mTORC1** (containing the regulatory protein raptor, regulatory associated protein of mTOR) interacting with Rheb, whereas the complex **mTORC2** (containing the regulatory protein rictor, rapamycin-insensitive companion of mTOR) activates small G-proteins of the type Rho and Rac controlling the dynamics of the cytoskeleton (and thus cell shape and cell movements). For a still-unknown reason, rapamycin inhibits mTOR only in mTORC1 but not in mTORC2. Probably further complexes and, therefore, further functions of mTOR will be discovered in the future.

9.4.2 Ribosomal S6 kinases: regulators of cell size and proliferation

In the preceding section, a ribosomal S6 kinase was introduced as a key substrate and effector protein of mTOR. Ribosomal S6 kinase is a collective term for two kinase families. It refers to the substrate protein S6, a component of the small ribosomal subunit, rather than to a location of these enzymes in ribosomes. In fact, the S6 kinases are found at membranes, in the cytoplasm, and in the nucleus and have functions going far beyond S6 phosphorylation.

Kinases p70^{S6K}

The ribosomal S6 kinase positioned downstream of mTOR belongs to the p70^{S6K} family. Mammalian cells express this type in two isoforms, with S6K1 being the S6 kinase proper. By up-regulating protein synthesis, the enzyme acts as an important positive controller of cell size and body growth. The corresponding gene-knockout animals are, indeed, significantly smaller than wild type. S6 kinase enhances translation by inactivating eEF2 kinase and activating the initiation factor eIF4B. Moreover, its many potential substrates include the insulin receptor substrate IRS1 (see above) as well as transcription factors and the pro-apoptotic protein Bad, indicating a much wider functional spectrum.

Until recently, phosphorylation of the S6 protein was thought to be essential for the translation of 5′-TOP-RNA required for ribosome synthesis. However, experiments with S6- and S6 kinase-knockout animals have raised doubts about this concept. As a control device for ribosome synthesis, the PI3K–PKB–mTOR cascade is, nevertheless, the prime candidate. However, the molecular mechanism behind this function is still largely mysterious, and the same holds true for the physiological role of S6 protein and the stimulatory effect of S6 phosphorylation on cell growth.

How is the activity of S6 kinase regulated? As we have seen, it has to be phosphorylated by mTOR to become active. This phosphorylation takes place in the N-terminal domain containing an mTOR signaling motif and abolishes the intrasterical auto-inhibition of the enzyme. However, since S6 kinases belong to the AGC kinase family, they also have to be phosphorylated at the activation loop. As for most other AGC kinases, this is catalyzed by the membrane-bound phospholipid-dependent kinase PDK1 (Section 4.6.4), encountered by S6K upon joining signal-transducing multiprotein complexes that assemble around transmembrane receptors. Thus, in principle the activation of S6 kinase resembles that of PKB/Akt as shown in Figure 4.43. Most interestingly, PDK1 and mTOR are S6 kinase substrates themselves. While the physiological consequences of mTOR phosphorylation are still elusive, PDK1 phosphorylation seems to speed up S6K activation by positive feedback.

Kinases p90^RSK

The second family of so-called S6 kinases are the ribosomal S6 kinases of the p90 type, p90^{RSK}. To distinguish them from S6Ks, they are called RSKs. Because they are also members of the AGC kinase superfamily, RSKs are related to the p70 S6Ks. For full activation, RSKs (like the p70 S6Ks) have to be phosphorylated by at least two different protein kinases: PDK1 and MAP kinases such as ERK or p38. Therefore, the RSKs are counted among the large group of MAP kinase-activated protein kinases (MAPKAPs) and are considered to be important links between the MAP kinase modules and the genome (for more details see Section 11.6). Like S6Ks, RSKs also phosphorylate PDK1 in a positive feedback loop and phosphorylate the S6 protein. However, S6 phosphorylation does not seem to be their primary function. Instead, RSKs act mostly at the level of gene transcription and their preferred substrates are transcription factors such as C/EBP, CREB, SRF, ATF1, and the estrogen receptor, as well as the transcriptional co-activator CBP/p300 and histone 3 (Figure 9.6). Moreover, RSK species counteract apoptosis by phosphorylating (like S6K) the pro-apoptotic protein Bad (Section 13.2.4), and they promote cell division by inactivating the cell cycle-inhibiting protein kinase MYT1 (Section 12.4). Finally, RSKs activate mTOR signalling, probably by inactivating (like PKB/Akt) the Rheb GAP TSC2. This provides an additional pathway along which growth factors stimulate protein synthesis to cope with cell proliferation.

9.4.3 AMP-dependent protein kinases: sensors of cellular energy status

In the preceding section it was shown that mTOR signaling is inhibited by 5′-AMP-activated protein kinase AMPK (not to be confused with cAMP-activated kinases!). This is a key event for translational control because the AMPK family has turned out to be a master regulator of energy metabolism in all eukaryotes studied thus far.

As we have repeatedly emphasized, movements, signal processing, and anabolic reactions such as protein biosynthesis are expensive cell physiological processes, critically depending on the energy status of the cell and specifically on the ATP

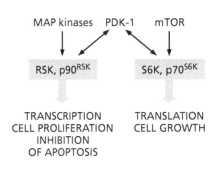

Figure 9.6 Activation and major functions of ribosomal S6 kinases
Double-headed arrows indicate mutual activation of the protein kinases PDK1 and RSK/S6K.

level. In fact, as long as it is alive, a cell maintains a ratio of ATP to ADP/AMP concentration that is several orders of magnitude above the chemical equilibrium, resembling a battery charged by chemical (ATP) energy (Section 2.1).

The state of charging is controlled by AMPK: the enzymes of this kinase family act as universal energy sensors, monitoring the ATP to 5′-AMP ratio and also the level of the energy store glycogen. When they recognize a decrease of ATP or glycogen accompanied by a corresponding increase of AMP, AMPKs turn down the ATP-consuming anabolic metabolism (synthesis of macromolecules, fatty acids, cholesterol, etc.) and, instead, stimulate ATP-delivering catabolic reactions (glycolysis, fatty acid oxidation, etc.) and – especially in muscle cells – glucose uptake. This occurs on the levels of transcription, translation, and metabolism.

Activation of AMP-dependent protein kinases

AMPKs are allosterically activated by 5′-AMP. They are heterotrimeric proteins consisting of a catalytic α-subunit and two regulatory subunits labeled β and γ (Figure 9.7). Humans possess two α-, two β-, and four γ-subunits that are tissue-specifically expressed. In addition, 12 AMPK-related protein kinases with mostly still ill-defined functions have been identified by screening of the human genome.

The activity of AMPK is regulated by an interaction between the three subunits. The α-subunit of AMPK harbors a typical Ser/Thr-specific protein kinase domain. It is activated by a combination of an allosteric effect with phosphorylation of the activation loop, a mechanism that is typical for most protein kinases. Activation loop phosphorylation occurs only in the presence of sufficient 5′-AMP, while in the absence of the nucleotide the kinase is rapidly dephosphorylated and blocked by auto-inhibitory sequences found both in the α- and γ-subunits. This blockade is relieved by a change of the quaternary structure induced when 5′-AMP binds to the γ-subunit. The protein kinase catalyzing activation loop phosphorylation is identical with the tumor suppressor protein LKB1 (see below).

ATP and glycogen prevent AMPK activation. In fact, AMPK (like many other protein kinases) combines the properties of a logical AND gate with those of a logical BUT NOT gate: it is activated by phosphorylation *and* binding of AMP *but not* by phosphorylation and an interaction with ATP or glycogen.

Multiple binding sites of 5′-AMP are located in the γ-subunit. At a high ATP:AMP ratio, ATP displaces AMP from these sites, thus blocking the activation of AMPK. A specific binding domain for glycogen is found in the β-subunit, which also serves as an adaptor for the two other subunits. Due to a pronounced interaction between the subunits, the enzyme is subject to efficient allosteric regulation, showing very steep sigmoidal kinetics and functioning as both a switch and a strong noise filter.

LKB1: a protein kinase suppresses tumor growth

The Ser/Thr-specific protein kinase LKB1, now considered to be the major receiver of input signals controlling AMPK activity, was discovered in the course of a genetic analysis of **Peutz–Jeghers syndrome**. Patients suffering from this relatively rare hereditary disease develop a large number of tumors in the gastrointestinal tract and in other organs that, although they are benign in the beginning, have a pronounced tendency to become malignant. Tumorigenesis was causally related to an *inactivation* of the *lkb1* gene, characterizing LKB1 as a tumor suppressor protein. This was a rather unexpected finding since protein kinases were thought to promote tumor formation through *overactivation* and mutation to hyperactive oncoproteins. LKB1 however inhibits cell proliferation, for instance, by inducing synthesis of the cell cycle inhibitor p21[WAF1] (Section 12.5). In

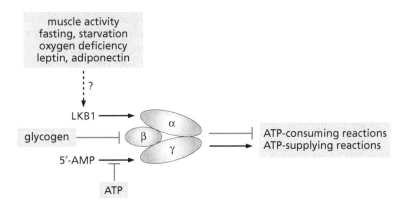

Figure 9.7 5'-AMP-dependent protein kinase AMPK Shown are the three subunits of AMPK, which respond to different input signals. AMPK inhibits ATP-consuming and stimulates ATP-supplying metabolic reactions. The mechanism of activation of the AMPK kinase LKB1 is not clear but may involve protein phosphorylation.

addition, LKB1 may stimulate the expression of the phosphatidylinositol 3-phosphatase PTEN, another abundant tumor suppressor protein that counteracts the effects of both PI3K and PKB/Akt. Such effects easily explain the tumor suppressor function of LKB1. Moreover, since LKB1 is an AMPK activator, its deletion leads to a disequilibrium between anabolic and catabolic metabolism. This situation results in a pathological overproduction of proteins and glycogen, a situation that promotes tumor development. The benign tumors of Peutz–Jeghers syndrome are indeed hamartoma-like with very large and overfed cells. Thus, the down-regulation of LKB1–AMPK signaling has the same consequence as an up-regulation of mTOR signaling: both produce tumors of a similar phenotype. Such an observation is understandable when the antagonism between LKB1–AMPK and mTOR signaling is considered.

LKB1 is only distantly related to other eukaryotic protein kinases. Not much is known about its regulation, in particular since it does not contain any of the known interaction domains. In fact, unlike most kinases, LKB1 appears to be constitutively active. Thus, AMPK phosphorylation by LKB1 most probably is controlled by 5'-AMP, which by inducing a conformational change converts AMPK into an LKB1 substrate. On the other hand, LKB1 becomes phosphorylated at several Ser and Thr residues, both by autophosphorylation and by kinases such as PKA, ERK, p90-RSK, and ATM (see Section 12.9); therefore, LKB1 might mediate certain anti-proliferative effects of these enzymes and the corresponding signaling pathways.

Apart from acting as a tumor suppressor, LKB1 plays an essential role in embryonic development, since mice carrying a homozygous knockout of the *lkb1* gene die before birth. Also in *Drosophila* and *C. elegans*, LKB-related kinases control developmental processes. In this context, regulation of cell polarity, such as the formation of axons and dendrites in neurons, is a major function of LKB1. Obviously, this effect is mediated by several AMPK-related kinases rather than by AMPK. Thus, LKB1 seems to be a master kinase controlling the activity of a whole series of protein kinases, a role reminiscent of that of the lipid-dependent kinase PDK1 with respect to the AGC kinases (see Section 4.6.4).

AMP-dependent protein kinase signaling in obesity, diabetes, and cancer

AMPK is an inhibitor of mRNA translation: it attenuates mTOR signaling by stimulating the Rheb GAP TSC2 and suppresses peptide chain elongation by activating eEF2 kinase. But AMPK can do much more. On the metabolic level, it phosphorylates and inhibits key enzymes of anabolic metabolism such as acetyl-CoA carboxylase 1 of fatty acid synthesis, 3-hydroxy-3-methylglutaryl-CoA reductase of cholesterol biosynthesis, glycerophosphate acyltransferase of triglyceride synthesis, and muscle glycogen synthase. Vice versa, AMPK activates proteins of catabolic metabolism such as the glucose transporters GluT1 and GluT4 and the enzyme phosphofructokinase. Translocation of GluT4 to the cell membrane is the key event in the regulation of glucose uptake by muscle and fat cells and a major target of insulin. As explained in Section 7.1.3, GluT4 translocation is prevented by the regulatory protein AS160, a GAP of a small G-protein of the Rab family. This inhibitory effect is abolished by phosphorylation of AS160, catalyzed either by insulin along the PI3K–PKB/Akt pathway or by AMPK. Thus, the effects of insulin and AMPK complement each other, and sufficient activation of AMPK (for instance by physical activity) may save insulin, an effect that is highly important for prevention of diabetes and other metabolic diseases (see below).

AMPK also regulates metabolism at the level of gene transcription. In yeast, for instance, the AMPK-homologous kinase SNF1 stimulates the transcription of genes that are repressed at a high supply of glucose (note the analogy to catabolite repression of prokaryotes), and in vertebrates no less than 370 genes have been identified whose transcription is influenced directly or indirectly by AMPK. In view of this functional versatility, it is conceivable that AMPK is stimulated by a wide variety of physiological and pathological stress factors such as fasting, starvation, oxygen deficiency, ischemia, intoxication, heat shock, hyperosmosis, bodily activity, and that disturbances of AMPK signaling are directly involved in metabolic disorders such as diabetes and obesity.

In this context, two proteohormones activating AMPK signaling deserve special attention. These are the so-called adipokinins **leptin** and **adiponectin**. Both are released by adipose tissue in response to excessive fat storage and play a significant role in the regulation of body mass. They do this by promoting the hypothalamus to take measures against an overload of the body's energy stores and by suppressing insulin secretion in the pancreas. Leptin, a 16 kDa peptide, has the structure of a cytokine and its receptors belong to the family of tyrosine (Tyr) kinase-coupled cytokine receptors of class I. As such, they induce various cellular reactions along the JAK–STAT and MAP kinase signaling pathways (Section 7.2.1). However, the mechanism by which leptin activates the LKB1–AMPK axis is still not known. The leptin receptors are expressed in many tissues, in particular, however, in certain cell types of the hypothalamus.

By stimulating hypothalamic functions, leptin suppresses the appetite, reduces the mass of adipose tissue, and activates the overall energy consumption of the body. To this end, it turns off the effects of the hypothalamus hormone neuropeptide Y that evokes the feeling of hunger and promotes anabolic metabolism. Moreover, leptin reduces the sensitivity of sweet-tasting sensory cells, thus rendering sweets less attractive (Section 15.1). Since sexual maturation depends on the state of nutrition, leptin has been proposed also to act as a triggering signal of puberty. In this respect it again functions as an opponent of neuropeptide Y, which retards sexual maturation by inhibiting gonadotropin release.

Disturbances of the leptin/adiponectin system are a major cause of **obesity**. In experimental animals, such a condition can be evoked by a loss-of-function mutation of leptin receptor genes, whereas in humans, an overload of the system due to excessive intake of high-energy food may result in leptin resistance due to adaptive down-regulation of the receptors. Since the cellular transduction of the leptin signal rather than leptin release seems to be impaired, the hope of curing obesity by leptin preparations has turned out to be somewhat premature.

Endogenous antagonists of leptin and adiponectin are the hormones **resistin** from fat cells, **ghrelin** produced by the digestive tract, and first and foremost, **insulin.** They all activate the anabolic metabolism, resulting in glycogen and fat deposition and an increase of protein synthesis and body mass (for a more detailed discussion of insulin signaling, see Section 7.1.3). It is assumed that leptin/adiponectin and insulin control each other by negative feedback along an "adipoinsulinic axis." For an obese person this has a serious consequence: the overproduction of leptin/adiponectin going hand in hand with adipokinin resistance dramatically diminishes both insulin production and insulin sensitivity.

Such a condition is **type 2 diabetes**, by far the most abundant metabolic disease, and its precursor stage, the so-called metabolic syndrome (in contrast, type 1 diabetes is caused by autoimmune destruction of insulin-producing β-cells in the pancreas). One cause of insulin resistance is considered to be a pathological deposition of fat in skeletal muscles and other organs, which massively impairs carbohydrate metabolism and causes down-regulation of insulin receptors. Since skeletal muscle is the tissue with the highest consumption of sugar, this lipotoxic effect has disastrous consequences.

Pathological fat and glycogen deposition is counteracted by the AMPK signaling system as well as by stress hormones. In fact, AMPK is activated by **metformin**, the most widely used antidiabetic drug. Moreover, as mentioned above, AMPK is stimulated not only by leptin but also by physical activity and other stress factors. In such situations the (nor)adrenaline level is also increased, speeding up lipolysis and glycogenolysis along the G_S–cAMP–protein kinase A pathway (see Section 5.3 for details). This easily explains why diabetes is promoted by obesity as well as by a lack of exercise, whereas physical activity and a reduction of body mass have a beneficial effect.

Diabetes and obesity clearly increase the risk of certain types of cancer such as mammary, colorectal, and prostate carcinomas, and again a diminished supply of calories as well as physical exercise have been unequivocally shown to exhibit a preventive effect. One of the reasons may be again the activation of AMPKs, which have been shown to suppress the formation of tumor-promoting growth factors such as insulin-like growth factor 1 (Section 7.1.3).

Diabetes and hexosamine signaling

The strong increase of blood glucose (hyperglycemia) in the course of type 2 diabetes evokes a series of feedback reactions that finally end up in insulin resistance of target cells, in particular in fat, liver, and muscle tissue, and as a consequence in long-term tissue damage. One reason for this glucose toxicity is the generation of highly reactive sugar metabolites that inactivate or even destroy essential cellular components (see Section 7.3.3). Another reason seems to be hyperactivation of a novel signaling pathway called hexosamine signaling, that was discovered in studies on **glucose-induced insulin resistance**. The key observation was that, in insulin target cells, this dreaded consequence of type 2 diabetes correlates with a characteristic post-translational modification of a wide variety of proteins: covalent binding of **N-acetylglucosamine (GlcNAc)**.

Sugar residues bound covalently at hydroxyl or amino groups of proteins are an abundant modification yielding glycoproteins. Most transmembrane proteins, for instance, are glycosylated at the extracellular domains. This is crucial for both their correct positioning and their function. The special thing about the binding of GlcNAc is that it fulfils the conditions of a signaling reaction possibly rivaling protein phosphorylation. Indeed, the reaction is a fully reversible energy-consuming switch that is operated by two enzyme families. These are the O-linked GlcNAc transferases, catalyzing the transfer of GlcNAc from UDP-GlcNAc to Ser and Thr residues of proteins, and the O-linked *N*-acetylglucosaminidases (O-GlcNAcases), which cancel the modification by hydrolysis (Figure 9.8). The human genome contains one O-linked GlcNAc transferase gene and one O-linked *N*-acetylglucosaminidase gene. Both give rise to a wide variety of isoenzymes that are generated by alternative pre-mRNA splicing and multiple start codons.

A large number or proteins are modified by hexosamine signaling. They include transcription factors, proteins of RNA translation, cytoskeletal and other structural proteins, signaling proteins, various enzymes, chaperones, proteasomes, and others. Our knowledge of the functional consequences of this posttranslational modification is still rather limited. Depending on the substrate protein, it

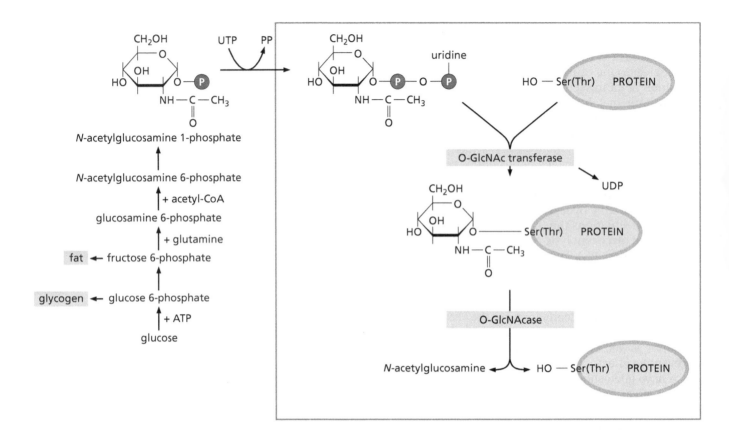

Figure 9.8 Protein modification by N-acetylglucosamine (hexosamine signaling)
N-Acetylglucosamine 1-phosphate is generated from glucose along the pathway shown to the left. It is activated by an O-glycosidic binding of a uridine diphosphate (UDP) residue and transferred to Ser or Thr residues by the enzyme O-N-acetylglucosamine transferase (O-GlcNAc transferase, box). The modification is reversed by O-linked N-acetylglucosaminidases (O-GlcNAcase), yielding the nonmodified protein and N-acetylglucosamine. Being a typical switch, hexosamine signaling requires energy supplied by the hydrolysis of one UTP molecule.

may block Ser and Thr residues, that is, counteract protein phosphorylation in logical BUT NOT gates, or it may cooperate with phosphorylation in logical AND gates. Blocking of phosphorylation may result in quite different effects including an inhibition of protein degradation along the ubiquitin proteasome pathway, which frequently depends on phosphorylation (in addition, proteasomes are directly inhibited by GlcNAc binding).

What is the present concept on the role of hexosamine signaling in insulin resistance? In insulin target tissues, glucose is sequestered as glycogen and transformed into fat, with hexosamine signaling being a branch of these two major pathways as depicted in Figure 9.8. In fact, hexosamine signaling has a regulatory function since it promotes the release of insulin antagonists, such as leptin and adiponectin, and stimulates fat and glycogen synthesis. In addition, the cell's insulin-dependent glucose transport system becomes gradually inactivated. Persisting hyperglycemia overactivates this feedback control, finally leading to insulin resistance and an impairment of insulin release along the adipoinsulinic axis mentioned above. This effect is amplified by down-regulation of the insulin receptor and other components of insulin signaling (see Section 7.1.3).

9.4.4 Glycogen synthase kinase 3: much more than a suppressor of glycogen deposition

Another signaling enzyme playing a key role in both translational and metabolic regulation is GSK3, an abundant eukaryotic Ser/Thr-specific protein kinase. As shown above, it suppresses mRNA translation by phosphorylating and inhibiting the initiation factor eIF2B (see Figure 9.2). Moreover, like several other protein kinases, GSK3 inhibits glycogen synthase by phosphorylation, thus suppressing the anabolic metabolism of glucose (Section 5.3). However, the activities of GSK are much more extensive. In fact, GSK3 is a multifunctional biochemical processor with a wide range of effects that in most cases have an inhibitory character.

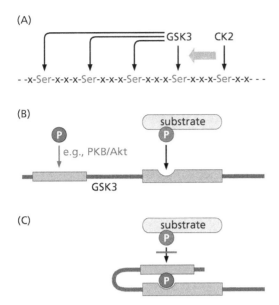

Figure 9.9 "Hierarchical" phosphorylation by and intrasteric inhibition of glycogen synthase kinase 3 (GSK3) (A) Hierarchical phosphorylation of a substrate protein (e.g., glycogen synthase) with casein kinase 2 (CK2) as a primer kinase. (B) Interaction of a prephosphorylated (primed) substrate protein with the catalytic domain (gray rectangle) of GSK3 and point of attack of an inhibitory phosphorylation, for instance by protein kinase B (PKB/Akt), in the N-terminal pseudosubstrate sequence (red rectangle). (C) In inactive GSK3, the phosphorylated pseudosubstrate sequence covers the substrate-binding site, hindering the access of the substrate.

Unusual mechanism of activation

In mammals GSK3 is expressed in two isoforms, α and β. Although these are closely related to MAP kinases, GSK3 does not require activation loop phosphorylation. Indeed, in contrast to most other protein kinases, GSK3 is constitutively active and, as a rule, becomes *inactivated* by input signals (an example of *activation* of GSK3 is discussed below). Such input signals are, first of all, transduced by Ser/Thr phosphorylation of GSK3 at sites outside the activation loop. These phosphorylations are catalyzed by several protein kinases.

A typical example is the inhibition of GSK3β by PKB/Akt, discussed in detail in Section 4.6.6. Here, the phosphorylation of a Ser residue modifies a pseudosubstrate sequence in such a way that it can cover the catalytic domain, which has a binding site for a phosphorylated amino acid sequence (Figure 9.9). This binding site plays a key role in substrate recognition, since GSK3 preferentially phosphorylates Ser and Thr residues placed C-terminally to a sequence of four amino acid residues with a phospho-Ser or phospho-Thr residue at the end; that is, the phosphorylation consensus sequence is Ser/Thr-x-x-x-Ser(P)/Thr(P) (with x standing for any amino acid and P for phosphate). When this peculiarity is considered, two conclusions can be drawn: one, to become recognized by GSK3, substrate proteins have to be prephosphorylated by another protein kinase, a reaction called "priming," and two, a GSK3 pseudosubstrate motif resembles a primed sequence lacking the upstream Ser/Thr residue to be phosphorylated by GSK3. This situation immediately suggests two mechanisms of GSK3 activation: the pseudosubstrate domain may be displaced from the catalytic center either by competition with the primed substrate protein or by dephosphorylation.

A protein kinase that frequently has been found to prime GSK3 is CK2 (see below). Provided a substrate protein has the corresponding primary structure, it may be phosphorylated by GSK3 in cooperation with such a primer kinase in a stepwise manner (see Figure 9.9). For an example (the hierarchical phosphorylation of glycogen synthase), the reader is referred to Section 5.3.1. As far as cellular data processing is concerned, GSK3-catalyzed hierarchical substrate phosphorylation resembles a logical AND gate combined with a digital–analog converter.

Negative key regulator of insulin and mitogenic signaling

The inhibitory effect of GSK3 on glycogen synthase is overcome by the above-mentioned phosphorylation of the pseudosubstrate domain, resulting in a suppression of GSK3 activity. Since PKB/Akt is a major player in this game, all signals

Figure 9.10 Three major pathways leading to an inhibition of glycogen synthase kinase 3 To the left are shown two signaling cascades that originate from Tyr kinase-coupled receptors. The Ras–RAF–ERK pathway results in activation of the protein kinase RSK1, while the phosphatidylinositol 3-kinase (PI3K) pathway results in activation of protein kinase B (PKB)/Akt. Both kinases inhibit GSK3 by phosphorylation of the pseudosubstrate domain, thus rescuing glycogen synthesis, protein synthesis and gene transcription from inhibition. To the right is shown the Wnt pathway, including the receptor Fz and the adaptor protein DSH. In the absence of Wnt, GSK3 (together with the primer kinase casein kinase 1, CK1) inhibits β-catenin activity by inducing proteolytic degradation. In the presence of Wnt, DSH inhibits GSK3, resulting in a stabilization of β-catenin. (For details see Section 5.9.)

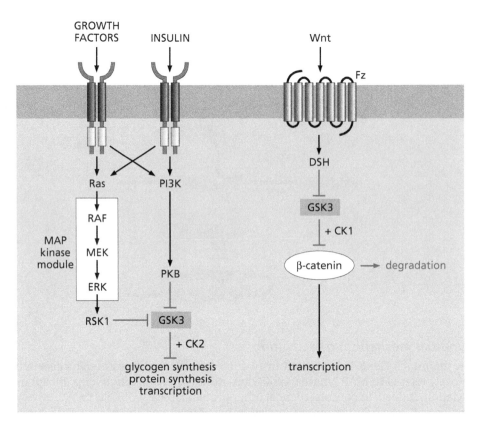

activating the PI3K–PKB/Akt pathway are expected to antagonize GSK3 (Figure 9.10). This holds true, in particular, for mitogens and insulin. In fact, GSK3 has turned out to be a central switching element in anabolic insulin effects such as glycogen deposition and protein synthesis. As already explained, insulin promotes protein synthesis by rescuing translation from the inhibitory phosphorylation of the initiation factor eIF2B catalyzed by GSK3 (Figure 9.2 and Section 9.3.3).

In addition, GSK3 is an inhibitory element in mitogenic pathways originating from Tyr kinase-coupled growth factor receptors. In such signaling cascades, GSK3 becomes switched off through pseudosubstrate phosphorylation either by PKB/Akt or by the ribosomal S6 kinase RSK1 that is (as explained above) positioned downstream of the Ras–RAF–ERK cascade (Figure 9.10).

The inhibitory effect of GSK3 on mitogenic signal transduction is due mainly to a blockade of translation and transcription. Transcription is suppressed by an inhibitory phosphorylation of mitogenic transcription factors such as cJun (see Section 11.7). Another example of a transcription factor that is switched off by GSK3 is β-catenin, which upon phosphorylation becomes a substrate of ubiquitin ligases priming the factor for degradation in proteasomes. Here the activity of GSK3 is controlled along the mitogenic Wnt signaling pathway with casein kinase CK1ε acting as a primer kinase (Figure 9.10; for details see Section 5.9). The formation of a complex with β-catenin and two adaptor proteins, axin and APC, probably renders GSK3 resistant to inactivation by PKB and other kinases. The breakdown of this regulatory mechanism is a major cause of colorectal cancer (Section 5.9).

Glycogen synthase kinase 3 and neurodegenerative diseases

Another aspect of GSK3 function should be mentioned, since it may turn out to be of considerable practical significance. This is the role of GSK3β in neurodegenerative diseases such as **Alzheimer's dementia.** GSK3β has been found to be hyperactive in impaired brain areas and to inactivate by phosphorylation some essential transcription factors besides β-catenin, such as NFAT (Section 14.6.4) and heat-shock factor HSF1. In this way, the vitality of nerve cells becomes diminished, culminating in apoptosis. Potentially even more dangerous seems to be the phosphorylation of another GSK3 substrate, the **microtubule-associated protein Tau**. By stabilizing microtubule tracks in axons, Tau provides the conditions for transport of vesicles filled with neurotransmitters to the synapse. Tau that is rendered insoluble by hyperphosphorylation aggregates into filamentous structures, which are unable to interact with microtubules. As a result, synapses become inactive, finally resulting in cell death. Tau aggregation is regarded as a major risk factor for Alzheimer's disease.

The overactivation of GSK3β in degenerating brain areas seems to be due to an impairment of blood supply, or ischemia. Ischemia is a strong promoter of neurodegeneration since it renders brain neurons sensitive to apoptotic signals. In such cells, GSK3 becomes activated by the β-amyloid peptide Aβ along an ill-defined pathway. Overproduction of Aβ and extracellular deposition of Aβ degradation products as so-called amyloid plaques is considered to be the major trigger of dementia (for details see Section 13.1.3). These findings nourish the hope that in the future GSK3 inhibitors may become suitable drugs for the prevention and treatment of neurodegenerative diseases, although in view of the multifunctionality of the enzyme, one certainly has to reckon with serious side effects.

9.4.5 "Casein kinases" CK1 and CK2: two multifunctional enzymes denying their names

As primer kinases of GSK3, the so-called casein kinases CK1 and CK2 were introduced in Section 9.4.4. However, this is by far not their only function. In fact, both represent two groups of Ser/Thr-specific protein kinases that exhibit a functional versatility comparable to that of GSK3. It is a curiosity that, despite their historic name, these kinases do not catalyze the phosphorylation of casein in vivo. Therefore, the acronyms CK1 and CK2 are preferred now. The structural similarity between the two enzymes is rather low. With GSK3 they share the property of being constitutively active; they are switched off rather than switched on by input signals.

CK1 and CK2 catalyze the phosphorylation of Ser and Thr residues localized adjacent to acidic amino acids (such as glutamic and aspartic acid) or phosphate. A phosphorylation consensus sequence of CK1 is, for instance, Ser(P)/Thr(P)-x-x-Ser/Thr. Therefore, like GSK3, CK1 and CK2 may phosphorylate proteins in a stepwise manner, cooperating with other protein kinases. By the way, their preference for a negatively charged phosphorylation consensus sequence distinguishes CK1 and CK2 as well as GSK3 from most other protein kinases that prefer positively charged target sequences.

Protein kinase CK1

Mammalian CK1 is expressed in at least eight isoforms, which usually are labeled by Greek letters. Since the enzyme inactivates itself by autophosphorylation, it requires a permanent contact with protein phosphatases to stay active. This situation provides a terminal for input signals that modulate phosphatase activity. CK1 has various functions including that of a primer kinase of AMPK and GSK3 (Figure 9.9). An important example is the control function of CK1 in the hedgehog and Wnt signaling cascades (Section 5.9). Recently its role in the regulation of **day–night rhythms** has found particular attention.

The cellular and organ functions of all eukaryotes and of certain prokaryotes (such as, for instance, the light-exploiting cyanobacteria) are adjusted to daytime.

347

Figure 9.11 Molecular control of day–night rhythm The diurnal rhythm is due to an autonomous oscillating signal system controlling the activity of a large number of genes. These genes encode and become activated by the heterodimeric CLOCK/MAL1 transcription factor (positive feedback loop, left side). Some of the control genes also give rise to the production of the repressor proteins PER and CRY that, upon phosphorylation by casein kinase 1 (CK1), inhibit CLOCK/MAL1; that is, they promote the inactivation of the control genes (negative feedback loop, right side). As a result, the undampened oscillation shown below is generated. CK1 is assumed to be controlled by exogenous signals.

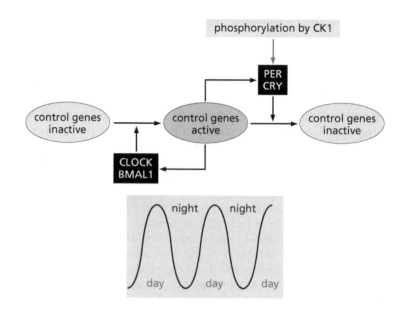

In vertebrates, the central oscillator consists of a group of neurons in the hypothalamus, called the nucleus suprachiasmaticus, which regulate subordinated feedback loops in the peripheral organs. Each air passenger travelling the Atlantic Ocean can tell a thing or two about the resistance of this internal clock! Although genetically fixed, the diurnal or 24-hour rhythm of physiological activities is additionally controlled by exogenous factors such as light, enabling adaptation to changing conditions. To this end, direct contact exists between the photoreceptor cells of the retina and the neurons of the nucleus suprachiasmaticus. In plants, the cryptochromes containing flavin and pterin chromatophores serve as light sensors. Homologous cryptochromes are also found in vertebrate cells. Whether or not they fulfil a similar purpose as the plant cryptochromes is not known.

Rhythmic oscillations such as the day–night rhythm are due to two-fold negative feedback or a combination of positive and negative feedback (see Section 1.9). As far as diurnal rhythms are concerned, such regulatory loops control the transcription of a large number of genes including three periodicity genes (*per1–3*) and two cryptochrome genes (*cry*). In the hypothalamus and other organs these genes are activated by the heterodimeric transcription factor **CLOCK/BMAL1**, the activity of which oscillates in a 24-hour rhythm due to a combination of negative and positive feedback. *Positive* feedback results from auto-induction of the genes encoding BMAL1, while *negative* feedback is due to production of the repressor proteins PER and CRY that, upon dimerization, migrate into the nucleus and inhibit CLOCK/BMAL1 (Figure 9.11).

Inhibition of the circadian genes correlates with multiple phosphorylation of the PER proteins that is catalyzed in particular by CK1. Studies on transgenic animals have shown this phosphorylation to be indispensable for a precise day–night rhythm. Phosphorylation by CK1 may have more than one effect. Thus, it promotes the ubiquitylation and degradation of PER in proteasomes, but in addition, it might also control the nuclear translocation of PER and its interaction with CLOCK/MAL1. The idea of CK1 as a transducer of exogenous signals modulating the day–night rhythm is easily at hand. However, the input signals controlling CK1 activity are still not known.

Protein kinase CK2

CK2 is a perfect example of a multifunctional protein kinase since more than 100 potential substrate proteins are known. The enzyme is found in all eukaryotes, where it is expressed in several isoforms. As a heterotetramer of two catalytic α-subunits and two regulatory β-subunits, CK2 exhibits a more complex quaternary structure than CK1. CK2 belongs to the very few protein kinases that are constitutively active even in the absence of all intercellular signals known. Activation loop phosphorylation does not seem to be necessary although phosphorylation by other kinases exhibits a modulatory effect on CK2 activity. CK2 is inhibited by acidic molecules such as heparin and is stimulated by basic polyamines. The physiological significance of these reactions is, however, as controversial as the interactions of CK2 with many other proteins.

The essential role of CK2 is emphasized by genetic knockout experiments ending in severe or even fatal developmental defects in such different organisms as yeast, slime molds, nematodes, and mice. However, the large number of potential substrate proteins has impeded most attempts to associate certain cellular events with CK2 activity. Among the putative functions, roles in cell proliferation and apoptosis have been studied in more detail. CK2 is required for an ordered course of the cell cycle and inhibits apoptosis by phosphorylating certain pro-apoptotic proteins such as Bid (Section 13.2.4). The function of CK2 as a primer kinase of GSK3 has been mentioned in Section 9.4.4, and an inhibitory effect on the transcription factor cJun will be discussed in Section 11.6.

Summary

Atypical protein kinase mTOR stimulates ribosomal protein synthesis in response to the nutrient and energy supply and to mitogenic and anabolic stimuli. Such stimuli activate the Ras-like G-protein Rheb, which in turn induces mTOR activity. The major target of stimulatory and inhibitory signals is the Rheb GTPase-activating protein (GAP) TSC2, a tumor suppressor protein deleted in hamartomas. The stimulatory effect of insulin on mTOR signaling is restricted by negative feedback. An overload of this control system may contribute to diabetes through overnutrition. Ribosomal S6 kinases S6K (p70^{S6K}) and RSK (p90^{S6K}) are multifunctional enzymes of the AGC kinase superfamily. Upon activation along the mTOR pathway, S6K phosphorylates the ribosomal protein S6 and other substrates, primarily controlling protein synthesis and cell size. RSKs are more involved in the regulation of cell proliferation and survival. To this end, they phosphorylate transcription factors regulating pro-mitogenic gene activity and inactivate pro-apoptotic proteins. AMP-dependent protein kinase (AMPK) is a cellular master regulator that adjusts energy-consuming anabolic metabolism (inhibited by AMPK) to energy-supplying catabolic metabolism (stimulated by AMPK). Since it is activated by 5′-AMP and inhibited by ATP and glycogen, AMPK monitors the energy status of the cell. Activation of AMPK requires activation loop phosphorylation by LKB1, a tumor suppressor protein. AMPK signaling is activated by factors that inhibit fat and glycogen deposition. On the other hand, disturbances of leptin–AMPK signaling are critically involved in metabolic disorders such as obesity and type 2 diabetes that increase cancer risk. Glycogen synthase kinase (GSK) is a constitutively active enzyme that becomes inactivated by input signals. To phosphorylate its many substrates, it frequently cooperates with primer kinases. GSK3 is a major negative regulator of insulin effects and mitogenic signaling and is thought to be critically involved in neurodegenerative diseases such as Alzheimer's dementia. "Casein kinases" CK1 and CK2 represent two families of widely distributed Ser/Thr-specific kinases that prefer acidic phosphorylation consensus sequences and are constitutively active. Among the many putative functions of these enzymes, their role as primer kinases of GSK3 and AMPK has gained attention. CK1, in particular, is involved also as a negative regulator in the control of diurnal rhythms.

9.5 Synopsis

Protein kinases that are major signal transducers in the network, which adjust cellular protein synthesis and growth (and proliferation) to the supply of food and energy, are summarized in Table 9.2.

Table 9.2 Protein kinases controlling the adjustment of cell growth to the supply situation

Kinase	Activation	Major substrates	Major functions
mTOR	PKB/Akt (or ERK) + amino acids, high ATP level	S6K, 4E-BP1, eEF2 kinase (?)	activation of translation increase of cell size
S6K	mTOR (or ERK) + PDK1	S6K, Bad, IRS1, eEF2 kinase	activation of translation and cell survival inhibition of insulin signaling
RSK	MAPK + PDK1	transcription factors, Bad	cell proliferation and survival
AMPK	5'-AMP + LKB1 (physical activity, stress)	metabolic enzymes proteins of mTOR signaling	activation of catabolic and inhibition of anabolic metabolism inhibition of translation
LKB1	? (constitutively active)	AMPK	AMPK kinase tumor suppressor
GSK3	constitutively active (inhibition by PKB/Akt, RSK, WNT)	glycogen synthase, RSK, transcription factors, Tau protein, eIF2B	inhibition of insulin signaling, cell proliferation, translation neurodegeneration
CK1	constitutively active	AMPK, GSK3, transcription factors	primer kinase of AMPK and GSK3 control of diurnal rhythms
CK2	constitutively active	GSK3 and many others	primer kinase of GSK3 cell proliferation and survival

Further reading

Ahmed K, Gerber DA & Cochet C (2002) Joining the cell survival squad: an emerging role for protein kinase CK2. *Trends Cell Biol.* 12, 226–230.

Alessi DF, Sakamoto K & Bayascas JR (2006) LKB1-dependent signaling pathways. *Annu. Rev. Biochem.* 75, 137–163.

Ali A, Hoeflich KP & Woodgett JR (2001) Glycogen synthase kinase-3: properties, functions, and regulation. *Chem. Rev.* 101, 2527–2540.

Andersen GR, Nissen P & Nyborg J (2003) Elongation factors in protein biosynthesis. *Trends Biochem. Sci.* 28, 434–441.

Bukau B, Weissman J & Horwich A (2006) Molecular chaperones and protein quality control. *Cell* 125, 443–451.

Carling D (2004) The AMP-activated protein kinase cascade – a unifying system for energy control. *Trends Biochem. Sci.* 29, 18–24.

Coll AP, Farooqi IS & O'Rahilly (2007) The hormonal control of food intake. *Cell* 129, 251–262.

Dever TE (2002) Gene-specific regulation by general translation factors. *Cell* 108, 545–556.

Frame S & Cohen P (2001) GSK3 takes centre stage more than 20 years after its discovery. *Biochem. J.* 359, 1–16.

Frühbeck G (2006) Intracellular signaling pathways activated by leptin. *Biochem. J.* 393, 7–20.

Gallego M & Virshup DM (2007) Post-translational modifications regulate the ticking of the circadian clock. *Nat. Rev. Mol. Cell Biol.* 8, 139–147.

Gething MJ (1999) Role and regulation of the ER chaperone BiP. *Cell Dev. Biol.* 10, 465–472.

Hardie DG (2005) New roles for the LKB1-AMPK pathway. *Curr. Opin. Cell Biol.* 17,167–173.

Hardie DG (2007) AMP-activated/SNF1 protein kinases: conserved guardians of cellular energy. *Nat. Rev. Mol. Cell Biol.* 8, 774–785.

Harding HP, Calfon M, Urano F et al. (2002) Transcriptional and translational control in the mammalian unfolded protein response. *Annu. Rev. Cell Dev. Biol.* 18, 575–599.

Harmer SL, Panda S & Kay SA (2001) Molecular basis of circadian rhythms. *Annu. Rev. Cell Dev. Biol.* 17, 215–253.

Harrington LS, Findlay GM & Lamb RF (2005) Restraining PI3K: mTOR signaling goes back to the membrane. *Trends Biochem. Sci.* 30, 35–42.

Hirsch C, Gauss R & Sommer T (2006) Coping with stress: cellular relaxation techniques. *Trends Cell Biol. 16,* 657–663.

Inoki K & Guan KL (2006) Complexity of the TOR signaling network. *Trends Cell Biol.*16, 206–212.

Jacinto E & Hall MN (2003) TOR signaling in bugs, brain and brawn. *Nat. Rev. Mol. Cell Biol.* 4, 117–126.

Jope RS & Johnson GVW (2004) The glamour and gloom of glycogen synthase kinase-3. *Trends Biochem. Sci.* 29, 95–102.

Kapp LD & Lorsch JR (2004) Molecular mechanics of eukaryotic translation. *Annu. Rev. Biochem.* 73, 657–704.

Kaufman RJ (2004) Regulation of mRNA translation by protein folding in the endoplasmic reticulum. *Trends Biochem. Sci.* 29, 152–158.

Kaufman RJ, Scheuner D, Schröder M et al. (2002) The unfolded protein response in nutrient sensing and differentiation. *Nat. Rev. Mol. Cell Biol.* 3, 411–421.

Keenan RJ, Freymann DM, Stroud RM et al. (2004) Translation factors: in sickness and health. *Trends Biochem. Sci.* 29, 25–31.

Knippschild U, Gocht A, Wolff S et al. (2005) The casein kinase 1 family: participation in multiple cellular processes in eukaryotes. *Cell. Signal.* 17, 675–689.

Lee C, Weaver DR & Reppert SM (2004) Direct association between mouse PERIOD and CKIε is critical for a functioning circadian clock. *Mol. Cell. Biol.* 24, 584–594.

Li Y, Corradetti MN, Inoki K et al. (2004) TSC2: filling the GAP in the mTOR signaling pathway. *Trends Biochem. Sci.* 29, 32–38.

Litchfield DW (2003) Protein kinase CK2: structure, regulation and role in cellular decisions of life and death. *Biochem. J.* 369, 1–15.

Love DC & Hanover JA (2005) The hexosamine signaling pathway: deciphering the "O-GlcNAc code". *Sci. STKE* 2005/312/re13.

Luo Z, Saha AK, Xiang X et al. (2005) AMPK, the metabolic syndrome and cancer. *Trends Pharmacol. Sci.* 26, 69–76.

Marshall S (2006) Role of insulin, adipocyte hormones, and nutrient-sensing pathways in regulating fuel metabolism and energy homeostasis: a nutritional perspective of diabetes, obesity, and cancer. *Sci. STKE* 2006/346/re7.

Meijer AJ & Dubbelhuis PF (2004) Amino acid signaling and the integration of metabolism. *Biochem. Biophys. Res. Commun.* 313, 397–403.

Moran O & Phillip M (2003) Leptin: obesity, diabetes and other peripheral effects – a review. *Pedriatr. Diabetes* 4, 101–109.

Muoio DM & Newgard CB (2006) Obesity-related derangements in metabolic regulation. *Annu. Rev. Biochem.* 75, 367–401.

Nielsen JN & Richter EA (2003) Regulation of glycogen synthase in skeletal muscle during exercise. *Acta Physiol. Scand.* 178, 309–319.

Nobukuni T, Kozma SC & Thomas G (2007) Hvps34, an ancient player, enters a growing game: mTOR complex 1/S6K1 signaling. *Curr. Opin. Cell Biol.* 19, 135–141.

Proud CG (2007) Signalling to translation: how signal transduction pathways control the protein synthetic machinery. *Biochem. J.* 403, 217–234.

Ron D (2002) Translational control in the endoplasmic reticulum stress response. *J. Clin. Invest.* 15, 1383–1388.

Rutter GA, Da Silva Xavier G & Leclerc I (2003) Roles of 5′-AMP-activated protein kinase (AMPK) in mammalian glucose homoeostasis. *Biochem. J.* 375, 1–16.

Ruvinsky I & Meyuhas O (2006) Ribosomal S6 phosphorylation: from protein synthesis to cell size. *Trends Biochem. Sci.* 31, 342–348.

Sakamoto K & Goodyear LJ (2002) Exercise effects on muscle insulin signaling and action. Invited review: Intracellular signaling in contracting skeletal muscle. *J. Appl. Physiol.* 93, 369–383.

Schröder M & Kaufman RJ (2005) The mammalian unfolded protein response. *Annu. Rev. Biochem.* 74, 739–789.

Soulard A & Hall MN (2007) Snap Shot: mTOR signaling. *Cell* 129, 434.

Wintermeyer W & Rodnina MV (2000) Translational elongation factor G: a GTP-driven motor of the ribosome. *Essays Biochem.* 35, 117–129.

Wullschleger S, Loewith R & Hall MN (2006) TOR signaling in growth and metabolism. *Cell* 124, 471–484.

Zabeau L, Lavens D, Peelman F et al. (2003) The ins and outs of leptin receptor activation. *FEBS Lett.* 546, 45–50.

Zhang K & Kaufman RJ (2003) Unfolding the toxicity of cholesterol. *Nat. Cell Biol.* 5, 769–770.

Signal Transduction by Small G-Proteins: The Art of Molecular Targeting

In addition to the heterotrimeric G-proteins (molecular masses between 85 and 90 kDa), which are almost exclusively coupled to transmembrane receptors, a large number of monomeric ("small") G-proteins with molecular masses between 16 and 36 kDa is known.

The prototype is the 21-kDa protein Ras. It has lent the collective term **Ras superfamily** to the whole group. In mammals this family consists of more than 150 isoforms, which are grouped into several subfamilies according to common functional properties (Table 10.1). These subfamilies have been found in all eukaryotes so far investigated in this respect.

A main function of small G-proteins is to organize the assembly of highly ordered multiprotein complexes of signal transduction, particularly at membranes. They do this as long as they are "switched on" by GTP binding. As explained in Section 2.4, the energy required for such signal-transducing processes is provided by exergonic reactions occurring in the protein complexes assembled, whereas GTP hydrolysis due to the intrinsic GTPase activity of the G-protein is a switching-off mechanism. However, some small G-proteins, in particular Ran, also use the energy of GTP hydrolysis to facilitate targeted transport processes, frequently against concentration gradients.

As is typical for G-proteins, signal transduction is stimulated by guanine nucleotide exchange factors (GEFs), which catalyze the exchange of bound GDP for GTP, and is terminated by GTP hydrolysis. For the reasons explained in Section 2.4, small G-proteins exhibit only a minute intrinsic GTPase activity. Therefore, the support of GTPase-activating proteins (GAPs) is indispensable. GEFs and GAPs are the major targets of signals controlling the function of small G-proteins. Both are expressed in a large variety of isoforms that connect G-protein signaling with almost any other pathway of signal transduction.

10.1 Ras proteins: generation of order in signal transduction

10.1.1 An abundant oncogene

The small G-protein Ras (p21^{Ras}) was discovered as an oncogenic mutant protein of an RNA virus (retrovirus) causing sarcomas in rats (Ras stands for Rat sarcoma). Later the nonmutated proto-oncogene product Ras was found to represent a highly conserved signaling protein of eukaryotes. Like all retroviral oncogenes, the *ras* proto-oncogene had been taken up by the virus from the host cell genome and undergone oncogenic mutation. For the *ras* gene it was shown for the first time that not only retroviruses but also tumor tissue from humans contained

Table 10.1 Ras superfamily of small G-proteins[a]

	Ras family		Rho family		Rab family			Arf family	Ran family
Mammals	Ha-Ras	Rheb	RhoA	Rnd2	Rab1A	Rab11A	Rab26	Arf1	Ran
	Ki-Ras	kB-Ras1	RhoB	Rho7	Rab1B	Rab11B	Rab27A	Arf2	
	N-Ras	kB-Ras2	RhoC	Tc10	Rab2	Rab12	Rab27B	Arf3	
	R-Ras		RhoD		Rab3A	Rab13	Rab28	Arf4	
	M-Ras		RhoE		Rab3B	Rab14	Rab29	Arf5	
	RalA		Rnd3		Rab3C	Rab15	Rab30	Arf6	
	RalB		Rho8		Rab3D	Rab16	Rab31	Sar1a	
	Rap1A		RhoG		Rab4	Rab17	Rab32	Sar1b	
	Rap1B		RhoH		Rab5A	Rab18	Rab33A	Arl1	
	Rap2A		TTF		Rab5B	Rab19	Rab33B	Arl2	
	Rap2B		Rac1		Rab5C	Rab20		Arl3	
	TC21		Rac2		Rab6	Rab21		Arl4	
	Rit		Rac3		Rab7	Rab22		Arl5	
	Rin		Cdc42		Rab8	Rab23		Arl6	
	Rad		Rnd1		Rab9	Rab24		Arl7	
	Kir/Gem		Rho6		Rab10	Rab25		Ard1	
Yeast	Ras1		Rho1		Ypt1	Ypt52		Arf1	Gsp1
	Ras2		Rho2		Sec4	Ypt53		Arf2	Gsp2
	Rsrl		Rho3		Ypt31	Ypt6		Arf3	
	Ycr7		Rho4		Ypt8	Ypt7		Sar1	
			Cdc42		Ypt32	Ypt10		Arl1	
			Yns0		Ypt9	Ypt11		Arl2	
					Ypt51			Cin4	
					Yps21				
	Signal transduction		**Signal transduction Cytoskeleton**		**Vesicle transport**			**Vesicle formation**	**Nuclear transport Mitosis**

[a]This selection of small G-proteins is grouped in subfamilies according to common structural features and cellular functions. From T. Matozaki et al. (2000) *Cell. Signal.* 12:515–524.

genes mutated to oncogenes. To prove this, NIH-3T3 fibroblasts, an immortal but non-tumorigenic cell line widely distributed in laboratories, was transfected with DNA from a human bladder carcinoma. The result was oncogenic transformation. As a cause, a mutated *ras* gene was identified that encoded a protein where a glycine residue in position 12 of normal Ras was replaced by another amino acid. Since then it has been shown that practically all oncogenic Ras proteins, including the viral ones, suffer from point mutations in positions 12, 13, and 61. However, in contrast to the situation in some animals, in humans the *ras* oncogene is generated by somatic mutation rather than being introduced by retroviral infection. Due to the oncogenic mutations Ras cannot interact any longer with the corresponding GTPase-activating proteins. In other words, the Ras switch is arrested in the ON position (see below). Being involved in about 30% of all human tumor diseases, *ras* mutations rank among the most frequent causes of cancer, particularly of the killer types such as pancreatic carcinoma (90%), colorectal carcinoma (50%), lung adenocarcinoma (30%), and myeloid leukemia (30%). In view of this situation it is easily conceivable that Ras is one of the best investigated proteins ever. It has turned out to be a central switching device of eukaryotic signal processing. Unfortunately, an effective approach to cancer therapy based on a pharmaceutical intervention in Ras signaling is still missing.

10.1.2 An all-around signal transducer

Mammalian Ras is expressed in 18 isoforms, constituting a Ras subfamily of its own within the Ras superfamily of small G-proteins (see Table 10.1). The prototype is **H-Ras** (referring to a Harvey rat sarcoma virus). Closely related are **K-Ras** (originally found in a Kirsten rat sarcoma virus) and **N-Ras** (N stands for Neuroblastoma). Within the Ras subfamily, these three isotypes are those undergoing oncogenic mutations.

Quite early, the transforming effect of Ras mutations raised the suspicion that the protein could play a central role in mitogenic signal transduction. In fact, proliferation of cells in culture could not be stimulated any longer by serum when the function of Ras had been suppressed, for instance by a monoclonal antibody. However, in the meantime the Ras signal has turned out to be ambiguous in that Ras not only controls cell proliferation but also may induce cell differentiation and even cell death depending on the cell type and the physiological conditions. The tumorigenic pheochromocytoma cell line P12, for instance, responds to epidermal growth factor (EGF) by proliferation but to nerve growth factor (NGF) by an outgrowth of neurites (that is, differentiation), although both factors activate the same Ras-controlled signaling cascade. The only differences are that the cells express many more EGF receptors than NGF receptors and that with NGF they have to stay in contact much longer than with EGF. Consequently, the two growth factors evoke different sets of interactions in the signal-processing protein network. When the NGF receptor is overexpressed, the cells indeed respond to NGF by proliferation rather than by differentiation.

How essential such cross talk may be is demonstrated by NIH-3T3 cells, where Ras induces cell proliferation only when the small G-protein Rho is activated simultaneously; otherwise Ras even exhibits an inhibitory effect (see below). Such contradictory findings are explained, of course, by the fact that principally signaling is an ambiguous process: the meaning of a signal is allocated by the receiver rather than being an intrinsic property of the signaling factor, which may be used for quite different purposes. This holds true both for inter- and intracellular signaling.

Ras is anchored at the inner side of the plasma membrane by two lipid side chains, a farnesyl and a palmitoyl residue (see Figure 4.8 for an illustration). Without such anchoring the protein is ineffective, because its main function (that it shares with other small G-proteins) is to recruit other signaling proteins to membranes. By this means Ras enables the formation of large signal-transducing complexes in the immediate vicinity of transmembrane receptors. Attempts to treat cancer with inhibitors of Ras farnesylation turned out to be unsuccessful, however.

10.1.3 Upstream of Ras: receptors and ion channels

Like all small G-proteins, Ras has an extremely low GDP–GTP exchange frequency and a minute GTPase activity. It is, therefore, entirely under the control of GEFs and GAPs, both serving as sensors of input signals.

Ras guanine nucleotide exchange factors: sensors of a wealth of input signals

Several types of Ras-GEFs connect Ras with practically all types of membrane receptors and ensure cross talk with other signaling pathways. The most prominent Ras-GEF is **mSOS** (mammalian Son Of Sevenless; for an explanation of the name see Sidebar 10.1) that is controlled by Tyr kinase- and G-protein-coupled receptors. Due to a SH3 domain interaction, mSOS forms a stable complex with the adaptor protein **Grb2** that contacts the autophosphorylated Tyr kinase-coupled receptors either directly (through a SH2 -domain) or by means of the scaffold protein **Shc** (Figure 10.1 and Section 7.1.2). Along the arrestin pathway mSOS is also activated by G-protein-coupled receptors (Figure 10.1 and Section 5.8). Thus mSOS is the target of a large number of input signals.

mSOS is widely distributed, whereas other Ras-GEFs seem to be more restricted to certain tissues, such as nerves and T-lymphocytes. These GEFs are **GRF** (Guanine nucleotide Releasing Factor) and **GRP** (Guanine nucleotide Releasing Protein). The peculiarity of these two GEFs is that they are activated by Ca^{2+} ions, and GRP in addition by diacylglycerol (DAG), thus connecting Ras with Ca^{2+}-transporting ion channels (Ca^{2+}-specific and nonspecific cation channels and cation-channel-coupled neurotransmitter receptors) as well as with phospholipase C (PLC) signaling elevating the cytoplasmic Ca^{2+} level along the inositol 1,4,5-trisphosphate ($InsP_3$) pathway and releasing DAG. A specialty of both mSOS and GRF is their integrating function: apart from Ras, both GEFs also activate the small G-proteins of the Rho family. This property is due to an additional Rho-GEF domain in the mSOS and GRF molecules (Section 10.3.1). In fact, as already mentioned, many cellular Ras effects depend on simultaneous activation of Rho-GTPases, providing a logical AND gate. A further Ras-GEF, **PLCε,** will be discussed in more detail in Section 10.1.4.

By increasing the cytoplasmic Ca^{2+} level, a large number of neurotransmitters can stimulate Ras activity in neurons. The same holds true for membrane depolarization and action potentials opening voltage-dependent Ca^{2+} channels (Section 14.5.2). Indeed, in brain neurons the Ras signaling pathway is involved in the regulation of long-term potentiation and synaptic plasticity. Long-term potentiation is understood as longer-lasting activation of synaptic signal transmission, while the term synaptic plasticity describes structural changes of frequently used (long-term potentiated) synapses. Both effects are thought to represent cellular equivalents of memory fixation. The concept of Ras playing a key role in these events is supported by observations on mice suffering from a

Figure 10.1 Upstream of Ras: signaling pathways leading to an activation of Ras guanine nucleotide exchange factors (GEFs) Activation of Ras-GEFs (colored box) and Ras is mediated by a wide variety of transmembrane receptors. From left to right: a phosphorylated (P) G-protein-coupled receptor activates, via the adaptor protein arrestin, a cytoplasmic tyrosine kinase (Src). By interacting with the adaptor Grb2 (and Shc), Src stimulates the Ras-GEF mSOS. mSOS is also activated by autophosphorylated tyrosine kinase-coupled receptors. The Ras-GEFs GRF and GRP are activated by an increase of the cytoplasmic Ca^{2+} concentration as brought about by voltage-dependent and receptor-coupled cation channels, or along the G-protein–phospholipase Cβ and Cε pathways. Moreover, PLCε exhibits Ras-GEF activity on its own. Not shown are signaling pathways that activate Ras through an inhibition of Ras-GAPs.

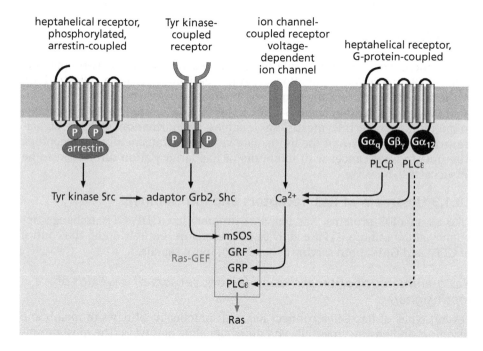

genetic GRF knockout: in such animals, learning capacity and memory are considerably curbed.

As Ca^{2+} sensors, GRF contains a Ca^{2+}/calmodulin interaction domain and GRP exhibits an EF-hand domain (more information about such domains is found in Section 14.6.1), while DAG is recognized by GRP through a C1 domain such as that found in DAG-dependent protein kinase C (see Section 4.6.3).

Ras GTPase-activating proteins

Like Ras-GEFs, Ras-GAPs (for humans more than five isoforms are known) are also controlled by various input signals. As discussed in Section 2.4.4, a GAP may be considered to be a regulatory subunit of a G-protein providing an arginine residue required for GTP hydrolysis, which thus becomes accelerated by a factor of 10^6. This mechanism does not work for Ras mutated to an oncoprotein because, due to the exchange of glycine 12 or 13 by another, larger amino acid residue, access of the Arg finger of GAP to the catalytic domain of Ras is hindered. Thus, the GTPase activity cannot be stimulated any more by GAP (mutation of glutamine 61 has a similar consequence because this residue stabilizes the transition state of GTP hydrolysis). Since the major role of GTP-loaded Ras is to organize signaling protein complexes at the membrane, the mutated Ras resembles a switch blocked in the ON position, or a valve that cannot be closed anymore.

The best-known Ras-GAP is **p120 GAP**. It is activated by Ras in a negative feedback loop but, as indicated by its SH2, SH3, and PH domains, must have additional possibilities for interaction that are not fully identified yet. Another Ras-GAP is **neurofibromin.** A loss-of-function mutation of this protein causes the Recklinghausen neurofibromatosis type 1, a hereditary disease characterized by multiple benign (sometimes also malignant) tumors of the nervous tissue appearing predominantly in the skin. A Ras-GAP activated by Ca^{2+} ions is **CAPRI** (**CA**lcium-**P**romoted **Ras I**nactivator). Its Ca^{2+}-binding sites are two C2 domains enabling Ca^{2+}-dependent binding to membrane phospholipids, thus bringing CAPRI into contact with membrane-bound Ras. Ca^{2+} ions, therefore, have a dual effect: on one hand Ras is activated by Ca^{2+}-dependent GEFs, while on the other hand it is inactivated by Ca^{2+}-dependent CAPRI. How the cell is able to reconcile these opposite effects is unknown.

10.1.4 Downstream of Ras: kinases and phospholipases

To perform its organizing effect on signal transduction Ras interacts with effector proteins, recruiting them to the membrane where they come in contact with receptors and other partners. Ras collects such proteins as long as it is in the active GTP-loaded state. The energy required for this organizing process is supplied by complex formation (the binding energies of the interacting partner molecules) and subsequent reactions within the complex rather than by Ras-catalyzed GTP hydrolysis. In fact, the latter only terminates the whole process. Of the various Ras effectors, p120 Ras-GAP has been mentioned already. Other effector proteins studied more thoroughly include protein kinases of the RAF family, phosphatidylinositol 3-kinases, PLCε, and GEFs of the Ras-like Ral proteins.

RAF kinases and the Ras–RAF–ERK cascade

Protein kinases of the RAF (**R**as-**A**ctivated **F**actor) family are the prototypical Ras effectors. They exist in three isoforms: A-RAF, B-RAF, and C-RAF. In addition, a retroviral v-RAF has been found. This oncoprotein is a truncated cellular RAF lacking an autoregulatory N-terminal domain, which normally inhibits RAF activity intrasterically, resembling a pseudosubstrate motif. The autoregulatory domain overlaps with two binding sites of Ras-GTP that ensure the membrane recruitment of RAF by Ras (Figure 10.2). Membrane-associated RAF is controlled by various input signals inducing a complex pattern of interactions with several protein kinases and scaffold proteins that has not yet been elucidated in every detail.

Interaction with Ras is a necessary but not sufficient condition for RAF activation. RAF has to be phosphorylated, in addition, in the C-terminal region. Two critical phosphorylation sites are Ser and a Thr residue in the activation loop that probably are autophosphorylated (Figure 10.2). To fully activate RAF, further phosphorylations at both sides of the active center are necessary: at a Ser residue (catalyzed by PKC) and a Tyr residue (catalyzed by Tyr kinases Src and JAK), and at a C-terminal Ser residue (probably catalyzed by a p21-activated protein kinase, PAK; see Sections 10.3.4 and 11.4 for more details about this kinase type). Stimulation by PKC of RAF activity and in turn of the mitogenic MAP kinase cascade is one reason for the frequently observed mitogenic effect of endogenous signals or drugs such as the tumor-promoting phorbol esters that activate PKC. PAK is stimulated by small G-proteins of the Rho/Rac family (see Section 10.3.2), indicating that RAF phosphorylation is one of the mechanisms ensuring the synergism of Ras and Rho/Rac signaling mentioned above.

RAF is also phosphorylated at Ras-binding sites. This phosphorylation is catalyzed either by protein kinase A (PKA) or by protein kinase B (PKB)/Akt and suppresses RAF activity because it hinders the interaction with Ras, instead enabling a binding of RAF to **14-3-3 proteins** (see below) that sequester the inactive kinase. 14-3-3 proteins also promote RAF effects by stabilizing the active form of RAF, provided the kinase had been phosphorylated before at the C-terminal Ser residue, the putative target of PAK (Figure 10.2). This apparently confusing and still incomplete picture of RAF regulation indicates that this protein kinase is connected with a wide variety of signaling pathways and provides a key element in Ras signaling. In fact, synthetic RAF inhibitors have proved to be promising anti-cancer drugs.

The only substrate proteins of RAF kinases identified with certainty are the RAF kinases themselves (becoming autophosphorylated) and the MAP2 kinases MEK1 and MEK2. Therefore, RAF fulfils the role of a MAP3 kinase within a MAP kinase module, with the MAP kinase ERK as a signaling outlet (see Sections 2.6.5 and 11.3). This module RAF–MEK (1 or 2)–ERK (1 or 2) has been called the mitogenic MAP kinase module because it predominantly (but not exclusively) processes mitogenic signals. It is controlled by all input signals of Ras and RAF. The (historically seen) canonical one among these signaling pathways starts from Tyr kinase-coupled growth factor receptors (Figure 10.3).

Figure 10.2 Complex regulation of the protein kinase C-RAF The kinase has two binding sites for Ras-GTP in the C-terminal region: a Ras-binding domain (RasBD) and a cysteine-rich domain (CRD). Phosphorylation by protein kinase A or B (PKA or PKB) of serine residues (S) in these domains hinders the interaction with Ras and induces sequestration by 14-3-3 protein; that is, phosphorylation has an inhibitory effect (red arrows). The kinase domain (gray rectangle), including the activation loop (AL), is localized in the C-terminal region. Phosphorylations of serine (S), threonine (T), and tyrosine (Y) residues in this region have a stimulatory effect on RAF activity (black arrows). The active form of RAF is stabilized by 14-3-3 proteins that recognize a C-terminal phosphoserine residue. Protein kinases found or assumed to catalyze the individual phosphorylations are shown. Auto-P, autophosphorylation.

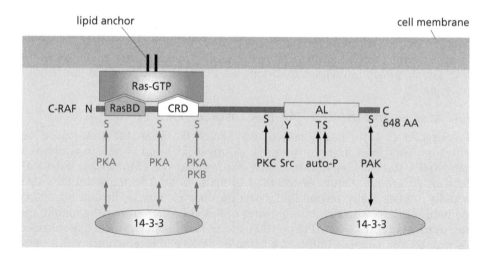

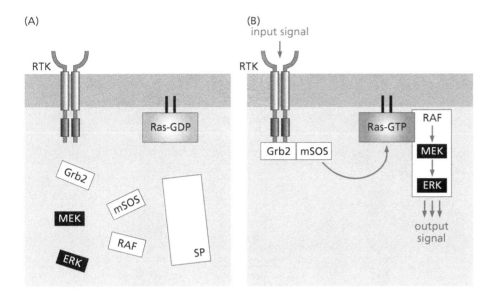

Figure 10.3 Activation of the mitogenic MAP kinase module by a receptor tyrosine kinase (RTK): an example of the organizing function of Ras (A) In the absence of an exogenous input signal, the components of the signal-processing protein complex are present in a disordered form in the cytoplasm while Ras is membrane-anchored and in the inactive GDP-loaded form. (B) By inducing autophosphorylation (P) of the receptor, an input signal (such as a growth factor) has triggered the organization of a highly ordered multiprotein complex bound at the surface of a scaffold protein (SP). Grb2 is an adaptor protein linking the activated receptor to the Ras-GEF mSOS that activates Ras and, as a consequence, the MAP kinase module consisting of the three protein kinases RAF, MEK, and ERK. As soon as the activation of the module has become stabilized by phosphorylation and binding of RAF to 14-3-3 proteins (not shown), Ras-GTP is free for a new cycle. The process is terminated by dephosphorylation of the receptor and GTP hydrolysis by a Ras/Ras-GAP complex.

The RAF isoform **B-RAF** (but not yet A-RAF and C-RAF) has been found to be mutated to an oncoprotein in several human malignancies such as malignant melanoma as well as thyroid, colorectal, and ovarian carcinoma. In most cases a valine residue in the activation loop is replaced by a glutamate residue. This mutation creates a negative charge that under normal conditions is introduced by reversible phosphorylation. The result is a constitutively active, nonregulated RAF mutant playing a role comparable to that of mutated Ras.

14-3-3 proteins: how signaling proteins are taken out of circulation

14-3-3 proteins are multifunctional components of cellular signal processing. Their strange name refers to their discovery: in an attempt to isolate proteins from bovine brain by chromatography and electrophoresis, in a fraction numbered 14-3-3, an acidic protein of about 30 kDa was found that later turned out to be representative of a whole family of highly conserved eukaryotic proteins. For humans, seven corresponding genes have been identified.

14-3-3 proteins are specialized for forming complexes with other proteins by binding to sequence motifs with phosphorylated Ser residues. In this respect they resemble proteins with SH2 and PTB domains that recognize motifs with phosphorylated Tyr residues.

The interaction of a 14-3-3 protein with its partner is due to a special structure: 14-3-3 proteins associate into homo- and heterodimers, forming a pocket of anti-parallel helices that fits two phosphorylated substrate proteins (or one protein with multiple phosphorylations). In most substrate proteins the 14-3-3-interacting domain is -Arg-Ser-X-X-Ser(P)-X-Pro- or -Arg-X-X-X-Ser(P)-X-Pro-, where Ser(P) is the phosphorylated Ser residue and X is any amino acid. The negative charge of the phosphate group is essential for the interaction; nonphosphorylated proteins are also bound by 14-3-3 proteins provided their interacting motif contains several acidic amino acids such as -Arg-Ser-Glu-Ser-Glu-Glu-. In some cases also a special zinc-finger structure does the job.

In some ways the function of 14-3-3 proteins resembles that of chaperones: they alter the conformation and, thus, the function of a protein, albeit dependent on whether or not this carries a negative charge at a specific site. As a consequence, 14-3-3 proteins may either inhibit or promote interactions between proteins (Figure 10.4). Inhibition is mostly due to sequestration of proteins for a definite period of time (until the effector protein becomes dephosphorylated, for instance). One example is provided by RAF kinase phosphorylated by PKA;

Figure 10.4 Mode of action of 14-3-3 proteins Dimeric 14-3-3 proteins bind other proteins, recognizing them by two phosphorylated serine (Ser) residues (or a corresponding negatively charged motif). (A) Sequestration of a two-fold phosphorylated protein is shown, resulting in either inactivation or activation. (B) The adaptor function of 14-3-3 is depicted, bringing together two interacting proteins upon specific Ser phosphorylation.

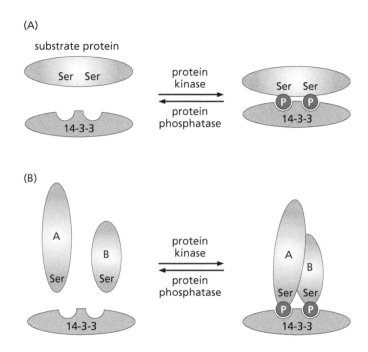

another example is the pro-apoptotic protein Bad that, upon phosphorylation by PKB/Akt, is taken out of circulation by 14-3-3 proteins (this represents one of the pathways along which PKB suppresses programmed cell death; see Sections 4.6.6 and 13.2.4).

Vice versa, 14-3-3 proteins may also facilitate protein interactions in that they act as adaptors, bringing together proteins or determining their intracellular localization, or they protect proteins from being inactivated by dephosphorylation or proteolysis (it should be remembered that ubiquitylation frequently requires a priming phosphorylation of the substrate protein). An example is provided by RAF kinases phosphorylated by PKC and PAK.

Phosphatidylinositol 3-kinases

These enzymes (PI3Ks) have been treated already in detail in Section 4.4.3. They catalyze the formation of membrane-bound second messengers such as phosphatidylinositol 3,4,5-trisphosphate, serving as docking sites for proteins with PH domains. The signal-processing complexes thus assembled promote subsequent reactions that are essential for most Ras effects. Certain (but not all) PI3Ks are activated by a combined action of Tyr kinase-coupled receptors and Ras in that the receptor inactivates an auto-inhibitory subunit p85 and Ras takes care of the correct membrane localization of the catalytic subunit (Figure 10.6).

Phospholipase Cε

A particularly interesting and versatile Ras effector is PLCε. The enzyme belongs to those type C phospholipases that hydrolyze phosphatidylinositol 4,5-bisphosphate, yielding the second messengers DAG and InsP$_3$. Thus, the role of PLCε in cellular signal transduction resembles that of PLCβ and -γ (Section 4.4.4). However in contrast to the latter, PLCε possesses two Ras-binding domains and a Ras-GEF domain; that is, it has two enzymatic activities (PLC and GEF). Therefore, the phospholipase is both a Ras effector and a Ras activator. The activating effect is amplified by the release of DAG and Ca^{2+} (via InsP$_3$; see Section 4.4.4) both activating the Ras-GEFs GRP and GRF (Figures 10.1 and 10.7). On the other hand, InsP$_3$ is integrated in *negative* feedback control of Ras by releasing the calcium that activates the Ras-GAP CAPRI (see above).

Sidebar 10.1 Ras and MAP kinase cascades in yeast and nonvertebrates Ras proteins have been found in all eukaryotes from yeast to humans.

In the fission yeast *Schizosaccharomyces pombe*, RAS1 (the only Ras form expressed in this organism) is a component of a signaling pathway leading from surface receptors to the genome and resembling a Ras–RAF–MAP kinase cascade (Figure 10.5). The activating signal is the mating pheromone of the haploid gametes that interacts with a G-protein-coupled receptor, activating the Ras-GEF Ste6 (the name refers to <u>ste</u>rile mutants) via the α-subunit of the trimeric G-protein GPA1 (see Section 5.2). The Ras-controlled MAP kinase module consists of the protein kinases Byr2 (a MAP3 kinase functionally resembling a RAF kinase), Byr1 (a MAP2 kinase resembling MEK), and Spk1 (a MAP kinase resembling ERK). The substrates of Spk1 are transcription factors of genes controlling the mating phenotype. Another Ras effector is a GEF of the small G-protein Cdc42 regulating cytoskeletal dynamics and cell shape. This pathway provides a clear analogy to vertebrate Ras signaling, which also cooperates with a Cdc42 homolog and other small G-proteins affecting the cytoskeleton. In fact, RAS1 of *S. pombe* is quite similar to H-Ras of mammals.

In multicellular nonvertebrates, Ras has various functions. Well-known is its role in morphogenetic processes. The classical example of a morphogenetic Ras–RAF–MAP kinase cascade is provided by the devel-opment of photoreceptor cells in the *Drosophila* eye. The ommatides of the insects' compound eyes are composed of eight different types of photoreceptor cells inducing each other during embryogenesis. So an R7 cell arises only when its precursor cell receives a specific differentiation signal from an adjacent R8 cell. This R8-derived signal is transmitted by a membrane-bound protein "Boss," a relative of the epidermal growth factor of vertebrates. By a juxtacrine interaction, Boss activates an EGF-receptor-related receptor Tyr kinase "Sevenless" exposed at the surface of R7 precursor cells (the name "Sevenless" refers to a loss-of-function mutation of the receptor resulting in animals lacking R7 cells). Upon autophosphorylation, the receptor interacts with an adaptor protein Drk (resembling the mammalian Grb2) and activates the Ras-GEF "son of sevenless (SOS)." The signaling cascade thus stimulated downstream provides an exact counterpart to the mammalian Ras–RAF–ERK cascade (Figure 10.5).

Quite a similar signaling cascade controls vulva development in *Caenorhabditis elegans*. As in the *Drosophila* eye, adjacent cells (called anchor cells) induce the differentiation of precursor cells by means of an EGF-related factor, here called Lin3 (Figure 10.5). Along analogous signaling pathways, further differentiation processes are controlled in the nematode. In the cases mentioned here, the targets of the MAP kinase cascades always are transcription factors of the Ets family (see Section 11.7.1).

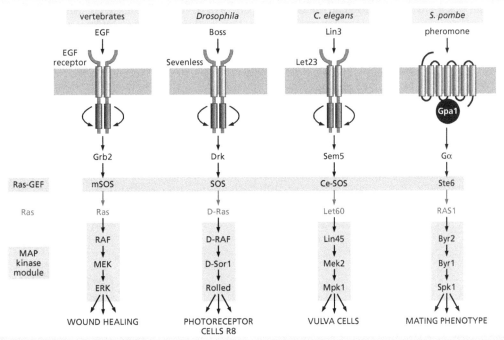

Figure 10.5 Ras-MAP kinase cascades Shown are two morphogenetic pathways in *Drosophila melanogaster* and *Caenorhabditis elegans* as compared with growth factor-stimulated signaling in vertebrates and pheromone signaling in *Schizosaccharomyces pombe*. Curved arrows symbolize the autophosphorylation of the receptor tyrosine kinases epidermal growth factor (EGF) receptor, Sevenless, and Let23. Yeast cells do not contain genes encoding receptor tyrosine kinases. Instead they activate the Ras-MAP kinase pathway by means of G-protein-coupled receptors. For other details see text.

Figure 10.6 Role of Ras in the activation of phosphatidylinositol 3-kinase type α For the activation of PI3Kα, Ras cooperates with a Tyr kinase-coupled receptor such as a growth factor receptor. PI3Kα consists of a regulatory subunit p85 and a catalytic subunit p110 (see Section 4.4.3). On the one hand, p85 inhibits enzymatic activity; on the other hand, it binds via an SH2 domain to the autophosphorylated receptor (P, phosphate group). As a result, the interaction with p110 is interrupted and the Ras-binding domain (RasBD) of p110 becomes exposed (red arrow), enabling p110 to bind to membrane-anchored Ras-GTP. The association with the membrane is reinforced by a Ca^{2+}–phospholipid-binding C2 domain. Thus the catalytic domain of p110 gains direct contact with its phospholipid substrates such as phosphatidylinositol 4- or 5-phosphate (PIP).

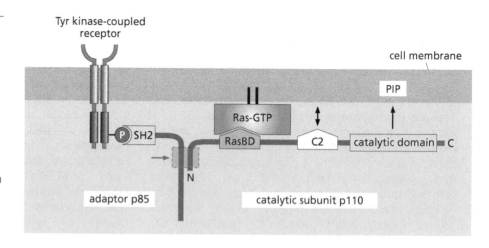

Figure 10.7 Integrating function of phospholipase Cε (A) Domain structure together with major interaction partners of PLCε is shown (GEF, guanine nucleotide exchange factor domain; PH, PIP_2-interacting pleckstrin homology domain; EFH and C2, Ca^{2+}- and phospholipid-binding domains; RasBD, Ras/Rap-binding domain; X and Y, catalytic domains). (B) PLCε is a Ras and Rap effector and, at the same time, both a positive and negative regulator of Ras and Rap activity. Stimulatory pathways are shown by black arrows; inhibitory pathways are shown in red. For other details see text.

This is not the end of the story, since PLCε has still more possibilities. It is also an effector and a GEF of the small G-proteins of the Rap type that alternatively are activated along cAMP-controlled pathways (see below). Moreover, PLCε has been reported to become activated by the βγ-subunits of trimeric G-proteins as well as by another small G-protein, Rho, that interacts with a special insert in the catalytic domain found only in PLCε but in no other C-type phospholipase. Most probably, the role of the G-proteins (Ras, Rap, Rho, Gβγ) is to direct the phospholipase to specific sites of the membrane where it becomes anchored through its PH, EF-hand, and C2 domains. As described in Chapter 4, reversible membrane anchoring is a critical step in phospholipase activation since it allows the enzyme to interact with its phospholipid substrates. Thus, the temporal and spatial control of PLCε activity provides an illustrative example of the targeting function of small G-proteins.

Although some PLCε effects still need final proof, at least as far as their physiological meaning is concerned, it is obvious that this phospholipase is able to connect various signaling pathways. The positive and negative feedback loops established by the enzyme might indicate an oscillating, frequency-modulated signaling system.

Summary

Transducing mitogenic, differentiation-controlling, and pro-apoptotic signals, the small G-protein Ras occupies a central position in cellular signal processing. Mutated Ras is an abundant oncoprotein. To perform its major function, recruiting signaling proteins at the membrane, Ras has to be GTP-loaded and anchored at the inner side of the plasma membrane. GTP-loading is catalyzed by various Ras-guanine nucleotide exchange factors (GEFs) that are targets of all kinds of input signals. Among the GEFs, mSOS is activated by Tyr kinase-coupled receptors and arrestin-coupled heptahelical receptors, while guanine nucleotide-releasing factor

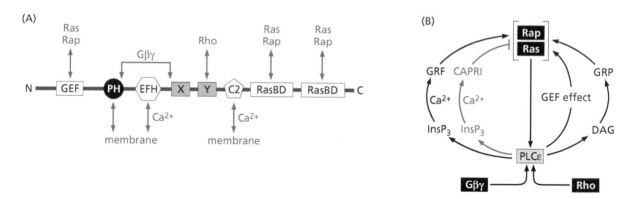

and protein (GRF and GRP) are stimulated by Ca^{2+} ions. The extremely low GTPase activity of Ras is up-regulated by GTPase-activating proteins (GAPs) that play an important role in feedback control and as receivers of input signals. In the Ras oncoprotein, interaction with GAPs is impaired due to point mutations. Thus, oncogenic Ras resembles a switch arrested in the ON position. Major effector proteins of Ras-GTP are RAF Ser/Thr kinases. As MAP3 kinases, they establish the contact between Ras and the mitogenic MAP kinase module RAF–MEK1,2–ERK. The isoform RAF-B is mutated to an oncogene in various human cancers. Ras recruits RAF to the membrane, where it becomes activated by protein kinase C, the Rho/Rac-activated kinase PAK, and cytoplasmic Tyr kinases. In contrast, protein kinases A and B inhibit RAF activity by interrupting Ras binding. RAF activity is regulated by 14-3-3 proteins. These resemble chaperones that modulate protein activity dependent on phosphorylation. Via specific recognition domains, they interact with proteins containing sequences with phosphorylated Ser residues. This may result either in temporal inactivation by sequestration or in a facilitated interaction with other proteins and protection from inactivation. Other important Ras effector proteins are PI3K and phospholipase Cε. The latter is both a Ras effector and a Ras activator. It connects Ras signaling with various other signaling pathways, thus promoting signaling cross talk.

10.2 Other G-proteins of the Ras subfamily: an unfinished story

As compared with the Ras proteins proper, particularly H-, K-, and N-Ras, much less is known about the functions of other members of the Ras subfamily. Many of them seem to be involved in the cross talk between pathways controlled by different small G-proteins.

Rap proteins are the closest relatives of Ras. They have been connected with regulation of the actin cytoskeleton and of intercellular adhesion. Indeed, Rap GTPases strengthen cellular contacts, particularly focal adhesions organized by integrins (see Section 7.2.3) and adherens junctions organized by cadherins (Section 7.3.1).To this end they recruit and activate proteins such as RapL (facilitating integrin aggregation at contact sites), Riam (linking integrins with the actin network), afadin (a multivalent adaptor for proteins in adherens junctions), and Tiam1 and Vav2, two GEFs of the small G-protein Rac, a master regulator of cytoskeletal dynamics (see Section 10.3).

Rap proteins have been proposed to also be Ras antagonists. In this context the activation of Rap by the Ras effector PLCε is of interest since it might provide a negative feedback control of Ras activity. Rap-GEFs are activated by the second messengers DAG, cAMP, and Ca^{2+}. One of the Rap-GEFs is, indeed, identical with the Ras-GEF GRP2, exhibiting binding sites for DAG and Ca^{2+}, while another Rap-GEF binds cAMP. The discovery of this factor, called **EPAC** (Exchange Protein Activated by Cyclic AMP), has broken the rule that in vertebrates only PKA and certain ion channels are effector proteins of cAMP.

Ral (Ras-like) is the common name of a group of small G-proteins (see Table 10.1) connecting Ras with Cdc42 signaling (see Section 10.3). Ral-GEF is stimulated by Ras and the Ral-GTP produced along this pathway is a negative regulator of Cdc42, exhibiting Cdc42-GAP activity. Almost nothing is known about other functions of Ral.

Only recently some information regarding the functions of other members of the Ras subfamily was obtained. The abundant GTPase **Rheb** has turned out to control the mTOR signaling pathway regulating ribosomal protein synthesis (Section 9.4.1), and the G-proteins **Rad** and **Kir/Gem** have been found to inhibit the activity of voltage-dependent Ca^{2+} channels and to suppress effects of Rho-type GTPases by inhibiting Rho-controlled protein kinases of the ROK family (see

Section 10.3). Finally, the Ras-like GTPase **TC21** has been identified as an oncoprotein, being able to transform cell lines derived from human tissues. Whether it is involved also in clinical forms of cancer remains an open question.

Summary

Only fragmentary information is available about the cellular functions of other members of the Ras subfamily. Some of them such as Rap, Ral, Rad and Kir/Gem seem to cooperate with other small G-proteins in the regulation of the actin cytoskeleton. Rheb plays a key role in the mTOR signaling pathway regulating ribosomal protein synthesis.

10.3 GTPases of the Rho family: master regulators of the actin cytoskeleton and more

Rho-type GTPases constitute a distinct family of small Ras-like G-proteins. They are found in vertebrates, invertebrates, yeast, and plants. In vertebrates this family, named after the protein Rho (Ras homology protein, discovered in 1985), consists of at least 22 GTPases grouped into several subfamilies. Among these, the **Rho** proteins proper and the **Rac** (an artificial acronym; note that in the literature Rac is used also for PKB/Akt, meaning "related to A- and C-kinases") and **Cdc42** proteins (originally discovered in a Cell-cycle defective cell type of *S. cerevisiae*) are best known. These proteins not only are key players in the regulation of cell shape, cell polarity, and cell motility but also are involved in enzyme regulation as well as in the transcriptional control of various genes. Moreover, they are essential for tumorigenesis caused by oncogenic Ras mutation, although they have not been found to undergo oncogenic mutations on their own. Nevertheless, cell cultures may become neoplastically transformed by an overexpression or overactivation of Rho GTPases.

Among the functions of Rho-type GTPases, their effects on the actin cytoskeleton have been investigated most thoroughly. The cytoskeleton is a highly ordered, although dynamic, structure responsible for the cell shape and all kinds of cellular movements as well as for intracellular vesicle trafficking, endocytosis, and cytokinesis. Moreover, together with the microtubular and intermediate filament skeletons and cellular membranes, it represents a supporting matrix of the data-processing protein network, thus playing a pivotal role in signal transduction (Section 1.5).

10.3.1 Upstream of Rho, Rac, and Cdc42: all kinds of transmembrane receptors

While Ras proteins are constitutively anchored in membranes, Rho-type GTPases shuttle between cytoplasm and membranes. The membrane anchor is a geranylgeranyl side chain. In the GDP-loaded, inactive Rho protein, this lipid residue is hidden in a complex with a specific **GDI** (GDP Dissociation Inhibitor) protein that renders the GTPase water-soluble and inhibits GDP–GTP exchange. How this inhibitory interaction is relieved, is not known. There is no doubt, however, that regulation of the Rho–GDI interaction plays a critical role in Rho signaling. Equally important as receivers of input signals are GEFs and GAPs.

Guanine nucleotide exchange factors and GTPase-activating proteins

More than 70 different GEFs activating Rho proteins have been identified. They constitute a very diverse family of proteins. The large number and high variability of GEFs indicates not only a wide variety of input signals (only a few of which have been identified yet with certainty) but also an involvement of Rho-type GTPases in many cellular processes.

The catalytic domain of Rho-GEFs is also known as DBL homology or DH domain (referring to an oncogenic Rho-GEF of diffuse B-cell lymphoma).

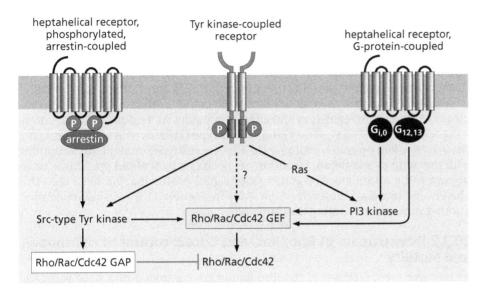

heptahelical receptor, phosphorylated, arrestin-coupled

Tyr kinase-coupled receptor

heptahelical receptor, G-protein-coupled

arrestin

Ras

Src-type Tyr kinase ⟶ Rho/Rac/Cdc42 GEF ⟵ PI3 kinase

Rho/Rac/Cdc42 GAP ⟶ Rho/Rac/Cdc42

Figure 10.8 Upstream of Rho, Rac, and Cdc42 To activate a Rho-, Rac-, or Cdc42 GEF, Tyr kinases of the Src family are assumed to cooperate with PI3 kinases. Src kinases are activated by arrestin-coupled heptahelical receptors or by autophosphorylated Tyr kinase-coupled receptors (P, phosphate). The latter also control, together with Ras, the activity of PI3 kinases of the types α, β, and δ, whereas the γ isoform is activated by G$\beta\gamma$-subunits mainly released from receptor-coupled G$_{i,0}$-proteins. In addition, at least some GEFs are activated by the α-subunits of trimeric G$_{12,13}$-proteins. A direct connection between Tyr kinase-coupled receptors and Rho/Rac/Cdc42 GEFs, indicated by the broken arrow with the question mark, is still a matter of debate. Not shown are pathways including Ras-GEFs with DBL homology domains, along which Rho, Rac, and Cdc42 in principle may become stimulated by nearly all signals that also activate Ras. Also not shown are signaling pathways involving Rho/Rac/Cdc42 GDIs and GAPs, some of which also become activated by Tyr phosphorylation.

Indeed, oncogenic mutations are known for several Rho-GEFs. Interestingly, in addition to their Ras-specific GEF domains, Ras-GEFs such as mSOS and GRF also contain DH domains, indicating again direct cross talk between Ras and Rho signaling.

Most Rho-GEFs have a PH domain C-terminal to the GEF domain. Certainly, the PH domain plays a role in the catalytic process; the mechanism of this effect, however, is still a matter of conjecture. Thus the obvious idea that the PH domain facilitates membrane anchoring of GEFs by binding to 3-phosphorylated phosphatidylinositols (see Section 4.4.5) is not convincingly supported by experimental data. To gain full activity, some Rho-GEFs must be phosphorylated, in addition, at Tyr residues, a reaction catalyzed by Tyr kinases of the Src family (Section 7.2) and others. Src kinase signaling is induced by Tyr kinase-coupled and heptahelical receptors that are arrestin-coupled (Figure 10.8).

Another signaling pathway leading to an activation of Rho-GEFs has the receptors of lysophospholipids, thrombin, endothelin, and thromboxane A$_2$ as starting points that may couple with trimeric G$_{12,13}$-proteins (Section 4.3.1). The G$\alpha_{12,13}$-subunits stimulate certain Rho-GEF isoforms directly, while the same Rho-GEFs turn up the GTPase activity of these Gα-subunits acting as regulators of G-protein signaling in a negative feedback loop (Section 4.3.2). This regulatory process provides an illustrative example of an interaction between large and small G-proteins.

Due to their highly variable and complex activation modes, Rho-GEFs provide perfect examples of logical gates interlinking many different signaling pathways. Considering the biological functions of Rho proteins, such interactions make sense, as for the majority of cell physiological processes (for instance, Tyr kinase-mediated cell proliferation), controlled changes of cell shape and motility are mandatory.

The extremely weak GTPase activity of Rho-type proteins requires stimulation by specific GAPs, which in mammals exist in about 80 isoforms. These are mostly large proteins with various interaction domains enabling intense cross-linking of Rho/Rac/Cdc42 signaling with other signaling pathways. As in the case of Rho-GEFs, input signals break up auto-inhibitory conformations of the GAPs. For this purpose, again the phosphorylation of Tyr residues seems to play an essential role.

Specificity by scaffold proteins

It has been found that individual GEFs and GAPS are coupled with distinct signaling pathways. Considering the large number and diversity of these proteins, how can cells select a specific GAP or GEF for a special purpose? The answer to this mystery is scaffold proteins that direct the GEFs and GAPs to distinct cellular sites and organize, as some kind of sockets, the formation of specific complexes with selected regulatory and effector proteins. As a result, scaffold protein A expressed in one cell type may couple Rho-type GTPases with gene expression, whereas scaffold protein B expressed in another cell type brings about a coupling with the actin cytoskeleton. Dictation of specificity by scaffold proteins is turning out to be a fundamental principle of signal processing. We have seen this already for protein kinases (Section 4.6). In Section 11.5 we shall encounter another example illustrating this principle.

10.3.2 Downstream of Rho, Rac, and Cdc42: control of cell shape and motility

As we have seen, GTPases of the Rho family are controlled by a wide variety of input signals. These correspond to an equally large number of effector proteins, indicating a wide array of functions that are still far from being known in all details. Among these functions, effects on the state of the actin cytoskeleton and its interaction with the motor protein myosin (see Sidebar 10.2) have been studied most thoroughly, revealing that many proteins involved in the regulation of these processes have interaction domains for the GTPases of the Rho family. In fact, such studies have ranked Rho, Rac, and Cdc42 among the key components of signaling pathways that control the architecture and motility pattern of cells as well as changes of cell shape occurring in the course of mitosis, endocytosis, exocytosis, and pinocytosis. A quick look at the extremely dense packing and branching of the actin fibers filling the cell (Figure 10.9) immediately demonstrates that such processes would be impossible without a rapid and continuous

Figure 10.9 Electron microscopic picture of the actin cytoskeleton
(A) Detail of a lamellopodium of a frog keratinocyte (the bar represents 0.5 µm). (B) Same detail at larger magnification (the bar represents 0.1 µm). (C–E) Actin cytoskeleton of a filopodium. Note the extremely dense packing and high degree of branching of the actin fibers. Note that the photographs represent snapshots of an ever-changing situation since the actin structures shown have a lifetime of minutes or less. (From T.H. Millard, S.J. Sharp & L.M. Machesky, *Biochem. J.* 380:1–17, 2004.)

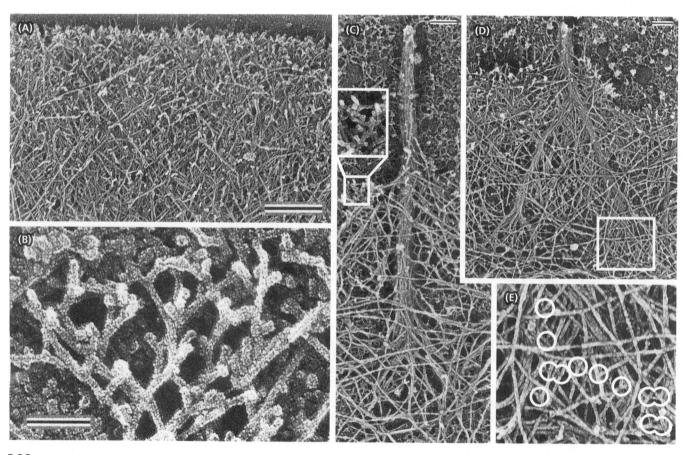

reorganization of this cytoskeleton, which has to be looked upon as a highly dynamic system rather than a stiff scaffold.

Originally the effects of Rho-type proteins were studied by use of fibroblasts grown in cell culture. Upon proper activation, these and some other cell types undergo characteristic morphological changes that have been found to depend on cooperation of Cdc42, Rac, and Rho proteins with each of them evoking a distinct cellular response (summarized in Figure 10.10).

When Cdc42 signaling is stimulated (for instance, along one of the pathways shown in Figure 10.8 or by a constitutively active Cdc42 mutant protein), the cells develop **filopodia.** These are fingerlike membrane extrusions possibly serving as some sort of antennas that help a migrating cell to reconnoiter the terrain. The characteristic cell morphological response to Rac activation consists of the formation of **lamellopodia** and **ruffled membranes**. Lamellopodia are fan-shaped mobile extrusions transiently developing at the front of migrating cells. Both filopodia and lamellopodia are pushed forward by targeted actin polymerization and are filled with either unbranched or branched actin fibers (Figure 10.9).

The cell responds to a selective stimulation of Rho by formation of **actomyosin stress fibers.** These are bundles of unbranched actin filaments decorated with myosin molecules that traverse the whole cell. Upon stimulation, these fibers contract under ATP consumption, causing changes of the cell shape and all kinds of movements. At the membranes they are anchored by special structures called **focal adhesion sites.** In a reversible way, focal adhesions sites fix the cell (during migration, particularly the membrane extrusions) with the supporting matrix (*in vitro* the Petri dish, *in vivo* connective tissue structures), thus providing a creeping (amoeboid) movement. In fact, the Rho-activating input signals frequently are derived from the extracellular matrix and are transmitted by integrins, a special type of Tyr kinase-coupled receptors concentrated at the focal adhesion sites. Combining the properties of highly dynamic cell adhesion molecules with those of signal receptors, integrins play an important role in the development and maintenance of tissues (for details see Section 7.2.3).

In a migrating cell, the morphological changes mentioned above occur in a strictly ordered sequence (Figure 10.10), indicating a high degree of cross talk and overlapping between Cdc42, Rac, and Rho signaling. In fact, sequential activation of Cdc42, Rac, and Rho in a small G-protein cascade has been repeatedly described, although the precise mechanism is still somewhat obscure.

Figure 10.10 Coordinated regulation of cell motility by Rho, Rac, and Cdc42: an overview (Top) Three phases of movement of a migrating cell. By actin polymerization the resting cell (1) develops lamello- and filopodia into the direction of movement (2). These extrusions become anchored to the substrate by new focal adhesions, enabling the cell to contract its rear part (3). (Bottom) Regulation of the motility apparatus and the cytoskeleton in different zones of a cell moving to the right, shown in a schematic and strictly simplified form. (A) Zone of actin polymerization induced by Rac and Cdc42 and resulting in the development of lamello- and filopodia; a re-contraction of these extrusions is prevented by Rac inhibiting myosin light-chain phosphorylation. (B) Zone of transient disintegration of the actin cytoskeleton due to an inactivation of Rho. (C) Zone of contraction; Rho induces the formation of actomyosin stress fibers and enhances the phosphorylation of myosin light chains.

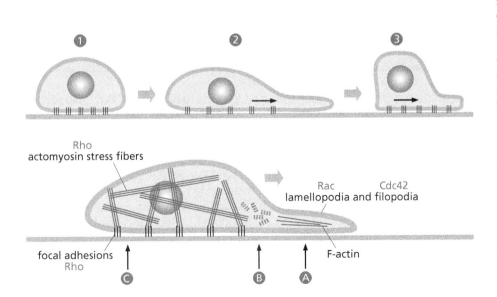

367

Sidebar 10.2 Rho, Rac, and Cdc42 control uptake of extracellular material Rho-type GTPases are typically involved in the uptake of extracellular material by phago- and pinocytosis and clathrin-dependent endocytosis. These processes are due to local changes of cell shape and motility.

Phagocytosis means uptake of larger particles. Some cell types such as macrophages and granulocytes are specialized in eliminating microorganisms by "professional" phagocytosis. They do this by clasping the foreign body with membrane extrusions called pseudopodia, thus forming an organelle, the phagosome. The process is triggered by certain receptors reacting with the complement factor C3b as well as with the Fc portion of immunoglobulins by which the invader has been "decorated." Via the corresponding GEFs, C3b receptors stimulate Rac and Cdc42, while Fc receptors activate Rho. In addition to phagocytes, other cells are qualified to respond by (in this case "nonprofessional") phagocytosis, for instance, in the final phase of programmed cell death (apoptosis) when the remnants of decayed cells are taken up by neighboring cells.

Pinocytosis, understood as the formation of vacuoles filled, for instance, with extracellular fluid or (in the case of antigen-presenting cells) antigens is also controlled by Rac and Cdc42. A special form of pinocytosis is **clathrin-mediated endocytosis** of transmembrane receptors or of material bound to such receptors (Section 5.8). In the regulation of this process, again GTPases of the Rho family are involved in a way still not entirely elucidated.

Like uptake reactions, also the secretory processes of **exocytosis** depend on reversible changes of the actin cytoskeleton that are controlled by Rho/Rac/Cdc42 proteins which here cooperate with Rab and Arf proteins (see next section).

In an intact tissue, the patterns of cell motility described here are probably less pronounced than in cell cultures. Nevertheless, cell migration is critical for a wide variety of physiological and pathological processes such as defense reactions, wound healing, and embryonic development as well as tumor invasion and metastasis. Moreover, by controlling the cytoskeleton, intracellular contact sites, and the motility apparatus, Rho-type GTPases dictate the shape and polarity of tissue cells, thus determining tissue morphology and function.

10.3.3 Downstream of Rho: protein kinases controlling dynamics of the actin cytoskeleton

A central question is how Rho, Rac, and Cdc42 signals are transduced to the molecular apparatus controlling the actin cytoskeleton and gene expression. In the following this question will be answered separately for Rho proteins proper and for Rac and Cdc42 proteins.

Among the large number of effector proteins of the Rho proteins proper, several inhibitors of actin polymerization and actin–myosin interaction are found. Rho inactivates these proteins, thus promoting the formation and contraction of actomyosin stress fibers. This effect is mediated by several Rho-controlled protein kinases, in particular ROK, LIM domain kinase, CRIK, and PRK.

Among these kinases, the RhO-activated Kinases **ROKα and ROKβ** (also known as ROCKs, RhO-Controlled Kinases) are particularly abundant. In addition to an amino-terminal kinase domain, these enzymes contain a Rho-binding domain and a PH domain (for membrane association and auto-inhibition) in the C-terminal part. ROKs belong to the family of AGC kinases and their activation follows the typical scheme explained for PKC-related kinases in Section 4.6.3: the auto-inhibition of the kinase brought about by binding of the C-terminal PH domain to the catalytic domain is relieved by interaction with membrane-bound Rho-GTP, enabling autophosphorylation of the activation loop and of a hydrophobic domain located near the C-terminus of the catalytic domain. As a result, the substrate-binding pocket of ROK becomes opened. [Note that kinase PDK1, which catalyzes activation loop phosphorylation in other AGC kinases (Section 4.6.4), does not seem to be involved in ROK activation.]

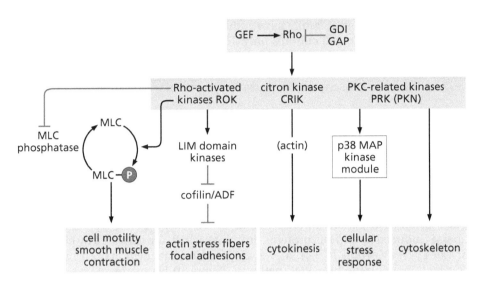

Figure 10.11 Effects of protein kinases controlled by Rho Rho is activated by guanine nucleotide exchange factors (GEF), inactivated by GTPase-activating proteins (GAP), and inhibited by GDP dissociation inhibitors (GDI). The effector protein kinases of Rho are shown in the colored box. MLC, myosin light chain; P, phosphate. For details, see text.

Among the many substrates of ROK, the regulatory light chains of myosin (MLC, myosin light chains) are of particular interest, since they serve as key regulators of actomyosin contraction by hindering the interaction between actin and myosin. This inhibitory activity is abolished by phosphorylation catalyzed by ROKs (or by a Ca^{2+}-dependent MLC kinase, see Section 14.6.2). To augment this effect, ROK simultaneously inactivates the MLC phosphatase (Section 14.6.2) that would reverse MLC phosphorylation. As a result, Rho increases the level of MLC phosphorylation, thus facilitating contraction of the actomyosin fibers. In parallel, ROK activates other protein kinases called **LIM domain kinases** (LIM stands for the proteins Lin11 of *C. elegans*, Isl1 of rats, and MEC3 of *C. elegans* where the LIM domain was found for the first time). The only substrates of the Ser/Thr-specific LIM domain kinases known are the proteins of the **cofilin/actin depolymerization factor** family. As indicated by their name, these proteins sever F-actin fibers, thus causing a breakdown of the cytoskeleton and providing monomeric actin for further polymerization (see Sidebar 10.3). This effect is prevented by LIM domain kinase-catalyzed phosphorylation, resulting in an increase and stabilization of polymerized actin as required for stress fiber formation. Since the promoting effect of Rho on the formation and contractility of actomyosin stress fibers depends entirely on ROK-catalyzed phosphorylation of MLCs, MLC phosphatase, LIM domain kinase, and some other cytoskeletal proteins, ROKs are considered to be key effectors of Rho (Figure 10.11).

As specialized as LIM domain kinase is the Ser/Thr-specific **citron kinase** or **CRIK,** another Rho (and Rac) effector protein (citron is the trivial name of a Rho-interacting protein). Citron kinase is distantly related to ROKs and contributes, in particular, to the formation of a contractile actomyosin ring that separates daughter and mother cells in the course of mitosis.

In contrast to the highly specific kinases LIMK and CRIK, the Protein kinase C-Related protein Kinases **PRK** are multifunctional Rho effectors. Effects of these kinases on the actin cytoskeleton, the intermediate filament cytoskeleton, and the microtubule cytoskeleton as well as on cell adhesion and vesicle transport have been repeatedly described, though the molecular details are not entirely known. As discussed in Section 4.6.3, PRKs are stimulated by phospholipid-dependent kinase 1 (PDK1)-catalyzed activation loop phosphorylation. The task of the membrane-anchored Rho-GTP is to bring PRK in contact with membrane-bound PDK1 and to change the PRK conformation in such a way that PDK1 gains access to the catalytic center. To this end Rho-GTP binds to a specific Rho interaction domain of PRK. Thus, in recruiting its effector kinase to the membrane, Rho plays a role resembling that of Ras in the activation of RAF and PI3K.

The enzymes of the PRK family have additional functions, which in most cases have been studied in cell cultures only. The results obtained indicate a role of PRKs as MAP4 kinases of stress-activated p38 MAP kinase modules (Figure 10.11 and Section 11.4). Isoform PRKα even translocates into the nucleus, possibly modulating gene transcription by phosphorylation of transcription factors.

Another enzyme stimulated by Rho and playing an important role in actin polymerization is **phosphatidylinositol 4-phosphate 5-kinase**. Its product, phosphatidylinositol 4,5-bisphosphate, not only is the precursor of the second messengers DAG and InsP$_3$ but also interacts with actin-binding proteins such as profilin, vinculin, and gelsolin, thus promoting actin polymerization at focal adhesion sites.

The formation of parallel unbranched actin fibers as they are typical for stress fibers is induced by Drf/Dia proteins (formins), which are activated by either Rho or Cdc42 (see Section 10.3.4). Moreover, tension of its own evoked by myosin contraction along the Rho–ROK pathway (Figure 10.11) seems to be a stimulus for stress fiber formation.

10.3.4 Rac and Cdc42 effects: regulation of actin polymerization and generation of reactive oxygen species

Rac- and Cdc42-controlled protein kinases

The number of Cdc42 and Rac effectors rivals that of Rho effectors. In Cdc42 and Rac signaling, three protein kinase families occupy positions resembling that of ROKs and PRKs in Rho signaling (Figure 10.12):

- Myotonin-Related Cdc42-binding Kinase **MRCK**

- p21-Activated Kinases **PAK** (p21 refers to the molecular mass of 21 kDa of the Rac protein; see also Section 11.4)

- Mixed-Lineage Kinases **MLK** (treated in more detail in Section 11.3.2)

As far as their cellular functions are concerned, these kinases have much in common with PRKs and ROKs, again indicating a close relationship and a high degree of overlap between Rho-, Rac-, and Cdc42-dependent signaling. They are recruited by the GTP-loaded small G-proteins to the inner side of the cell

Figure 10.12 A selection of pathways downstream of Cdc42 and Rac Both Cdc42 and Rac are selectively activated by guanine nucleotide exchange factors (GEF) and inhibited by GTPase-activating proteins (GAP) and GDP dissociation inhibitors (GDI). Double-headed arrows indicate reciprocal activation of Cdc42, Rac, and Rho in small G-protein cascades. Direct effector proteins of Rac and Cdc42 are shown in the colored box (MLKs, mixed-lineage kinases; PAKs, p21-activated protein kinases; MRCK, myotonin-related Cdc42-binding kinase; WASP, Wiscott–Aldrich syndrome proteins; WAVE, WASP/verprolin-homologous proteins). From the many putative pathways, only a few supported by strong experimental evidence are shown. For other details see text.

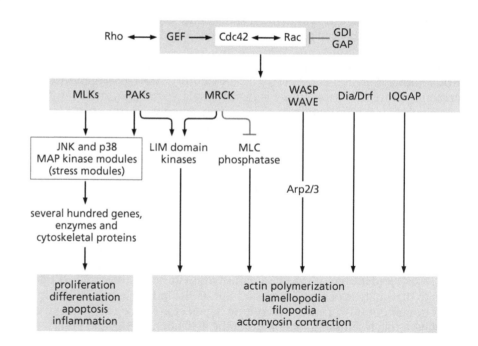

membrane where PAKs become phosphorylated and activated by PDK1 while MLK and MRCK undergo autophosphorylation. Experiments showing that in the absence of Cdc42 and Rac certain membrane components such as sphingolipids and gangliosides may stimulate PAK phosphorylation by PDK1 support the concept of a membrane-recruiting effect of Cdc42 and Rac.

Moreover, PAKs, MRCK, and MLKs share several substrate proteins with ROKs and PRKs. Like ROKs, MRCK stimulates actomyosin contraction by inactivating the phosphatase via dephosphorylation of the myosin light chains. Another example of overlapping functions is provided by the stimulatory effects of PRKs, PAKs, and MLKs on stress-activated MAP kinase modules, whose activation results in a wide variety of cellular responses including gene expression (for details see Sections 11.3 and 11.4).

Like ROKs, MRCK, and PAKs stimulate LIM domain kinase, thus promoting the formation of actomyosin stress fibers. However, in contrast to ROKs, PAKs *suppress* myosin light chain phosphorylation and actomyosin contraction. To this end they phosphorylate and inactivate the Ca^{2+}-dependent myosin light chain kinase. It has been speculated that, through this effect, a retraction of Rac-induced lamellopodia is prevented.

Signaling networks controlling formation and outgrowth of actin fibers

As mentioned above, activation of LIM domain kinase hinders the cofilin-induced *breakdown* of actin fibers, thus stabilizing the actin cytoskeleton and promoting the formation of actomyosin stress fibers. In addition, Rho, Rac, and Cdc42 promote the *outgrowth* of new actin fibers. The latter effect is mediated by actin-associated regulatory proteins such as WASP, WAVE, Drf/Dia, Arp2/3, and IQGAP, the activities of which are controlled by the small G-proteins (Figure 10.13).

As far as their function is concerned, WAVE, Drf/Dia, and WASP belong to a family of factors that control the formation of new actin fibers in a process called nucleation (see Sidebar 10.3). **WASPs** got their name from **Wiscott–Aldrich syndrome**; this is a severe albeit rare hereditary immune deficiency caused by a loss-of-function mutation of the protein-encoding gene. Consequently, cells of the immune system (T-lymphocytes, macrophages, and monocytes) are restricted in their mobility, rendering them unable to perform movements in response to chemotactic signals.

WASPs are scaffold proteins that form functional complexes of nucleation-controlling proteins. As in many other signal-transducing proteins, WASP activity

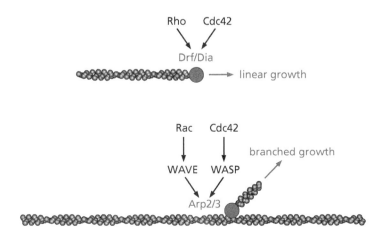

Figure 10.13 Linear and branched growth of actin fibers induced by Rho-type GTPases Rho or Cdc42 activates Drf/Dia proteins (formins) to induce linear growth at barbed ends, whereas Rac or Cdc42 stimulates WAVE or WASP proteins to form complexes with Arp2/3, thus triggering the outgrowth of a branch from a preexisting actin fiber.

Sidebar 10.3 A quick glance at dynamics of the actin cytoskeleton Actin is one of the most abundant and most conservative proteins of the eukaryotic cell. It forms tracks for motor proteins and contractile protein complexes such as actomyosin, and with microtubules and intermediate filaments, it constitutes the cytoskeleton. To be particular, the term cytoskeleton is somewhat misleading because the actin network is not a rigid scaffold but a highly dynamic structure that transforms chemical energy (from ATP) into mechanical energy and undergoes continuous reorganization, thus determining the ever-changing shape and motility of cells.

Actin exists as a monomer (globular or G-actin, molecular mass 42 kDa) and as a polymer (fibrillar or F-actin) and is subject to continuous polymerization and depolymerization depending on the cellular conditions and input signals. G-actin polymerizes spontaneously unless it is hindered by other proteins. In the starting phase, however, spontaneous polymerization is kinetically blocked because actin dimers and trimers are unstable. Once the trimeric state is surpassed, polymerization proceeds rapidly, yielding long filaments. Therefore, the trimers are called nuclei and the process of trimer formation is known as nucleation. In the cell nucleation is controlled (promoted or prevented) by several regulatory proteins, which are targets of signaling pathways. Due to a head-to-tail arrangement of the asymmetric actin monomers, F-actin filaments show a polarity with pointed and barbed ends.

Although actin binds ATP with high affinity and exhibits ATPase activity, filament formation is a spontaneous exergonic reaction and does not require ATP hydrolysis. Logically, disassembly of the filaments is endergonic and depends on ATP hydrolysis catalyzed by the intrinsic ATPase activity of actin. In other words: actin polymerization has an installed timer. The rates of filament assembly and disassembly are different at both ends of the filament, resulting in a continuous treadmilling of actin monomers from the barbed end, where they become bound as ATP complexes, to the pointed end, where they become released as ADP complexes. As a result, the growing filament appears to move forward and generates pressure that may push forward cell membranes; that is, ATP energy is transformed into mechanical energy. Treadmilling is controlled by regulatory proteins modulating both ADP–ATP exchange at freshly released monomers (such an exchange factor is profilin) and ATPase activity in the polymer (note that the activity of actin is regulated in a similar manner as that of G-proteins). The same holds true for branching and debranching of the filaments, which in addition to treadmilling is essential for the protrusion of membrane processes such as lamello- and filopodia and microvilli. Debranching is driven by ATP hydrolysis, catalyzed by those regulatory proteins (such as Arp2/3) that serve as templates for branching.

In the formation and breakdown of the actin cytoskeleton, a large number of regulatory proteins (more than 50) is involved, exhibiting (among others) the following functions. [Examples of corresponding proteins are given in parentheses; for details the reader is referred to textbooks of cell biology. A complete list of actin regulators is provided by A.D. Siripala and M.D. Welch, *Cell* 128:626, 1041, 2007.]

- nucleation of new actin fibers, in particular at branching points (Arp2/3)
- nucleation of unbranched actin filaments (Spir, Drf/Dia)
- nucleation of branched actin filaments (WASP, WAVE)
- binding and inhibition of spontaneous aggregation of G-actin (profilin, β-thymosine)
- depolymerization, debranching, severing of the polymer (cofilin, gelsolin)
- capping of the growing end of the polymer (tropomodulin, IQGAP)
- cross-linking of polymers (α-actinin, fascin)
- membrane anchoring (annexin, spectrin, ezrin, moesin, radixin)
- motor protein (myosin)
- cell–extracellular matrix junctions (talin, tensin, dystrophin, vinculin)
- regulation of actin–myosin interaction (troponin, tropomyosin, caldesmon, myosin light chains)

In addition to generating the power for membrane protrusion, actin is involved in still another energy-transforming process: by binding to actin, the ATPase function of myosin is activated. As a result, the actomyosin complex contracts due to a conformational change. In eukaryotes this reaction is the basis of most cellular movements ranging from muscle contraction to amoeboid creeping and changes of the cell shape by stress fiber contraction. Actomyosin is also involved in intracellular cargo transport. Triggered by Ca^{2+}, the interaction between actin and myosin is evoked by all input signals inducing an increase of the cytoplasmic Ca^{2+} concentration. The classical example is provided by muscle contraction (see Section 14.6.2). In parallel, actin dynamics and actin–myosin interactions are controlled by signals transmitted along the Rho/Rac/Cdc42 pathways. As a rule, the input signals affect regulatory proteins rather than actin or myosin themselves.

(the ability to bind actin and other proteins) is suppressed by an auto-inhibitory conformation that is broken up by an input signal. In the case of WASPs, this is the interaction with membrane-bound GTP-loaded Cdc42. To this end Cdc42 binds to a specific site of the WASP molecule known as the CRIB (Cdc42 and Rac Interactive Binding) or GTPase binding domain that is found in many Rac/Cdc42 effector proteins. In addition, WASPs have proline-rich domains that interact with SH3 domains of adaptor proteins, especially those of the MAP kinase pathways (such as Grb2 and Nck). Moreover, WASPs are activated by Tyr phosphorylation catalyzed by Src-type kinases. These alternative mechanisms of WASP activation resemble a logical AND gate that links mitogenic stimulation (for instance, by growth factors) with a reorganization of the cytoskeleton. Changes of the cell shape such as rounding-off are, indeed, typical features of cell division.

In an analogous way, albeit along another pathway, Rac activates **WAVEs** (WASP/VErprolin-homologous proteins), a group of scaffold proteins related to A-kinase anchor proteins (Section 4.6.1), which like WASPs assemble nucleation-promoting multiprotein complexes. Since WAVEs do not contain a CRIB domain, Rac interacts with accessory proteins, which in turn transform the WAVE protein proper into an active conformation. WAVEs are enriched particularly in lamellopodia. Like WASPs, the WAVE proteins are under the control of various signaling pathways that either up- or down-regulate their activities. So WASPs and WAVEs may be looked upon as hubs in a network of interactions controlling cell shape and motility.

To induce the outgrowth of actin filaments, WASPs and WAVEs have to form complexes with actin trimers or higher oligomers. Kinetically the formation of such oligomers is highly unfavorable and requires the assistance of regulatory proteins that bypass this obstacle (see Sidebar 10.3). Here the so-called nucleator proteins such as **Arp2/3** (Actin-related protein) enter the stage. Arp2/3, a complex of seven polypeptides, mimics a *stable* actin dimer interacting with both WASP (or WAVE) and actin. In the absence of an input signal, its activity is blocked by an auto-inhibitory conformation. When it is activated along the Cdc42–WASP (or Rac–WAVE) pathway, Arp2/3 binds to the side of actin filaments and induces the polymerization of an F-actin side chain (Figure 10.13), thus playing a key role in the formation of *branched* actin fibers and their outgrowth to a branched lattice (as shown in Figure 10.9). In the cell such branches have a lifetime of only a few minutes before they disintegrate, reflecting the dynamics of the actin cytoskeleton. The process of debranching requires energy provided by ATP hydrolysis, which is catalyzed by the ATPase function of Arp2/3.

In contrast to WAVE and WASP, the **Drf** or **Dia proteins** (Diaphanous-related formins, referring to the homologous *Drosophila* protein Diaphanous that belongs to a family of proteins with formin domains), also called **formins,** do not interact with Arp2/3. Instead they interact with the barbed end of F-actin fibers and induce the outgrowth of parallel bundles of unbranched actin filaments as found in filopodia and stress fibers (Figure 10.13). Drf/Dia proteins are abundant in animal cells. They are bound and activated by either Rho or Cdc42.

The outgrowth of actin fibers is also under *negative* control by what is called capping. Capping is understood as a binding of special regulatory proteins at the growing end of the fiber, the further polymerization of which is thus prevented (see Sidebar 10.3). Rac and Cdc42 suppress capping and promote the outgrowth of actin fibers. This effect is thought to be mediated by proteins of the **IQGAP** family (the name refers to the original identification of those proteins as Ras-GAPs, which later turned out to be a misconception), although IQGAPs do not contain CRIB domains. IQGAPs are multipurpose scaffold proteins. They contain various interaction domains that bind other cytoskeletal proteins as well as receptors, protein kinases, and proteins involved in cell adhesion. This indicates a wide variety of IQGAP functions, the physiological relevance of which is still known only fragmentarily. Moreover, it should be mentioned that the role of

IQGAPs as Rac/Cdc40 effector proteins is still somewhat controversial since other data indicate a function *upstream* of these G-proteins; for instance, by inhibition of their GTPase activity, thus increasing the concentration of GTP-loaded forms at the leading edge of migrating cells. Probably IQGAPs exhibit both effects depending on the cell type and the physiological context.

Generation of reactive oxygen species

With the control of actin dynamics, the potencies of Rho, Rac, and Cdc42 are by no means exhausted. In fact, cooperating, as mentioned, with Ras along the MAP kinase pathways, they modulate the effects of a large number of proteins and genes.

Among the special functions of Rac, regulation of **NADPH oxidase (NOX)** activity has found particular attention. Using nicotinamide adenine dinucleotide phosphate (NADPH) as a reductant, this enzyme (or more precisely a whole electron transport chain) catalyzes the reduction of molecular oxygen to superoxide anion radicals and hydrogen peroxide. NOX is a multienzyme complex consisting of Rac, the Rac effector protein p67^{Phox} (Phox stands for Phagocyte NADPH oxidase), the protein p47^{Phox}, and a membrane-bound flavoprotein–cytochrome b complex (gp91^{Phox}). Fulfilling the typical task of a small G-protein, Rac promotes the assembly of this complex at the membrane.

NADPH oxidase is expressed predominantly by phagocytes (monocytes and neutrophilic granulocytes). Committing some sort of chemical warfare, these cell types use the reactive oxygen species produced by NOX as well as even more aggressive derivatives thereof to kill microorganisms taken up by phagocytosis. The reaction proceeds in an explosive manner called oxidative burst and represents an essential component of nonspecific defense and inflammation. NOX isoforms probably serving additional purposes are found in practically every tissue. In fact, the concept of reactive oxygen species providing intracellular signals is gaining steadily increasing support (for details and examples see Sections 2.2, 7.1.1, and 13.2.4). These signaling pathways are expected, therefore, to be under the supervision of Rac.

10.3.5 Signals controlling microtubule and intermediate filament cytoskeletons

The cytoskeleton of eukaryotic cells consists of three major structural components: actin fibers (microfilaments), microtubules, and intermediate filaments. Although in contrast to the actin filaments they are found to be connected only loosely with small G-proteins, microtubules and intermediate filaments are briefly reviewed here as far as their role in cellular signal processing is concerned. For detailed information the reader may consult textbooks of cell biology.

Microtubules: tracks and scaffolds

Microtubules are polymers of α,β-tubulin dimers. In determining cell shape and polarity, building up the mitotic spindle, and providing tracks for intracellular trafficking, they cooperate with actin filaments. Like the actin cytoskeleton, the microtubule cytoskeleton is a highly dynamic structure undergoing continuous reorganization driven by polymerization and depolymerization. Again these processes are controlled by various regulatory proteins called **MAPs** (Microtubule-Associated Proteins; not to be confused with MAP kinases), which are targets of signaling cascades. One of these proteins is Tau, the hyperphosphorylation of which has a bad reputation as one major cause of Alzheimer's dementia (see Chapter 13).

Beside other kinases, those activated by Rho-type GTPases have been observed to catalyze MAP phosphorylation, although as compared with the role of small G-proteins in actin dynamics this field of research is still in its infancy. For instance, microtubules disassemble when certain polymerization-promoting

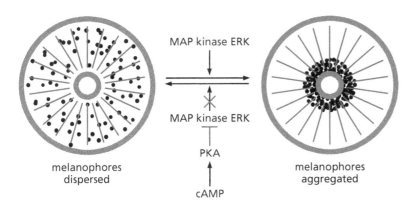

MAP kinase ERK

MAP kinase ERK

PKA

cAMP

melanophores dispersed

melanophores aggregated

Figure 10.14 Signal-controlled cargo transport along microtubule tracks Reversible transport of pigment granules (melanophores) along microtubule tracks (red lines) in melanocytes is shown. Phosphorylation by the MAP kinase ERK of the motor proteins dynein and kinesin promotes aggregation in the cell center and is inhibited by protein kinase A (PKA). As a consequence, cAMP promotes melanophore dispersion.

regulatory proteins (stathmins and collapsin response mediator protein 2) become phosphorylated and inactivated along the Rac/Cdc42–PAK and Rho–ROK pathways. Moreover, the Rac/Cdc42 effector protein IQGAP seems to regulate the interaction of microtubules with membrane proteins.

Microtubules serve as tracks for **cargo transport** driven by kinesin and dynein motor proteins. A peculiarity of the microtubule cytoskeleton is that it forms a radial network of filaments with the fast-growing (plus) ends directed to the cell periphery and the slow-growing (minus) ends facing the cell center. Depending on extracellular signals and the dynein and kinesin type, cargo may be transported in either direction.

To study this process, melanocytes have turned out to be the model of choice. In these cells pigment granula, or melanophores, are found either dispersed over the whole cell or concentrated in the cell center, depending on environmental stimuli. The central aggregation is promoted by phosphorylation of the motor proteins catalyzed by the MAP kinase ERK. As explained in Chapter 11, ERK activity is attenuated by PKA. In fact, an increase of cAMP activating PKA correlates with dispersion of melanophores (Figure 10.14).

Due to the short signaling pathways the redistribution of pigment granula occurs rapidly, explaining the immediate changes of color pattern in the skin of certain animals such as octopusses. Provided these findings can be generalized, microtubule-directed cargo transport must be assumed to be under the control of a wide variety of extracellular signals that activate the Ras–RAF–ERK cascade and cAMP production.

Intermediate filaments: control through phosphorylation

Beside actin-derived microfilaments and tubulin-derived microtubules, intermediate filaments (IFs) are the third cytoskeletal structure found in all eukaryotic cells (except yeast). They are formed by a wide variety of proteins that assemble into insoluble homo- and heteropolymers, filling the cell with a spongelike structure. Members of the IF protein family are (among others) epithelial keratins, mesenchymal vimentins, muscular desmins, neuronal neurofilaments, and the abundant laminins of the nuclear lamina. In other words, in contrast to the chemical composition of microfilaments and microtubules, that of intermediate filaments is highly variable and tissue-dependent.

While actin fibers and microtubules are anchored to focal adhesion sites and adherens junctions, the contact sites connecting the intermediate filaments of adjacent cells are known as **desmosomes**. Moreover, via **hemidesmosomes**, intermediate filaments also anchor cells to connective tissue structures such as the basal lamina of epithelia. Intermediate filaments, therefore, decisively determine the mechanical properties of tissues.

Figure 10.15 Schematic structure and sites of post-translational modifications of an intermediate filament protein For details see text.

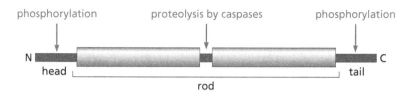

What is known about interactions of the data-processing apparatus with this cytoskeletal structure? A common structural feature of IF proteins is a helical rod module flanked by an N-terminal head and a C-terminal tail domain (Figure 10.15). Via a coiled-coil structure of their rod modules, IF proteins form dimers, which then polymerize to filaments. Thus, while the rod module determines the biomechanical properties of the IF protein, the head and tail domains are targets of regulatory signaling pathways. To this end they contain a high number of Ser, Thr, and Tyr phosphorylation sites. A major difference between microfilaments and microtubules on the one hand and intermediate filaments on the other is indeed that polymerization of the former is regulated by ATP and GTP hydrolysis controlled by accessory proteins, while for the latter direct phosphorylation is the control mechanism of choice. In fact, intermediate filaments rank among the most abundant phosphoproteins, and their phosphorylation is a highly dynamic process that in a reversible way controls their grade of polymerization and their intracellular distribution.

Like the actin and microtubule cytoskeletons, the IF cytoskeleton is highly dynamic, rapidly becoming polymerized and depolymerized on demand. Depending on the grade of polymerization, IF proteins shuttle between a soluble cytoplasmic pool and an insoluble cytoskeletal pool, thus resembling actin and tubulin. Multiple phosphorylation promotes depolymerization (or hinders polymerization). It increases during mitosis (when disintegration of the nuclear lamina and of the cytoskeleton is mandatory) and upon tissue injury. In the latter case, phosphorylation certainly is needed to prime cells for cell division aiming at wound repair. In addition, however, intermediate filaments seem to provide a phosphate sink, dampening the activity of pro-apoptotic protein phosphorylation that is stimulated via the MAP kinase stress modules. Moreover, phosphorylation protects IF proteins from degradation by apoptotic caspases or proteasomes. Thus, counterproductive cell death is prevented in the damaged tissue.

Phosphorylation also controls the intracellular compartmentalization of intermediate filaments, for instance, in neurons and polarized epithelial cells. As far as the role of IF proteins in cellular stress protection is concerned, an exciting new aspect has been added by studies on wounded nerve cells: this is the role of IF proteins in **intracellular cargo transport**. At the site of damage, MAP kinase cascades (and other signaling pathways) become activated and the MAP kinases migrate into the nucleus to modulate gene expression. When the damage has occurred, for instance, in the axon near to the synaptic terminal, the sheer size of a neuron raises the question as to how the large distance (up to 1 meter) between the nerve ending and the nucleus can be bridged in a reasonable time. The most probable mechanism is active transport by motor proteins such as kinesins and dyneins. In fact, in neurons a nonfilamentous form of the IF protein vimentin has been found to serve as a scaffold, anchoring activated (phosphorylated) MAP kinases to the dynein motor and protecting them from inactivation (by dephosphorylation) during their long journey to the nucleus. This special vimentin type is produced from conventional vimentin by limited proteolysis catalyzed by the Ca^{2+}-dependent protease calpain (note that the cytoplasmic Ca^{2+} level rises at sites of cell damage). The idea is easily at hand that the role of IF proteins in cellular transport is not restricted to damaged neurons but may represent a general mechanism of long-range signaling. The large number of different IF proteins expressed in mammalian cells might provide this process with a high degree of selectivity.

A wide variety of protein kinases have been found to catalyze IF protein phosphorylation. These include PKC, PKB/Akt, RAF, CDK1, CDK5, ERK-type MAP kinases, and (last but not least) ROKs. In addition, Tyr phosphorylation has been observed. Therefore, intermediate filaments are targets of numerous signaling cascades, and pronounced cross talk between the control of intermediate filament dynamics and pathways regulating cell proliferation, cell differentiation, cell shape, and cell motility has to be postulated.

Summary

Rho, Rac, and Cdc42 couple various cell physiological processes such as proliferation, tissue-specific function, stress responses, and endo- and exocytosis with changes of cell shape, cell motility, and selective gene expression. In this context a major function of Rho/Rac/Cdc42 is to control the dynamics of the actin cytoskeleton. Guanine nucleotide exchange factors (GEFs) of Rho, Rac, and Cdc42 proteins constitute a highly diverse group of proteins with various interaction motifs. For activation they require an association with membrane-bound phosphatidylinositol 3-kinase (PI3K) products and, in addition, phosphorylation of Tyr residues catalyzed by Src-type Tyr kinases. Effects of Rho on the cytoskeleton are mediated by a series of protein kinases that directly interact with Rho-GTP. Rho-activated kinases (ROKs) rescue the contractile actomyosin system from the blockade by myosin light chains (MLCs) while at the same time promoting formation of focal adhesion sites and stress fibers; citron kinase (CRIK) is specifically involved in cytokinesis; and protein kinase C-related kinases (PRKs) transduce stress signals to the genome via the p38 MAP kinase module. Stress fiber formation at focal adhesion sites is also facilitated by other Rho effectors and by tension. Like Rho, Cdc42 and Rac control the dynamics of the actin cytoskeleton by activation of certain protein kinases. These include myotonin-related Cdc42-binding kinases (MRCKs) that, as activators of MLC phosphorylation and actin polymerization, functionally resemble ROKs and p21-activated kinases (PAKs), as well as mixed-lineage kinases (MLKs) that stimulate stress-processing MAP kinase modules. Moreover, by inhibition of MLC phosphorylation, PAKs are probably involved in negative feedback control of actomyosin contraction. Cdc42 and Rac effector proteins that control the nucleation and polymerization of actin fibers are WASP, WAVE, Drf/Dia, and IQGAP. They play a key role in actin-dependent cellular events such as motility, intracellular vesicle trafficking, exo- and endocytosis, phagocytosis, and outgrowth of membranous structures such as neurites. Rac transduces signals to NADPH oxidase, thus stimulating the generation of reactive oxygen species used either as antimicrobial and pro-inflammatory substances or as messengers in intracellular signaling. Another component of the cytoskeleton is microfilaments. The polymerization and depolymerization of these tubulin aggregates is regulated by a series of MAPs, the activities of which are controlled by protein phosphorylation rather than by Rho/Rac/Cdc42 directly. Cargo transport along microtubule tracks is regulated by phosphorylation of the associated motor proteins (kinesins and dyneins). Intermediate filaments represent the third type of cytoskeletal structure, formed by the spontaneous aggregation of several quite diverse proteins. Their polymerization is controlled predominantly by protein phosphorylation catalyzed by ROKs and other kinases.

10.4 Arf and Rab proteins: control of vesicle transport

10.4.1 Intracellular traffic of secretory vesicles

A characteristic feature of eukaryotic cells is compartmentalization: the separation by membranes into organelles and other areas of individual biochemical activities. Like the plasma membrane, the intracellular membranes consist of lipid bilayers that are impermeable for hydrophilic substances such as proteins and nucleic acids. Such material is packed, therefore, in membrane-covered vesicles that travel between the different compartments. The principle of this

Figure 10.16 Transport of cargo-filled vesicles between cellular compartments

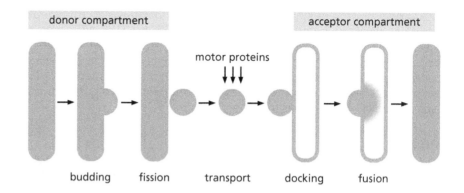

macromolecular trafficking is shown in Figure 10.16: a cargo-filled vesicle is released from the membrane of a donor compartment and transported by means of ATP-driven motor proteins along microtubule tracks to an acceptor compartment, where it fuses with the membrane and delivers its content.

This principle holds true for import of extracellular material (**endocytosis**), secretion of intracellular substances (**exocytosis**), and intracellular traffic between different compartments. Endo- and exocytosis are essential for communication between the cell and the environment. By exocytosis, signaling factors such as hormones and neurotransmitters are released and receptors and other proteins are exposed at the cell surface, and the reverse holds true for endocytosis (Figure 10.17).

Exocytosis of a new protein starts at the site of biosynthesis (the endoplasmic reticulum), proceeds via the different Golgi compartments (cis-, medial-, and trans-Golgi), and ends with the emptying of the secretory vesicle at the cell surface. For endocytosis, the material bound to receptors at the cell surface assembles in membrane invaginations (coated pits and caveolae), which become separated and released as vesicles into the cytoplasm. The vesicles are taken up by larger organelles, the endosomes, which transfer the material either to lysosomes (to become degraded) or back into the exocytotic cycle (for additional details on endocytosis, see Section 5.8).

Figure 10.17 Secretion by exocytosis The principle is shown for a glandular cell (left). The secretion is produced at the endoplasmic reticulum, filled into vesicles, and transported via the Golgi apparatus to the cell surface to be released. A cellular structure specialized for secretion, the presynaptic terminal of a chemical synapse, is shown on the right. The secretory vesicles filled with neurotransmitter migrate along cytoskeletal tracks (neurofilaments) from the cell body of the neuron (not shown) to the end of the axonal nerve fiber to become emptied into the synaptic cleft, that is, the gap between pre- and postsynaptic cells. (Adapted from R.V. Krstic, Ultrastruktur der Säugetierzelle, Berlin: Springer-Verlag, 1976.)

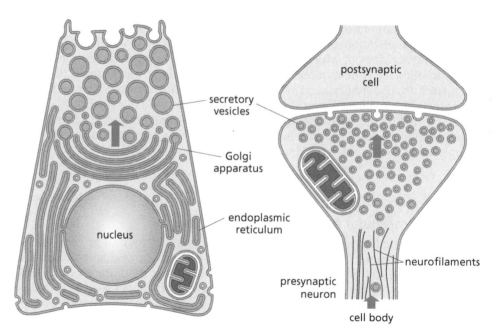

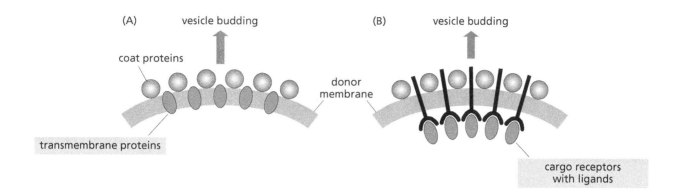

Figure 10.18 Formation of secretory vesicles Cargo molecules (colored ellipses) are concentrated in the donor membrane by interacting with vesicular coat proteins. (A) Transmembrane proteins as cargo. (B) Soluble proteins as cargo; for uptake into the vesicle, they are bound by vesicular target receptors.

To become separated from the donor membrane, the vesicles are covered by a tangle of COat Proteins (COPs), producing a conformational tension that leads to a budding of the membrane. Three types of coat protein complexes, called coatomeres, are known, the functions of which depend on the donor compartment:

- **COPI** controls early phases of secretion, in particular at the Golgi complex

- **COPII** is required for vesicle formation at the endoplasmic reticulum

- **clathrin adaptor protein** (Section 5.8) ensures the endocytotic separation of vesicles from the plasma membrane and vesicle release from the trans-Golgi compartment

The protein coat promotes both the loading and the release of vesicles. If the cargo is membrane proteins, such as receptors, these make contact with coat proteins via their cytoplasmic domains and become concentrated at the site of vesicle that is budding. In contrast, soluble cargo, such as hormones, binds to transmembrane cargo receptors that interact with coat proteins (Figure 10.18). Prior to fusion with the acceptor membrane, the vesicle has to be uncoated. Vesicle traffic is strictly controlled by the signal-processing protein network. As in most targeted processes, G-proteins are the essential switches, working according to principles conserved from yeast to humans.

10.4.2 Fusion machinery

A vesicle spontaneously fuses with an acceptor membrane when they come in close contact. The surfaces are covered, however, by water molecules, which hinder direct contact between the lipid bilayers. To overcome this obstacle, complementary surface proteins interact with each other, displacing the water coat and pulling the opposite membranes together. This process is strictly controlled in order to prevent chaotic fusions. In other words, the proteins of the fusion machinery become active only when they receive a proper input signal.

Fusion is a two-step event. In the first step, the vesicle becomes fixed by so-called tethering proteins at the site of fusion, and in the second step, specific protein machinery promotes the fusion of the membranes (Figure 10.19).

The first fusion-competent protein identified was found by a classical biochemical approach. It was observed that *N*-ethylmaleimide (NEM), a customary enzyme inhibitor that alkylates SH groups in proteins, prevented the fusion of vesicles with membranes. The effect was traced back to a protein inactivated by the agent. This NEM-Sensitive Factor **NSF**, an ATPase of the AAA+ protein family (see Section 2.5.2), has turned out to be required for every step of exocytosis. NSF binds to fusing membrane areas, but it needs the support of adaptors, Soluble NSF-Attachment Proteins or **SNAPs**. The SNAPs make contact with the NSF effector proteins called SNAP REceptors or **SNARE**, which are the fusion proteins proper. Mammalian cells express various SNARE isoforms that are differently

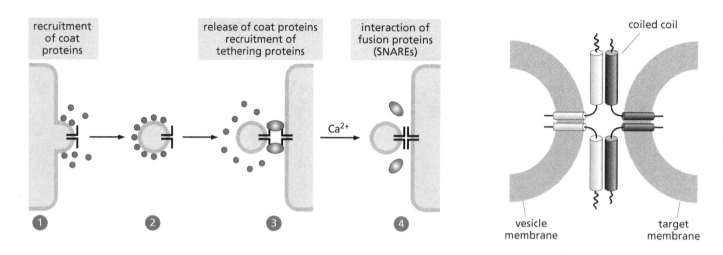

Figure 10.19 Steps of vesicle formation and fusion (1) Coat proteins are assembled at the site of vesicle germination; fusion proteins are symbolized by L-shaped hooks. (2) The vesicle separated from the donor membrane is transported (by motor proteins, not shown) to the target compartment. (3) The vesicle becomes uncoated and recruits tethering proteins (red ellipses) making contact between vesicle and acceptor membrane. (4) The Ca^{2+}-dependent interaction between complementary fusion proteins (SNAREs) triggers membrane fusion. Binding and dissociation of the coat and tethering proteins is facilitated by small G-proteins of the Arf and Rab families. Complementary binding between v-SNAREs (light gray cylinders) and t-SNAREs (dark gray cylinders) through coiled-coil formation of the cytoplasmic domains is shown schematically at the right.

distributed among tissues and intracellular compartments (see textbooks of cell biology for more details on this subject). Most SNAREs are bound to vesicular and target membranes by a single transmembrane domain. Their cytoplasmic extensions contain helical SNARE motifs that have a strong propensity to develop coiled-coil structures. This mechanism enables SNAREs to form tight complexes with each other. It is generally assumed that, to accomplish fusion, SNAREs bound to vesicle membranes (v-SNAREs) interact with complementary SNAREs on the target membrane (t-SNAREs).

In the course of endo- and exocytosis, more and more proteins of the fusion apparatus inevitably accumulate in the target membrane and must be recycled. This retrograde transport process proceeds along the COPI pathway. To this end the v-SNARE/t-SNARE complexes have to become disintegrated, which is the job of the ATPase NSF, supplying ATP energy for the endergonic complex dissociation.

Neurotransmitter release: a perfect example of signal-controlled exocytosis

Input signals triggering the SNARE interaction are transmitted by Rab proteins (see below) and Ca^{2+} ions. This mechanism has been studied most thoroughly for neurotransmitter exocytosis at synapses. The synaptic vesicles filled with neurotransmitter are docked at active zones of the inner side of the presynaptic membrane (see Figure 10.17), a process controlled by the small G-protein Rab3. To be able to fuse and release their content into the synaptic cleft, they require a brief increase of cytoplasmic Ca^{2+} from 0.1 to 1.20 µM triggered by membrane depolarization (or an action potential). In this way, voltage-dependent Ca^{2+} channels in the postsynaptic membrane are opened, allowing Ca^{2+} to flow in from the outside (for details see Section 14.5.2). These channels are components of the fusion machinery.

The Ca^{2+} ions penetrating the cell bind to **synaptotagmin**, a transmembrane protein of the vesicles. Synaptotagmins (nine mammalian isoforms are known) harbor two C2 domains (Section 4.6.3), each binding 2–3 Ca^{2+} ions in its cytoplasmic extension (Figure 10.20). This results in a conformational change exposing hydrophobic areas of the protein. Now synaptotagmin is able with its cytoplasmic domain to dip into the adjacent synaptic membrane, where the C2 domains become fixed by phospholipids. By this means it makes contact with the t-SNARE **syntaxin**, a transmembrane protein, and **SNAP25** (SyNaptosome-Associated Protein of 25 kDa, which must not be confused with the NSF adaptor SNAP), a soluble SNARE protein. Through its two symmetrically arranged SNARE motifs, SNAP25 forms a complex with both syntaxin and the corresponding v-SNARE **synaptobrevin**, thus ensuring membrane fusion. Subsequently the NSF–SNAP complex catalyzes the dissociation of the SNARE complex.

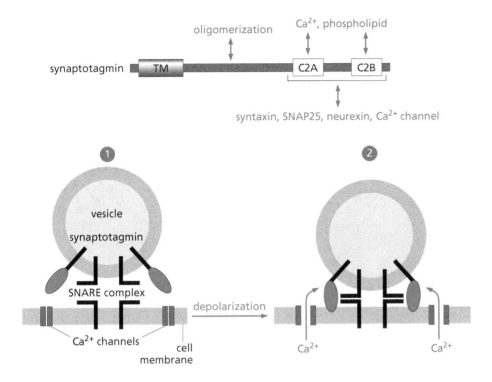

Figure 10.20 Model of Ca²⁺-dependent neurotransmitter release (Top) Simplified domain structure and interaction partners (red) of synaptotagmin (TM, transmembrane domain). (Bottom) Proposed role of synaptotagmin in vesicle fusion. (1) The vesicle has made contact with the inner side of the presynaptic membrane. (2) Depolarization opens voltage-dependent Ca²⁺ channels. Ca²⁺ binds to the C2 domains in the cytoplasmic extensions of synaptotagmin (colored ellipses), which upon a conformational change dip into the cell membrane, reinforcing the interaction between the proteins of the SNARE complex: synaptobrevin, syntaxin, and SNAP25. As a result, membrane fusion is induced (not shown). For the sake of clarity, only two of many synaptotagmin, SNARE, and channel proteins surrounding the site of membrane fusion are depicted. Moreover, additional regulatory proteins not shown here are involved in membrane fusion.

The synaptic proteins are targets of extremely powerful bacterial toxins. The neurotoxins* of **Clostridium botulinum** (cause of meat poisoning) and **Clostridium tetani** (cause of tetanus) rank among the most dangerous biogenic substances. Their average lethal dose (LD_{50}) is less than 0.0001 mg/kg of body weight (for comparison, the LD_{50} is 0.3 mg/kg for cobra toxin and 6.5 mg/kg for sodium cyanide). The clostridium toxins are zinc-containing proteases that specifically destroy synaptic SNARE proteins.

Neurexin, another t-SNARE-related protein of the presynaptic membrane, is attacked by α-latrotoxin, a poison of the **black widow spider** (*Latrodectus tredecimguttatus*) causing a catastrophic emptying of synaptic vesicles (LD_{50} = 0.6 mg/kg). A similar effect is observed upon interaction of the spider poison with latrophilin, a heptahelical receptor with unknown ligand(s) found in presynaptic membranes.

10.4.3 Arf proteins and vesicle formation

To assemble proteins into large complexes at membranes is one of the main functions of small G-proteins. Vesicle formation requires such protein recruitment, in this case of coat proteins. This is managed by the small G-proteins of the **Arf/Sar1 family**. Arf proteins were discovered by their property of increasing the ADP-ribosyltransferase activity of cholera toxin (Arf stands for ADP ribosylation factor). However, this reaction, the physiological significance of which is unknown, has nothing to do with the role of Arf in vesicle formation. Arf proteins have been found in all eukaryotes from yeast (seven isoforms) to humans (15 isoforms). They have been grouped into three subfamilies: Arf proteins proper, Sar (Secretion-associated Ras-like) proteins, and Arf-like (Arl) proteins (Table 10.1).

Vesicle formation is controlled by four Arf proteins (Arf1, -2, -3, and -6) while the others have different effects. The functions of the eight Arf proteins are known

---

*Neurotoxins are only one ingredient of a whole cocktail of *Clostridium* toxins. Other components include ADP-ribosyltransferases, which selectively attack actin, and inhibitors of small G-proteins of the Rho family, which cause actin depolymerization and a breakdown of the cytoskeleton.

Figure 10.21 Role of Arf (or Sar1) in vesicle formation (1) Arf-GEF receives an input signal and activates a cytoplasmic Arf protein by GDP–GTP exchange. This causes a conformational change exposing a membrane anchor (black horizontal bar) that was hidden before in a hydrophobic pocket of the Arf-GDP complex. By means of the anchor, the Arf-GTP complex associates with the donor membrane. (2) Arf-GTP recruits the coat proteins and the coated vesicle is released from the donor membrane. (3) A vesicle-bound Arf-GAP (which also may receive input signals) inactivates Arf by GTP hydrolysis, inducing uncoating of the vesicle and the release of Arf-GDP. For the sake of clarity, only one of the many Arf molecules associating with the vesicle is shown.

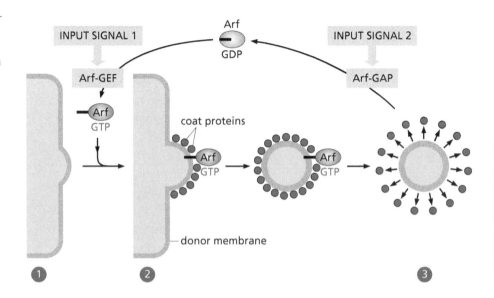

only incompletely. Some of them seem to be involved in vesicle and membrane trafficking within the Golgi apparatus, probably by promoting the association with tethering proteins. The main function of Arf1, -2, -3, and -6 as well as Sar1 is to recruit coat proteins at the site of vesicle budding. The steps of this process are depicted in Figure 10.21.

In the cell, the individual G-proteins have different sites of action. Sar1 controls the formation of vesicles with a COPII coat at the endoplasmic reticulum; and Arf1, of COPI-coated vesicles in the Golgi apparatus. Arf1–3 and probably also Arf6 regulate, in addition, the formation of clathrin-coated endocytotic vesicles at the plasma membrane.

The central role played by Arf in secretory processes is underlined by inhibitor experiments using, for instance, a dominant negative mutant Arf protein with a defective GTPase function. Under such conditions the coat proteins cannot be recruited any longer at the donor membrane, the Golgi apparatus disintegrates, and secretion and endocytosis are blocked almost completely.

Arf and Sar1 are active only when associated with membranes. For association Arf has two membrane anchors, a tetradecanoyl residue bound N-terminally as an amide and a hydrophobic helix structure, while Sar1 lacks the fatty acid side chain. The membrane anchoring must be reversible, since Arf leaves the donor membrane together with the vesicle and has to be released into the cytoplasm to return and to be used for a new cycle of vesicle formation. This is achieved by a conformational change rendering Arf more hydrophilic: upon GTP hydrolysis a hydrophobic pocket opens in the Arf protein, hiding the lipid anchor.

Guanine nucleotide exchange factors and GTPase-activating proteins: receivers of input signals controlling Arf activity

Arf is active only in the GTP-loaded form. Signaling inputs are therefore transmitted by GDP–GTP exchange factors. These **Arf-GEFs** (mammalian cells express 14 isoforms) catalyze the nucleotide exchange only when bound to membranes; they are either membrane-integrated or associate upon activation with membranes. Some Arf-GEFs are selectively inhibited by the mold toxin **brefeldin A** (from *Penicillium brefeldianum*) used experimentally for a blockade of intracellular vesicle trafficking.

As is typical for small G-proteins, the GTPase activity of Arf is extremely weak and has to be stimulated by a GAP. Arf-GAPs are a large heterogeneous protein family. The factors exhibit numerous interaction domains linking Arf activity with other signaling pathways. In the course of vesicle formation, Arf-GAPs associate with the vesicular membrane to induce the dissociation of coat proteins by inactivating Arf. To prevent premature uncoating, however, this activity remains inhibited by specific protein interactions until the vesicle has been completely released.

10.4.4 Rab proteins and vesicle trafficking

Like vesicle budding, vesicle targeting and fusion with the acceptor membrane also requires the assembly of proteins at the vesicle membrane. This process is controlled by the small G-proteins of the Rab family that, like Arf proteins, are found throughout the eukaryotic kingdom and are indispensable for exo- and endocytotic vesicle transport.

Saccharomyces cerevisiae expresses 11 isotypes, with 10 of them called **YPT proteins** (Yeast Protein Transport) and one called **SEC4** (referring to a secretion-defective mutant). The corresponding mammalian homologs were isolated originally from a rat brain cDNA library by polymerase chain reaction with YPT-analogous oligonucleotides. With more than 60 isoforms, these Rat brain (Rab) proteins constitute by far the largest subfamily of small G-proteins. They are, of course, not restricted to brain but found in all cell types. Rab and YPT proteins are related to such an extent that they may replace each other in experimental approaches. Therefore, for the investigation of Rab functions, yeast provides an excellent model.

Each step of vesicle transport is controlled by Rab (as well as, of course, by Arf) GTPases. Within the Rab family the responsibilities are clearly distributed. Rab1 and Rab2 control, for instance, vesicular traffic from the endoplasmic reticulum to the Golgi apparatus; Rab6, inter-Golgi traffic; Rab8, transport from the trans-Golgi apparatus to the plasma membrane; and Rab4, -5, and -9 regulate endocytosis. Finally, Rab7 is specialized for recycling of down-modulated membrane receptors (Section 5.8). Moreover, many Rab isoforms exhibit more tissue-specific functions. An example is provided by Rab3 that controls neurotransmitter release at synapses.

Like Arf proteins, Rab GTPases are active only when they are GTP-loaded and membrane-bound, shuttling between cytoplasm and vesicle membranes. Membrane association is achieved by two geranylgeranyl side chains bound to C-terminal cysteine residues as thioethers (more details on lipid anchors are found in Section 4.3.1). This membrane anchoring is controlled by two proteins called **GDI** (GDP Dissociation Inhibitor; such a factor was discussed earlier with regard to the control of Rho proteins, Section 10.3.1) and **GDF** (GDI Dissociation Factor). By hiding the membrane anchor of Rab in a lipophilic pocket, GDI forms a soluble complex with Rab-GDP that is released into the cytoplasm. In this complex the GDP form of Rab is stabilized (explaining the name GDI). For reassociation of Rab with membranes the complex with GDI has to be dissolved. This job is done by GDF targeting its Rab isoform to a site in the vesicle membrane where it becomes activated by a membrane-bound Rab-GEF (Figures 10.22 and 10.23). Among the G-proteins, this sophisticated mechanism of reversible membrane anchoring is restricted to Rab and Rho proteins. It offers additional opportunities for control, because GDIs and GDFs may be targets of input signals.

There is no final answer to the question as to where exactly Rab proteins are involved in vesicle trafficking and whether there is at all a common mode of Rab action. Generally it is assumed that Rab recruits the tethering proteins for the initial contact between vesicle and target compartment. Thus, Rab would play a role comparable to that of Arf in vesicle coating and that of Ras in the formation of signaling protein complexes. However, besides tethering proteins, many additional effector proteins of Rab have been found, exhibiting functions that go far beyond vesicular transport. Such observations indicate further functions of Rab, in particular in the organization and dynamics of the cytoskeleton.

Figure 10.22 Control of Rab activity The active Rab–GTP complex (shown in color) is anchored in the vesicle membrane by two geranylgeranyl side chains (horizontal bars) and able to interact with its effector proteins. A Rab GTPase-activating protein (GAP) catalyzes the inactivation to Rab-GDP, which is released from the membrane upon complex formation with GDP dissociation inhibitor (GDI), hiding the membrane anchors of Rab. A signal-activated Rab GDI dissociation factor (GDF) displaces GDI, bringing back Rab to a membrane where it may become activated again by a Rab guanine nucleotide exchange factor (GEF). Most probably, the regulatory proteins GDI, GDF, GEF, and GAP are all controlled by input signals.

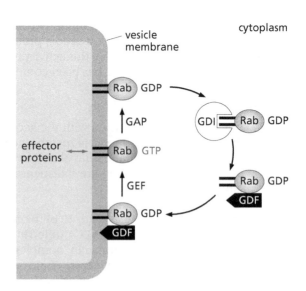

Figure 10.23 Targeting of vesicular transport and induction of fusion by Rab GTPases The figure illustrates a current model which, however, is still partially hypothetical. (1) The vesicle loaded with active Rab-GTP (red) approaches the acceptor membrane. (2) Rab-GTP recruits tethering proteins (colored ellipses) for the initial contact between vesicle and acceptor membrane. (3) Interaction between the SNARE proteins prepares the membrane for fusion (priming step). (4) Fusion is induced by Ca^{2+} ions (see Figure 10.20), and the NSF–SNAP complex catalyzes the ATP-consuming dissociation of the SNARE complexes. During steps 3 and 4, Rab is inactivated by a GTPase-activating protein (GAP; probably bound to the acceptor membrane) and released from the vesicle by binding to GDP dissociation inhibitor (GDI) to be re-used upon interaction with GDF and activation by guanine nucleotide exchange factor (GEF; probably bound to the vesicular membrane). Upon inactivation of Rab, the tethering proteins are released. For the sake of clarity, only one (or two) of many Rab, SNARE, and tethering proteins involved in the process is shown.

Summary

Vesicle trafficking between cellular compartments is a targeted process that consumes energy stored in ATP and GTP and proceeds in several steps. To be loaded with cargo and released from target membranes, vesicles are coated by coat proteins (COPs) and clathrin adaptor proteins that have to be removed prior to fusion with the target membrane. Tethering proteins enable the initial contact with the target membrane. This contact is reinforced by SNARE proteins found in the vesicle and target membranes. Ca^{2+} triggers their interaction, which leads to membrane fusion. Subsequently the SNARE complex is dissociated under ATP consumption catalyzed by the AAA+ protein NSF. The exocytotic release of neurotransmitter, the central event in neuronal signal transmission, provides the classical example of an exocytotic process. Here the Ca^{2+} ions required for membrane fusion enter the cell through voltage-dependent Ca^{2+} channels and enable a vesicular membrane protein (synaptotagmin) to reinforce the contact between vesicular and cellular membrane. The most powerful neurotoxic poisons, such as

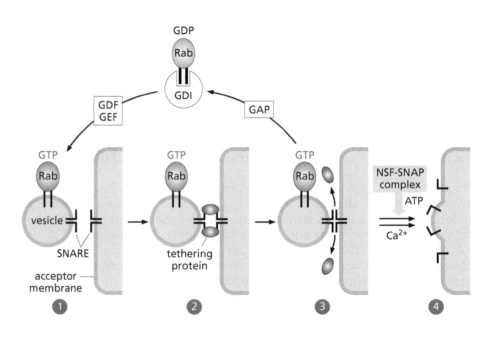

botulinum and tetanus toxin, are proteases that destroy the synaptic SNAREs. Vesicle formation is regulated by the small G-proteins of the Arf family. Arf-GTP bound to the budding vesicle recruits the coat proteins, while uncoating of the vesicle occurs upon GTP hydrolysis. Both guanine nucleotide exchange factors (Arf-GEFs, catalyzing GTP loading) and GTPase-activating proteins (Arf-GAPs, catalyzing GTP hydrolysis) are targets of input signals controlling vesicle formation. Membrane binding of Arf-GTP is due to a lipid anchor that in Arf-GDP is hidden inside the protein molecule. This renders the protein water-soluble, enabling release from the vesicular membrane into the cytoplasm as required for recycling. Fusion of secretory vesicles with target membranes is under the control of Rab proteins, which constitute the largest family of small G-proteins. Rab-GTP bound to the vesicular membrane recruits the tethering proteins for the initial contact with the target membrane. Upon SNARE interaction, Rab catalyzes GTP hydrolysis and in the GDP-loaded form becomes released into the cytoplasm. By binding to a protein called GDP dissociation inhibitor (GDI), Rab-GDP is stabilized and rendered water-soluble. For reactivation of Rab-GDP, GDI is replaced by a GDI dissociation factor (GDF) that facilitates contact of Rab with a membrane-bound Rab-GEF.

10.5 Ran, nuclear transport, and mitosis

Ran (**Ra**s-like **n**uclear GTPase) is the most abundant among the G-proteins of the Ras superfamily. It differs from the other species in two points.

(1) It does not associate with membranes.

(2) Its major function is to organize the traffic of macromolecules (proteins and nucleic acids) between cytoplasm and nucleus rather than to assemble signal-transducing multiprotein complexes at membranes. GTP hydrolysis provides the energy that gives the transport a defined direction, frequently against concentration gradients.

RNA and ribosomes are exported from the nucleus while chromatin proteins and transcription factors are imported. The traffic proceeds through several thousands of sophisticated structured pores in the nuclear membrane that allow passage of more than a million macromolecules per minute (for more information about nuclear pores, the reader may consult textbooks of cell biology). As a rule, the export occurs against a concentration gradient.

Most proteins and RNAs cannot travel unaccompanied between cytoplasm and nucleus but have to be carried by transport proteins, the **exportins** and **importins** (collectively called β-karyopherins). These carriers recognize their cargo by short amino acid sequences: the Nuclear Localization Sequence (or Signal) **NLS** and the corresponding Nuclear Export Sequence **NES** (an additional Nuclear Retention Signal **NRS** is found in proteins preventing the nuclear export of immature RNA).

Importins and exportins are expressed in several isotypes exhibiting different specificities. In the most abundant "classic" import pathway, cytoplasmic proteins carrying a classic NLS bind importin α, an adaptor protein, together with importin β, the carrier protein proper. Probably by interacting with pore proteins (nucleoporins), importin β directs this ternary complex through the nuclear pore. Upon arrival in the nucleoplasm, the complex meets with GTP-loaded Ran that binds to the importins, displacing the cargo protein from the complex. By means of a special exportin (called CAS), Ran-GTP then carries the importins back to the cytoplasm. Since this occurs against a concentration gradient, energy is required; this is supplied by hydrolysis of the Ran-bound GTP, catalyzed by a Ran-GAP bound at the cytoplasmic side of the nuclear membrane. This reaction is supported by **Ran-binding proteins** acting as adaptors and catalyzing sumoylation (see Section 2.8.3), which is essential for the correct localization of Ran-GAP at the mouth of the nuclear pore. As a consequence of GTP hydrolysis, the Ran–importin complexes dissociate and the transport process is rendered irreversible. Ran-GDP is transported back into the nucleus by the protein nuclear transport factor 2 and becomes recharged with GTP by means of a nuclear Ran-GEF localized in the nucleoplasm.

For the export of other macromolecules, mainly ribonucleoproteins, additional exportins interacting with Ran-GTP are available. The Ran cycle is schematically depicted in Figure 10.24. Note that export, not import, is the Ran-controlled step.

Ran-GEF is also known as **Regulator of Chromosome Condensation 1 (RCC1)**, referring to an additional function of Ran: to promote the polymerization of the mitotic spindle. Again the importins α and β are involved in that they sequester spindle assembly factors, which are proteins required for spindle formation. Similar to its role in nuclear transport, Ran-GTP catalyzes the dissociation of such complexes. During mitosis, Ran-GEF (RCC1) is bound to histones H2A and

Figure 10.24 Ran cycle Protein with a nuclear localization sequence [(NLS)protein] is imported into the nucleus; protein with a nuclear export sequence [(NES)protein] is exported from the nucleus. Only export depends on Ran and GTP hydrolysis. NTF2 (nuclear transport factor 2) is a carrier protein for re-import of Ran-GDP into the nucleus; Ran-GEF is a guanine nucleotide exchange factor (also known as regulator of chromosome condensation 1, RCC1); Ran-GAP is a GTPase-activating protein.

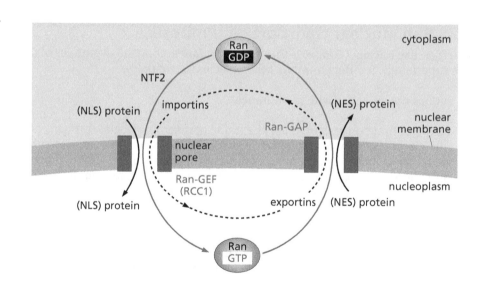

H2B in chromatin, ensuring a high local concentration of Ran-GTP. At the end of mitosis, the chromatin-associated Ran also seems to promote the formation of the new nuclear membrane.

Summary

Cycling between the cytoplasm and the nucleus, Ran, the most abundant small G-protein, controls the targeted transport of macromolecules across the nuclear membrane. The importins promoting the influx of proteins into the nucleus are re-exported against a concentration gradient to the cytoplasm under GTP hydrolysis catalyzed by Ran. Ran-GDP migrates back passively into the nucleus, where it becomes activated by GDP–GTP exchange. Ran is involved also in the formation of the mitotic spindle.

Further reading

Bai J & Chapman ER (2004) The C2 domains of synaptotagmin – partners in exocytosis. *Trends Biochem. Sci.* 29, 143–151.

Barr F (2000) Vesicular transport. *Essays Biochem.* 36, 37–48.

Bokoch GM (2003) Biology of the p21-activated kinases. *Annu. Rev. Biochem.* 72, 743–781.

Bos JL (2005) Linking Rap to cell adhesion. *Curr. Opin. Cell Biol.* 17, 123–128.

Bos JL, Rehmann H & Wittinghofer A (2007) GEFs and GAPs: critical elements in the control of small G proteins. *Cell* 129, 865–877.

Brown MD & Sacks DB (2006) IQGAP1 in cellular signaling: bridging the GAP. *Trends Cell Biol.* 16, 242–249.

Bunney TD & Katan M (2006) Phospholipase Cε: linking second messengers and small GTPases. *Trends Cell Biol.* 16, 640–648.

Burridge K & Wennerberg K (2004) Rho and Rac take center stage. *Cell* 116, 167–179.

Chapman ER (2002) Synaptotagmin: a Ca^{2+} sensor that triggers exocytosis. *Nat. Rev. Mol. Cell Biol.* 3, 498–507.

Chen YA & Scheller RH (2001) SNARE-mediated membrane fusion. *Nat. Rev. Mol. Cell Biol.* 2, 98–106.

Collins RN (2003) Rab and ARF GTPase regulation of exocytosis. *Mol. Membr. Biol.* 20, 105–115.

Cullen PJ & Lockyer PJ (2002) Integration of calcium and Ras signaling. *Nat. Rev. Mol. Cell Biol.* 3, 239–248.

Dasso M (2002) The Ran GTPase: theme and variations. *Curr. Biol.* 12, R502–R508.

Deneka M, Neeft M & van der Sluijs P (2003) Regulation of membrane transport by Rab GTPases. *Crit. Rev. Biochem. Mol. Biol.* 38, 121–142.

Goley ED & Welch MD (2006) The Arp2/3 complex: an actin nucleator comes of age. *Nat. Rev. Mol. Cell Biol.* 7, 713–725.

Gruenberg J (2001) The endocytotic pathway: a mosaic of domains. *Nat. Rev. Mol. Cell Biol.* 2, 721–730.

Hall A (ed) (2000) GTPases. Oxford University Press.

Hancock JF (2003) Ras proteins: different signals from different locations. *Nat. Rev. Mol. Cell Biol.* 4, 373–385.

Helfand BT, Chou YH, Shumaker DK et al. (2005) Intermediate filament proteins participate in signal transduction. *Trends Cell Biol.* 15, 568–570.

Helmreich EJM (2004) Structural flexibility of small GTPases. Can it explain functional versatility? *Biol. Chem.* 385, 1121–1136.

Hermann H, Bär H, Kreplak L et al. (2007) Intermediate filaments: from cell architecture to nanomechanics. *Nat. Rev. Mol. Cell Biol.* 8, 562–573.

Jaffe AB & Hall A (2005) Rho GTPase: biochemistry and biology. *Annu. Rev. Cell Dev. Biol.* 21, 247–269.

Jope RS & Johnson GV (2004) The glamour and gloom of glycogen synthase kinase-3. *Trends Biochem. Sci.* 29, 95–102.

Kashina A & Rodionov V (2005) Intracellular organelle transport: few motors, many signals. *Trends Cell Biol.* 15, 396–398.

Koh TW & Bellen HJ (2003) Synaptotagmin I. A Ca^{2+} sensor for neurotransmitter release. *Trends Neurosci.* 26, 413–422.

Kuersten S, Ohno M & Mattaj IW (2001) Nucleocytoplasmic transport: Ran, beta and beyond. *Trends Cell Biol.* 11, 497–503.

Marinissen MJ & Gutkind JS (2005) Scaffold proteins dictate Rho GTPase-signaling specificity. *Trends Biochem. Sci.* 30, 423–426.

Millard TH, Sharp SJ & Machesky LM (2004) Signaling to actin assembly via the WASP (Wiscott–Aldrich syndrome protein)-family proteins and the Arp2/3 complex. *Biochem. J.* 380, 1–17.

Omary MB, Ku NO, Tao GZ et al. (2006) "Heads and tails" of intermediate filament phosphorylation: multiple sites and functional insights. *Trends Biochem. Sci.* 31, 384–394.

Pfeffer S & Aivazian D (2004) Targeting Rab GTPases to distinct membrane compartments. *Nat. Rev. Mol. Cell Biol.* 5, 886–896.

Pollard TD & Borisy GG (2003) Cellular motility driven by assembly and disassembly of actin filaments. *Cell* 112, 453–465.

Qualmann B & Mellor H (2003) Regulation of endocytotic traffic by Rho GTPases. *Biochem. J.* 371, 233–241.

Randazzo PA & Hirsch DS (2004) ARF-GAPs: multifunctional proteins that regulate membrane traffic and actin remodelling. *Cell Signal.* 16, 401–413.

Ridley A (2006) Rho GTPases and actin dynamics in membrane protrusions and vesicle trafficking. *Trends Cell Biol.* 16, 522–528.

Ridley AJ (2001) Rho family proteins: coordinating cell responses. *Trends Cell Biol.* 11, 471–476.

Riento K & Ridley AJ (2003) ROCKs: multifunctional kinases in cell behaviour. *Nat. Rev. Mol. Cell Biol.* 4, 446–455.

Rizo J, Chen X & Arac D (2006) Unraveling the mechanisms of synaptotagmin and SNARE function in neurotransmitter release. *Trends Cell Biol.* 16, 339–350.

Robinson MS (2004) Adaptable adaptors for coated vesicles. *Trends Cell Biol.* 14, 169–175.

Rossman KL, Der CJ & Sondek J (2005) GEF means go: turning on Rho GTPases with guanine nucleotide-exchange factors. *Nat. Rev. Mol. Cell Biol.* 6, 167–180.

Segev N (2001) Ypt/Rab GTPases: regulators of protein trafficking. *Sci. STKE* 2001/100/re11.

Sorkin A & von Zastrow M (2002) Signal transduction and endocytosis: close encounters of many kinds. *Nat. Rev. Mol. Cell Biol.* 3, 600–613.

Stewart M (2007) Molecular mechanism of the nuclear protein import cycle. *Nat. Rev. Mol. Cell Biol.* 8, 195–207.

Symons M & Rusk N (2003) Control of vesicular trafficking by Rho GTPases. *Curr. Biol.* 13, R409–R418.

Takenawa T & Suetsugu S (2007) The WASP–WAVE protein network: connecting the membrane to the cytoskeleton. *Nat. Rev. Mol. Cell Biol.* 8, 37–48.

Turton K, Chaddock JA & Acharya KR (2002) Botulinum and tetanus neurotoxins: structure, function and therapeutic utility. *Trends Biochem. Sci.* 27, 552–558.

Tzivion G & Avruch J (2002) 14-3-3 proteins: active cofactors in cellular regulation by serine/threonine phosphorylation. *J. Biol. Chem.* 277, 3061–3064.

Vartiainen MK & Machesky LM (2004) The WASP-Arp2/3 pathway: genetic insights. *Curr. Opin. Cell Biol.* 16, 174–181.

Walker SA, Lockyer PJ & Cullen PJ (2003) The Ras binary switch: an ideal processor for decoding complex Ca^{2+} signals? *Biochem. Soc. Trans.* 31, 966–969.

Wehrle-Haller B & Imhof BA (2002). The inner lives of focal adhesions. *Trends Cell Biol.* 12, 382–389.

Weis K (2002) Nucleocytoplasmic transport: cargo trafficking across the border. *Curr. Opin. Cell Biol.* 14, 328–335.

Wellbrock C, Karasarides M & Marais R (2004) The RAF proteins take centre stage. *Nat. Rev. Mol. Cell Biol.* 5, 875–885.

Westermann S & Weber K (2003) Post-translational modifications regulate microtubule function. *Nat. Rev. Mol. Cell Biol.* 4, 938–947.

Wieland F & Harter C (1999) Mechanisms of vesicle formation: insights fom the COP system. *Curr. Opin. Cell Biol.* 11, 440–446.

Yaffe MB (2002) How do 14-3-3 proteins work? Gatekeeper phosphorylation and the molecular anvil hypothesis. *FEBS Lett.* 513, 53–57.

Zhao Z & Manser E (2005) PAK and other Rho-associated kinases – effectors with surprisingly diverse mechanisms of regulation. *Biochem. J.* 386, 201–214.

Mitogen-activated Protein Kinase and Nuclear Factor κB Modules

Eukaryotic cells contain switching elements integrated into modules consisting of several interacting signaling proteins. To a certain extent, such modules resemble microprocessors. Even more, one is tempted to refer to them as "molecular ganglions". This means that they collect and bundle the signals transmitted by the "sensory organs" of the cell (the signal receptors at the cell surface) and transduce them to the "cellular brain" (the data-processing molecular network, including the genomic data bank), thus controlling cellular behavior as well as the metabolic apparatus and activity of the genome. Two of these modules, the mitogen-activated protein (MAP) kinase module and the nuclear factor (NF)κB module, have been investigated in great detail; they provide especially clear insights into the mode of action of such "nanoganglions."

Before entering this chapter, the reader may be warned: the bewildering variety of input signals flowing through the "bottleneck" of the MAP kinase modules, as well as the likewise bewildering number of output signals generated by the modules, creates a didactic problem, as does the confusing nomenclature of the components. However, to grasp the central position of these modules in cellular data processing, we have to clear a path through this thicket of details.

11.1 MAP kinase modules: universal relay stations of eukaryotic signal processing

MAP kinase modules are complexes of three protein kinases that are interconnected in series: they are known as MAP kinase (MAPK, MK), MAP2 kinase (MAP2K, MK kinase), and MAP3 kinase (MAP3K, MKK kinase) (Figure 11.1). Each of these kinases exists in various isoforms that are integrated in individual modules. The core of each module is the pair of MAP kinase and MAP2 kinase, which interact with each other in a highly specific manner. In contrast, the interaction between MAP2 kinases and MAP3 kinases is much less specific: many protein kinases may function as MAP3 kinases, thus transmitting a wide variety of input signals to the module. The same holds true for MAP kinases, which, admittedly, are firmly wired with the MAP2 kinases, but are not choosy as far as their substrate proteins are concerned. Thus, MAP kinase modules are "bottleneck" devices that transform a wide variety of input signals into a comparatively large variety of output signals. As such, they play a key role in cellular data processing ranging from yeast to humans (no such modules have been found yet in prokaryotes). While the input signals are mainly derived from signal receptors, the output signals address metabolic reactions, the architecture and mobility of cells, and gene transcription.

What is the secret of this successful model of evolution? Most probably it is the special connection of three protein kinases that provides a switch with a powerful

Figure 11.1 MAP kinase module: signal processor and noise filter
(A) Arrangement of the three protein kinases and phosphorylations within the module. The colored and gray boxes symbolize scaffold proteins holding together the module. Upon activation of the module, the strong interaction between MAP2K and MAPK may become weakened, enabling the activated MAPK to translocate, for instance from the cytoplasm into the nucleus. (B) Relationship between input and output signal with (solid line) and without (broken line) modular organization. Note the sharp switch-like response of the module. The area of noise suppression is shown in color.

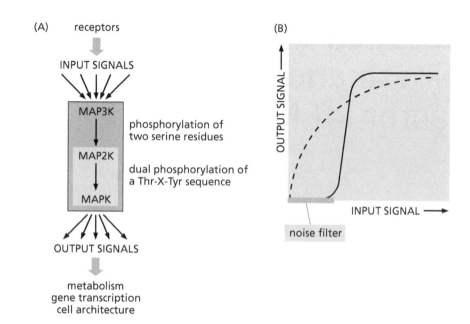

noise filter, and in some cases an amplifier, and has – at least theoretically – the properties of an oscillator, allowing the generation of frequency-modulated signals. Moreover, MAP kinase modules are perfect logical gates because they are able to process a large variety of additive, synergistic, and antagonistic signals.

Summary

MAP kinase modules comprising three protein kinases interconnected in series are central relay stations of cellular signal transduction that seem to be distributed all over the eukaryotic kingdom. They transform a wide variety of input signals, mostly transmitted by receptors, into a large number of output signals addressing metabolism, gene transcription, and cell architecture.

11.2 MAP kinases and MAP2 kinases: the core of MAP kinase modules

MAP kinases (<u>M</u>itogen-<u>A</u>ctivated <u>P</u>rotein kinases) is the collective term for the serine/threonine (Ser/Thr)-specific protein kinases generating the output signals of the MAP kinase modules. The historically based name is somewhat misleading since MAP kinase modules by no means are activated only by mitogenic signals but also are involved in cell differentiation and cell death. The mammalian (human) genome encodes 14 MAP kinases and eight MAP2 kinases. The MAP kinases have been arranged in several subfamilies that are known by historical names such as:

- the **ERK1,2 family** (<u>E</u>xtracellular signal-<u>R</u>egulated <u>K</u>inases), including isoforms ERK1 and ERK2

- the **JNK family** (c<u>J</u>un <u>N</u>-terminal <u>K</u>inases, also called SAPK, <u>S</u>tress-<u>A</u>ctivated <u>P</u>rotein <u>K</u>inases; cJun is a transcription factor), including isoforms JNK1–3, which due to alternative splicing exist in 11 subspecies

- the **p38 family** (molecular mass of 38 kDa), with four isoforms p38α, -β, -γ, and -δ

- the **ERK3,4 family** and the **ERK5 family** (also called big MAP kinases)

Each family, probably even each isoform, organizes a module of its own, with the ERK1,2, JNK, and p38 modules being the best-known ones (Figure 11.3). The

Sidebar 11.1 MAP kinase modules of yeast and plants
MAP kinase modules have been found in all eukaryotes studied so far. The baker's yeast *Saccharomyces cerevisiae* expresses six different modules, and the fission yeast *Schizosaccharomyces pombe* has a similar number. Some of the yeast modules are shown in Figure 11.2. As can be seen, these modules are coupled to quite different receptors and signaling devices, while downstream they phosphorylate transcription factors, thus controlling the function of various genes. The genome of the plant *Arabidopsis thaliana* encodes no less than 20 MAP kinases, 10 MAP2 kinases, and 17 putative MAP3 kinases. Their cellular functions are widely unknown.

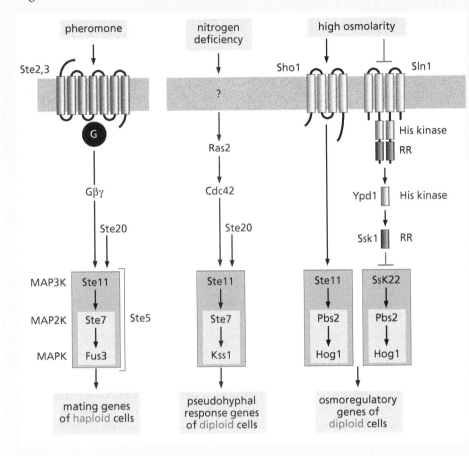

Figure 11.2 Signaling pathways including MAP kinase modules in *Saccharomyces cerevisiae* The MAP kinase modules (colored boxes) are controlled by various exogenous stimuli along quite different signaling pathways, leading to individual cell-physiological effects. Note that the two modules shown to the left differ only in their MAP kinases (Fus3 versus Kss1) addressing different downstream effectors, whereas the two modules shown to the right have different MAP3 kinases (Ste11 versus SsK22), thus responding to different input signals but generating identical output signals. Ste2 and Ste3, G-protein-coupled receptors; Sln1, sensor histidine kinase of a two-component system; RR, response regulator of a two-component system; Ste20, Ser/Thr-specific protein kinase; Ste5, scaffold protein of the pheromone module (see Sections 11.4 and 11.5). In the fission yeast *Schizosaccharomyces pombe*, pheromones also interact with a G-protein-coupled receptor and the corresponding MAP kinase module Byr2–Byr1–Spk1 is activated by the Gα-subunit together with a Ras-homologous protein (Section 10.1.4).

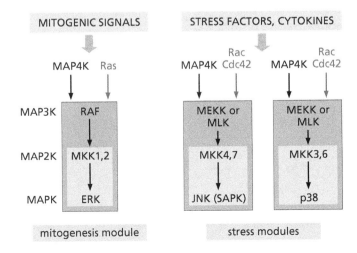

Figure 11.3 Three best-known MAP kinase modules of vertebrates In most cases, the modules are activated by different MAP4 kinases (MAP4K) together with small G-proteins of the Ras and Rac/Cdc42 families. By convention the MAP2 kinases frequently are called MKK (MAPK kinases) or MEK (MAPK/ERK kinases). Other abbreviations are MEKK for MEK kinase and MLK for mixed-lineage kinase (see Section 11.3). Further MAP3 kinases are described in the text.

ERK1,2 modules, in particular, respond to mitogenic signals such as transmitted by growth factor receptors (and have been called, therefore, mitogenesis modules), whereas the JNK and p38 modules predominantly (but not exclusively) process stress and pro-inflammatory cytokine signals (therefore also called stress modules). MAP kinases catalyze the phosphorylation of Ser and Thr residues adjacent to proline residues and have been called **proline-directed protein kinases**. As mentioned in Section 8.5, such kinases cooperate with Pin1 peptidyl-prolyl *cis–trans*-isomerases in modulating the conformation and function of signaling proteins.

Normally MAP kinases are found in the cytoplasm, where they phosphorylate cytoplasmic proteins. However, upon activation they may become translocated into the nucleus to phosphorylate transcription factors (Section 11.7). The intracellular distribution depends on specific domains in the MAP kinase molecule. ERK, for instance, has both a nuclear localization sequence (NLS) and a cytoplasmic retention sequence (CRS). Via the CRS domain, ERK associates with its MAP2 kinase and is hindered from nuclear translocation unless the module dissociates when activated by an input signal.

11.2.1 MAP2 kinases and phosphatases: activation and inactivation of MAP kinases

Each MAP kinase is activated by a firmly attached MAP2 kinase, also called MKK (MAP kinase kinase), catalyzing in an absolutely specific way the dual phosphorylation of a Thr-X-Tyr sequence in the activation loop. Since from module to module these sequences differ in the amino acid X, each module is equipped with its own specific MAP2 kinases. The Thr-X-Tyr sequences are:

- Thr-Glu-Tyr (TEY) in ERK1, -2, and -5, phosphorylated by MAP2 kinases MKK1 and MKK2
- Thr-Gly-Tyr (TGY) in p38, phosphorylated by MAP2 kinases MKK3 and MKK6
- Thr-Pro-Tyr (TPY) in JNK, phosphorylated by MAP2 kinases MKK4 and MKK7

MAP2 kinases belong to the small group of dual-specific kinases phosphorylating both Thr and Tyr residues. MAP kinases are their only substrates. As shown below, all MAP2 kinases are activated by MAP3 kinases, which catalyze the phosphorylation of two adjacent Ser and Thr residues in the activation loop. The nomenclature of MAP2 kinases is somewhat confusing (Table 11.1).

Antagonists of the MAP2 kinases are the **dual-specific MAP kinase phosphatases** (DS-MKP) that dephosphorylate the Thr-X-Tyr sequences of MAP kinases. These enzymes are as selective as MAP2 kinases: one distinguishes between at least 13 isoforms exhibiting different MAP kinase specificities. DS-MKPs are interconnected with MAP kinases in several regulatory loops. Many

Table 11.1 Nomenclature of MAP2 kinases[a]

MAP2 kinase	Synonyms
MKK1	MEK1 (MAPK/ERK kinase)
MKK2	MEK2
MKK3	SKK1 (SAPK kinase)
MKK4	SKK2, SEK1 (SAPK/ERK kinase), JNKK1 (JNK kinase)
MKK6	SKK3
MKK7	JNKK2

[a]Frequently, but not logically, MEK is used as a collective term for all MAP2 kinases.

MAP3 kinases: sensors and logical gates of MAP kinase modules

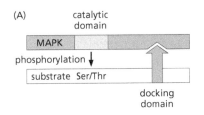

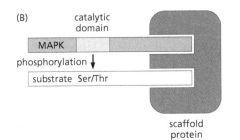

Figure 11.4 Selection and reinforcement of the interaction between MAP kinase and substrate Kinase specificity is due to (A) specific interactions with docking domains or (B) a specific arrangement of the partners by scaffold proteins.

DS-MKPs are encoded, for instance, by immediate early response genes induced by transcription factors, which are under the control of MAP kinases. Moreover, DS-MKPs require phosphorylated MAP kinases as allosteric activators, and finally, the lifespan of the rather labile DS-MKPs is prolonged by MAP kinase-catalyzed phosphorylation. On the other hand, DS-MKPs function as anchor proteins, sequestering MAP kinases in the nucleus for dephosphorylation. Thus, MAP kinases, MAP2 kinases, and the related phosphatases are interlinked by several positive and negative feedback loops, indicating an oscillating system.

11.2.2 MAP kinase specificity: role of docking motifs and scaffold proteins

The phosphorylation consensus sequences of MAP kinases are quite abundant: there are many potential substrate proteins, including protein kinases, transcription factors, and cytoskeletal proteins as well as signal-generating and signal-terminating enzymes. Nevertheless, the individual MAP kinases are surprisingly selective, recognizing only a limited set of substrate proteins. This selectivity is due to additional, highly specific protein–protein interactions that considerably restrict the number of enzyme–substrate interactions and are achieved by **docking motifs** (or D-domains) that are not parts of the catalytic substrate binding domain. By means of these motifs, the kinases selectively bind to complementary motifs of individual substrate proteins (Figures 11.4A and 11.5). The principle is reminiscent of a bank safe that is unlocked by two keys. Another tool bringing together MAP kinases and selected substrate proteins are **scaffold proteins** (Figure 11.4B). Docking domains and scaffold proteins are also used to establish contact between MAP kinases and upstream kinases such as MAP2 and MAP3 kinases, to hold together the whole module (see Section 11.5).

Summary

Mitogen-activated protein (MAP kinases) are Pro-directed Ser/Thr-specific protein kinases that are activated by dual phosphorylation of a Thr-X-Tyr sequence. Among the various isoforms, extracellular signal-regulated kinases (ERK1,2), cJun N-terminal kinases, also known as stress-activated protein kinases (JNK/SAPK), and p38 kinases are the most prominent ones. While ERK1,2 are incorporated in Ras-controlled mitogenesis modules, JNK/SAPK and p38 are components of stress modules. Upon receiving an upstream signal, the highly specific MAP2 kinases activate MAP kinases by dual phosphorylation of a Thr-X-Tyr sequence: X = Glu for ERK1,2 modules, X = Gly for JNK/SAPK modules, and X = Pro for p38 modules. The phosphorylations are reversed selectively by dual-specific MAP kinase phosphatases. In principle, MAP kinases are rather nonspecific. However, their specificity and intracellular localization are improved considerably by sequences (docking motifs) and scaffold proteins, enabling more selective interaction with target proteins.

11.3. MAP3 kinases: sensors and logical gates of MAP kinase modules

To become active, MAP2 kinases have to be phosphorylated at two Ser or Thr residues in the activation loop. A rather large and heterogeneous group of

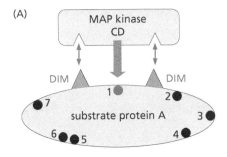

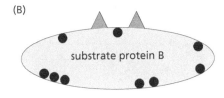

Figure 11.5 Accuracy of MAP kinase–substrate interactions Despite the abundant phosphorylation motif, Ser/Thr-Pro, MAP kinases phosphorylate substrate proteins selectively rather than randomly. This specificity is achieved by domain-interacting motifs (DIM, triangles) of the substrate protein binding to the docking domains of the kinase. (A) Two interacting motifs direct the catalytic domain (CD) of the kinase selectively to one of seven potential phosphorylation sites (black dots; the number is chosen arbitrarily). (B) If the docking motifs are positioned differently, the protein cannot interact with this particular kinase but possibly with another kinase isoform.

protein kinases, collectively named MAP3 kinases, can do this job. MAP3 kinases are the signal receivers of MAP kinase modules; individual modules are associated with distinct MAP3 kinase types rather than with individual enzymes. These types are:

- RAF kinases of ERK1,2 modules, and the related mixed-lineage kinases (MLKs) of JNK and p38 modules

- MEK kinases (MEKKs), mainly of JNK and p38 modules

- a heterogenous collection of other kinases named ASK1, TAK1, TPL2, NIK, and TAO

The enormous versatility of MAP3 kinases and the remarkable variety of their interaction domains reflect the integration of MAP kinase modules in almost any signaling pathway. As shown in Figures 11.2 and 11.3, MAP3 kinases are activated both by small G-proteins such as Ras, Rac, and Cdc42 and by MAP4 kinases. For this reason, most MAP3 kinases have small-G-protein-interacting domains and essential phosphorylation sites.

11.3.1 RAF kinases: MAP3 kinases of mitogenesis modules

The most prominent representatives of MAP3 kinases are the three protein kinases A-, B-, and C-RAF (also called MEKK- or MKKK-1, -2, and -3) that selectively activate the MAP2 kinases MKK1 and MKK2 and, thus, the MAP kinases ERK1 and ERK2. RAF was discovered as a retroviral oncoprotein v-Raf isolated from a murine sarcoma virus (the acronym RAF appears to be artificial but may be understood as Ras-activated factor). The complex regulation of RAF kinases has been discussed already in detail in Section 10.1.4. Critical steps are the interaction with GTP-loaded Ras via a Ras-binding domain and a cysteine-rich domain and several phosphorylations within and around the activation loop of RAF (see Figure 10.2). In an exemplary manner, RAF shows how MAP kinase modules process input signals by interacting with other protein kinases (for instance, protein kinase C, PKC) acting as MAP4 kinases as well as with small G-proteins. On the other hand, phosphorylation of the Ras-interacting domains by protein kinase A (PKA) or B (PKB/Akt) inhibits RAF. Thus RAF, like other MAP3 kinases, is the prototype of a logical gate enabling the connected MAP kinase module to perform all operations required for data processing (Figure 11.6).

11.3.2 Mixed-lineage kinases: MAP3 kinases of stress modules

What RAF kinases are for the ERK modules, mixed-lineage kinases (MLKs) are for the JNK and p38 modules. The strange name of this enzyme family is based on an error: due to structural peculiarities, they were assumed to be a lineage of

Figure 11.6 MAP3 kinase RAF as a logical gate RAF is activated by both Ras and protein kinase C (PKC), resembling an AND gate, or by Ras and p21-activated kinase (PAK), resembling an OR gate, but not by Ras and protein kinase A (PKA) or protein kinase B (PKB)/Akt, resembling a NOR (or BUT NOT) gate.

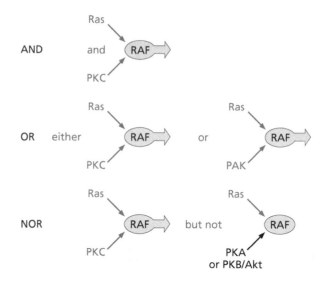

MAP3 kinases: sensors and logical gates of MAP kinase modules

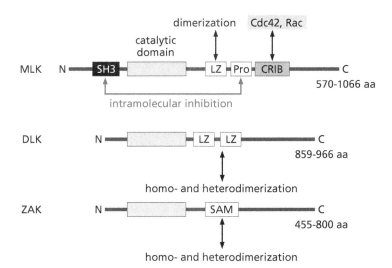

Figure 11.7 Domain structures of mixed-lineage kinases MLK, mixed-lineage kinase; SH3, Src homology domain 3; LZ, leucine zipper motif; Pro, proline-rich domain interacting with SH3 domains; CRIB, Cdc42/Rac-interactive binding domain; DLK, double leucine-zipper kinase; ZAK, zipper sterile-α kinase; SAM, zipper sterile α motif. For further details, see text.

"mixed-functional" kinases phosphorylating both Ser/Thr and Tyr residues (called dual-specific kinases today). Since then, mixed-lineage kinases have turned out to be bona fide Ser/Thr kinases. Phylogenetically they are distant relatives of the RAF kinases.

In mammalian cells, nine MLK isoforms have been found. Five of them constitute the MLK family proper, while two are known as DL kinases, referring to a characteristic double (D) leucine (L)-zipper domain, and two as ZA kinases, containing a so-called zipper (Z) sterile α (A) motif (see Figure 11.7). These nine enzymes are components of JNK modules; isoforms MLK3 and DLK in addition are found in p38 modules.

LZ and sterile α motif (SAM) domains play a regulatory role: to become active, mixed-lineage kinases have to dimerize via these domains, thus enabling trans-autophosphorylation of the activation loops. This dimerization is promoted by upstream regulatory proteins.

To activate MLKs proper, intramolecular inhibition due to an interaction between a Pro-rich and a SH3 domain must be relieved. This is achieved by small G-proteins of the type Cdc42 or Rac, which here play a role resembling that of Ras for RAF activation. Both interact in their GTP-loaded forms with a special binding domain CRIB (Cdc42/Rac Interactive Binding domain; Figure 11.7) thus inducing a conformational change by which the Pro-rich domain is displaced from the SH3 domain. Probably, this provides one of the pathways along which Cdc42 and Rac stimulate the JNK and p38 modules (see Figure 11.3; a parallel pathway is described in Section 11.4).

Together with their partner molecules, mixed-lineage kinases are fixed on special scaffold proteins. These JNK-interacting proteins or JIPs (see Section 11.5) bind, in addition, other signal-transducing proteins such as Rac GTP–GDP exchange factors (Rac-GEFs), which load Rac with GTP. Inside the cell, such complexes may be transported as cargo by kinesin motor proteins. By this means, probably whole sets of signal-processing proteins are rearranged, resulting in mechanical changes of the "hardware components" aiming at an adaptation of the network's switching pattern.

Presently mixed-lineage kinases are gaining considerable medical interest, since apoptosis is induced in neurons along the MLK–JNK signaling pathway, in particular in the course of **neurodegenerative disorders** such as Alzheimer's dementia, Huntington's and Parkinson's disease, and old-age deafness. In fact, relatively specific MLK inhibitors have been found to exhibit neuroprotective

Figure 11.8 MEK kinases (A) Association of the MEK kinases MEKK1–4 with individual MAP kinase core modules and upstream signaling proteins. A broken arrow indicates that the physiological significance of the corresponding interaction is still debated. (B) Domain structures of MEKK1 and MEKK2. MEKK4 has a similar structure as MEKK1 except that a Cdc42/Rac-binding (CRIB) domain replaces the Ras-binding domain (RasBD). MEKK3 closely resembles MEKK2. Pro, proline-rich domain interacting with SH3 domains; PH, pleckstrin homology domain interacting with membrane phospholipids and Gβγ-subunits.

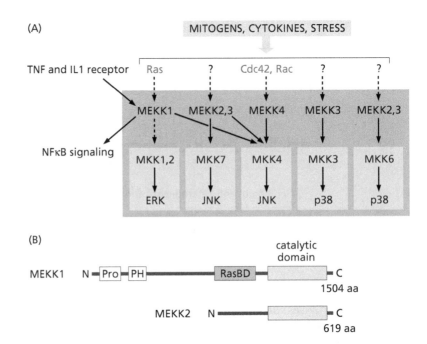

effects in animal experiments and are in clinical evaluation. An example is provided by CEP-1347, a semisynthetic derivative of an indocarbazole compound K252 from actinomycetes. Interestingly, these bacteria also produce other K252-related kinase inhibitors such as staurosporine that are used widely as experimental tools.

11.3.3 MEK kinases: MAP3 kinases and more

Another heterogeneous family of MAP3 kinases are the MEK kinases or MKK kinases (MEKK, MKKK). In contrast to MLKs and RAF kinases, they are phylogenetic descendants of the yeast enzyme Ste11 (Figure 11.2), and many of them were indeed isolated by the polymerase chain reaction (PCR) with Ste11 probes. There are four closely related isoforms MEKK1–4 (Figure 11.8) and the more distantly related kinases ASK1, TAK1, and TPL2 that are discussed separately below.

The isoform most thoroughly investigated is **MEKK1**. In addition to a membrane-interacting pleckstrin homology (PH) domain and a Pro-rich motif interacting with SH3 domains, this enzyme has a Ras-binding domain and activates the ERK1,2 modules in exactly the same way as RAF kinases. In addition, MEKK1 may serve as a scaffold protein for a RAF–MKK2–ERK2 module. Moreover, together with the MAP2 kinase MKK7 and the MAP kinase JNK, MEKK1 is fixed also on another scaffold protein known as JSAP1 (JNK/SAPK-Associated Protein 1; see Section 11.5). Thus, MEKK1 is a MAP3 kinase in both the ERK and JNK pathways. Under physiological conditions, however, the enzyme seems to be mainly integrated in JNK modules. But this is not the end of the story, since the kinase is also a component of signaling cascades controlled by cytokine receptors that lead to activation of the transcription factor NFκB (Chapter 6 and Section 11.8). In this case, MEKK1 is activated along alternative pathways, for example, by the proteolytic destruction of an auto-inhibitory peptide sequence.

Like MEKK1, other MEK kinases also are mainly associated with JNK and p38 modules. Thus, **MEKK2** binds, for instance, kinases MKK7 and JNK1 and acts as both a MAP3 kinase and a scaffold protein of this particular module. Like mixed-lineage kinases, **MEKK4** contains a CRIB domain functioning as an additional link between the small G-proteins Cdc42/Rac and JNK modules. The interaction between Cdc42/Rac and MEKK4 seems to be mediated, in addition, by a protein kinase, most probably the p21-activated kinase PAK (see also Section 10.3.4).

MAP3 kinases: sensors and logical gates of MAP kinase modules

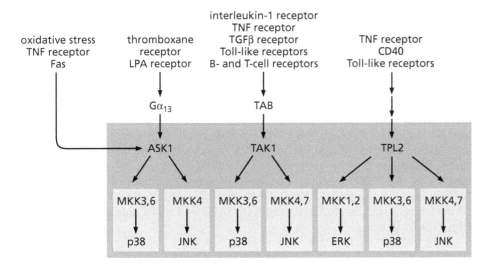

Figure 11.9 Signal transduction by MAP3 kinases Apoptotic signal-related kinase 1 (ASK1), transforming growth factor β-activated kinase (TAK1), and kinase "tumor progression locus 2" (TPL2) are depicted. For details, see text. TAB, TAK1-binding protein. Fas is a death receptor inducing apoptosis. CD40 is a member of the TNF receptor family expressed by antigen-presenting cells.

How MEKK2 and -3 become activated is not known. Since these isoforms do not contain small-G-protein-interacting motifs, stimulatory pathways differing from the canonical mechanism of MAP3 kinase activation have to be postulated. Some such alternative pathways are discussed in the following section. They render MAP kinase modules even more versatile than hitherto assumed.

11.3.4 Other MAP3 kinases: processors of stress signals

Other MAP3 kinases that are activated mainly by stress signals but do not belong to the MEKK and MLK subfamilies are mostly restricted to certain tissues (Figure 11.9). An interesting example is provided by **ASK1** (Apoptotic Signal-regulated Kinase 1). This ubiquitously expressed enzyme phosphorylates the MAP3 kinases MKK3, -4, and -6, thus stimulating the p38 and JNK signaling modules. The kinase is activated along quite different pathways, either by the Gα-subunit of the large G13-protein or by tumor necrosis factor α (TNFα) and death receptor signaling or by oxidative stress. In fact, by promoting stress module-dependent transcription of pro-apoptotic genes, ASK-1 plays an important role as downstream mediator of stress- and receptor-transmitted signals that cause programmed cell death (for details see Section 13.2.4).

Another MAP3 kinase thought to be involved in the processing of cellular stress is **TAK1**. This abbreviation stands for transforming growth factor β- (TGFβ-) activated kinase, indicating that the enzyme is integrated in the TGFβ signaling pathway (Section 6.2). However, TAK1 is also a component of signaling cascades originating from TNFα and interleukin-1β receptors with specific TAK-binding proteins (TAB1 and -2) serving as adaptors between the kinase and the receptor complexes (see Section 6.3.1). Like ASK1, TAK1 catalyzes the phosphorylation of the MAP2 kinases MKK3, -4, and -6 (and -7), thus stimulating the p38 and JNK modules as well as NFκB signaling (Section 11.8.2).

A functionally related MAP3 kinase, **TPL2,** was identified originally as a proto-oncogene product encoded by a Tumor Progression Locus 2. Its oncogenic mutation causes T-cell lymphomas in rats, whereas a role in human cancer is still obscure. TPL2 activates all MAP kinase modules as well as the NFκB module. Found primarily in macrophages and B-lymphocytes, TPL2 seems to play a role in defense reactions, in particular since it has been found to be stimulated along signaling pathways activated by bacterial lipopolysaccharides and T-lymphocytes.

Finally, the **NFκB-inducing kinase NIK** and the **TAO kinases** (referring to the Thousand And One amino acids of the first characterized TAO kinase) remain to be mentioned. The fact that NIK is a component of the NFκB module (Section

11.8.2) as well as a MAP3 kinase in p38 and JNK modules indicates intense cross talk between these signaling modules. TAO kinases serve as MAP3 kinases in p38 modules except for a splice variant in prostate tumors, which stimulates JNK activity.

Summary

MAP3 kinases, the signal receivers of MAP kinase modules, constitute a large and heterogeneous group of Ser/Thr-specific protein kinases that catalyze the activation loop phosphorylation of MAP2 kinases. Up to a certain extent, distinct subfamilies can be assigned to individual modules: Ras-dependent RAF kinases to the ERK1,2 modules and mixed-lineage kinases (MLKs) and most MEK kinases to the JNK/SAPK and p38 modules. Other MAP3 kinases mainly process stress signals activating the JNK/SAPK and p38 modules and, in addition, NFκB signaling.

11.4 MAP4 kinases and G-proteins: lessons learned from yeast

As we have seen, MAP kinase modules are under the control of a wide variety of input signals, which are received by the MAP3 kinases. Frequently such input signals are transmitted by protein kinases known as MAP4 kinases in collaboration with G-proteins. An example of this canonical pathway is shown in Figure 11.6, in which the activation of the RAF–ERK module by Ras is in collaboration with protein kinase C acting as MAP4 kinase. However, the precise mechanisms of these interactions are not always clear.

The system best understood is the pheromone signaling cascade of *S. cerevisiae*. Here the MAP3 kinase Ste11 is activated by the Gβγ-subunit of the pheromone receptor-coupled G-protein in cooperation with the Ser/Thr-specific protein kinase Ste20 acting as a MAP4 kinase. This is considered to represent a standard model of MAP3 kinase activation consisting of two steps.

(1) To become activated by a membrane receptor, the MAP kinase module must be assembled at the inner side of the cell membrane by interacting, according to cell type and species, with either a membrane-bound G-protein subunit or a membrane-bound GTP-loaded small G-protein.

(2) The module thus primed is activated by phosphorylation of the MAP3 kinase through a MAP4 kinase that also has been recruited to the membrane by a GTP-loaded small G-protein. Due to the interaction with the G-protein, the conformation of the MAP4 kinase is changed in such a way that an activation loop phosphorylation (by autophosphorylation or by another protein kinase) becomes possible.

In Figure 11.10, this two-step process is shown for the pheromone module of yeast and applied to the ERK1 module stimulated by a growth factor receptor in vertebrate cells. Note that in the vertebrate cell the organizing role of the Gβγ-subunit is taken over by Ras-GTP. Another example we came across already is the activation of the p38 module by a MAP4 kinase of the PKC-related kinase family cooperating with the small G-protein Rho (Section 10.3.3).

The mammalian counterpart of the yeast MAP4 kinase Ste20 is PAK2. This kinase is a member of the **p21-Activated protein Kinase** family (referring to the molecular mass of 21 kDa of the small G-protein Rac) discussed already in Section 10.3.4. The PAKs are evolutionary descendants of yeast Ste20. Characteristic structural features of both PAK and Ste20 are a C-terminal kinase domain and a domain, such as CRIB, for membrane binding and activation by small G-proteins. PAK2 is one out of six PAK isoforms (in humans). In contrast to the MAP4

MAP4 kinases and G-proteins: lessons learned from yeast

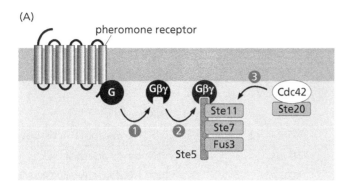

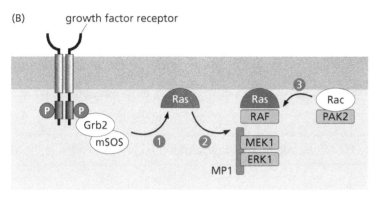

Figure 11.10 Activation of MAP kinase modules by G-proteins cooperating with MAP4 kinases of the Ste20 group The common theme is recruitment of the signal-transducing apparatus to the membrane, where it comes in close contact with receptors of input signals. (A) Activation of the pheromone module in *S. cerevisiae*. (Step 1) Activated pheromone receptor induces the release of a Gβγ-subunit from the receptor-coupled trimeric G-protein. (Step 2) Membrane-anchored Gβγ-subunit recruits the pheromone module to the membrane by interacting with the scaffold protein Ste5 (gray bar). (Step 3) MAP4 kinase Ste20, bound to the membrane by the small membrane-anchored G-protein Cdc42 (or by the Gβγ-subunit) and activated by (auto)phosphorylation, catalyzes the phosphorylation of the MAP3 kinase Ste11 and thus activates the pheromone module. (B) Analogous activation of the ERK1 module in mammalian cells; cooperation of the two G-proteins Ras and Rac is assumed. (Step 1) Activated and autophosphorylated growth factor receptor Tyr kinase stimulates, via the adaptor protein Grb2 and the Ras-GEF mSOS, the small G-protein Ras. (Step 2) Membrane-anchored active Ras-GTP complex binds the MAP3 kinase RAF, thus recruiting the ERK1 core module (bound to the scaffold protein MP1) to the membrane. (Step 3) MAP4 kinase PAK2, bound to the membrane by the small membrane-anchored G-protein Rac and activated by autophosphorylation, phosphorylates RAF, thus activating the ERK1 module. Rac is activated along the Ras signaling pathway or other pathways (see Section 10.3).

kinase PAK2, the other PAKs are not very efficient in activating MAP kinase modules. Instead, they preferentially phosphorylate proteins controlling the dynamics of the cytoskeleton (see Section 10.3).

Another family of mammalian protein kinases related to yeast Ste20 are the **germinal-center kinases (GCK)**. The strange name is explained by the original discovery of these enzymes in B-cell germinal centers, sites of B-cell maturation. However, these kinases by no means are restricted to lymphoid tissue. Mammalian GCKs exist in 20 isoforms (Table 11.2 shows a selection). They differ from PAKs in the position of the kinase domain (located in the N-terminal region) and a pronounced variability of the regulatory domains.

Experimental data showing selective activation of the JNK modules by GCKs confirm observations indicating that they are predominantly involved in the processing of stress signals. In addition, these kinases mediate signals that induce the differentiation of lymphoid and hematopoietic tissues. Such signals are, for instance, transmitted by cytokines of the TNF family. In fact, a characteristic property of GCKs is that by interacting with TRAF (TNF receptor-associated factor) adaptor proteins they make contact with and are activated by cytokine receptors of the TNFα receptor family (Section 6.3.1). By adaptors containing SH3 domains such as Nck, Grb2, Crk, etc., some GCKs are linked also to Tyr kinase-coupled receptors transmitting growth factor and cytokine signals (Section 7.1). Thus, GCKs transmit a series of different exogenous signals to the JNK modules. Whether or not they are bona fide MAP4 kinases or interact more indirectly with the modules is not entirely clear.

A special member of the GCK family is the **JNK-inhibiting kinase JIK** that competes with MAP2 kinases for JNK. While MAP2 kinases activate JNK, phosphorylation by JIK, occurring at another site of the JNK molecule, has an inhibitory effect. This inhibition is relieved by growth factors such as epidermal growth factor. It has been proposed that epidermal growth factor, although it normally activates only the ERK module, may stimulate the JNK modules indirectly along this pathway.

Table 11.2 Selection of putative MAP4 kinases of the mammalian GCK family

		Activation	MAPK module
GCK	germinal-center kinases	TNFα	JNK
GCKR	GCK-related kinases	TNFα, UV	JNK
GLK	GCK-like kinases	TNFα	JNK
HPK1	hematopoietic progenitor kinase	TNFα, Tyr kinases	JNK; hematopoietic tissues
NIK[a]	Nck-interacting kinase	TNFα	JNK; binds adaptor Nck
TNIK	TRAF- and Nck-interacting kinase	TNFα	JNK; binds adaptors TRAF and Nck
NRK	NIK-related kinase	TNFα	JNK
SOK1	Ste20-like oxidant stress-activated kinase	oxidative stress	?
KRS1	kinase responsive to stress	extreme stress	?
MST	mammalian Ste 20-like kinases	extreme stress, Fas	JNK, p38
LOK	lymphocyte-oriented kinase	extreme stress	?; in lymphocytes only
JIK	JNK-inhibiting kinase	?	inhibition of JNK

[a]The MAP4 kinase NIK must not be confused with the MAP3 kinase NIK (NFκB-inducing kinase), a signal transmitter in the NFκB system (see Sections 11.3.4 and 11.8.2).

Other Ste20-homologous mammalian kinases related to both GCKs and MLKs are the above-mentioned MAP3 kinases of the TAO kinase family.

Summary

The standard mechanism of MAP3 kinase activation is protein phosphorylation catalyzed by a wide variety of MAP4 kinases. Most of them are related to the yeast kinase Ste20, which transduces pheromone signals. For activation they require an interaction with small G-proteins of the Rho/Rac/Cdc42 family, which also induces membrane attachment of the kinases and activation loop (auto)phosphorylation. Other MAP4 kinases become activated along tumor necrosis factor (TNF) receptor-controlled signaling pathways. With the exception of PKC functioning as a MAP4 kinase of the ERK modules, the MAP4 kinases known thus far preferentially stimulate the stress modules.

11.5 Organization of MAP kinase modules by scaffold proteins

Scaffold proteins are molecular sockets that bind several signal-transducing proteins in such a way that the signaling interactions are facilitated and rendered more specific. Somehow, such complexes are reminiscent of integrated electronic circuits. For the assembly and function of MAP kinase modules, scaffold proteins play a key role.

The prototype of such a scaffold protein is **Ste5** of *S. cerevisiae*. It organizes the pheromone-activated Fus3 module (see Figures 11.2 and 11.11) and is indispensable for the module's function. Quite a similar module, differing from the pheromone-activated module in the MAP kinase only (Kss1 instead of Fus3), controls an entirely different physiological event: the invasive growth of diploid and filamentous yeast cells in response to nitrogen deficiency (Figure 11.2). This module is not fixed on Ste5 but possibly bound by another, still-elusive scaffold protein. In an additional yeast module, the Hog1 module of osmolarity adaptation, the MAP2 kinase Pbs2 serves as a scaffold protein (Figure 11.11).

The MAP kinase modules of mammalian cells are hold together by similar scaffold proteins. The proteins **JIP** binding the JNK modules have been mentioned already. They are active only as dimers, thus probably facilitating a trans-autophosphorylation between the kinases of two modules. **KSR**, Kinase Suppressor of Ras, has a similar function for the ERK1,2 modules. KSR is a RAF-related pseudokinase that, like RAF, interacts with the MAP2 kinase MKK1, at the same time being a substrate of the MAP4 kinase TAK1. [Pseudokinases are proteins with an incomplete and, therefore, inactive protein kinase domain that are assumed to be involved in the control of active protein kinases; the human genome harbors about 50 pseudokinases.] When fully phosphorylated, the KSR–MKK1 complex is sequestered by a 14-3-3 protein and cannot organize the Ras-activated ERK modules (explaining its name). Only upon partial dephosphorylation is KSR able to bind RAF and ERK. A collection of scaffold proteins of mammalian MAP kinase modules is shown in Table 11.3.

Scaffold proteins increase the specificity or targeting and the efficacy of signal transduction. Moreover, they assemble "their" signal-transducing complexes at distinct cellular sites, shielding them from disturbing influences. This occurs, however, at the expense of signal amplification, which is not possible in a module with a 1:1 ratio of components. For this reason, MAP kinase modules *without* scaffold proteins may also exist, enabling a sequential signal amplification, albeit now at the expense of specificity (Figure 11.12).

Summary

MAP kinase modules frequently are organized on scaffold proteins, increasing the specificity and efficiency of signal transduction. However, this is achieved at the expense of signal amplification, which can occur only in the absence of scaffolding.

11.6 Downstream of MAP kinase modules: MAP kinase-activated protein kinases

MAP kinases have various substrates, such as the cytoplasmic phospholipase A_2 (the key enzyme of arachidonic acid metabolism), the Ras-GEF mSOS, and the above-mentioned MAP kinase phosphatases, as well as transcription factors that will be discussed below. Moreover, MAP kinase-derived signals are transduced and distributed by protein phosphorylation. In other words, with the series

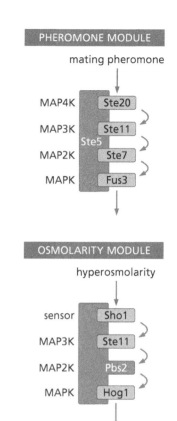

Figure 11.11 Scaffold proteins of MAP kinase modules in *S. cerevisiae* The pheromone module has a separate scaffold protein (Ste5) organizing the ordered interaction of four kinases. In the osmolarity module, the MAP2 kinase Pbs2 serves as a scaffold protein on its own, holding together the whole module. Both scaffold proteins represent standard types. Analogous proteins are found in vertebrate cells. Red curved arrows symbolize activation steps (phosphorylations).

Table 11.3 Scaffold proteins of mammalian MAP kinase modules[a]

Scaffold protein	MAP kinase module	Remarks
KSR	RAF–MKK1–ERK1	RAF-related pseudokinase
MP1	MKK1–ERK1	
JIP1,2	MLK–MKK7–JNK	interacts with Rho-GEF p190
JSAP1 (JIP3)	MEKK1–MKK4–JNK	
MEKK1	RAF–MKK2–ERK2	interacts with cytoskeleton
MEKK2	MEKK2–MKK7–JNK1	
filamin	TRAF2–MKK4–JNK	interacts with actin filaments
β-arrestin 2	ASK–MKK4–JNK3	activated by angiotensin II

[a]KSR, kinase suppressor of Ras; MP, MEK partner; JIP, JNK-interacting protein; JSAP, JNK/SAPK-associated protein; TRAF2, TNF receptor-associated factor 2, an adaptor protein bringing together the TNF receptor with JNK and NFκB modules (see Section 6.3.1).

Figure 11.12 Signal amplification by a MAP kinase module not fixed on a scaffold protein Upon activation by an input signal, MEKK1 phosphorylates many MKK4 molecules, each of which then phosphorylates many JNK molecules (red arrows; P, phosphate group). The cascade is interrupted only by dephosphorylation of MKK4 and inactivation of MEKK1. An asterisk marks the active MEKK1.

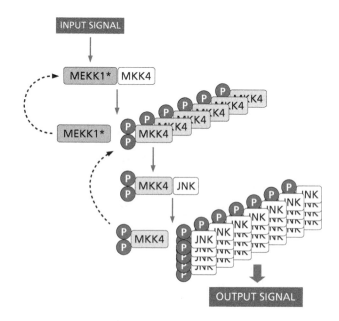

MAP4K–MAP3K–MAP2K–MAPK the phosphorylation cascade associated with MAP kinase modules is by no means at an end but may continue downstream to MAP kinase-activated protein kinases (MAPKAP kinases). In mammalian cells, 11 structurally related types of such kinases have been found:

- ribosomal S6 kinases (RSK), also known as p90^{RSK} or MAPKAP kinases 1 (expressed in four isoforms)

- <u>M</u>itogen- and <u>S</u>tress-activated <u>K</u>inases MSK (two isoforms)

- <u>M</u>APK-i<u>N</u>teracting <u>K</u>inases MNK (two isoforms)

- MK2, -3, and -5 (unfortunately the acronym MK is also used for MAP kinases, which must not be confused with these MAPKAP kinases)

By means of corresponding docking motifs, MKs associate with MAP kinases, in particular ERK and p38, and become activated by phosphorylation of Ser and Thr residues adjacent to Pro residues. A major function of MAPKAP kinases is activation of proliferation and survival programs in response to mitogenic and stress signals (see Figure 11.13). To evoke such responses, further protein kinases or even kinase cascades as well as transcription factors and other regulatory

Figure 11.13 Effects of MAP kinase-activated protein kinases In response to mitogens and stress factors transmitted by MAP kinase modules, the protein kinases shown in the colored box modulate proliferation and activate survival programs. To this end, they stimulate the transcription of early response and survival genes and inactivate cell cycle inhibitors such as p27^{KIP} and MYT1 but also the cell cycle stimulator CDC25. At the same time, they promote mRNA translation by inactivating the translational inhibitor eEF2 kinase and stimulating the translation factor eIF4E. Inhibition of the Ras-GEF mSOS by p90 RSK provides a negative feedback control of the Ras–RAF–ERK cascade.

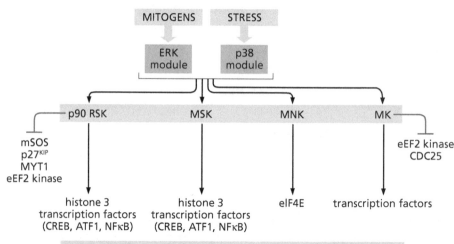

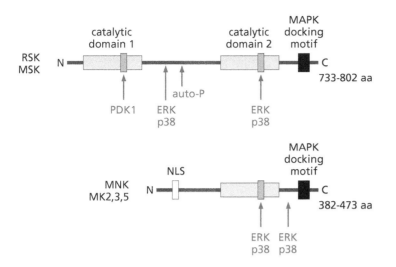

Figure 11.14 Basic structure of MAP kinase-activated protein kinases (Top) A peculiarity of ribosomal S6 kinases (RSK) and mitogen- and stress-activated kinases (MSK) are two catalytic domains: an N-terminal protein kinase C-related domain 1 and a C-terminal calmodulin-dependent kinase-related domain 2. (Bottom) Other MAPK-interacting kinases (MNK) and MAP kinase-activated protein kinases (MK) contain only the calmodulin-dependent kinase-related domain 2. Activation loops are symbolized by colored bars. Arrows indicate the sites phosphorylated by individual MAP kinases and phospholipid-dependent kinase 1 (PDK1). Auto-P, autophosphorylation; NLS, nuclear localization sequence.

proteins become phosphorylated and activated by MAPKAP kinases. Thus, in mammalian cells phosphorylation cascades may easily comprise more than half a dozen steps. Yeast cells express almost no MAPKAP kinases, indicating that (as expected) their data-processing network is organized more simply.

11.6.1 RSK and MSK subfamilies: activators of survival programs

The **ribosomal S6 kinases (RSK)** got their name from the protein S6 of the small ribosomal subunit, which is, however, not their major substrate. [RSK or p90^{RSK} must be distinguished from the S6 kinases proper (S6K or p70^{RSK}), which are not MAP kinase substrates (for details see Section 9.4.2)]. They are characterized by two functional kinase domains, a PKC- and a calmodulin kinase-related one (Figure 11.14).

Upon activation by ERK1,2 or p38, RSKs accumulate in the nucleus to phosphorylate transcription factors. Together with ERK and JNK, RSKs stimulate the transcription of immediate early response genes as a rapid reaction to mitogenic signals and stress factors. Typical RSK substrates are the transcription factors SRF (serum response factor), CREB (cAMP response element binding protein), cFos, and cJun, as well as factors controlling the biosynthesis of ribosomal RNA. Moreover, RSKs protect cells from apoptosis by inactivating pro-apoptotic proteins such as Bad and stimulating the transcription of anti-apoptotic genes. Finally, RSKs play an important role in the up-regulation of the cell cycle: they inactivate the cell cycle inhibitors p27^{KIP} and protein kinase MYT1, thus shortening the G1 phase and promoting the cell's progression into the S phase (see Section 12.5 for a more in-depth discussion of this subject). More details on RSK and further substrates thereof, such as glycogen synthase kinase 3 (GSK3) and protein kinase LKB1, are found in Section 9.4.

The widely distributed kinases of the **MSK subfamily** are close relatives of RSKs and (like these) are substrates of ERK and p38 (Figure 11.14). They are found mainly in the nucleus where they phosphorylate and activate transcription factors such as CREB, NFκB, and ATF1 (activating transcription factor 1), as well as histone H3. Because it weakens the histone–DNA interactions in nucleosomes, H3 phosphorylation is an important step in transcriptional stimulation such as that of immediate early response genes (Section 8.2.5).

11.6.2 MNK and MK subfamilies of MAPKAP kinases: up-regulation of protein synthesis and cell proliferation

Like RSK and MSK, kinases of type **MNK** are activated by both ERK1,2 and p38, whereas type MK seems to be a p38 substrate only. The best-known function of isoform MNK1 is phosphorylation of the translation factor eIF4E, resulting in increased translational efficiency (see Section 9.3.3). Such an up-regulation of protein synthesis supports transcriptional activation induced along the MAP kinase cascades.

The MAPKAP kinases of the **MK** subfamily arrive at the same result by phosphorylating and inactivating the eEF2 kinase, a powerful translational repressor (Section 9.3.4). In addition, they interrupt the cell cycle by phosphorylating the cyclin-dependent kinase phosphatase CDK25, stabilizing mRNA, and (like RSK and MSK) activating transcription factors controlling immediate early response genes. Taken together, these reactions (summarized in Figure 11.13) are essential for the cellular stress response.

Summary

MAP kinase-activated protein kinases (MAPKAP kinases) transduce signals received from the MAP kinase modules. One of their major functions is to promote cell proliferation and to prevent apoptosis. To this end, they modulate the activities of regulatory proteins and stimulate mRNA translation and gene transcription, particularly of immediate early response genes. A well-known subfamily of MAPKAP kinases is the ribosomal S6 kinases p90^{RSK}.

11.7 Downstream of MAP kinase modules: transcription factors

As we have seen, MAP kinase modules are central relay stations of signal transduction from the periphery to the genome. In fact, all MAP kinases, from yeast to humans, catalyze the phosphorylation of transcription factors either directly or via MAPKAP kinases, thus controlling the transcriptional activity of a large number of genes.

Pivotal targets are the immediate early response genes. As the name says, such genes are regulated by signaling cascades directly, without any detour (for more details see Section 12.6). The transcription factors associated with such genes have (apart from MAP kinase phosphorylation sites with the characteristic Ser/Thr-Pro core) docking domains that render the kinase–substrate interaction specific and direct the phosphorylation to selected sites in the protein molecule. The principle of this mechanism is illustrated in Figure 11.5. MAP kinase-catalyzed phosphorylation may either activate or inactivate transcription factors by regulating one or more of the following features:

- DNA binding
- transactivating or transrepressing efficacy
- interactions with transcriptional co-activators or co-repressors
- nuclear import or export
- ubiquitin-mediated degradation

Which of these effects predominates depends on the type of transcription factor, the MAP kinase isoform, and the cell-physiological context. In the following, some transcription factors will be introduced that have turned out to be standard substrates of MAP kinases.

11.7.1 Transcription factors of the Ets family

Ets factors are of critical importance for the formation, maintenance, and function of tissues. Their name refers to an erythroblastosis virus E26 (<u>E</u> <u>t</u>wenty-<u>six</u>) where one of these factors was found to be encoded by an oncogene. Nevertheless, the Ets family is probably restricted to multicellular organisms. In mammals, it comprises around 30 members grouped into 10 subfamilies.

Ets factors are characterized by an **Ets domain** that exhibits a helix–loop–helix structure serving both DNA binding and interactions with other proteins. Like most transcription factors, Ets factors are activated by input signals that change the conformation from an inactive resting state into an active state. These input signals consist of protein–protein interactions and protein phosphorylation uncovering the Ets domain. Other helix–loop–helix proteins containing inhibitory domains may compete with Ets domains for DNA binding sites, thus acting as negative regulators of Ets signaling (Figure 11.15).

Depending on the particular isoform, Ets factors are selectively phosphorylated by individual MAP kinases, in some cases also simultaneously by more than one MAP kinase. Mostly, the phosphorylation has a stimulatory effect on Ets activity,

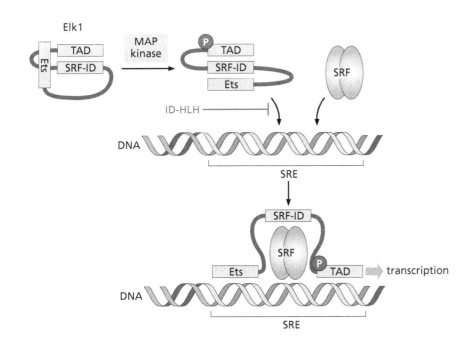

Figure 11.15 Activation of a ternary-complex factor By MAP kinase-catalyzed phosphorylation (P), the conformation of an Ets factor (here the isoform Elk1) is changed in such a way that the "hidden" Ets domain is exposed, enabling Elk1 to form a complex with the serum response factor SRF (via an SRF interaction domain, SRF-ID) and to bind to a serum response element (SRE) in a corresponding gene promoter. In turn, SRF further unfolds the conformation of Elk1 so that its transactivating domain (TAD) may now recruit the initiation complex of transcription. Helix–loop–helix proteins with inhibitory domains (ID-HLH) may compete with Elk1 for DNA binding, suppressing gene transcription. The oncogenic Ets protein of erythroblastosis virus has a constitutively active conformation and does not require phosphorylation as a stimulatory input signal.

although there are exceptions to this rule (see below). Moreover, some Ets factors are transcriptional repressors, becoming inactivated by phosphorylation. Like other transcription factors, Ets proteins are active only as dimers, recruiting not only transcriptional co-modulators such as the histone acetyltransferases CBP/p300 but also other transcription factors.

An example demonstrating the interaction between Ets and other factors is provided by the **ternary-complex factors (TCF)**, an Ets subfamily comprising the isoforms Elk1 (Ets-like) and SAP1,2 (Serum response factor-Accessory Proteins). These factors control the transcription of various immediate early response genes by binding to a serum response element located in the gene promoter. (Serum response elements were discovered in experiments with cell cultures, which require serum, a cocktail of growth factors, for proliferation.) To acquire this activity the TCFs have to interact with another transcription factor, the homodimeric **serum-response factor (SRF)**. As a result, a ternary complex consisting of Elk1, SRF, and serum response element is established, which gave rise to the name ternary-complex factor. This complex is activated by MAP kinase-catalyzed phosphorylation of the Ets component. While Elk1 is a substrate of ERK, JNK, and p38, SAP1 and -2 are phosphorylated by p38 only (interestingly, SAP2 is inhibited by JNK-catalyzed phosphorylation).

Ets factors are phosphorylated not only by MAP kinases but also by various other protein kinases and, of course, are substrates of protein phosphatases terminating the Ets signal. Moreover, like Ets, SRF also is regulated by protein phosphorylation as catalyzed, for instance, by the MAPKAP kinase RSK. Thus, the control of ternary complex factor activity shows in a particularly clear manner that transcriptional regulation represents a combinatorial process where the resulting effect depends on the local state of excitation of the protein network rather than on a single signaling event.

Still, the physiological functions of Ets factors are not completely understood in all details. Relatively well-known is their role in developmental and regenerative processes such as the formation of blood vessels (angiogenesis) and in hematopoiesis as well as in neuronal differentiation. At least in cell cultures, Ets factors are controlled also by proliferative signals transmitted by growth factors and serum and mediated by the ERK1,2 modules. This has been thought to indicate a role in wound repair and explains the tumor-promoting efficacy of Ets, proven by numerous experiments.

11.7.2 Transcription factors of the AP1 family

The second large family of transcription factors directly controlled by MAP kinase-catalyzed phosphorylation is collectively termed AP1 (Activator Proteins 1). Each AP1 factor is a homo- or heterodimer composed of proteins belonging to the subfamilies Jun, Fos, ATF2 (Activating Transcription Factor 2), and MAF (Musculo-Aponeurotic Fibrosarcoma factor). These factors exist in 20 isoforms at least. Dimerization is achieved by leucine zipper motifs. This combinatorial principle gives rise to a large number of transcription factors differing in gene specificity. Jun and Fos were discovered originally as oncoproteins of animal sarcoma and osteosarcoma viruses (vJun and vFos); the cellular proto-oncogene products are differentiated with the prefix c (cJun and cFos).

The efficacy of AP1 factors is regulated by both *de novo* synthesis and phosphorylation. Fos factors are encoded by typical immediate early response genes, the transcription of which ranks among the most rapid responses of cells to stress and proliferation signals transmitted by MAP kinase modules. Like most eukaryotic gene promoters, the *fos* promoter contains several response elements that interact with different input signals. In other words, the transcription of the *fos* genes is activated along several signaling pathways (Figure 11.16) including the Ras–RAF–ERK–TCF cascade, cAMP-controlled pathways, and the JAK–STAT

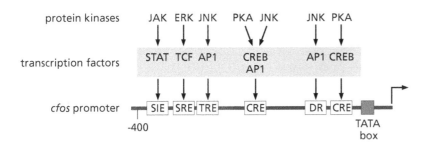

Figure 11.16 *cfos* gene promoter
This promoter provides a striking example of a gene-regulatory sequence with various signaling inputs rendering it an efficient logical gate. Shown are the different response elements with the associated transcription factors (colored box) and protein kinases. The protein kinases represent a selection: as described for CREB in Section 8.6, most transcription factors are substrates of more than one kinase. Protein kinases: JAK, Janus kinase; ERK, extracellular signal-related kinase; JNK, c-Jun N-terminal kinase; PKA, protein kinase A. Transcription factors: STAT, signal transducer and activator of transcription; TCF, ternary-complex factor; AP1, activator protein 1; CREB, cAMP response element binding protein. Response elements: SIE, STAT-inducible element; SRE, serum response element; TRE, TPA response element; CRE, cAMP response element; DR, direct repeat. For additional details, see text.

cascade with various cytokine receptors as starting points (Section 7.2.1). This situation resembles a multifunctional logical gate.

In nonstimulated cells, the cFos partner cJun is phosphorylated in the carboxy-terminal domain, preventing its interaction with the gene promoter. This phosphorylation is catalyzed by GSK3β or casein kinase 2 (CK2) (see Section 9.4.5). When GSK3β is inactivated by mitogenic signals along the PI3K–PKB/Akt pathway (Section 4.6.6), cJun becomes dephosphorylated by omnipresent protein phosphatases. To become active, cJun as well as ATF2 (the two AP1 factors most thoroughly studied in this respect) have to be phosphorylated at two Ser residues in their amino-terminal transactivating domains. These phosphorylations are catalyzed by JNK (for cJun, explaining the name Jun N-terminal kinase) or p38 (for ATF2), and as in the case of Ets, their specificity is enhanced by docking domains. Together with the *de novo* synthesis of cFos, this represents the major pathway along which stress factors, cytokines, and growth factors change gene expression.

cJun homodimers and cJun/cFos heterodimers preferentially bind to so-called **TPA response elements** in gene promoters. TPA (12-*O*-tetradecanoylphorbol 13-acetate), a poison from the spurge *Croton tiglium*, has become most prominent as a potent tumor promoter in animal experiments. It mimics the effects of the second messenger diacylglycerol (for details see Section 4.6.5) and is a strong activator of genes, though the molecular mechanism of this effect is not entirely clear. The corresponding TPA response element is found also in the *cfos* and many other gene promoters including the *cjun* promoter, enabling cJun to stimulate its own biosynthesis and that of cFos in a positive feedback loop.

Different from cJun/cFos, the cJun/ATF heterodimers recognize their target genes by cAMP response elements, which were detected originally as binding sites of the CREB transcription factors. The latter are activated along the cAMP–PKA cascade and other signaling pathways (Sections 4.6.1 and 8.6). Still other response elements mediate the interaction of cJun/MAF heterodimers with their target genes.

The effects of the individual MAP kinase modules on Ets/TCF as well as on some other transcription factors are summarized in Figure 11.17. However, this is not the end of the story, because as we have seen, ERK and p38 (at least) may activate transcription factors also via protein kinases positioned downstream of the modules, such as RSK, MSK, MNK, and MK. This situation and additional secondary reactions make plausible the fact that the activities of more than 1000 genes are changed upon stimulation of the Ras signaling pathway. In other words and to repeat what has been stated above: exogenous signals do not activate linear biochemical reactions but stimulate large areas of the data-processing protein network, thus generating a diffuse excitation pattern similar to that seen in the brain's neuronal network upon sensory stimulation (see Figure 1.13).

Summary

Prototypes of transcription factors controlled by MAP kinase-catalyzed phosphorylation are the Ets factors (including ternary-complex factors, TCFs). Upon

Figure 11.17 Direct control of transcription factor activity by MAP kinase modules To the left and right of the three major MAP kinase modules (colored boxes), a selection of transcription factors phosphorylated by MAP kinases is shown. **MEF2** (Myocyte Enhancer Factor 2) controls genes of muscle differentiation and stimulates the transcription of the *cjun* gene. **CHOP** (CREB-HOmologous Protein) is involved in the genotoxic stress response. Together with other factors, it induces cell cycle arrest at the G1–S transition to provide time for DNA repair (see Section 12.9). **NFAT** (Nuclear Factor of Activated T cells) predominantly controls interleukin genes, playing an important role in inflammatory reactions and in the immune response. Phosphorylation by JNK prevents its translocation from the cytoplasm into the nucleus (an inhibitory effect; Section 14.6.4). **MAX** is a binding partner and activator of the transcription factor **cMyc** that controls many genes for proliferation, apoptosis, and cell differentiation and may undergo oncogenic mutation. Not shown are transcription factor phosphorylations, catalyzed by MAP kinase-activated protein kinases (see Figure 11.13). PI3K, phosphatidylinositol 3-kinase; ATF2, activating transcription factor 2. To the right, phosphorylation of cJun is shown (activation is symbolized by black arrows; inactivation, by red arrows). As can be seen, cJun works as a logical BUT NOT gate. JNK, c-Jun N-terminal kinase; GSK3β, glycogen synthase kinase 3β; CK2, casein kinase 2.

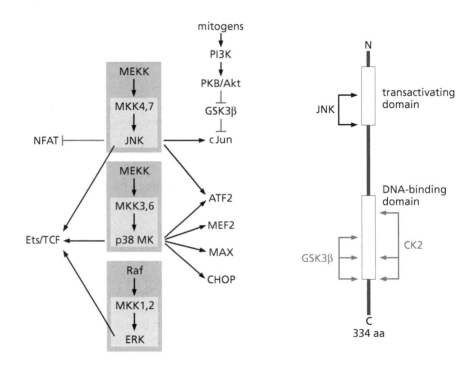

phosphorylation they form complexes with other transcription factors (such as serum-response factor, SRF). These complexes modulate the transcriptional activity of various immediate early response genes involved in developmental and regenerative processes. Transcription factors of the activator protein 1 (AP1) family, comprising a large number of homo- and heterodimers of the proteins cJun, cFos, ATF2, and MAF, are controlled by JNK- and p38-catalyzed phosphorylation. In combination with other transcription factors, they regulate the activities of a wide variety of genes.

11.8 NFκB signaling pathway

A major "spinal cord" of cellular data processing, which rivals the MAP kinase modules in its versatility and complexity, is the signaling pathways that control the NFκB (nuclear factor κB) family of transcription factors.* These factors rank among the most versatile regulators of gene transcription. In fact, the NFκB cascades have been estimated to connect more than 200 exogenous stimuli with about the same number of different genes. By binding to specific DNA sequences (κB elements) in gene promoters and enhancers, NFκB transcription factors can either stimulate or inhibit gene transcription.

Like MAP kinase modules, NFκB signaling is involved in an immense number of physiological processes dealing with organ development, cell proliferation, cell death, and cell differentiation. A major task of the NFκB system is to process signals that control innate and acquired immune defense. Thus it plays a central role in inflammatory reactions (see also Section 6.3). For inflammation and defense, NFκB factors regulate the transcription of genes encoding pro-inflammatory cytokines and chemokines, cell adhesion molecules, and enzymes such as NO synthase and cyclooxygenase that catalyze the production of inflammatory mediators. Certain cytokines induced along the NFκB pathways, such as TNFα and interleukin 1, activate the *de novo* synthesis of NFκB through positive

---

*Originally NFκ B was found to bind to the enhancer of the immunoglobulin κ gene in B-lymphocytes and was thought to stimulate antibody formation. Today we know that this is not a major function of the factor. Moreover, it is not exclusively nuclear but is found in the cytoplasm as well.

Sidebar 11.3 Control of gene transcription by frequency-modulated signals? Although appearing to be quite simple at a first glance, the modulation of gene activity by signals such as phosphorylation harbors a fundamental problem, at least as the situation in a single cell is concerned. In fact, genes respond to input signals in a variable, graded manner. However, the law of mass action describing stochastically graded chemical reactions is not applicable here due to the small number of reaction partners (two gene copies only). Therefore, variable gene activity should obey other rules than those of random distribution. One possibility certainly consists of the interaction of a gene with more than one stimulatory and inhibitory signal, rendering it a stepwise-regulated logical gate. Another quite appealing and not mutually exclusive mechanism resulting in a graded transcriptional response would be frequency modulation of signals. Indeed, due to its particular properties and diverse intrinsic feedback loops, the MAP kinase module, in particular, may fulfil the conditions of a chemical oscillator (Section 1.9). This would mean that the activity of a transcription factor is not controlled according to a rigid ON–OFF scheme but instead by a discrete sequence of phosphorylating and dephosphorylating impulses, the frequencies of which depend on the intensity of the particular input signal. Such a mechanism would enable a graded transcriptional response and may explain why a single input signal, although processed by the same MAP kinase module, may exhibit different or even opposite effects depending on the concentration of the signaling factor and the duration of action. It is conceivable that this might be one reason for the triumphant progress of the MAP kinase modules in evolution. As a restriction, it must be emphasized, however, that currently the concept of frequency-modulated transcriptional regulation is still widely hypothetical, mainly due to a lack of suitable experimental methods.

feedback. Such a system may easily run out of control, resulting in severe disorders such as rheumatoid arthritis, bronchial asthma, chronic inflammatory bowel disease, multiple sclerosis, and cancer as well as in death by shock. Therefore, NFκB signaling is under strict and expensive control to ensure that the factors become active only when several safety devices have been unlocked.

11.8.1 NFκB module and its components

The basic structure of an NFκB module is shown in Figure 11.18. It consists of a protein kinase that senses input signals and a transcription factor that transmits output signals. The transcription factor is composed of a catalytic subunit and a regulatory subunit, which in the absence of an input signal inhibits the catalytic subunit. This inhibition is relieved by phosphorylation and subsequent degradation of the regulatory subunit. In other words: the module works according to the principle of double negation (minus times minus is equal to plus).

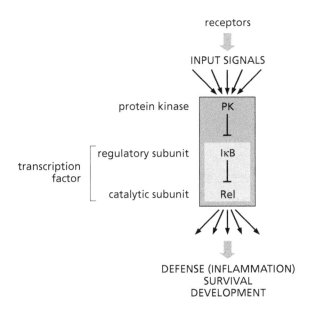

Figure 11.18 NFκB module The module functions according to the principle of double negation. The core of the module is a transcription factor consisting of a regulatory subunit (IκB) that, in the absence of an input signal, inhibits the catalytic subunit (Rel). Input signals activate a protein kinase (PK) that phosphorylates IκB, priming it for degradation. When it is relieved from inhibition, Rel activates a series of genes involved in defense reactions (inflammation), cell survival, and tissue development.

The active transcription factors of the NFκB family are homo- or heterodimers of the proteins cRel, RelA, RelB, p50 (NFκB1), and p52 (NFκB2) (the term Rel refers to a retroviral oncogene vRel derived from cRel and found in avian reticuloen-dotheliosis virus). In common these proteins have N-terminal **Rel homology domains (RHD)** for dimerization, nuclear targeting, and DNA binding (a closely related Rel similarity domain is also found in NFAT transcription factors that address another set of genes; see Section 14.6.4). Upon activation by input signals, the dimers accumulate in the nucleus and bind to the above-mentioned κB elements in the promoter and enhancer regions of the target genes. The Rel and NFκB transcription factors may dimerize to any possible combination except for the RelB homodimer, which has not been found *in vivo*. The 15 dimers thus generated exhibit individual but overlapping gene specificities. Among them, the RelA–p50 heterodimer is the most abundant one, representing the prototypical NFκB protein. Principally, the Rel factors activate transcription whereas p50 and p52 have inhibitory effects when they occur as *homodimers* (whereas heterodimers with cRel, RelA, or RelB are stimulatory). The reason for this is that p50 and p52 are truncated forms that lack a transactivating domain (see Figure 11.19) and thus are unable to recruit the protein complex of the transcriptional machinery, though via their RHDs they bind like Rel to DNA. Moreover, in the presence of p50–p50 or p52–p52 homodimers, transcriptional co-repressors such as histone deacetylases are assembled at the gene promoter. Thus, p50 and p52 homodimers are powerful competitive inhibitors of NFκB signaling that resemble dominant-negative mutants.

Rel/NFκB-homologous transcription factors are also found in invertebrates. *Drosophila*, for instance, expresses three isoforms called Dif, Dorsal, and Relish. They are essential for certain stages of embryonic development as well as for the defense against pathogenic organisms. Some examples are discussed in more detail in Section 6.3.2. In agreement with its functions in defense and development, the NFκB system probably entered the stage of evolution with the appearance of multicellular eukaryotes, since it is not found in unicellular organisms such as yeast (the lack of NFκB in *C. elegans* is explained by secondary gene loss).

11.8.2 Signal processing by NFκB modules: canonical and noncanonical pathways

A characteristic feature of Rel and NFκB transcription factors is that in most cell types they shuttle between cytoplasm and nucleus. In the absence of input signals (that is, in nonstimulated cells), they are mainly concentrated in the cytoplasm where they exist as complexes with their inhibitory subunits called **IκBs**.

Figure 11.19 Modular structures of NFκB subunits Rel proteins are catalytically active transcription factors that contain a Rel homology domain (RHD) for DNA binding, nuclear translocation (NLS = nuclear localization sequence), dimerization, and interaction with IκB, and a transactivating domain (TAD) for binding of transcriptional co-activators such as CBP/p300. Proteins NFκB1 and NFκB2 are truncated forms of Rel that lack TAD. NFκB1 is released from the protein p100 by limited proteolysis. p100 and p105 contain ankyrin repeats that block the RHD, thus preventing nuclear localization and DNA binding. The same effect is achieved by the inhibitory subunits IκB when they interact with Rel or NFκB proteins.

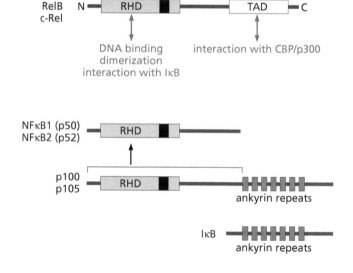

IκBs are expressed in several isoforms, with IκB-α, -β, and -ε being the species most studied. Two other proteins, p100 and p105, have a C-terminal IκB domain linked with an N-terminal RHD (Figure 11.19). These domains are also known as IκBγ and -δ, respectively. The inhibitors block the RHD either intermolecularly or, in the case of p100 and p105, intramolecularly. Important for this inhibitory effect are ankyrin repeats (see Section 13.1) that also mask the nuclear localization sequences of the associated Rel/NFκB proteins, which thus are hindered from entering the nucleus. At least for IκBα, the major partner of the prototypical RelA–p50 heterodimer, this effect is augmented by a nuclear export signal.

It is easily conceivable that a key event in NFκB signaling is the inactivation of IκBs. This is achieved by phosphorylation, priming the inhibitors for destruction along the ubiquitin–proteasome pathway (see Section 12.16). Input signals that stimulate this reaction are pro-inflammatory cytokines, antigens, bacterial components, and stress factors like UV light, hypoxia, oxidative stress, and ionizing radiation, as well as factors that control embryonic development. Upon interaction, either with cellular receptors or directly, these signals activate a complex of two protein kinases, **IK kinases** IKK1 (alias IKKα) and IKK2 (alias IKKβ), held together by the scaffold protein NEMO (<u>N</u>FκB <u>E</u>ssential <u>MO</u>dulator, formerly known as IKKγ although it lacks kinase activity) that at the same time acts as an additional sensor for upstream signals. In this complex, the phosphorylation of IκB is catalyzed by IKK2 (Figure 11.20).

Upon being rescued from the inhibitor, Rel/NFκB accumulates in the nucleus to stimulate the transcription of its target genes, including that of IκBα. Thus, like MAP kinases (which induce the production of MAP kinase phosphatases), NFκB also controls its own activity via a negative feedback loop. Another analogy with MAP kinase modules is that the NFκB module is controlled by upstream protein kinases that catalyze activation loop phosphorylation of IKKs (see Section 11.8.3). Most of these IKK kinases are members of the MAP3 kinase family such as MEKK1, MEKK3, NIK, NAK, and TAK1. NFκB and MAP kinase modules are also stimulated by PKC. This means that frequently both modules are activated simultaneously. Another IKK kinase is the DNA checkpoint kinase ATM linking the NFκB system with DNA damage repair (see Section 12.9). Moreover, certain

Figure 11.20 Major pathways of NFκB signaling (Left) Canonical NFκB module, including phosphorylation (P) and proteolytic degradation of the inhibitor IκB. The factor Rel thus activated forms a heterodimer, for instance with p50 as shown here, that accumulates in the nucleus. The complex of IKK1 and -2 and IκB assembled by the scaffold protein NEMO has been called a signalosome. (Right) Noncanonical NFκB module. Upon phosphorylation (P), the protein p100 is partially degraded along the ubiquitin–proteasome pathway and releases active p52, which as a heterodimer with RelB accumulates in the nucleus. In a similar way, p50 is released from a precursor protein p105 (not shown). In addition to these major pathways, some atypical pathways of NFκB signaling exist. Upon UV irradiation, IκB is phosphorylated by casein kinase CK2 instead of by IKK, releasing the Rel–p50 complex, while under genotoxic stress NEMO forms a complex with the checkpoint kinase ATM (Section 12.9) that activates the canonical module (not shown). TNFR, tumor necrosis factor receptor; IL1R, interleukin 1 receptor; TLR, Toll-like receptor of bacterial components; BCR and TCR, B- and T-cell antigen receptors; NIK, NFκB-inducing kinase; MHC, major histocompatibility complex.

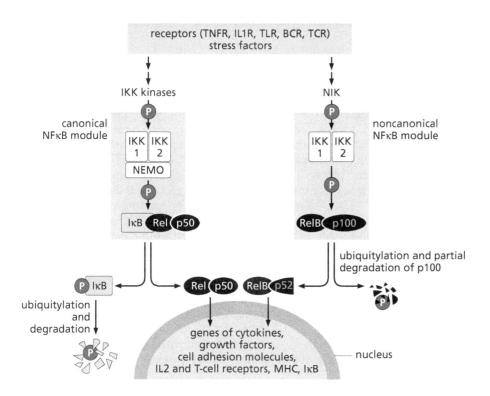

protein kinases can replace the IKKs in NFκB activation. For example, casein kinase CK2 is stimulated upon UV irradiation and directly phosphorylates IκBα and -β and primes them for proteolytic degradation, in this case by the protease calpain rather than by proteasomes.

Along these pathways the **canonical** module, which releases an active RelA–p50 heterodimer, is activated. The input signals that typically stimulate this module are transmitted by receptors of pro-inflammatory cytokines such as TNFα and interleukin 1, as well as bacterial components interacting with Toll-like receptors. The response is an activation of innate immunity, resulting in an inflammatory reaction (see Section 6.3 for details). Another NFκB signaling pathway, termed **noncanonical**, leads to the exclusive formation of RelB–p52 heterodimers. This pathway starts from certain TNF receptor types and from the T- and B-cell antigen receptors, which activate the MAP3 kinase NIK (NFκB-inducing kinase) that in turn phosphorylates IKK1. The substrate of the NIK-activated IKK1 is not IκB but an inactive complex of a Rel protein and p100, which here functions as a regulatory subunit. Upon phosphorylation, the IκBγ domain of p100 is degraded to release the active p52/NFκB2 which then accumulates in the nucleus together with RelB. This pathway plays a role in the regulation of acquired immunity: it leads to B-cell maturation and triggers the development of secondary lymphoid organs such as lymph nodes. IKK1 supports the effect of NFκB in still another way: by phosphorylating the Ser 10 residue in histone 3. As explained in Section 8.2.5, this reaction primes the histone for subsequent acetylation, thus playing a key role in transcriptional activation.

Unlike p100, the IκB isoform p105 is phosphorylated by IKK2 rather than by IKK1 resulting in ubiquitylation and complete degradation. In fact, p50 (NFκB1) is generated directly by translation rather than by partial precursor degradation.

The multiple pathways of NFκB signaling show that each dimer addresses an individual set of genes. This considerably increases the versatility of the system.

11.8.3 A complex system of signal processing

What we have seen up to now is nothing but the bare skeleton of NFκB signaling. In reality the NFκB system ranks among the most complex and most interlinked systems of signal processing known. Like the MAP kinase modules, it is connected directly or indirectly with numerous signaling pathways, and we are still far from giving a reliable estimate of the range of this circuitry.

Cross talk with other transcription factors

First of all, the functional repertoire of Rel/NFκB factors is far from exhausted by dimerization with each other. In fact, NFκB binding sites are positioned on gene promoters in such a way that the individual Rel/NFκB monomers or dimers may associate with other transcription factors to form signaling complexes, each of which addresses one or more selected genes and either promotes or suppresses transcription. Such transcription factors include STAT (Section 7.2.1), nuclear receptors (Section 8.3), ATF and CREB (Section 8.6), cJun and cFos (Section 11.7), C/EBPβ (CCAAT/enhancer-binding protein beta, an abundant transcription factor with special functions in the inflammatory response and immune reactions), and others, which are activated along quite varied signaling pathways. Thus, when stimulated through TNFα or interferon γ, RelA and STAT1 may form a heterodimeric transcription factor that controls genes encoding certain pro-inflammatory mediators, while upon stimulation of the Toll-like receptor TLR4, the pair RelA–IRF3 (interferon-regulated factor 3; see Section 6.3.2) induces the transcription of genes with an interferon response element. The well-known anti-inflammatory effect of glucocorticoids is explained at least partially by a direct interaction with the nuclear glucocorticoid receptor, resulting in sequestration and inhibition of NFκB, an effect which is called transrepression.

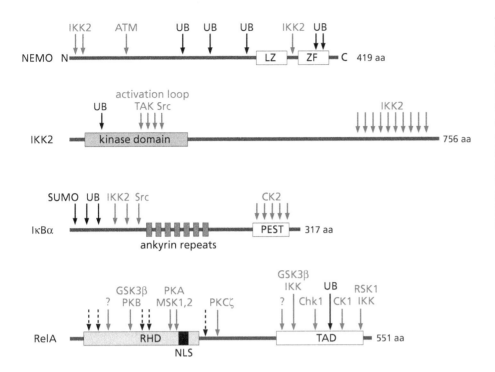

Figure 11.21 Multiple post-translational modifications of selected proteins of the NFκB module Red arrows symbolize phosphorylation (with the corresponding protein kinases), black arrows symbolize ubiquitylation (UB) or sumoylation (SUMO), and broken black arrows symbolize acetylation. RHD, Rel homology domain; TAD, transactivating domain; LZ, leucine zipper domain; ZF, zinc finger domain; PEST, domain containing the sequence Pro-Glu-Ser-Thr.

Post-translational modifications

The components of the NFκB modules provide a most instructive example of how the activity of signaling proteins is modulated by a large number of post-translational modifications. However, a closer look reveals an extremely complex and somewhat confusing picture that is not yet understood in every detail. Therefore, only some basic facts are described in the following. An overview is found in Figure 11.21.

Let us start with the protein NEMO, the scaffold and sensor of the canonical NFκB modules. Here signal transduction to the IKK complex requires ubiquitylation by polyubiquitin chains linked via Lys63. As explained in Section 2.8, this modification does not trigger protein degradation in proteasomes but provides interaction sites for other proteins. The importance of NEMO ubiquitylation is underlined by the fact that different stimuli activate different E3 ubiquitin ligases that catalyze the ubiquitylation of individual Lys residues in NEMO. Vice versa, de-ubiquitylating enzymes may attenuate NFκB signaling. Among the ubiquitin ligases targeting NEMO, we have encountered the TRAFs as transducers of TNFα and interleukin 1 signals (Section 6.3).

The key event of the canonical pathway is activation loop phosphorylation of IKK2 at two Ser residues, and a major function of ubiquitylated NEMO is to establish the contact with the IKK2 kinases mentioned above. Frequently this requires a cooperation with other poly-ubiquitylated adaptor proteins. An example is given in Section 6.3.2: as shown in Figures 6.8 and 6.9, both the activated interleukin 1 receptor and Toll-like receptors, via the TRAF6 adaptor protein, stimulate the protein kinase TAK1 complexed with the adaptor proteins TAB1 and TAB2. TAK1 is an efficient IKK2 kinase. Its binding to NEMO depends on a Lys63-linked poly-ubiquitylation of the TAB proteins and of TRAF6. Equally complicated is the process by which genotoxic stress signals activate NFκB modules: it includes sumoylation and ubiquitylation of NEMO and its phosphorylation by the stress-activated protein kinase ATM (see Section 12.9), which then phosphorylates IKK2. Interestingly, activation loop phosphorylation of IKK2 can alternatively occur at two Tyr residues. Another target of Tyr phosphorylation is IκBα, which thus becomes either degraded or separated from NFκB. In both cases

NFκB signaling is stimulated. These reactions open a route connecting the NFκB modules with Tyr kinase-coupled receptors.

The signaling sensor of the noncanonical NFκB module is the protein kinase NIK (Figure 11.20). It catalyzes activation loop phosphorylation of IKK1 independently of NEMO.

Let us now turn to the NFκB transcription factors. Both their nuclear translocation and their transactivating efficacy are regulated by a wide variety of post-translational modifications. Again phosphorylation plays a major role. Upon stimulation by input signals, phosphorylation occurs both in the cytoplasm and in the nucleus. While cytoplasmic phosphorylation controls dimerization and nuclear translocation and primes the transcription factor for DNA binding, nuclear phosphorylation affects DNA binding, transactivating efficacy, and lifetime. Various protein kinases (such as PKA, IKK1 and IKK2, ribosomal S6 kinases, checkpoint kinase Chk1, MSK1, GSK3β, and PKCζ) catalyze Rel phosphorylation, linking NFκB signaling with many other signaling pathways. Phosphorylation may have opposite effects depending on the phosphorylation site: phosphorylation of the transactivating domains generally enhances transcription factor activity, whereas phosphorylation of the DNA binding domain is inhibitory. Moreover, phosphorylation of distinct sites may prime Rel/NFκB proteins for ubiquitylation and subsequent degradation in proteasomes. Such modifications terminate the NFκB signal.

Like many other gene-regulatory proteins, RelA is acetylated at several Lys residues by CBP/p300 acetyltransferases, the canonical transcriptional co-activators. As a result, DNA binding becomes reinforced and acetylated peptide sequences are generated that serve as binding sites for proteins with Bromo domains to form a multiprotein signaling complex. Such proteins, including CBP/p300 (which in this way promotes its interaction with RelA), are involved in transcriptional activation (Section 8.2.2). Acetylation also hinders the interaction of Rel with IκBα that would promote its export from the nucleus.

Transcriptional co-repressors exhibiting histone deacetylase activity catalyze the deacetylation of Rel and attenuate its effects on gene transcription. However, it must not be overlooked that, like phosphorylation, acetylation may either stimulate or inhibit transcription factor activity, depending on the site of the protein molecule where it occurs. Thus the intensity as well as the duration of the Rel signal is regulated by acetylation in a complex manner.

11.8.4 NFκB signaling: a decision between life and death

In most tissues, NFκB activates genes that promote the survival of cells by encoding anti-apoptotic and pro-mitogenic proteins. A classical example is provided by peripheral B-lymphocytes, where along the NFκB pathways antigens stimulate the transcription of survival genes. Signaling starts from the B-cell receptor (Section 7.2.2), which upon activation interacts with cytoplasmic Tyr kinases such as Btk and Syk. These kinases phosphorylate and stimulate phospholipase Cγ, thus releasing the second messengers diacylglycerol and inositol 1,4,5-trisphosphate. This leads, in turn, to an activation of PKCβ, which here acts as an IKK kinase, promoting both canonical and noncanonical NFκB signaling. As a final outcome, genes encoding anti-apoptotic proteins such as the caspase inhibitors IAP and FLIP and proteins of the Bcl2 family are activated (for details see Sections 13.2.2 and 13.2.4).

NFκB proteins also guarantee the survival of immature T-lymphocytes whereas in mature T cells they mediate apoptosis by inducing the synthesis of death receptors such as Fas and TRAIL-R, the corresponding ligands, pro-apoptotic Bcl2 proteins, and pro-apoptotic transcription factors such as p53 and cMyc. Since the T-cell receptor resembles the B-cell receptor in many aspects, signal transduction may be similar also.

A pro-apoptotic effect of NFκB is not restricted to mature T cells but also has been observed in other tissues. Frequently it may be an outcome of inflammation, which in its final stages leads to collateral cell death caused by cytotoxic mediators such as reactive oxygen species and pro-apoptotic cytokines released from white blood cells.

NFκB and disease

Cancer is characterized by disequilibrium between cell gain and cell loss due to decreased apoptosis and increased cell proliferation. Thus, an overactivation of NFκB is thought to be oncogenic. This is indeed the case as demonstrated by the retroviral vRel oncogene, which causes lymphomas in birds. However, an oncogenic mutation of an NFκB factor has not yet been found in human tumors. Here NFκB signaling plays the role of a powerful tumor-promoting mechanism. A hallmark of tumor promotion is chronic inflammation. Since NFκB is a key mediator of the inflammatory response, constitutive overactivation of NFκB signaling goes hand in hand with many **inflammatory diseases** such as rheumatoid arthritis, asthma, chronic gastritis, hepatitis, atherosclerosis and multiple sclerosis. Inflammatory diseases such as *Helicobacter pylori*-associated gastritis, Crohn's colitis, and viral-induced hepatitis are preliminary stages of cancer. In fact, an overactivation of NFκB signaling is associated with a wide variety of clinical and experimental neoplasias, whereas pharmacological or genetic inhibition of NFκB signaling has a protective effect.

Tumor promotion by NFκB signaling is easily explained by a constitutive over-stimulation of anti-apoptotic and pro-mitogenic effects. This may occur at any level of the NFκB complex. For instance, a hyperactive IKKε (an IKK isoform found in a subset of NFκB modules) has turned out to be an oncogenic factor in breast cancer. Moreover, along the NFκB-controlled pathways, reactive oxygen species are produced that are important mediators of innate defense, but as endogenous genotoxic agents they may also contribute to the genetic instability of tumor cells (see Section 12.1 for more details).

Apoptosis not only prevents neoplastic growth but also, by removal of infected cells, provides one of the body's defense strategies against pathogenic viruses. Many virus types are able, however, to switch off the apoptotic machinery by overactivating NFκB signaling. Human viruses misusing NFκB signaling are directly or indirectly involved in tumorigenesis. These are **hepatitis C virus**, **herpes simplex virus**, human immunodeficiency virus (**HIV**) and human T-cell leukemia virus (**HTLV-1**). For example, HIV activates NFκB via various signaling proteins in its host cells and harbors an NFκB-responsive element in its own genome.

With the involvement of NFκB signaling in so many diseases, it is clear that strong efforts are being made to apply corresponding inhibitors in the clinic. This has turned out, however, to be a nontrivial task due to the variability, versatility, and redundancy of this signaling system.

Summary

The transcription factors of the NFκB family participate in the control of a wide variety of genes, particularly those controlling specific and nonspecific defense reactions that result in inflammation. The active forms are homo- or heterodimers of the proteins RelA and cRel and heterodimers of these proteins with the truncated Rel proteins p50 and p52, respectively. Since p50 and p52 lack a transactivating domain, their homodimers are transcriptional repressors. Rel factors are kept in the cytoplasm in an inactive state by binding to inhibitory subunits. Upon activation, these subunits become phosphorylated and degraded by proteolysis, enabling the oligomerization and nuclear translocation of Rel factors. The catalytic and inhibitory subunits of the transcription factors and activating kinases are combined into module-like "signalosomes." The signaling versatility of Rel factors is increased by interactions with other transcription factors. Moreover, the activities of NFκB modular components are controlled and fine-tuned by post-translational modifications such as ubiquitylation, phosphorylation, and acetylation, indicating interactions with a wide variety of signaling events. In most cases the NFκB modules process survival (anti-apoptotic) signals. However, depending on cell type and physiological context, cell death may also be induced. This ambivalence plays an important role in the homeostasis of the immune system. An overstimulation of NFκB signaling may promote inflammatory diseases and tumor growth. Several highly pathogenic viruses ensure the survival of their host cells by overstimulating anti-apoptotic NFκB signaling.

Further reading

Beinke S & Ley SC (2004) Functions of NFκB1 and NFκB2 in immune cell biology. *Biochem. J.* 382, 393–409.

Buchwalter G, Gross C & Wasylyk B (2004) Ets ternary complex transcription factors. *Gene* 324, 1–14.

Chang L & Karin M (2001) Mammalian MAP kinase signaling cascades. *Nature* 410, 37–40.

Chen LF & Greene WC (2004) Shaping the nuclear action of NFκB. *Nat. Rev. Mol. Cell Biol.* 5, 392–401.

Chong H, Vikis HG & Guan KL (2003) Mechanisms of regulating the RAF kinase family. *Cell. Signal.* 15, 463–469.

Dan I, Watanabe N & Kusumi A (2001) The Ste20 group kinases as regulators of MAP kinase cascades. *Trends Cell Biol.* 11, 220–230.

Davis RJ (2000) Signal transduction by the JNK group of MAP kinases. *Cell* 103, 239–252.

Dunn C, Wiltshire C & MacLaren A (2002) Molecular mechanism and biological functions of c-Jun N-terminal kinase signaling via the c-Jun transcription factor. *Cell. Signal.* 14, 585–593.

Eferl R & Wagner EF (2003) AP-1: double-edged sword in tumorigenesis. *Nat. Rev. Cancer* 3, 859–869.

Escarcega RO, Fuentes-Alexandro S, Garcia-Carrasco M et al. (2007) The transcription factor nuclear factor κB and cancer. *Clin. Oncol.* 19, 154–161.

Farooq A & Zhou MM (2004) Structure and regulation of MAPK phosphatases. *Cell. Signal.* 16, 769–779.

Gallo KA & Johnson GL (2002) Mixed-lineage kinase control of JNK and p38 MAPK pathways. *Nat. Rev. Mol. Cell Biol.* 3, 663–672.

Gilmore TD (2006) Introduction to NFκB : players, pathways, perspectives. *Oncogene* 25, 6680–6684.

Gosh S & Karin M (2002) Missing pieces in the NFκB puzzle. *Cell* 109, S81–S96.

Gustin MC, Albertyn J, Alexancer M et al. (1998) MAP kinase pathways in the yeast *Saccharomyces cerevisiae*. *Microbiol. Mol. Biol. Rev.* 62, 1264–1300.

Hagemann C & Blank JL (2001) The ups and downs of MEK kinase interactions. *Cell. Signal.* 13, 863–875.

Harper SJ & LoGrasse P (2001) Signaling for survival and death in neurons: the role of stress-activated kinases, JNK and p38. *Cell. Signal.* 13, 299–310.

Hazzalin CA & Mahadevan LC (2002) MAPK-regulated transcription: a continuously variable gene switch? *Nat. Rev. Mol. Cell Biol.* 3, 30–41.

Hoffmann A & Baltimore D (2006) Circuitry of nuclear factor κB signaling. *Immunol. Rev.* 210, 171–186.

Kucharczak J, Simmons MJ, Fan Y et al. (2003) To be, or not to be: NFκB is the answer—role of Rel/NFκB in the regulation of apoptosis. *Oncogene* 22, 8961–8982.

Kyriakis JM & Avruch J (2001) Mammalian mitogen-activated protein kinase signal transduction pathways activated by stress and inflammation. *Physiol. Rev.* 81, 807–869.

Lewis SZ, Shapiro PS & Ahn NG (1998) Signal transduction through MAP kinase cascades. *Adv. Cancer Res.* 74, 49–139.

Millar JBA (1999) Stress-activated MAP kinase pathways of budding and fission yeast. *Biochem. Soc. Symp.* 64, 49–62.

Ono K & Han J (2000) The p38 transduction pathway: activation and function. *Cell. Signal.* 12, 1–13.

Perkins ND (2006) Post-translational modifications regulating the activity and function of the nuclear factor κB pathway. *Oncogene* 25, 6717–6730.

Perkins ND (2007) Integrating cell-signaling pathways with NFκB and IKK function. *Nat. Rev. Mol. Cell Biol.* 8, 49–61.

Raabe T & Rapp UR (2002) KSR: a regulator and scaffold protein of the MAP kinase pathway. *Sci. STKE Perspect.* 2002/136/pe28.

Raman M & Cobb MH (2003) MAP kinase modules: many roads home. *Curr. Biol.* 13, R886–R888.

Ravid T & Hochstrasser M (2004) NFκB signaling: flipping the switch with polyubiquitin chains. *Curr. Biol.* 14, R898–R900.

Roux PR & Blenis J (2004) ERK and p38 MAPK-activated protein kinases: a family of protein kinases with diverse biological functions. *Microbiol. Mol. Biol. Rev.* 68, 320–344.

Scheidereit C (2006) IκB kinase complexes: gateways to NFκB activation and transcription. *Oncogene* 25, 6685–6705.

Sharrocks AD (2001) The Ets domain transcription factor family. *Nat. Rev. Mol. Cell Biol.* 2, 827–836.

Tanoue T & Nishida E (2003) Molecular recognitions in the MAP kinase cascades. *Cell. Signal. 15*, 455–462.

Ubersax JA & Ferrell JE Jr (2007) Mechanisms of specificity in protein phosphorylation. *Nat. Rev. Mol. Cell Biol.* 8, 530–541.

van Drogen F & Peter M (2002) MAP kinase cascades: scaffolding signal specificity. *Curr. Biol.* 12, R53–R55.

Viatour P, Merville MP, Bours V et al. (2005) Phosphorylation of NFκB and IκB proteins: implications in cancer and inflammation. *Trends Biochem. Sci.* 30, 44–51.

Wagner EF (Ed) (2001) AP-1: Introductory remarks. *Oncogene* 20, 2334–2335.

Wang LH, Besirli CG & Johnson EM Jr (2004) Mixed-lineage kinases: a target for the prevention of neurodegeneration. *Annu. Rev. Pharmacol. Toxicol.* 44, 451–474.

Weston CR & Davis RJ (2002) The JNK signal transduction pathway. *Curr. Opin. Genet. Dev.* 12, 14–21.

Widmann C, Gibson S, Jarpe MB et al. (1999) Mitogen-activated protein kinase: conservation of a three-kinase module from yeast to human. *Physiol. Rev.* 79, 143–180.

Wietek C & O'Neill LAJ (2007) Diversity and regulation in the NFκB system. *Trends Biochem. Sci.* 32, 312–319.

Yamamoto Y & Gaynor RB (2004) IκB kinases: key regulators of the NFκB pathway. *Trends Biochem. Sci.* 29, 72–79.

Yang SH, Sharrocks AD & Whitmarsh AJ (2003) Transcriptional regulation by the MAP kinase signaling cascades. *Gene* 320, 3–21.

Yoshioka K (2004) Scaffold proteins in mammalian MAP kinase cascades. *J. Biochem. (Tokyo)* 135, 657–661.

417

Cancer and Regulation of Cell Division

Cell proliferation is mandatory for reproduction and for development, repair, and maintenance of tissues. In fact, the apparatus of cell division is an ultimate target of most of the signal-processing reactions explained in the previous chapters. In the following, we shall have a closer look at pathways that connect this apparatus with the cell's signal-processing network. The investigation of these interactions has tremendous practical consequences because defective transduction of signals controlling cell proliferation is the major cause of cancer.

12.1 Cancer: a "cellular psychosis"

Tissue cells are permanently threatened with death from apoptosis, terminal differentiation, senescence, and injury. In fact, their lifespan generally is much shorter, typically weeks to months, than that of the whole organism. To compensate this never-ending cell loss, continuous replacement of cells by division of stem cells is indispensable; the rate of which differs from tissue to tissue. That means that, in the adult organism, the constant mass (and thus the constant function) of a tissue is due to a precisely controlled steady state between the birth rate and the decay rate of cells. This steady state is called **tissue homeostasis**. There may be a functional requirement for a temporary overshooting proliferation: for instance, wound repair or in the course of hormone-regulated physiological cycles. When the tissue moves away permanently from homeostasis, it either wastes away (atrophy) or hyperproliferates (hyperplasia), possibly ending up as a tumor.

Tissue homeostasis is not maintained automatically but is controlled by an intricate interplay of positive and negative signals that may be both local and systemic. Survival signals such as growth factors and hormones stimulate cell proliferation, whereas death signals such as the Fas ligand and related cytokines as well as a wide variety of stress factors exert opposite effects (Figure 12.1). In the state of tissue homeostasis, the effects of death and survival signals compensate each other. A tumor develops when the birth rate permanently surpasses the decay rate. This occurs when cells respond more strongly to survival signals than to death signals. In other words, cancer is a disease of cellular signal processing. If one prefers the metaphor of the "cellular brain" for typifying the signal-transducing protein network, one may speak of a "cellular psychosis."

The major causes of tumors are mutations of genes encoding certain types of signal-processing proteins. To emphasize their role in cancer development (oncogenesis), these genes are called proto-oncogenes and tumor suppressor genes (Figure 12.2). The proteins encoded by **proto-oncogenes** are components of survival networks that process proliferative and anti-apoptotic signals. Upon mutation of the corresponding proto-oncogenes to oncogenes, these proteins

Figure 12.1 Tissue homeostasis and tumor growth (A) A tissue is depicted schematically, where the birth rate of cells is matched by the death rate because the positive and negative effects of survival signaling are precisely balanced out by those of death signaling and vice versa (homeostasis). (B) In a tumor, gene mutations have led to down-regulation of death signaling and up-regulation of survival signaling, causing a permanent disequilibrium of the rates of cell birth and cell death (arrows indicate activation; pistils indicate inhibition).

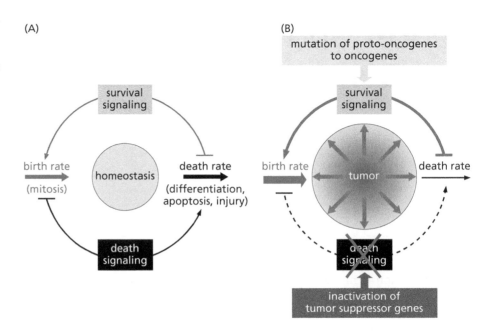

become hyperactive and control-resistant (this type of gene mutation is called, therefore, a gain-of-function mutation). In contrast, the proteins encoded by **tumor suppressor genes** are integrated in networks that process anti-proliferative, pro-apoptotic, and differentiation-promoting signals. In this case, tumor growth may be stimulated by loss-of-function mutations (gene mutations that result in inactivity of the encoded proteins) of the tumor suppressor genes. In most cancer cells, several proto-oncogenes and tumor suppressor genes have undergone mutations. If one defines cell division as forward movement, a healthy cell would resemble a car with a properly working gas pedal (proto-oncoproteins) and brakes (tumor suppressor proteins), whereas in a car resembling a tumor cell, the gas pedal is sticking and the brakes are defective.

Tumorigenic mutations of proto-oncogenes or tumor suppressor genes are caused mostly by environmental factors. These are certain viruses, genotoxic chemicals, UV light, and ionizing radiation. In most cases environmental carcinogens are not sufficient to evoke cancer but have to synergize with endogenous components such as certain hormones, pro-inflammatory mediators, reactive oxygen species, and defects of DNA repair. Due to a hereditary predisposition, the sensitivity to both exogenous and endogenous factors may be increased considerably (as found in so-called cancer families). When this predisposition is a deletion of one of the two alleles of a tumor suppressor gene, as is frequently the case, such a hereditary effect is easily understood. As long as the

Figure 12.2 Examples of signal-processing proteins the mutation of which may cause cancer

FROM PROTO-ONCOGENES	
ErbB	EGF receptor (receptor tyrosine kinase)
Ras	small G-protein
PKB/Akt	Ser/Thr kinase
Src	Tyr kinase
Jun, Fos	transcription factors

FROM TUMOR SUPPRESSOR GENES	
PTEN	PIP-3 phosphatase
APC	inhibitor of β-catenin signaling
Rb	transcriptional repressor
p53	transcription factor
RolB	phospho-Tyr phosphatase (*Agrobacterium rhizogenes*)

second allele is intact, the functional suppressor protein is produced, albeit perhaps at a diminished rate. However, only one more mutagenic hit is needed to delete the second allele and fully ruin the cellular brake (a prominent example of this mechanism is provided by the retinoblastoma gene discussed in Section 12.7). Other reasons for a hereditary predisposition to cancer may be phenotypic features (such as poor sun protection in fair-skinned people of Irish descent) as well as inborn defects of DNA repair, of chromatin modifications controlling the constancy of tissue differentiation (see Section 8.2.8), of enzymes metabolizing chemical carcinogens (see Section 8.4), of endogenous pathways leading to potentially carcinogenic metabolites, and of other mechanisms that would promote genetic damage.

Cancer develops in a stepwise manner over many years. In general, it starts with a single genetic defect that has escaped DNA repair in a single cell; that is, most tumors are monoclonal. For a malignant tumor, additional gene mutations are required, frequently amounting to about half a dozen defects. The probability is practically zero that such a sequence of specific genetic alterations occurs in a single cell. This suggests that, besides gene mutations, there is an additional process involved in cancer development called tumor promotion. Tumor promotion is a selective mitogenic stimulation of pre-malignant cells, that is, cells that have undergone only one or a few of the genetic alterations required to become malignant. The larger the clone of such cells, the higher the probability of further mutations occurring accidentally.

Another important effect of tumor promotion is an up-regulation of metabolic pathways generating metabolites such as reactive oxygen species, organic free radicals, and other compounds that damage DNA by acting as endogenous mutagens. Moreover, the cellular mechanisms of DNA repair and elimination of mutated cells may be impaired. As a result, the tissue acquires a mutator phenotype characterized by a strongly increased probability of further mutations, a situation that has been called **genetic instability**.

Tumor promotion is due to a long-lasting stimulation of cellular signal processing rather than to gene mutations. Classical examples of tumor-promoting effects are repeated wounding or permanent activation of protein kinase C caused by the diterpene ester 12-O-tetradecanoylphorbol 13-acetate mimicking the effect of the second messenger diacylglycerol (Section 4.6.5). Under pathophysiological conditions, hormones such as prolactin, estradiol, and testosterone as well as bile acid metabolites, long-wave UV light (as used in tanning salons), a wide variety of chemicals, and chronic inflammation due to continuous irritation or infections can exert tumor-promoting effects. Tobacco smoke contains both mutagenic and tumor-promoting components, and lung cancer develops against a background of chronic irritation.

Normal noncancerous cells also respond to tumor promoters by hyperproliferation due to an increased rate of cell division, a decreased death rate (or both), and increased genotoxic metabolism. As compared with genetically damaged precancerous cells, however, they are less sensitive and,-even more importantly, may rapidly adapt to the tumor-promoting stimulus. Therefore, the mutated cells gain a selective advantage upon prolonged contact with the tumor promoter.

Summary
Cancer results mainly from gene mutations (predominantly caused by environmental factors) that lead to a continuous deregulation of cellular signal processing and, as a consequence, to an imbalance between the rates of cell division and cell death. The progression from a single mutated cell to a malignant tumor requires additional genetic and epigenetic alterations. It is promoted by non-mutagenic endogenous and exogenous stimuli (tumor promoters) that cause a

long-lasting stimulation of mutagenic and anti-apoptotic signaling and, in addition, may stimulate the production of genotoxic metabolites, thus creating a state of genetic instability.

12.2 The cell cycle

The first and foremost symptom of cancer is overshooting cell proliferation due to defective communication between the signal-processing network and the apparatus of cell division. This apparatus is a particularly exciting cellular device: working with overwhelming accuracy, it manages one of the most elaborate processes in nature, the birth of a new cell.

Proliferating cells pass through a sequence of phases presented as a cyclic process. This cell cycle is conserved in all eukaryotes. It provides a perfect example of how a highly complex cellular event is controlled temporally and spatially by signaling reactions that are tightly interlinked and feedback-controlled to ensure that the genetic material has been correctly copied and evenly distributed to the daughter cells. Pioneering work on cell cycle regulation was honored by the Nobel Prize in Physiology or Medicine for 2001, awarded to Leland H. Hartwell, R. Timothy Hunt, and Paul M. Nurse.

The production of intact and genetically identical daughter cells presupposes a precise sequence of the cycle phases. For instance, separation of chromosomes must not occur prior to complete chromosome condensation, which requires DNA replication to be finished in order to proceed correctly. Therefore, the cell cycle is interrupted at least three times by **checkpoints** (Figure 12.3) that give the data-processing network an opportunity to monitor the precision of DNA replication and chromosome segregation and, if necessary, to launch repair measures or, in the case or irreparable damage, to promote cell death (for details see Section 12.9). By this means, dissemination of harmful mutations is prevented. However, as is typical for data processing, this control is not perfect but may be evaded in the course of tumorigenesis.

The clockwork-like precision of the cell cycle is guaranteed by control elements meshing like gears. In this mechanism, strictly regulated formation and degradation of regulatory proteins together with precisely timed protein phosphorylation play a key role. The master regulators of cell cycle progression are the **cyclin-dependent protein kinases (CDKs)**. Their importance was discovered

Figure 12.3 Mammalian cell cycle
Phases of the cell cycle: S, synthesis phase, period of DNA replication; M, mitosis phase, period of chromosome separation and cytokinesis; G1 and G2, so-called gap phases, where the cell realizes specific programs of differentiation and prepares for DNA replication and mitosis, respectively; G0, resting phase; upon mitogenic stimulation the cell may return to the cell cycle. The three major checkpoints are shown in red.

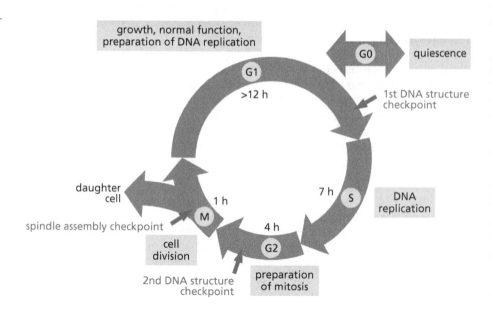

(A)

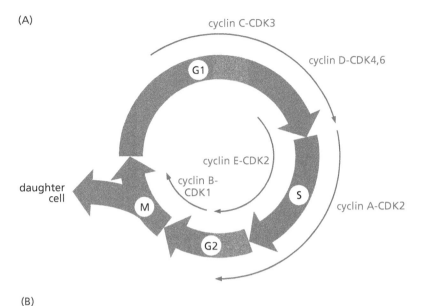

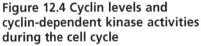

(B)

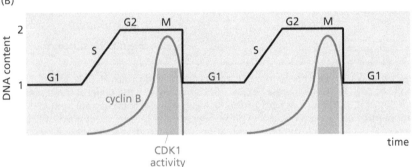

Figure 12.4 Cyclin levels and cyclin-dependent kinase activities during the cell cycle
(A) Cyclin–CDK activities associated with the individual phases.
(B) Oscillation of the cyclin level (red curve) and the CDK activity (colored bars) depending on the cell cycle phase as represented by the DNA content (black line and ordinate). As an example, the cyclin B–CDK1 pair has been chosen.

originally by studies on yeast mutants with defective or overexpressed *cdc* (*cell division cycle*) genes* and on starfish and amphibian oocytes.

To make a fertilized egg, the cell cycle of oocytes is arrested at the G2–M transition. This arrest could be overcome by progesterone or by a maturation-promoting factor (MPF) isolated from dividing frog cells. MPF was identified as a protein kinase in the late 1980s. A homologous kinase was found to be encoded by the yeast *cdc* gene number 2. Both kinases were originally called Cdc2 but today are known as CDK1. They turned out to be the catalytic subunits of a complex with a regulatory protein called **cyclin B**. Cyclin B appeared in the G2 phase to stimulate CDK1 and then disappeared abruptly at the end of the M phase. In contrast, Cdc2 (CDK1) was always present but was active only in the presence of cyclin B (Figure 12.4).

CDK1 and cyclin B are representatives of larger protein families. The human genome, for instance, encodes 20 CDKs and CDK-like proteins, and 29 cyclins or cyclin-like proteins (proteins with an amino acid sequence known as the cyclin box; see below). Only six CDKs and seven cyclins have been shown unequivocally to be involved in cell cycle regulation, while the rest exhibit other or unknown functions (Table 12.1).

*The yeast species *Saccharomyces cerevisiae* and *Schizosaccharomyces pombe* have been extremely valuable models for genetic and biochemical analysis of the cell cycle. Conditional mutations, for instance, those that become effective only above a distinct temperature, have enabled the identification of various cell division cycle-controlling genes (*cdc*) and their products, significantly facilitating the search for homologous vertebrate proteins.

Table 12.1 Cyclin-dependent protein kinases of mammalian cells

Kinase	Regulatory protein	Major function
CDK1	cyclin B	cell cycle control
CDK2	cyclin A and E (D, B)	cell cycle control
CDK3	cyclin A, E, C	cell cycle control
CDK4	cyclin D	cell cycle control
CDK5	p35 and p39	neuronal differentiation
CDK6	cyclin D	cell cycle control
CDK7	cyclin H	CDK-activating kinase
		transcriptional control[a]
CDK8	cyclin C	transcriptional control[a]
		inhibition of cyclin H–CDK 7
CDK9	cyclin K and T	transcriptional control[a]
CDK10	?	transcriptional control
CDK11	cyclin L	transcriptional control[a]
		pre-mRNA splicing

[a]CDK7– 9 and -11, together with transcription-associated cyclins H, C, K, L, and T, form complexes with RNA polymerase II. CDK7 (a component of the general transcription factor TFIIH), CDK8 (a component of the RNA polymerase holoenzyme), and CDK9 (a component of an elongation factor) phosphorylate the C-terminal domain of RNA polymerase II, thus participating in regulation of transcriptional elongation (see Section 8.1). The CDK11–cyclin L complex coordinates transcription with RNA splicing (Section 8.9).

Summary

The cell cycle is divided into several phases through which a proliferating cell must pass. The correct sequence of phases depends on the periodic formation and degradation of regulatory proteins known as cyclins. At several checkpoints, cell cycle passage is slowed down, giving the data-processing network an opportunity to monitor the precision of DNA replication and chromosome segregation and, if necessary, to launch repair measures or to promote cell death.

12.3 Cyclins: cell cycle regulators and beyond

Most but not all CDKs require cyclins for activation. Cyclins are heterogeneous proteins that share a so-called cyclin box: a domain consisting of a helix bundle that enables them to interact with the corresponding CDKs. The amino acid sequence of the cyclin box distinguishes the cell cycle-associated cyclins A, B, D, and E from the transcription-associated cyclins C, H, K, L, and T and from other cyclins with still-unknown functions (Table 12.1).

Cell cycle-associated cyclins are periodically synthesized during individual cell cycle phases. They initiate the entry of the cell into a phase and, at the same time, inhibit the transition into the next phase. In other words, as long as a particular cyclin is available, the cell stays in the corresponding phase. Therefore, cyclins are synthesized at the beginning and must be destroyed at the end of a particular phase. Cyclin synthesis is induced by a cyclin–CDK pair from the previous phase of the cell cycle. Cyclin degradation proceeds along the ubiquitin–proteasome pathway and is triggered by ubiquitin ligases that bind to N-terminal destruction domains of cyclins. The cell cycle is dominated by two types of ubiquitin ligases: SCF ubiquitin ligases induce the degradation of G1- and S-phase cyclins, whereas APC/C ubiquitin ligases are more specialized for G2- and M-phase cyclins (more information about ubiquitin ligases is found in Sections 12.11 and 12.16).

Summary

Cyclins are regulatory proteins that activate cyclin-dependent kinases (CDKs). They are periodically produced and degraded in the course of the cell cycle. Each cell cycle phase is associated with and controlled by a distinct cyclin–CDK pair. In addition to cell-cycle associated cyclins, other cyclins are known that, together with corresponding CDKs, participate in the control of gene transcription and pre-mRNA splicing.

12.4 Cyclin-dependent protein kinases: dual control by phosphorylation and dephosphorylation

CDKs are typical eukaryotic Ser/Thr-specific protein kinases that are intrasterically inhibited in the absence of cyclin. As shown in Table 12.1, enzymes of this family are involved not only in the control of cell division but also in other processes, in particular gene expression. While each cell cycle phase has been assumed to be controlled by a distinct CDK isoform, this concept has been challenged recently by experiments with genetically manipulated mice. These data indicate that CDK2, -4, and -6 are dispensable for the basic cell cycle or at least replaceable by other CDKs such as CDK1 in most (but not all) tissues.

A characteristic structural feature of the cell cycle-associated CDKs is the PSTAIRE (or a closely related) sequence localized in a helix of the ATP-binding site. By means of this sequence, CDK binds to the cyclin box of the corresponding cyclin and is relieved from auto-inhibition. By means of X-ray crystallography this interaction has been studied in detail (Figure 12.5). It has been shown that, in the absence of cyclin, the activation loop of CDK buries the ATP binding site and hinders access of the substrate protein to the catalytic center. Due to a conformational change induced by cyclin binding, the activation loop turns aside, opening the entrance of the catalytic cleft. As in most protein kinases, this open conformation is stabilized by phosphorylation of a strategic threonine residue in the activation loop. This addition of phosphate occurs either via autophosphorylation or via a **CDK-activating kinase**, which is itself a cyclin H–CDK7 complex.

At least two cell cycle-associated kinases, CDK1 and CDK2, are arrested in an inactive state by dual phosphorylation of an N-terminal Thr-Tyr sequence catalyzed by two **CDK-inactivating kinases**: **MYT1** [Membrane-associated tyrosine (Y) and threonine (T)-specific kinase], which predominantly phosphorylates the Thr residue, and **WEE1**, a Tyr kinase (the name refers to *WEE1* deletion mutants of fission yeast, characterized by particularly small cells due to an accelerated

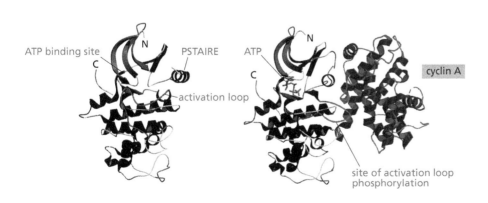

Figure 12.5 Interaction of CDK2 with cyclin A (Left) In the absence of cyclin A, the activation loop (red) closes the catalytic cleft including the ATP-binding site of CDK2. (Right) Cyclin A (red) interacts with the PSTAIRE sequence of CDK2 and induces a conformational change that leads to an opening of the catalytic cleft, allowing ATP and substrate protein binding. The active conformation is stabilized by activation loop phosphorylation.

425

Sidebar 12.1 CDK-activating kinase: more than a cell cycle regulator The functions of CDK-activating kinase go beyond the control of CDK activity, since as a subunit of the general transcription factor TFIIH, the enzyme also phosphorylates RNA polymerase II. This reaction, providing an example of CDK functions outside the cell cycle, is important for the transition from the initiation into the elongation phase of transcription (Section 8.1). CDK-activating kinase is activated either by a regulatory protein or by phosphorylation catalyzed by another cyclin–CDK pair, cyclin C–CDK8, which is also involved in transcriptional regulation rather than in cell cycle control.

cell cycle). To activate a CDK, both phosphate groups have to be removed by a highly specialized dual-specific protein phosphatase of the **Cdc25** family, which in humans comprises three isoforms A, B, and C. For the CDK1–cyclin B complex as an example, these relationships are summarized in Section 12.8, Figure 12.9.

CDK inhibitors MYT1 and WEE1 and CDK activator Cdc25 are controlled by input signals. In this context, the role of **checkpoint kinases** has been studied in detail. With Cdc25, phosphorylation by checkpoint kinases has an inhibitory effect since it labels Cdc25 for an interaction with 14-3-3 proteins, hindering translocation into the nucleus, and enhances Cdc25 degradation by the ubiquitin–proteasome pathway. On the contrary, MYT1 and WEE1 are stimulated by phosphorylation.

In other words, checkpoint kinases bring the cell cycle to a halt. Since they become active upon DNA damage (Section 12.9), they provide one of the mechanisms stopping cell proliferation at the DNA structure checkpoints in the late G1 phase and late G2 phase to enable DNA repair or, as a final consequence, to drive an irreparable cell into apoptosis. Another protein kinase inhibiting Cdc25 is the MAP kinase p38-activated kinase 2, which connects checkpoint control with the general stress response.

Summary

Cyclin-dependent protein kinases (CDKs) are the master regulators of the individual cell cycle phases. For activation they require, in addition to cyclin binding, activation loop phosphorylation (either by autophosphorylation or by the cyclin

Sidebar 12.2 The outsider CDK5 Although closely related to CDK1 and CDK2, the enzyme CDK5 is an outsider of the CDK family, since (despite its name) it is not regulated by cyclins but by other proteins called p35 and p39 that are expressed exclusively in brain. Moreover, CDK5 does not require activation loop phosphorylation and does not respond to MYT1 and WEE1. Most probably, the enzyme does not participate in cell cycle control. Instead, it plays an important role in nonproliferating nerve cells, being essential for **brain development** since it controls the migration and correct positioning of neurons in the embryo.

Exhibiting similar substrate specificity as CDK1, CDK5 phosphorylates many proteins. By controlling cytoskeletal dynamics, the enzyme regulates cell motility, cell shape (such as the outgrowth of neurites), intercellular contacts (for instance at synapses), and neurotransmitter release. CDK5 cooperates with focal adhesion kinases (Section 7.2.3), which are CDK5 substrates.

Considering the multiple functions, it is conceivable that a deregulation of CDK5 may result in serious consequences. So a hyperactive CDK5 due, for instance, to a truncation of the regulatory subunit p35 (to p25) by the Ca^{2+}-dependent protease calpain (Section 14.6.5) has been related to neurodegenerative diseases such as **amyotrophic lateral sclerosis** (a fatal destruction of neuromuscular contacts), **Parkinson's disease**, and **Alzheimer's dementia**. A pathological consequence of a hyperactive CDK5 is an excessive phosphorylation of the microtubule-associated protein Tau resulting in massive deformations of the cytoskeleton and finally in cell death (for details about Alzheimer's disease see Section 13.1.3). The reader is reminded that Tau is also phosphorylated by glycogen synthase kinase 3 (GSK3; Section 9.4.4).

H–CDK7 complex) and dephosphorylation of an N-terminal Thr-Tyr sequence (by the dual-specific phosphatase Cdc25). Threonine kinase MYT1 and tyrosine kinase WEE1 catalyze rephosphorylation of this sequence, resulting in CDK inactivation. These kinases and Cdc25 are targets of signaling reactions.

12.5 Cyclin-dependent kinase inhibitors: keeping the cell cycle under control

Strict regulation of CDK activity, as required for precise cell cycle progression, depends not only on inhibitory phosphorylation but also on proteins that specifically antagonize the stimulatory effects of cyclins. These important regulators of cell cycle progression exist in two families, called INK4 (INhibitors of CDK4) and CIP or KIP (CDK- or Kinase-Inhibiting Proteins).

The **INK4** family of CDK inhibitors includes the proteins $p15^{INK4}$, $p16^{INK4}$, $p18^{INK4}$, and $p19^{INK4}$. They specifically inhibit CDK4 and CDK6, hindering the access of cyclin D by means of ankyrin repeats (Section 13.1). $p16^{INK4}$ is a tumor suppressor protein inactivated, for example, in malignant melanoma. Transcription of the $p15^{INK4}$ gene is induced by transforming growth factor β, and this effect contributes to the strong anti-proliferative effects of TGFβ, especially on epithelial cells. Probably INK4 proteins are also involved in cell aging or senescence.

The **CIP/KIP** family comprises the inhibitor proteins $p21^{KIP}$, $p27^{KIP}$, and $p57^{KIP}$ (found in only a few tissues). They are less selective than the INK4s and block various CDKs. $p27^{KIP}$ plays a particularly important role in the control of tissue homeostasis: it keeps the cell in a nonproliferative state of quiescence called G0 phase and is a primary target of mitogenic signals (see Section 12.6). In contrast, $p21^{KIP}$ permanently regulates the cell cycle progression of proliferating cells. Its *de novo* synthesis is strongly induced by the transcription factor **p53**. Since p53 is activated by checkpoint kinases (that is, upon DNA damage), this response provides, in addition to the above-mentioned activation of WEE1 and MYT1 and inactivation of Cdc25, a second mechanism to interrupt the cell cycle at the DNA structure checkpoints and to trigger apoptosis if necessary.

Like cyclins, CDK inhibitors are short-lived proteins that are rapidly ubiquitylated and degraded in proteasomes. To be recognized by ubiquitin ligases (of the SCF type, see Section 12.16), the CDK inhibitors have to be phosphorylated. This is accomplished mostly by cyclin–CDK complexes, thus allowing the cell to enter the next phase of the cell cycle. The factors controlling CDK activity are summarized in Figure 12.6.

Summary

The activities of cyclin-dependent kinases are specifically controlled by inhibitory proteins comprising the two families INK4 (inhibitors of CDK4) and CIP/KIP (CDK- or kinase-inhibiting proteins). They are short-lived: produced only on demand to be rapidly destroyed afterwards.

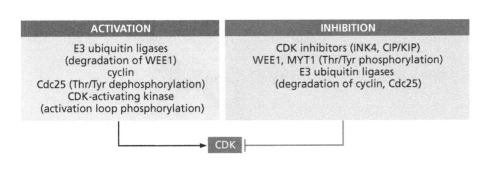

Figure 12.6 Activation and inhibition of cell cycle-associated cyclin-dependent kinases: an overview

12.6 G0 cells, restriction points, and the effect of mitogenic signals

Inhibition of cell proliferation does not lead automatically to apoptotic cell death. Instead, cells may remain in a resting state, the G0 phase, to return after some time into the cell cycle. For example, liver cells can leave and re-enter before they finally lose their proliferative capacity by differentiation or senescence.

For a cell entering the G0 phase, the cell cycle must be brought to a halt in the G1 phase by CDK inhibitors and a shortage of G1 cyclins such as cyclins C and D. **Cyclin D** has been investigated most thoroughly in this respect. It is tissue-specifically expressed in three variants that, depending on the cell type, interact with either CDK4 or CDK6 (Figure 12.7). These complexes control the transition of the cell into the S phase.

The G0 blockade can be overcome only by a synergistic effect of several mitogenic signals (such as hormones and growth factors) acting on the cell for several hours until the cell cycle begins to progress independently of exogenous signals. This occurs at the so-called **restriction point** in the late G1 phase. The reason is a sequential activation and repression, respectively, of genes.

In the beginning (0-4 h) the triggering signal, for instance a growth factor, directly controls (via particular signaling cascades) the transcription of immediate early proliferation genes. Later, when the restriction point has been passed, the products of these genes stimulate delayed early genes, the products of which finally turn on late genes in the early S phase. Over 100 early genes encode, for instance, transcription factors such as cJun, cFos, and cMyc, growth factors and their receptors, as well as proteins of the cytoskeleton and the extracellular matrix. Concomitantly with the activation of such genes, the genes of the CDK inhibitors are repressed. One of the delayed genes encodes cyclin D, and the late genes are responsible for events involved in DNA replication.

Figure 12.7 Mitogenic stimulation of resting (G0) cells
The figure shows three phases of gene expression and the activation of the E2F transcription factors due to phosphorylation of the negative regulator Rb by cyclin C–CDK3 at the G0–G1 transition and cyclin D–CDK4 (or -6) as well as cyclin E–CDK2 at the G1–S transition (colored boxes). The dotted red line represents the time a mitogen has to be present until the restriction point has been passed.

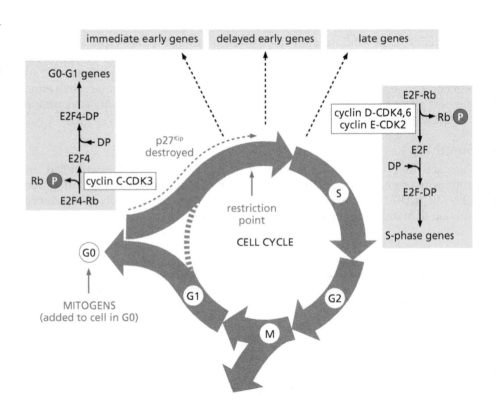

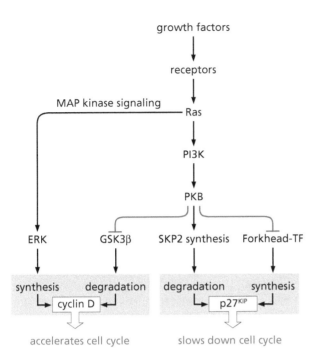

growth factors

receptors

Ras

PI3K

PKB

MAP kinase signaling

ERK   GSK3β   SKP2 synthesis   Forkhead-TF

synthesis   degradation   degradation   synthesis

cyclin D   p27^{KIP}

accelerates cell cycle   slows down cell cycle

Figure 12.8 Effects on cell cycle-regulatory proteins of mitogenic signaling cascades The scheme shows signaling pathways activated by growth factors and controlling the *de novo* synthesis (via gene transcription) and the degradation (via the ubiquitylation–proteasome pathway) of the CDK activator cyclin D and the CDK inhibitor p27^{KIP}. These signaling cascades frequently undergo oncogenic mutations. PI3K, phosphatidylinositol 3-kinase; PKB, protein kinase B/Akt; ERK, extracellular signal-regulated kinase; GSK3β, glycogen synthase kinase 3β; SKP2, subunit of the ubiquitin ligase SCF; TF, transcription factor. For details see text.

Among the CDK inhibitors, p27^{KIP} is responsible primarily for keeping the cell in the G0 phase. Appropriately, this factor must be inactivated for mitogenic stimulation. To this end, the synthesis of p27^{KIP} is inhibited and its degradation is promoted, while the reverse holds true for cyclins C and D (Figure 12.8). Both occur via mitogenic signaling cascades, frequently containing the small G-protein **Ras** as an essential switching device that transduces signals from the periphery to the genome along the MAP kinase and phosphatidylinositol 3-kinase–protein kinase B (PKB)/Akt pathways (Section 10.1). In fact, most growth factors as well as many hormones and cytokines activate Ras via their receptors.

A particularly significant role is played in this context by PKB/Akt, which relieves the cell cycle from the restraints caused by CDK inhibitors. For this purpose, PKB phosphorylates and inactivates the inhibitors p21^{KIP} and p27^{KIP} and, in addition, prevents their *de novo* synthesis by phosphorylating the corresponding transcription factors, such as those of the Forkhead family that control the *p27^{KIP}* gene. Upon phosphorylation, these factors are sequestered in the cytoplasm by 14-3-3 proteins and, thus, are hindered from entering their sites of action in the nucleus (Section 10.1.4).

To reinforce these effects, PKB activates, in addition, transcription of the gene encoding subunit SKP2 of ubiquitin ligase SCF, the enzyme triggering the degradation of p27. Here PKB cooperates with CDK2, since to be recognized by SCF and degraded, p27 must first be phosphorylated at a distinct Thr residue by the cyclin E–CDK2 complex. Consequently, a positive feedback loop becomes established since CDK2 activity increases to the same extent as the level of p27 drops. Another enzyme priming p27 for ubiquitylation and degradation is ERK, the executor kinase of the mitogenic MAP kinase module (Section 11.1).

Along these and other survival pathways (such as the NFκB cascade), transcription of the cyclin C and D genes is also turned on. Formation of these cyclins is also promoted at the level of mRNA translation since the protein kinase MNK1, an effector enzyme of MAP kinases, activates the translation factor eIF4E by phosphorylation (Section 9.3.3). This response is antagonized by a phosphorylation through the ubiquitous GSK3β (Section 9.4.4) that primes the cyclins for ubiquitylation and degradation. PKB, on the other hand, inactivates GSK3β and thus stabilizes the cyclins. These complicated relationships are depicted schematically in Figure 12.8.

Cultivated cells enter the resting phase not only at a growth factor deficiency but also at an increasing population density. This phenomenon, known as **contact inhibition**, similar to bacterial quorum sensing, is caused by interactions of cell adhesion molecules, such as cadherins (Section 7.3), and dramatically lowers the cell's sensitivity for proliferative signals.

Tumor cells have little or no tendency to undergo contact inhibition. In addition, they may supply themselves by growth factors acting in an autocrine manner. In fact, a typical sign of the **transformed state** is that the cells grow in Petri dishes one on top of another to generate a criss-cross pattern, and transformed cells do not need serum as a source of growth factors. As mentioned, neoplastic transformation requires oncogenic mutations of signal-transducing proteins such as Ras, PKB, growth factor receptors, and transcription factors as well as deletion of tumor suppressor proteins such as the PKB antagonist PTEN or the retinoblastoma protein pRb. Together such disturbances result in excessive production of cyclin D and C, as is found in a wide variety of cancer cells.

Summary

During the G1 phase, cells may leave the cell cycle, entering a G0 phase. This resting state is associated with a surplus of CDK inhibitors (in particular p27^{KIP}) and a shortage of G1 cyclins (such as cyclin D). Vice versa, mitogenic signals trigger a return into the proliferative state by inducing down-regulation of CDK inhibitors and up-regulation of G1 cyclins. The signaling pathways leading to cell proliferation are targets of carcinogens.

12.7 Retinoblastoma proteins: tumor repressors and master regulators of the cell cycle

For a resting cell, the synthesis of cyclins C and D is a signal to enter the proliferative phase. As a key event, the associated CDKs (CDK3, -4, and -6, see Figure 12.7) activate the **E2 transcription factors (E2F)** through phosphorylation. Upon heterodimerization with factors DP1 and DP2 (DNA-binding Proteins), the E2 factors control the genes of the G0–G1 transition and practically all late genes required for the entry into the S phase and DNA replication (including genes encoding DNA polymerase, factors of the DNA pre-replication complex, ribonucleotide reductase, thymidylate synthase, thymidine kinase, and proteins of DNA repair).

Outside of the G0–G1 and G1–S transitions, the activation of such genes would lead to a chaotic situation. To avoid this, in nonproliferative cells the E2F–DP complex is inhibited by the **Rb proteins** p107 or p130 (also called pocket proteins because of a characteristic shape of the E2F binding domain). These transcriptional repressors hinder the interaction of E2F–DP dimers with the transcriptional machinery, in particular by recruiting transcriptional co-repressors such as histone deacetylases. Rb proteins become inactivated by multiple phosphorylations catalyzed by cyclin C–CDK3 at the G0–G1 transition and by cyclin D–CDK4 (or CDK6) at the G1–S transition (Figure 12.7). As a result, E2F is released from the inhibitory complex and the co-repressors are displaced by transcriptional co-activators such as histone acetyltransferases.

The prototype of Rb proteins, providing their name, is the **retinoblastoma protein pRb**. It was the first tumor suppressor protein identified. A loss-of-function mutation of both alleles causes tumor development, in particular of retinoblastomas in the eye. In an exemplary fashion, this neoplastic disease has shown how a hereditary defect of a tumor suppressor gene combines with environmental influences. Predisposition is due to a recessive deletion of one *rb* allele and a tumor arises only when the second allele is inactivated by an exogenous factor. The risk of such a second hit is rather high, as indicated by an early occurrence of retinoblastoma in children.

As expected, the Rb–E2F system occupies a key position in proliferation control. This is underlined by observations showing inactivation of Rb through excessive phosphorylation in almost *all* types of tumors. Major causes of this deregulation are a pathologically increased expression of cyclins (along the signaling pathways shown in Figure 12.8) together with an inactivation or complete loss of CDK inhibitors.

The Rb–E2F complex is a point of attack of **DNA tumor viruses**. Viral components such as the SV40 antigen, the E7 protein of certain papilloma viruses, and the E1A protein of adenoviruses inactivate Rb (in contrast, RNA tumor viruses induce tumorigenesis by means of oncogenes derived from their host cells). Considering these tumorigenic effects, one might be surprised at first glance that the Rb–E2F signaling pathway also may lead to apoptosis. One reason is that E2F controls the gene encoding the factor ARF that stabilizes the pro-apoptotic protein p53 (Section 12.9.3). This effect appears to be restricted, however, to the isoform E2F-1. In neoplastically transformed cells, E2F-1 is hyperactive. This seems to be the mechanism by which oncogenes may interrupt the cell cycle by triggering cell death. By this means, the organism tries to prevent a dissemination of oncogenic mutations.

The genes activated by E2F-DP1 include those of cyclins E and A. Together with CDK2, these cyclins control S-phase progression. While cyclin E–CDK2 speeds up the cell cycle via increased Rb phosphorylation, cyclin A–CDK2 has a dampening effect in that it inhibits DP1 by phosphorylation. The result is a precisely balanced equilibrium of positive and negative feedback. To avoid a disturbance of S-phase progression by cyclin D, the latter is phosphorylated by GSK3β and thus primed for ubiquitylation and proteosomal degradation at the end of the G1 phase. The same happens to cyclin E in the late S phase and to cyclin B in the mitosis phase.

Summary

Transcription of the genes promoting the progression from G0 phase to and through the phase of DNA replication is stimulated by E2 transcription factors. This effect is opposed by Rb (retinoblastoma) proteins acting as transcriptional repressors. Rb proteins are inactivated by phosphorylation catalyzed by cyclin C–CDK3 and cyclin D–CDK4/6. Excessive Rb phosphorylation is a hallmark of tumorigenesis. Moreover, the pRb isoform is the epitome of a tumor suppressor protein undergoing carcinogenic loss-of-function mutations.

Sidebar 12.3 Cdc7, a protein kinase for DNA replication DNA replication is a complex event controlled by a large number of regulatory proteins. A discussion of details would go far beyond the purpose of this book; instead, the reader is referred to textbooks of cell and molecular biology. Nevertheless, Cdc7 (not to be confused with CDK7!), a ubiquitous eukaryotic Ser/Thr kinase, deserves to be mentioned. Like a CDK, this protein kinase is activated by regulatory subunits, called Dbf4 and Drf1, that are produced during a certain cell cycle phase, that is, at the G1–S transition, and degraded again by means of the ubiquitin ligase APC/C and proteasomes at a later time point. In this respect the regulatory subunits of Cdc7 resemble the otherwise unrelated cyclins. Cdc7 is required for the initiation of DNA replication cooperating with cyclin E–CDK2. An essential substrate of both kinases is the MCM protein complex that initiates DNA replication at the starting sites with partially melted DNA double strands. As a result, the double strand becomes unwound, provided that certain MCM proteins, most probably DNA helicases, had been phosphorylated before. Cdc7 seems to be involved also in the genotoxic stress response induced by DNA damage.

Figure 12.9 Regulation of cyclin-dependent kinase 1 In the G2 phase, CDK1 forms a complex with newly synthesized cyclin B. At this time point, however, the kinase activity of CDK1 is still suppressed by the phosphorylation (P) of an adjacent Thr-Tyr pair catalyzed by the Ser/Thr kinase MYT1 and the Tyr kinase WEE1. At the G2–M transition, these residues become dephosphorylated by the phosphatase Cdc25, and a Thr residue in the activation loop of CDK1 (red dot) is phosphorylated by CDK-activating kinase, which is a cyclin H–CDK7 complex. Due to strictly controlled *de novo* synthesis of cyclin H, CDK-activating kinase becomes active exactly at this time point. The now active cyclin B–CDK1 complex phosphorylates various proteins required for the G2–M transition and progression of the M-phase. In the telophase, cyclin B is ubiquitylated by the ubiquitin ligase APC/C and degraded in proteasomes. CDK1 is released and is inactivated by dephosphorylation of the activation loop (catalyzed by the protein phosphatase PP2A) and rephosphorylation of the N-terminal Thr-Tyr pair (catalyzed by MYT1 and WEE1). The process includes several feedback loops, which for the sake of clarity are not shown. Cyclin B–CDK1 stimulates Cdc25 and inhibits WEE1, thus increasing the efficacy and punctuality of CDK1 activation by positive feedback. Moreover, the two inhibitory kinases MYT1 and WEE1, as well as their opponent Cdc25, are under the control of upstream signaling pathways. DNA damage and incompletely replicated DNA, respectively, induce the genotoxic stress response, preventing CDK1 activation and interrupting the cell cycle at the late G2 phase (also known as antephase) prior to chromosome condensation.

12.8 Regulation of G2 phase and G2–M transition: precise like clockwork

At the transition from the S to the G2 phase, CDK2 is replaced by CDK1 as a partner of cyclin A. The cyclin A–CDK1 complex then induces the synthesis of cyclin B, which subsequently interacts with CDK1, controlling the transition into the mitosis phase. In an exemplary fashion, the **cyclin B–CDK1 complex** shows the sophisticated control of a CDK by the signal-processing network, ensuring a to-the-point activation in the correct time window.

Cyclin B–CDK1 controls the function of proteins needed by the cell to progress from G2 phase into M phase, and from prophase into metaphase of mitosis. These CDK-regulated proteins include laminin of the nuclear membrane (that disintegrates upon laminin phosphorylation), condensin proteins and histone 1 (the phosphorylation of which initiates chromosome condensation), and nucleolin (its phosphorylation leads to the breakdown of the nucleolus). Moreover, transcription is inhibited by phosphorylation of RNA polymerase II, and the suppression of protein synthesis observed at the G2–M transition is thought to be due to phosphorylation and inactivation by CDK1 of the translation factor eEF1B.

Finally, microtubule-associated proteins required for the formation of the mitotic spindle become phosphorylated, and the actin and intermediary filament cytoskeleton is rearranged, enabling the dividing cell to detach from the support and to round up. At the end of metaphase, CDK1 is inactivated by cyclin degradation, dephosphorylation of the activation loop, and rephosphorylation of the N-terminal Thr-Tyr pair catalyzed by MYT1 and WEE1 (Figure 12.9).

However, let us back up in the cell cycle. During late G2 phase, immediately prior to chromosome condensation, the cell once more slows down cell cycle progression to inspect the integrity of the genetic material. This second DNA structure checkpoint, also known as **antephase checkpoint**, offers a last chance to repair defects in one DNA strand using the intact strand as a template, or to commit suicide in the case of irreparable damage.

Summary

The progression from G2 phase to mitosis is controlled by cyclin B–CDK1 activated precisely at the entry into M phase. At the end of metaphase, CDK1 is inactivated by cyclin degradation, dephosphorylation of the activation loop (which was phosphorylated by CDK-activating kinase), and rephosphorylation of the N-terminal Thr-Tyr pair catalyzed by MYT1 and WEE1.

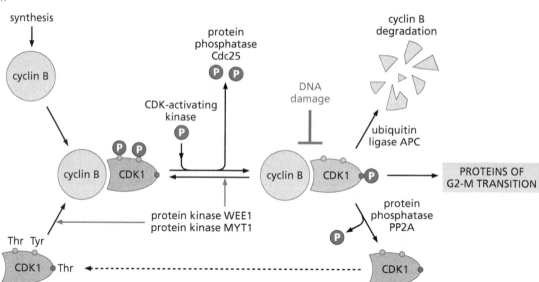

12.9 Genotoxic stress response: a matter of life and death

The stability of the genetic material is jeopardized permanently by exogenous influences such as the UV radiation of the sun; ionizing radiation of minerals, nuclear waste, or from space; genotoxic chemicals (which may also be of endogenous origin); and viruses. Moreover, DNA replication does not work without error. All these factors damage DNA and may cause gene mutations, resulting in immediate cell death or in diseases such as cancer. The ability to cope with genotoxic stress is, therefore, a fundamental condition of health and survival.

As already mentioned, somatic cells have a dual strategy: they try to repair the damage or, if this turns out to be impossible, to commit suicide by undergoing apoptosis. This must be done prior to DNA replication and chromosome segregation, that is, at the DNA structure checkpoints in G1 and G2 phase. At these checkpoints, a sophisticated data-processing network is activated to inspect and evaluate the damage and to launch countermeasures if necessary. Although this network still is the subject of ongoing research, most agree that signal transduction through protein phosphorylation plays an essential role. In general, the DNA break repair response involves three stages:

- recognition of the damage by sensor proteins
- activation of transducer protein kinases
- activation of downstream effector proteins controlling the repair process, cellular stress response, cell cycle progression, and if necessary apoptosis

12.9.1 DNA-dependent protein kinases: gatekeepers of gene stability

The machinery of DNA repair is activated within a few minutes after DNA damage. In this process, transducer protein kinases known as DNA-dependent protein kinases play a key role by monitoring damage and inducing repair measures. They are members of a larger family of eukaryotic Ser/Thr-directed kinases called atypical since they are not closely related to the conventional eukaryotic protein kinases (see Section 2.6.3). A characteristic structure found in many atypical kinases is a lipid kinase domain. Indeed, some members of the family such as phosphatidylinositol 3- and 4-kinases are specialized for the phosphorylation of phospholipids rather than proteins. However, DNA-dependent kinases are not thought to phosphorylate lipids; they are bona fide protein kinases and therefore are subsumed under the collective term **phosphatidylinositol 3-kinase-like kinases (PIKK)**. With up to 4100 amino acids, the PIKKs are unusually large proteins. On the basis of structural variations, they are divided into the subfamilies TOR, DNA kinases, and ATM/ATR.

The enzymes of the TOR group have been examined already in Section 9.4.1. They are involved in the adjustment of ribosomal protein synthesis and cell growth to hormonal stimulation and the food supply rather than in DNA repair. In contrast, DNA kinases and ATM/ATR are engaged in genetic recombination and DNA repair.

Enzymes of the **DNA kinase** subfamily transduce signals triggering the repair of DNA double-strand breaks by illegitimate recombination and V(D)J-recombination (which is the mechanism that regulates, through gene rearrangements, the diversity of antibodies and T-cell receptors). The enzyme is attached to DNA by a regulatory subunit consisting of the two proteins Ku70 and Ku80 serving as sensors in that together they recognize free ends of double-stranded DNA (as generated by double-strand breaks) and other structural irregularities of the DNA double helix. As a result the Ku70/80 heterodimer is allosterically activated and stimulates DNA kinase activity. In certain autoimmune diseases, such as lupus erythematosus, Ku70/80 acts as an auto-antigen.

Scid mice, which suffer from a loss-of-function mutation of DNA kinase, provide an animal model of immune deficiency diseases. A role of DNA kinase in genotoxic stress response is still a matter of debate.

Genuine stress enzymes that are activated in emergency situations such as DNA damage are the DNA kinase-homologous protein kinases **ATM** and **ATR**. Their names come from the hereditary disease **ataxia telangiectasia** (ATM stands for Ataxia Telangiectasia Mutated and ATR stands for ATM- and Rad3-related; Rad3 is a DNA-dependent protein kinase of yeast). This disease is characterized by genomic instability leading to brain defects, increased radiation sensitivity, immune deficiency, and a high cancer risk, particularly of leukemia. The causes are homozygous loss-of-function mutations of the genes encoding ATM and ATR. Both kinases respond to genotoxic stress: while ATM primarily is a sensor of DNA defects caused by ionizing radiation, ATR is more specialized for damage from UV light and inhibitors of DNA replication (e.g., cytostatics). For the integrity of DNA, ATM and ATR are of uppermost importance. Therefore, they have been called guardians at the gateway to genomic stability.

The factors monitoring DNA damage and recruiting ATM and ATR to the sites of damage are large protein complexes. These complexes are multifunctional stress sensors that, in addition to DNA strand breaks and UV-induced defects, recognize a blocked DNA replication fork, and respond to oxidative and nitrosative stress, starvation, and heat shock.

The core components of the complexes are ATR-interacting protein (for ATR) and MRN (for ATM). MRN is a complex of the endo- and exonuclease MRE11 (referring to Meiotic REcombination), the adaptor protein Rad50, and the protein NBS1. A mutation of NBS1 causes the **Nijmegen breakage syndrome**, an autosomal recessive disorder characterized by increased radiation sensitivity and cancer incidence. Like ATM, NBS1 is an essential protein since corresponding gene knockout mice are not viable. Additional components of the complexes include Rad and Hus proteins (these terms refer to radiation-sensitive and hydroxyurea-sensitive yeast mutants, where the proteins were discovered; hydroxyurea is an artificial inhibitor of DNA replication) and **BRCA2** (BReast CAncer protein 2).BRCA2 is one of the two tumor suppressor proteins causally related to the hereditary or familiar form of breast cancer. The other one, BRCA1, is a type 3 ubiquitin ligase that also is part of the DNA repair complex (see below and Section 2.8).

When they are bound to the site of damage, ATM and ATR become activated by autophosphorylation and assemble substrate and other proteins such as enzymes of DNA repair, histone variant H2AX that recruits substrate proteins of ATM/ATR, histone deacetylases to stop transcription during DNA repair, transcription factor E2F1 (which is inhibited by ATM but, on the other hand, stimulates transcription of the *atm* gene), Tyr kinase cAbl, and others. A major function of such proteins is to connect the machinery of DNA repair with other signaling pathways the activation of which is essential for a proper cellular stress response. Such pathways include the MAP kinase and NFκB cascades, which occupy a central position in stress adaptation (note that two of the MAK kinases, JNK and p38, had been discovered as stress-activated protein kinases; see Section 11.2). Activation of MAP kinase modules by ATM and ATR possibly involves the Tyr kinase cAbl (see Section 7.1.1), whereas the NFκB complex is activated through phosphorylation of the scaffold protein NEMO, which upon ubiquitylation and association with IKK1 and IKK2 triggers the degradation of the NFκB inhibitor IκBα (Section 11.8).

12.9.2 Reversible inhibition of the cell cycle at DNA structure checkpoints

As far as the effects of DNA damage on cell cycle progression are concerned, two ATR/ATM substrates occupy a key position. These are the **checkpoint kinases**

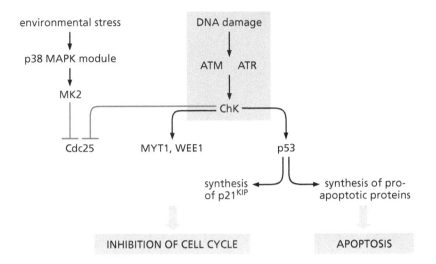

Figure 12.10 Events occurring at the DNA structure checkpoints
Damaged DNA sequentially activates the protein kinases ATM, ATR, and ChK. By phosphorylation, ChK primes Cdc25 for degradation and activates the inhibitory kinases MYT1 and WEE1 as well as the transcription factor p53, stimulating the synthesis of CDK inhibitor p21^{KIP} and of pro-apoptotic proteins. Cdc25 is also inhibited by the protein kinase MK2 that is activated by the p38 MAP kinase module. Along this pathway, the cell cycle is slowed down by a variety of environmental stress factors including genotoxic radiation.

ChK1 and **ChK2** that transduce the signals leading to cell cycle arrest and apoptosis when they are activated by ATM/ATR.

DNA repair is possible only when the cell cycle is transiently halted. To this end, CDKs are inhibited by:

- degradation of cyclins
- activation of the CDK-deactivating kinases WEE1 and MYT1
- destruction of the CDK-activating phosphatases of the Cdc25 family
- *de novo* synthesis of the CDK inhibitor p21^{KIP}

These reactions are controlled by the checkpoint kinases: by phosphorylating specific Ser residues, they activate WEE1 and MYT1, and prime the stimulatory CDK phosphatases of the Cdc25 type for ubiquitylation and proteolytic degradation. To induce p21^{KIP} synthesis, the checkpoint kinases stabilize the corresponding transcription factor p53. This is also the way leading to apoptosis if DNA repair turns out to be impossible (Figure 12.10).

Once the integrity of the genome has been successfully examined or reestablished, the blockade of the cell cycle must be lifted. This **checkpoint recovery** is achieved mainly by protein phosphatases that reactivate Cdc25 and inactivate the checkpoint kinases as well as MYT1, WEE1, and p53, and by a ubiquitin transferase that triggers the degradation of proteins of the DNA damage sensor complexes and of WEE1. The type 3 ubiquitin ligase in charge belongs to the SCF family. As is typical for these enzymes, their substrate proteins must be phosphorylated to be recognized (see Section 12.16). At least in the G2 phase of the cell cycle, this phosphorylation is catalyzed by the Polo-like kinase Plk1 (Section 12.12). How Plk1 senses the end of the repair process remains a mystery.

SCF-type ubiquitin ligases are multiprotein complexes. Using alternative substrate-binding subunits they can specifically recognize individual phosphorylation sites in substrate proteins. This may explain why the ubiquitin ligase required for checkpoint recovery has a bivalent function since it is also involved in checkpoint initiation where it triggers the proteosomal *destruction* of the phosphatase Cdc25. Cdc25 is labeled for ubiquitylation by phosphorylation via ChK1 and this phosphorylation is recognized by another substrate-binding subunit of the ubiquitin ligase than that catalyzed in WEE1 by PlK1. In other words, the same ubiquitin ligase may control different or even antagonistic signaling pathways (Figure 12.11). This case provides a particularly clear example of the key role played by the cooperation of protein phosphorylation with protein ubiquitylation in cellular signal processing.

Figure 12.11 Reversible stop of cell cycle progression at the DNA damage checkpoint in the G2 phase The key role of a specific SCF-type ubiquitin (Ub) ligase exhibiting a bivalent effect is shown. Upon DNA damage, the cell cycle activator CDK phosphatase Cdc25A becomes phosphorylated along the ATR–ChK1 pathway and labeled by ubiquitylation for proteasomal degradation. When the damage has been repaired, a still-unknown signal activates the Polo-like kinase Plk1, which labels the cell cycle inhibitor CDK kinase WEE1 and other DNA damage sensor proteins for ubiquitylation and proteasomal destruction.

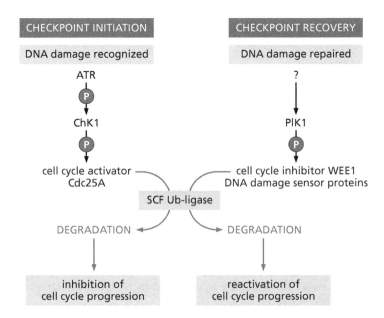

Figure 12.12 Domain structure, interactions, and modifications of transcription factor p53 Through its DNA-binding domain (DNA-BD), p53 interacts with the regulatory sequences of genes encoding the cell cycle inhibitor p21^{KIP} and pro-apoptotic proteins, and recruits via its transactivating domain (TD) transcriptional co-activators such as the acetyltransferase CBP/p300, which in addition to histones also acetylates the C-terminal region of p53. The transactivating domain is also a binding site of the ubiquitin ligase MDM2 preparing p53 for proteolytic degradation. The ubiquitylation of the C-terminal domain may be competitively inhibited by sumoylation preventing degradation. In another pathway leading to a stabilization of p53, the protein ARF interrupts the interaction with MDM2. OD, oligomerization domain (in the figure this domain is shown to be flanked by nuclear import and export sequences); RD, regulatory domain (containing nuclear import and export sequences). Phosphorylation sites (serine residues) are marked by red numbers.

12.9.3 Tumor suppressor p53: "guardian of the genome"

p53 is a multifunctional transcription factor, exhibiting the typical domain structure of such proteins (Figure 12.12). In the human genome, more than 4000 potential p53-binding sites have been found, leaving open the question as to whether or not all of them are bona fide target genes of p53.

A major function of p53 is to protect the genome from harmful mutations, that is, to prevent genetic instability, justifying the name "guardian of the genome." Due to this property p53 is a tumor suppressor protein, the loss-of-function mutation of which ranks among the most frequent causes of human neoplasia, affecting more than 50% of human cancers. p53 has two close relatives, the transcription factors **p63** and **p73**. They have not yet been associated with tumorigenesis. Instead, they suppress mutations leading to embryonic malformations.

Characteristic effects of p53 are:

- to arrest the cell cycle
- to induce apoptosis
- to promote DNA repair
- to inhibit angiogenesis (particularly in tumors)

For these purposes p53 activates the genes of the cell cycle inhibitor p21^{KIP} and of the pro-apoptotic proteins Bax, Noxa, Puma, APAF1, and Fas (see Section 13.2.4), as well as genes encoding inhibitors of angiogenesis such as thrombospondin and

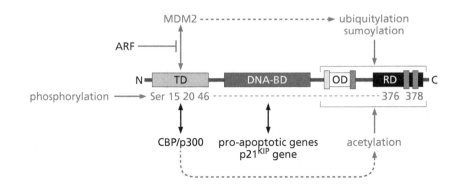

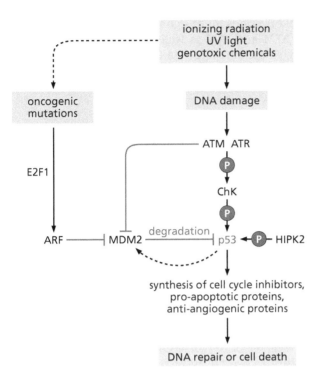

Figure 12.13 Activation of the p53 signaling pathway by genotoxic stress and oncogenic mutations DNA damage leads to a stabilization of p53 due to ChK-catalyzed phosphorylation of p53 and inactivation of the ubiquitin ligase MDM2 that primes p53 for degradation. The transcription factor E2F1, rendered hyperactive in response to oncogenic mutations, activates the gene of the MDM2 inhibitor ARF, thus also stabilizing p53. HIPK2, homeodomain-interacting protein kinase 2; P, phosphorylation. More details are found in the text.

maspin. In addition, p53 induces the synthesis of the phospholipid phosphatase PTEN, a tumor suppressor protein blocking the anti-apoptotic phosphatidylinositol 3-kinase–PKB/Akt signaling pathway (Section 4.6.6). At the same time, transcription of genes encoding anti-apoptotic and pro-angiogenic factors is suppressed. These p53-regulated processes are induced either by oncogenic alterations and excessive mitogenic stimulation or by DNA damage (Figure 12.13).

Interestingly, cell cycle arrest by p53 is only partially mediated by p21^{KIP} synthesis. Along a parallel pathway, p53 stimulates the transcription of genes encoding specific **microRNA** species, which in turn repress via RNA interference (see Section 8.8) the formation of a series of cell-cycle-promoting factors including CDK4, cyclin E, and E2F.

Experimental animals exhibiting excessive p53 activity are more or less protected from becoming ill with cancer, but they pay for this benefit by a shortened lifespan. This may explain why oxidative stress and the loss of telomeres are life-restricting factors since both stimulate p53 signaling. Therefore, an early death may be the price to be paid for protection against cancer.

Thus p53 is under strict control in order to bring the two risks, death by cancer versus death by aging, into a tolerable relationship. p53 activity is regulated at several levels, providing a perfect example of the complex interconnection of a transcription factor with the signal-processing network resembling a series of logical gates. In nondamaged cells p53 is prone to rapid proteolytic degradation, so its cellular level is very low. In fact, p53 provokes its own destruction by inducing the synthesis of a specific **ubiquitin ligase MDM2** (the name refers to Mouse Double Minutes, extrachromosomal pairs of aggregates consisting of amplified genes found in particular in tumor cells). MDM2, together with an isoform MDMX, not only primes p53 for proteolysis in proteasomes but also binds to the transactivating domain, thus directly inhibiting the interaction of p53 with the transcriptional machinery. MDM2 knockout mice show once more how important it is to keep p53 activity at a low level under normal conditions: already as embryos, these animals die from excessive apoptosis in all tissues unless their p53 gene has been deleted too.

MDM2 is a target of various input signals. Among them, phosphorylation by the protein kinases ATM and cAbl as well as interaction with the protein **ARF** or p14^{ARF} have been studied in more detail. Both interrupt the interaction of MDM2 with p53 thus stabilizing p53 and turning on p53 signaling. These effects explain the tumor suppressor function of MDM2, whereas ARF may mutate to an onco-protein. [The name ARF indicates that the protein is encoded by an Alternative open Reading Frame of the gene for the CDK inhibitor p16^{INK4A}. ARF should not be confused with the ADP-ribosylation factor Arf, a small G-protein controlling vesicle formation (see Section 10.4.3).]

While ATM responds primarily to DNA damage, ARF is rather activated by onco-genic signals (Figure 12.13) such as overexpression or mutation of β-catenin of the Wnt signaling pathway (Section 5.9), mutated Ras, overexpression of the cMyc transcription factor, or deletion of the retinoblastoma gene. As explained above, such alterations lead to strong activation of the transcription factor E2F1, resulting in excessive transcription of the gene encoding ARF.

The activity of p53 itself is regulated by post-translational modifications in a complex manner, determining whether apoptosis or another process is stimu-lated. p53 has numerous phosphorylation sites concentrated, in particular, in the transactivating domain, the phosphorylation of which results in p53 activa-tion, and in the regulatory domains, the phosphorylation of which leads to p53 inactivation (Figure 12.12). Phosphorylation of several Ser and Thr residues in the transactivating domain is catalyzed by the checkpoint kinases that (as we have seen) are under control of the kinases ATM and ATR. It disrupts the interac-tion with MDM2 (which, in addition, is inactivated by ATM/ATR-catalyzed phos-phorylation), thus increasing the stability of p53. Moreover, the activity of p53 is also intensified because phosphorylation promotes the interaction with the transcriptional co-activator CBP/p300. In addition to histones, this acetyltrans-ferase also acetylates p53 at several Lys residues in the regulatory domain, thus further stimulating p53 activity (Figure 12.12). Such acetylation is observed, in particular, after DNA damage and is accompanied by removal of inhibitory phos-phate residues from the regulatory domain. The phosphorylation of a special Ser residue (Ser46 in humans) in the transactivating domain of p53 is a signal trig-gering apoptosis. p53 thus modified forms a complex with the transcriptional co-activator PML (ProMyelocytic Leukemia protein, a tumor suppressor) and stimulates, among others, the transcription of the gene encoding apoptosis-inducing protein 1, which changes the mitochondrial membrane to evoke release of the apoptotic trigger cytochrome *c* (Section 13.2.4). The protein kinase catalyzing the pro-apoptotic phosphorylation of p53 is called **HIPK2** (Figure 12.13). HIPK stands for Homeodomain-Interacting Protein Kinase; it was discov-ered as a co-repressor of genes controlled by transcription factors with homeo-domains. The enzyme is a member of the so-called **DYRK family**. These "dual-specific Y(Tyr)-phosphorylated and -regulated kinases" belong to the rare protein kinases that phosphorylate Ser, Thr, and Tyr residues. DYRKs are involved in the differentiation of the nervous system. An isoform called Minibrain C (referring to a corresponding *Drosophila* mutant) is overexpressed in the brain of patients suffering from **Down syndrome** and is thought to be partially responsi-ble for pathological alterations.

Summary

At DNA structure checkpoints prior to DNA replication and chromosome con-densation, the cell cycle is slowed down to enable DNA repair or elimination of an irreversibly damaged cell by apoptosis. Damage sensors are the DNA-depend-ent atypical protein kinases ATM (ataxia telangiectasia mutated) and ATR (ATM- and Rad3-related) that are recruited by Rad and Hus proteins. The kinases assemble and phosphorylate a large number of proteins required for inhibition of cell cycle progression, DNA repair, and apoptosis. To inhibit cell cycle progres-sion and trigger apoptosis, the damage-activated kinases ATR and ATM stimulate checkpoint kinases, which in turn up-regulate inhibitory CDK phosphorylation. By activating the transcription factor p53, checkpoint kinases induce the synthesis of CDK inhibitors and pro-apoptotic proteins. Transcription factor p53 is the supervisor of the DNA structure checkpoints. Becoming activated and rescued from proteolytic degradation along the ATM/ATR–checkpoint kinase pathway or because of oncogenic mutation, p53 activates genes encoding cell cycle inhibitors (such as p21^{KIP} and microRNA), pro-apoptotic proteins, and proteins inhibiting angiogenesis. Whether p53 activation results in cell cycle inhibition or apoptosis depends on specific post-translational modifications. p53 is the tumor suppressor protein most frequently deleted in human cancers.

12.10 Phases of mitosis

If at the second DNA structure checkpoint (G2–M checkpoint) the genome is found to be intact, the cell is allowed to enter the phase of mitosis. As one of the most complex cellular processes, mitosis – the symmetric distribution of genetic material among daughter cells – is controlled by a highly sophisticated signaling apparatus that works with extreme precision and monitors again at several checkpoints. In analogy to other phases of the cell cycle, reversible protein phosphorylation and the periodic formation and degradation of control proteins provide the key regulatory events and the corresponding enzymes – kinases, phosphatases and ubiquitin ligases – are the major targets of control signals.

Mitosis is divided into the following phases:

- **Prophase:** chromatin condenses to chromosomes, the centrosomes begin to separate, and the mitotic spindle begins to grow

- **Prometaphase:** the cell rounds off, the nuclear membrane disintegrates, the centrosomes are separated, and the mitotic spindle continues to grow

- **Metaphase:** the chromosomes are aligned in the equatorial plain (metaphase plate) and recruit in their central region (the centromere) the kinetochore protein complex, which anchors the microtubule fibers that link the centrosomes at the spindle poles with the chromosomes

- **Anaphase:** the sister chromatids separate to become daughter chromosomes, which are pulled to the spindle poles by motor proteins and by shortening of the microtubules; and in preparation for the division into daughter cells through cytokinesis, a contractile microfilament ring forms around the cell equator

- **Telophase:** the new nuclear membrane is formed, the chromosomes de-condense, and the daughter cells become separated by contraction of the equatorial microfilament ring

The central event of mitosis is the separation of the sister chromatids. It requires the formation of the bipolar mitotic spindle consisting of the centrosomes with the attached microtubules and various regulatory and motor proteins. Only a spindle working with flawless accuracy guarantees correct alignment of the chromosomes in metaphase and their proper segregation during anaphase. If half the chromosomes do not end up in each daughter cell (**aneuploidy**), developmental disorders such as Down syndrome and tumorigenesis result. In fact, a correlation exists between the degree of aneuploidy and the malignancy of cancer. To avoid such accidents, the cell cycle is interrupted in late metaphase at a **spindle assembly checkpoint** (Figure 12.14) until the spindle–chromosome complex has taken on the correct structure (see below). Only then does anaphase begin.

Organizers of the spindle are the **centrosomes**. These small nuclear organelles consist of two centrioles each serving as microtubule-organizing centers. The

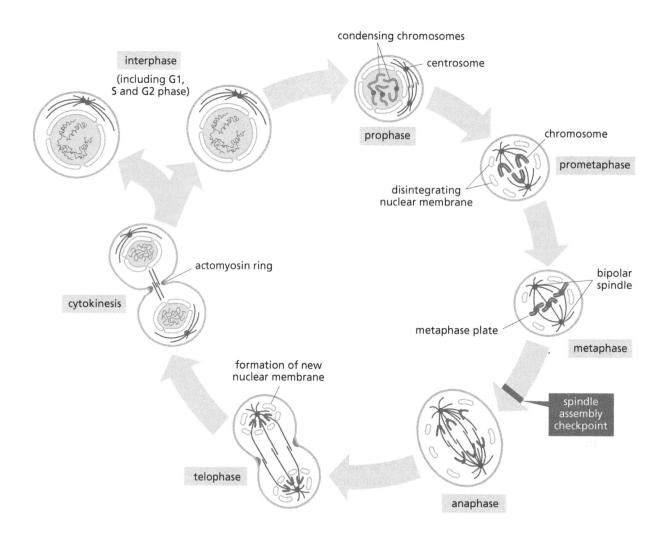

Figure 12.14 Mitotic cycle and spindle assembly checkpoint For details, see text.

microtubules radiating from the centrosomes are anchored at the kinetochores, which are localized in specific chromosomal structures, the centromeres. Kinetochores are large complexes of structural, motor, and regulatory proteins. To form a bipolar spindle, the centrosomes have to duplicate. This occurs in the S phase together with DNA replication (it has been postulated that, prior to the S phase, centrosome duplication is prevented by a protein "nucleophosmine" that becomes inactivated by cyclin E–CDK2-catalyzed phosphorylation at the G1–S transition). The two daughter centrosomes do not segregate immediately after duplication but are held together until prophase, when they migrate to the cell poles. Disturbances of centrosome duplication are one cause of genetic instability promoting cancer development. These occur, for instance, when DNA damage is not repaired due to loss of p53 activity, and give rise to multipolar mitotic spindles.

Summary

The central event of mitosis is condensation of the chromosomes and separation of the sister chromatids directed by the centrosomes that generate the mitotic spindle. To avoid aneuploidy and tumor development, correct distribution of the genetic material is controlled at a spindle assembly checkpoint prior to chromosome segregation.

12.11 Ubiquitin ligase APC/C: giving the beat of mitosis

As we have seen, to execute cell division in a flawless manner, a cell must both activate (produce) and inactivate (destroy) regulatory proteins such as CDKs and cyclins at the appropriate time. This holds true also for mitosis, which like other cell cycle phases is controlled by cyclin–CDK complexes. Thus cyclin A–CDK1 is active until the prometaphase, when cyclin A is degraded by proteolysis. The complex controls the early phases of chromosome condensation and, together with cyclin B–CDK1, the breakdown of the nuclear envelope. Cyclin B–CDK1 remains active up to late metaphase and regulates chromosome condensation and spindle assembly. CDK activity follows a strict time course: it must be high in the beginning of mitosis and be turned down at the onset of anaphase since otherwise a separation of the sister chromatids is not possible. In fact, the switch from high to low CDK activity brought about by cyclin degradation is a key event that determines the onset of anaphase. The elimination of cyclins (and other proteins regulating mitosis) is triggered by E3 ubiquitin ligases, in particular those of types **SCF** and **APC/C** [Anaphase-Promoting Complex/Cyclosome; this notation is used to avoid confusion with the adaptor protein APC of the Wnt signaling pathway (Section 5.9)], which tag the proteins for degradation in proteasomes. These ligases are essential: without them, the cell cannot divide and arrests at the end of metaphase. SCF ubiquitin ligases preferentially attack cyclins and other proteins involved in regulation of G1 and S phases, whereas APC/C enzymes control later phases of mitosis.

A key event is the separation of sister chromatids in anaphase. To this end, a specific endopeptidase, **separase**, destroys a subunit of the protein cohesin, which is the "glue" holding together the chromatids at the centromeres. To prevent premature separation before anaphase, this enzyme is inhibited by binding to a protein called **securin**. Coincident with cyclin degradation and the onset of anaphase, securin is eliminated with the help of APC/C, rendering separase active. To perform this job, APC/C must be phosphorylated by CDK1–cyclin B. This explains why the activity of the kinase must be high in late metaphase.

Phosphorylation determines the substrate specificity of APC/C in that it controls interactions with various regulatory subunits. In fact, the ligase is a large multi-protein complex consisting of more than a dozen subunits (see Figure 12.20, Section 12.16) that resembles in size other molecular machines such as ribosomes, proteasomes, or chaperonins, thus justifying the name cyclosome.

An important regulatory principle of the cell cycle is that APC/C has the choice between two major substrate-recognizing subunits, known as **Cdc20** and **Cdh1**, which confer different specificities. While Cdc20 mediates the degradation of securin and some prometaphase proteins such as cyclin A and (at least partially) cyclin B, Cdh1 is associated with the elimination of proteins that control the onset of anaphase and later phases of mitosis. These include cyclin B, mitotic protein kinases (see Section 12.12), and subunit Cdc20. The degradation of such proteins is essential for mitotic exit and a clearing of the cell for the subsequent G1 phase. Since APC–Cdh1 also induces the degradation of cyclin A, which is required for DNA replication, the ubiquitin ligase becomes inactivated at the end of the G1 phase by the combined effects of an inhibitory protein EMI1 (which is produced just in time along the E2F pathway) and of cyclin A–CDK2. The latter phosphorylates Cdh1, priming it for ubiquitylation by the E3 ligase SCF and degradation in proteasomes. Thus, like cyclins, subunits Cdc20 and Cdh1 are synthesized only during definite periods of the mitotic cycle to become sequestered or degraded afterwards, and the regulation of APC/C activity appears to be as sophisticated as that of CDK activity.

Which of the subunits interacts with APC/C depends on the state of APC/C phosphorylation: APC/C must be phosphorylated at several sites to bind Cdc20 but dephosphorylated to interact with Cdh1 (Figure 12.15). APC/C phosphorylation is catalyzed predominantly by CDK1–cyclin B and occurs as long as enough cyclin B is available. It provides the triggering signal for chromosome separation in anaphase. In contrast, APC/C–Cdh1 becomes active when cyclin B disappears, giving protein phosphatases the chance to dephosphorylate APC/C

Figure 12.15 Reprogramming of ubiquitin ligase APC/C during the cell cycle At the beginning of mitosis, APC/C is phosphorylated by the CDK1–cyclin B complex. In the phosphorylated state, APC/C binds the subunit Cdc20 and induces the transition from metaphase to anaphase by priming securin for degradation. By this means, the separation of sister chromatids is triggered. In dephosphorylated APC/C, Cdc20 is replaced by the subunit Cdh1, rendering the complex specific for an alternative set of substrates such as cyclin B, Cdc20, and other mitotic proteins. Their degradation triggers cytokinesis and "clears" the cell for the subsequent G1 phase. At the end of G1 phase, an inhibitory protein (EMI1) generated along the E2F pathway suppresses APC/C–Cdh1 activity, and cyclin A–CDK2 phosphorylates Cdh1, priming it for ubiquitylation by the E3 ligase SCF and degradation in proteasomes. The diagram to the left shows in a schematic manner the activities of the two APC/C complexes during different cell cycle phases.

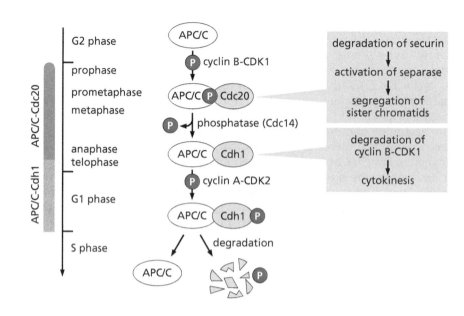

(Figure 12.16). By eliminating cyclins and mitotic protein kinases as well as Cdc20, the APC/C–Cdh1 complex speeds up its own formation in an autocatalytic process. This reprogramming of the ubiquitin ligase takes place in late anaphase.

Summary

The E3 ubiquitin ligase anaphase-promoting complex/cyclosome (APC/C) induces two key events of mitosis: in early anaphase, it induces proteolytic degradation of securin, an endogenous inhibitor of separase, the protease that causes sister chromatid separation; and in metaphase, it induces degradation of

Figure 12.16 Correlation of anaphase transit with declining cyclin B–CDK1 activity and a change of substrate specificity of APC/C For details, see text.

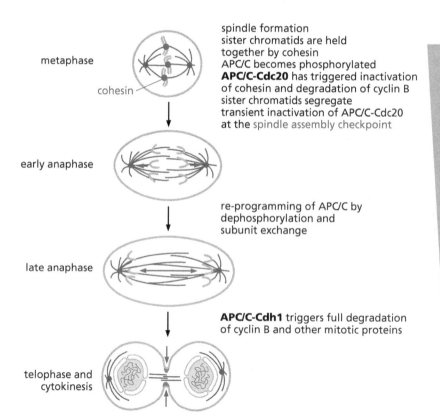

spindle formation
sister chromatids are held together by cohesin
APC/C becomes phosphorylated
APC/C-Cdc20 has triggered inactivation of cohesin and degradation of cyclin B
sister chromatids segregate
transient inactivation of APC/C-Cdc20 at the spindle assembly checkpoint

re-programming of APC/C by dephosphorylation and subunit exchange

APC/C-Cdh1 triggers full degradation of cyclin B and other mitotic proteins

cyclin B and other proteins, thus enabling anaphase and telophase. The switch in function of APC/C is due to reprogramming by dephosphorylation and an exchange of regulatory subunits.

12.12 Mitotic protein kinases: formation of the mitotic spindle

Mitosis is controlled by a network of protein kinases that includes, in addition to CDK1–cyclin B, the following families of highly specialized Ser/Thr-directed kinases:

- Aurora kinases
- Polo-like kinases
- NIM-A-related kinases
- Bub kinases
- Mps kinases

Like cyclins, mitotic kinases are expressed only for a short time interval. Most of them are present from the G2–M transition up to the end of the M phase, when they undergo degradation triggered by the APC/C–Cdh1 complex. Major targets of the mitotic kinases are proteins of centrosomes, of sister chromatids (such as securin), and of the centromere–kinetochore complex.

Mitotic kinases require activation loop phosphorylation. It is not always clear whether this is catalyzed by other kinases or is due to autophosphorylation, particularly because the input signals activating these kinases are known only incompletely. Moreover, the way in which these kinases share the job of controlling chromosome segregation and cytokinesis is still not fully understood.

Aurora kinases got their name from a *Drosophila* mutant (although initially they were discovered in *Xenopus* eggs). Mammalian cells express three isoforms called Aurora A, B, and C. While isoforms A and B are widely distributed, isoform C is restricted to sperm cells. Aurora kinases are key regulators of cell division that cooperate but differ in substrate specificity and cellular localization. Thus, Aurora A and C are enriched at centrosomes, while Aurora B is enriched at chromosomes. **Aurora A** appears at the G2–M transition and is degraded at the end of mitosis. By phosphorylating several centrosomal proteins in the prometaphase, it regulates centrosomal functions that are essential for the formation of a bipolar spindle. Therefore, Aurora A controls chromosome segregation.

The same holds true for **Aurora B**, albeit at chromosomes. Together with some other proteins, Aurora B is a component of a chromosomal passenger complex that stays with the chromosomes up to the end of metaphase, becoming more and more concentrated at the centromeres. During anaphase and telophase, it migrates to the equatorial zone that finally develops into the cleavage furrow of the dividing cell. Inhibitor and knockout experiments show that the complex is essential for spindle formation, correct alignment of chromosomes, spindle assembly quality control, formation of a central spindle during anaphase, and cytokinesis. Substrates of Aurora B are various centromere and kinetochore proteins. Their phosphorylation is required for spindle formation but also loosens the contact between kinetochores and microtubules. In this way, Aurora B might correct defaults in the links between chromosomes and spindle.

Other substrates are histones, particularly histone 3, the phosphorylation of which is essential for mitosis (Section 8.2.5). A pathological overexpression of Aurora kinases is characteristic for a variety of human tumors. Moreover, in cell culture Aurora overexpression causes aneuploidy and genetic instability resulting in a transformed phenotype. Therefore, Aurora kinases are potential targets of novel anti-cancer strategies.

Polo-like kinases (Plk), which owe their name to the "Polo" mutant of *Drosophila*, have been found in all eukaryotes studied in this respect. Among the four isoforms of the mammalian cell, at least Plk1 and Plk4 play a significant role as mitotic regulators. Thus, the duplication of centrioles in the S phase critically depends on **Plk4**: inhibition or elimination of Plk4 correlates with severe spindle anomalies, defective embryonic development, and increased cancer susceptibility. Probably, the Plk4 substrates are structural or regulatory proteins required for centriole duplication.

In controlling spindle formation, Plk4 cooperates with the isoform **Plk1**. Like Plk4 and Aurora kinases, Plk1 is enriched in centrosomes and kinetochores. Major substrates are microtubule-associated proteins, the phosphorylation of which triggers spindle formation at the centrosomes. As explained above, Plk1 is also involved in the recovery from G2 checkpoint control when DNA damage has been successfully repaired. To this end, the kinase phosphorylates key regulatory proteins, priming them for ubiquitylation and degradation. To augment this effect, Plk1 also phosphorylates the CDK-activating phosphatase Cdc25, whose activity thus becomes increased (through this reaction, progesterone stimulates meiosis of amphibian oocytes; see Section 12.2). This reaction provides an instructive example of the ambiguity of the phosphorylation signal: while Cdc25 is primed for ubiquitylation and degradation by phosphorylation of a Ser residue in the "ubiquitylation degron" (see Section 12.16) catalyzed by checkpoint kinase 1, phosphorylation of another Ser residue by Plk1 has a stimulatory effect.

Like CDK1–cyclin B, Plk1 also may phosphorylate the ubiquitin ligase APC/C, thus promoting the entry into anaphase. Vice versa, APC/C complexed with Cdh1 primes Plk1 for degradation at the end of mitosis.

Plk1 is under the control of other protein kinases such as **SLK** (S̲te20-L̲ike K̲inase, referring to the protein kinase Ste20 of yeast) found to be widely distributed in eukaryotes. SLK is activated by auto(?)phosphorylation and reaches maximal activity at the G2–M transition.

The name of the **NIM-A-related kinases (Nrk)** comes from the NIM (N̲ever I̲n M̲itosis) mutant of the mold *Aspergillus nidulans*, which exhibits a blockade of the G2–M transition. The mold enzyme NIM-A is active only during prophase and metaphase and is degraded afterwards.

Mammals have seven genes encoding Nrk isoforms, which here are called Nek1 to Nek7. **Nek2** is associated with centrosomes, where it phosphorylates a centrosomal Nek-associated protein (C-Nap1) that (like cohesin) probably adheres centrosome pairs. Since the glue effect is lost upon phosphorylation of C-Nap1, Nek2 can cause centrosome separation and spindle formation. Almost nothing is known about the functions of the other Nek isoforms.

Additional mitotic protein kinases found in all eukaryotes are associated with the kinetochores. They act as **spindle checkpoint kinases** (see below) and are known as **Bub**, **BubR** (Bub-related), and **Mps** kinases. Bub (B̲udding u̲ninhibited by b̲enzimidazole) refers to yeast mutants that, due to a defective spindle checkpoint, are unable to interrupt the cell cycle in response to the cytotoxic agent benzimidazole. Mps (M̲ono̲polar s̲pindle) refers to a yeast mutant with disturbed centrosome duplication. At least *in vitro*, Mps kinases exhibit rare dual specificity: the ability to phosphorylate Ser, Thr, and Tyr residues. The yeast enzyme has also been shown to regulate centrosome duplication in G1 phase (in animals this function is attributed to Plk4). These effects indicate that proteins of the spindle checkpoint and of centrosomes might be Mps kinase substrates.

Summary

Apart from cyclin B–CDK1, several other Ser/Thr-directed protein kinases control the passage of the cell through the mitotic cycle. These mitotic protein kinases (Plk, SLK, Nrk, Aurora, Bub, and Mps) are thought to be involved in control of centrosome separation and function, spindle formation, histone phosphorylation, regulation of APC/C activity, and mitotic checkpoints.

12.13 Spindle assembly checkpoint: last chance to correct mistakes

During metaphase, free microtubules of the mitotic spindle search around until they have "captured" the kinetochores (the microtubule-binding protein complexes associated with sister chromatids). This stochastic process is prone to mistakes. Therefore, a control mechanism is essential to ensure the correct bipolar attachment of sister chromatids (where each pair of sister chromatids is attached by a microtubule to each centrosome). In fact, spindle quality control is extremely important for the well-being of the cell: if it does not work properly, the risk of malignant transformation is strongly increased. The control mechanism becomes active at the spindle assembly checkpoint, where the separation of the chromosomes is delayed until it is guaranteed that each sister chromatid is attached correctly to prevent aneuploidy. To this end, the degradation of securin is interrupted by a transient inhibition of the ubiquitin ligase APC/C–Cdc20.

The checkpoint control is triggered by several proteins, forming a mitotic checkpoint complex. However, it is still not entirely clear how this complex inhibits APC/C–Cdc20 until all chromatids have become precisely aligned and fixed at the spindle. One concept, the mechanism of reversible Cdc20 inactivation, is shown in Figure 12.17.

According to this hypothesis, the complex binds to kinetochores that lack a contact with the spindle or whose microtubules are incorrectly aligned in relationship to the spindle pole. The kinesin motor protein **centromeric protein E (CENP-E)** transiently associates with kinetochores in the metaphase and recognizes such defects and recruits the **checkpoint kinase BubR1**, which becomes activated.

A major substrate of BubR1 is a complex of the proteins **Mad1** and **Mad2** (the name refers to <u>M</u>itotic <u>a</u>rrest-<u>d</u>efective yeast mutants). Upon phosphorylation,

Figure 12.17 Mitotic checkpoint complex: a still somewhat speculative model As the key mechanism of anaphase arrest, a transient inactivation of the ubiquitin ligase APC/C–Cdc20 complex is postulated that becomes reversed when the kinetochore has been found by a microtubule or incorrect microtubule–kinetochore binding has been corrected. (A) The motor protein CENP-E (dark gray horizontal bar) recognizes a kinetochore lacking a microtubule and recruits the mitotic checkpoint complex consisting of several protein kinases (shown in color) and the regulatory proteins Mad1 and Mad2. (B) One (or several) of the protein kinases phosphorylates Mad1 (P, phosphate residue), disrupting the interaction with Mad2. Mad2 thus released displaces the regulatory subunit Cdc20 from the ubiquitin ligase APC/C, thus inactivating APC/C. This results in a transient stop of the mitotic cycle. (C) The kinetochore is correctly contacted by a microtubule. Probably due to a conformational change of CENP-E, the mitotic checkpoint complex dissociates from the kinetochore and disintegrates. By dephosphorylation of Mad1, the Mad1–Mad2 complex is reestablished. As a result, APC/C is relieved from inhibition and can now induce the metaphase–anaphase transition. The kinetochore is depicted here as a single entity; in reality, it is a large multiprotein complex. This rather simplified scheme is based on assumptions, some of which are awaiting final experimental proof.

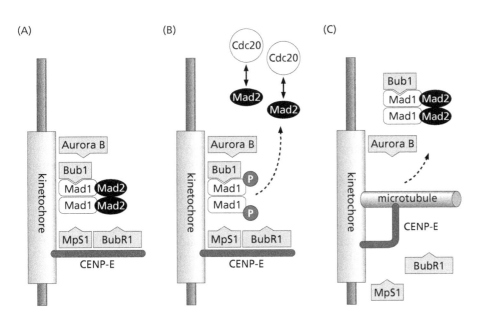

this complex displaces subunit Cdc20 from the ubiquitin ligase APC/C, rendering APC/C unable to induce securin degradation and sister chromatid segregation. As a result, the mitotic cycle is blocked in late metaphase. This blockade is overcome only when a microtubule from the spindle pool displaces the BubR1–Mad complex from the kinetochore and a mechanical tension develops between the kinetochore pair of adjacent sister chromatids.

Transient inhibition of APC/C–Cdc20 is only one aspect of spindle quality control. In addition, incorrectly attached microtubules must be released from the kinetochore. This step depends on **Aurora B**-catalyzed phosphorylation of proteins that control the mechanical tension between microtubules and kinetochores (incorrect attachment changes this tension). Aberrant tension is detected by accessory proteins that form the chromosomal passenger complex with Aurora B, as mentioned above. Experiments with mutants and inhibitors have shown that, in addition to Bub and Aurora B kinases, spindle control also requires the kinase Msp. How Msp shares the job with the other kinases remains an open question.

Summary

At the spindle assembly or anaphase checkpoint, separation of the chromosomes is interrupted until it is guaranteed that all sister chromatids are bound by microtubules from the opposite spindle poles to ensure symmetric segregation of the genetic material among the daughter cells. The checkpoint control is due to a sophisticated (not yet fully understood) interaction between proteins that recognize spindle defects and enzymes that catalyze reversible phosphorylation of regulatory proteins to induce a transient stop of the mitotic cycle and allow time for repair measures.

12.14 Cytokinesis: how a new cell is born

Cytokinesis is the final stage of cell division, where the daughter cells form. As far as biochemical control mechanisms are concerned, it is the cell cycle phase least understood. By means of functional genomics and proteomics, several hundred proteins and low-molecular weight factors have been found to be potentially involved without obtaining a clear picture of their interactions.

The major event of cytokinesis is the contraction of the cell perpendicular to the spindle by a microfilament ring formed in the equatorial plain. Until complete separation, both daughter cells stay connected by the **central spindle** that consists of microtubules not linked to kinetochores and arises when the chromosomes are pulled to the spindle poles. The central spindle is essential for correct progress of cytokinesis since it assembles regulatory proteins in the region of membrane constriction. How this occurs is not known. One possibility would be a mechanical transport by motor proteins along the microtubule "tracks" of the spindle.

As we have seen, a critical condition of cytokinesis and mitotic exit is to inhibit cyclin B–CDK1 activity, since this activity would suppress continuation of M-phase events such as formation of a new nuclear envelope and cytokinesis. Therefore, the destruction of cyclin B is initiated in late metaphase by the ubiquitin ligase APC/C–Cdh1 complex (Figure 12.16). Additional control elements are considered to be the protein kinases Aurora B and Plk since their inhibition retards cytokinesis. Moreover, Aurora B dissociates from chromosomes during anaphase and telophase to assemble in the central part of the spindle and in the zone of membrane constriction (midzone), where it probably is bound by specific anchoring proteins. However, for a better understanding of the mechanisms involved, identification of the elusive kinase substrates is mandatory.

The cleavage itself is controlled by several GTPases, which become concentrated in the equatorial plain of the central spindle. Among them, the small G-protein

Rho (Section 10.3) promotes the aggregation of actin fibers to the contractile ring. A Rho GTP–GDP exchange factor (GEF) and a Rho GTPase-activating protein (GAP), both regulating Rho activity, are bound by adaptor proteins in the central spindle plain. Rho-GEF can activate Rho only when cyclin B has been degraded, since otherwise it would be inhibited by CDK1-catalyzed phosphorylation. Such an inhibitory effect seems to be supported additionally by activation of Rho-GAP through phosphorylation by Aurora B.

Further GTPases participating in cell division are the **dynamins** (see Section 5.8) and the **septins**. Septins are large proteins found in all eukaryotes except plants (in the human genome, seven septin genes have been identified). When activated by GTP, they aggregate to short filaments associating with the actin fibers of the contractile ring where they might recruit additional proteins required for cytokinesis. Like dynamins, septins are also involved in exocytosis.

Contraction of the microfilament ring depends on a local increase of the cytoplasmic Ca^{2+} concentration activating the Ca^{2+}/calmodulin-regulated **myosin light chain (MLC) kinase**. This enzyme phosphorylates the short myosin chains enabling myosin to interact with actin and thus induce contraction (Section 14.6.2). The increase in cytoplasmic Ca^{2+} is also associated with destruction of cyclin B and onset of anaphase. In fact, CDK1–cyclin B phosphorylates and activates **MLC phosphatase**, which efficiently counteracts the effect of MLC kinase unless it becomes dephosphorylated. Inactivation of MLC phosphatase is enhanced by Rho activating the protein kinase ROK, which inhibits MLC phosphatase activity (Section 10.3.3). [Note that the dual control of MLC phosphatase resembles that of Cdc25: depending on the target site, phosphorylation either promotes or inhibits enzyme activity.] As mentioned above, Rho becomes active only upon degradation of cyclin B. So, as noted, the removal of this cyclin initiated by the ubiquitin ligase APC/C provides a critical step in mitosis (see Figure 12.16).

Summary

Mechanisms regulating cytokinesis, the mechanical formation of daughter cells, are only incompletely understood. A critical step is the inactivation of cyclin B–CDK1 through degradation of cyclin B triggered by the ubiquitin ligase APC/C–Cdh1 complex. It provides the condition for cell separation by the contractile microfilament ring the formation and function of which is controlled by G-proteins of the Rho, dynamin, and septin families as well as by myosin light chain phosphorylation.

12.15 Mitotic exit network of yeast

To finish mitosis and to open the gate to a new interphase, the signaling system that initiated M phase must be switched off completely. A significant step is the reprogramming of the ubiquitin ligase APC/C by subunit exchange, which renders the enzyme specific for anaphase and telophase proteins including cyclin B (see Figure 12.15). As shown for *S. cerevisiae*, to this end a signaling complex is activated in late anaphase. The major executor of this mitotic exit network (MEN) is the dual-specific **protein phosphatase Cdc14** that counteracts the mitotic CDKs and promotes the reprogramming of APC/C via replacement of the subunit Cdc20 by the subunit Cdh1 (see Figure 12.15). At least in *S. cerevisiae*, this reaction clears the way for the degradation of Clb2 (the yeast homologue of cyclin B), resulting in inactivation of the kinase Cdc28 (the yeast homologue of CDK1). To reinforce this effect, Cdc14 also activates a CDK inhibitor Sic1.

Punctual activation of Cdc14 is a critical condition of exact regulation of the mitotic exit. In fact, the enzyme is sequestered in the nucleolus and blocked by a protein complex known as Fob1–Net1 until it becomes activated by the MEN signaling system and transiently released into the cytoplasm during anaphase. The MEN complex of yeast is composed of a small Ras-like G-protein Tem1, the

Figure 12.18 Mitotic exit network of *S. cerevisiae* The signaling system depicted in a simplified form is composed of the proteins shown in the large gray box. Its activation in late anaphase is probably due to a rearrangement of the components resulting in conformational changes. The black solid arrows symbolize activating reactions, while the broken arrows indicate (putative) positive feedback loops based on a dephosphorylation of Lte1 and Cdc15 that both are inhibited by CDK1-catalyzed phosphorylation. These and other inhibitory effects are shown in red. For more details, see text.

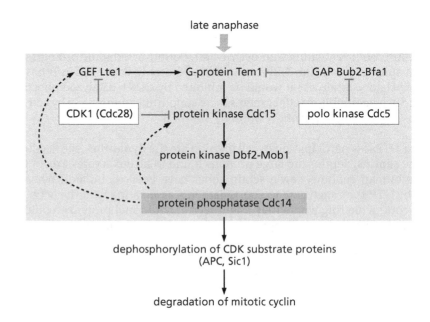

associated GEF (Lte1) and GAP (a heterodimer of the proteins Bub2 and Bfa), and the protein kinases Dbf2–Mob1, Cdc15, and Cdc5 (a Polo-like kinase). These proteins are interconnected to a signaling cascade as depicted in Figure 12.18. Since Tem1 and Cdc15 are inactivated by CDK-catalyzed phosphorylation and become reactivated by the phosphatase Cdc14, the mitotic exit network seems to enhance its efficacy by positive feedback.

To get the network in full swing in *late* anaphase, Cdc14 becomes partially activated in *early* anaphase. This is brought about by another protein network called **FEAR** (Cdc **F**ourteen **E**arly **A**naphase **R**elease) that involves the inactivation of the Cdc14 inhibitor Fob1–Net1 by the combined action of the protease separase (here called Esp1) and CDK1 (here called Cdc28). The clockwork-like interaction of several regulatory circuits guarantees flawless temporal coordination between chromosome segregation, cytokinesis, and termination of mitosis.

Cdc14 is a highly conserved protein found from yeast to humans. Moreover, homologues of the other MEN proteins are distributed equally widely. Therefore,

Sidebar 12.4 NDR kinases: cell cycle inhibitors and tumor suppressors The protein kinase Dbf2 of the mitotic network of yeast is a member of a larger family of related Ser/Thr-directed protein kinases that probably occur in all eukaryotes, called NDR kinases (**N**uclear **D**bf2-**R**elated). They represent a subfamily of the AGC kinases (Section 2.6.3) and are activated typically by (auto)phosphorylation of the activation loop and phosphorylation in the C-terminal hydrophobic region. The mammalian enzymes bind, in addition, to a calmodulin-containing Ca^{2+} sensor protein of the S100 family that stimulates the activating (auto)phosphorylation depending on the cellular Ca^{2+} level. The role of NDR kinases in cell cycle control has been studied most thoroughly for yeast. Whether or not the homologous enzymes of higher eukaryotes fulfill similar purposes is still an open question.

One of the mammalian enzymes has become known as the **tumor suppressor protein LATS** (**LA**rge **T**umor **S**uppressor), the deletion of which causes tumor growth in multiple organs in experimental animals. LATS is a negative regulator of cell division, and its activity and cellular localization strictly depend on the cell cycle phases. Active LATS displaces cyclin B from CDK1, thus inhibiting the G2–M transition. Moreover, by inducing the synthesis of the pro-apoptotic protein Bax, LATS promotes cell death. The activities of the two additional NDR kinases of mammals, NDR1 and NDR2, seem to be regulated also in correlation with the cell cycle. Hyperactive NDR, due to an overexpression of the regulatory S100 protein, has been repeatedly found in malignant melanoma.

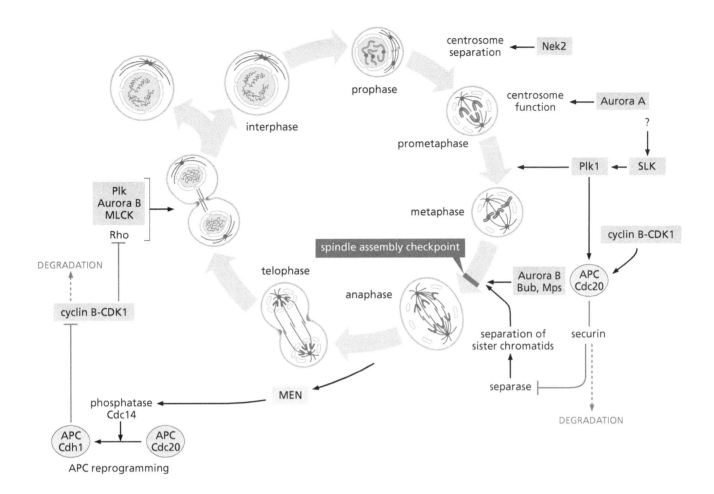

Figure 12.19 Signaling network controlling mitosis: a preliminary summary of the state of the art
The scheme summarizes the interactions of protein kinases (gray boxes) and of the ubiquitin ligase APC/C as far as they are known at present. MEN, mitotic exit network of yeast.

one might expect mammalian cells to have as much complexity in regulation of the late mitotic phase as yeast. It must be emphasized, however, that for metazoans the mechanism of mitotic exit differs from that of yeast in a way that is still not understood.

In Figure 12.19, an attempt is made to summarize schematically the complexity of signaling pathways controlling mitosis, with particular emphasis laid on protein kinases. It must not be overlooked, however, that this scheme provides only a preliminary image of the state of the art, since this field of research still is under intense investigation.

Summary

In yeast cells, a mitotic exit network activated in anaphase is thought to provide the conditions for entry into the G1 phase of the cell cycle. A key event is dephosphorylation (by the phosphatase Cdc14) of CDK1 substrate proteins including the ubiquitin ligase APC/C, which are reprogrammed to trigger the elimination of cyclins and other mitotic proteins.

12.16 Ubiquitin ligases

The cell cycle provides a particularly instructive example of signal-controlled protein degradation to ensure a precise timing of events. The key reaction of protein degradation consists of tagging the protein with a Lys48-linked polyubiquitin chain that marks a protein for proteolysis in proteasomes. This posttranslational modification terminates a signaling event by destruction of a required protein and is of central importance for cell cycle regulation. However, as we have seen in Section 2.8, proteasome targeting is only one among many other functions of ubiquitylation.

As explained in Section 2.8, among the three types of enzymes required for ubiquitylation, the ubiquitin ligases E3 determine the substrate specificity of the reaction. In cell cycle control, these are the enzymes SCF and APC/C, which represent a large protein family. In fact, the striking diversity of E3 ligases is reflected by the number of their genes—more than 500 in humans. In addition, the significance of these enzymes is underscored by the finding that quite a number of them are tumor suppressors. However, others may mutate to dominant-negative fragments that act as oncoproteins by antagonizing the nonmutated forms.

E3 ubiquitin ligases recognize specific sequences, known as "degrons," in their substrate proteins. A typical degron is the destruction motif of cyclins. Frequently, degron recognition depends on other signaling reactions, particularly on phosphorylation. Characteristic degron-interacting domains define two E3 families: HECT domain ligases and RING domain ligases.

12.16.1 Ubiquitin ligases with HECT domains

The C-terminal HECT domain characterizing these enzymes contains a strategic cysteine (Cys) residue where the ubiquitin obtained from the ubiquitin-conjugating enzyme E2 is bound transiently as a thioester and transferred subsequently to the substrate bound at the N-terminal domain of the ligase (see Section 2.8).

The prototypical enzyme of this family is **E6-AP** (Figure 12.20) that is induced by the papilloma virus E6 (HECT stands for Homologous to E6-AP C-Terminus). By priming p53 for degradation, this enzyme allows passage of the DNA structure checkpoint and prevents apoptosis of damaged cells to ensure virus replication. Another HECT domain ligase triggers the destruction of the SMAD transcription factors that mediate the effects of TGFβ and related tissue hormones (see Section 6.2).

12.16.2 Ubiquitin ligases with RING domains

The ubiquitin ligase **MDM2**, discussed as a negative regulator of p53-dependent cell cycle arrest and apoptosis at the DNA structure checkpoints, as well as the

Figure 12.20 E3 ubiquitin ligases
E3 ubiquitin ligases play the role of scaffold proteins that bring together the ubiquitin-conjugating enzyme E2 (shown in color) with selected substrate proteins. Substrate-binding domains/subunits of E3 are depicted in black. E2 catalyzes the transfer of the ubiquitin residue (UB) from the ubiquitin-activating enzyme E1 to the substrate (gray). The figure shows some representative complexes of E2 with HECT and RING domain enzymes, that is, the major families of E3 ubiquitin ligases. HECT, RING, ROC1, and Apc11 are E3 domains/subunits with E2-binding sites; cullin, SKP1, elongins, and Apc proteins (except Apc11) are scaffold and adaptor proteins. For details, see text.

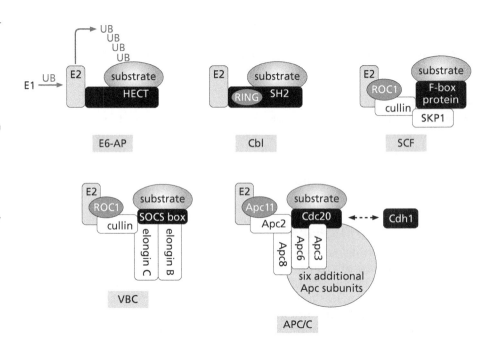

TNF receptor-associated factors involved in cytokine signaling (Section 6.3.1), are representatives of the **RING domain proteins**. The RING domain (named after a Really Interesting New Gene) contains a group of eight Cys and histidine (His) residues arranged in a fingerlike structure, the "RING finger," that is stabilized by two zinc ions. As the active center of the E3 ligase, this domain is responsible for bringing about the specific interaction between the E2 enzyme and the substrate of ubiquitylation. Thus, in contrast to HECT domain ligases, the RING domain enzymes do not covalently bind the ubiquitin residue.

An example of a relatively simple RING ubiquitin ligase is provided by **Cbl** (Section 7.1.1). Here the RING domain and the substrate-binding domain share the same polypeptide chain. This simple structure appears to be an exception rather than the rule, since in most other RING domain ligases the two domains are located on separate subunits. Through an SH2 domain, Cbl binds to Tyr-phosphorylated proteins such as autophosphorylated growth factor receptors, which become ubiquitylated and are taken up by endocytosis. Cbl thus triggers the adaptive down-regulation of such receptors and of cytoplasmic Tyr kinases (see Section 7.1.1). Deletion of the RING domain transforms Cbl into an oncoprotein that competitively inhibits the ubiquitin ligase activity of the intact enzyme derived from the nonmutated allele. In contrast, another RING ubiquitin ligase, the breast cancer protein **BRCA1**, is a tumor suppressor: upon a loss-of-function mutation, the ability of cells to adapt to proliferative signals becomes impaired.

RING domain ligases have a tendency to form large multiprotein complexes held together by scaffold proteins of the **cullin** family (comprising seven members in humans) and related adaptor proteins. The regulation of such complexes seems to be quite sophisticated. For instance, the function of the RING domain ligases is enhanced by reversible covalent binding of the ubiquitin-like peptide NEDD8 (Neuronal precursor cell-Expressed Developmentally Down-regulated protein 8) at a specific Lys residue of cullin. In contrast to poly-ubiquitylation, "**neddylation**" does not trigger proteolytic degradation (thus resembling sumoylation).

A clear advantage of the multisubunit structure of most RING-domain ligases is that it offers the chance to choose between different substrate-binding or regulatory subunits, thus conveying a particularly high degree of versatility. Prominent examples of the more complex RING ubiquitin ligases are SCF, VBC, and APC/C.

SCF-type enzymes are composed of at least four subunits: an adaptor protein SKP1, the scaffold protein Cullin, the RING domain ligase Roc1/Rbx1, and a substrate-binding F-box protein (Figure 12.21). An **F-box*** is a sequence motif of about 50 amino acids that, from yeast to humans (it is estimated that 38–100 F-box genes may exist), is found in various proteins, involved in protein–protein interactions. In humans, F-box proteins differing in their substrate specificities have been identified as subunits of E3 ubiquitin ligases. By choosing between different F-box proteins, an SCF-type ligase may tag different sets of proteins for destruction, for instance, during cell cycle progression.

To become recognized by an F-box, the degron of the substrate protein must be phosphorylated (Figure 12.21). This is because the F-box contains a phospho-amino acid-docking domain such as WD40 [a propeller-shaped phospho-Thr-specific sequence of about 40 amino acids ending with tryptophan (W) and aspartate (D)]. This substrate phosphorylation provides a particularly clear example of the reversible ad hoc generation of a docking motif and explains why

---

*The name F-box goes back to **cyclin F**, which, however, is not a ubiquitin ligase. Instead, it binds to cyclin B–CDK1, probably transporting this complex into the nucleus. In fact, cyclin F has several nuclear localization sequences that cyclin B–CDK1 is lacking.

Figure 12.21 Cooperation between phosphorylation and ubiquitylation: degradation of cyclin E In the active cyclin E–CDK2 complex, cyclin E is phosphorylated at a Thr residue (P). The phosphorylated sequence is recognized by the WD40 domain of an SCF-type E3 ubiquitin ligase. Upon binding by SCF, cyclin E becomes poly-ubiquitylated and, in turn, undergoes degradation in proteasomes. CDK2 is not ubiquitylated and degraded.

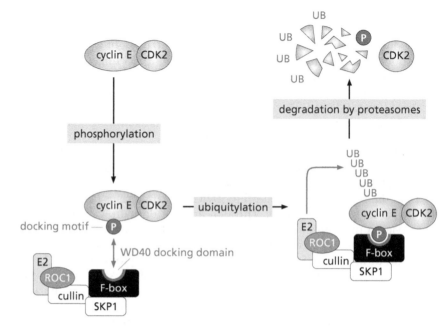

ubiquitylation frequently goes along with phosphorylation, thus representing a logical AND gate.

Cooperating in such a way with various protein kinases, SCF-type ligases prime numerous proteins for degradation in proteasomes. These substrates include the cyclins of the G1 phase, CDK phosphatases (Cdc25), CDK inhibitors (such as p21^{KIP} and p27^{KIP}, and the inactivating kinase WEE1), transcription factors such as E2F, β-catenin and the hedgehog-controlled factor Ci/Gli (Section 5.9), Notch (Section 13.1.1), and the inhibitor IκB of the transcription factor NFκB (Section 11.8). When the variety of SCF substrates is considered, it becomes clear that these ligases play a key role in signal termination and adaptation.

SCF-type ubiquitin ligases are particularly abundant in plants (in *Arabidopsis thaliana*, about 700 F-box genes have been found). Some of these enzymes have been identified as receptors for the phytohormones auxin and gibberellic acid that, upon ligand binding, trigger the proteolytic degradation of transcriptional repressors, thus releasing hormone-responsive gene transcription from inhibition (see Section 13.1).

Another subfamily of RING ubiquitin ligases are the **VBC complexes**. They got their name from the ligase <u>V</u>HL and the subunits elongin <u>B</u> and elongin <u>C</u>. The ubiquitin ligase VHL is a tumor suppressor deleted in <u>V</u>on <u>H</u>ippel–<u>L</u>indau disease (see Section 8.7) and in kidney carcinomas. Elongin B and C were described originally as components of the elongation complex of transcription. Both are adaptor proteins that, together with cullin, organize the VBC complex by binding the E2 enzyme and a substrate-binding protein (Figure 12.20). The interaction domain of the latter is a **SOCS box** that is structurally and functionally closely related to the F-box (the name SOCS refers to the <u>S</u>uppressors <u>O</u>f <u>C</u>ytokine <u>S</u>ignaling that also contain this motif; see Section 7.2.1). VBC ubiquitin ligases, particularly VHL, are involved in the cellular response to hypoxic stress (Section 8.7). Human cells express about 50 different SOCS box proteins that combine with ubiquitin ligases, providing substrate specificity and versatility.

As shown above, **APC/C** is a ubiquitin ligase that plays an essential role in controlling mitosis. This rather complex enzyme is composed of at least 12 different subunits. Subunits Apc11 (a RING domain protein) and Apc2 (a cullin analog) are essential for ubiquitin ligase activity, while subunits Apc3, -6, and -8 are SKP1-like adaptors with mostly unknown functions. Cdc20 and Cdh1 act as two different substrate-binding proteins that are exchanged during mitosis (Figure 12.21). The functions of the other subunits are largely unknown.

Summary

More than 500 different mammalian E3 ubiquitin ligases render ubiquitylation specific by coupling E2 ubiquitin-conjugating enzymes with selected substrate proteins. Depending on E2-binding domains, one distinguishes between HECT domain and RING domain ligases. In many ligases, the substrate-binding domain is localized on a separate subunit bound to the E2-binding RING domain subunit by adaptor and scaffold proteins. The role played by E3 ubiquitin ligases in signal transduction rivals that of protein kinases.

Further reading

Bartek J, Falck J & Lukas J (2001) CHK2 kinase – a busy messenger. *Nat. Rev. Mol. Cell Biol.* 2, 677–687.

Bartek J, Lukas C & Lukas J (2004) Checking on DNA damage in S phase. *Nat. Rev. Mol. Cell Biol.* 5, 792–803.

Bollen M & Beullens M (2002) Signaling by protein phosphatases in the nucleus. *Trends Cell Biol.* 12, 138–145.

Bracken AP, Ciro M, Cocito A et al. (2004) E2F target genes: unraveling the biology. *Trends Biochem. Sci.* 29, 409–417.

Carcinogenesis (2000) A collection of review articles in *Carcinogenesis* 21 (3).

Cardozo T & Pagano M (2004) The SCF ubiquitin ligase: insights into a molecular machine. *Nat. Rev. Mol. Cell Biol.* 5, 739–751.

Chan GK, Liu ST & Yen TJ (2005) Kinetochore structure and function. *Trends Cell Biol.* 15, 589–598.

Chan HM, Shikama N & LaThangue NB (2001) Control of gene expression and the cell cycle. *Essays Biochem.* 37, 87–96.

Cleveland DW, Mao Y & Sullivan KF (2003) Centromeres and kinetochores: from epigenetics to mitotic checkpoint signaling. *Cell* 112, 407–421.

Cheeseman IM & Desai A (2008) Molecular architecture of the kinetochore-microtubule interface. *Nat. Rev. Mol. Cell Biol.* 9, 33-46.

Coqueret O (2002) Linking cyclins to transcriptional control. *Gene* 299, 35–55.

Dai W, Wang Q & Traganos F (2002) Polo-like kinases and centrosome regulation. *Oncogene* 21, 6195–6200.

D'Amours D & Amon A (2004) At the interface between signaling and executing anaphase: Cdc14 and the FEAR network. *Genes Dev.* 18, 2581–2596.

De Gramont A & Cohen-Fix O (2005) The many phases of anaphase. *Trends Biochem. Sci.* 30, 559–568.

Downs JA & Jackson SP (2004) A means to a DNA end: the many roles of Ku. *Nat. Rev. Mol. Cell Biol.* 5, 367–378.

Durocher D & Jackson SP (2001) DNA-PK, ATM and ATR as sensors of DNA damage: variations on a theme? *Curr. Opin. Cell Biol.* 13, 225–231.

Eggert US, Mitchison TJ & Field CM (2006) Animal cytokinesis: from parts list to mechanisms. *Annu. Rev. Biochem.* 75, 543–566.

Fry AM (2002) The Nek2 protein kinase: a novel regulator of centrosome structure. *Oncogene* 21, 6184–6194.

Gonzalez C (2003) Cell division: the place and time of cytokinesis. *Curr. Biol.* 13, R363–R365.

Gutierrez GJ & Ronai Z (2006) Ubiquitin and SUMO systems in the regulation of mitotic checkpoints. *Trends Biochem. Sci.* 31, 324–332.

Hansen DV, Hsu JY, Kaiser BK et al. (2002) Control of centriole and centrosome cycles by ubiquitylation enzymes. *Oncogene* 21, 6209–6221.

Harper JV & Brooks G (2004) The mammalian cell cycle: an overview. *Methods Mol. Biol.* 296, 113–153.

Hichman HS, Moroni MC & Helin K (2002) The role of p53 and pRB in apoptosis and cancer. *Curr. Opin. Genet. Dev.* 12, 60–66.

Humbert PO, Brumby AM, Quinn LM et al. (2004) New tricks for old dogs: unexpected roles for cell cycle regulators revealed using animal models. *Curr. Opin. Cell Biol.* 16, 614–622.

Jackson PK, Eldridge AG, Freed E. et al. (2000) The lord of the RINGs: substrate recognition and catalysis by ubiquitin ligases. *Trends Cell Biol.* 10, 429–439.

Karlsson-Rosenthal C & Millar JBA (2006) Cdc25: mechanisms of checkpoint inhibition and recovery. *Trends Cell Biol.* 16, 285–292.

Lew DJ & Burke DJ (2003) The spindle assembly and spindle position checkpoints. *Annu. Rev. Genet.* 37, 251–282.

Malumbres M & Barbacid M (2005) Mammalian cyclin-dependent kinases. *Trends Biochem. Sci.* 30, 630–641.

Mayo LD & Donner DB (2002) The PTEN, Mdm2, p53 tumor suppressor–oncoprotein network. *Trends Biochem. Sci.* 27, 462–467.

McGowan CH & Russell P (2004) The DNA damage response: sensing and signaling. *Curr. Opin. Cell Biol.* 16, 629–633.

Melino G, DeLaurenzi V & Vousden KH (2002) p73: Friend or foe in tumorigenesis. *Nat. Rev. Cancer* 2, 605–815.

Melino G, Lu X, Crook T et al. (2003) Functional regulation of p73 and p63: development of cancer. *Trends Biochem. Sci.* 28, 663–670.

Melo J & Toczyski D (2002) A unified view of the DNA-damage checkpoint. *Curr. Opin. Cell Biol.* 14, 237–245.

Meraldi P, Honda R & Nigg EA (2004) Aurora kinases link chromosome segregation and cell division to cancer susceptibility. *Curr. Opin. Genet. Dev.* 14, 29–36.

Michael D & Oren M (2002) The p53 and Mdm2 families in cancer. *Curr. Opin. Genet. Dev.* 12, 53–59.

Miele L (2004) The biology of cyclins and cyclin-dependent protein kinases: an introduction. *Methods Mol. Biol.* 285, 3–21.

Millband DN, Campbell L & Hardwick KG (2002) The awesome power of multiple model systems: interpreting the complex nature of spindle checkpoint signaling. *Trends Cell Biol.* 12, 205–209.

Murray AW (2004) Recycling the cell cycle: cyclins revisited. *Cell* 116, 221–234.

Musacchio A & Salmon ED (2007) The spindle checkpoint in space and time. *Nat. Rev. Mol. Cell Biol.* 8, 379–392.

Nevins JR (2001) The Rb/E2F pathway and cancer. *Hum. Mol. Genet.* 10, 699–703.

Nigg EA (2001) Mitotic kinases as regulators of cell division and its checkpoints. *Nat. Rev. Mol. Cell Biol.* 2, 21–32.

Nigg EA (2007) Centrosome duplication: of rules and licenses. *Trends Cell Biol.*17, 215–221.

Nikolic M (2004) The molecular mystery of neuronal migration: FAK and Cdk5. *Trends Cell Biol.* 14, 1–5.

Norbury CJ & Hickson ID (2001) Cellular responses to DNA damage. *Annu. Rev. Pharmacol. Toxicol.* 41, 367–401.

Nurse P (2002) Cyclin dependent kinases and cell cycle control (Nobel Lecture). *Chembiochem.* 3, 596–603.

O'Connell MJ, Krien MJE & Hunter T (2003) Say never: the NIMA-related protein kinases in mitotic control. *Trends Cell Biol.* 13, 221–228.

Osborn AJ, Elledge SJ & Zou L (2002) Checking on the fork: the DNA-replication stress-response pathway. *Trends Cell Biol.* 12, 509–516.

Peters JM (2006) The anaphase promoting complex/cyclosome: a machine designed to destroy. *Nat. Rev. Mol. Cell Biol.* 7, 644–656.

Petroski MD & Deshaies RJ (2005) Function and regulation of cullin-RING ubiquitin ligases. *Nat. Rev. Mol. Cell Biol.* 6, 9–20.

Pines J (2006) Mitosis: a matter of getting rid of the right protein at the right time. *Trends Cell Biol.* 16, 55–63.

Pinsky BA & Biggins S (2005) The spindle checkpoint: tension versus attachment. *Trends Cell Biol.* 15, 486–493.

Robinson DN & Spudich JA (2004) Mechanics and regulation of cytokinesis. *Curr. Opin. Cell Biol.* 16, 182–188.

Ruchaud S, Carmena M & Earbshaw WC (2007) Chromosomal passengers: conducting cell division. *Nat. Rev. Mol. Cell Biol.* 8, 798–812.

Sage J (2004) Cyclin C makes an entry into the cell cycle. *Dev. Cell* 6, 607–616.

Sears RC & Nevins JR (2002) Signaling networks that link cell proliferation and cell fate. *J. Biol. Chem.* 277, 11617–11620.

Sherr CJ (2000) The Pezcoller Lecture: Cancer cell cycles revisited. *Cancer Res.* 60, 3689–3695.

Sherr CJ (2001) The *INK4a/ARF* network in tumour suppression. *Nat. Rev. Mol. Cell Biol.* 2, 731–737.

Shiloh Y (2006) The ATM-mediated DNA-damage response: taking shape. *Trends Biochem. Sci.* 31, 402–410.

Smith DS & Tsai LH (2002) Cdk5 behind the wheel: a role in trafficking and transport? *Trends Cell Biol.* 12, 28–36.

Sullivan M & Morgan DO (2007) Finishing mitosis, one step at a time. *Nat. Rev. Mol. Cell Biol.* 8, 894–903.

Sumara I, Maerki S & Peter M (2008) E3 ubiquitin ligases and mitosis: embracing the complexity. *Trends Cell Biol.* 18, 84–94.

Tyers M (2004) Cell cycle goes global. *Curr. Opin. Cell Biol.* 16, 602–613.

Vargas DA, Takahashi S & Ronal Z (2003) Mdm2: a regulator of cell growth and death. *Adv. Cancer Res.* 89, 1–34.

Vodermaier HC (2004) PC/C and SCF: controlling each other and the cell cycle. *Curr. Biol.* 14, R787–796.

Vousden KH & Lu X (2002) Live or let die: the cell's response to p53. *Nat. Rev. Cancer* 2, 594–605.

Winey M & Huneycutt BJ (2002) Centrosomes and checkpoints: the MPS1 family of kinases. *Oncogene* 21, 6161–6169.

Yang J, Yu Y, Hamrick HE et al. (2003) ATM, ATR and DNA-PK: initiators of the cellular genotoxic stress responses. *Carcinogenesis* 24, 1571–1580.

Signal Transduction by Proteolysis and Programmed Cell Death

Proteolysis plays a most critical role in cellular data processing since it terminates signaling by proteins and peptides in a controlled way. The major player in this process of signal extinction is the ubiquitin–proteasome system. Various examples of this pathway are found throughout this book.

Besides ubiquitin-triggered protein degradation, proteases are involved in signal processing in two additional ways.

(1) Proteases are components of proteolytic switches standing at the top of signaling cascades (Section 2.11). They are coupled with signal receptors in such a way that, upon activation, they release from proteins peptide fragments that act as second messengers. The energy needed for signal processing stems from hydrolysis of the peptide bond. A perfect example of this signaling principle is the notch system.

(2) Proteases also stand at the bottom of signaling cascades. The most prominent examples are the proteases that, upon activation by death signals, drive cells into suicide.

13.1 Secretase-coupled receptors: generation of peptide second messengers

13.1.1 Notch: a key signal of animal development

The notch system represents a proteolysis-driven mechanism of signal transduction across the cell membrane, or a proteolytic switch, in a particularly clear form. This system enables communication between cells of developing tissues and, therefore, is of central importance for the embryogenesis of metazoans. Mutations of notch proteins cause malformation and cancer.

Notch (the name refers to a *Drosophila* mutant with notched wings) is a dimeric transmembrane receptor that is expressed in four isoforms in mammals. Due to a specific proteolytic cut, each subunit consists of two parts linked by noncovalent binding at the cell surface. The extracellular domain contains 36 so-called epidermal growth factor (EGF)-like repeats (see Sidebar 13.1) that are essential for ligand binding. The high variability of notch effects results from post-translational modifications of the protein. Thus, the ligand and cell-type specificity of notch is regulated by the glycosyltransferase "fringe" that attaches *N*-acetylglucosamine residues to the EGF-like repeats (for more details on this post-translational protein modification see Section 9.4.3).

Figure 13.1 Notch signaling cascade The transmembrane receptor notch consists of two proteins linked by Ca²⁺-dependent noncovalent binding. Upon interaction with the membrane-bound (or proteolytically released) ligand DSL notch is cleaved in two steps by (1) ADAM protease and (2) the intramembrane protease γ-secretase. The cytoplasmic fragment NICD (notch intracellular domain) thus generated activates the transcription factor CSL by heterodimerization via ankyrin repeats (see inset; α-helices are symbolized by cylinders, β-sheets by arrows). In reality notch exists as a homodimer rather than (as shown here for the sake of simplicity) as a monomer.

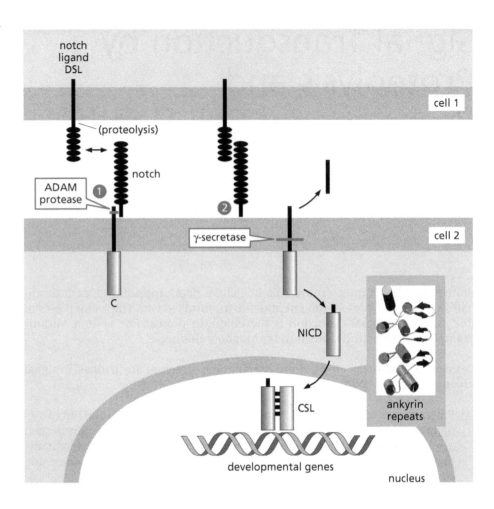

The notch ligands are called **DSL proteins**, referring to the *Drosophila* mutants Delta and Serrate and the *Caenorhabditis elegans* mutant Lag-2. They also are transmembrane proteins bound at the surface of nearby cells and acting as juxtacrine signals. On the other hand, their extracellular domains, containing (like notch) EGF-like repeats, may be released proteolytically into the extracellular medium. By this "ectodomain shedding" peptides are generated which interact with notch in a paracrine mode.

DSL induces a two-step proteolytic activation of notch (Figure 13.1). In the first step, known as S2 cleavage, a transmembrane protease of the ADAM family (see below) clips the extracellular domain of notch, releasing it from the cell surface in a reaction called **juxtamembrane proteolysis**. Into the transmembrane fragment of notch thus generated, a second membrane-integrated protease makes a single cut. By this **regulated intramembrane proteolysis** or S3 cleavage, the cytoplasmic portion of notch, NICD, (Notch IntraCellular Domain), becomes detached into the cytoplasm, acting as a second messenger in that it binds the NFκB -related transcription factor CSL. As a result, CSL is set free from a complex with transcriptional co-repressors and now can stimulate the expression of a number of genes involved in developmental processes. NICD and CSL interact with each other through so-called ankyrin repeats (see Sidebar 13.1).

The intramembrane protease that catalyzes the release of NICD is called **γ-secretase**. This enzyme is a membrane-bound multiprotein complex consisting of a catalytic subunit known as presenilin and three regulatory transmembrane proteins, called nicastrin, APH1 (originally discovered in a maldeveloped Anterior PHarynx of *C. elegans*), and PEN2 (Presenilin ENhancer 2). The catalytic subunit has eight transmembrane domains, with domains 6 and 7 containing the

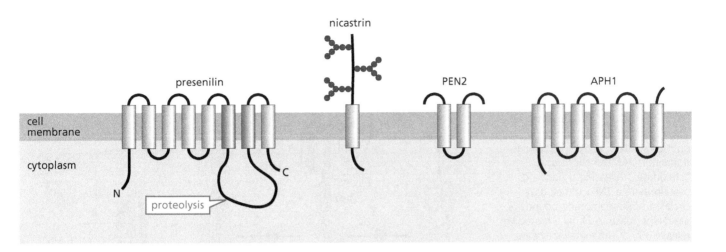

Figure 13.2 γ-Secretase complex
The membrane-bound protease γ-secretase is composed of the catalytic subunit presenilin (with the site of proteolytic activation in the third cytoplasmic loop and the catalytic center shown in red), the regulatory subunits nicastrin and PEN2, and the multipass transmembrane protein APH1, which probably functions as a scaffold. The extracellular domain of nicastrin is heavily glycosylated, as indicated by the dotted structures.

catalytic center (see Figure 13.2). Like most proteases, γ-secretase exists as an inactive zymogen that, upon demand, is activated by limited proteolysis, in this case by a cut in the third cytoplasmic loop. In transmembrane domains 6 and 7 this cleavage exposes two catalytically competent aspartate residues that were hidden before inside the protein. The reaction seems to be catalyzed by a still ill-defined protease called presenilinase and critically depends on the above-mentioned regulatory proteins, in the absence of which presenilin becomes rapidly degraded. The signaling pathways (including protein phosphorylations and ubiquitylations) along which γ-secretase activity is fine-tuned are still not fully understood.

13.1.2 ADAM proteins

ADAM proteins are membrane-bound **proteases**. Two of them, ADAM10 and ADAM17, have been shown to catalyze the S2 cleavage of notch. They represent a larger family of related enzymes (with more than 30 isotypes) found in all animal species but not in plants, yeast, and prokaryotes. All are characterized by a single transmembrane domain and an extracellular disintegrin and metalloprotease (ADAM) domain. The disintegrin domain, a sequence of about 70 amino acids, interacts with integrins of neighboring cells and provides ADAM proteases with the properties of **cell adhesion molecules** (originally, this domain was found in certain snake venoms, which impair blood clotting by blocking integrin interactions required for platelet aggregation). This role of ADAM proteins is best exemplified by the sperm-derived protein fertilin (Figure 13.3 and Section 7.2.3).

Another major function of ADAM proteases is **ectodomain shedding.** This is the proteolytic detachment of peptides and proteins from the extracellular domains of membrane-bound precursor proteins (Figure 13.3). Ectodomain shedding is emerging as an important step in intercellular communication. Typical examples are provided by the proteins of the EGF, ephrin, tumor necrosis factor α (TNFα), and notch ligand families that, upon detachment from membrane-bound precursors, mediate paracrine effects. Moreover, the extracellular domains of many transmembrane receptors are released, as we have seen for notch. While notch is activated this way, other receptors may become inactivated. Moreover, upon detachment their extracellular domain may function as a decoy, intercepting the ligand. Ectodomain shedding is specifically catalyzed by individual ADAM proteases such as ADAM10 and ADAM17 (as "TNFα-converting enzyme", ADAM17 catalyzes, for instance, the release of the cytokine TNFα from a membrane-bound precursor protein).

Like most proteases, ADAM proteins are under strict control to avoid potentially dangerous proteolysis by mistake. To this end, the enzymes are produced as zymogens that are kept silent by an intramolecular interaction of a cysteine

Figure 13.3 Activation and mode of action of ADAM proteases The protease activity of ADAM proteins is stimulated by the proteolytic detachment (P1) of an inhibitory prodomain (shown in color) converting pro-ADAM into ADAM (reaction 1). Upon additional activation by phosphorylation of the intracellular domain, ADAM functions as a "sheddase," releasing a signaling protein (here a TNFα trimer, gray circles) from a membrane-bound precursor (reaction 2). By a proteolytic release (P2) of the metalloprotease domain MpD, certain ADAM isoforms may become cell adhesion molecules, such as the sperm-derived fertilin (an ADAM1–ADAM2 heterodimer, reaction 3) that, upon dimerization, interacts with integrins of neighboring cells (left; see Section 7.2.3). Whether the removal of MpD is a general condition of ADAM to act as a cell adhesion molecule or is restricted to certain subtypes is not known. MpD (gray cylinder), metalloprotease domain with an essential zinc ion; Pro (colored cylinder), inhibitory prodomain); DisD (gray ellipse), disintegrin domain; C, cysteine-rich domain; E, EGF-like repeats; T, transmembrane domain.

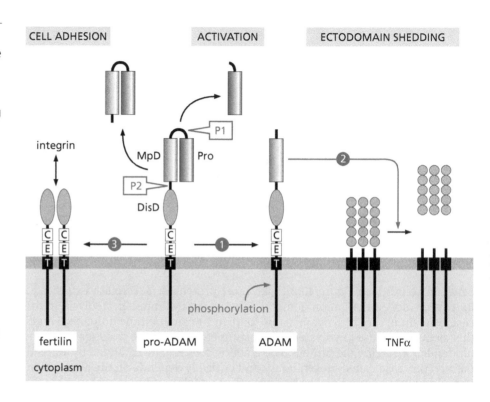

residue in the N-terminal prodomain with an essential zinc ion in the catalytic sequence (known as a cysteine switch). For activation the prodomain is removed by autoproteolysis or by a specific protease or proprotein convertase. In addition, a series of inhibitory proteins known as tissue inhibitors of metalloproteases have been identified. Finally, the cytoplasmic tails of ADAM proteins contain various recognition motifs for intracellular signals that most probably are involved in the regulation of protease and cell adhesion activities. In fact, ADAM-catalyzed ectodomain shedding is induced by a large number of factors that activate protein kinase C, MAP kinase cascades, and other signaling pathways. An important example of this inside-out signaling has been discussed in Section 5.8.3: transactivation of EGF receptors by G-protein-coupled receptors that activate ADAM proteases, releasing EGF from membrane-bound precursor proteins.

Sidebar 13.1 EGF-like and ankyrin repeats These structural elements have been found in many proteins where they mediate homo- and heterotypical protein–protein interactions.

EGF (Epidermal Growth Factor)-like repeats are sequences of around 40 amino acids arranged in series. Each repeat is characterized by a conserved group of six cysteine residues linked by three disulfide bonds. Originally discovered in EGF, such repeats have been found in the extracellular domains of many receptors and cell adhesion molecules as well as in cytokines and other intercellular signaling molecules (for examples see Chapter 6).

Ankyrin repeats are composed of a series of sequences of 33 amino acids originally found in ankyrin, a cytoskeletal protein of the erythrocyte membrane. They rank among the most abundant sequence motifs and are characterized by a folding structure consisting of stacked α-helices linked by β-sheets, which are aligned on one side of the stack (Figure 13.1). These structural elements interact with spatially fitting structures of a partner molecule according to the zipper principle. In other words: in contrast to most other interaction domains, ankyrin repeats recognize spatial arrangements rather than specific amino acid sequences. The interaction of the notch effector protein NICD with the transcription factor CSL, as well as the binding of the transcription factor NFκB to the inhibitor protein IκB (Section 11.8.2) and of the INK4 inhibitor proteins to cyclin-dependent protein kinases (Section 12.5), provide well-known examples of this type of interaction.

13.1.3 Protease-mediated signaling and Alzheimer's disease

γ-Secretase is a key enzyme not only of notch signaling but also of Alzheimer's disease. Due to an ever-increasing lifespan, this age-dependent dementia has become a medical and social problem of top priority, at least in highly developed countries. Research on the relatively rare hereditary or familial form of Alzheimer's disease, comprising about 5% of all Alzheimer cases, has led to the characterization of two genes encoding two closely related proteins called **presenilins**. Overactivation of these proteins due to gene mutation is the primary cause of the familial form of the disease. It is now generally accepted that presenilins represent two isoforms of the catalytic subunit of γ-secretase (see Figure 13.2).

What is the role of γ-secretase in Alzheimer's dementia? In the course of the disease, brain neurons die because insoluble protein aggregates accumulate between the cells. As shown in Figure 13.4, these **β-amyloid plaques** are generated by a two-step proteolytic cleavage of the membrane protein APP (β-Amyloid Precursor Protein; its mutation also may cause the familial form of the disease). In the first step, a transmembrane protease called **β-secretase** removes the N-terminal extracellular domain of APP, while in a second step, the membrane-bound fragment becomes dissected by γ-secretase into two additional fragments. One of these fragments, APP intracellular domain, closely resembles the notch fragment NICD in that it heterodimerizes with two transcription factors, Fe65 and Tip60, thus controlling a series of genes such as that of the transcription factor NFκB, which becomes repressed. However, in contrast to the notch system, a primary signal inducing the proteolytic cleavage of APP (that is, the APP ligand) is not known.

The second fragment produced by γ-secretase, called **β-amyloid peptide** or **APβ**, is released into the extracellular space, where it may aggregate into the dreaded β-amyloid plaques. An especially aggressive variant, APβ42, with 42 instead of 40

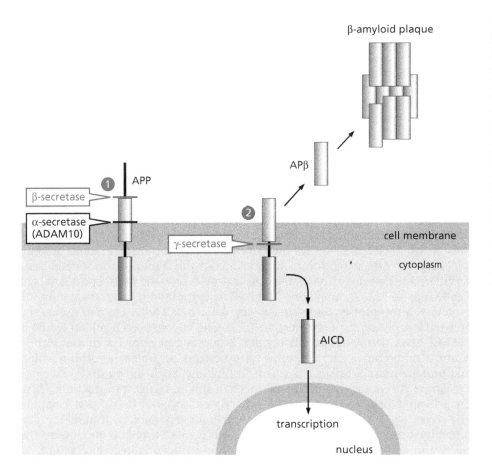

Figure 13.4 Formation of β-amyloid plaques in the course of Alzheimer's disease The transmembrane protein APP (β-amyloid precursor protein) is degraded in two subsequent steps by β- and γ-secretases (reactions 1 and 2). The cytoplasmic fragment AICD (APP intracellular domain) acts as a transcriptional regulator, whereas the extracellular fragment APβ (β-amyloid peptide) may aggregate into β-amyloid plaques, causing neuronal cell death and Alzheimer's dementia. α-Secretase (an ADAM10 protease) cleavage of APP within the APβ -domain (gray cylinder) prevents β-amyloid formation.

amino acids, predominates in the brain of Alzheimer patients; because of a more hydrophobic C-terminus, this variant has a greater tendency to oligomerize.

An important difference from notch is that β-secretase is not an ADAM-type protease. In contrast to β-secretase, ADAM proteases (also called α-secretases) have been shown to cleave APP *within* the APβ sequence and thus preclude β-amyloid formation. In fact, experimental overexpression of ADAM10, the major α-secretase in brain, protects mice from amyloid deposition and neurological disorders resembling Alzheimer's disease. This puts both ADAM10 and β-secretase into the position of key factors in Alzheimer's dementia and in therapeutic strategies. Yet little is known about their regulation and, in particular, about the physiological role of β-secretase.

In the course of Alzheimer's disease, insoluble protein aggregates are deposited both outside and inside the cell. The intracellular deposits consist of bundled microfibrils called degeneration fibrils or neurofibrillar tangles that lead to deformations of the cytoskeleton and, thus, massively impair secretory processes such as neurotransmitter release as well as cell motility. The production of these aggregates is due to a malfunction of one of the many microtubule-associated proteins (MAPs, not to be confused with MAP kinases) that regulate the formation and degradation of the microtubule cytoskeleton. This MAP species, called **Tau protein**, has been found to be hyperphosphorylated and interlinked by disulfide bonds in neurons of Alzheimer patients. One Tau kinase responsible for the pathological phosphorylation is identical with the cyclin-dependent kinase CDK5 (see Section 12.4). Under the conditions of Alzheimer dementia and other neurodegenerative diseases, this enzyme is hyperactive since its regulatory subunits p35 and p39 have become inactivated by partial proteolysis. Another Tau kinase involved in Alzheimer's dementia is glycogen synthase kinase 3 (see Section 9.4.4). Such kinases catalyze proline-directed phosphorylation of serine and threonine residues. Interestingly, the effect of Tau (and also of APP) depends on the conformation of the peptidyl-prolyl bonds adjacent to the phosphorylation sites: only when these bonds are *cis* do pathological protein aggregates form, whereas the *trans* proteins fulfill normal physiological purposes. One reason for this is that phosphorylated *cis*-Tau is resistant to dephosphorylation by phosphatases such as PP2A that are *trans*-specific. As explained in Section 8.5.1, peptidyl-prolyl bonds adjacent to phosphorylation sites undergo *cis–trans* isomerization catalyzed by the peptidyl-prolyl c*is/trans* isomerase PIN1. This enzyme is down-regulated in neurons of Alzheimer patients. In contrast, experimental PIN1 overexpression protects animals from amyloid deposition and brain damage. Such findings not only emphasize the important role of enzymes like PIN1 in cellular signal processing but also indicate new opportunities for diagnosis and therapy.

13.1.4 Other examples of protease-mediated signaling

Notch and APP signaling are representative for other examples of signal transduction by γ-secretase. Always the enzyme cleaves a transmembrane protein, releasing the intracellular domain that functions as a modulator of gene transcription. An especially interesting case is provided by the partial proteolysis of **ErbB4** (Figure 13.5). A member of the EGF receptor family, this transmembrane protein is a receptor tyrosine kinase (also known as HER4, Human Epidermal growth factor Receptor 4; see Section 7.1.1). The endogenous ligand of ErbB4 is the EGF-related growth factor heregulin. Like other receptor Tyr kinases, ErbB4 controls numerous signaling pathways by means of protein phosphorylation and protein–protein interactions. For ErbB4 it has been shown that a receptor Tyr kinase still possesses additional mechanisms of signal transduction. Thus, the activated (liganded and autophosphorylated) receptor becomes a substrate of γ-secretase. Its cytoplasmic Tyr kinase domain generated by intramembrane proteolysis migrates into the nucleus, where it most probably modulates the activity of transcription factors by phosphorylation or direct interaction. In

Sidebar 13.2 Protein misfolding disorders and protein deposition diseases Fibrillary protein deposits resulting from protein misfolding and misassembly and resembling the β-amyloid plaques of Alzheimer's disease are found in the final stages of a variety of diseases that are extremely difficult to treat. Like β-amyloid, the aggregates are insoluble, have a steel-like strength, and are widely resistant to proteolytic degradation. They strongly impair cellular functions, finally leading to cell death. Such amyloids [the name refers to a starch (amylose) -like appearance of the aggregates] can derive from quite different proteins, mostly because of an acquired or genetically predetermined misfolding. Since protein misfolding and misassembly normally are prevented by molecular chaperones; the corresponding diseases are certainly due to severe disturbances of the cellular chaperone system such as the unfolded protein response (Section 9.3.2).

Examples of mutated proteins involved in familial (hereditary) forms of disease are huntingtin, a neuronal protein of unknown function forming aggregates in brain neurons of patients suffering from **Huntington's disease**; superoxide dismutase 1, a redox enzyme found as amyloid in motor neurons of patients afflicted with **amyotrophic lateral sclerosis**; and cystic fibrosis transmembrane conductance regulator (CFTR), a chloride channel found to be aggregated in certain types of **cystic fibrosis**.

Examples of proteins undergoing an acquired misfolding and aggregation are β-amyloid and α-synuclein, a cytoplasmic neuronal protein found to form aggregates (called Lewy bodies) in the degenerating brain areas of patients suffering from **Parkinson's disease**. Moreover, for a century amyloid deposits in the pancreas have been known to be a hallmark of **type 2 diabetes**. They are due to an acquired misfolding of **amylin**, a hormone-like peptide that is co-secreted with insulin, suppressing the emptying of the stomach and producing a feeling of satiety and thirst. As one of the pathogenic factors of diabetes, amylin presently gains considerable interest in basic and applied medical research.

The mechanism of spontaneous aggregation has been studied in detail for the prion protein PrP, providing another example of acquired misassembly. Prions are infectious proteins causing **transmissible spongiform encephalopathies**. The bovine form BSE, better known as mad cow disease, and its human counterpart, **Creutzfeldt–Jacob disease (kuru)**, have become notorious in the past.

The correctly folded cellular protein PrP^C is abundant in the central nervous system and in hematopoietic cells, where it is found to be bound to membranes via a glycosylphosphatidylinositol anchor (see Section 4.3.1). Its physiological function is still widely elusive. Under certain conditions it can change its helical structure into a conformation characterized by stacks of β-sheets. This form, called PrP^{SC} (SC stands for scrapie, the corresponding disease in sheep), has a high tendency to form amyloid aggregates. Moreover, PrP^{SC} catalyzes the conformational change of native PrP^C; it is an infectious agent. The autocatalytic process, resembling crystallization, starts with a seed of a few aggregated PrP^{SC} molecules, which serve as a nucleus, inducing misfolding and aggregation of PrP^C.

In addition to pathological formation, functional amyloid formation is also known. It facilitates the generation of an extracellular matrix in bacterial colonies, mating of yeast gametes, spore formation in fungi, blood clotting and melanin biosynthesis in humans, and protection of eggs of insects and fishes. Functional amyloidogenesis in humans provides a serious obstacle in the development of anti-amyloidogenic drugs.

---

analogy to notch and APP, a shedding of the extracellular domain precedes the intramembrane proteolysis of ErbB4. The protease catalyzing this step is ADAM17. Vice versa, the ligand binding domain of a receptor may become detached and intercept its diffusible ligand, thus preventing the ligand from interacting with its receptor. Along such a pathway, the activity of growth hormone is attenuated, for instance.

Signaling peptides are also released from proteins integrated in *intracellular* membranes. An example is provided by cholesterol metabolism. Upon a shortage of cholesterol, genes encoding key enzymes of **cholesterol biosynthesis** become activated. In their promoters these genes have a sterol response element (SRE) that interacts with a transcription factor SREBP (SRE Binding Protein). An inactive precursor of this factor is a protein with two membrane-spanning domains, which at sufficient cholesterol supply is stored in the endoplasmic reticulum together with another transmembrane protein acting as a cholesterol sensor. A shortage of cholesterol causes these proteins to translocate to the Golgi

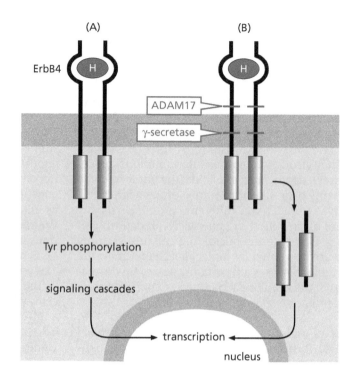

Figure 13.5 Signal transduction by the receptor tyrosine kinase ErbB4/HER4 Upon activation by the ligand heregulin (H), the homodimeric receptor ErbB4 controls gene transcription either (A) along a phosphorylation-dependent pathway or (B) by a proteolytic release of the cytoplasmic domains by ADAM17 and γ-secretase; the released domains then migrate into the nucleus. Colored cylinders symbolize the Tyr kinase domains. For more details, see text.

membranes, where the active SREBP becomes released by juxta- and intramembrane proteolysis catalyzed by the proteases S1P (site-1 protease) and S2P (site-2 protease) and migrates into the nucleus (Figure 13.6). In principle this reaction resembles the activation of notch, with S1P doing the job of ADAM and S2P that of γ-secretase. SREBP activates gene transcription with the consequence that the cholesterol level increases, the process of SREBP activation becomes stopped, and upon phosphorylation by glycogen synthase kinase 3, SREBP (together with key enzymes of cholesterol biosynthesis) becomes degraded along the ubiquitin–proteasome pathway. In other words: cholesterol formation is under strict feedback control. It is easily seen that SREBP signaling is of great clinical interest

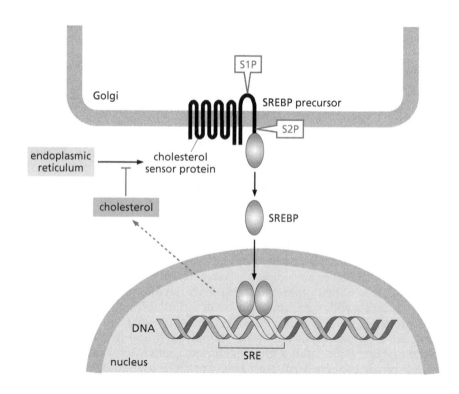

Figure 13.6 Regulation of cholesterol synthesis Transcription of genes containing a sterol-responsive element (SRE) and encoding enzymes of cholesterol synthesis is stimulated by the dimeric transcription factor SREBP. In the presence of cholesterol, SREBP and a cholesterol sensor protein are sequestered in the endoplasmic reticulum. Shortage of cholesterol causes a translocation of the protein complex to the Golgi apparatus, where SREBP is released from the membrane by the site-1 and site-2 proteases S1P and S2P. Cholesterol inhibits the translocation (red pistil).

Sidebar 13.3 Plant hormones regulate gene expression along proteolytic pathways The plant hormones auxin and gibberellic acid are potent stimulators of growth and differentiation. Although they have been known for more than 50 years, their molecular mechanism of action remained obscure until recently. It has now been shown that both hormones activate discrete sets of genes along a quite direct but rather unusual signaling pathway. The key event consists of a proteolytic degradation of transcriptional repressors known as IAA/AUX proteins in the case of auxin and DELLA domain proteins in the case of gibberellic acid. This occurs along the canonical proteasome pathway controlled by SCF-type E3 ubiquitin ligases. As explained in Section 12.16, these enzymes are multiprotein complexes with F-box subunits serving as specific binding sites for the substrate proteins to become ubiquitylated (Figure 13.7). In the SCF ubiquitin ligases of plants involved in hormone effects, the F-box subunits interact not only with the substrate proteins (the transcriptional repressors mentioned above) but also with auxin or gibberellic acid. Moreover, the hormones strongly stimulate substrate ubiquitylation. In other words: these enzymes resemble hormone receptors with an intrinsic ubiquitin E3 ligase activity. The *Arabidopsis* genome seems to encode about 700 F-box proteins, which is about 10 times more than in animal cells. Thus, it is easy to imagine that the role of receptor ubiquitin ligases is not restricted to auxin and gibberellic acid but may be a more general mechanism of signal transduction in plants. Indeed, jasmonic acid, a plant wound hormone, may use such a pathway.

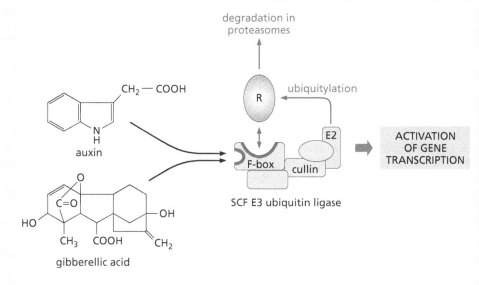

SCF E3 ubiquitin ligase

Figure 13.7 Mechanism of gene activation by the plant hormones auxin and gibberellic acid The hormones stimulate gene expression by triggering proteolytic degradation of transcriptional repressors (R, gray ellipse). To this end they interact with individual F-box subunits of SCF-type E3 ubiquitin ligases (multiprotein complex shown in gray; cullin is an adaptor protein), which specifically promote the ubiquitylation of R catalyzed by the ubiquitin transferase E2 (shown in color). The figure shows a generalized scheme. In reality, each hormone has its own set of E3 ligases and repressors.

since defects of this control mechanism may result in a pathological overproduction of cholesterol.

In quite a similar manner, the **transcription factor ATF6** controlling certain chaperone genes is activated. In this case the trigger is an accumulation of incorrectly or incompletely folded proteins in the endoplasmic reticulum. These proteins become rearranged into their native conformations by their chaperones to be newly synthesized. The signaling pathway, called unfolded protein response, is discussed in detail in Section 9.3.2. Taken together, these examples show that controlled intramembrane proteolysis provides a highly efficient and versatile mechanism of signal transduction, the potencies of which cannot yet be estimated down to the last detail.

Summary

The receptor-dependent release of peptide second messengers seems to be a widespread mechanism of signal transduction. The prototype of this signaling system is represented by notch, a transmembrane receptor the intracellular domain of which is released as a transcriptional activator upon receptor activation by ligands of the DSL family. The reaction is catalyzed by an ADAM protease and the intramembrane protease γ-secretase. Related enzymes play a key role also in the pathogenesis of Alzheimer's dementia, where they catalyze the formation of plaque-forming β-amyloid peptide. Another pathogenic process involved in Alzheimer's disease is the deposition of insoluble aggregates of cytoskeletal proteins due to a dysfunction of Tau protein, a microtubule-associated protein that becomes hyperactive through phosphorylation. γ-Secretase and related proteases also mediate the release of transcription factors such as the intracellular domain of receptor Tyr kinases, sterol-response element binding proteins, and ATF6 inducing the unfolded protein response. In plants, the hormones auxin and gibberellic acid induce the proteolytic degradation of transcriptional repressors by activating specific ubiquitin ligases.

13.2 Apoptosis: a signal-controlled suicide of cells

Cell death is a physiological process where input signals activate proteolysis that results in a complete degradation of proteins rather than in the release of signaling peptide fragments. It plays a central role in the development, maintenance, and repair of tissues. As a rule, the lifespan of somatic cells is much shorter than that of the whole organism. Therefore, a constant tissue mass (as a condition of constant tissue function) is due to a dynamic equilibrium between cell gain and cell loss adjusting the birth rate of cells (amounting to approximately 10^{12} cells per day in adult humans) precisely to the death rate. This equilibrium becomes transiently shifted in the course of developmental and growth processes as well as during tissue repair, with each change being controlled by a complex system of stimulatory and inhibitory signals. As explained in Chapter 12, such control mechanisms are permanently disturbed in tumors.

Cells die in four different ways, necrosis, autophagy, terminal differentiation, and apoptosis, with fluid transitions between these processes. **Necrosis** is understood to be due to nonspecific destruction of tissue cells caused primarily by external influences but also by chronic inflammation and poor blood supply. In contrast to the pathological necrosis resembling tissue damage, autophagy, terminal differentiation, and apoptosis are strictly regulated physiological events. The term **autophagy** describes the self-eating of cells, where cellular components including organelles are digested by lysosomal enzymes. By providing luxury material for vital metabolic processes, it is a survival strategy in stress situations such as starvation. Autophagy also serves the elimination of invading microorganisms. Only in the extreme case does autophagy result in cell death. Cell death by **terminal differentiation** occurs, for instance, in the epidermis, where cells develop into dead horny scales, or in the hematopoietic system, where erythrocytes and thrombocytes undergo maturation to nonreproductive cells of limited lifespan. While terminal differentiation may be understood as a consequence of cell aging, **apoptosis** is a signal-controlled cellular suicide occurring in cells of any developmental stage and age.

Microscopic examination helps to identify the course of cell death. For instance, in contrast to terminal differentiation and apoptosis, necrosis generally is accompanied by inflammatory processes. Apoptotic cells, on the other hand, exhibit characteristic features such as membrane blebbing, shrinking, condensation of the nucleus, fragmentation of chromosomes, and formation of cytoplasmic vacuoles. The cell fragments generated by apoptosis are taken up and eliminated by neighboring cells and phagocytes.

The signal-controlled cellular suicide is of central importance for the maintenance and functionality of tissues. As a kind of a cleansing operation, apoptosis removes potentially dangerous cells, such as virus-infected or cancer cells, as well as cells that have become superfluous for the organism or, for any possible reason, have escaped from their tissue or have lost their function otherwise. Likewise apoptosis is important for tissue formation, since in the course of embryonic organ development normally more cells become produced than are finally needed in the mature tissue. The excess dies by apoptosis.

An instructive example is provided by the nematode *C. elegans,* a favored model of apoptosis research. During its development to the mature animal consisting of 959 somatic cells, exactly 131 cells undergo apoptosis under the control of 10 different *ced* (cell death) genes. Another example is insect metamorphosis, but vertebrate development may also be considered: for instance, the metamorphosis of amphibians or the formation of hand and foot limbs. A particularly high excess of cells is required for the morphogenesis of the nervous system in order to establish synaptic contacts. As soon as this has been accomplished, more than 50% of the embryonic neurons become superfluous and die by apoptosis. Especially stricken by death are those cells that failed to made contact with their target tissue, which is a source of survival factors, the neurotrophins. Such a mechanism prevents incorrect "wiring" of the nervous system.

These studies have shown that apoptosis is triggered either by specific death signals or by a lack of survival signals or trophic factors as well as by related stress situations. In its strongest form, the hypothesis of trophic factors postulates somatic cells to undergo suicide automatically in the absence of survival signals, when they become neglected.

13.2.1 Caspases: the executors of cell death

Apoptosis is due to a destruction of DNA and vital proteins resulting in the decay of the cell. The enzymatic apparatus for the execution of apoptosis is found in a latent form in every cell. Upon demand, it becomes activated by exogenous and endogenous signals that interact either with membrane receptors (death receptors) or mitochondria. The major executors of apoptosis are several proteases, which are interconnected in series to form a proteolytic cascade (Figure 13.8). They are called caspases (<u>c</u>ysteine/<u>asp</u>artate-specific prote<u>ases</u>).

Caspases constitute a family of endopeptidases (13 members in humans) showing a peculiarity: in contrast to most other proteases, they hydrolyze peptide bonds C-terminally to aspartate residues, thus degrading a large number of vital proteins including those responsible for signal processing, cellular structures

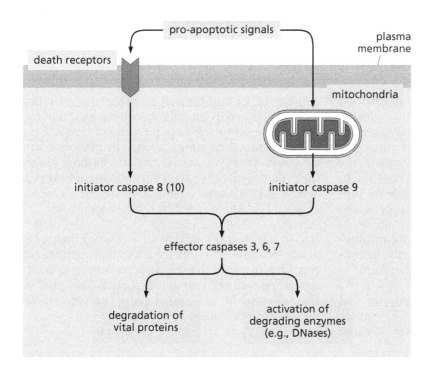

Figure 13.8 Two major pathways leading to apoptosis Apoptotic signals either stimulate membrane-bound death receptors along an extrinsic pathway or damage mitochondria along an intrinsic pathway. Both pathways lead to an activation of caspase cascades. The destruction of vital cellular components, such as proteins and DNA, finally results in cell death.

(for example, the nuclear membrane), DNA repair, and transcriptional regulation. To complete the arsenal of cell degradation, caspases stimulate other hydrolytic enzymes such as the **Caspase-Activated DNase (CAD)** that cuts DNA between the nucleosomes, yielding fragments of $n \times 190$ nucleotides that can be visualized by gel electrophoresis as a characteristic apoptotic ladder. As a potentially highly dangerous enzyme, CAD is always expressed together with an inhibitor protein, **ICAD**, that forms a complex with CAD. Caspases destroy ICAD, thus releasing the active DNase, which is caspase-resistant.

Potentially as dangerous as CAD are the caspases themselves. Therefore, they are subject to strict control, and several safeguards against accidents are installed in the system. In the absence of input signals, all caspases are present as inactive zymogens or procaspases, and in addition, they are blocked by specific caspase inhibitors (see Section 13.2.2).

When apoptosis is triggered, the caspases start to activate each other by partial proteolysis, forming a **caspase cascade** that to some extent resembles the blood-clotting and complement cascades. Along such proteolytic cascades the initial signal becomes amplified enormously.

Like ADAM proteases, procaspases are kept silent by the interaction of the N-terminal prodomain with the catalytic center. For activation, this domain must be removed proteolytically. At the same time, a second proteolytic cut is made in the C-terminal half of the molecule, yielding a small and a large fragment (Figure 13.9). The active caspase is a heterotetramer composed of two small and two large fragments.

Caspase activation follows the mode of trans-autoactivation: each enzyme activates the next one. Such a mechanism requires many enzyme molecules to join and is called, therefore, **proximity-induced autoactivation**. This is an important principle of signal transduction, being valid also for other processes such as phosphorylation cascades. The aggregation of the proteins is facilitated by corresponding interaction domains as well as by adaptor and scaffolding proteins.

Aggregation of caspases is induced by pro-apoptotic signals acting on procaspases 2, 8, 9, and 10 (Figure 13.8). Among the procaspases, only these species have in their prodomains sequences such as DED (Death Effector Domain, in procaspases 8 and 10) or CARD (CAspase-Recruiting Domain, in procaspases 2 and 9) capable of interaction with receptor and sensor proteins of pro-apoptotic

---

Sidebar 13.4 Inflammatory caspases Not all caspases are involved in apoptosis. Therefore, one distinguishes between apoptotic caspases (caspase 3 family), including the isoforms 2, 3, 6, 7, 8, 9, and 10, and inflammatory caspases (caspase 1 family), including the isoforms 1, 4, 5, 11 (not in humans), and 12. A major function of inflammatory caspases is to release pro-inflammatory key mediators such as the cytokines interleukin 1 and 18 from their precursor proteins. In fact, caspase 1 was originally discovered as an interleukin1β-converting enzyme (ICE). Among the inflammatory caspases, it is the best investigated isoform.

Caspase 1 is strongly expressed by monocytes and macrophages and becomes activated by a wide variety of microbes and toxic substances. The key reaction consists of an autoproteolytic processing of procaspase 1 catalyzed by an inflammosome, a protein complex that resembles the apoptosome discussed below. Inflammosomes become activated by microbial structures and other cytotoxic factors such as those released from damaged cells taken up by macrophages (see Section 6.3.2 for more details). Like apoptotic caspases, caspase 1 is held in check by inhibitory proteins.

It should be mentioned that an increasing body of evidence indicates both inflammatory and apoptotic caspases are involved in other cellular processes such as differentiation, cell-fate determination, and proliferation. *Drosophila* has turned out to provide an ideal model for such studies.

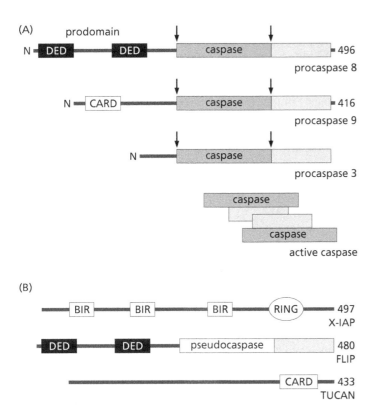

(A)

(B)

Figure 13.9 Domain structure of apoptosis-regulating caspases and their endogenous inhibitors
(A) The initiator or upstream procaspases 8 and 9 as well as the effector or downstream procaspase 3 and an active caspase heterotetramer are shown (colored rectangle, large catalytic subunit; gray rectangle, small catalytic subunit; arrows, sites of proteolysis). The structure of procaspase 2 resembles that of procaspase 9. (B) The inhibitors X-IAP (of caspases 3 and 7), the dominant-negative "mutants" FLIP (of caspases 8 and 10), and TUCAN (of caspase 9) are depicted. For further details, see text. DED, death effector domain; CARD, caspase-recruiting domain; BIR, baculovirus IAP repeat.

signals (Figure 13.9). Called initiator (or upstream) caspases, caspases 2, 8, 9, and 10 are placed, therefore, at the top of the caspase cascade, inducing the activation of the effector caspases 3, 6, and 7 placed downstream.

13.2.2 Caspase inhibitors: how to protect cells from suicide by mistake

A second safeguard against an errant caspase activation consists of several endogenous caspase inhibitors. Effector caspase 9 is competitively inhibited by the protein **TUCAN** (Tumor-Up-regulated CARD-containing Antagonist of caspase Nine). Like caspase 9, TUCAN has a CARD sequence; however, it lacks the caspase domain (Figure 13.9), thus functionally resembling a dominant-negative mutant of caspase 9. According to the same principle, the protein **FLIP** (Fas-ligand-activated ICE-Like Inhibitory Protein) inhibits the initiator caspases 8 and 10 of receptor-induced apoptosis (see Section 13.2.3). FLIP shares with the two caspases the interaction domains but contains an enzymatically inactive pseudocaspase domain (Figure 13.9).

Both initiator and effector caspases are inhibited, in addition, by the proteins of the **IAP family** (Inhibitors of Apoptotic Proteins). Among the eight IAP isoforms known, X-IAP (X-chromosome-linked IAP) is the most efficient one (Figure 13.9). By means of BIR domains (from Baculovirus IAP Repeat; the inhibitor was originally discovered in this insect virus), X-IAP hinders substrate proteins from access to the active center of caspases 3, 7, and 9. X-IAP has, in addition, a RING domain with ubiquitin ligase activity, probably labeling the caspases for degradation in proteasomes (see Section 12.16).

In cells undergoing apoptosis, X-IAP is neutralized by the protein **SMAC** (Second Mitochondrial Activator of Caspases; also called DIABLO, Direct IAP-Binding protein with LOw p*I* value). Quite a similar effect is shown by the IAP inhibitor **OMI/HtrA2** (an Inhibitor released upon permeation of the Outer Mitochondrial membrane; HtrA refers to a homologous high temperature requirement protein A of bacteria). Both inhibitors are stored in mitochondria, becoming released into the cytosol and activated by partial proteolyis upon induction of apoptosis. OMI/HtrA2 is a trypsin-related serine protease itself, which can also undergo autoactivation and induce apoptosis independently of caspases.

In some tumor types the IAP proteins are overexpressed, protecting the tumor cells from apoptosis. Some virus species such as the cowpox virus and the baculovirus of insects also produce caspase inhibitors to hinder apoptosis of the infected host cells. Additional safeguards against accidental apoptosis are installed in death receptor complexes and the mitochondrial membrane.

13.2.3 Receptor-induced apoptosis: the kiss of death

A number of intercellular signal proteins are specialized to induce apoptosis. They are counted among the TNF family of cytokines, acting in both a paracrine and a juxtacrine manner. Their transmembrane receptors constitute a subfamily of TNF receptors (see Section 6.3.1). These death receptors are characterized by death domains (DDs; see Sidebar 13.5) localized in their cytoplasmic portion.

Thus far six such receptors have been identified in mammals. Their prototype is **Fas**, also known as CD95 or Apo-1. Fas is strongly expressed in lymphocytes but is also found in other tissues. Its ligand, **FasL**, is a homotrimeric transmembrane protein expressed in particular by cytotoxic T-lymphocytes and natural killer cells, which in this way kill Fas-positive cells.* This mechanism plays an important role for elimination of virus-infected and tumor cells as well as in the autoregulation of the immune system. Autoimmune diseases are prevented and excessive antigen-specific T cells are destroyed in the subsiding phase of an immune reaction, preserving a small population of FasL-resistant memory cells.

Fas activates the caspase cascade via the adaptor protein **FADD** (Fas-Associated Death Domain protein) recruited by a homotypical interaction between DD motifs (Figure 13.10). FADD is the only protein that contains, in addition to DD, a DED motif (see Sidebar 13.5) that interacts homotypically with the corresponding domain of initiator procaspase 8 or 10, the two caspases that contain DEDs. [Caspase 8 is also known in the older literature as FLICE, Fas-Ligand-activated ICE-like protease, referring to its relationship with ICE (caspase 1). Caspase 10 is restricted to humans, having developed only recently by a doubling of the caspase 8 gene.]

Both procaspases thus become activated and trigger the caspase cascade. For this purpose a rather large **death-inducing signaling complex (DISC)** of many receptor molecules is formed, assembling enough molecules of procaspase 8 or 10 for proximity-induced autoactivation. DISC formation is facilitated by the clustering of membrane lipids in rafts, with ceramides playing an important role in this process. This may explain, at least partially, the pro-apoptotic effect of such lipids, as mentioned in Section 4.4.5. For certain cell types (type I cells) DISC formation is sufficient to induce apoptosis, whereas in other cell types (type II cells) it has to be amplified by the mitochondria-based mechanism. In the latter case, caspase 8 activates the protein Bid by partial proteolysis (see Section 13.2.4).

The DED–DED interaction between FADD and procaspase 8 (or 10) is controlled by inhibitor proteins of the type **FLIP** (also known as I-FLICE). As already mentioned, these proteins resemble procaspase 8 with the enzymatic domain

---

*Cytotoxic T cells and natural killer cells may kill virus-infected cells not only along the Fas pathway but also by injecting the content of secretory granules. A component of the secretion, the protein **perforin**, forms channels in the cell membrane, thus causing a breakdown of the membrane potential and the osmotic pressure. Another component, the protease **granzyme B**, triggers (like an initiator caspase) the caspase cascade; it activates, in addition, the pro-apoptotic protein Bid and destroys the DNase inhibitor ICAD.

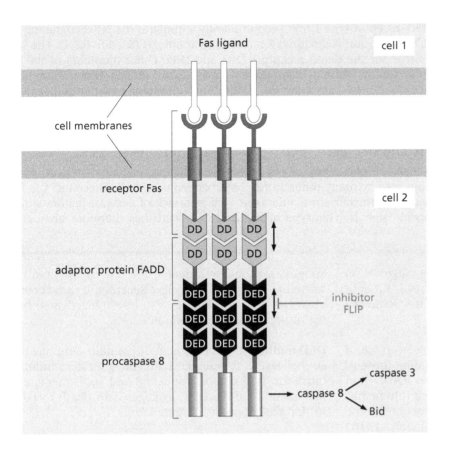

Figure 13.10 Induction of apoptosis by the death receptor Fas/Apo1 The trimeric receptor is activated by interaction with a trimeric ligand exposed at the surface of a neighboring cell and stimulates, via the adaptor protein FADD , procaspase 8 and, thus, the caspase cascade leading to cell death. The direct downstream substrates of caspase 8 are caspase 3 and the mitochondrial "BH3-only" protein Bid. The inhibitor FLIP, a pseudoprocaspase 8, competes with procaspase 8 for the binding sites of FADD. Note the homotypical, noncovalent interactions between the death domains (DD) and the death effector domains (DED).

replaced by an inactive pseudocaspase domain, thus acting as dominant-negative "mutants". By expressing a viral FLIP, certain virus species such as herpesvirus may alleviate the death receptor-induced defense reaction of the organism.

Other death receptors, including those for ligands called TRAILs (TNF-Related Apoptosis-Inducing Ligands), trigger the caspase cascade via the same mechanism as Fas, whereas an additional adaptor protein called TRADD (TNF Receptor-Associated Death Domain) connects FADD with the TNF receptor type 1 (see Section 6.3.1).

As mentioned above, apoptosis plays an important role in the embryonic development of the nervous system. Here cell death is induced by (among others) the **low-affinity p75-nerve growth factor receptor**; however, this receptor stimulates the *de novo* synthesis of pro-apoptotic proteins rather than triggering the caspase cascade directly. Ligands of the receptor are neurotrophins such as the nerve growth factor, a protein that does not belong to the TNF family. How can a growth factor, typically a survival signal, induce cell death? The answer lies in the expression of two different receptor types: in contrast to other TNF receptors, the p75-nerve growth factor receptor is able to establish a heteromeric complex with a receptor Tyr kinase, TrkA (see Section 7.1.1). Only via this **high-affinity nerve growth factor receptor** does nerve growth factor function as a survival signal. Opposing effects have been described also for other death signals and receptors. So, depending on the physiological context, Fas may either trigger apoptosis or co-stimulate mitogenic activation of T cells, thus providing an especially clear example of the principle that the meaning of a signal always depends on the functional state of the receiver cell (for mechanistic details on the mitogenic effect of death receptors, see Section 6.3.1).

For some receptors of the TNF receptor family, the pro-apoptotic effect even recedes into the background when compared with other effects. An example is the TNF receptor type 1 that predominantly stimulates the NFκB signaling cascade, a typical survival pathway, rather than apoptosis (Section 6.3.1). The same holds true for the death receptors DR3 and DR6. Other members of the TNF receptor family contain incomplete or no death domains and are unable to induce apoptosis. They may act similarly to dominant-negative mutants and dampen pro-apoptotic signaling.

13.2.4 Mitochondria-induced apoptosis: the consequences of stress and neglect

Apoptosis is induced not only by endogenous signals reacting with death receptors but also by many other, rather heterogeneous stimuli. These include DNA damage and further stress effects as well as a lack of survival factors such as mitogenic signals. This type of induced cell death has therefore been called

---

Sidebar 13.5 Death domains. Protein aggregation required for the autoactivation of caspases is brought about by homotypical interactions of the sequence motifs DD, DED, CARD, and PYRIN.

DD motif. This Death Domain proper has been found to be encoded in 33 human genes including those of receptors and adaptor proteins such as FADD and TRADD. A major function of the DD motif is to recruit multiprotein complexes at death receptors and some other receptors of the TNF receptor family. **FADD** exclusively transduces apoptotic signals, whereas **TRADD** controls two potentially opposing signaling pathways: the pro-apoptotic cascade, when it interacts with FADD, and the mitogenic NFκB cascade, when it interacts with the adaptor protein RIP2 (see Section 6.3.1). Under normal conditions, the receptors coupled with TRADD activate preferentially the NFκB pathway. Only when the latter is blocked may apoptosis be induced.

Alternatively to FADD, another adaptor protein, **RAIDD** (the acronym indicates that RAIDD associates with the adaptor RIP2 and is homologous to the ICE homology protein 1, which is identical with caspase 2), may interact with TRADD. RAIDD is the only protein that exhibits both DD and CARD motifs. Along this pathway, procaspase 2 becomes activated. A further adaptor protein with a DD motif is **MyD88**, which couples the protein kinases of the interleukin 1 receptor-activated kinase family (also DD proteins) and the associated NFκB module with Toll-like receptors of bacterial surface components (details in Section 6.3.2).

Other protein kinases with DD motifs are the **DAP kinases** (Death-Associated Protein kinases). These enzymes induce apoptosis and, therefore, are tumor suppressor proteins. DAP kinases exist in several isoforms. Some of them are similar to myosin light-chain kinase, becoming activated by Ca^{2+}/calmodulin and phosphorylating myosin light chains. This type of DAP kinase is assumed to be involved in Ca^{2+}-induced

apoptosis, probably inducing the typical membrane blebbing of apoptotic cells. Recently it has been observed that DAP kinases also become activated by the pro-apoptotic lipid ceramide.

DED motif. This Death Execution Domain, a bundle of six helices, is found in 13 human genes, including those for the initiator caspases 8 and 10. It mediates the assembly of the DISC aggregates for death receptor-induced apoptosis (Figure 13.9).

CARD motif. This CAspase-Recruiting Domain is related to the DED motif and has been found in 22 human proteins including caspases 1, 2, 4, 5, 9, 11, and 12. In the course of mitochondria-induced apoptosis (see Section 13.2.4), APAF1 and procaspase 9 aggregate via CARD motifs (Figure 13.11).

For nonspecific immunosurveillance, the **NOD proteins** (Nucleotide-binding Oligomerization Domains) resemble the APAF1 proteins (Section 6.3.2). Their CARD motifs enable them to recruit the CARD protein **RIP2**, thus activating the pro-inflammatory NFκB signaling cascade (Section 11.8). In addition, RIP2 may become activated along a pathway including the inflammatory caspase 1, leading to a release of the pro-inflammatory cytokine interleukin-1β.

PYRIN motif. Close relatives of the CARD motifs are the PYRIN domains. They are restricted to vertebrates (with 19 human isoforms) and are involved in inflammatory processes in particular. The PYRIN domain (derived from the Greek prefix *pyro* for fire) was initially discovered in the pyrin protein that is produced by white blood cells and keeps the inflammatory process under control by a still ill-defined mechanism. A mutation of the corresponding gene is the cause of familial Mediterranean fever, a hereditary disease. Poxviruses paralyse the host's immune system by using PYRIN sequences found in their genome. As a result, defense reactions (inflammation) are suppressed and infected cells cannot be removed by apoptosis.

"apoptosis by neglect" as a counterpart to the receptor-mediated "apoptosis by instruction."

The apoptosome: a molecular killing machine

The primary points of attack of apoptosis by neglect are the mitochondria. In fact, the regulation of programmed cell death is one of their major functions, and for this purpose they are equipped with a complex machinery of interacting proteins. In a way still not fully understood, all kinds of stress factors bring mitochondria to degenerate, rendering the outer membrane permeable for macromolecules. One of the proteins released into the cytoplasm from damaged mitochondria is the respiratory enzyme **cytochrome *c*** that is considered to be the actual intracellular death signal.

Once in the cytoplasm, cytochrome *c* binds to **APAF1** (Apoptotic Proteases-Activating Factor), a large chaperone-like protein of the NOD family (Section 6.3.2). APAF1 contains a CARD motif and a nucleotide-binding oligomerization domain (NOD) and exhibits ATPase activity. The interaction with cytochrome *c* induces a conformational change that triggers the hydrolysis of dATP (or ATP) bound to NOD and enables the protein to oligomerize to a wheel-like structure with seven spokes. This complex is called an **apoptosome** (Figure 13.11). It represents a molecular machine for the activation of caspase 9: via their CARD motifs, several procaspase molecules are homotypically bound to the CARD motifs of the apoptosome to become transformed into active caspase 9 by proximity-induced autoproteolysis. Being an initiator caspase, caspase 9 then triggers the caspase cascade leading to cell death. In other words: apoptosomes play the same role for mitochondria-induced apoptosis as the DISC aggregates for receptor-mediated apoptosis.

Other proteins released from degenerating mitochondria are a DNA-degrading endonuclease G that cooperates with the caspase-activated DNase CAD, the IAP-inhibitors SMAC/DIABLO and OMI/Htr2a already mentioned, as well as an apoptosis-inducing factor (AIF) that triggers apoptosis independently of caspases in a way still not fully understood. AIF is a flavoprotein with NADH oxidase

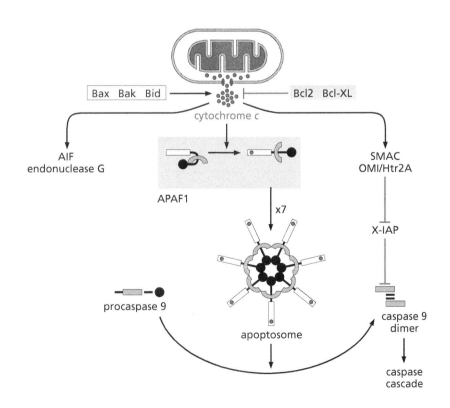

Figure 13.11 Mitochondria-mediated apoptosis Mitochondria permeabilized by the stress-sensitive proteins Bax, Bak, and Bid release several pro-apoptotic proteins. One of them, cytochrome *c* (colored dots), binds to the scaffold protein APAF1, causing a conformational change by which CARD motifs (black dots) and nucleotide-binding oligomerization domains (NODs, gray curved shapes) become exposed. Upon ATP-hydrolysis (not shown) the activated APAF1 oligomerizes to a wheel-like structure, the apoptosome. This "machine" recruits (via its CARD motifs) procaspase 9 molecules, which undergo proximity-induced autoproteolysis to yield active caspase 9 dimers (catalytic domains are colored). Further pro-apoptotic proteins released from mitochondria are SMAC and OMI/Htr2A, which sequester the caspase inhibitor X-IAP, as well as apoptosis-inducing factor (AIF) and endonuclease G. The anti-apoptotic proteins Bcl2 and Bcl-XL (shown in the colored box) protect the mitochondria from permeabilization.

activity and has, like cytochrome *c*, a physiological function in intact mitochondria, particularly in oxidative phosphorylation and detoxification of reactive oxygen species.

Bcl2 proteins: key regulators of mitochondria-mediated apoptosis

The structural integrity of mitochondria and the permeability of their outer membrane is controlled by Bcl2 proteins. This protein family, with about 25 members in humans, has been named after the <u>B</u>-cell <u>l</u>ymphoma gene <u>2</u>, which by a chromosomal translocation may become a hyperactive oncogene protecting the cell from apoptosis. For the first time this oncogenic mutation has shown that cancer can be due not only to uncontrolled cell proliferation but also to a suppression of natural cell death. As expected, the Bcl2 protein encoded by this gene inhibits apoptosis, and the same holds true for the related Bcl-XL protein, which also may mutate to an oncoprotein.

Other proteins of the Bcl family *stimulate* apoptosis. Such death proteins are Bax, Bak, Bad, Bid, Bim, Noxa, and Puma. The only structural element common to pro- and anti-apoptotic Bcl2 proteins is the **BH (<u>B</u>cl2 <u>H</u>omology) domain**, which is expressed in four isoforms and through which the proteins interact with each other (Figure 13.12). In addition, many Bcl2 proteins have a transmembrane domain for anchoring in the outer mitochondrial membrane. Such a membrane interaction enables Bax, Bak, and Bim to form pores large enough for macromolecules such as cytochrome *c* to pass. However, normally permeabilization is prevented by the anti-apoptotic proteins Bcl2 and Bcl-XL, which are also membrane-bound and antagonize the effects of Bax, Bak, and Bim.

Pro-apoptotic signals overcome this control by interacting with Bid, Bad, Bim, Noxa, or Puma. Since these proteins contain only the BH isoform BH3, they are called "**BH3-only proteins.**" They transduce apoptosis-inducing input signals downstream to the "death proteins" Bax and Bak, which are activated, and to the "survival proteins" Bcl2 and Bcl-XL, which are competitively inhibited (Figure 13.13). Moreover, Bcl2 and Bcl-XL are also turned off by active Bax and Bak (and vice versa). Thus, the competition between pro- and anti-apoptotic proteins finally decides between life and death for a cell.

How are Bax and Bak activated by BH3-only proteins? Inactive Bax is found in the cytosol. Interaction with a BH3-only sensor protein causes a conformational change and the release of a Bax inhibitor protein, thus exposing the transmembrane domain that was hidden before inside the Bax molecule. As a result, Bax oligomerizes and forms membrane pores in mitochondria. In contrast to Bax, Bak is a constitutive component of the outer mitochondrial membrane, which upon stimulation by a BH3-only sensor protein oligomerizes directly to membrane pores.

Figure 13.12 Proteins with BH domains BH, B-cell lymphoma gene 2 (Bcl2) homology; gray rectangles, transmembrane domains. For details, see text.

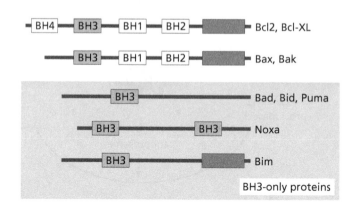

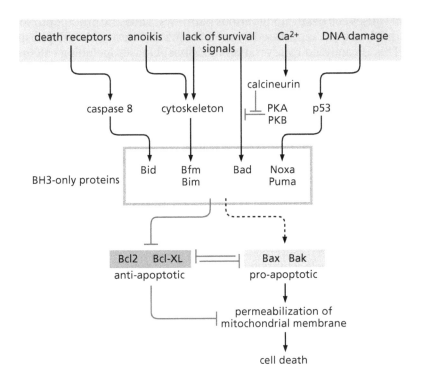

Figure 13.13 Control of apoptosis by proteins of the Bcl2 family
Various stress signals interact directly or indirectly with the sensor proteins of the BH3-only subfamily, which then inhibit the anti-apoptotic Bcl2 proteins (colored box), thus leading to an activation of the pro-apoptotic proteins Bax and Bak. Bax and Bak permeabilize (probably together with the BH3-only protein Bid) the outer mitochondrial membrane, thus triggering apoptosis. Bad is inactivated by phosphorylation catalyzed by protein kinase A or B (PKA, PKB). An antagonist of these kinases is the Ca^{2+}-dependent protein phosphatase calcineurin, which mediates apoptosis by Ca^{2+} stress, for instance, in the course of stroke and other neurodegenerative diseases (Sections 14.6.4 and 16.4.2). The transcription factor p53 becomes active upon DNA damage, stimulating the *de novo* synthesis of Noxa and Puma (as well as of Bax). For additional details, see text.

BH3-only proteins receive their input signals along various pathways that are activated in stress situations. As we have seen, Bid is stimulated by death receptors via caspase 8-catalyzed partial proteolysis. The sensor proteins Noxa and Puma are up-regulated at the level of gene transcription: their genes are turned on by the transcription factor p53, which becomes activated upon DNA damage (Section 12.9.3). Changes of the microtubule cytoskeleton cause a release of active Bim, while a more severe breakdown of the cytoskeleton activates the pro-apoptotic sensor protein, Bfm. Such a situation, called **anoikis** (from the Greek word for "homeless"), occurs when the cell's contact with the extracellular matrix and with neighbor cells is interrupted and, consequently, the cell is in danger of release from the tissue. This protective mechanism against tissue disintegration is repealed in infiltrating and metastasizing tumors.

An important mediator of extracellular signals is Bad. Bad is inhibited by phosphorylation catalyzed by **protein kinase B/Akt (PKB)** as well as by protein kinase A and MAP kinases. PKB prevents apoptosis in still another way: it phosphorylates and inhibits the Forkhead/FOXO transcription factors that promote the transcription of the Bim gene. When these anti-apoptotic effects of PKB are considered, it is not surprising that hyperactive PKB functions as an oncoprotein, Akt, preventing the elimination of tumor cells by Bad- and Bim-mediated apoptosis (see also Section 4.6.6). PKB and the other kinases mentioned above are normally stimulated by a wide variety of growth factors, cytokines, and hormones. Upon a lack of these survival factors, the kinase activity ceases and consequently Bad becomes dephosphorylated, now stimulating apoptosis by neglect.

Ca^{2+}-induced apoptosis

Calcium, the most abundant and most versatile second messenger, may promote either cell survival or cell death. Whether or not it kills a cell depends on the dose and duration of the Ca^{2+} signal: low doses and short pulses facilitate cell survival whereas any signal inducing a strong or long-lasting increase of the cytoplasmic Ca^{2+} concentration may trigger apoptosis. The deadly effect of Ca^{2+} ions is assumed to play an important role in stroke, where brain neurons are flooded by Ca^{2+} due to an overactivation of cation channels in the plasma membrane (details in Section 16.4.2).

Calcium causes cell death along various pathways. Thus, the pro-apoptotic effect of Bad that had been suppressed by protein kinases is restored by dephosphorylation. This is catalyzed by several protein phosphatases including **calcineurin**, which contains the calcium-binding subunit calmodulin and is activated by Ca^{2+} (Section 14.6.4).

However, calcineurin does more: it also activates the NFAT transcription factors (see Section 14.6.4 for details), which in turn stimulate the transcription of death receptor genes. In addition to calcineurin, other Ca^{2+}-dependent enzymes, including proteases such as calpain, endonucleases and enzymes producing stress signals such as reactive oxygen species and NO, become activated along signaling pathways that lead to an increase of the cytoplasmic calcium concentration. These all may contribute to the pro-apoptotic effect.

Finally, a group of Ca^{2+}/calmodulin-dependent Ser/Thr-specific protein kinases known as **DAP kinases** (Death-Associated Protein kinases) should be mentioned (see also Sidebar 13.5). Their activation by various apoptotic signals, including the pro-apoptotic lipid ceramide, evokes characteristic morphological changes of dying cells such as a loss of cell adhesion, membrane blebbing, and the formation of autophagosomes. DAP kinases are closely related to myosin light-chain kinases, and most of their functions may be explained by effects on the cytoskeleton. DAP kinases are tumor suppressor proteins the deletion of which may promote neoplastic growth.

The expression of DAP kinases is controlled by **p53**. As discussed in detail in Section 12.9.3, this transcription factor controls a wide variety of genes that encode (in addition to DAP kinases) proteins of cell cycle regulation, DNA repair, cell differentiation, and senescence. p53 also stimulates the expression of other pro-apoptotic proteins such as Bax, Noxa, Puma, APF-1, and death receptors. Appropriately, p53 is a tumor suppressor protein because its loss of function protects tumor cells from dying by apoptosis.

Cell death by oxidative stress

The transcription of genes relevant for apoptosis is controlled by still other signals including those transduced by the stress modules of the MAP kinase complexes. These modules may be stimulated by the MAP3 kinase **ASK1** (Apoptotic Signal-regulated Kinase 1; see Section 11.3) with the input signals transmitted by receptors of the TNF receptor family. Moreover, ASK1 also responds to oxidative stress, since in its inactive form it is complexed with the redox sensor thioredoxin (details on redox signaling are found in Section 2.2). Reactive oxygen species act as redox signals by oxidizing the dithiol form of thioredoxin to the disulfide form, thus causing the complex to disintegrate and release active ASK1 (Figure 13.14).

Figure 13.14 Cell death by oxidative stress: role of the oxygen-sensitive protein kinase ASK1 In its inactive form, the homodimeric apoptotic signal-regulated kinase 1 (ASK1) exists as a complex with reduced thioredoxin (TRX). Reactive oxygen species (ROS) oxidize TRX promoting the dissociation of ASK1, which activates itself by trans-autophosphorylation (P). As a MAP3 kinase, ASK1 activates stress-sensitive MAP kinase modules, thus leading to expression of pro-apoptotic genes. MKK, MAP2 kinases; JNK, c-Jun N-terminal kinase.

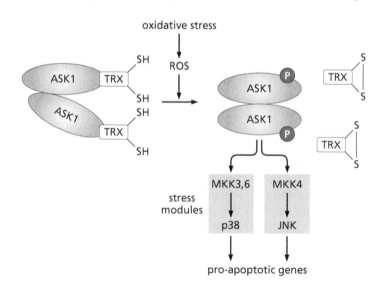

Another target of oxidative stress is the NFκB module that processes survival signals by stimulating the transcription of genes for the anti-apoptotic proteins Bcl2, Bcl-XL, IAP, and FLIP. The transcription factor NFκB becomes inactivated by reactive oxygen species due to an oxidation of essential cysteinyl-thiol groups to intramolecular disulfide groups. The oxidized factor is reduced and reactivated by reduced thioredoxin.

Summary

Programmed cell death (apoptosis) is a signal-controlled physiological process. The death signals are transduced by either membrane-bound death receptors or mitochondrial proteins. Apoptosis is executed primarily by a series of cysteine/aspartate-specific proteases called caspases that kill cells by destroying vital proteins and activating additional macromolecule-degrading enzymes such as caspase-activated DNase (CAD). Caspases are strictly controlled. In the absence of death signals they exist as inactive procaspases. Activation occurs through controlled proteolysis, progressing in a cascade of interconnected caspases. It starts with an autoactivation of initiator procaspases 2, 8, 9, and 10, which possess domains that enable specific interaction with death signals. The initiator caspases then activate the effector caspases 3, 6, and 7. Apart from autoinhibition, caspases are controlled by specific inhibitor proteins. In cells undergoing apoptosis, the activities of these inhibitors are neutralized by regulatory proteins released upon death signaling. Endogenous signals, inducing apoptosis, are members of the tumor necrosis factor (TNF) family of cytokines. The corresponding trimeric transmembrane receptors have cytoplasmic death domains enabling a homotypic interaction with the Fas-associated death domain (FADD) adaptor protein that, in turn, activates initiator procaspase 8 (by a homotypic interaction with death effector domains). Alternatively, death receptors may stimulate survival pathways of signaling by interacting with another adaptor protein, TNF receptor-associated death domain protein (TRADD). Apart from death receptors, mitochondria are the major targets of apoptosis-inducing signals. Various stress factors including a lack of survival signals (such as growth factors) cause damage to the outer mitochondrial membrane. As a result pro-apoptotic factors become released, primarily cytochrome c. In the cytoplasm, cytochrome c induces the aggregation of multiprotein complexes, called apoptosomes, that facilitate the activation of initiator procaspase 9. The structural integrity of the mitochondrial membrane (and thus the release of pro-apoptotic factors) is under the control of the proteins of the Bcl2 family that may either stimulate or inhibit apoptosis. Their activities are regulated by a wide variety of extra- and intracellular signals (mostly stress factors), which are transduced by BH3-only proteins, a subfamily of Bcl2.

Further reading

Adrain C, Brumatti G & Martin SJ (2006) Apoptosomes: protease activation platforms to die from. *Trends Biochem. Sci.* 31, 243–247.

Apoptosis (2003) *Immunol. Rev.* 193.

Barnhart BC, Lee JC, Alappat EC et al. (2003) The death effector domain protein family. *Oncogene* 22, 8634–8644.

Bengoechea-Alonso MT & Ericsson J (2007) SREBP in signal transduction: cholesterol metabolism and beyond. *Curr. Opin. Cell Biol.* 19, 215–222.

Bialik S & Kimchi A (2006) The death-associated protein kinases: structure, function, and beyond. *Annu. Rev. Biochem.* 75, 189–210.

Blobel CP (2005) ADAMs: key components in EGFR signalling and development. *Nat. Rev. Mol. Cell Biol.* 6, 32–42.

Boatright KM & Salvesen GS (2003) Caspase activation. *Biochem. Soc. Symp.* 70, 233–242.

Bray SJ (2006) Notch signalling: a simple pathway becomes complex. *Nat. Rev. Mol. Cell Biol.* 7, 678–689.

Breckenridge DG & Xue D (2004) Regulation of mitochondrial membrane permeabilization by BCL2 family proteins and caspases. *Curr. Opin. Cell Biol.* 16, 647–652.

Burgering BMT & Klops GJPL (2002) Cell cycle and death control: long live forkheads. *Trends Biochem. Sci.* 27, 352–359.

Chiti F & Dobson CM (2006) Protein misfolding, functional amyloid, and human disease. *Annu. Rev. Biochem.* 75, 333–366.

Chow B & McCourt P (2007) Plant hormone receptors: perception is everything. *Genes Dev.* 20, 1998–2008.

Cory S, Huang DCS & Adams JM (2003) The Bcl2 family: roles in cell survival and oncogenesis. *Oncogene* 22, 8590–8607.

Danial NN & Korsmeyer SJ (2004) Cell death: critical control points. *Cell* 116, 205–219.

Fortini ME (2002) γ-Secretase-mediated proteolysis in cell-surface-receptor signaling. *Nat. Rev. Mol. Cell Biol.* 3, 673–683.

Garrido C & Kroemer G (2004) Life's smile, death's grin: vital functions of apoptosis-executing proteins. *Curr. Opin. Cell Biol.* 16, 639–646.

Gelinas C & White E (2005) BH3-only proteins in control: specificity regulates MCL-1 and BAK-mediated apoptosis. *Genes Dev.* 19, 1263–1268.

Golde TE & Eckman CB (2003) Physiological and pathological events mediated by intramembranous and juxtamembranous proteolysis. *Sci. STKE* 2003, 1–14.

Haass C & Selkoe DJ (2007) Soluble protein oligomers in neurodegeneration: lessons from the Alzheimer's amyloid β-peptide. *Nat. Rev. Mol. Cell Biol.* 8, 101–112.

Haass C & Steiner H (2002) Alzheimer disease γ-secretase: a complex story of GxGD-type presenilin proteases. *Trends Cell Biol.* 12, 556–562.

Hong SJ, Dawson TM & Dawson VL (2002) Nuclear and mitochondrial conversations in cell death: PARP-1 and AIF signaling. *Trends Pharmacol. Sci.* 25, 159–265.

Huovila APJ, Turner AJ, Pelto-Huikko M et al. (2005) Shedding light on ADAM metalloproteinases. *Trends Biochem. Sci.* 30, 413–422.

Igney FH & Krammer PH (2002) Death and anti-death: tumor resistance to apoptosis. *Nat. Rev. Cancer* 2, 277–288.

Iwatsubo T (2004) The γ-secretase complex: machinery for intramembrane proteolysis. *Curr. Opin. Neurobiol.* 14, 379–383.

Jiang X & Wang X (2004) Cytochrome *c*-mediated apoptosis. *Annu. Rev. Biochem.* 73, 87–106.

Kaufmann SH & Hengartner MO (2001) Programmed cell death: alive and well in the new millennium. *Trends Cell Biol.* 11, 526–534.

Krammer PH, Kaminsky M, Kiessling M et al. (2007) No life without death. *Adv. Cancer Res.* 97, 110–132.

Kuranaga E & Miura M (2006) Nonapoptotic functions of caspases: caspases as regulatory molecules for immunity and cell-fate determination. *Trends Cell Biol.* 17, 135–144.

Lord SJ, Rajotte RV, Korbutt ES et al. (2003) Granzyme B: a natural born killer. *Immunol. Rev.* 193, 31–38.

Modjtahedi N. Giodarnetto F, Madeo F et al. (2006) Apoptosis-inducing factor: vital and lethal. *Trends Cell Biol.* 16, 264–272.

Nadiri A, Wolinski MK & Saleh M (2006) The inflammatory caspases: key players in the host response to pathogenic invasion and sepsis. *J. Immunol.* 177, 4239–4245.

Nunan J & Small DH (2002) Proteolytic processing of the amyloid-β protein precursor of Alzheimer's disease. *Essays Biochem.* 38, 37–49.

Orrenius S, Zhivozovsky B & Nicotera P (2003) Regulation of cell death: the calcium–apoptosis link. *Nat. Rev. Mol. Cell Biol.* 4, 552–564.

Rathmell LC & Thompson CB (2002) Pathways of apoptosis in lymphocyte development, homeostasis, and disease. *Cell* 109, S97–S107.

Reed J (ed) (2003) Apoptosis. *Oncogene* 22 (53), Review Issue 7.

Reed JC, Doctor KS & Godzik A (2004) The domains of apoptosis: a genomics perspective. *Sci. STKE* 2004, re9.

Riedl SJ & Shi Y (2004) Molecular mechanisms of caspase regulation during apoptosis. *Nat. Rev. Mol. Cell Biol.* 5, 897–907.

Salvesen GS (2002) Caspases and apoptosis. *Essays Biochem.* 38, 9–19.

Seals DF & Courtneidge SA (2003) The ADAMs family of metalloproteases: multidomain proteins with multiple functions. *Genes Dev.* 17, 7–30.

Selkoe D & Kopan R (2003) Notch and presenilin: regulated intramembrane proteolysis links development and degeneration. *Annu. Rev. Neurosci.* 26, 565–597.

Shiozaki EN & Shi Y (2004) Caspases. IAPs and Smac/DIABLO: mechanisms from structural biology. *Trends Biochem. Sci.* 39, 486–494.

Stennicke HR, Ryan CA & Salvesen GS (2002) Reprieval from execution: the molecular basis of caspase inhibition. *Trends Biochem. Sci.* 27, 94–101.

Thorburn A (2004) Death receptor-induced cell killing. *Cell. Signal.* 16, 139–144.

Vaux DL & Silke J (2005) IAPs, RINGs and ubiquitylation. *Nat. Rev. Mol. Cell Biol.* 6, 287–297.

Wajant H (2002) The Fas signaling pathway: more than a paradigm. *Science* 296, 1635–1636.

Wang S & El-Deiry WS (2003) TRAIL and apoptosis induction by TNF-family death receptors. *Oncogene* 22, 8628–8633.

Weihofen A & Martoglio B (2003) Intramembrane-cleaving proteases: controlled liberation of proteins and bioactive peptides. *Trends Cell Biol.* 13, 71–78.

Weng AP & Aster JC (2004) Multiple niches for Notch in cancer: context is everything. *Curr. Opin. Genet. Dev.* 14, 48–54.

White JM (2003) ADAMs: modulators of cell–cell and cell–matrix interactions. *Curr. Opin. Cell Biol.* 15, 598–606.

Zhivotovsky B (2003) Caspases: enzymes of death. *Essays Biochem.* 39, 25–40.

Signal Transduction by Ions

In the previous chapters, mechanisms of signal processing have been discussed that derive their energy from biochemical reactions such as the hydrolysis of energy-rich bonds or redox processes. In parallel, all cell types, from the simplest prokaryotes to the most complex brain neurons, use the electrical energy stored in the membrane potential (like in a charged battery) for data processing and information transfer. The membrane potential is due to a disequlibrium of ions generated and maintained by ATP-consuming ion pumps. To significantly change the potential it is sufficient to transport only a minute proportion of ions across the membrane (Section 2.10.1). Therefore, most signals using the membrane potential as transducer interact with ion channels rather than with ion pumps. In fact, ion channels are controlled by a wealth of extra- and intracellular stimuli, and are able to regulate almost any cellular function including gene expression. This is achieved by connecting the influx and efflux of ions with the standard pathways of biochemical signal transduction.

Ion channels are transmembrane proteins. They are composed of subunits with several transmembrane domains, mostly α-helices, that form a central pore for ion flux. Like most signal-transducing proteins, ion channels exist in at least two conformations, an active one (the channel is open) and an inactive one (the channel is closed). These conformations are in an allosteric equilibrium that provides a signal-operated switch: channel-gating signals stabilize the open form, while channel-closing signals stabilize the closed form. These signals may consist of electricity, controlling voltage-operated ion channels; molecules, interacting with ligand-operated ion channels; and mechanical pressure, acting on mechanosensitive ion channels. Moreover, temperature-sensitive ion channels are known (see Chapter 15). The ions transported by such channels have the following functions:

- H^+: control of pH value

- Na^+: change of membrane potential (depolarization)

- K^+: change of membrane potential (hyperpolarization)

- Cl^-: change of membrane potential (de- or hyperpolarization)

- Ca^{2+}: second messenger, sometimes also change of membrane potential

In addition, monovalent cations and anions determine the osmotic pressure inside the cell. The families of anion and cation channels are strictly separated.

Basic information on the membrane potential and ion channels is found in Section 2.10. Here and in Chapters 15 and 16 we will discuss the structural and chemical features of individual ion channel families and their role in cellular signal processing. A detailed discussion of (electro) physiological phenomena goes beyond the scope of this book; for that the reader is referred to textbooks of neurosciences and cell biology.

14.1 Cation channels: prototypical structures and gating mechanisms

A typical cation channel is composed of at least four α-subunits that associate with regulatory subunits labelled by additional Greek letters (β, γ, δ, etc.). Each α-subunit contains up to six transmembrane helices or segments and a P-loop that dips into the membrane surface. The pore consists of a selectivity filter determining the ion specificity of the channel, a water-filled central cavity for ion enrichment, and a gate that opens or closes in a reversible manner upon signal reception (Figure 14.1). While the gate and central cavity are lined by transmembrane segments, the selectivity filter is formed by the P-loops (see Section 2.10.3 for more details on ion selectivity).

One distinguishes between two prototypes of cation channels that have not changed very much from prokaryotes to humans. They differ in the architecture of the subunits (Figure 14.2).

(1) Prototype 1: Each subunit contains six transmembrane segments, numbered S1–S6, and one P-loop between S5 and S6. This architecture is characteristic for prokaryotic and eukaryotic voltage-gated cation channels, cyclic nucleotide-gated (CNG) cation channels, transient receptor potential (TRP) channels, and the ryanodine and inositol 1,4,5-trisphosphate (InsP$_3$) receptor calcium channels of the endoplasmic reticulum (ER).

(2) Prototype 2: Each subunit contains only two transmembrane segments, S5 and S6, and one P-loop in between. This simple and probably more ancient architecture is characteristic for the epithelial Na$^+$ channel (ENaC) family and for inwardly rectifying potassium channels (K$_{ir}$). Immediate descendants are the two-P-domain potassium channels (K$_{2P}$) where two of the subunits are covalently linked.

Figure 14.1 Principles of cation channel structure Cation channels are transmembrane proteins composed of at least four subunits that constitute a central pore. The pore is lined by transmembrane helices and contains a selectivity filter that determines ion specificity, a central cavity, and a gate that reversibly opens or closes in a signal-controlled manner. (A) Molecular model of a channel pore based on structure analysis of the bacterial potassium channel KcsA. Shown are two of the four subunits. Each subunit consists of two tilted transmembrane helices and one P-loop. The P-loops form the selectivity filter (red) that strips the ions (black balls) from water molecules (red dots). In the water-filled central pore, the ions become rehydrated. The gate is formed by the inner helices. (B) Space-filling model of a cation channel in closed and open states. Three of the four subunits of a prototypical channel are shown; the positions of the individual pore components may differ from channel type to channel type. Moreover, both the extracellular and intracellular domains may contain ligand binding sites. (A, adapted from fig.cox.miami.edu/.../150/memb/ion channels/htm. B, adapted with modifications from M. Schumacher and J. P. Adelman, *Nature* 417: 501–502, 2002.)

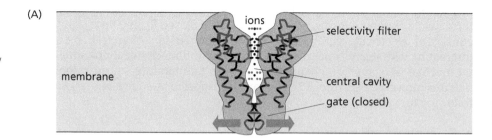

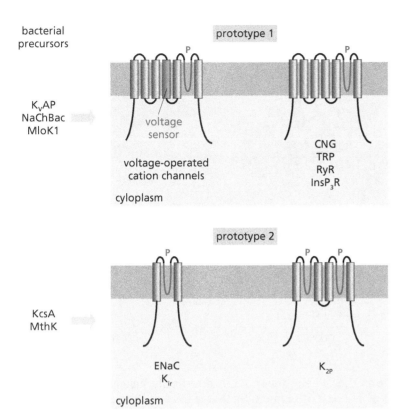

bacterial precursors

$K_V AP$
NaChBac
MloK1

prototype 1

voltage sensor

voltage-operated cation channels

cyloplasm

CNG
TRP
RyR
$InsP_3R$

KcsA
MthK

prototype 2

ENaC
K_{ir}

K_{2P}

cyloplasm

Figure 14.2 Prototypes of cation channels Sodium, potassium, and calcium channels have two prototypes. These are characterized by the structure of the subunits, shown here in schematic form. Nonspecific cation channels coupled with neurotransmitter receptors are distantly related (not shown). CNG, cyclic nucleotide-gated channel; ENaC, epithelial Na^+ channel; $InsP_3R$, inositol 3,4,5-trisphosphate receptor; KcsA, K^+ channel of *Streptomyces lividans;* K_{ir}, inwardly rectifying K^+ channel; K_{2P}, 2P-domain K^+ channel; $K_V AP$, voltage-gated K^+ channel of *Aeropyrum pernix*; MloK1, cyclic nucleotide-gated K^+ channel of *Mesorhizobium loti*; MthK, Ca^{2+}-controlled K^+ channel of *Methanobacterium thermoautotrophicum*; NaChBac, voltage-gated Na^+ channel of *Bacillus halodurans*; RyR, ryanodine receptor; TRP, transient receptor potential channel. The cylinders represent transmembrane domains.

Most of these ion channels are rather selective, transporting either sodium, potassium, or calcium ions. Other, more unspecific cation channels are distantly related to these prototypes or to other prokaryotic precursors. They are coupled with neurotransmitter receptors and are discussed in Chapter 16. As shown in Section 14.7, anion channels may exhibit a completely different architecture.

Voltage-gated ion channels open upon membrane depolarization. They represent the majority among the nearly 500 different ion channel proteins of humans. From prokaryotes to humans, all voltage-gated ion channels are thought to operate according to the same principle based on the characteristic molecular architecture resembling prototype 1 (Figure 14.3). In the most ancient species, the voltage-gated potassium channels, the four subunits are separated, whereas in the evolutionarily more modern sodium and calcium channels the four subunits share one large polypeptide chain. Voltage-gated ion channels are ultra-high-speed switches that easily put into shadow technical devices such as transistors. Therefore, they are predestined for rapid signal propagation in the nervous system. To this end they have a structural element that, upon depolarization, changes its position or conformation in such a way that the channel gate opens. The core of this structure is transmembrane segment S4: it carries a strong positive charge due to a clustering of arginine residues and is lipophilic enough to move within the lipid bilayer of the membrane. Mutation analyses have confirmed the concept of S4 acting as voltage sensor. The precise mechanism of voltage-dependent gating remained, nevertheless, a mystery for half a century. Recently this situation has changed dramatically, particularly since it became possible to crystallize a series of (mostly prokaryotic) ion channels enabling X-ray analysis and other studies of protein structure. The results thus obtained seem to have superseded or at least relativized older hypotheses on channel gating. Such a putatively outdated theoretical concept, still found in most textbooks, is the sliding helix model. It implies that, upon depolarization, helix S4 makes a screw turn in the direction of the cell surface, resulting in an iris-like opening of the channel pore. Detailed studies on the voltage-gated potassium channel $K_V AP$ of the archaebacterium *Aeropyrum pernix* indicate that the truth might be much simpler. They have shown that, in contrast to previous

Figure 14.3 Gating mechanism of a voltage-dependent ion channel
(A, B) Molecular model of the archaebacterial potassium channel KvAP. (A) View from above, showing the four subunits with six transmembrane helices each (numbered S1–S6) and the central pore occupied by an ion (red dot). (B) Vertical section, showing two of the four subunits and a closed ion gate. Note the perpendicularly bent hairpin loops consisting of the voltage-sensitive S3b and S4 segments (paddles). (C) Model of channel gating. Upon depolarization, the positively charged paddles turn upward, opening the ion gate. (A and B, adapted from www.aups.org.au/Proceedings/37/15–27/.)

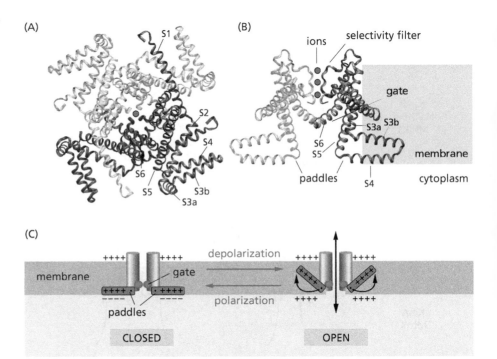

assumptions, the S4 helix does not move vertically. Instead, it forms a hairpinlike structure with a part of segment S3 that is fixed at the basis of the channel protein like a paddle at its axis of rotation. A change of the membrane potential lets the paddle swing within the membrane's interior: in the hyperpolarized state, its position is almost perpendicular to the channel closing the gate, while upon depolarization it swings upward, opening the gate (Figure 14.3). Certainly, many variations of this basic theme may exist. Thus, in voltage-gated potassium channels of animals, the S4 helix seems to undergo both tilting and sliding.

Voltage-dependent ion channels interact with a wide variety of other signal-transducing components that adjust their effectiveness to the given physiological context. Thus, their activities are fine-tuned by protein phosphorylation. To this end the cytoplasmic domains of the α-subunits harbor several phosphorylation sites. An example for the consequences of ion channel phosphorylation is found in Section 14.5.2.

In contrast to the voltage-gated channels, most of the other channels of prototype 1 are controlled by intra- or extracellular ligands rather than by the membrane potential (some channels also by both). The reason for voltage insensitivity is an insufficient positive charge of the S4 segment. The ligands interact with extended intra- or extracellular structures of the receptors (for an example, see Section 14.4.1).

Channels of prototype 2 do not contain an S3–S4 paddle or a comparable voltage sensor. Nevertheless, some of them may respond to voltage but in a reversed manner: they are closed upon depolarization. To this end the channel is plugged by positively charged polyamines and Mg^{2+} ions, a mechanism that differs entirely from that of voltage-gated channels (see Section 14.4.2). Other channels of prototype 2 either are consistently open or are controlled by ligands.

For their pioneering work on the structure and function of transmembrane channels Roderick MacKinnon and Peter Agre were awarded the Nobel Price in Chemistry in 2003.

Summary

Cation channels are composed of at least four subunits forming a central pore. Two basic subunit structures are distinguished that can be traced back to prokaryotic precursors. The simplest structure consists of two transmembrane segments (helices) lining the pore and a P-loop forming the selectivity filter. Voltage-gated channels contain four additional segments. Here a hairpin structure of segment 3 with the positively charged segment 4 (paddle) acts as a voltage sensor that opens or closes the channel depending on the membrane potential. Related channels with an insufficiently charged segment 4 are not gated by depolarization. A special channel type is closed upon depolarization due to a plugging of the pore by positively charged polyamines and Mg^{2+} ions.

14.2 Voltage-gated Na⁺ channels: masters of the action potential

Sodium ions have two major functions: they control the membrane potential and they determine the osmotic pressure. Accordingly, Na⁺-selective channels exist in two basic forms for these different purposes: voltage-gated channels for changing the membrane potential and voltage-independent channels for regulating electrolyte transport. As mentioned before, in the voltage-dependent sodium channel the four groups of transmembrane helices characteristic for prototype 1 channels share together as subdomains one polypeptide chain, a 260 kDa protein (Figure 14.4). The structure of this α-subunit, a rather late development of evolution found only in multicellular organisms, has been investigated in detail. With four influx and four efflux openings, it is surprisingly complex (Figure 14.5). The pore is formed by the hydrophobic S6 transmembrane helices and, at a central bottleneck, by the selectivity filter of the P-loops, while the S4 voltage sensors mark the boundaries of the influx and efflux openings.

Although the experimental expression of the pore-forming α-subunits is sufficient to generate a functional channel, in most voltage-dependent channels the α-subunits are surrounded by several regulatory subunits (Figure 14.4). Thus, the α-subunit of the voltage-dependent Na⁺ channel forms a complex with three different β-subunits, β₁, β₂, and β₃. These are single-pass transmembrane proteins that exhibit structural features of cell adhesion molecules of the immunoglobulin (Ig)-like superfamily (Section 7.3.1). As such, they probably control the interaction of the channel with the extracellular matrix, adjacent cells (such as glial

Figure 14.4 Voltage-controlled axonal Na⁺ channel (A) Transmembrane topography of the α-subunit with the voltage sensors S4 (red cylinders) and the P-loops (shown in red). The inactivation domain localized in an intracellular loop between transmembrane subdomains III and IV blocks the gated channel, rendering it transiently refractory (see also Figure 14.11). (B) Arrangement of the transmembrane helices and P-loops (P) seen from above. (C) Model of the subunit architecture of the complete channel, showing the four subdomains of the α-subunit (red) surrounded by the regulatory transmembrane subunits β₁, β₂, and β₃. (B, adapted from S.F. Stevens *Nature* 349:657–658, 1991.)

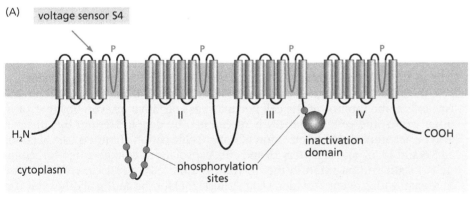

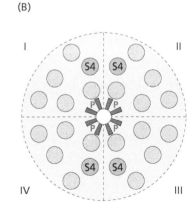

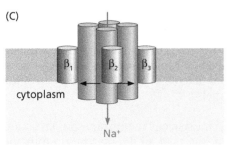

Figure 14.5 Architecture of the axonal Na⁺ channel (α-subunit)
The central pore of the channel has four influx and four efflux openings lined by the S4 voltage sensors, the S6 transmembrane domains, and the P-loops. The model is based on electron microscopy, X-ray diffraction, and nuclear magnetic resonance spectroscopy. (From W.A. Catterall *Nature* 409:988–990, 2001.)

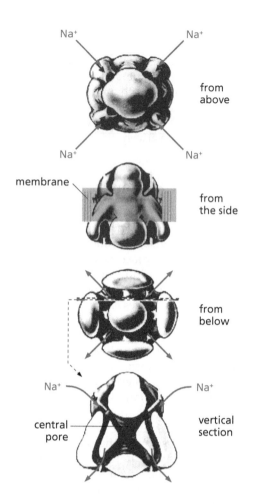

cells), and the cytoskeleton. Such interactions play an important role in the regulation of channel activity and in the clustering of sodium channels along myelinated axons at the nodes of Ranvier. The saltatory propagation of action potentials along nerve fibers depends on this clustering (see textbooks of neurobiology). The important role of the β-subunits is underlined by the fact that point mutations of β1 cause hereditary epilepsy. As may be expected from their putative functions, the β-subunits are a late development of evolution, found only in vertebrates.

Due to the low cytoplasmic Na⁺ concentration and negative charge inside the cell, the gating of voltage-dependent sodium channels always results in a rapid Na⁺ influx: the interior of the cell becomes less negative or even positive, or in other words, the cellular battery is discharged by depolarization. But this can also be accomplished by activation of nonspecific cation channels (see Sections 14.5.6 and 14.5.7 and Chapters 15 and 16). A specialty of voltage-gated Na⁺ channels is signal propagation in the nervous system and in muscle cells, because here each gating event provides a triggering signal for the gating of other voltage-dependent ion channels in the neighborhood. As a result, an action potential migrating across the cell membrane is triggered.

As explained in Section 2.10.2, an action potential is defined as a complete pulse-like depolarization of the cell. This is caused by a rapid discharge and recharge of the membrane battery due to a controlled influx of Na⁺ ions followed by an efflux of K⁺ ions through voltage-operated channels.

To migrate like a wave along neuronal and muscle cell membranes for an action potential two conditions must be fulfilled: the peak of depolarization must be

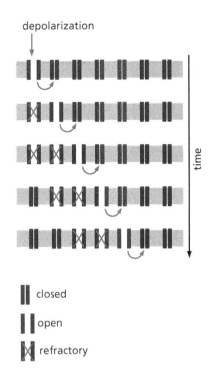

depolarization

time

closed

open

refractory

Figure 14.6 Unidirectional spreading of a wave of membrane depolarization propagated by cooperating Na⁺ channels The membrane depolarization caused by channel opening is sufficient to open the next channel. Since, after activation, a channel does not return immediately into the closed state but becomes refractory for a certain period of time, the wave can spread in one direction only (here from left to right). A second wave may arise when the refractory channels have again taken on the closed conformation.

sharp (transient) and it must be propagated. The sharpening is due to a rapid inactivation of voltage-gated sodium channels that terminates the depolarizing phase of the action potential, enabling voltage-gated potassium channels, by hyperpolarization, to reset the membrane potential to the resting value (see Section 2.10.1). For the propagation of action potentials, adjacent voltage-gated Na⁺ channels cooperate in such a way that the depolarization caused by one channel triggers the gating of the next channel and so on. Such a unidirectional propagation of the action potential, resembling a burning fuse, is possible only when, after activation, the channel becomes refractory for a certain period of time, ensuring that it is not gated again by the depolarization caused by the next channel (Figure 14.6). Here again, rapid channel inactivation is necessary. The inactive state is thought to be due to a cytoplasmic globular domain of the α-subunit blocking the channel pore until it has taken on the closed conformation (ball-on-a-chain model, see Section 14.4.1).

An inhibition by drugs and toxic compounds of voltage-dependent sodium channels interrupts impulse propagation in the nervous system. This is the mechanism of **local anesthetics**, such as lidocaine and tetracaine, but also of highly toxic neurotoxins that are applied as experimental probes. Both famous and infamous is **tetrodotoxin**, the poison of the puffer fish (Figure 14.7). Its half-maximal lethal dose (LD_{50}) for humans is 8 μg/kg of body weight (though botulinus toxin, at 0.005 μg/kg, is far more toxic). In the puffer fish, as "Fugu" a particular highlight of Japanese cuisine, the toxin is concentrated in inner organs such as the liver. These have to be removed carefully but not entirely, since the gourmet is said to experience a slight paralysis of the tongue. In spite of intense training and strict examinations of the kitchen staff, about 200 people are said to fall victim to the delicacy per year.

Of comparable toxicity are the secretions of skin glands of South American colored frogs. They contain highly active Na⁺ channel blockers such as **histrionico-toxin** (Figure 14.7) and are used as arrow poisons by native tribes. Particularly infamous is **saxitoxin**, the poison of the dinoflagellate *Gonyaulax catenella*. The explosive multiplication of these algae causes the ecologically disastrous red tide. Saxitoxin is a major cause of mussel poisoning.

Figure 14.7 Toxins, of the puffer (Fugu) fish and of colored frogs, that block the axonal Na⁺ channel

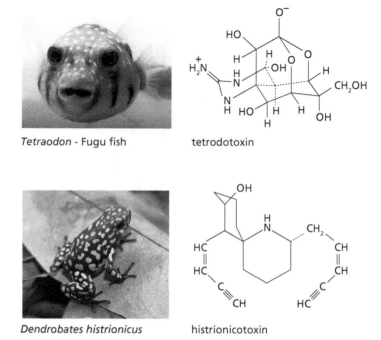

Tetraodon - Fugu fish   tetrodotoxin

Dendrobates histrionicus   histrionicotoxin

Summary

Voltage-gated Na⁺ channels propagate action potentials in excitable cells. They are complexes consisting of at least four proteins. The channel-forming protein, the α-subunit, comprises four groups of six transmembrane helices and one P-loop each, sharing together one large polypeptide chain. It is associated with

Sidebar 14.1 Investigation of ion channels by the patch-clamp technique A detailed investigation of the pharmacological and molecular properties of ion channels has become possible by the patch-clamp technique developed by Erwin Neher and Bert Sakmann (awarded the Nobel Prize in Physiology or Medicine for 1991). By means of this method, a tiny piece of the cell membrane (the patch) is isolated that contains a few channels (or even a single one), the conductivity of which may be measured without confounding interactions. For this purpose, a small-diameter glass pipette filled with a salt solution and connected with a measuring instrument is pressed onto the cell surface and a membrane patch (about 1 μm diameter) sealed with the mouth of the pipette by suction. Depending on whether the patch is left on the cell or separated, channel conductivity may be measured in a whole-cell environment or in an isolated preparation (Figure 14.8).

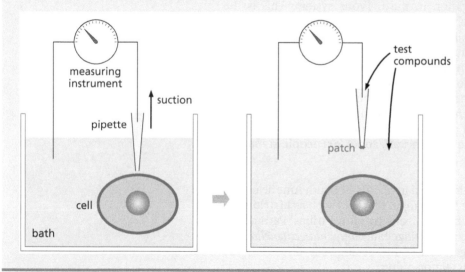

Figure 14.8 Patch-clamp method A small piece (patch) of a cell membrane sealing the tip of a micropipette is separated by suction. An electrical circuit is established by generating a potential difference between bath and pipette. In this circuit, the patch provides resistance. Test compounds added either to the bath or to the salt solution within the pipette may change the conductivity by interacting with the extra- or intracellular domains of the ion channel(s) in the patch.

regulatory transmembrane proteins, the β-subunits. After gating, the channel transiently becomes blocked by a cytoplasmic domain of the α-subunit, sharpening action potentials and enabling their targeted propagation. Axonal Na⁺ channels are inhibited by local anesthetics and some powerful animal neurotoxins.

14.3 Epithelial Na⁺ channels: regulation of electrolyte balance and induction of action potentials

Epithelial Na⁺ channels (ENaCs) constitute the other large family of sodium-specific ion channels. With two transmembrane helices and a P-loop but lacking a voltage sensor, an ENaC subunit resembles prototype 2 of channel structure. The subunits are expressed in several isoforms that associate into a heterotetramer to form a channel. ENaCs are widely distributed in multicellular organisms. The homologous channels of *Caenorhabditis elegans* are called degenerins; therefore, an ENaC/degenerin superfamily is sometimes cited. Since ENaCs are specifically blocked by the diuretic amiloride, they are also frequently called **amiloride-sensitive sodium channels**. Although some of them manage the directed transport of Na⁺ ions through epithelia, the adjective "epithelial" is somewhat misleading, since channels of this family are found in many other tissues.

Those ENaCs that transport sodium ions through epithelia play a key role in regulating the body's salt and water balance. Many epithelia (in the kidney, intestine, glands, and respiratory tract) take up Na⁺ from the urine, contents of the bowel, and secretions and supply them to the blood. The best-known example is Na⁺ reabsorption in kidney tubules. This process is stimulated by hormones such as vasopressin and aldosterone and turned down by ANP, the natriuretic peptide hormone from atrial heart cells. The effects of the hormones are partially mediated by ENaCs: while vasopressin and aldosterone induce the *de novo* synthesis of ENaC proteins and probably activate latent channels, ANP exhibits an opposite effect.

To do their job, ENaCs have to cooperate with ion pumps and K⁺ channels. Directed Na⁺ transport requires an asymmetrical distribution of these components in epithelial cells: the Na⁺/K⁺-dependent ATPase (ion pump) is localized in the basolateral (inner) cell membrane, together with K⁺ channels, and produces an ion gradient promoting the influx of Na⁺ through ENaCs that are components of the apical (outer) membrane (Figure 14.9).

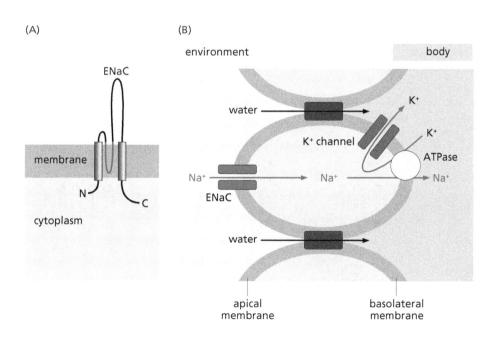

Figure 14.9 Mode of action of epithelial Na⁺ channels (ENaC)
(A) Subunit of the channel that exhibits two transmembrane helices (shown as cylinders) and a large extracellular loop that dips into the membrane, forming a P-loop (red). Each channel consists of at least four subunits. (B) Epithelial cell connected to neighboring cells by tight junctions (dark gray bars) that seal the body against the environment. Active ion transport (against a concentration gradient) accomplished by the apical ENa channel (red) and the basolateral Na⁺/K⁺-dependent ATPase is accompanied by a passive water flux. K⁺ taken up by the cell through the ATPase is extruded again through basolateral K⁺ channels. For chloride, the counterion of Na⁺, a separate transport system exists (see Section 14.7).

Other ENaCs are gated by sensory stimuli and trigger action potentials that are propagated by voltage-gated Na+ channels and migrate to the brain. They are found in mechanosensitive and pain receptors of the skin and in certain taste cells (see Chapter 15). Channels of a subfamily called **acid-sensing ion channels (ASICs)** open at an acidic pH value. They respond to sour food but also to tissue acidification resulting from inflammation, injury, and diminished blood supply (for instance, in the course of angina pectoris) and evoke pain sensations. Acid-sensing sodium channels are also expressed in brain neurons. Gene-knockout experiments, indicate that they are involved in learning and memory fixation. This concept is supported by observations showing pronounced pH fluctuations and waves of acidification to spread in active brain areas.

Summary

Epithelial Na+ channels (ENaCs) are composed of four subunits with two transmembrane helices and one P-loop each. ENaCs regulate the osmolarity of body fluids by transporting ions through epithelia (e.g., salt reabsorption in the kidney). In certain sensory cells the voltage-independent gating of ENaCs triggers action potentials. Some ENaC isoforms (ASICs) preferentially found in neurons are acid-sensitive, responding to tissue acidification.

14.4 K+-selective ion channels: regulators of hyperpolarization and osmotic pressure

K+-selective channels constitute the largest ion channel family and have a long evolutionary history. Their variabilty and versatility, resulting both from numerous genes and from alternative splicing of pre-mRNA, is amazing. Figure 14.10 provides an overview of the K+ channel families of vertebrates.

Due to the high intracellular K+ concentration, the gating of K+ channels mostly results in an efflux of K+ ions against the voltage gradient. This exerts a hyperpolarizing effect (the electrical charge inside the cell becomes more negative) that

Figure 14.10 K+-selective ion channels of vertebrates Shown are the pore-constituting α-subunits (with the numbers of α-subunits forming a channel). From left to right: voltage-controlled channel Kv; voltage- and Ca2+-controlled channel BK; Ca2+/CaM (calmodulin)-controlled channel SK; inwardly rectifying channel Kir; Gi-protein-controlled inwardly rectifying channel GIRK interacting with Gβγ; and 2P-domain channel. Each channel subunit exists in various isoforms. Voltage-sensing helices are colored. The upper diagram shows the influence of K+ channels on the action potential: the influx of K+ lowers the resting potential and terminates the phase of depolarization, causing repolarization and afterhyperpolarization.

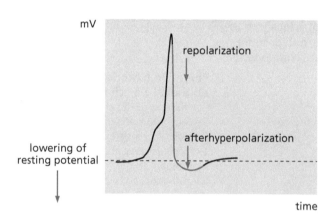

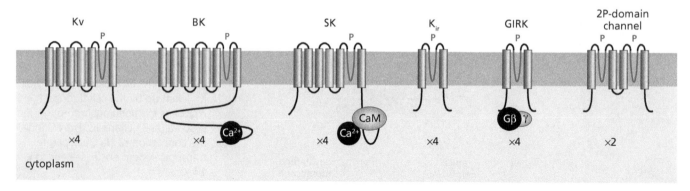

has two major consequences: it lowers the resting potential, thus hindering depolarization, and it terminates depolarization waves such as action potentials. Thus, an activation of K⁺ channels attenuates and blocks depolarization-dependent events such as neurotransmitter release and muscle contraction, whereas vice versa, channel closing or blockade may facilitate depolarization. In addition, K⁺ channels help to control the osmotic pressure of cells.

14.4.1 Voltage-controlled K⁺ channels: termination of action potentials

The first voltage-dependent potassium channel discovered was the product of a gene responsible for the Shaker mutation of the *Drosophila* fly (mutant animals lacking this channel exhibit shaking under anesthesia). In excitable cells, the voltage-dependent potassium channel K_v is the direct opponent of the voltage-dependent Na^+ channel. Since it opens upon depolarization, its major effect is to terminate the action potential by negative feedback (Figure 14.10). The K_v channel has prototype 1 architecture and resembles the voltage-dependent Na^+ channel, albeit with the variation that the four α-subunits are separated rather than joined on a single polypeptide chain. This constructive principle renders K_v channels extremely variable: in mammals, the α-subunits are encoded by more than 50 genes and, according to a unit construction system, the isoforms assemble into homo- and heterotetramers. The different combinations are tissue-specifically expressed and respond individually to changes of the membrane potential. A Kv isoform that has gained considerable medical interest is derived from a human ether-a-go-go-related gene and called hERG. The strange name refers to a *Drosophila* mutant that when anesthesized by ethyl ether starts to shake its legs like a go-go-dancer. The major function of hERG is to shape and terminate the action potential that controls the heartbeat. This occurs in two steps: immediately after the peak of depolarization and at the end when the resting potential is to be reached again. In between the channel becomes desensitized for a short period of time. This results in a plateau phase of partial repolarization that is important for supplying the heart muscle with Ca^{2+} (a typical action potential of the heart is shown in Figure 14.33). Defects of hERG function – due to point mutations or intoxication (for instance by anti-psychotic drugs) – are the reason for life-threatening cardiac arrhythmias. Therefore, the channel is a target of corresponding medical drugs.

In the vertebrate K_v channel, each α-subunit associates with a β-subunit (also expressed in various isoforms). These K_v β-subunits share only the name but neither structure nor function with the β-subunits of the voltage-dependent Na^+ channels. They are not membrane-spanning cell adhesion molecules of the Ig-like family. Instead they are cytoplasmic proteins that play the role of a door-keeper: by means of a positively charged globular structure fixed to a mobile polypeptide chain, they reversibly plug the channel pore (ball-on-a-chain model; see Figure 14.11). This effect resembles that of the inactivation domain in the sodium channel's α-chain. As we have seen, such a short-term desensitization of ion channels is essential for rapid termination as well as for unidirectional spreading of the action potential.

Certain K_v channels, particularly those at neuromuscular synapses, are blocked by **dendrotoxins**. These proteins are components of mamba snake venom. They are widely used as tools for the pharmacological investigation of potassium channels. Other K_v channel blockers employed for research are tetraethylammonium salts and 4-aminopyridine. Derivatives of the latter are clinically used for the treatment of multiple sclerosis.

Calcium-sensitive potassium channels

Action potentials induce secretory processes, such as neurotransmitter release at synapses, and trigger muscle contraction. Both responses depend on an increase of the cytoplasmic Ca^{2+} concentration. To this end, voltage-dependent calcium

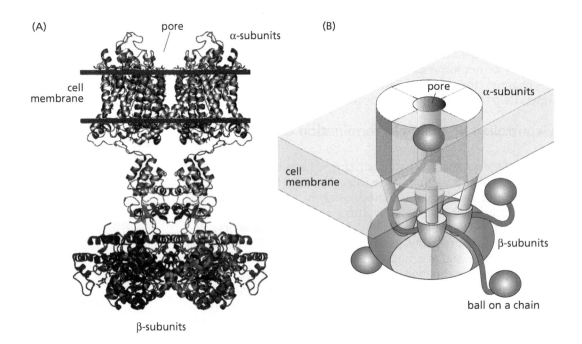

(A) pore α-subunits
cell membrane
β-subunits

(B) pore α-subunits
cell membrane
β-subunits
ball on a chain

Figure 14.11 Model of the voltage-controlled potassium channel K$_V$ (A) Molecular model showing α-subunits that form the membrane pore (black). At the cytoplasmic side, they are associated with the β-subunits (red) exhibiting mobile globular domains (ball on a chain) (B) that fit the central pore and, thus, may interrupt the ion flux. The molecular model is based on X-ray crystallography. (B, adapted from R.W. Aldrich, *Nature* 411:643–644, 2001.)

channels in the presynaptic membranes of neurons and in muscle cell membranes become gated by depolarization. This effect is, of course, opposed by K$_V$ channel gating. In addition, the resulting Ca^{2+} influx is regulated by a special group of Ca^{2+}-sensitive potassium channels via negative feedback. These K$^+$ channels are structurally related to K$_V$ channels; two types are distinguished by their electrophysiological features.

BK channels (Big conductance K$^+$ channels) function as coincidence detectors resembling logical AND gates: though they are voltage-controlled, gating is not evoked by depolarization alone but requires, in addition, an increase of the cytoplasmic Ca^{2+} level. Their role in a particular physiological process, the relaxation of smooth muscles, is described in Section 14.6.2. Unlike K$_V$ channels, the α-subunits of BK channels have a unique seventh transmembrane segment S0 at the N-terminus and an extended cytoplasmic Ca^{2+} sensor domain (Figure 14.10).

SK channels (Small conductance K$^+$ channels) become gated by Ca^{2+} alone, interacting with the calcium sensor protein calmodulin, which is a channel subunit. The α-subunits resemble those of K$_V$ channels, although SK channels are not voltage-controlled since they do not contain a sufficient number of positively charged amino acid residues in the S4 helix to function as a voltage sensor. SK channels are (at least partially) responsible for the afterhyperpolarization phase of an action potential, when the membrane potential falls below the resting potential for a short time (Figure 14.10).

Calcium-sensitive potassium channels are targets of **apamin**, a component of bee venom, and of the scorpion poison **charybdotoxin**. These peptides block the channels and cause pathological hyperactivation of the nervous system. The consequences are extreme pain and other severe dysfunctions.

14.4.2 Inwardly rectifying K$^+$ channels: setting the resting potential

A large group of potassium channels was discovered due to their apparently strange electrophysiological properties: in contrast to K$_V$ channels, they became gated upon hyperpolarization and closed upon depolarization. In humans the pore-forming α-subunits of these channels are encoded by at least 15 genes giving rise to a high structural and functional variability. The individual channel isoforms are tissue-specifically expressed, in particular in the central nervous system, the musculature, endo- and exocrine glands, and epithelia.

Hyperpolarization-gated potassium channels belong to channel prototype 2. As such they lack an intrinsic voltage sensor such as the S3–S4 hairpin domain. How can they be voltage-dependent in a reverse manner from K_V channels? The answer is that upon depolarization the cytoplasmic side of the ion gate becomes plugged by positively charged polyamines like spermine and by Mg^{2+} ions.

The gating of K_V channels always results in an efflux of K^+ ions. In contrast, K^+ channels that are gated by hyperpolarization may allow a K^+ *influx*, resulting in depolarization when (depending on the membrane potential and the intracellular K^+ concentration) the electrical force of the inward-directed voltage gradient exceeds the chemical force of the outward-directed concentration gradient. This effect, which is essential for setting the resting potential, led to the name inwardly rectifying potassium channels (K_{ir} channels). However, when the potential becomes less negative, such channels begin to propagate K^+ efflux, causing hyperpolarization. This would impair or even prevent depolarization-dependent processes. Therefore, the channels must be closed upon depolarization.

K_{ir} channels that regulate the resting potential are consistently open at sufficiently negative potentials (leak channels). They are of utmost importance for the generation and shaping of action potentials: in the initial phase their closure facilitates depolarization, and in the subsiding phase (when the repolarizing K_V channels are already closed again) they open again and complete the repolarization of the cell, thus terminating the action potential (Figure 14.10). This holds true, in particular, for long-lasting excitations such as those required for heart contraction that critically depend on the correct function of the leak channels.

The manipulation of K_{ir} channels offers a possibility to dampen processes triggered by action potentials: all that has to be done is to keep the channel open by appropriate signals in spite of depolarization. In fact, signal-regulated K_{ir} channels are major negative control devices. This is shown by the following examples.

G_i-protein-regulated K^+ channels

This channel type is termed $K_{ir}3$ or better known as GIRK (<u>G</u>-protein-gated <u>I</u>nwardly <u>R</u>ectifying <u>K</u>$^+$ channel). Its depolarization-dependent closure is prevented by Gβγ-subunits that in sufficient concentrations are released from the particularly abundant $G_{i,0}$-proteins and bind to the four channel subunits (Figure 14.12). Open GIRK channels lower the resting potential and impair the generation of an action potential. Together with inhibition of cAMP production, this is the effect of all extracellular signals that interact with $G_{i,0}$-protein-coupled receptors, provided the corresponding cell expresses GIRK channels (as is the case in many organs, particularly in brain, heart, and endocrine glands). $G_{i,0}$-coupled receptors activating GIRK channels are widely distributed. In particular, they mediate inhibitory effects of neurotransmitters (see Table 16.1 in Section 16.1).

Thus, this channel type plays an important role in the nervous system and the target organs of neuronal signaling. Again the effects focus on Ca^{2+}-dependent events: as explained above, hyperpolarization primarily turns down the influx of Ca^{2+} ions through voltage-operated Ca^{2+} channels. This results in an attenuation of Ca^{2+}-dependent processes such as muscle cell contraction and neurotransmitter secretion. We have already encountered such GIRK effects: the parasympathetic down-regulation of heart activity (Section 5.5) and the $α_2$-adrenergic feedback inhibition of noradrenaline release at sympathetic synapses (Section 5.3.3). Other examples are found in Chapter 16. Gβγ-subunits also inhibit

Figure 14.12 G$_i$-protein-regulated K$^+$ channel GIRK Gβγ-subunits, mainly released from G$_{i,0}$-proteins upon activation of a G-protein-coupled receptor (GPCR), prevent closure of the K$_{ir}$ channel and induce hyperpolarization, while the corresponding Gα$_i$-subunits inhibit adenylate cyclase. As a result, voltage-, Ca^{2+}- and cAMP-dependent signaling is inhibited.

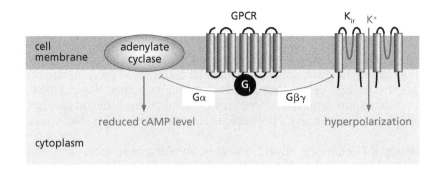

voltage-operated Ca^{2+} channels directly, thus augmenting the effect of hyperpolarization (Section 14.5.2).

G$_{i,0}$-dependent signaling opposes G$_s$- and G$_{q,11}$-controlled signaling. Vice versa, GIRK channels are inhibited along the G$_{q,11}$-pathway, most probably by protein kinase C-catalyzed phosphorylation.

ATP-regulated potassium channel K$_{ATP}$

This channel provides a further example of a ligand-controlled K$_{ir}$ channel (Figure 14.13). Although closely related to the GIRK channel, it forms a complex with the ATP/ADP-binding transmembrane protein **SUR** (**SulfonylUrea Receptor**) rather than with Gβγ-subunits. SUR belongs to the family of ATP-binding cassette (ABC) transporters (Section 3.2.1). As a specific target of antidiabetic drugs of the sulfonylurea type, this protein is of eminent medical importance. In fact, the K$_{ATP}$–SUR team participates in the control of insulin secretion by the β-cells of the islets of Langerhans. Here SUR acts as an energy sensor, regulating channel conductance in accordance with the intracellular ATP/ADP ratio that directly depends on the blood sugar level: the cells respond to rising blood sugar by increased uptake of glucose through membrane-bound glucose transporters. This leads to an increased formation of ATP that is bound by both SUR and the channel subunits. As a result, the channel is closed and the resting potential becomes less negative, facilitating membrane depolarization and insulin release. Since the latter is an exocytotic process, it is triggered by Ca^{2+} ions entering the cell through voltage-dependent Ca^{2+} channels that sensitively respond to any change of the membrane potential. A lowering of the blood sugar level causes a fall of the intracellular ATP level and the ATP bound to SUR becomes displaced

Figure 14.13 Regulation of insulin secretion by the K$_{ATP}$ channel (A) The channel consists of four pore-forming K$_{ir}$ subunits and four regulatory subunits SUR (sulfonylurea receptor). SUR has the structure of an ABC transporter with two groups of six transmembranes each (TMD1 and -2, light gray) and two cytoplasmic ATP/ADP-binding sites. In addition, it contains a group of five N-terminal transmembrane domains TMD0 (dark gray). At low ATP levels, the channel is open and thus inhibits insulin secretion. ATP produced in response to glucose uptake binds to both the channel and SUR. As a result, the channel closes and depolarization and insulin secretion are facilitated. (B) Arrangement of the subunits (K$_{ir}$, colored) as seen from above (based on X-ray crystallography).

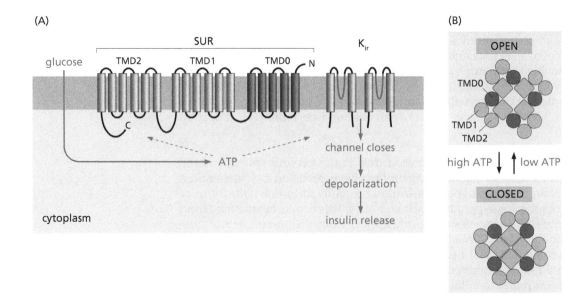

by ADP. Since under this condition the K$_{ATP}$ channel remains open, the cell becomes hyperpolarized and insulin secretion decreases. By stabilizing the ATP complex of SUR, sulfonylurea drugs promote insulin release from the pancreas.

As we have seen, K$_{ATP}$ channels control the membrane potential depending on the metabolic state of the cell. They do this not only in the pancreas but also in other tissues including heart and vascular smooth muscle, skeletal muscle and brain. In a specialized set of hypothalamus neurons, a fall of the blood sugar level activates K$_{ATP}$ channels. In turn a neuronal impulse is sent to the pancreatic α-cells that release the sugar-mobilizing hormone glucagon. Another important response to channel gating is the widening of blood vessels because hyperpolarization hinders smooth muscle contraction. The brain and the heart are protected in this way from the harmful consequences of diminished blood supply (resulting in decreased ATP production) in the course of stroke and myocordial infarction. Such effects are targets of additional pharmaceutical drugs. As a more curious example, minoxidil may be mentioned: by stimulating potassium transport through K$_{ATP}$ channels, it lowers the blood pressure and is said to prevent hair loss by promoting blood circulation in the skin.

14.4.3 2P-domain channels: background channels, pressure relief valves, temperature sensors, and pain relievers

A third family of K$^+$-selective ion channels is closely related to K$_{ir}$ channels. These channels consist of two α-subunits: each has four transmembrane helices and two P-loops and may be looked upon as two covalently linked subunits of a K$_{ir}$ channel (Figure 14.10). The P-loop tandem led to the name 2P-domain or K$_{2P}$ channels. Since, like other K$^+$ channels, 2P-domain channels exhibit only one pore, the name twin pore channels (sometimes found in the literature) is misleading.

Fifteen human isoforms have been found. On the basis of structural parameters, they are subdivided into six subfamilies denoted by four-letter acronyms beginning with T (Tandem of P-loops), such as TWIK (T in a Weak Inwardly rectifying K$^+$ channel) or TREK (TWIK-RElated K$^+$ channel). In contrast to K$_{ir}$ channels, K$_{2P}$ channels do not fully close upon depolarization. This qualifies them as background channels, setting the resting potential at different levels and generally dampening excitability. While some are constitutively open, others are gated by a variety of physical or chemical stimuli.

Such a physical stimulus is mechanical pressure from inside the cell. Thus, some K$_{2P}$ channels are cellular pressure relief valves, opening when the cell membrane is under tension. Such a situation occurs in a hypotonic environment. To prevent a rupture, the cell has to lower the concentration of osmotically effective molecules, mainly by releasing electrolytes, such as K$^+$, the cytoplasmic concentration of which is particularly high. To retain electric neutrality, pressure-sensitive chloride channels open simultaneously with potassium channels (see Section 14.7.1). As a result, a KCl solution is poured out by the cell. Another physical stimulus is temperature. Temperature-sensitive K$_{2P}$ channels are expressed, for instance, in sensory neurons and in the hypothalamus. They are thought to function as temperature sensors because they close in response to cooling, thus facilitating the generation of action potentials. Certain hypothalamus neurons use this channel type to control body temperature. This effect is mediated by prostaglandin E$_2$ which, by interacting with a G$_s$-protein-coupled receptor, induces cAMP formation in neurons. As a result, protein kinase A becomes activated and phosphorylates a particular K$_{2P}$ channel. Due to this modification, the channel becomes blocked, rendering the resting potential less negative and facilitating depolarization. This leads to an increased activity of those hypothalamus neurons that cause a rise in body temperature. Since **fever** develops along the same route, drugs that inhibit prostaglandin synthesis, such as acetylsalicylic acid (aspirin), exhibit an antipyretic effect (Section 4.4.5).

Other 2P-domain channels are gated by intracellular signaling molecules such as arachidonic acid and lysophospholipids. These lipid mediators are released in the brain in response to disturbances of the blood supply. The corresponding 2P-channels probably participate in neuronal protection from damage caused by **stroke**: by suppressing a pathological overstimulation of voltage-dependent Ca^{2+} channels, which is a characteristic feature of stroke (Section 16.4.2), they prevent a fatal flooding of the cell by Ca^{2+} ions.

Some K_{2P} channels are of great medical interest because they play an important role in the action of **gaseous narcotics** such as chloroform, halothane, and diethyl ether that open the channels, thus hindering the generation of action potentials in those neurons that transmit pain signals to the brain.

Summary

The highly variable voltage-dependent K^+ channels (K_V) are the counterparts of the voltage-dependent Na^+ channels. Upon gating by depolarization, they cause hyperpolarization, thus terminating action potentials. The pore-forming α-subunits of K_V channels are composed of four separate polypeptides, each containing six transmembrane helices and one P-loop. The central pore is transiently blocked by a globular domain of cytoplasmic β-subunits. A subfamily of K_V-related channels is regulated by Ca^{2+} ions and is important for negative feedback control of Ca^{2+}-dependent processes such as smooth muscle contraction and neurotransmitter release. Inwardly rectifying potassium channels (K_{ir} channels) are open at negative membrane potentials and become closed upon depolarization. They are important for setting the resting potential and shaping action potentials. The K_{ir} channel GIRK, gated by $G\beta\gamma$-subunits released from $G_{i,0}$-coupled receptors, is a key mediator of inhibitory neurotransmitter effects and plays an important role in the negative feedback control of neurotransmitter release at synapses and in the parasympathetic attenuation of heart activity. Another K_{ir} channel is under the control of the ABC transporter protein SUR that closes the channel upon ATP binding, thus facilitating events depending on depolarization (such as insulin secretion and smooth muscle contraction), whereas ATP shortage has an opposite effect. SUR is a target of antidiabetic drugs. A third family of K^+ channels characterized by a 2P-domain structure also controls the resting potential. In addition, it is specialized for osmotic pressure relief to prevent a rupture of the cell membrane. For this reason, these channels have to cooperate with chloride channels. 2P-domain channels expressed in neurons control body temperature and transmit pain sensations. Their inactivation by phosphorylation occurring along the prostaglandin–G_s–cAMP–PKA pathway may result in fever. Certain 2P-domain channels are gated directly by gaseous narcotics and indirectly by nonsteroidal anti-inflammatory drugs that prevent prostaglandin formation.

14.5 Calcium ions: the most versatile cellular signals

Ca^{2+} ions occupy a central position in cellular data processing. Without any exaggeration, it may be stated that there is practically no cell physiological process that is not influenced and regulated directly or indirectly by calcium. For calcium signaling, cells keep the cytoplasmic calcium concentration at a level that is about 4 orders of magnitude lower than in the extracellular medium. As a consequence, gating of Ca^{2+} channels automatically leads to a calcium influx that provides the triggering signal for cellular responses. This gating is induced and modulated by exogenous and intracellular input signals. In most vertebrate cells, calcium influx does not change the membrane potential significantly because the Ca^{2+} concentration is 2–4 orders of magnitude lower than that of sodium, potassium, and chloride ions. Instead, Ca^{2+} ions function as second messengers, being recognized (specifically bound) by effector proteins that, upon interaction with Ca^{2+}, undergo conformational and functional changes, thus producing output signals. This is in clear contrast to the monovalent ions Na^+, K^+, and Cl^-, which primarily control the membrane potential.

14.5.1 Generation and termination of Ca^{2+} signals

As will be shown below, cells modulate the Ca^{2+} signal in an almost unlimited manner, with respect to intracellular localization, intensity (or amplitude), and temporal fluctuation (or frequency). Moreover, they are able to generate stationary and spreading Ca^{2+} waves as well as spatial concentration patterns. These sophisticated processes require precise control of both the rise and the decay of Ca^{2+} signals because otherwise the cell would be flooded by Ca^{2+} and become paralyzed and eventually killed. Therefore, we first shall address the question of how the cellular Ca^{2+} level is regulated.

Calcium pumps

Like other second messengers, Ca^{2+} acts as an allosteric modulator of protein functions. In the cytoplasm of nonstimulated cells the concentration of free Ca^{2+} ions is kept at a level of about 0.1 μM, which is too low to significantly influence effector proteins (which respond only at 1–10 μM Ca^{2+}). To achieve this low concentration, Ca^{2+} ions are pumped out of the cell and sequestered in the lumen of the ER, in the mitochondrial matrix, and by cytoplasmic calcium-binding buffer proteins. The concentration of free Ca^{2+} in the extracellular fluid, ER, and mitochondria is about 1 mM, or about 4 orders of magnitude higher than in the cytoplasm. Thus, to generate and maintain the low cytoplasmic Ca^{2+} level, the ions must be transported against a steep concentration gradient. This requires energy derived from ATP hydrolysis, catalyzed by **Ca^{2+}-dependent ATPases**. These enzymes are integrated in the ER and plasma membranes. ATP hydrolysis causes a conformational change of the ATPase that conveys a noncovalently bound Ca^{2+} ion from the inside to the outside of the cell. Thus, these ATPases are calcium pumps that keep the resting cytoplasmic calcium level low and, upon gating of Ca^{2+} channels, ensure a rapid termination of the Ca^{2+} signal, preventing the fatal consequences of calcium overload.

Similar or even stronger pumps are the membrane-bound **Na^+–Ca^{2+} exchangers** or transporters. They compensate the efflux of one Ca^{2+} ion by an influx of three Na^+ ions, thus exhibiting (in contrast to the ATPase) a depolarizing effect. The energy is derived from the Na^+ gradient across the cell membrane; that is, it is ultimately based on the activity of the Na^+/K^+-dependent ATPase generating this gradient. Due to its depolarizing action, the exchanger not only lowers the cytoplasmic calcium level but also raises the resting potential and facilitates the subsequent generation of an action potential. This effect plays an important role in the periodic firing of synapses and in continuously oscillating heart muscle cells, which both depend on Ca^{2+} influx provoked by depolarization.

An electrogenic Na^+–Ca^{2+} exchanger is also found in mitochondrial membranes. Together with a neutral H^+–Ca^{2+} exchanger, it catalyzes the efflux of mitochondrial calcium into the cytoplasm in a strictly controlled way against a steep charge gradient across the membrane (180mV, inside negative) that is generated by the respiratory electron transport chain and also drives mitochondrial ATP synthesis. Mitochondria contain, in addition, nonselective Ca^{2+} pores that open upon Ca^{2+} overload. These permeability transition pores penetrating the outer and inner mitochondrial membranes are probably also responsible for the release of cytochrome *c* and other pro-apoptotic proteins from mitochondria in the course of programmed cell death (Section 13.2.4).

While the ER takes up cytoplasmic Ca^{2+} by means of the Ca^{2+}-dependent ATPase SERCA (Sarco/Endoplasmic Reticulum Calcium-dependent ATPase), mitochondria absorb cytoplasmic Ca^{2+} through a special Ca^{2+} channel. This **uniporter** is localized in the inner mitochondrial membrane and allows the influx of Ca^{2+} from the cytoplasm into the mitochondrial matrix. It works only at rather high Ca^{2+} concentrations of 0.01 mM and more and is assumed, therefore, to be positioned in the immediate neighborhood of Ca^{2+} channels in the plasma membrane and the ER, where the Ca^{2+} concentration is particularly high. The pathways and devices controlling the cytoplasmic calcium level are summarized in Figure 14.14.

Figure 14.14 Regulation of the cytoplasmic Ca²⁺ level Extracellular Ca²⁺ is taken up through voltage- and ligand-dependent membrane channels (red) that are opposed by a Ca²⁺-dependent ATPase (PMCA, plasma membrane Ca²⁺-dependent ATPase) and an electrogenic Na⁺–Ca²⁺ exchanger (EX$_{Na}$). Intracellular calcium stores are the endoplasmic reticulum (ER) and the mitochondria. The ER membrane contains ligand-controlled Ca²⁺ channels (red) opposed by a Ca²⁺-dependent ATPase (SERCA, sarco/endoplasmic reticulum Ca²⁺-dependent ATPase). The mitochondria deliver Ca²⁺ to the cytoplasm by an electrogenic Na⁺–Ca²⁺ exchanger (EX$_{Na}$), a H⁺–Ca²⁺ exchanger (EX$_{H}$), and permeability transition pores (not shown). Ca²⁺ influx into the mitochondrial matrix occurs through a uniporter (UP) along a charge gradient that is generated by the respiratory electron transport chain. The resting concentrations of Ca²⁺ in the individual compartments are given in red.

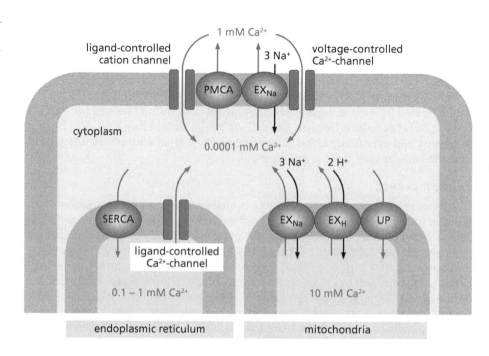

Calcium channels

Calcium channels are the opponents of calcium pumps. The cell membrane contains both ligand- and voltage-controlled types, whereas in the ER, only channels gated by intracellular messengers are found. A third channel type of the plasma membrane, called a store-operated calcium channel, responds to the Ca²⁺ concentration in the lumen of the ER, which beside mitochondria is the major cellular calcium store. Finally, Ca²⁺ is also transported together with Na⁺ and K⁺ by nonspecific cation channels, particularly those controlled by neurotransmitters (Chapter 16). The different types of calcium channels are summarized in Table 14.1. They will be discussed in detail below.

Table 14.1 Ca²⁺-transporting membrane channels: an overview

In the plasma membrane
- voltage-operated Ca²⁺-selective channels (VOCCs)
- storage-operated Ca²⁺-selective channels (SOCCs)
- arachidonic acid-regulated Ca²⁺-selective channels (ARCCs)
- nonselective cation channels of the TRP family
- cyclic nucleotide-gated nonselective cation channels
- ionotropic neurotransmitter receptors with intrinsic nonselective cation channel activity

In the ER and in the outer nuclear membrane
- InsP₃/Ca²⁺-controlled Ca²⁺-selective channels
- ryanodine receptor Ca²⁺-selective channels

In other organelles
- NAADP– controlled channels
- sphingosine 1-phosphate-controlled channels
- mitochondrial uniporters
- mitochondrial permeability transition pores

14.5.2 Voltage-operated calcium channels

The plasma membrane contains voltage-operated Ca^{2+} channels (VOCCs) that are opened by depolarization. They have been found in all eukaryotic cell types examined thus far, but not in prokaryotes. Because they connect changes of the membrane potential with Ca^{2+}-dependent cellular processes such as secretion and movements, they are major players in neurotransmitter release at synapses and in cellular motility, including muscle contraction. In fact, action potentials generated by nerve and muscle cells primarily serve the gating of VOCCs.

Basic structures and types of voltage-operated calcium channels

The molecular architecture of VOCCs is similar to that of the voltage-dependent Na^+ channel. The structural core is a large pore-forming protein with 4×6 transmembrane helices and four P-loops. It is called α_1-subunit and is flanked by additional subunits (α_2, β, γ, δ) that exhibit stabilizing and regulatory functions (Figure 14.15). The β-subunit, a cytoplasmic protein, controls the exposure of the α_1-subunit at the cell membrane by masking sequences that otherwise would hinder membrane trafficking. In addition, it modulates the electrophysiological properties of the channel and interacts (via an SH3 domain) with other signaling proteins. Thus, it differs entirely from the β-subunits of voltage-dependent Na^+- and K^+-channels. The role of the δ-subunit, a single-pass transmembrane protein, is that of an adaptor for the extracellular α_2-subunit that modulates the efficacy of the channel. Both subunits are encoded by the same gene and are connected to each other by disulfide bonds. The role of the γ-subunit, a protein with four transmembrane domains, is not well understood. Whether this subunit is common to all VOCCs or restricted to the cytoskeletal type is an open question. An additional cytoplasmic subunit of most VOCCs is the Ca^{2+}-binding protein calmodulin (CaM). It mediates feedback inhibition of channel gating (see below).

Humans possess 22 genes encoding α_1-subunits. Together with four $\alpha_2\delta$, four β, and eight γ genes, they could assemble theoretically into more than 1000 different combinations. How many of these are realized is an open question, however.

On the basis of pharmacological data, VOCCs are grouped in several subfamilies labeled L, N, Q, R, and T and differing from each other primarily by the α_1-isoforms.

Apart from being expressed in certain brain regions, **L-type channels** (Long-lasting) are typical for skeletal, smooth, and heart muscle cells, where they are involved in stimulus–contraction coupling (see below). They are known also as **dihydropyridine receptors** in reference to a family of medical drugs that selectively block L-type channels and are used for the treatment of angina pectoris, cardiac arrhythmia, and high blood pressure. L-type channels open gradually when the membrane potential becomes less negative. At maximal depolarization (at the peak of an action potential), they are opened to about 70%. By such a graded response, the digital all-or-nothing signals of action potentials are transformed into analog signals.

VOCCs that induce neurotransmitter release across presynaptic neuronal membranes belong to the **N** (Neural), **P, Q**, and **R subfamilies**. They respond to moderate to strong potential changes, as are typical for action potentials, and they are targets of highly toxic animal poisons. Examples are ω-conotoxin and ω-agatoxin. **Conotoxins** are a group of peptides found in the venom of marine cone snails (genus *Conus*). They block various ion channels; the ω-form in particular inhibits the N-type VOCC. This blockade exerts an analgesic (pain-relieving) effect much stronger than that of morphine. In fact, synthetic derivatives of ω-conotoxin are used in pain therapy.

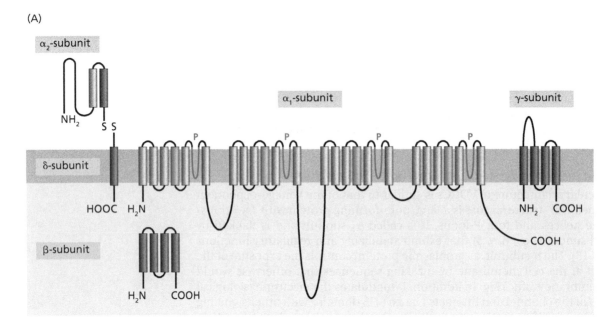

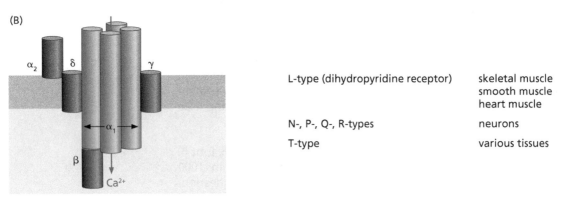

L-type (dihydropyridine receptor)	skeletal muscle
	smooth muscle
	heart muscle
N-, P-, Q-, R-types	neurons
T-type	various tissues

Figure 14.15 Voltage-operated Ca²⁺-channels: basic architecture and major occurrences Shown are the transmembrane architecture (A) and the arrangement of the subunits (B) of an L-type channel. P, P-loops of α_1-subunit. In panel A, the S4 voltage sensors are colored. (Adapted from Z.W. Hall Molecular Neurobiology. Sunderland, MA: Sinauer Associates, 1992.)

The peptide **ω-agatoxin** is a component of the venom of funnel web spiders (family *Agelenidae)*. It blocks P and Q channels. The **T-type channels** (<u>T</u>ransient) expressed by neurons and other tissues (such as heart muscle) are gated, in contrast to other VOCCs, upon a minute increase of the resting potential and become refractory above –40 mV. Due to this peculiar property, they play an important role in periodic changes of the membrane potential such as the oscillations occurring in the pacemaker cells of the heart (Section 14.6.2).

Regulation of voltage-operated calcium channels

The key role of VOCCs in cellular signal transduction and the potentially dangerous effects of Ca²⁺ require precise regulation of channel gating (Figure 14.16). Like Na⁺ and K⁺ channels, VOCCs undergo auto-inactivation, ensuring a sharpening of the Ca²⁺ signal and preventing fatal flooding of the cell by Ca²⁺. For this voltage-dependent desensitization, a cytoplasmic domain of the α_1-subunit is thought to plug the pore, resembling the ball-on-a-chain mechanism of Na⁺ and K⁺ channel inactivation. Ca²⁺ itself also blocks the channel by interacting with the calcium sensor protein CaM, which is a separate subunit of the channel noncovalently bound at the C-terminal domain. Transient channel refractoriness is also a condition for unidirectional propagation of calcium waves (see below).

In contrast, phosphorylation catalyzed by PKA or Ca²⁺/calmodulin-dependent protein kinase II (CaMKII) facilitates and prolongs channel gating, causing, for instance, sympathetic and stress effects in the heart (Section 14.6.2). Neuronal

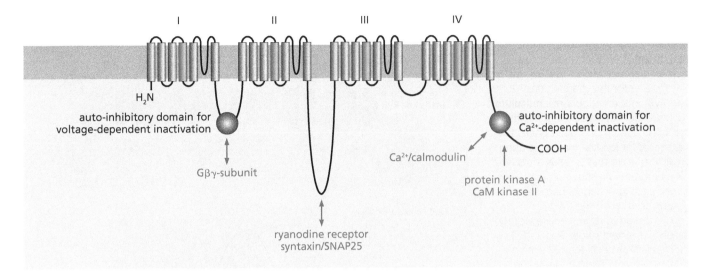

Figure 14.16 Regulation of voltage-operated Ca^{2+} channels
Resembling voltage-dependent Na^+ channels, the Ca^{2+} channels undergo desensitization by an auto-inhibitory domain positioned between subdomains I and II. A C-terminal domain interacts with a cytoplasmic complex of Ca^{2+} and the major calcium-binding protein calmodulin (providing negative feedback control of channel permeability). Moreover, Ca^{2+} channels are inhibited by interaction with $G\beta\gamma$-subunits (mainly derived from $G_{i,0}$-proteins). In the presynaptic neuronal membrane, an interaction with syntaxin/SNAP25 triggers exocytosis of secretory vesicles. The L-type channel couples with ryanodine receptors in skeletal muscle and is activated by C-terminal phosphorylation catalyzed by protein kinase A or CaM kinase II in heart muscle (sympathetic effect; see Section 5.5).

calcium channels, in particular, are blocked by $G\beta\gamma$-subunits mainly released from receptor-coupled $G_{i,0}$-proteins. As mentioned above, $G\beta\gamma$-subunits, in addition, activate potassium channels of the GIRK type, rendering the resting potential more negative. In other words, $G\beta\gamma$-subunits suppress the generation of a Ca^{2+} signal at two points: membrane depolarization and Ca^{2+} influx. This reaction plays a major role in negative feedback control of neurotransmitter release (Section 16.3). As an example we have encountered the α_2-adrenergic inhibition of noradrenaline release in sympathetic neurons discussed in Section 5.3.3. Moreover, along these pathways, certain neurotransmitters and sedative drugs attenuate neuronal activity by interacting with $G_{i,0}$-coupled receptors (Chapter 16).

14.5.3 Calcium channels of the endoplasmic reticulum

The lumen of the ER provides the intracellular calcium store primarily used by the signal-processing apparatus (whereas mitochondria regulate the overall calcium homeostasis of the cell). The calcium store of the ER cooperates with the cellular environment in that it is refilled by extracellular Ca^{2+} (see Section 14.5.5). In fact, the Ca^{2+} concentration in the ER lumen (about 1 mM) resembles that of the extracellular space. The ER membrane contains Ca^{2+}-selective ion channels, which at the cytoplasmic side become gated by ligand binding. Two such ion channel families, the **inositol 1,4,5-trisphosphate (InsP$_3$) receptors** and the **ryanodine receptors**, have been investigated in detail. These families are closely related and consist of four subunits each. Each subunit has six transmembrane domains, a P-loop, and an extended N-terminal (cytoplasmic) domain that includes the ligand-binding site. Thus, in their membrane topology these ligand-dependent calcium channels resemble the α-subunit of the voltage-dependent potassium channel as depicted in Figure 14.10. However, as compared with the K^+ channel subunits their membrane integration is upside-down, with the ligand binding sites facing the cytoplasm. Moreover, they are not voltage-gated because the positive charge of their S4 helices is not high enough. Both receptor types exist in various isoforms that are tissue-specifically expressed and associate into hetero-oligomers.

Strictly speaking, InsP$_3$ and ryanodine receptors are not only Ca^{2+}-selective channels but also Ca^{2+} sensors. In fact, their calcium permeability is regulated by cytoplasmic Ca^{2+} according to a biphasic mode: the channels open at Ca^{2+} concentrations below 300 nM (positive feedback) and close above 300 nM Ca^{2+} (negative feedback). Channel gating by Ca^{2+} is known as **Ca^{2+}-induced Ca^{2+} release** from the ER (Figure 14.17). As shown below, this provides the basic mechanism of intracellular calcium signaling.

Figure 14.17 Ca²⁺-induced Ca²⁺ release from the endoplasmic reticulum Upon sensitization by intracellular messengers (cADPR, InsP₃, Ca²⁺), receptor ion channels (InsP₃R and RyR) in the endoplasmic reticulum (ER) membrane release Ca²⁺ into the cytoplasm. Each Ca²⁺ pulse opens the next receptor ion channel, generating a calcium wave that spreads across the ER membrane. As shown in the lower panel, each ion channel becomes desensitized when the Ca²⁺ concentration in its immediate vicinity exceeds a certain value. This mechanism protects the cell from a fatal increase of the Ca²⁺ level and renders a Ca²⁺ wave unidirectional, somewhat resembling an action potential.

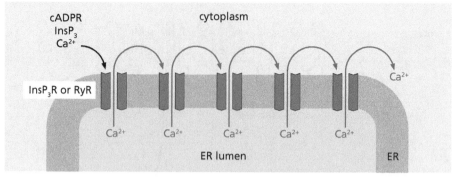

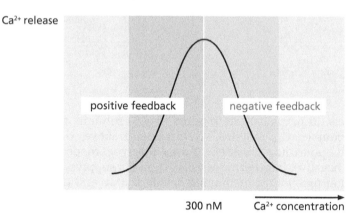

The effect of InsP₃, the Ca²⁺-mobilizing second messenger of a wide variety of extracellular signals (Chapter 4), is to increase the InsP₃ receptor's sensitivity for Ca²⁺-induced Ca²⁺ release. To this end, up to four InsP₃ molecules bind specifically to the four receptor subunits.

Ryanodine receptors serve the same purpose. They were discovered as cellular binding proteins of the plant alkaloid and insecticide ryanodine. Their endogenous ligand, functionally resembling InsP₃, has been sought for many years. Although not yet entirely finished, this task has led to the identification of **cyclic ADP-ribose (cADPR)** as the most probable candidate. This compound is released from nicotinamide adenine dinucleotide (NAD⁺) by the abundant enzyme ADP-ribosyl cyclase, which due to an intrinsic hydrolase activity also inactivates cADPR (Figure 14.18). The signaling pathways leading to cADPR generation and the mechanism of action of cADPR are not fully understood. Probably cADPR interacts with associated proteins rather than with the ryanodine receptor directly. There is also evidence that it stimulates the calcium pump SERCA, thus ensuring that the ER always contains enough Ca²⁺ for Ca²⁺-induced Ca²⁺ release.

Another reaction catalyzed by ADP-ribosyl cyclase is the deamination of NADP⁺. The **nicotinic acid adenine dinucleotide phosphate (NAADP)** thus generated also induces a Ca²⁺ influx into the cytoplasm; however, the source is lysosome-like organelles rather than the ER. The content of NAD⁺ and NADP⁺ is a measure of the metabolic (redox) state of the cell, which via cADPR and NAADP might be linked with Ca²⁺ signaling.

In certain tissues, a release of Ca²⁺ from intracellular stores is triggered, in addition, by **sphingosine 1-phosphate** cooperating with InsP₃. The mechanism of this effect is still widely unknown.

The calcium channels of the ER are interlinked with various signaling pathways. InsP₃ receptors are inhibited by phosphorylation catalyzed by Ca²⁺-dependent

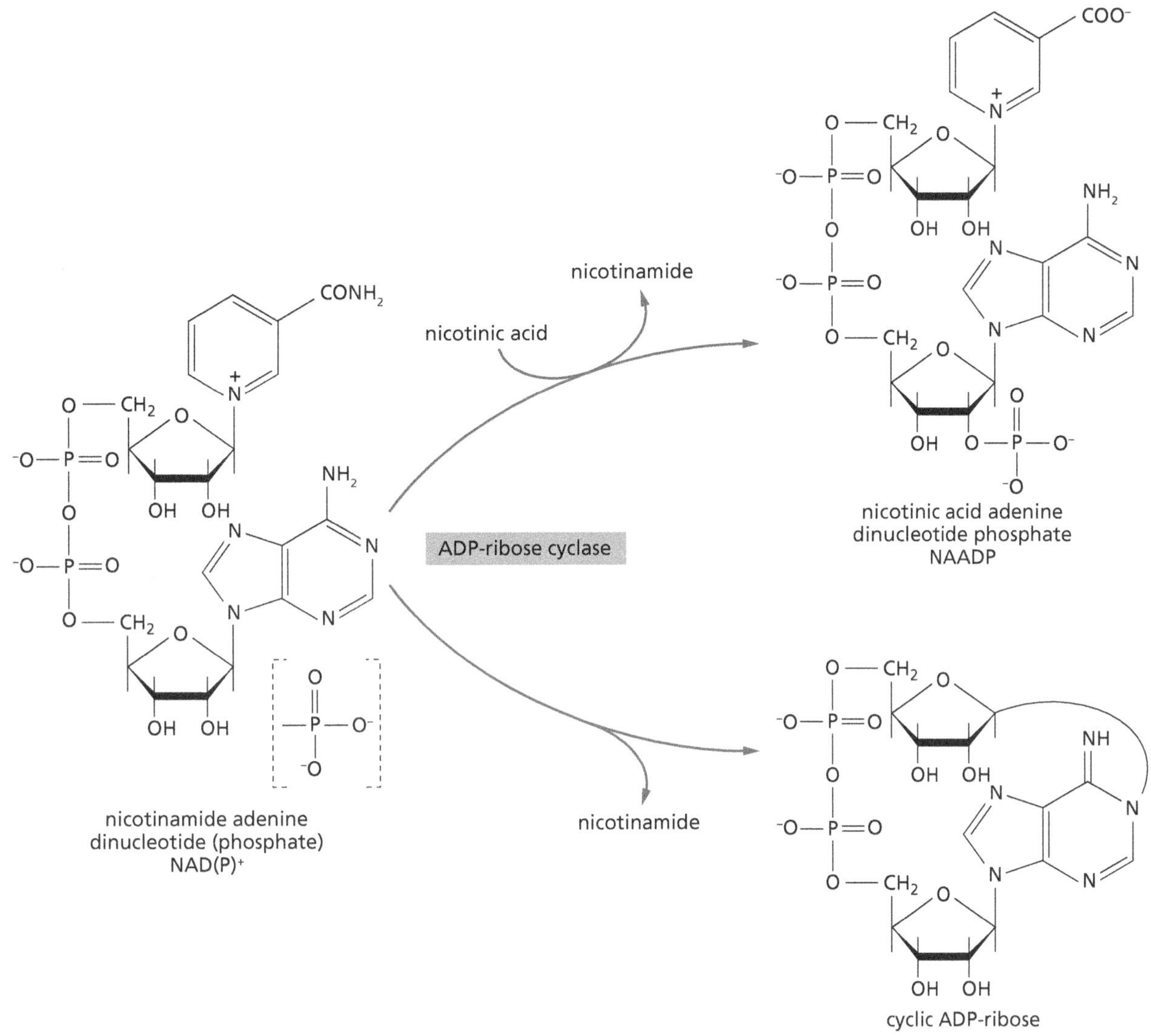

nicotinamide

nicotinic acid

nicotinamide

ADP-ribose cyclase

nicotinamide adenine
dinucleotide (phosphate)
NAD(P)$^+$

nicotinic acid adenine
dinucleotide phosphate
NAADP

cyclic ADP-ribose

protein kinases such as PKC and CaMKII. The same effect is seen for protein kinase G, the activator of which (cGMP) is produced along a pathway involving the Ca^{2+}-dependent synthesis of nitrogen monoxide, NO (see Section 16.5.1). Thus, by attenuating Ca^{2+} release, these kinases are components of negative feedback loops controlling Ca^{2+}-dependent pathways.

Phosphorylation by PKA, in contrast, stimulates InsP$_3$ receptors and ryanodine receptors, and the same holds true for phosphorylation by tyrosine-specific kinases of the Src family. The effect of the cAMP-activated PKA provides another example of complex feedback control, since both the generation and the breakdown of cAMP are either stimulated or inhibited by Ca^{2+}, depending on the adenylate cyclase or phosphodiesterase isoform involved (see Sections 4.4.1 and 4.4.2).

14.5.4 Waves and patterns: the beauty of frequency modulation

Like neuronal data processing, cellular data processing is assumed to be dominated by the principles of spatial privacy and temporal transitoriness. This has been convincingly shown for subcellular Ca^{2+} signals, which may be either restricted to very small microdomains or spread as waves across the whole cell or even within a syncytium coupled by gap junctions.

When visualized by specific dyes (see below), local Ca^{2+} signals appear as narrow (1–3 µm) and short-lived (less than 1 s) sparks or as somewhat more extended

Figure 14.18 Intracellular messengers inducing Ca^{2+} release from internal stores ADP-ribosyl cyclase catalyzes the formation of cyclic ADP-ribose from NAD$^+$ and of NAADP from NADP$^+$. Other Ca^{2+}-releasing messengers are inositol 1,4,5-trisphosphate (see Figure 4.18, Section 4.4.4) and sphingosine 1-phosphate (see Figure 4.25, Section 4.4.5).

Figure 14.19 Ca²⁺ waves An input signal induces a local accumulation of intracellular messenger molecules (such as InsP₃, cADPR, or Ca²⁺), which at a "hot spot" triggers Ca²⁺-induced Ca²⁺ release from the endoplasmic reticulum. This effect may be locally restricted, spread as a wave, or develop into a stationary pattern. The frequency of the oscillating Ca²⁺ signal depends on the intensity of the input signal (upper panel).

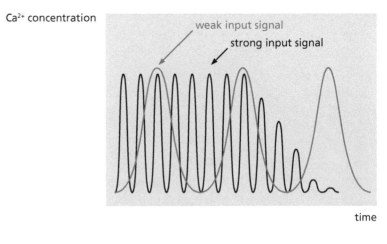

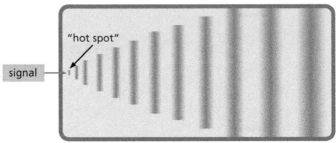

and longer-lasting puffs and spikes. The smallest sparks develop at the cytoplasmic mouth of Ca²⁺ channels; larger ones and spikes are found at so-called "hot spots": where receptors of extracellular input signals and sensitized InsP₃ and ryanodine receptors are concentrated (Figure 14.19).

Since, at higher concentrations, calcium causes apoptosis (Section 13.2.4), a strict limitation and temporal oscillation of Ca²⁺ signals is primarily a question of life and death. This essential control is guaranteed by a strong inhibitory field consisting of a high Ca²⁺-buffering capacity of the cytoplasm, which is brought about by the cooperation of Ca²⁺-binding proteins with the membrane-bound calcium pumps and uniporters mentioned above. As a result, the diffusion rate of Ca²⁺ in the cytoplasm (which is a semisolid medium due to the high concentration of macromolecules) is reduced to a value of 10 $\mu m^2/s$, or about 2% of the rate in water. Therefore, we are dealing with a system where the stimulatory events (such as Ca²⁺-induced Ca²⁺ release) are short-range and controlled by local autofeedback as well as by feedback with the much more extended inhibitory field. This provides the conditions for Ca²⁺ signals to be intimate (restricted to very small areas of the cell) or, on the other hand, to oscillate generating waves as well as complex patterns of excitation that may be looked upon as interferences of stationary waves (Figure 14.20).

In fact, sparks as well as hot spots are starting points of calcium waves migrating across the ER. Such waves are generated by a cooperation of many channels. The underlying mechanism is Ca²⁺-induced calcium release. It enables adjacent calcium channels to activate each other (Figure 14.17). The unidirectional wavelike propagation of Ca²⁺ release requires the activated channel to become refractory for a short period of time. This desensitization is achieved by Ca²⁺-dependent channel blocking at concentrations above 300 nM (Figure 14.17). Such a situation resembles the spreading of action potentials (Figure 14.6). The ER has been called, therefore, an excitable medium, though the chemical calcium waves are much slower than the electrochemical action potentials. However, this apparent drawback is compensated by the short distances within the cell as compared with the nervous system.

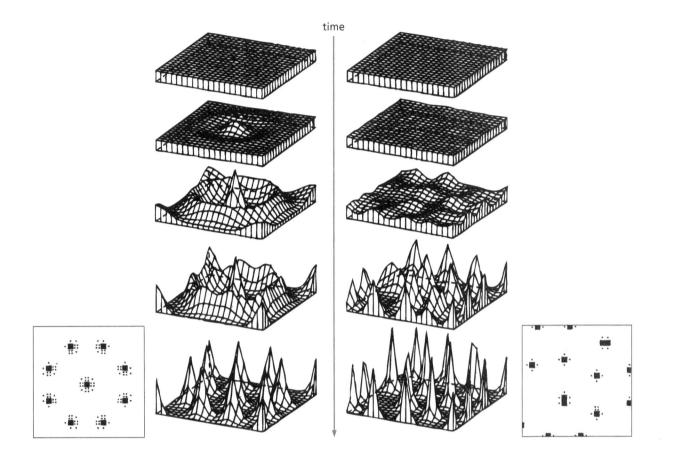

time

Figure 14.20 Pattern formation A computer-generated pattern in a feedback system with a short-range stimulatory and a long-range inhibitory effect is shown. To the left, the pattern was induced by a single input signal and has developed in a highly regular fashion. To the right, the result is shown for statistically fluctuating input signals inducing an irregular pattern. (From A. Gierer, Die Physik, das Leben und die Seele. München and Zürich: Piper-Verlag, 1985.)

Like any wave, a calcium wave is subject to modulation of both amplitude and frequency. By frequency-dependent modulation, the efficacy of information processing is increased tremendously [one may consider, for instance, the large difference between a black-and-white photograph (being mainly amplitude-modulated) and a color picture, or between a series of drumbeats and a symphony].

The wavelength of an oscillating calcium signal may fluctuate between milliseconds and several minutes. It directly depends on the nature and the intensity of the input signal: the stronger this is, the more $InsP_3$ receptors (or their relatives) become activated, the more sparks are generated at a hot spot, and the higher is the frequency of the resulting calcium wave.

The Ca^{2+} effector proteins located downstream respond to the different frequencies in an individual manner. In other words, depending on the frequency of the wave, different sets of proteins are addressed by calcium. A prototype of a protein acting as a frequency detector is CaM kinase II, which is discussed in detail in Section 14.6.3.

How to prove frequency modulation of signals

To reiterate, practically all cellular processes are modulated directly or indirectly by Ca^{2+}. Therefore, one may start from the assumption that, via signaling cross talk, the oscillating pattern of Ca^{2+} signals is transduced to other signals, which, vice versa, may modulate calcium signaling. This has been studied in detail for the interaction between Ca^{2+} and cAMP signaling.

Investigation of frequency-modulated signals requires the monitoring of rapid changes of concentrations in living cells. Optical methods have turned out to be particularly suitable since they work almost without delay, enabling an exact localization of signals. Among the most advanced techniques are those

Figure 14.21 Assay of Ca²⁺ and cAMP by FRET analysis (A) A fusion protein consisting of calmodulin (CaM) and the CaM-binding domain of myosin light-chain kinase (MLCK) is linked on both sides with green fluorescent protein (GFP) derivatives 1 and 2 emitting different-colored fluorescent light upon excitation by UV light. Binding of Ca²⁺ to CaM induces a conformational change bringing 1 and 2 close together, thus enabling a fluorescence resonance energy transfer (FRET). As a result, the intensity of the fluorescent signal of 1 decreases by the same value as the intensity value for the signal of 2 increases. (B) For a cAMP assay, the catalytic subunits (CU) and the regulatory subunits (RU) of protein kinase A are fused with different-colored GFP derivatives 1 and 2, which in the absence of cAMP are close enough to enable a FRET from 1 to 2. Upon binding of cAMP, the complex dissociates and the FRET signal is diminished.

employed for an analysis of calcium. For instance, organic dyes that undergo a change in their absorption spectra upon Ca²⁺ binding are used. Such probes are applied by microinjection or as membrane-penetrating derivatives that release the dye upon enzymatic hydrolysis in the cell. More advanced approaches make use of bioluminescence proteins. An example is the protein aequorin from the luminous jellyfish *Aequorea victoria*: it is activated by Ca²⁺ to emit light the intensity of which depends on the Ca²⁺ concentration. An even better yield of light is obtained with the **green fluorescent protein (GFP)**. The fluorescence of GFP is activated by aequorin in the jellyfish, but it can also be excited in the laboratory by UV light (see also Section 5.2). Moreover, GFP has the advantage that derivatives obtained by an exchange of certain amino acid residues or by molecular rearrangements emit light of different colors. However, GFP does not interact directly with Ca²⁺. Therefore, by means of genetic manipulations, fusion proteins have been constructed that, in addition to GFP, contain CaM as a calcium sensor and the CaM-binding motif of the myosin light-chain kinase (MLC kinase) as an effector domain. In the presence of Ca²⁺, CaM induces a conformational change of this effector domain, which is transduced to the GFP moiety, changing its fluorescence activity. Based on this approach, several Ca²⁺-indicators have been developed that may be expressed deliberately in cells, enabling a very sensitive assay of temporal and spatial fluctuations of Ca²⁺ concentrations.

An interesting variant of this analytical technique is offered by fusion proteins where the CaM/MLC kinase effector domain pair is flanked on both sides by different-colored GFP derivatives. These so-called chameleons make use of the quantum-mechanical phenomenon of fluorescence resonance energy transfer

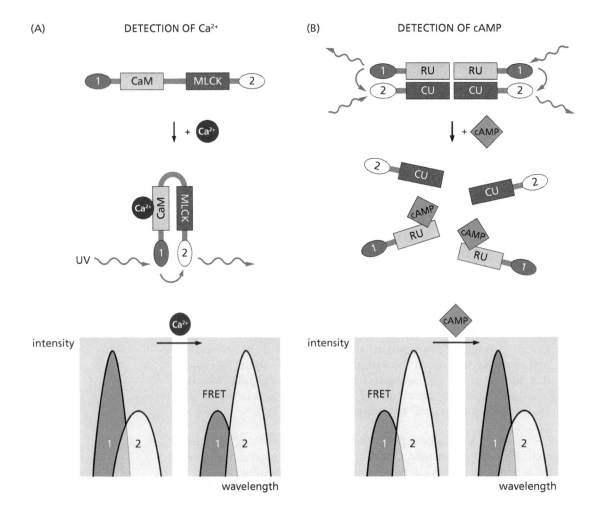

(FRET; for details see Section 5.2) where one of the chromophores activates the other one to fluoresce, provided they come together very closely. The latter condition is fulfilled only in the presence of Ca^{2+} when the fusion protein changes from a stretched into a folded conformation (Figure 14.21A).

Such methods can also be employed to study other signaling molecules. An illustrative example is provided by cAMP (Figure 14.21B). Here one uses cAMP-binding PKA as a sensor, making use of a characteristic property of this enzyme: as described in Section 4.6.1, the catalytic and regulatory domains of PKA are located on separate subunits, which in the absence of cAMP are firmly associated. Binding of cAMP at the regulatory subunits results in a dissociation of the complex. When by genetic manipulations the subunits have become fused with different-colored fluorescent proteins, the concentration of cAMP can be quantitatively assayed by FRET. Since a maximal FRET signal is obtained only when both subunits are completely associated (in the intact complex), cAMP diminishes the FRET signal in a concentration-dependent manner. By employing this and related approaches, a tight coupling between Ca^{2+} and cAMP oscillations has been found that is due, for instance, to the activities of Ca^{2+}-controlled adenylate cyclases (Section 4.4.1) and cAMP phosphodiesterases (Section 4.4.2). Moreover, cAMP oscillations may occur independently of Ca^{2+} due to the oscillator-type construction of cAMP response modules as shown in Section 4.6.1.

Since feedback interactions enabling undamped oscillations of output signals are a hallmark of most signal-processing protein complexes, the frequency modulation of signaling certainly is not restricted to Ca^{2+}- and cAMP-driven systems but may be assumed to be a (or even *the*) major mode of cellular data processing.

14.5.5 Store-operated calcium channels recharge the cell's calcium battery

Longer-lasting waves and patterns require a continuous replenishment of the Ca^{2+} store in the ER because the calcium-dependent ATPases and ion exchangers of the plasma membrane permanently pump cytoplasmic calcium into the extracellular medium. To replenish the store, the ER is in contact with a distinct class of Ca^{2+} channels in the plasma membrane called **store-operated calcium channels (SOCCs)**, which are opened in response to store depletion in the ER. These channels cooperate with the ER-bound ATPase SERCA, which pumps the Ca^{2+} entering the cell through the SOCCs into the lumen of the ER. Thus, like electrical capacitors, the intracellular calcium stores are discharged and recharged; the replenishment of the stores has therefore been called **capacitative Ca^{2+} influx**. It is essential for any long-lasting or oscillating calcium signal. The nature and mode of operation of SOCCs are still a matter of debate. It seems certain, however, that tissues express different channel types with different mechanisms of action.

As far as the function and the structure are concerned, the channel **CRAC**, a subtype of SOCCs found especially in lymphocytes and some other cell types, has been studied most thoroughly. It causes an electrophysiologically detectable Calcium Release-Activated calcium Current that is highly ion-selective. The CRAC channel is composed of subunits called **Orai**, reminiscent of the mythological gatekeepers of the ancient Greek heaven. The Orai protein is expressed in several isoforms. It consists of four transmembrane helices with the C- and N-termini facing the cytoplasm, but it does not exhibit any structural relationship with other Ca^{2+} channels (Figure 14.22). Although in comparison with VOCCs the conductivity of the CRAC channel is small, it is sufficient to guarantee a continuous Ca^{2+} influx and replenishment of the ER store.

The Ca^{2+} load of the ER lumen is monitored by a sensor protein. This protein was discovered in quite another context, as a STromal Interaction Molecule (**STIM**)

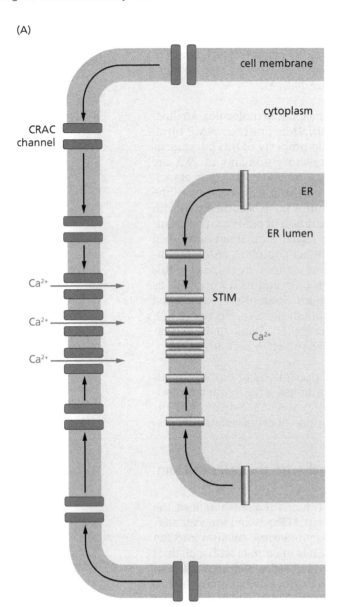

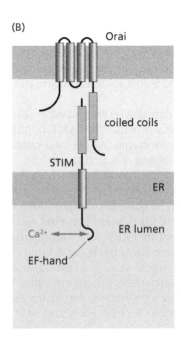

Figure 14.22 Model of the capacitative Ca²⁺ influx (A) Upon Ca²⁺ depletion of the endoplasmic reticulum (ER) lumen, the calcium-sensing protein STIM (stromal interaction molecule; shown as gray cylinders) in the ER membrane aggregates opposite to clusters of the calcium-release-activated Ca²⁺ channel (CRAC), activating Ca²⁺ influx from the cellular environment. Ca²⁺ entering the cytoplasm is pumped into the ER lumen by the Ca²⁺-dependent ATPase SERCA (not shown). (B) STIM, a single-pass transmembrane protein of the ER that exhibits a Ca²⁺-binding EF-hand domain, is shown to interact via coiled-coil domains with the CRAC subunit Orai, a cell membrane protein with four transmembrane helices. Whether such interaction controls the gating of the CRAC channel is not clear yet.

and putative tumor suppressor in a splenic stromal cell line. Later STIM was identified as a protein involved in capacitative Ca²⁺ influx. Integrated in the ER membrane by a single transmembrane domain, STIM exhibits a Ca²⁺-binding EF-hand domain (Section 14.6.1) that serves as a Ca²⁺ sensor at the lumeinal side. At the cytoplasmic side STIM has an extended coiled-coil structure and interaction domains (Figure 14.22). Upon store depletion, STIM aggregates into patches concentrating opposite assemblies of CRAC channels and other SOCCs in the plasma membrane (Figure 14.22). The mechanism by which the signal is transduced from STIM to the channels is still not known. Evidence exists for both a direct interaction with channel proteins and a diffusible messenger molecule.

In addition to Orai channels, a series of **TRP channels** (Transient Receptor Potential; see Section 14.5.6) participates in store-dependent Ca²⁺ influx. In fact, almost all TRP channels of the TRPC subfamily seem to function as SOCCs, rivaling or replenishing (probably in a tissue-specific way) the effect of Orai channels. Moreover, like Orai, the TRP channels bind to and are regulated by STIM proteins, indicating that this type of interaction represents a universal signal reporting the charge of the ER battery. Nevertheless, other Ca²⁺-binding ER proteins such as calreticulin and calsequestrin have been proposed also to monitor store depletion.

Store-controlled channels cooperate with arachidonic acid-controlled calcium channels to regulate the amplitude and frequency of Ca^{2+} waves

As shown by inhibitor experiments, cytoplasmic Ca^{2+} oscillations not only depend on channels like CRAC and on the charge of the ER store but also are regulated by novel ligand-controlled Ca^{2+} channels in the plasma membrane. The gating ligand is arachidonic acid and the channel is called, therefore, **ARCC** (<u>A</u>rachidonic acid <u>R</u>egulated <u>C</u>alcium <u>C</u>hannel). ARCC plays a crucial role since calcium waves frequently break down when this particular channel is blocked. In the cell, ARCC and SOCC exhibit different functions: while SOCC predominantly controls the amplitude, ARCC regulates the frequency of Ca^{2+} waves. Moreover, Ca^{2+}-dependent adenylate cyclases are activated exclusively by CRAC, probably because they form complexes with this channel.

Arachidonic acid is released from membrane phospholipids by Ca^{2+}-dependent phospholipase A_2 (PLA_2), a reaction controlled by a wide variety of extracellular signals (Section 4.4.5). Its gating effect on ARCC shows that, in addition to being the precursor of eicosanoids, arachidonic acid is a bona fide second messenger (Figure 14.23).

Other ligand-controlled Ca^{2+} channels of the plasma membrane

In contrast to ARCC, which transports exclusively Ca^{2+}, other ligand-controlled "calcium channels" of the plasma membrane are nonspecific cation channels rather than being Ca^{2+}-selective. Nevertheless, some of them exhibit some preference for calcium. Such channels are controlled by intra- and extracellular signals. Prominent examples are the nicotinic acetylcholine receptors, the P2X purine receptors, and the ionotropic glutamate receptor NMDA. These receptor-coupled ion channels are discussed in Chapter 16. For another type of ligand-controlled cation channels, the cyclic nucleotide-gated channels, see Section 14.5.7.

14.5.6 TRP ion channel family: epithelial Ca^{2+} channels and more

As mentioned above, several SOCCs are members of the channel family TRP, which will be briefly reviewed here. The term transient receptor potential refers

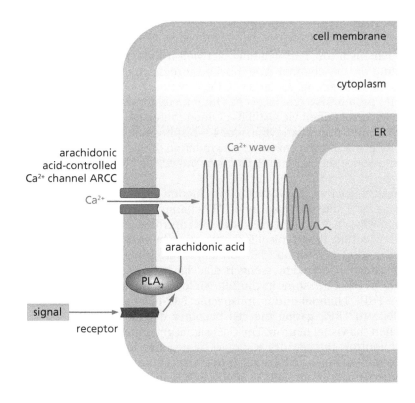

Figure 14.23 Arachidonic acid-controlled Ca^{2+} influx A signal (hormone, neurotransmitter, cytokine, or environmental stimulus) stimulates the receptor-dependent release of arachidonic acid by phospholipase A_2 (PLA_2). Arachidonic acid opens the Ca^{2+} channel ARCC, and the Ca^{2+} entering the cell induces and modulates Ca^{2+} oscillations generated at the ER according to the mechanism of Ca^{2+}-dependent Ca^{2+} release.

to the discovery of this channel type in the course of electrophysiological studies on *Drosophila* mutants exhibiting a defect of light adaptation (see Section 15.5).

In the meantime the TRP family has turned out to be quite large, including no less than 28 genes in humans. They are classified into six subfamilies (a comprehensive list of mammalian TRP channels is provided by D.E. Clapham, *Cell* 129:220–221, 2007):

- TRPC (canonical)

- TRPV (V refers to the vanilloid receptor explained in Section 15.3)

- TRPM (related to melastatin, which was discovered as a tumor suppressor protein)

- TRPP (related to polycystins, proteins that are mutated in autosomal dominant polycystic kidney disease; see Section 15.1 for details)

- TRPML (related to mucolipin, a protein associated with the disease mucolipidosis)

- TRPA or ANKTM1 (ankyrin-like transmembrane protein 1)

Due to alternative splicing, the TRP genes give rise to a large number of proteins that associate into homo- and hetero-oligomers. They constitute one of the most versatile ion channel families, all the manifold functions of which are still the subject of ongoing research. Relatively well-known is the key role TRP channels play in **sensory signal processing** (Chapter 15).

TRP and TRP-related channels are active in all multicellular eukaryotes examined so far. They transport cations: some of them in a rather nonspecific manner, others with a strong preference for calcium. TRP channels belong to the cation channel prototype 1 (see Figure 14.2). They are composed of four subunits with six transmembrane helices and a P-loop each (Figure 14.24). With the exception of the heat sensor TRPV1, the cold sensor TRPM8, and the taste-sensitive ion channel TRPM5 (see Chapter 15), TRP channels are not voltage-gated because the positive charge of transmembrane domain S4, the voltage sensor in potential-dependent channels, is insufficient.

The mechanism of TRP channel activation is not fully understood yet. Depending on the channel type, obviously more than one signaling pathway leads to channel gating. As far as SOCC function is concerned, an interaction with STIM proteins is a crucial event. Other mediators of channel activation may be diacylglycerol (albeit via a still ill-defined route), cADPR (Section 14.5.3), and Ca^{2+} ions. Since phosphatidylinositol 4,5-bisphosphate inhibits the opening of many TRP channels, the enzymatic degradation of this membrane-bound phospholipid by phospholipase C (PLC) might provide an additional gating signal.

Regardless of whether diacylglycerol production or phospholipid degradation is the triggering signal, the majority of TRP channels seems to be activated by the PLC isoenzymes β and γ that are under the control of G-protein- and Tyr kinase-coupled receptors. Prototypical is the situation in the **insect eye**. Here the light-sensitive rhodopsin (Section 15.5) stimulates via a trimeric $G_{q,11}$-protein a β-type PLC, releasing the second messengers diacylglycerol and InsP3. InsP3 triggers the depletion of the Ca^{2+} store in the ER, while in the plasma membrane a Ca^{2+}-selective TRPC channel and a nonspecific TRP-like cation channel (TRPL) are gated. Due to TRPL gating the cell becomes depolarized, evoking an action potential in the visual neuron. The Ca^{2+} signal generated in parallel by the TRPC channel controls the sensory response by negative feedback and, in addition, serves to refill the intracellular calcium stores. Other examples of TRP channels with known functions are TRPC1, both a SOCC and a putative mechanoreceptor

(Section 15.2), and a corresponding receptor NOMPC (NO Mechanoreceptor Potential Channel) of insects, both of which respond to touch stimuli, as well as TRPP1 and 2, which sense fluid currents in kidney cilia. A particular subfamily of sensory TRP channels is the **vanilloid receptors** that belong to the TRPV type. These nonspecific cation channels expressed by sensory neurons respond to temperature stimuli. They are discussed in more detail in Section 15.3.

A channel of special interest is TRPC2 that is expressed in most vertebrates with the exception of primates. The gating of TRPC2 has been found to induce the so-called **acrosomal reaction** in mouse sperm. This Ca^{2+}-dependent exocytotic process is a key event of reproduction: it promotes the release of lytic enzymes that enable the sperm cell to penetrate the *zona pellucida* wrapping the oocyte. TRPC2 is a Ca^{2+}-transporting channel that is activated by proteins of the *zona pellucida* along the PLCβ pathway but also by store depletion. The channel is expressed also in the cells of the vomeronasal organ, the sensory organ for (sex) pheromones in vertebrates (Section 15.4). In primates, the *trpc2* gene has degenerated to an inactive pseudogene and the vomeronasal organ is probably out of function. Whether or not the sperm channel TRPC2 is replaced here by another channel remains an open question.

In vertebrates the **epithelial Ca^{2+} channels (ECaC)**, found particularly in the epithelia of kidney and small intestine, have been identified as the highly Ca^{2+}-selective TRP channels TRPV5 and -6. Their function resembles that of the nonrelated ENaCs: being located in the apical membrane of the epithelial cell, ECaCs are the mediators of vitamin D-controlled Ca^{2+} (re)absorption. In the basolateral membrane two calcium pumps, a Ca^{2+}-dependent ATPase and a Na^+–Ca^{2+} exchanger, convey the Ca^{2+} that has entered the cell through ECaC into the bloodstream (Figure 14.24). TRPV5 and -6 are leaky channels: they are permanently open. This creates a problem, because a continuous Ca^{2+} influx would lead to undesirable second messenger effects and eventually culminate in cell death. To avoid this, in the cell Ca^{2+} is transiently sequestered by a Ca^{2+}-buffering protein, calbindin, which hands over Ca^{2+} to the ion pumps. The transcription of the genes encoding TRPV5, TRPV6, and calbindin is promoted by calcitriol (1,25-dihydroxyvitamin D_3). The latter is the active form of vitamin D_3 that binds to the vitamin D receptor VDR, a transcription factor (Section 8.3). Along this route, D vitamins stimulate the uptake of calcium from food and promote bone formation. Vitamin D deficiency is the cause of rickets.

Figure 14.24 Mode of operation of the epithelial Ca^{2+} channel ECaC (A) One of the four subunits forming a channel. (B) Epithelial cell with an apical ECaC, enabling passive Ca^{2+} influx from the environment (food, urine), and a basolateral ATPase/exchanger complex, pumping Ca^{2+} into the blood. To avoid undesirable second messenger effects, the Ca^{2+} to be transported is transiently bound by the cytoplasmic protein calbindin.

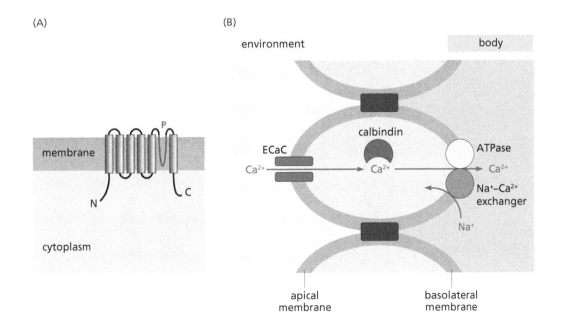

(A)

(B)

507

Figure 14.25 Gating of a cyclic nucleotide-regulated cation channel by cAMP The model, based on X-ray crystallographic studies, shows the conformational change of ion channel proteins induced by binding of cAMP (red) to the cyclic nucleotide-binding domains (cNBD). Transmembrane helices S5 and S6, which form the membrane pore, are turning aside and the positively charged bundle of transmembrane helices S1–4 are turning upward, opening the channel. (Adapted with modifications from H. Rehmann, A. Wittinghofer, and J.L. Bos, *Nat. Rev. Mol. Cell Biol.* 8:70, 2007.)

A TRP channel probably functioning as an osmosensor is expressed by heart muscle, liver, and kidney cells and is thought to play an important role in salt and water balance and, thus, in blood pressure control. A similar function is attributed to the channel TRPC4 in endothelial cells. By transporting Ca^{2+} into the cell, this channel triggers the synthesis and release of vasodilatory mediators such as nitrogen monoxide (generated by a Ca^{2+}-dependent NO synthase) and the arachidonic acid derivative prostacyclin (the synthesis of which is triggered by a Ca^{2+}-dependent PLA_2 providing the precursor lipid arachidonic acid). In contrast, Ca^{2+}-selective TRP channels in smooth muscle cells induce vasoconstriction by triggering muscle contraction.

14.5.7 Cyclic nucleotide-gated ion channel family

Another family of ligand-dependent nonspecific cation channels with a high permeability for Ca^{2+} ions is the cyclic nucleotide-gated (CNG) channels. The best-known representatives are the cGMP- or cAMP-controlled ion channels of vertebrate photoreceptor and odorant receptor cells (Chapter 15). Upon gating, a Na^+ efflux induces depolarization while Ca^{2+} enters the cell along a steep concentration gradient. The ligand specificity of the CNG channels is variable: for

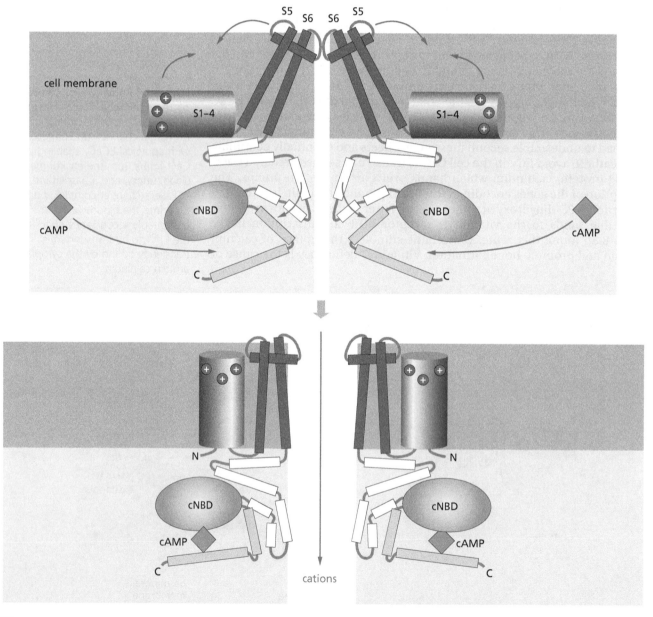

instance, the channel in odorant receptor cells does not distinguish between cAMP and cGMP, whereas the channel of photoreceptor cells exhibits a strong preference for cGMP.

CNG channels exist as heterotetramers consisting of two A- and two B-subunits. Each subunit has a cyclic nucleotide binding site as well as six transmembrane helices and a P-loop (Figure 14.25); thus the architecture of a CNG channel resembles that of voltage-dependent potassium channels (prototype 1, see Figure 14.2). However, with the exception of the HCN subfamily (see below), CNG channels are not voltage-controlled because the transmembrane helix S4, which normally serves as a voltage sensor, lacks some of the positively charged amino acid residues required for voltage responsiveness (see Section 14.2). Another component of the CNG channel complex is the Ca^{2+}-binding protein calmodulin. It mediates a blockade of the channel by calcium ions and plays a key role in sensory adaptation (Section 15.6).

CNG channels and their relatives are widely distributed among eukaryotes. They have been found not only in sensory cells but also in other tissues, where their function remains more or less unknown. An exception may be provided by sperm cells. From experiments on sea urchins, it is known that peptide factors released by the egg stimulate motility, energy metabolism, and the acrosomal reaction of sperm cells. To this end, these peptides interact with a receptor guanylate cyclase on the sperm membrane. The cGMP thus generated induces a Ca^{2+} influx into the cytoplasm, which acts as a trigger of sperm activation. In mammalian sperm cells (unfortunately not yet in sea urchin sperm), the corresponding cGMP-sensitive Ca^{2+} channel has been identified as a member of the CNG family.

A subfamily of CNG channels is the **HCN channels** (Hyperpolarization and Cyclic Nucleotide-gated channels). These voltage-dependent cation channels are unique in that they are already gated by depolarization at extremely negative resting potentials and are, in addition, activated by cAMP and cGMP. As pacemaker channels, they play a central role in oscillating processes such as the heartbeat (for details see Section 14.6.2).

Summary

By means of Ca^{2+}-dependent ATPases and Na^+ gradient-driven exchangers, cells generate and maintain a steep gradient between extracellular and cytoplasmic Ca^{2+} concentrations (1 mM versus 0.1 μM). Gating of corresponding membrane channels automatically leads to an influx of Ca^{2+} into the cytoplasm. In most cell types, such an influx does not cause a significant change of the membrane potential. Instead, Ca^{2+} has the effect of a second messenger interacting with a wide variety of proteins. Voltage-operated Ca^{2+} channels (VOCCs) gated by depolarization are found in all cell types. They are the ultimate targets of action potentials. Their overall structure resembles that of voltage-dependent Na^+ channels but the subunit composition is more complex. Several such channel types are distinguished that are tissue-specifically expressed. The L-type channel of heart muscle is an important target of pharmaceutical drugs used for the treatment of high blood pressure and cardiac arrhythmia. VOCCs are stimulated by protein kinase A- and CaM kinase II-catalyzed phosphorylation (sympathetic effect) and are inhibited by Ca^{2+} and by βγ-subunits of $G_{i,o}$-proteins. Two major types of ligand-operated Ca^{2+} channels are found in the ER: inositol 1,4,5-trisphosphate ($InsP_3$) receptors and ryanodine receptors. Structurally they are closely related and resemble the upside-down membrane topology of the voltage-dependent potassium channel. Both are gated by Ca^{2+}. This Ca^{2+}-induced Ca^{2+} release from the ER into the cytosol is augmented by the ligands $InsP_3$ and cyclic ADP-ribose (cADPR) and blocked by Ca^{2+} at concentrations above 300 nM. Other intracellular messengers that induce Ca^{2+} release from internal stores are nicotinic acid adenine dinucleotide phosphate (NAADP) and sphingosine 1-phosphate. Ca^{2+}-

induced Ca^{2+} release through $InsP_3$ receptors and ryanodine receptors provides a means for the generation of frequency-modulated Ca^{2+} signals, appearing as both propagating waves and stationary patterns. The wavelength and amplitude depend on the intensity of input signals inducing at "hot spots" the release of $InsP_3$ and cADPR, respectively. Because of the cross talk between individual signal-transducing pathways, the frequency modulation of Ca^{2+} signals can be transferred to other cellular signals. Storage-operated calcium channels are connected with Ca^{2+}-sensor proteins in the ER and become gated when the Ca^{2+} level of the internal store falls below a critical value. The capacitative calcium influx thus induced is required for a continuous reloading of the internal store, being a condition of long-lasting Ca^{2+} oscillations. Intracellular Ca^{2+} oscillations also depend critically on arachidonic acid-controlled calcium channels. TRP channels are cation channels with some preference for Ca^{2+} that exhibit the membrane topography of voltage-dependent K^+ channels although most of them are not voltage-sensitive. They play an important role in the processing of sensory stimuli and are involved in capacitative calcium influx, trans-epithelial calcium transport, acrosomal reaction of sperm cells, vasoconstriction, osmosensation, and in NO and prostacyclin synthesis by endothelial cells. Depending on the isotype, TRP channels are gated along various signaling pathways, most of which include an activation of C-type phospholipases. Cyclic nucleotide-gated channels are nonspecific cation channels with some preference for Ca^{2+} that are activated by cAMP or cGMP. They are widely distributed in eukaryotes and play a particular role in sensory signal processing and sperm activation. Their overall architecture resembles that of voltage-dependent potassium channels. However, with the exception of the subtype HCN, which constitutes the pacemaker channels in the heart, cyclic nucleotide-gated channels are not voltage-sensitive.

14.6 Downstream of Ca^{2+} signals

As a chemical signal, Ca^{2+}, like other second messengers, interacts with proteins. Such calcium-sensitive proteins are expressed by cells in large numbers. Some of them fulfill the role of buffer proteins: in cooperation with ion channels, ion pumps, and ion exchangers, they control the level of free Ca^{2+} in the cytoplasm and in organelles by binding and releasing Ca^{2+} in a concentration-dependent manner. Examples are calreticulin and calsequestrin in the endo(sarco)plasmic reticulum and calbindin and parvalbumin in the cytosol.

Most Ca^{2+}-binding proteins, however, are components of signaling cascades. For them Ca^{2+} is an input signal modulating their conformation and function and prompting them to emit an output signal. Some of these effector proteins directly interact with Ca^{2+}, whereas others receive the input signal from a separate Ca^{2+}-sensor protein. Ca^{2+}-effector proteins have Ca^{2+}-recognition domains, with the EF-hand domain being the most abundant one, followed by the C2 and annexin domains. A selection of Ca^{2+}-binding proteins is given in Table 14.2.

14.6.1 Calmodulin and other calcium sensor proteins

Proteins with EF-hand domains constitute the largest and most heterogeneous family of calcium-binding proteins. The prototypical **EF-hand protein** is calmodulin (CaM). This universal Ca^{2+}-sensor protein of eukaryotic cells transduces the Ca^{2+} signal to a large number of effector proteins. The mammalian form consists of 148 amino acids and has a molecular mass of 16.7 kDa. Two helix–loop–helix motifs at the amino terminus and two at the carboxy terminus are denoted as EF-hands (Figure 14.26). These are the binding sites of four Ca^{2+} ions that interact with the protein in a cooperative manner. The interaction changes the conformation in such a way that complementary structures of effector proteins fit a hydrophobic pocket of CaM. As a result, both the conformation and the function of the effector protein are modulated. CaM itself is a substrate of numerous protein kinases that change its activity by phosphorylation.

Table 14.2 A selection of Ca^{2+}-binding proteins

Proteins that buffer and sequester Ca^{2+}

    calbindin

    calsequestrin

    calreticulin

    parvalbumin

Ca^{2+} sensor proteins

    calmodulin

    troponin C

    S100 proteins

    neuronal Ca^{2+}-sensor proteins (recoverin, GCAP)

    synaptotagmins

Enzymes

    protein kinase C (c- and n-subfamilies)

    calpain (protease)

    phospholipase A$_2$

    phospholipases C

    NO synthases (constitutively expressed isoforms eNOS and nNOS)

Receptors and ion channels

    InsP$_3$ receptors

    ryanodine receptors

    Ca^{2+}-activated K$^+$ and Cl$^-$ channels

    Ca^{2+}-dependent ATPases

Cytoskeletal proteins

    α-actinin

    gelsolin

    caltractin

Cell adhesion and other proteins

    annexins

    cadherins

Transcription factors

    DREAM

Factors controlling G-protein activity

    GRF (a Ras-GEF)

    GRP (a Ras-GEF)

    CAPRI (a Ras-GAP)

In Table 14.3, some enzymes are summarized that contain CaM as a regulatory subunit or the activity of which is controlled by Ca^{2+}/CaM. They are discussed in detail elsewhere.

An isoform of CaM, **troponin C**, is specialized to interact with the troponin–tropomyosin complex of skeletal muscle cells (see below). The **S100 proteins** (100% soluble in ammonium sulfate solution) are a group of at least 20

Figure 14.26 Calmodulin Upon binding of Ca²⁺ (red spheres), the inactive stretched form of calmodulin (left) develops a central helical structure (middle) that, bends into a hydrophobic cavity that fits corresponding domains of partner proteins (red, right). This rearrangement is supported by hydrophobic interactions between the N- and C-terminal EF-hand domains. EF-hands are helix–loop–helix structures binding one Ca²⁺ ion each at aspartate and glutamate residues in the central loop. The name refers to the arrangement of the two helices E and F to a handlike structure with stretched thumb and forefinger.

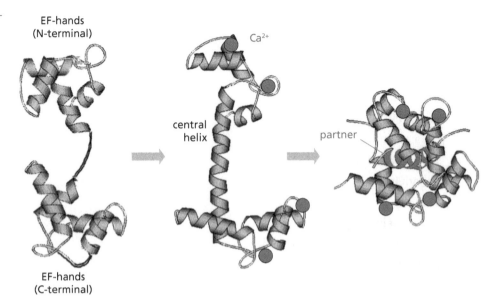

EF-hand proteins that, like CaM, interact with a large number of effector proteins, but in contrast to CaM, they are tissue-specifically expressed. In the clinic they are used as markers for certain tumor types and for brain damage by accidents or stroke. Much remains to be learned about the physiological role of this protein family. Other Ca²⁺-sensor proteins with EF-hands have been found particularly in nerve cells. Well-known examples are recoverin and guanylate cyclase-activating proteins (GCAPs), the expression of which is restricted to retina cells, where they are involved in light adaptation (see Sections 15.5 and 15.6). Both are representatives of the large family of **neuronal Ca²⁺-sensor (NCS) proteins**. An interesting member of this family, the transcriptional repressor DREAM, will be discussed in Section 14.6.4. NCS proteins may shuttle between cytoplasm and membranes. For this reason, most of them contain a tetradecanoyl (myristoyl) residue covalently bound at the N-terminus that functions as a membrane anchor. Depending on the Ca²⁺ signal, this lipid moiety either is buried inside the protein, unable to interact with membranes, or is exposed at the outside (Figure 14.27). By means of this **Ca²⁺–myristoyl switch** such proteins can be anchored reversibly at membranes,

Table 14.3 Ca²⁺/calmodulin-dependent enzymes: a selection

Enzyme	Major function
CaMKI	regulation of gene transcription (Section 14.6.3)
CaMKII	regulation of synaptic plasticity (Section 14.6.3)
CaMKIV	regulation of gene transcription (Section 14.6.3)
eEF2 kinase (CaMKIII)	eEF2-specific; inhibition of translation (Section 9.3.4)
MLC kinase	MLC-specific; contraction of the actin–myosin complex (Section 14.6.2)
phosphorylase kinase	specific activation of glycogen phosphorylase (Sections 5.3.1 and 14.6.3)
CaMK kinases α and β	activation of CaMKI and -IV (Section 14.6.3)
calcineurin (protein phosphatase 2B)	protein dephosphorylation (Section 14.6.4)
Ca²⁺-dependent ATPase	Ca²⁺ export (Section 14.5.1)
constitutive NO synthases eNOS and nNOS	synthesis of nitrogen monoxide (Section 16.5.1)
adenylate cyclases 1, 3, and 8	synthesis of cAMP (Section 4.4.1)
cAMP/cGMP phosphodiesterases type 1	degradation of cAMP and cGMP (Section 4.4.2)

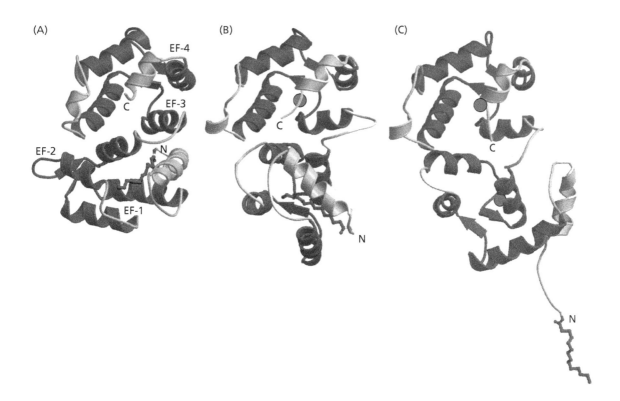

providing an important condition for a controlled interaction with other signaling proteins (the reader may recall that the intracellular localization of small G-proteins of the Arf family is regulated by an analogous mechanism; see Section 10.4.3).

NCS proteins have various functions including modulation of voltage-dependent calcium and potassium channels, of neurotransmitter receptors, and of pro- and anti-apoptotic effects. As indicated by experiments with gene-manipulated animals, they are involved in learning and memory fixation. In addition, they seem to play a role in psychiatric diseases. Similar functions have been attributed to other Ca^{2+}-binding proteins of the EF-hand family.

The second frequent Ca^{2+}-binding motif is the **C2** or **Ca^{2+}–lipid-binding domain**. As indicated by the name, through this domain proteins are reversibly attached to membranes, where Ca^{2+} interacts electrostatically with acidic phospholipids such as phosphatidylserine. NCS proteins of this family are the **synaptotagmins** that control neurotransmitter release at synapses (as discussed in detail in Section 10.4.2). In addition, a series of enzymes with C2 domains are known, for instance, conventional and new PKCs (in fact, C2 stands for conserved domain number 2 of these enzymes), several phospholipases, and the Ras GTPase-activating protein CAPRI (Section 10.1.3).

The **annexin domain**, the third frequent Ca^{2+}-binding motif, shares with the C2 domain the property of binding proteins via Ca^{2+} ions to acidic membrane phospholipids. It is particularly found in the annexin proteins, which are widely distributed among eukaryotes. By interacting in a Ca^{2+}-dependent manner with membranes, annexins form pseudocrystalline aggregates, thus stabilizing individual membrane regions. Moreover, they function as scaffolds and interact with numerous cytoplasmic proteins, thus constituting signal-transducing multiprotein assemblies involved, for instance, in exo- and endocytosis and in the organization of ion channel complexes.

14.6.2 Muscle contraction

The role of the Ca^{2+} signal in stimulus–secretion and stimulus–contraction coupling in nerve and muscle cells, respectively, has been scrutinized in detail.

Figure 14.27 The Ca^{2+}-myristoyl switch of recoverin (A) Recoverin in the absence of Ca^{2+}. The structure consisting of four EF-hands is folded in such a way that the myristoyl residue (red) is buried inside the protein, rendering the protein water-soluble. (B, C) Binding of Ca^{2+} (colored circles) at two EF-hands causes an unfolding of the protein, exposing the myristoyl residue that now anchors the protein in lipid bilayers of membranes. (Modified from O.H. Weiergräber & K.W. Koch, *Biospektrum* 11:174–175, 2005.)

Figure 14.28 Stimulation of muscle contraction by Ca²⁺: the principle Contraction is due to a conformational change and activation of the ATPase myosin triggered by an interaction with F-actin and resulting in a relative shift of the proteins against each other (see textbooks of cell biology for details). This interaction is blocked by the protein tropomyosin and by myosin light chains. Ca²⁺ overcomes this blockade. To this end, it binds in skeletal and heart muscle to the sensor protein troponin C, which, through the adaptor proteins troponin I and troponin T (not shown), interacts with tropomyosin. In other cells, including smooth muscle cells, calmodulin serves as a Ca²⁺ sensor that activates caldesmon and the myosin light-chain kinase (MLCK). While caldesmon, like troponin C, displaces tropomyosin from the actomyosin complex, MLCK inactivates the myosin light chains by phosphorylation. The latter reaction is reversed by a myosin light-chain phosphatase (not shown). In smooth muscle and other tissues, phosphorylation of myosin light chains is essential for contraction and movements, whereas in skeletal muscle it has a modulatory function only. Input signals control the channels and enzymes of cellular calcium homeostasis.

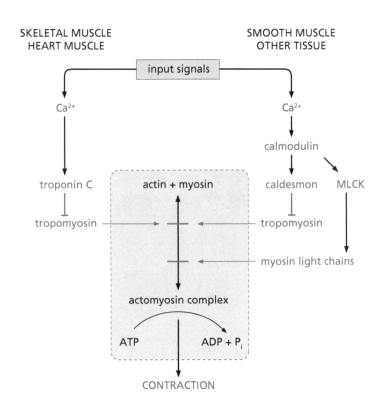

Several different calcium channels cooperate in these processes. This becomes particularly clear for muscle contraction.

Muscle contraction is due to an interaction between the proteins actin and myosin. The extended conformational change thus induced in myosin generates a mechanical tension, resulting in a shift of myosin relative to actin. The energy is derived from ATP hydrolysis catalyzed by myosin, which functions as an actin-stimulated ATPase. In the absence of input signals, the interaction between actin and myosin is blocked by inhibitory proteins such as myosin light chains or tropomyosin (Figure 14.28). Ca²⁺ entering the cytoplasm upon stimulation relieves this blockade according to the principle of double negation (inhibition of an inhibitor). In the three major types of muscles – skeletal muscle, smooth muscle, and heart muscle – Ca²⁺ influx and contraction are triggered by membrane depolarization gating VOCCs of the L-type. These channels cooperate with other channels according to a pattern that depends on the muscle type.

Skeletal muscle

The contraction of skeletal muscle cells is triggered by action potentials that are generated by acetylcholine interacting with nicotinic receptor cation channels (Section 16.2). In turn, L-type Ca²⁺ channels located in the immediate vicinity become gated. These channels are coupled with ryanodine receptors in the adjacent sarcoplasmic reticulum, which become activated (Figure 14.29). The Ca²⁺ thus released into the cytoplasm is bound by troponin C, which induces a conformational change of tropomyosin. This reaction clears the way for the interaction of actin and myosin, resulting in ATP hydrolysis and contraction. Such an arrangement of interacting proteins is based on a special cellular morphology and enables particularly rapid signal processing [one should bear in mind that more than 100 single contractions per second (100 impulses of signal generation and signal extinction) are taking place in a continuously working muscle].

Smooth muscle

Smooth or vegetative muscles surround the gastrointestinal tract, respiratory tract, blood and lymph vessels, uterus, oviduct, and vagina, as well as the urinary

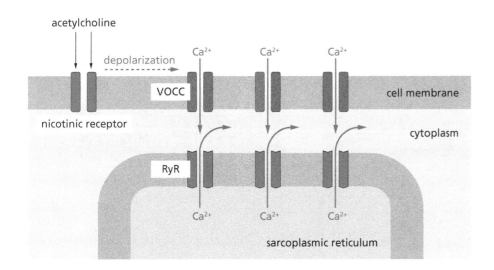

Figure 14.29 Signal transduction in skeletal muscle contraction By interacting with a nicotine receptor cation channel, acetylcholine induces depolarization. Consequently, voltage-operated Ca²⁺ channels (VOCC) of the L-type become gated. Each of these channels activates an associated ryanodine receptor (RyR) in the sarcoplasmic reticulum, both by direct protein–protein interaction and by Ca²⁺ entering the cell through the L-type channel.

bladder, gall bladder, and many other tissues. They control the state of contraction and thus the function of those organs and are major targets for the signals of the vegetative nervous system.

While skeletal muscle contraction is triggered by an action potential, smooth muscle cells respond to subthreshold depolarizations such as those induced by ligand-operated cation channels, in particular by the ionotropic P2X purine receptor gated by ATP (Section 16.6). As a result, adjacent L-type channels, which as we have seen gradually respond to depolarization, become partially gated and contraction is induced. This process is supported by additional mechanisms enabling a more subtle control, in particular by the vegetative nervous system. For example, β-adrenergic **sympathetic signals** (as transmitted by noradrenaline) evoke dilatation of most smooth muscles by activating the cAMP–PKA cascade. As a result the Ca²⁺ signal is suppressed, mainly due to phosphorylation and activation of the hyperpolarizing K⁺ channel BK and of the Ca²⁺ pumps in the cell membrane and the ER. A similar effect is produced by cGMP released in smooth muscle cells by nitrogen monoxide, the most effective vasodilatory agent of all (Section 16.5.1).

In contrast, **parasympathetic signals** transmitted by $G_{q,11}$-coupled muscarinic acetylcholine receptors support smooth muscle contraction by triggering InsP₃-promoted Ca²⁺ release from the ER and, in turn, activation of ryanodine receptors by Ca²⁺. On the other hand, Ca²⁺ released from the ER may also evoke smooth muscle dilatation. Such apparently contradictory observations find an explanation in interactions of Ca²⁺ with other ion channels such as the Ca²⁺-activated K⁺ channel BK and a Ca²⁺-stimulated chloride channel of the ClC family (Figure 14.30; see also Sections 14.4.1 and 14.7.1). While the gating of the BK channel hinders voltage-dependent Ca²⁺ influx by hyperpolarization, the chloride channel evokes contraction by depolarization due to the relatively high chloride concentration in smooth muscle cells. By such positive and negative feedback control, the tone of the smooth muscle cell is fine-tuned.

Ca²⁺ evokes smooth muscle contraction by stimulating the phosphorylation of the myosin light chains that results in a deblockade of the actin–myosin interaction (Figure 14.28). This reaction is catalyzed by **myosin light-chain kinase (MLC kinase)**, a Ca²⁺-dependent enzyme with CaM as a regulatory subunit. In addition to MLC kinase, Ca²⁺/CaM activates the protein caldesmon that displaces tropomyosin from the actin–myosin complex. In other words, in smooth muscle (and other cells), the function of the CaM/caldesmon pair resembles that of troponin C in skeletal muscle.

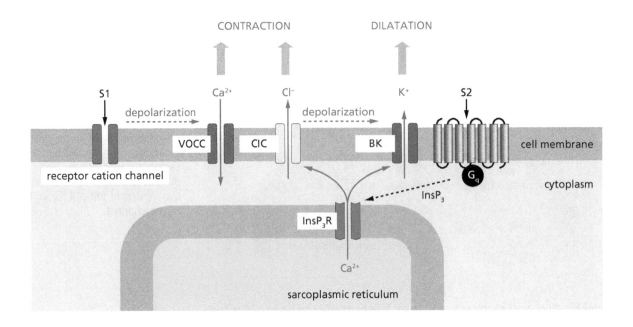

Figure 14.30 Signal transduction in smooth muscle contraction An input signal S1 (left), for instance ATP, activates a receptor cation channel, for instance the P2X purine receptor. By the depolarization thus induced, a voltage-operated Ca^{2+} channel (VOCC) of the L-type is gated and contraction is triggered by Ca^{2+} influx. This mechanism is completed and supported by signals S2 interacting with $G_{q,11}$-coupled receptors and inducing InsP$_3$ release (right). By interacting with its receptor (InsP$_3$R), InsP$_3$ turns up Ca^{2+}-dependent Ca^{2+} release from the sarcoplasmic reticulum. The increase of cytoplasmic Ca^{2+} leads to an activation of BK potassium channels evoking hyperpolarization and muscle dilatation, and depolarization by chloride channels (ClC) supporting contraction. The BK channel is activated, in addition, by depolarization as well as by cAMP- and cGMP-dependent protein phosphorylation (not shown).

An alternative mechanism by which input signals control smooth muscle contraction is based on the *dephosphorylation* of the myosin light chains by which their inhibitory effect on the actin–myosin interaction is reestablished. The reaction is catalyzed by a specific **myosin light-chain phosphatase (MLC phosphatase)**. The enzyme is a heterodimer consisting of the phosphoserine/phosphothreonine-directed phosphatase PP1 (Section 2.6.4) and a myosin phosphatase target subunit (MYPT1). MLC phosphatase is inactivated by protein kinase ROK, which is controlled by the small G-protein Rho (Section 10.3.3; Figure 14.31). To this end, ROK phosphorylates and activates a specific inhibitor protein CPI-17 that interrupts the interaction between the phosphatase subunits. At the same time ROK phosphorylates and inactivates myosin light chains, an effect that resembles that of MLC kinase. Along this pathway, contraction can be evoked without a further supply of Ca^{2+}. Stimulation of MLC phosphorylation by Rho/ROK is essential for cytoskeletal rearrangements and motility changes controlled by small G-proteins of the Rho family, not only in smooth muscle but in any other cell types (for details see Section 10.3). A well-known example of a universal Rho-controlled process is cytokinesis.

The reactions required for smooth muscle activation (release of InsP$_3$ and phosphorylation of myosin light chains and MLC phosphatase) are much slower than a voltage-stimulated Ca^{2+} influx and the noncovalent interactions between troponin C and tropomyosin in skeletal muscles. These mechanistic features and anatomical peculiarities explain the relative sluggishness of smooth muscles and their ability to execute long-lasting contractions.

Heart muscle

As in skeletal muscle, in heart muscle the gating of L-type VOCCs triggers Ca^{2+} influx that results in a Ca^{2+}/troponinC-dependent deblockade of the actin–myosin complex and muscle contraction. However, differing from skeletal muscle, here L-type Ca^{2+} channels communicate with ryanodine receptors through Ca^{2+} signals alone rather than by direct protein–protein interaction. By activating a Ca^{2+}-induced Ca^{2+} release at the mouth of the ryanodine receptor, Ca^{2+} entering the cell from outside generates calcium sparks. For this purpose each L-type channel is associated with a battery of 4–5 ryanodine receptors constituting a structure known as a "calcium synapse" (Figure 14.32). Such "synapses" work independently from each other with the result that the stronger the depolarizing input signal, the more "synapses" become activated and the more powerful is the contraction. This mechanism demonstrates in a particularly clear way how, by cooperation of a large

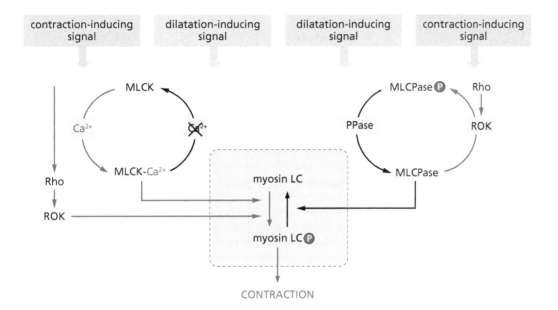

Figure 14.31 Control of smooth muscle tone by myosin light-chain phosphorylation
Phosphorylation (P) of myosin light chains (LC) results in contraction. Of the two kinases catalyzing this modification, myosin light-chain kinase (MLCK) is activated by an increase and inactivated by a decrease of the cytoplasmic Ca^{2+} concentration, while the kinase ROK is activated by small G-proteins of the Rho family (left). Vice versa, myosin light-chain phosphatase (MLCPase) causes smooth muscle dilatation. The phosphatase is phosphorylated and inactivated by ROK and re-activated by a protein phosphatase (PPase) (right).

number of independent binary units, a digital mode of signal processing based on the binary all-or-nothing code of an action potential is transformed into an analog mode enabling continuous modulation of the amplitude of the output signal (heart muscle contraction). This regulation is superimposed by a frequency modulation due to the positive and negative chronotropic effects of the vegetative nervous system on the pacemaker cells (see below).

Autonomous oscillations control the heartbeat

The heart muscle is the prototype of a biological oscillator. Its rhythmic activity is brought about by a periodic Ca^2 signal. As a peculiarity, the oscillator and the effector systems are located in different cell types. The periodic signals are generated by a special type of heart muscle cells in the sinoatrial knot called pacemaker cells. As action potentials, they spread across the entire heart muscle, evoking a synchronous contraction of the atrial and ventricular muscle cells, which are electrically coupled by gap junctions. A system of positive and negative feedback loops based on the cooperation of several ion channels renders pacemaker cells autonomous oscillators (Figure 14.33).

Each action potential and thus each heartbeat is started by gating a special type of cation channels called pacemaker or **HCN channels** (see Section 14.5.7).

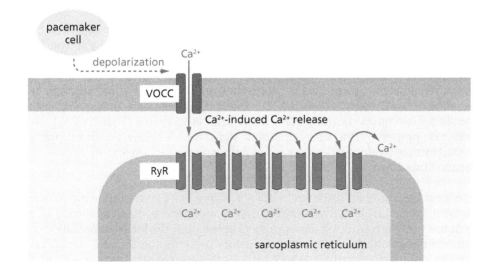

Figure 14.32 Signal transduction in heart muscle contraction
Through the action potential generated by pacemaker cells, voltage-operated calcium channels (VOCC) of the L-type are gated. Ca^{2+} ions entering the cell stimulate a Ca^{2+}-induced Ca^{2+} release from a battery of cooperating ryanodine receptors (RyR).

Figure 14.33 Generation of pacemaker potentials (A) Oscillations of the membrane potential in pacemaker cells. In the final phase of each action potential, HCN (hyperpolarization- and cyclic nucleotide-activated) channels become gated, causing slow depolarization (shown in red). This results in a gradual activation of voltage-operated T-type (T) and L-type (L) Ca^{2+} channels, ryanodine receptors (RyR) and $Na^+–Ca^{2+}$ exchangers (EX), leading finally to a new action potential. The repolarization is initiated by voltage-operated K^+ channels (K_v). (B) Rhythmic fluctuations of the cytoplasmic Ca^{2+} concentration in nonstimulated (solid line) as compared with sympathetic- and stress-activated pacemaker cells (dotted line), where the HCN channels are stimulated by cAMP.

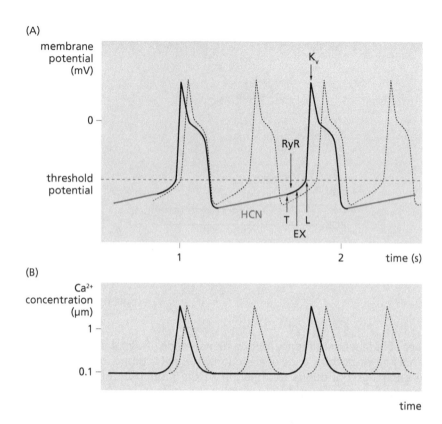

These channels are voltage-dependent but, in contrast to most voltage-operated channels, they become gated at an extremely negative potential, as occurs at the end of an action potential. By the influx of Na^+ ions through the HCN channels, the cell becomes slowly depolarized. Beyond a threshold potential, VOCCs, first T-type and later L-type, take over the role of the HCN channels, which more and more become inactive. The influx of Ca^{2+} contributes to depolarization and, at the same time, activates ryanodine receptors and depolarizing $Na^+–Ca^{2+}$ exchangers. Due to this positive feedback, the Ca^{2+} influx and the depolarization become accelerated, triggering a new action potential when the threshold of about –35 mV is reached. Repolarization occurs in two steps through a gating of voltage-dependent potassium channels of the type hERG (Section 14.4.1). This gives the action potential the characteristic shape with a plateau phase of partial repolarization between the peak of depolarization and terminal repolarization (Figure 14.33).

HCN channels belong to the family of cyclic nucleotide-gated ion channels (see Section 14.5.7). In fact, cAMP and cGMP are ligands that specifically activate the channels in addition to depolarization. This has immediate consequences for the heart frequency: upon excitation and stress, sympathetic neurons release noradrenaline that evokes cAMP formation in pacemaker cells. Through the binding of cAMP to HCN channels, the interval between two subsequent action potentials becomes shortened and the heartbeat increases (Figure 14.33). This effect is opposed by the parasympathetic signal acetylcholine, which suppresses cAMP formation along a $G_{i,o}$-controlled pathway (see Section 5.5). As a result, the heart calms down.

Under stress, the heart beats faster but also stronger. This amplitude modulation is again caused by cAMP and in this case is mediated by PKA. By means of an adaptor protein AKAP (A-Kinase Anchoring Protein), the kinase associates with protein phosphatases and several regulatory proteins and forms a signal-transducing complex with the ryanodine receptor, which becomes sensitized by

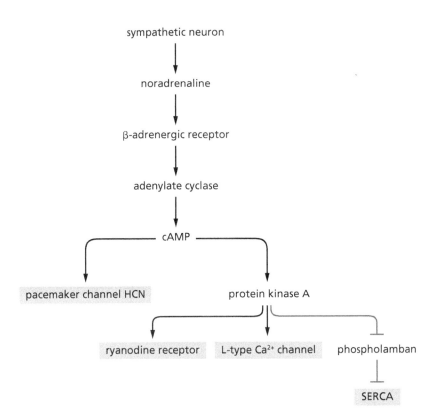

Figure 14.34 Sympathetic- and stress-activated signaling cascade in heart muscle cells For details, see text. HCN, hyperpolarization- and cyclic nucleotide-activated channel; SERCA, sarco/endoplasmic reticulum Ca²⁺-dependent ATPase.

phosphorylation. Moreover, as mentioned above, PKA phosphorylates and sensitizes the L-type Ca²⁺ channel, thus augmenting the calcium signal.

Another stimulatory pathway leads to the Ca²⁺-dependent ATPase SERCA. In the sarcoplasmic reticulum of nonstimulated heart muscle cells, this enzyme is inhibited by an interaction with the small transmembrane protein **phospholamban**. Phospholamban again is a substrate of PKA; upon phosphorylation, it loses its ability to interact with SERCA. As a result, cytoplasmic Ca²⁺ is pumped more rapidly into the lumen of the sarcoplasmic reticulum and the "charge" of the calcium battery is continuously kept at a high level. This context is depicted schematically in Figure 14.34.

Caffeine enhances the sympathetic effect on the heart by inhibiting phosphodiesterase-catalyzed cAMP degradation and sensitizing ryanodine receptors (for more details see Section 16.9.1). This is the reason for a lively heartbeat following excessive consumption of coffee.

14.6.3 How Ca²⁺ controls protein phosphorylation

Several protein kinases contain CaM as a regulatory subunit that, upon binding of Ca²⁺, displaces an auto-inhibitory pseudosubstrate domain from the catalytic center, thus activating the kinase. We have already encountered members of this enzyme family: phosphorylase kinase (Section 5.3.1), eEF2 kinase (also known as CaMKIII; Section 9.3.4), and MLC kinase. These three kinases are highly specific for their particular substrates. In contrast, other Ca²⁺/CaM-dependent kinases are less choosy: CaMKI, -II, and -IV and CaMK kinases α and β. They are structurally related to each other and to MLC kinase.

CaMKI is a widely distributed enzyme, whereas its isoform **CaMKIV** seems to be restricted to a very few tissues. Both phosphorylate a large number of proteins. Among the substrates are several transcription factors, such as CREB (cAMP response element binding protein) and its co-activator CBP/p300 (Section 8.6) as well as the myocyte enhancer factor MEF2, which are activated by phosphorylation. Thus the two kinases play an important role in Ca^{2+}-controlled gene transcription, in particular since they have been shown to shuttle between cytoplasm and nucleus.

For activation of CaMKI and -IV, interaction with Ca^{2+}/CaM is a necessary but not sufficient condition. In addition, like most protein kinases, both enzymes must be phosphorylated at a strategic threonine (Thr) residue in the activation loop. This phosphorylation is catalyzed by the **CaMK kinases** in a so-called CaMK cascade. Like CaMKs, CaMK kinases are activated by Ca^{2+}/CaM.

In addition to CaMKs, the CaMK kinases activate other substrates such as protein kinase B (PKB)/Akt. Since PKB/Akt inhibits apoptosis, this is probably a protective mechanism that hinders normal fluctuations of the calcium level from driving cells to death (note that cells undergo programmed suicide when flooded by Ca^{2+} for a *longer* time period; see Section 13.2.4).

Calmodulin kinase II: a biochemical frequency detector with memory

By far the best-studied Ca^{2+}/CaM-dependent kinase is CaMKII. The reason for this interest is obvious: this extraordinary enzyme is crucially involved in learning and memory fixation (Section 16.4.1).

Among the four isoforms of the kinase, CaMKIIα and -β predominantly are expressed in neurons, whereas CaMKIIγ and -δ are found in nearly all tissues. Each of the four isoforms exists in various splice variants. Although closely related to CaMKI and CaMKIV, CaMKII has an additional oligomerization domain in the C-terminal region that enables the enzyme to aggregate into star-shaped hexamers, two of which are stacked (Figure 14.35). The outstanding cellular functions of CaMKII critically depend on this quaternary structure, which seems to be unique in the world of protein kinases.

To trigger the activation of CaMKII, an interaction with Ca^{2+}/CaM is necessary. By this means an auto-inhibition through a pseudosubstrate motif is overcome. To stabilize the active conformation, phosphorylation is also required. To this end, the monomer initially activated by Ca^{2+}/CaM phosphorylates the adjacent monomer at a Thr residue in the regulatory domain (provided this subunit has also bound Ca^{2+}/CaM) (Figure 14.35). In other words: in contrast to CaMKI and CaMKIV, which require the assistance of a CaMK kinase, CaMKII is activated by Ca^{2+}-dependent trans-autophosphorylation proceeding in a stepwise manner. The autophosphorylation has an important consequence: on the one hand it potentiates the affinity of the monomers for Ca^{2+}/CaM several thousandfold; on the other hand it renders the kinase activity autonomous (it does not depend any longer on Ca^{2+}/CaM).

The stronger the Ca^{2+} signal, the more monomers of the complex become occupied by Ca^{2+}/CaM and autophosphorylated until finally the entire complex is activated. Due to this stepwise stimulation, the CaMKII oligomer functions as a detector, monitoring duration, intensity (amplitude), and frequency of a Ca^{2+} signal and staying active even when the signal has vanished. The duration of this "molecular memory" depends only on the rate of CaMKII dephosphorylation.

CaMKII comes close to Hans von Foerster's model of a nontrivial calculator: a signal-processing device that is able to learn by adapting its algorithm to the previous operations (see Section 1.6). As shown in Section 16.4.1, this property of CaMKII is highly significant for long-term potentiation of synaptic signal transmission, which is thought to resemble a learning process.

In fact, CaMKII phosphorylates many proteins involved in neurotransmission. These include ryanodine receptors, excitatory glutamate receptors (Section

16.4), and various transcription factors (such as CREB and the serum response factor), which are activated, as well as the neuronal NO synthase (Section 16.5.1) and the neuronal Ras-GAP SynGAP, which are both inactivated. To recognize individual substrate proteins and to become targeted to individual cellular sites, CaMKII (like other protein kinases) needs the assistance of various adaptor and scaffold proteins.

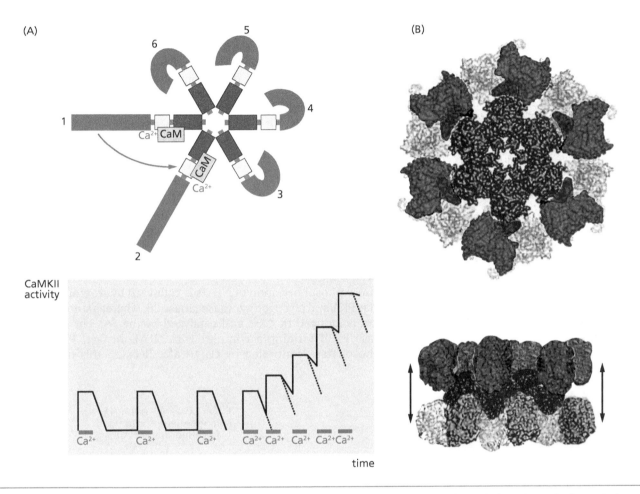

Figure 14.35 Activation of CaMKII (A) Six enzyme molecules are shown that aggregate through their C-terminal association domains (dark gray), forming a six-armed starlike structure. Two of these aggregates are stacked (for the sake of clarity, only one aggregate is shown here). In the absence of an input signal, the kinase activity is suppressed as shown for monomers 3–6: a pseudosubstrate motif located in a central regulatory domain (light gray rectangle) blocks the N-terminal catalytical domain (shown in color). In monomers 1 and 2, this blockade has been relieved by Ca²⁺/calmodulin (CaM). The conformational change resulting from this interaction has exposed a strategic Thr residue in the regulatory domain that becomes phosphorylated by the adjacent monomer (red arrow). This trans-autophosphorylation renders the kinase activity of this monomer autonomous, or Ca²⁺/CaM-independent. With an increasing concentration of Ca²⁺/CaM, the remaining kinase monomers (3–6) also become activated by their neighbors in a stepwise manner. Thus, the degree of phosphorylation and the overall kinase activity of the complex directly depend on the *intensity* of the Ca²⁺ input signal. In the panel below, the activity of the multimeric complex is depicted as a function of the *frequency* of the (oscillating) Ca²⁺ signal (red). At a low frequency, Ca²⁺/CaM has enough time to dissociate from the complex between the single pulses (left side; the phosphorylation is reversed by protein phosphatases). At higher frequencies this is not possible any longer. Instead, the individual phosphorylations summarize and the overall activity increases in a sigmoidal mode (right side). A similar situation arises when the *amplitude* of the Ca²⁺ signal increases. In this case the red bars in the figure would become longer (not shown). Note that the *rate* of inactivation (the control value of which is indicated by the dotted lines) slows down with the number of kinase monomers activated since the binding affinity for calmodulin increases with each step. (B) Model of the inactive CaMKII based on X-ray crystallography. One recognizes the two stacked star-shaped oligomers composed of six monomers each are shown from above (upper structure) and from the side (lower structure, pulled apart a little). The association domains are black; the catalytic domains containing the Ca²⁺–CaM binding sites are colored (upper oligomer) or gray (lower oligomer). (B, adapted with modifications from O.S. Rosenberg, S. Deindl, R.J. Sung, et al., *Cell* 123:849–860, 2005.)

14.6.4 How calcium regulates gene transcription

As explained above, Ca^{2+}-dependent protein kinases, primarily CaMKI, -II, and –IV, phosphorylate various transcription factors, thus stimulating the transcription of numerous genes. A well-known example is provided by the transcription factor CREB, which is phosphorylated by CaMK at the same site as by PKA (see Section 8.6 for details).

NFAT: nuclear factor of activated T cells and more

Another group of transcription factors called **NFAT** is activated by Ca^{2+}-dependent *dephosphorylation* (Figure 14.36). The name Nuclear Factors of Activated T-cells refers to the initial discovery of these proteins. However, NFATs by no means are restricted to T-lymphocytes but are found in many cell types, where they control a wide variety of physiological processes. For gene activation, NFATs cooperate with other transcription factors, the type of which depends on the tissue. In T-lymphocytes, for instance, the partners are Jun–Fos dimers of the AP1 transcription factor family that are activated along MAP kinase pathways (Section 11.7.2). Like most other genes, therefore, NFAT-controlled genes are logical gates that are controlled by the combined effect of at least two different input signals.

Five NFAT isoforms are known and are tissue-specifically expressed. In the absence of Ca^{2+} signals, they are phosphorylated at 10 serine residues at least. Since phosphorylation blocks both the DNA-binding domain and the nuclear localization sequence, the factors cannot translocate into the nucleus and interact with DNA unless they are dephosphorylated.

The inhibitory phosphorylation of NFATs is catalyzed by several protein kinases, in particular PKA and glycogen synthase kinase 3β, whereas stimulatory dephosphorylation is induced by Ca^{2+} and catalyzed by the Ser/Thr-directed protein phosphatase PP2B forming a complex with NFAT. In fact, PP2B is the only protein phosphatase controlled by Ca^{2+}/CaM. Because this otherwise widely

Figure 14.36 Calcineurin–NFAT signaling pathway Various signals activating receptors and membrane channels induce an increase of cytoplasmic Ca^{2+} that, upon binding to calmodulin (CaM), activates the protein phosphatase calcineurin. Calcineurin dephosphorylates the transcription factor NFAT (nuclear factor of activated T cells), which thus becomes active and translocates into the nucleus to stimulate, in cooperation with other transcription factors (here AP-1), numerous genes. NFAT is inactivated by phosphorylation catalyzed by kinases such as protein kinase A (PKA) and glycogen synthase kinase 3β (GSK3β). DSCR-1 protein, overexpressed in neurons of Down syndrome patients and immunosuppressive drugs inhibit calcineurin.

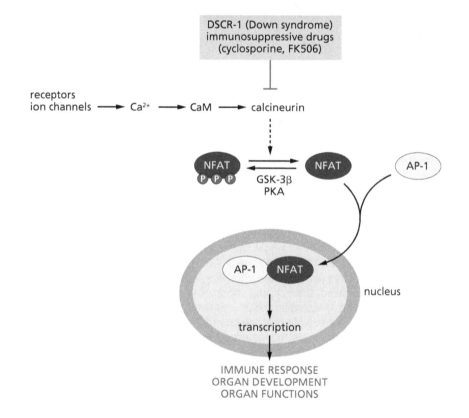

distributed enzyme is strongly expressed in neurons, it is known also as **calcineurin**. The phosphatase is a heterodimer consisting of a catalytic subunit A and a regulatory subunit B with four EF-hand motifs. In the absence of a calcium signal, the enzymatic activity is suppressed by an auto-inhibitory domain. This blockade is relieved partially by an interaction with Ca^{2+}/CaM and completely by additional binding of Ca^{2+} to the regulatory domain. In other words, like CaMKII, calcineurin is activated by Ca^{2+} in a *stepwise* manner, although according to a different mechanism.

Apart from NFATs, calcineurin dephosphorylates other proteins. Its strong expression in the brain indicates that it participates in learning and memory fixation, a concept that is supported by a large body of evidence (see Section 16.4.1). A permanent (genetically fixed) insufficiency of calcineurin seems to be one among several causes of schizophrenia.

In principle, calcineurin and NFAT are activated by every signal inducing an increase of the cytoplasmic calcium concentration. In T-lymphocytes, antigens that stimulate Ca^{2+}-release from the ER by activating the T-cell receptor PLCγ–InsP₃ cascade transmit such signals. Along this pathway, antigens turn up the transcription of genes controlling T-cell proliferation and the immune response, including genes of the T-cell mitogen interleukin 2 and its receptor; interleukins 3, 4, 5, and 8; interferon γ; Tyr kinase Lck (Section 7.2.2); death receptor Fas and its ligand (Section 13.2.3); and prostaglandin synthase COX-2 (Section 4.4.5).

Due to its key function in T-cell activation, the NFAT signaling pathway has found considerable interest in the clinic, in particular since it has turned out to be a target of **immunosuppressive drugs** such as cyclosporine and FK506, which have revolutionized transplantation medicine. These drugs bind to cellular proteins called immunophilins (such as cyclophilin and the FK506-binding protein; for details see Section 8.5) that block the interaction between calcineurin and NFAT thus suppressing the immune response.

In addition to their effects on the immune system, NFAT transcription factors exhibit other functions. They participate in the development of the central nervous system, cardiovascular system, skin, and skeletal muscles. Moreover, they regulate the adaptation of organ functions to environmental conditions. An example with considerable medical consequences is provided by the hypertrophy of the heart muscle in response to continuous strain (the athletic heart). The underlying mechanism is thought to consist of a chronic activation of postsynaptic Ca^{2+} channels (L-type) in neuromuscular synapses. As a result, the calcineurin–NFAT system becomes overstimulated and activates genes of muscle proteins.

Sidebar 14.2 Calcineurin and Down syndrome
Affecting one out of 700 newborns, Down syndrome is by far the most frequent birth defect caused by chromosomal damage. The disease is due to a tripling of chromosome 21 (trisomy 21) resulting in a pathological overexpression of approximately 300 genes. One of these genes encodes a protein that inhibits calcineurin (and thus NFAT) signaling. It is called **calcipressin** or **DSCR1** (Down Syndrome Critical Region of chromosome 1). Because the synthesis of this inhibitor is induced by NFAT, it seems to be a component of a negative feedback mechanism controlling the calcineurin–NFAT signal. Another gene that maps to a Down syndrome region of chromosome 21 is the unique cell adhesion molecule and receptor **DS-CAM** (Section 16.1.5). Because these proteins play a central role in signaling pathways controlling embryonic development, continuous overexpression of DSCR1 and DS-CAM may contribute significantly to the organic and mental symptoms of Down syndrome patients.

Figure 14.37 Transcriptional repressor DREAM As a tetramer, DREAM (downstream regulatory element antagonist) binds to downstream regulatory elements (DRE) of certain genes inhibiting transcription (top panel). Upon binding of Ca^{2+} the tetramer dissociates into two dimers, which leave the DNA giving way to the RNA polymerase II complex (bottom panel).

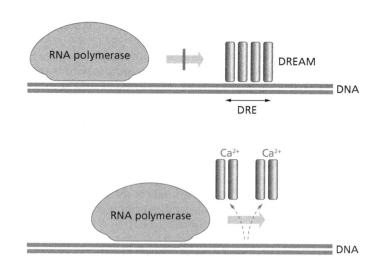

The DREAM of pain research

Presently only one transcription factor is known that *directly* interacts with Ca^{2+}. This is **DREAM**, <u>D</u>ownstream <u>R</u>egulatory <u>E</u>lement <u>A</u>ntagonist. The name of this protein refers to its discovery as a factor bound by a regulatory sequence of the prodynorphin gene located downstream of the transcription start point. DREAM has four EF-hand motifs and belongs to the same family of neuronal Ca^{2+}-sensor proteins as the retinal protein recoverin.

Figure 14.38 Pain relief by Ca^{2+} signals (A) A peripheral pain sensor activates (via the neurotransmitter substance P) a spinal neuron that transmits a high-frequency burst of action potentials (red circle) as a pain signal to the brain. (B) A spinal interneuron simultaneously activated by substance P decreases the firing frequency of the pain neuron by means of dynorphin. At the level of transcription, the synthesis of dynorphin is inhibited by DREAM (A) and this inhibition is overcome by Ca^{2+} released in the course of an action potential (B). (Modified from B.A. Vogt, *N. Engl. J. Med.* 347: 362–364, 2002.)

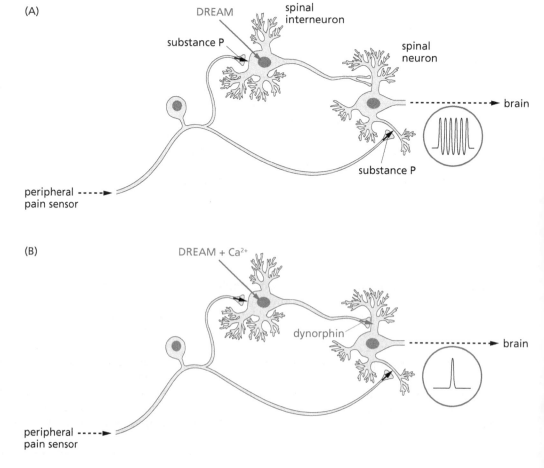

In the absence of a Ca^{2+} signal, the factor inhibits transcription, acting as a gene repressor. Upon binding of Ca^{2+}, DREAM dissociates from the DNA, allowing access by the RNA polymerase complex and transcriptional activators (Figure 14.37). In addition to its repressor functions, DREAM seems to have other effects. It has been independently characterized as "potassium channel interacting protein 3," binding to voltage-dependent K^+ channels, and as "calsenilin," interacting with the protein presenilin-2 associated with Alzheimer's disease (Section 13.1). In both cases, DREAM modulates the functions of its binding partners. Thus, DREAM is a multifunctional transcription factor and in this respect resembles, for instance, p53 and β-catenin.

DREAM has caused some stir in the context of **pain research**. Mice carrying an inactive *dream* gene have been found to be widely insensitive to pain, probably due to an overproduction of **dynorphin** in the spinal cord. This endogenous "opiate" interacts with the $G_{i,0}$-coupled opioid receptor κ (see Section 16.8) and inhibits the generation of action potentials and the transmission of pain signals to the brain, apparently without development of tolerance and addiction (Figure 14.38). The name DREAM thus takes on a second meaning, as it may represent a "dream target" in the battle against chronic pain. On the other hand, DREAM may also *enhance* pain by inhibiting neurons that dampen the activity of pain neurons.

In addition to CREB, NFAT, and DREAM, there are many other transcription factors the activity of which is regulated by Ca^{2+} signals, albeit in an indirect way. An overview is given in Figure 14.39. The individual signaling pathways are discussed elsewhere in this book.

14.6.5 Calcium and cell death

Uncontrolled flooding of the cytoplasm by Ca^{2+} inevitably kills the cell by causing apoptosis and necrosis. This is the reason for tissue destruction resulting from stroke or myocardial infarction. In theses cases Ca^{2+} enters the cell from the outside or is released by damaged mitochondria or by the ER (for a more detailed discussion of apoptosis, see Section 13.2).

The apoptotic signal is transduced by several Ca^{2+}-dependent proteins. One of these, calcineurin, acts as an antagonist of the anti-apoptotic PKB/Akt pathway by dephosphorylating and activating the pro-apoptotic protein Bad. Moreover, NO synthases and oxygenases of arachidonic acid metabolism activated along the Ca^{2+}-dependent PLA_2 pathway catalyze the generation of cytotoxic metabolites such as peroxynitrite and reactive oxygen species that damage mitochondria, thus triggering apoptosis.

Figure 14.39 Ca²⁺-regulated gene transcription: an overview Shown are signaling pathways leading to an activation of transcription factors or to a removal of repressors. These factors are shown in the oval. For details see text. CaM, calmodulin; cPKC, conventional protein kinase C; GRF, guanine nucleotide releasing factor, and GRP, guanine nucleotide releasing protein (both Ras-guanine nucleotide exchange factors; see Section 10.1.3); MEKK1, MAP2 kinase 1; MAPK, mitogen-activated protein kinase; PKA, protein kinase A; CaMKK, calmodulin-activated kinase kinase; CaMK, calmodulin-activated kinase; DREAM, downstream regulatory element antagonist; AP-1, activator protein 1; NFκB, nuclear factor κB; Ets/TCF, erythroblastosis virus E26/ternary-complex factors; CREB, cAMP-response element binding protein; SRF, serum response factor; NFAT, nuclear factor of activated T cells.

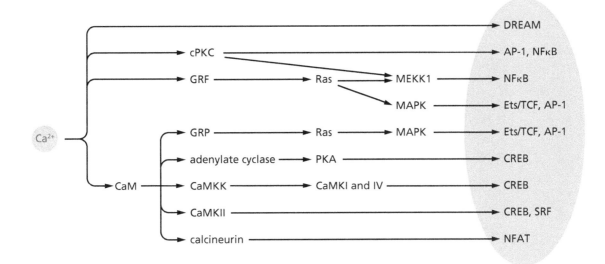

525

Sidebar 14.3 Calpains: calcium-dependent proteases
Calpains are cysteine proteases with EF-hand domains. They become activated by relatively high Ca^{2+} concentrations and are inhibited by specific proteins, the calpastatins. Since calpains have a wide spectrum of substrate proteins, they have been connected with various cellular processes, although their physiological role is still not fully understood.

An effect with significant consequences for signal transduction consists of the proteolytic removal of auto-inhibitory domains from signal-processing proteins, which thus may become constitutively active. As examples, the continuous activation of PKC by destruction of the pseudosubstrate motif, and of the neuronal protein kinase CDK5 by partial proteolysis of the regulatory subunits p35 or p38 have been mentioned already (Sections 4.6.3 and 12.4). Though the physiological significance of such effects is still a matter of debate, an increasing body of evidence indicates a role of permanent CDK5 activation in neurodegenerative diseases such as Alzheimer's dementia.

Other processes in which calpains are thought to be involved are:

- the cell cycle, where under certain circumstances they support protein degradation (for instance of cyclins) by proteasomes
- cell movements, where they disintegrate contacts between integrins and the cytoskeleton
- synaptic long-term potentiation
- necrosis and certain types of apoptosis

In several diseases, accompanying disturbed Ca^{2+} homeostasis, calpains contribute to functional disorders and cell death. Examples include cataracts, muscular dystrophy, myocardial infarction as well as stroke, multiple sclerosis, Alzheimer's dementia, and other neurodegenerative disorders.

Ca^{2+}-dependent proteases participating in apoptotic and necrotic protein degradation are the **calpains** (see Sidebar 14.3). These enzymes, for instance, destroy anti-apoptotic proteins such as Bcl2 and Bcl-XL and activate pro-apoptotic proteases of the caspase family. More recently, a family of **death-associated protein kinases** has been discovered that promotes apoptosis. Because these Ser/Thr-specific enzymes are closely related to MLC kinase, most of them are activated by Ca^{2+}/CaM. Like cyclin-dependent kinases, they also require dephosphorylation to become fully active (see Section 13.2.4 for more details).

Summary

Ca^{2+} ions represent the most versatile second (or third) messengers. By interacting as input signals with a wide variety of proteins, they participate in the control of almost any cellular process. Ca^{2+}-binding sensor proteins transduce Ca^{2+} signals to effector proteins. The most abundant Ca^{2+}-sensor proteins are the EF-hand proteins calmodulin (found in all cell types) and its close relative troponin C (found in muscle cells). Neuronal Ca^{2+}-sensor (NCS) proteins constitute another large group of EF-hand proteins. A second Ca^{2+}-binding motif known as C2 is found in synaptotagmins and various enzymes such as protein kinases C and phospholipases. A third Ca^{+}-binding motif of numerous proteins is the annexin domain. A major function of Ca^{2+} ions is stimulus–contraction coupling in muscle cells. Skeletal muscle contraction is triggered by nicotine receptors that induce an influx of Ca^{2+} into the cytoplasm through coupled L-type Ca^{2+} channels and ryanodine receptors. For the control of smooth muscle activity, various ion channels interact with each other and with G-protein-coupled receptors. While membrane depolarization triggers Ca^{2+} influx through L-type channels, $InsP_3$ released by parasympathetic signals along $G_{q,11}$-controlled pathways promotes Ca^{2+} release from the ER and thus contraction. Ca^{2+}-dependent chloride channels facilitate and Ca^{2+}-dependent K^+ channels hinder membrane depolarization. Protein phosphorylation induced by sympathetic signals and by NO causes dilatation, mainly due to an activation of hyperpolarizing BK channels and Ca^{2+} pumps. Smooth muscle activity depends on the phosphorylation status of myosin light chains, controlled by Ca^{2+}/CaM-activated MLC kinase (triggering contraction) versus MLC phosphatase (triggering dilatation). Signals stimulating MLC phosphorylation and inhibiting the dephosphorylation (transmitted by Rho-activated protein kinase) facilitate contraction, and the reverse holds true for dilatation. Heart muscle contraction is initiated by oscillating action potentials generated by pacemaker cells. The Ca^{2+} entering the cell through an L-type channel triggers Ca^{2+}-induced Ca^{2+} release through a battery of 4–5 ryanodine receptors. The graded activation of such "calcium synapses" enables a continuous response of contraction power to modulatory signals. Periodic generation of action potentials by pacemaker cells is based on cooperation of several types of ion channels. Hyperpolarization and cyclic nucleotide-gated (HCN) channels opening at extreme hyperpolarization start the depolarization. Subsequently several types of voltage-operated Ca^{2+} channels, ryanodine receptors, and an electrogenic Na^+–Ca^{2+} exchanger are activated, accelerating depolarization that culminates in an action potential and finally is terminated and reversed by voltage-dependent K^+ channels and Ca^{2+} pumps. The pacemaker channel HCN belongs to the family of cyclic nucleotide-gated cation channels. Its sensitization by cAMP explains the sympathetic and stress effects on the heart frequency. Through protein kinase A-catalyzed phosphorylation of the ryanodine receptor, the L-type channel, and the inhibitor protein phospholamban, cAMP also increases the power of heart contraction. Caffeine, an inhibitor of cAMP degradation, intensifies heart activity. Several protein kinases are activated by Ca^{2+}–CaM. These include highly specific enzymes such as glycogen phosphorylase kinase, EF2 kinase, and MLC kinase as well as the less specific CaMKI, -II, and -IV. CaMKI and -IV participate in gene activation. They require activation loop phosphorylation by CaMK kinases, which also are Ca^{2+}/CaM-controlled. CaMKII is a frequency detector. Based on a unique oligomeric structure that allows stepwise activation, this kinase monitors the duration, amplitude, and frequency of oscillating Ca^{2+} signals and thus provides a molecular short-term memory store. CaMKII modulates the activity of various receptors, transcription factors, and other signal-transducing proteins. In the brain, the enzyme plays a crucial role in long-term potentiation of synaptic activity and thus in learning and memory fixation. The transcription factor NFAT is activated by dephosphorylation catalyzed by the Ca^{2+}/CaM-dependent protein phosphatase PP2B (calcineurin), and thus by all input signals that evoke an increase of cytoplasmic Ca^{2+}. In antigen-exposed lymphocytes, NFAT promotes the transcription of genes controlling the immune response. Immuno-suppressive drugs inhibit the interaction between calcineurin and NFAT. Overexpression of an endogenous calcineurin inhibitor (DSCR1) seems to be involved in Down syndrome. The NFAT–calcineurin system is also thought to play a critical role in acquired heart insufficiency. The only transcription factor known to interact *directly* with Ca^{2+} is the repressor DREAM; it is inactivated by Ca^{2+} binding. Among others, DREAM inhibits the expression of genes encoding the opioid dynorphin, an endogenous painkiller; thus it may be involved in the processing of pain sensations. At higher concentrations Ca^{2+} is a strong inducer of apoptosis by activating Ca^{2+}-dependent enzymes such as calcineurin, proteases of the calpain family, NO synthases, enzymes generating active oxygen species, and other proteins that trigger pro-apoptotic signal transduction. Ca^{2+}-induced cell death plays a crucial role in stroke, myocardial infarction, and neurodegenerative diseases such as Alzheimer's dementia.

14.7 Anion channels

Anion channels are found in all organisms from the simplest prokaryotes to mammals (Section 3.5.7). Nevertheless, considering their investigation, they are the Cinderellas among the ion channels. Only recently this situation has changed – in particular due to structural studies – and now anion channels are gaining steadily increasing attention. In addition to chloride, they transport bromide, iodide, nitrate, and other negatively charged ions; however, in animal cells these ions are much less abundant than chloride. Therefore, anion channels are usually called chloride channels. The gating of chloride channels may have either a depolarizing or a hyperpolarizing effect depending on the steepness of the ion gradient across the cell membrane. When the concentration of chloride in the cytoplasm is much lower than in the extracellular space, chloride ions will enter the cell through an open channel despite the negative charge inside. Consequently, the intracellular charge becomes more negative: the cell hyperpolarizes, and events depending on depolarization are impaired. This is the case, for instance, in skeletal muscle cells and in most neurons. In other cell types, such as smooth muscle cells, the concentration gradient is not steep enough to overcome the electric force of the membrane potential. Here the opening of an anion channel results in an efflux of chloride, causing depolarization (see also Section 14.6.2).

In vertebrates there are three families of chloride channels, illustrated in Figure 14.40:

- the ClC (Cl Channel) family
- the CFTR (Cystic Fibrosis Transmembrane conductance Regulator) family
- the family of GABA (Gamma-AminoButyric Acid) and glycine receptor anion channels

The major functions of these channels are control of cell volume and osmolarity, trans-epithelial electrolyte transport, regulation of the intracellular pH value, setting of the resting potential, and (mostly inhibitory) neurotransmission. In this chapter, channels of the ClC and CFTR types will be discussed; the receptor anion channels used for neurotransmission are explained in Section 16.3.

Figure 14.40 Families and major functions of chloride channels Helices are symbolized by cylinders. The number of subunits per channel is shown in parentheses. ATP-binding sites are depicted as colored rectangles. ClC, chloride channel; R, regulatory domain; CFTR, cystic fibrosis transmembrane conductance regulator; GABA, γ-aminobutyric acid. For more details, see text.

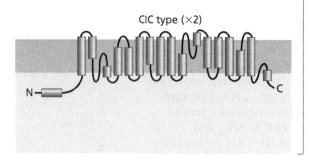

ClC type (×2)

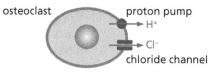

acidification (e.g., for bone resorption)

osteoclast — proton pump — H⁺ — Cl⁻ — chloride channel

control of the cell volume
NaCl reabsorption in the kidney
control of the resting potential

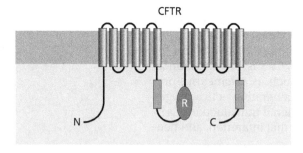

CFTR

trans-epithelial salt and water transport

GABA and glycine receptor (×5)

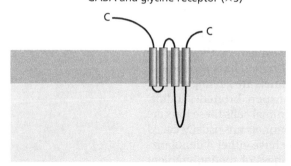

inhibitory neurotransmission

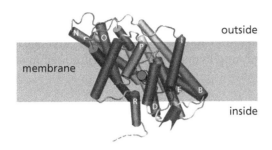

Figure 14.41 Three-dimensional structure of a chloride channel of the type ClC The figure shows one of the two subunits of the channel. Note that the transmembrane helices (cylinders) are arranged in a twisted and tilted rather than in a parallel and perpendicular manner. The red sphere in the central pore symbolizes the chloride ion. (Adapted from T.J. Jentsch, *Nature* 415:276–277, 2002.)

14.7.1 Anion channels of type ClC

This ancient channel type is found in prokaryotes, most probably participating in regulation of the cell volume (Section 3.5.7). Mammals express at least nine different isoforms. Their overall structure, derived from the bacterial channel, is not related to that of cation channels. Instead, it comprises no less than 18 α-helices; 10–12 of these are transmembrane domains, whereas the others dip into the membrane only partially or not at all (Figures 14.40 and 14.41). Each channel is composed of two such subunits and exhibits two pores (see Figure 3.34 in Section 3.5.7).

Most chloride channels are gated by low pH, high Cl⁻ concentration and membrane depolarization. Their voltage-sensitive domain is not related to the S3–S4 voltage sensor of cation channels and the mechanism of voltage-dependent gating is not understood. The individual ClC types differ in their functions and tissue distributions. Type **ClC1**, for instance, is more or less restricted to skeletal muscle cells, whereas the related isoform **ClC2** is broadly expressed. [The first channels of the ClC type to be isolated were those of the electric organ of the electric ray *Torpedo californica*. These so-called ClC0 channels resemble the type ClC1 of skeletal muscles, which are the evolutionary precursors of electric organs.] While ClC2 participates in the regulation of the intracellular pH value and the cell volume, ClC1 controls the resting potential, which in skeletal muscle cells is set mainly by chloride channels rather than by potassium channels as in neurons. Consequently, the channel is gated by depolarization. Defects of the *clc1* gene cause severe neuromuscular diseases such as congenital myotony.

Another highly specialized chloride channel is **ClC-K**, expressed predominantly in the inner ear and kidney. In the kidney, the channel is involved in NaCl reabsorption (Figure 14.42). ClC-K forms a complex with a regulatory subunit with

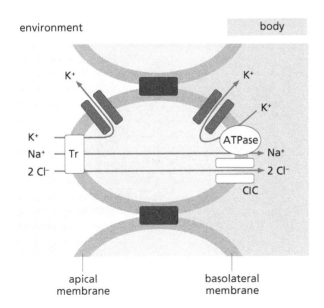

Figure 14.42 NaCl reabsorption in Henle's loop of the kidney By means of a Na⁺/K⁺/2 Cl⁻ co-transporter (Tr), the epithelial cell (shown as an oval) takes up NaCl and KCl from the primary urine (environment). At the basolateral side (body), Na⁺ is pumped into the blood by a K⁺/Na⁺-dependent ATPase with chloride following passively through a ClC channel. Excessive K⁺ leaves the cell through apical and basolateral potassium channels of the type K_{ir} (Figure 14.10). In an alternative transport system, epithelial cells take up Na⁺ by Na⁺ channels (ENaC, Figure 14.9).

two transmembrane domains called barttin. This name alludes to the **Bartter syndrome** that is characterized by a severe defect of salt balance. The reason is loss-of-function mutations of the barttin and ClC-K genes, leading to massive salt loss in the kidney and to deafness since the production of endolymph in the inner ear is impaired.

The channel **ClC7** is involved primarily in the **resorption of bone material**. This is the task of a phagocyte-like cell type, the osteoclast, which produces acidic hydrolases that dissolve the bone substance. The acidic milieu required is generated by H^+-dependent ATPases. They pump protons from the intra- to the extracellular space (Figure 14.40). To attain electric neutrality, this proton efflux is matched by a chloride efflux through ClC7 channels; that is, the cell releases hydrochloric acid. Bone resorption is of utmost importance for the development of the skeleton and the repair of bone damage. Indeed, a deletion of the *clc7* gene seems to be a cause of juvenile osteoporosis, a severe developmental disorder. Acidification is also important for the function of certain intracellular organelles such as endosomes, lysosomes, and secretory vesicles, the membranes of which also contain H^+-dependent ATPases and chloride channels.

Some pressure-sensitive ClC types play a role as **cellular pressure relief valves**. In cooperation with 2P-domain potassium channels, they enable the cell to adapt to hypotonic stress by promoting an efflux of KCl (see Section 14.4.3).

14.7.2 Chloride channels of type CFTR

The CFTR channel plays an essential role in the hormonal regulation of mucous epithelia. This was discovered during studies on the genetic background of **cystic fibrosis** (or mucoviscidosis), a severe and fatal hereditary disease that is characterized by massive disturbances of water and electrolyte balance. As a consequence, mucous membranes suffering from desiccation produce a highly viscous secretion that strongly impairs their function and promotes bacterial infection, in particular in the respiratory and gastrointestinal tracts. The reason is a loss-of-function mutation of the *cftr* gene resulting in reduced CFTR channel permeability.

Figure 14.43 Domain structure and regulation of the CFTR chloride channel To open the channel, the regulatory domain R must be phosphorylated by either cAMP- or cGMP-dependent kinases and ATP must be hydrolyzed by the intrinsic ATPase domains (shown as colored rectangles). The figure shows two (patho)physiologically important signaling pathways leading to channel gating. AC, adenylate cyclase; PKA, protein kinase A; PKG, protein kinase G; GC, guanylate cyclase.

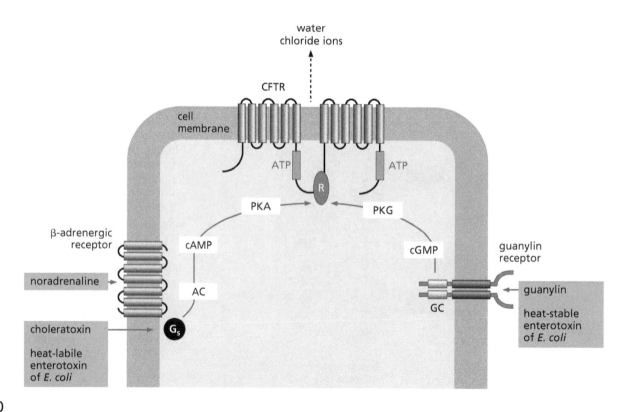

Among the ion channels, CFTR is unique because it belongs to the family of ABC proteins, which normally transport organic molecules rather than ions (see Section 3.2.1). Like most ABC transporters, CFTR has 12 transmembrane domains arranged in two groups of six each, as well as two ATPase domains in the cytoplasmic region (Figure 14.43). These structural elements are completed by a regulatory domain (R in Figure 14.43) that contains various phosphorylation sites and is not found in other human ABC proteins. The regulatory domain is the target of hormonal signals controlling CFTR permeability.

Gating of the channel resembles a logical AND gate because it requires both phosphorylation of the regulatory domain and ATP binding. Upon ATP hydrolysis by the intrinsic ATPase activity (as well as upon dephosphorylation), the channel is closed. This built-in timer function resembles that of G-proteins. Phosphorylation is required because the unphosphorylated regulatory domain blocks the channel pore. It is mainly catalyzed by either cAMP- or cGMP-dependent protein kinases. The cAMP-dependent pathway is activated, for instance, by (nor)adrenaline as part of the systemic stress response; the cGMP-controlled pathway is stimulated by the peptide hormone **guanylin** produced by the small intestine and regulating the electrolyte balance. While (nor)adrenaline interacts with a G_s-protein-coupled β-adrenergic receptor, guanylin stimulates a receptor with intrinsic guanylate cyclase activity (this receptor type is discussed in detail in Section 4.5).

While the inactivation of the CFTR channel causes desiccation of mucous membranes, an overactivation has the opposite result. This explains the effect of some prominent bacterial toxins that attack CFTR-related signal transduction. The G_s-protein is a specific target of **cholera toxin** and **heat-labile enterotoxin** of *Escherichia coli*; they cause a hyperphosphorylation and continuous activation of CFTR (Section 4.3.3), which leads to a catastrophic loss of salt and water, particularly in the intestinal epithelium. The **heat-stable enterotoxin** of bacteria (such as certain *E. coli* strains) causing diarrhea has a similar effect by overstimulating the guanylin receptor.

Summary

ClC-type channels are complex transmembrane proteins with prokaryotic ancestry. They are unrelated to other ion channels. Depending on the steepness of the Cl$^-$ gradient, their gating has either a hyperpolarizing or a depolarizing effect. Individual members of this channel family participate in control of the resting potential, regulation of the osmotic pressure, trans-epithelial electrolyte transport, and bone resorption. CFTR-type channels belonging to the ABC transporter family are essential for the function of mucous membranes. They are gated by a combined effect of ATP hydrolysis and protein phosphorylation, catalyzed by either cAMP- or cGMP-dependent protein kinases. Cholera toxin and heat-stable enterotoxin of colibacteria deregulate the function of the channel, while a loss-of-function mutation causes cystic fibrosis.

Further reading

Alvarez de la Rosa D, Canessa CM, Fyfe GK et al. (2000) Structure and regulation of amiloride-sensitive sodium channels. *Annu. Rev. Physiol.* 62, 573–594.

Ambudkar IS, Ong HL, Liu X et al. (2007) TRPC1: the link between functionally distinct store-operated calcium channels. *Cell Calcium* 42, 213–223.

Anderson PAV & Greenberg RM (2001) Phylogeny of ion channels: clues to structure and function. *Comp. Biochem. Physiol., Part B: Biochem. Mol. Biol.* 129, 17–28.

Arikkath J & Campbell KP (2003) Auxiliary subunits: essential components of the voltage-gated calcium channel complex. *Curr. Opin. Neurobiol.* 13, 298–307.

Ashby MC & Tepikin AV (2002) Polarized calcium and calmodulin signaling in secretory epithelia. *Physiol. Rev.* 82, 701–734.

Bezanilla F (2008) How membrane proteins sense voltage. *Nat. Rev. Mol. Cell Biol.*, 9, 323-332.

Berger F, Ramirez-Hernandez MH & Ziegler M (2004) The new

life of a centenarian: signaling functions of NAD(P). *Trends Biochem. Sci.* 29, 111–118.

Berridge JR, Bootman MD & Roderick HL (2003) Calcium signaling: dynamics, homeostasis and remodelling. *Nat. Rev. Mol. Cell Biol.* 4, 517–529.

Berridge MJ (2003) Cardiac calcium signaling. *Biochem. Soc. Trans.* 31, 930–933.

Berridge MJ, Lipp P, Bootman MD (2000) The versatility and universality of calcium signaling. *Nat. Rev. Mol. Cell Biol.* 1, 11–21.

Bezprozvanny I (2005) The inositol 1,4,5-trisphosphate receptors. *Cell Calcium* 38, 261–272.

Blaustein MP & Golovina VA (2001) Structural complexity and functional diversity of endoplasmic reticulum Ca^{2+} stores. *Trends Neurosci.* 24, 602–608.

Bootman MD, Lipp P & Berridge MJ (2001) The organisation and functions of local Ca^{2+} signals. *J. Cell Sci.* 114, 2213–2222.

Burgoyne RD (2007) Neuronal calcium sensor proteins: generating diversity in neuronal Ca^{2+} signaling. *Nat. Rev. Neurosci.* 8, 182–193.

Campbell JD, Sansom MS & Ashcroft FM (2003) Potassium channel regulation. *EMBO Rep.* 4, 1038–1042.

Carafoli E (2004) Calcium-mediated cellular signals: a story of failures. *Trends Biochem. Sci.* 29, 371–379.

Carmeliet E (2004) Intracellular Ca^{2+} concentration and rate adaptation of the cardiac action potential. *Cell Calcium* 35, 557–573.

Catterall WA (2000) From ionic currents to molecular mechanisms: the structure and function of voltage-gated sodium channels. *Neuron* 26, 13–15.

Catterall WA (2000) Structure and regulation of voltage-gated Ca^{2+} channels. *Annu. Rev. Cell. Dev. Biol.* 16, 521–555.

Chang W & Shoback D (2004) Extracellular Ca^{2+}-sensing receptors – an overview. *Cell Calcium* 35, 183–196.

Chin D & Means AR (2000) Calmodulin: a prototypical calcium sensor. *Trends Cell Biol.* 10, 322–328.

Costigan M & Woolf CJ (2002) No DREAM, no pain: closing the spinal gate. *Cell* 108, 297–300.

Deisseroth K, Mermelstein PG, Xia H et al. (2003) Signaling from synapse to nucleus: the logical behind the mechanisms. *Curr. Opin. Neurobiol.* 13, 354–365.

Dodson PD & Forsythe ID (2004) Presynaptic K^+-channels: electrifying regulators of synaptic terminal excitability. *Trends Neurosci.* 27, 210–216.

Estevez R & Jentsch TJ (2002) CLC chloride channels: correlating structure with function. *Curr. Opin. Struct. Biol.* 12, 531–539.

Fleig A & Penner R (2004) The TRPM ion channel subfamily: molecular, biophysical amd functional features. *Trends Pharmacol. Sci.* 25, 633–639.

Foskett KJ, White C, Cheung KH et al. (2007) Inositol trisphosphate receptor Ca^{2+} release channels. *Physiol. Rev.* 87, 593–658.

Gadsby DC, Vergani P & Csanady L (2006) The ABC protein turned chloride channel whose failure causes cystic fibrosis. *Nature* 440, 477–483.

Galione A & Ruas M (2005) NAADP receptors. *Cell Calcium* 38, 273–280.

George AL, Bianchi L, Link EM et al. (2001) From stones to bones: the biology of ClC chloride channels. *Curr. Biol.* 11, R620–R628.

Giamarchi A, Padilla F, Coste B et al. (2006) The versatile nature of the calcium-permeable cation channel TRPP2. *EMBO Rep.* 7, 787–793.

Gill DL & Patterson RL (2004) Toward a consensus on the operation of receptor-induced calcium entry signals (and additional articles related to this topic). *Sci. STKE* 2004/243/pe39.

Goldin AL (2003) Mechanisms of sodium channel inactivation. *Curr. Opin. Neurobiol.* 13, 284–290.

Goll DE, Thompson VF, Li H et al. (2003) The calpain system. *Physiol. Rev.* 83, 731–801.

Gormley K, Dong Y & Sagnella GA (2003) Regulation of the epithelial sodium channel by accessory proteins. *Biochem. J.* 371, 1–14.

Guse H (2004) Regulation of calcium signaling by the second messenger cyclic adenosine diphosphoribose (cADPR). *Curr. Mol. Med.* 4, 239–248.

Hamilton SL (2005) Ryanodine receptors. *Cell Calcium* 38, 253–260.

Hanson CJ, Bootman MD & Roderick HL (2004) Cell signaling: IP$_3$ receptors channel calcium into death. *Curr. Biol.* 14, R933–R935.

Hewavitharana T, Deng X, Soboleff J et al. (2007) Role of STIM and Orai proteins in the store-operated calcium signaling pathway. *Cell Calcium* 42, 173–182.

Hofer AM & Brown EM (2003) Extracellular calcium sensing and signaling. *Nat. Rev. Mol. Cell Biol.* 4, 530–538.

Hogan PG & Rao A (2007) Dissecting I_{CRAC}, a store-operated calcium current. *Trends Biochem. Sci.* 32, 235–245.

Honoré E (2007) The neuronal background K_{2P} channels: focus on TREK 1. *Nat. Rev. Neurosci.* 8, 251–261.

Hook SS & Means AR (2001) Ca^{2+}/CaM-dependent protein kinases: from activation to function. *Annu. Rev. Pharmacol. Toxicol.* 41, 471–505.

Hosley V & Pavlath GK (2002) NFAT: ubiquitous regulator of cell differentiation and adaptation. *J. Cell Biol.* 156, 771–774.

Hudmon A & Schulman H (2002) Neuronal Ca^{2+}/calmodulin-dependent protein kinase II: the role of structure and autoregulation in cellular function. *Annu. Rev. Biochem.* 71, 473–510.

Hudmon A & Schulman H (2002) Structure-function of the multifunctional Ca^{2+}/calmodulin-dependent protein kinase II. *Biochem. J.* 364, 593–611.

Ikura M, Osawa M & Ames JB (2002) The role of calcium-binding proteins in the control of transcription: structure to function. *Bioessays* 24, 625–636.

Jentsch TJ, Stein V, Weinreich F et al. (2002) Molecular structure and physiological function of chloride channels. *Physiol. Rev.* 82, 503–568.

Jentsch TJ, Neagoe I & Scheel O (2005) ClC chloride channels and transporters. *Curr. Opin. Neurobiol.* 15, 319–325.

Koh TW & Bellen HJ (2003) Synaptotagmin I. A Ca^{2+} sensor for neurotransmitter release. *Trends Neurosci.* 26, 413–422.

Kortvely E & Gulya K (2004) Calmodulin, and various ways to regulate its activity. *Life Sci.* 74, 1065–1070.

Krishtal O (2003) The ASICs: signaling molecules? Modulators? *Trends Neurosci.* 26, 472–482.

Lee HC (2001) Physiological functions of cyclic ADP-ribose and NAADP as calcium messengers. *Annu. Rev. Pharmacol. Toxicol.* 41, 317–345.

Lemmon MA (2008) Membrane recognition by phospholipid-binding domains. *Nat. Rev. Mol. Cell Biol.*, 9, 99-111.

Lewis RS (2003) Calcium oscillations in T-cells: mechanisms and consequences for gene expression. *Biochem. Soc. Trans.* 31, 925–929.

Macian F, Lopez-Rodriguez C & Rao A (2001) Partners in transcription: NFAT and AP-1. *Oncogene* 20, 2476–2489.

MacLennan DH & Kranias EG (2003) Phospholamban: a crucial regulator of cardiac contractility. *Nat. Rev. Mol. Cell Biol.* 4, 566–577.

Meissner G (2004) Molecular regulation of cardiac ryanodine receptor ion channel. *Cell Calcium* 35, 621–628.

Miller C (2006) ClC chloride channels viewed through a transporter lens. *Nature* 440, 484-489.

Minke B (2006) TRP channels and Ca^{2+}-signaling. *Cell Calcium* 40, 261–275.

Minke B & Cook B (2002) TRP channel proteins and signal transduction. *Physiol. Rev.* 82, 429–472.

Montell C (2003) Thermosensation: hot findings make TRPNs very cool. *Curr. Biol.* R476–R478.

Nilius B (2003) From TRPs to SOCs, CCEs and CRACs: consensus and controversies. *Cell Calcium* 33, 293–298.

Nichols CG (2006) K$_{ATP}$ channels as molecular sensors of cellular metabolism. *Nature* 440, 470-476.

Nilius B, Owsianik G, Voets T et al. (2007) Transient receptor potential cation channels in disease. *Physiol. Rev.* 87, 165–217.

Orrenius S, Zhivozovsky B & Nicotera P (2003) Regulation of cell death: the calcium–apoptosis link. *Nat. Rev. Mol. Cell Biol.* 4, 552–564.

Patel AJ & Honore E (2001) Properties and modulation of mammalian 2P domain K$^+$ channels. *Trends Neurosci.* 24, 339–347.

Patterson RL, Boehning D & Snyder SH (2004) Inositol 1,4,5-trisphosphate receptors as signal integrators. *Annu. Rev. Biochem.* 73, 437–465.

Petersen OH, Michalak M & Verkhratsky A (eds) (2005) Frontiers in calcium signalling. *Cell Calcium* 38, 161–446.

Riordan JR (2005) Assembly of functional CFTR chloride channels. *Annu. Rev. Physiol.* 67, 701–718.

Rooshild TP, Le KT & Choe S (2004) Cytoplasmic gatekeepers of K$^+$-channel flux: a structural perspective. *Trends Biochem. Sci.* 29, 39–45.

Roux B (2002) Theoretical and computational models of ion channels. *Curr. Opin. Struct. Biol.* 12, 182–189.

Rudolf R, Mongillo M, Rizzuto R et al. (2003) Looking forward to seeing calcium. *Nat. Rev. Mol. Cell Biol.* 4, 679–686.

Rusnak F & Mertz P (2001) Calcineurin: form and function. *Physiol. Rev.* 80, 1483–1521.

Sadja R, Alagem N & Reuveny E (2003) Gating of GIRK channels: details of an intricate, membrane-delimited signaling complex. *Neuron* 39, 9–12.

Sanguinetti MC & Tristani-Firouzi (2006) hERG potassium channels and cardiac arrhythmia. *Nature* 440, 463-469.

Sansom MSP & Shrivastava IH (2002) Ion channels: frozen motion. *Curr. Biol.* 12, R65–R67.

Santella L, Lim D & Moccia F (2004) Calcium and fertilization: the beginning of life. *Trends Biochem. Sci.* 29, 400–407.

Santoro B & Baram TZ (2003) The multiple personalities of h-channels. *Trends Neurosci.* 26, 550–554.

Shibasaki F, Hallin U & Uchino H (2002) Calcineurin as a multifunctional regulator. *J. Biochem. (Tokyo)* 131, 1–15.

Shieh CC, Coghlan M, Sullivan JP et al. (2000) Potassium channels: molecular defects, diseases, and therapeutic opportunities. *Pharmacol. Rev.* 52, 557–593.

Shuttleworth TJ & Mignen O (2003) Calcium entry and the control of calcium oscillations. *Biochem. Soc. Trans.* 31, 916–919.

Sigworth FJ (2003) Life's transistors. *Nature* 423, 21–22.

Soderling TR & Stull JT (2001) Structures and regulation of calcium/calmodulin-dependent protein kinases. *Chem. Rev.* 101, 2341–2351.

Stotz SC & Zamponi GW (2001) Structural determinants of fast inactivation of high-voltage Ca^{2+}. *Trends Neurosci.* 24, 176–181.

Taylor CW, da Fonseca PCA & Morris EP (2004) IP$_3$ receptors: the search for structure. *Trends Biochem. Sci.* 29, 210–219.

Tengholm A (2007) Cyclic AMP: swing that message! *Cell. Mol. Life Sci.* 64, 382–385.

West AE, Chen WG, Dalva MB et al. (2001) Calcium regulation of neuronal gene expression. *Proc. Natl Acad. Sci. U.S.A.* 98, 11024–11031.

Worley PF, Zeng W, Huang GN et al. (2007) TRPC channels as STIM1-regulated store-operated channels. *Cell Calcium* 42, 205–211.

Yellen G (2002) The voltage-gated potassium channels and their relatives. *Nature* 419, 35–42.

Yu FH, Yarov-Yarovoy V, Gutman GA & Catterall WA (2005) Overview of molecular relationships in the voltage-gated ion channel superfamily. *Pharmacol. Rev.* 57, 387–395.

Yule DI, Straub SV & Bruce JIE (2003) Modulation of Ca^{2+} oscillations by phosphorylation of $Ins(1,4,5)P_3$ receptors. *Biochem. Soc. Trans.* 31, 954–957.

Zaccolo M & Pozzan T (2003) cAMP and Ca^{2+} interplay: a matter of oscillation. *Trends Neurosci.* 26, 53–55.

Sensory Signal Processing

Sensory cells translate environmental signals into the universal language of the central nervous system: action potentials, the frequency of which is interpreted by the brain as a sensation. To this end they express specific receptors for light, temperature, touch, sound waves, gravity, smell, and taste (or, more precisely, for the corresponding chemicals) as well as, in certain animals, for electricity and magnetism.

In most cases the activation of a sensory receptor results in a depolarization of the sensory cell. This so-called receptor potential induces the release of neurotransmitter at the synaptic contact with the postsynaptic neuron. The neurotransmitter interacts with corresponding receptors at the postsynaptic membrane, causing an action potential in the sensory neuron. The depolarization of sensory cells is due to a gating of cation channels (in some cases a blockade of hyperpolarizing K^+ channels produces the same result). Most of these channels are cation-nonspecific and belong to the transient receptor potential (TRP) and cyclic nucleotide-gated (CNG) channel families (Sections 14.5.6 and 14.5.7). Depending on the environmental stimulus, the ion channels are receptors themselves or are controlled by other sensory receptors via second messengers (Figure 15.1). Sensory cells are highly sensitive. To respond over a large range of signal intensity, they make use of very efficient adaptive mechanisms, thus adjusting their input sensitivity to the strength of the stimulus.

15.1 Taste

A fundamental condition of life is that organisms must be able to check the quality of food in order to distinguish between edible and inedible and to avoid poisoning. Therefore, taste probably represents the most ancient of all senses. Taste receptors are chemoreceptors that recognize environmental key compounds. In bacteria, for instance, certain amino acids and sugars signal food. We can discriminate between at least five basic taste qualities: sour, salty, sweet, bitter, and "umami." Recently evidence has been provided for a sixth sense recognizing "fatty taste." Other variants of tasting are mediated by smelling. In mammals, taste cells are found predominantly in the taste buds of the oral epithelium. Like most other sensory cells they exhibit a bipolar shape: their apical membrane, directed towards the oral cavity, is divided into numerous microvilli and densely occupied by taste receptors, while the basolateral membrane forms a synaptic contact with a sensory neuron (Figure 15.2). Each taste quality is thought to have its own cell and receptor type; a simplified overview of taste signal transduction is shown in Figure 15.3.

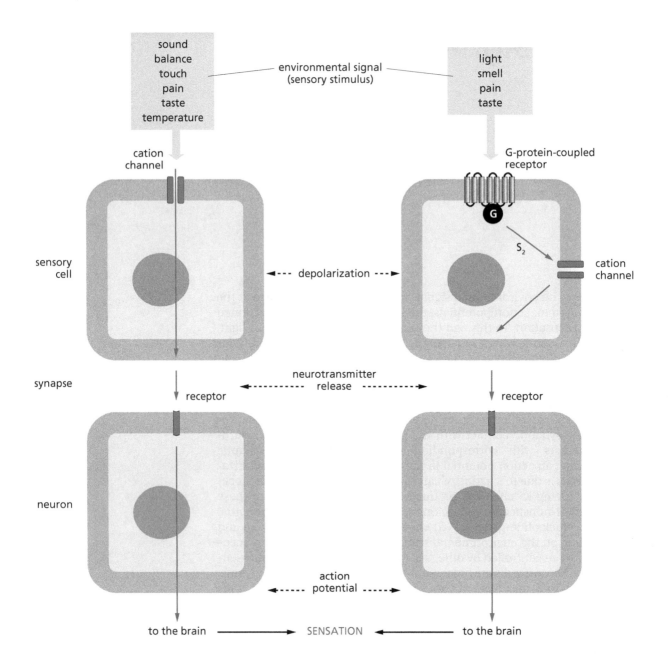

Figure 15.1 Two pathways for the transformation of sensory stimuli into action potentials In sensory cells, environmental signals activate cation channels either directly (left side) or via a second messenger S_2 (right side), the formation of which is mediated by a G-protein-coupled receptor. In both cases the sensory cell depolarizes and releases a neurotransmitter, which via a postsynaptic receptor induces an action potential in the neuron connecting the sensory organ with the brain. Most of the S_2-activated cation channels belong to the TRP family (Section 14.5.6).

15.1.1 Salty and sour tastes: mediated directly by ion channels

The **salty taste** is important for the controlled intake of electrolytes. It is caused mainly by sodium chloride, or more precisely by Na^+ ions. The corresponding "receptors" are epithelial Na^+ channels (ENaC family; see Section 14.3) that depolarize the cell via Na^+ influx. As is typical for most ENaCs, their expression is induced by aldosterone. In fact, this steroid hormone, which is secreted by the adrenal cortex as a response to salt deficiency, stimulates adaptive responses that are mediated by ENaCs: the salty taste sensation and Na^+ reabsorption in kidney tubules and other epithelia.

The **sour taste** first of all is a warning signal indicating poisonous chemicals, unripe fruits, or food that has gone bad. It is evoked by protons, which activate depolarizing ion channels. In some species these are ENaCs of the acid-sensing ion channel (ASIC) type (see Section 14.3). Another widespread acid-sensitive cation channel involved in sour taste processing is the P2-type TRP channel (also called PKD2L1). Channels of this type are also known as type 2 **polycystins** (also called PKD2). Together with type 1 polycystins (PKD1) they

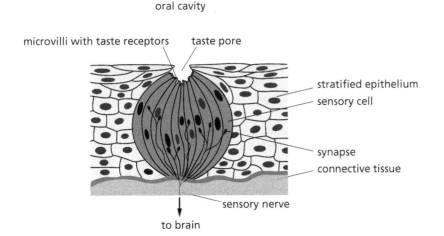

oral cavity

microvilli with taste receptors   taste pore

stratified epithelium
sensory cell

synapse
connective tissue

sensory nerve

to brain

Figure 15.2 Schematic diagram of a taste bud Taste buds are sensory organs at the surface of the tongue. Each taste bud contains about 50 sensory cells (red) that are in synaptic contact with the fibers of the sensory nerves. The cells' microvilli lining the taste pore are equipped with taste receptors. (Adapted from B. Alberts et al. Molecular Biology of the Cell. New York: Garland Science, 1983.)

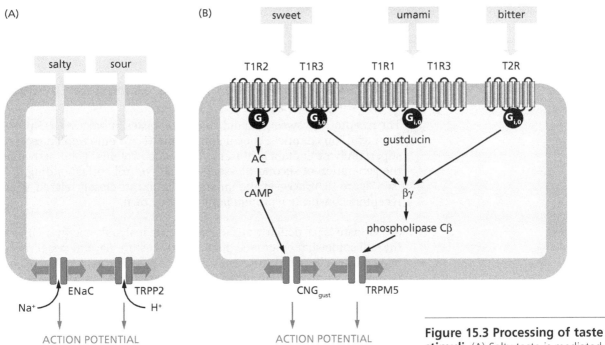

(A)

(B)

salty   sour

sweet   umami   bitter

T1R2   T1R3   T1R1   T1R3   T2R

G_s $G_{i,0}$ $G_{i,0}$ $G_{i,0}$

gustducin

AC

cAMP

βγ

phospholipase Cβ

ENaC   TRPP2

Na^+ H^+

CNG$_{gust}$ TRPM5

ACTION POTENTIAL

ACTION POTENTIAL

Figure 15.3 Processing of taste stimuli (A) Salty taste is mediated by a sodium channel of the epithelial Na^+ channel (ENaC) family, sour taste by the proton-sensitive transient receptor potential, polycystin-related (TRPP2) cation channel. (B) The $G_{i,0}$-protein gustducin couples with sweet, umami, and bitter receptors activating phospholipase Cβ via Gβγ-subunits. In turn the ion channel TRPM5 (transient receptor potential, melastatin-related) is gated. Sweet taste may also be mediated by a cyclic nucleotide-gated cation channel CNG$_{gust}$ along the G_s–adenylate cyclase (AC)–cAMP pathway. For other details see text. Note that the figure represents a simplified scheme since in reality each of the five taste qualities is processed by a unique cell type.

have been found to be mutated in autosomal dominant polycystic kidney disease (PKD, a severe genetic disorder ending up in renal failure). [Polycystic kidney disease proteins (PKD1 and -2) should not be confused with the entirely unrelated protein kinases D (PKD; see Section 4.6.5).] Type 1 and 2 polycystins have only the name in common. Structurally they differ completely: type 2 polycystins constitute the TRPP2 subfamily of TRP cation channels (Section 14.5.6), whereas type 1 polycystins are proteins with 11 membrane-spanning domains, an extended extracellular N-terminus, and a G-protein-activating cytoplasmic domain.

The precise function of PKD1 is unknown except that to constitute an acid-sensitive cation channel, PKD1 and PKD2 must interact via their intracellular C-terminal domains. The heterodimers thus formed have been found to be expressed selectively in distinct taste receptor cells of the mammalian tongue as well as in neurons thought to monitor the acidity of the cerebrospinal fluid. How the channel senses acidity is not clear. The same holds true for its role in polycystic kidney disease.

537

15.1.2 Sweet, umami, and bitter tastes: mediated by G-protein-coupled receptors

The receptors for sweet, umami, and bitter tastes interact with G-proteins rather than with ion channels; they belong to the G-protein-coupled receptor (GPCR) superfamily of heptahelical transmembrane proteins. Their activation leads to the generation of second messengers that control corresponding ion channels (see Figure 15.3B). Sweet and umami receptors are closely related, whereas bitter receptors constitute a protein family of their own.

Sweet taste is particularly attractive because it signals energy-rich food (as does the still somewhat enigmatic quality "fatty taste"). Several sweet receptors have been found. They constitute the T1R (Taste 1 Receptors) family, are coupled with G_S-proteins, and respond primarily to sugars but also to some other natural and artificial sweeteners. In the plasma membrane, receptor isoforms combine into heterodimers of the type T1R2–T1R3. Along the G_S-controlled signaling pathway, these dimeric receptors stimulate cAMP production and inhibit (probably through protein kinase A-catalyzed phosphorylation) potassium channels, at the same time gating the cAMP-controlled cation channel CNG_{gust} (Section 14.5.7). This reaction cascade resulting in depolarization is activated by sugars, whereas artificial sweeteners seem to open the $G_{q,11}$–phospholipase Cβ pathway of signaling. The question as to how this pathway leads to depolarization has not been finally settled. On the one hand, a hyperpolarizing potassium channel has been reported to become blocked by protein kinase C-catalyzed phosphorylation. On the other hand, the cation channel TRPM5 (Section 14.5.6) has recently moved into the center of interest since it seems to respond to many if not all taste stimuli. Like some other TRP channels, TRPM5 becomes gated by phospholipase C. The effect is mediated primarily by a $G_{i,0}$-protein called gustducin in analogy to the transducin of photoreceptor cells via Gβγ-subunits. As explained in Section 14.5.6, the mechanism of TRP channel activation along the phospholipase C pathway is still a matter of debate. It may include both degradation of phosphoinositides, blocking TRP activity, and phosphorylation of TRPs by protein kinase C and Ca^{2+}/calmodulin-activated kinases. In the case of TRPM5 the membrane potential also may play a role because this channel belongs to the few voltage-sensitive TRP ion channels.

Most interestingly, the adipokinetic hormone **leptin** lowers the sensitivity of sensory cells for signals transducing sweet taste. As explained in Section 9.4.3, leptin prevents excessive weight gain by suppressing the feeling of hunger and by stimulating energy consumption through effects on the hypothalamus, as well as by inhibiting insulin secretion in the pancreas. In parallel, the taste quality of sweetness obviously becomes less attractive. Upon starvation, the secretion of leptin is turned down, resulting in the well-known craving for sweets. Inborn or acquired defects of cellular mechanisms processing leptin signals are some of the main causes of obesity.

The amino acid L-glutamate is a key signal indicating protein-rich food. The taste sensation, typical of chicken soup, is called **umami** after *umai*, the Japanese word for "delicious." Umami receptors belong to the same T1R family as the sweet receptors. However, their functional form is a T1R1–T1R3 heterodimer that, like the T1R2–T1R3 sweet receptors, activates the TRPM5 cation channel. T1 receptors share the selectivity for glutamate with the metabotropic glutamate receptors of the nervous system. Both receptors are active only as heterodimers and quite similar to each other in that they have an extended extracellular domain that contains the ligand binding site, originating probably from a bacterial binding protein for sugars and amino acids (details in Section 16.4).

Bitter taste mostly indicates that a food that may be inedible or even toxic. Humans have 26 and mice have 33 genes encoding putative bitter receptors. Like T1 receptors, they are G-protein-coupled, although they belong to another family that is derived from rhodopsin and called T2R (see also Section 5.2). Their great variability reflects the large number of putative ligands with bitter taste. The trimeric G-protein coupled with T2 receptors is gustducin. Upon activation by bitter taste receptors, gustducin appears to stimulate two different types of cation channels: its Gβγ-subunit activates phospholipase Cβ2 and in turn the TRPM5 cation channel, while the α-subunit stimulates a cAMP/cGMP-specific phosphodiesterase of the PDE1 family that gates a cyclic nucleotide-suppressed cation channel by degrading cyclic nucleotides. Gene knockout experiments indicate that the phospholipase Cβ2–TRPM5 pathway is the major route of bitter taste signal processing.

Summary

Sensory receptors transform environmental stimuli into action potentials that are transmitted to the brain. They are parts of depolarizing ion channels or are G-protein-coupled heptahelical transmembrane proteins that induce channel gating along intracellular signaling pathways. There are at least five taste qualities: salty, sour, sweet, bitter, and umami. Salty taste is evoked directly by depolarizing epithelial sodium channels (ENaCs); sour taste is evoked by various pH-sensitive ion channels of the ASIC and TRPP families. The receptors for sweet, bitter, and umami-tasting chemicals are G-protein-coupled and activate, along cellular signaling cascades, depolarizing cation channels of the transient receptor potential (TRP) and cyclic nucleotide-gated (CNG) superfamilies.

15.2 Mechanical stimuli: touch and sound

Mechanical stress and fluctuations of osmotic pressure are experienced by even the most primitive cells. Therefore, all cells have ion and water channels for osmotic pressure equalization that respond to deformations of the cell membrane. In the MscL channel of *Escherichia coli* we have already seen such a pressure valve (Section 3.5.7). The corresponding channels of eukaryotic cells belong, however, to another protein family. In fact, mechanosensitive channels seem to have developed independently several times in the course of evolution. Accordingly, this receptor family is quite varied, including nonspecific cation channels as well as selective Na^+, K^+, and anion channels. In addition, many neurotransmitter receptors (Chapter 16) respond to mechanical stress. The mechanosensitive ion channels of certain neurons in the hypothalamus, for instance, are sensors of blood osmolarity. When the osmotic pressure rises, they facilitate (through depolarization) the secretion of the neurosecretory hormone vasopressin (also called adiuretin) that inhibits the excretion of water by the kidneys. Particularly sensitive for mechanical stimuli of blood flow and blood pressure are vasoendothelial cells, which respond by production of blood pressure-regulating factors such as prostacyclin, nitric oxide, endothelin, ATP, growth factors, and other signaling substances.

Numerous **sensors of touch and movement** note every muscle and tendon tension and every organ movement. They are concentrated in special sensory organs such as muscle spindles and Ruffini, Meissner, and Pacini bodies of skin.

Most probably, all these cell types possess mechanosensitive ion channels. The touch-sensitive organs in the skin of insects and nematodes have been studied most thoroughly in this respect. In the case of *Drosophila* they belong to the TRP family of cation channels and are connected with a flexible bristle. Any touch, even the slightest air breeze, moves the bristle and generates shearing forces between the bristle shaft and a sensory neuron connected with the bristle. Because the channels in the neuronal membrane are fixed by anchor proteins to the bristle outside and to the actin cytoskeleton inside the cell, they are opened by these forces and the neuron is depolarized, sending an action potential to the brain. Similar organs, albeit without bristles, are found in the skin of *Caenorhabditis elegans*, Here the mechanosensitive channels belong to the degenerin/ENaC family (Section 14.3). Like the TRP channels of insects, they are gated by shearing forces, which depolarize the cell (Figure 15.4). The mechanoreceptors of vertebrate skin seem to belong also to a particular type of ENaCs.

In the inner ear, highly specialized mechanosensors are found that, as stato-acoustic apparatus, respond to sound waves, gravity, and acceleration forces. Acoustic signals are processed by the sensory cells of the cochlea. In reference to their dense bundles of apical stereocilia, these cells are called hair cells. They are located on the basal lamina of the organ of Corti and anchored to the overlying tectorial membrane by the tips of the stereocilia. Sound causes vibrations of the membrane, which are transduced to the stereocilia bundles. Consequently, the mechanosensitive cation channels of stereocilia are opened. Each channel is anchored elastically at both sides: inside the cell to the actin cytoskeleton and

Figure 15.4 Cellular processing of mechanical stimuli (A) Schematic depiction of a mechanosensory organ in the cuticle of *Drosophila melanogaster*. The organ monitors the slightest air movements. It consists of a flexible bristle that is connected to the dendrite of a sensory neuron (red). The dendrite contains several mechanosensitive cation channels of the TRP family and is embedded in a sac filled with endolymph. (B) Model of a mechanosensitive cation channel. The channel is connected with the cytoskeleton and an extracellular anchoring structure in such a way that it becomes gated by any movement. (A, adapted from P.G. Gillespie and R.G. Walker, *Nature* 413:194–202, 2001.)

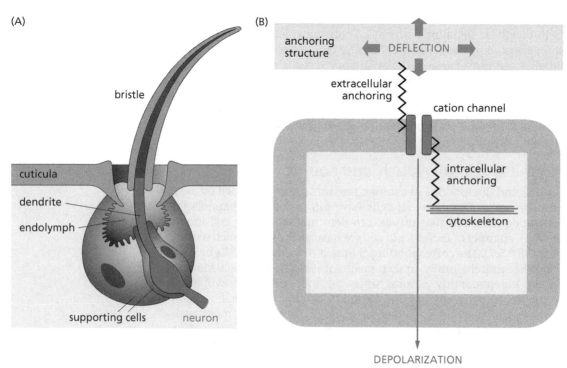

(A)

bristle

cuticula

dendrite

endolymph

supporting cells

neuron

(B)

anchoring structure

DEFLECTION

extracellular anchoring

cation channel

intracellular anchoring

cytoskeleton

DEPOLARIZATION

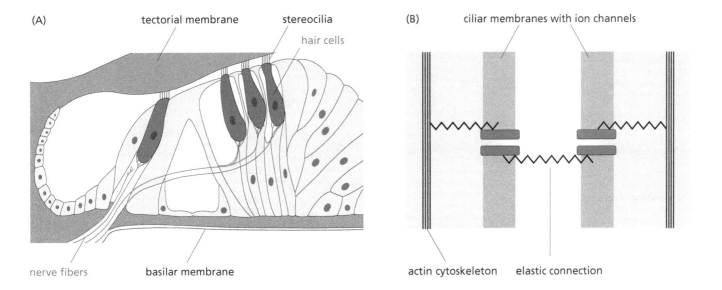

(A) tectorial membrane   stereocilia   hair cells

nerve fibers   basilar membrane

(B) ciliar membranes with ion channels

actin cytoskeleton   elastic connection

outside the cell to the neighboring stereocilium (Figure 15.5). The identity of the channels is not entirely clear. Probably they belong to the TRP superfamily.

The closely related sensory cells of the organ of balance are localized in the semicircular canals or labyrinth of the inner ear. They inform the organism about its spatial position and all kinds of acceleration. The inert mass of this system is increased by an incorporation of small limestones (statoconia) in the tectorial membrane. These resemble the gravity stones or statoliths in the organs of balance of many invertebrates.

Figure 15.5 Processing of mechanical stimuli in the inner ear (A) Anatomical scheme of the organ of Corti. (B) Detail of two neighboring stereocilia. For further details see text.

Summary

Mechanosensors are cation channels firmly anchored to intra- and extracellular structures. They are gated by mechanical forces (exerted by movements, touch, pressure, changes of osmotic pressure, gravity, and sound waves) that induce membrane depolarization.

15.3 Temperature and pain

Painful pressure, such as that due to an excessively filled urinary bladder or flatulence, is mediated by mechanoreceptors localized at the surface of neighboring neurons. In contrast, another receptor type responds to painful heat and cold. These thermoreceptors have been discovered in nociceptor (pain-sensitive) neurons in the course of studies with natural substances evoking heat, cold, or pain sensations in skin.

A prototype of such a substance is **capsaicin,** the hot ingredient of red pepper (Figure 15.6). Capsaicin stimulates a receptor that normally responds to temperatures above 43°C (the pain limit of heat in humans). This receptor, called TRPV1, belongs to the family of the depolarizing TRP cation channels (Section 14.5.6). Because of the vanilloid-related structure of capsaicin and other hot natural substances, it is also known as **vanilloid receptor 1**. [Interestingly, the TRPV1 receptor of birds does not respond to capsaicin and related chemicals, which allows the animals to consume hot foods such as peppers.] Related receptors exhibit other temperature sensitivities. So TRPV2 becomes active from 52°C, TRPV3 (known as the camphor receptor) from 31°C and TRPV4 from 27°C upward.

The pain limit of cold is around 15°C. The corresponding receptors, called TRPM1, TRPA1, and TRPM8, have been identified by using cooling substances such as menthol or eucalyptol as artificial ligands. They are known as **menthol**

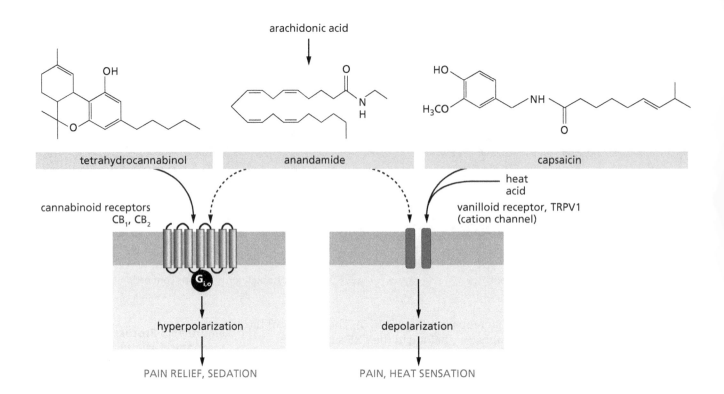

arachidonic acid

tetrahydrocannabinol

anandamide

capsaicin

heat
acid

cannabinoid receptors
CB₁, CB₂

vanilloid receptor, TRPV1
(cation channel)

hyperpolarization

depolarization

PAIN RELIEF, SEDATION

PAIN, HEAT SENSATION

Figure 15.6 Signal transduction by the vanilloid receptor TRPV1 as compared with cannabinoid receptors The receptor ion channel TRPV1 (transient receptor potential, vanilloid type, right) is activated by heat, acid, capsaicin, and the endogenous ligand anandamide, an ethanolamide of arachidonic acid. As an endocannabinoid, anandamide also activates the cannabinoid receptors, exerting an opposite effect (left).

receptors and also belong to the TRP cation channel family. In contrast to most other receptors of the TRP family, TRPV1 and TRPM8 are weakly voltage-dependent, becoming more readily opened upon depolarization. This effect is modulated by the temperature. If it is assumed that the receptors exist in two conformations, OPEN and CLOSED, the conformation OPEN (for heat-activated receptors) and the conformation CLOSED (for cold-activated receptors) may become stabilized with increasing temperature.

Why does the binding of capsaicin, menthol, and other hot and cold natural products to TRP channels produce a "wrong" temperature sensation? These compounds induce a change of protein conformation. As a result the channel is gated at room temperature instead of at its threshold temperature. In physiology the phenomenon of increased temperature sensitivity is called thermal hyperalgesia.

At least for TRPV1 and TRPV4 receptors an endogenous ligand has been identified. Surprisingly, this substance turned out to be identical with the endocannabinoid **anandamide,** an arachidonic acid derivative (Figure 15.6). Endocannabinoids are endogenous neurotransmitter-like substances that interact with the cellular receptors for cannabis drugs. However, these cannabinoid receptors are not identical with TRPV ion channels but belong to the G-protein-coupled receptors (Section 16.7). Their activation has a pain-relieving effect. So, like many other neurotransmitters, endocannabinoids interact with both G-protein-coupled and ion channel-coupled receptors, eliciting opposite responses and dampening their own effects.

Other endogenous stimulants of TRPV receptors are substances that are released from wounded or infected tissue, and provoke the characteristic heat sensation of an inflammatory reaction. Inflammation causes an acidification of the tissue, and the protons thus produced activate the TRPV1 receptor as well as ASIC receptors of the ENaC family (Section 14.3).

Pro-inflammatory mediators mediating pain sensations include the tissue hormone **bradykinin**, the neuropeptide **substance P** (a member of the so-called tachykinin family), and **prostaglandin E**. They all interact with G_q-protein-

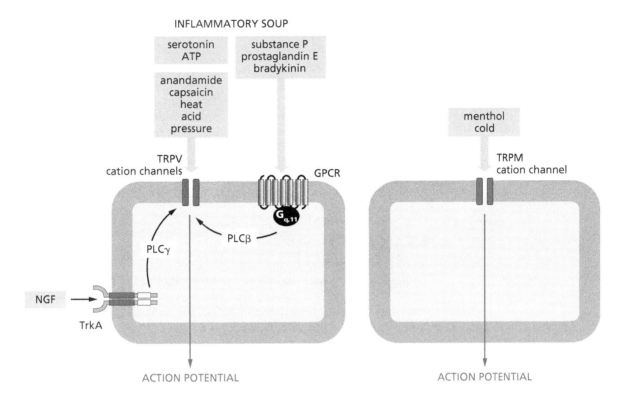

coupled heptahelical receptors (BK2-R for bradykinin, TAC1-R for substance P, and EP1-R for prostaglandin E) that activate phospholipase Cβ and, in turn, the TRPV1 channel (see Figure 15.7). The formation of prostaglandin E catalyzed by cyclooxygenases is inhibited by nonsteroidal anti-inflammatory drugs (NSAIDs), which include most important pain relievers such as acetylsalicylic acid, better known as aspirin (Section 4.4.5). Other ingredients of the inflammatory soup such as ATP and serotonin facilitate TRPV-dependent depolarization by gating receptor-coupled cation channels (see Table 16.1 in Chapter 16).

An endogenous factor whose role in pain has become obvious only recently is the **nerve growth factor**, **NGF**. Originally, this tissue hormone was discovered as a neurotrophic protein: an endogenous mediator that promotes the embryonic development of the nervous system, particularly of sensory and sympathetic neurons (for details see Chapter 16.1). This is, in fact, its predominant function in the prenatal phase, whereas in the adult organism NGF is a major pain mediator that becomes strongly expressed in wounded and inflamed tissues. NGF interacts with two quite different types of receptors: a receptor tyrosine kinase of the Trk family (Section 7.1) and a p75 neurotrophin receptor (p75^{NTR}) belonging to the tumor necrosis factor receptor family (Section 6.3.1). Like other members of this family, p75^{NTR} may induce apoptosis under distinct conditions; for instance, when the NGF-Trk pathway is inhibited or down-regulated.

To evoke pain sensations NGF has to bind to type A Trk receptors that are expressed in sensoric neurons. As a result, TRPV1 ion channels become sensitized. As in the case of bradykinin and other pain mediators, this may occur through an activation of phospholipase C (here of the γ-isoform; see Section 4.4.4). In addition, the expression of the TRPV1 channel and the acid-sensitive cation channel ASIC3 (Section 14.3) has been found to be induced by TrkA receptors along MAP kinase-dependent signaling cascades stimulating gene transcription.

Since in humans NGF is involved in several painful diseases, which sometimes are resistant to conventional pain relievers such as NSAIDs and opiates (including chronic headaches and arthritis as well as bladder and prostate

Figure 15.7 Cellular processing of temperature and pain signals
Shown are effects on heat and pain receptors (left) and on cold receptors (right). The inflammatory soup is a cocktail of endogenous pain-inducing mediators that become locally released upon wounding and inflammation. While anandamide directly interacts with TRPV (transient receptor potential, vanilloid type) channels, serotonin and ATP activate other cation channels, thus facilitating depolarization by TRPV. Substance P, prostaglandin E, and bradykinin bind to G$_{q,11}$-protein-coupled receptors (GPCR), probably activating TRPV channels via phospholipase C (PLC) signaling. The same holds true for NGF (nerve growth factor) that interacts with the receptor tyrosine kinase TrkA, thus activating PLCγ. TRPM, transient receptor potential, melastatin-related.

inflammation), the development of drugs that block NGF signaling has great therapeutic potential.

Summary

Pain receptors respond to high and low temperature, acid, corrosive and irritating chemicals, mechanical forces, and endogenous factors released in the course of inflammatory processes. The receptors are depolarizing cation channels that either interact directly with the stimulants or become gated along intracellular signaling pathways originating from G-protein- or tyrosine kinase-coupled receptors of pro-inflammatory mediators. Major players in this game are the vanilloid receptors belonging to the TRPV cation channel family. These are temperature sensors that are sensitized by hot- and cold-tasting food ingredients.

15.4 Smell

Smell is processed by the most versatile chemical sense. Vertebrates have several million olfactory cells that form a neuroepithelium in the nasal mucosa. These sensory neurons are bipolar, with strongly branched dendrites packed with receptors and extending to the surface of the mucosa, and an axon leading to the brain at the opposite pole (Figure 15.8).

Due to the pioneering work of Linda B. Buck and Richard Axel (who were awarded the 2004 Nobel Prize in Physiology or Medicine), we know that smell or odorant receptors constitute by far the largest subfamily of G-protein-coupled receptors and are encoded by one of the largest gene families (constituting up to 4% of a mammalian genome). The nematode *C. elegans* has around 500 genes for putative odorant receptors; these are not related to the vertebrate receptors, though they exhibit the same heptahelical architecture characteristic for G-protein-coupled transmembrane proteins.

For mammals around 1000 odorant receptor genes have been found, 600 of which have regressed to inactive pseudogenes in humans. This number of receptors is sufficient to discriminate between almost countless smell variants, because the nose works similarly to the eye, which distinguishes any number of color varieties. Although each sensory neuron seems to express only one receptor type, an odorant may react with several receptor types or a single receptor may respond to several scents. Therefore, a smell sensation is assumed to result from a combined excitation pattern involving many different receptors and neurons.

As G-protein-coupled receptors, odorant receptors have seven transmembrane domains. The individual receptor types differ from each other primarily in the composition of the transmembrane domains 3, 4, and 5 forming the binding pocket for the ligand (see Section 5.2). The associated olfactory G-protein G_{olf}

Figure 15.8 Schematic diagram of the olfactory epithelium The diagram shows a section of the olfactory epithelium consisting of sensory neurons and supporting epithelial cells. The neurons have direct contact with the brain. The cilia of the neurons directed into the nasal cavity and bathed in nasal mucus contain the olfactory receptors. The stem cells serve the regeneration of sensory neurons, which have a limited lifespan of a few weeks. (From B. Alberts et al., Molecular Biology of the Cell, New York: Garland Science, 1983.)

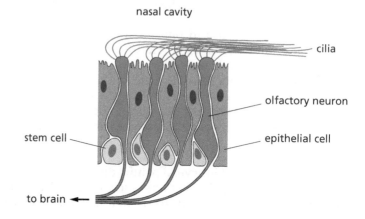

nasal cavity

cilia

olfactory neuron

stem cell

epithelial cell

to brain ←

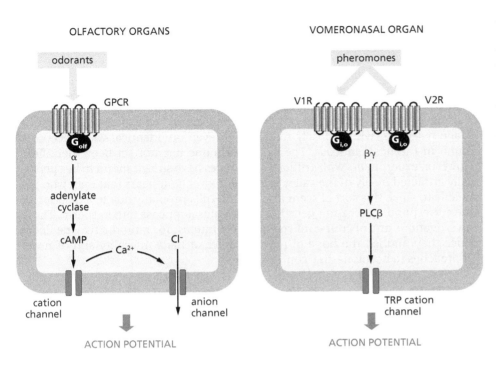

OLFACTORY ORGANS

odorants

GPCR

G_olf
α

adenylate
cyclase

cAMP   Ca²⁺   Cl⁻

cation
channel

anion
channel

ACTION POTENTIAL

VOMERONASAL ORGAN

pheromones

V1R   V2R

G_i,o   βγ   G_i,o

PLCβ

TRP cation
channel

ACTION POTENTIAL

Figure 15.9 Cellular processing of smell signals in the olfactory system and the vomeronasal organ For details, see text. GPCR, G-protein-coupled receptor; V1R and V2R, vomeronasal receptor isoforms 1 and 2; PLCβ, phospholipase Cβ; TRP, transient receptor potential.

belongs to the trimeric G_S proteins with its α-subunit activating adenylate cyclase type III (Figure 15.9).

The action potential required for signal transduction to the brain is generated by cAMP that binds to and opens nonspecific cation channels of the CNG channel family (Section 14.5.7). This triggering event is accompanied by a tremendous amplification of the signal: a single activated receptor stimulates at least 10 G-protein molecules, one after the other, thus releasing more than 10,000 cAMP molecules, three of which are sufficient to open an ion channel. In addition, olfactory neurons possess another obviously unique mechanism of amplification. In contrast to most other nerve cell types, they are able – by activating the corresponding ion pumps – to generate a steep charge gradient of chloride ions from the inside to the outside. When Ca^{2+} ions enter the cell through the cAMP-gated cation channels, they stimulate chloride channels, thus inducing an efflux of chloride ions and facilitating depolarization. On the other hand, calcium ions block the cation channels; that is, they exhibit an inhibitory feedback effect in addition to their amplifying effect. The inhibitory effect is important for a rapid adaptation of the sensory apparatus (Section 15.6).

Many vertebrates have a second olfactory system, the **vomeronasal organ** or Jacobson's organ in the paranasal sinus (Figure 15.9). Its sensory neurons respond specifically to pheromones, signal substances that are released by individual organisms and induce species-specific behavior. Best known are the sexual pheromones that attract sexual partners and stimulate the endocrine system in order to induce mating readiness. The cells of the vomeronasal organ express two families of G-protein-coupled receptors – V1 and V2 receptors, each consisting of approximately 150 receptor isoforms. V1 receptors are closely related to conventional odorant receptors, whereas V2 receptors have an extended extracellular domain, structurally resembling metabotropic glutamate receptors of the central nervous system (Section 16.4) and T1 receptors that sense the umami taste. These receptor types are coupled with $G_{i,0}$-proteins that activate phospholipase Cβ via their β,γ-subunits. Along this pathway, V1 and V2 receptors open the cation channel TRPC2, thus generating an action potential.

In primates, the vomeronasal organ regresses after birth. Moreover, with one exception, all vomeronasal receptor genes are inactive pseudogenes. Thus, humans obviously recognize sexual attractants through the normal olfactory system in addition to other signals indicating mating readiness.

Recently, this story has been enriched by another aspect. It is a general biological principle that, in addition to individuals, gametes also attract each other by means of pheromones. This holds true also for human **sperm**, which responds to a scent released from the egg cell. In fact, the sperm cell expresses transmembrane receptors, which closely resemble, or even are identical with, receptors found in the nasal mucosa. The egg pheromone has not yet been identified. However, experiments with artificial substances have shown sperm to be specifically attracted by lily of the valley scent. Moreover, it appears that men who are unable to smell this sort of scent also are infertile, probably due to a deletion of the corresponding receptor gene both in the nasal mucosa and in sperm. These investigations are, of course, of considerable interest for reproductive medicine since they nourish the hope of new therapies of male infertility and of novel approaches to hormone-free contraception.

Summary

Odorants are environmental chemicals that interact with odorant or smell receptors. These are G-protein-coupled heptahelical transmembrane proteins, constituting by far the largest subfamily of this receptor type. There are two groups of odorant receptors: those expressed in olfactory organs and those expressed in the vomeronasal organ. Closely related receptors have been found in sperm. Through receptor activation, signaling cascades are stimulated, leading to membrane depolarization due to a gating of cation channels and depolarizing anion channels.

15.5 Vision

Among the mechanisms of sensory signaling, visual signal transduction has been investigated most thoroughly. In the lenticular eye – common to all vertebrates – the light-sensitive layer of the retina is composed of two types of photoreceptor cells, rods and cones. The highly light-sensitive rods (several million in the human eye) are specialized for the reception of dim light, while the less sensitive cones (in humans around 6 million) deal with bright daylight and colors (according to the RGB model, colors are processed by three types of cones that are red-, green-, or blue-sensitive). The light-sensitive membrane surface of a photoreceptor cell is extremely enlarged by invaginations (in cone cells) or separation of up to 1000 stacked discs (in rod cells). These membrane structures contain the light-sensitive pigment rhodopsin (in rods) or the closely related color-specific iodopsins (in cones) forming a highly sensitive light conductor in the outer segment of the cell. The inner segment resembles the cell body containing the organelles and establishing synaptic contacts to the bipolar cells, another type of neurons in the retina (see below). Outer and inner segments are connected by a thin ciliar structure (Figure 15.10).

In the compound eyes of invertebrates, another type of photoreceptor cell is found. Its light-sensitive membrane is enlarged by a brushlike structure of tightly packed microvilli called a rhabdomere. Again, rhodopsin is the light-sensitive pigment.

The rhodopsin structure is an ancient "invention" of evolution, found even in archaea. However, although they resemble each other in their 3D structures, prokaryotic and eukaryotic rhodopsin probably developed independently. Rhodopsin is composed of the G-protein-coupled heptahelical receptor protein opsin and its ligand, vitamin A aldehyde or retinal. It is inserted as a homodimer into the disc membrane of rod cells (or the membrane invaginations of cone cells). The retinal binding site is localized in a pocket formed by the seven transmembrane domains on the inner side of the disc (that resembles the extracellu-

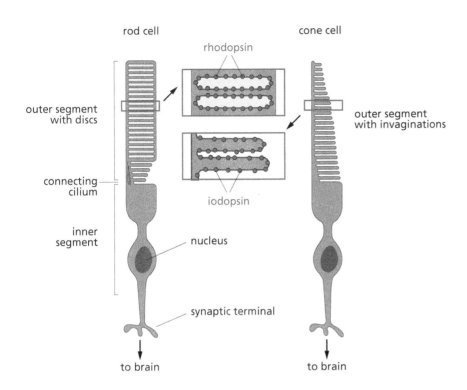

rod cell

cone cell

rhodopsin

outer segment
with discs

outer segment
with invaginations

connecting
cilium

iodopsin

inner
segment

nucleus

synaptic terminal

to brain

to brain

Figure 15.10 Schematic diagram of photoreceptor cells For details see text. (Adapted from J.G. Nicholls, A.R. Martin, and B.G. Wallace, From Neuron to Brain, Sunderland, MA: Sinauer Associates, 1992.)

lar space). In the absence of light, retinal exists as an 11-*cis* isomer that, similar to the 13-*cis*-retinal of bacterial rhodopsin (Section 3.3.5), is bound to a lysine residue (Lys269 in humans) as a protonated Schiff base. Upon illumination, retinal rearranges to the *all-trans* configuration. As a result, the covalent Schiff base bond is hydrolyzed and retinal finally dissociates from opsin, diffusing into the disc lumen where it becomes enzymatically re-isomerized.

Hydrolysis of the covalent bond between opsin and retinal causes an allosteric change of opsin conformation: the transmembrane helices are shifted in such a way that the active form (called metarhodopsin II) can interact with the trimeric G-protein transducin (G_t) on the cytoplasmic side of the disc membrane (Figure 15.11; see also Section 5.1). Transducin belongs to the $G_{i,0}$-protein family but differs from the other members of this family because it interacts with a unique effector protein, a cGMP-specific phosphodiesterase of the type PDE6 (see Section 4.4.2). By promoting the dissociation of an inhibitory subunit (the γ-subunit of PDE6), the α-subunit of transducin activates this enzyme.

PDE6 catalyzes the hydrolysis and inactivation of cyclic GMP, the key signal of light processing in photoreceptor cells. cGMP opens nonspecific cation channels in the cell's outer segment, thus inducing depolarization. Like the channels of odor-sensing cells, these cation channels are members of the CNG channel family (Section 14.5.7). Rod cells express the channel isoform CNG1, while cone cells express the channel isoform CNG6. In dark-adapted rods, the constitutively active guanylate cyclases E and F produce cGMP in a concentration sufficient for maximal gating of the CNG1 channels. Both cyclases are receptorlike transmembrane enzymes. Their ligands, if they exist at all, are unknown.

As a consequence of light-induced cGMP degradation, the CNG ion channels are closed and the membrane *hyperpolarizes*. This is because the membrane potential of a photoreceptor cell depends on an equilibrium between two opposing processes: the influx of cations through cGMP-gated ion channels and the removal of those ions by a constitutively active K^+–Na^+–Ca^{2+} exchanger in the membrane. In contrast, light may also result in a *depolarization* of the photoreceptor cell in nonvertebrates.

Figure 15.11 Processing of light signals in a rod cell of the vertebrate retina Light activates, via rhodopsin and the α-subunit of the G-protein transducin (G_t), phosphodiesterase 6 (PDE 6), which hydrolyzes cGMP. As a result, the cyclic nucleotide gated (CNG) cation channel becomes closed and the cell hyperpolarizes. More details are found in the text.

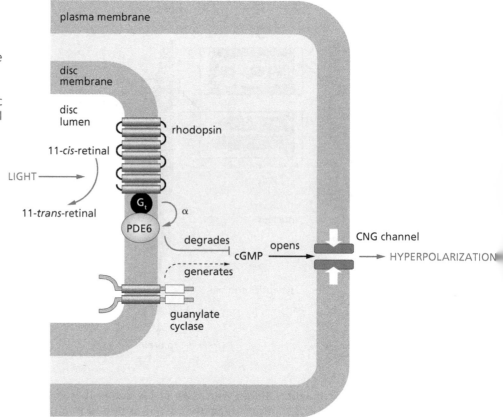

The system has an enormous amplification rate. At complete darkness, a single photon has been found to activate up to 500 molecules of transducin, resulting in the degradation of 250,000 molecules of cGMP. Under this condition half-maximal hyperpolarization is induced by only 30 photons. Without adaptation, such an extreme sensitivity is incompatible with efficiency, or the ability to respond over a wide range of signal intensity. At least as impressive, therefore, is the system's ability to adapt, enabling the cell to process between one and 1 million photons per second. The mechanisms of adaptation will be explained in Section 15.6.

Upon illumination, a photoreceptor cell transmits an action potential to the brain. How is this possible, although light induces *hyperpolarization*? The answer is found in the principle of double negation ("minus times minus is equal to plus") and a corresponding neuronal wiring in the retina with several cell types communicating with each other (Figure 15.12). In the dark, the photoreceptor cells are depolarized and release the neurotransmitter glutamate at their synaptic contacts with the bipolar cells. These cells exist in two different types. One type expresses *inhibitory* glutamate receptors, rendering the cell inactive upon interaction with glutamate released from the photoreceptor cells. When, upon illumination, the photoreceptor cells are hyperpolarized and the glutamate release ceases, these bipolar cells are relieved from inhibition; they depolarize and release their own neurotransmitter, which again is glutamate. As a result, an action potential is evoked in the ganglion cell connected downstream and equipped with excitatory glutamate receptors. Since these ganglion cells respond to a switching on of light, they are called ON cells. The other type of bipolar cells expresses *excitatory* glutamate receptors. These cells are, therefore, stimulated by the glutamate released from photoreceptor cells and are depolarized in the dark. They induce an action potential in connected ganglion cells, which is switched off upon illumination (OFF cells; see Figure 15.13).

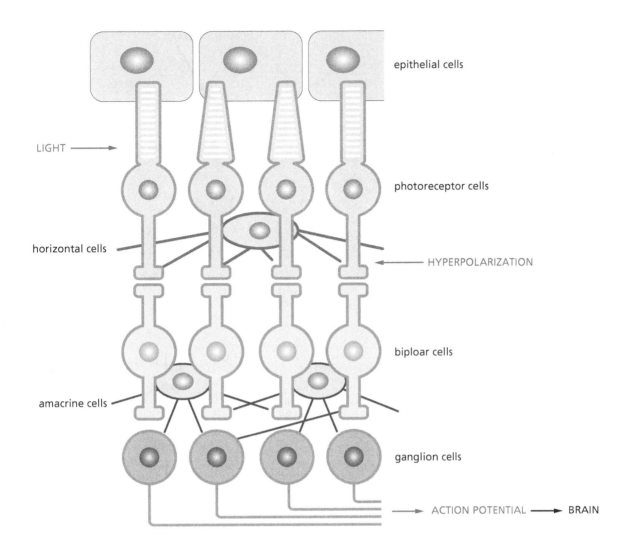

epithelial cells

LIGHT →

photoreceptor cells

horizontal cells

← HYPERPOLARIZATION

biploar cells

amacrine cells

ganglion cells

→ ACTION POTENTIAL → BRAIN

The photoreceptor cells of the compound eyes of invertebrates process light signals along another pathway (Figure 15.14). In the *Drosophila* eye, retinal is replaced by 3-hydroxyretinal, which also undergoes a *cis–trans* rearrangement upon illumination. However, in contrast to vertebrates, the *all-trans* form remains bound to opsin and becomes re-isomerized to the 11-*cis* form by longer-wavelength (red) light alone, without an enzyme. This allows an unparalleled temporal discrimination of light signals, up to 10 times faster than in the vertebrate eye and corresponding with the high-speed lifestyle of insects.

In invertebrate eyes, light always induces a *depolarization* of the photoreceptor cell. In *Drosophila* this occurs via gating of a nonspecific TRP-like cation channel (TRPL). Simultaneously a Ca^{2+}-selective TRP channel is opened that is essential for adaptation (see below). Both these channels are unrelated to the CNG channels of the vertebrate eye; they belong to the TRP ion channel family, to which they have given the name (Section 14.5.6). Like most other TRP channels they are regulated along the phospholipase C signaling pathway. To this end, light-activated metarhodopsin stimulates a trimeric $G_{q,11}$-protein that activates phospholipase Cβ type 4 (also called NorpA for *Drosophila*).

Summary

Light receptors are heptahelical proteins of the rhodopsin family incorporated in the disc membrane of retina cells and coupled to transducin. Their ligand is retinal; upon illumination, it undergoes a *cis–trans* rearrangement resulting (in vertebrates) in activation of a phosphodiesterase that degrades the second

Figure 15.12 Wiring of the vertebrate retina The photoreceptor cells transmit their signals via intermediate bipolar cells to the ganglion cells of the visual nerve. Horizontal and amacrine cells make horizontal connections, serving to fine-tune the response.

Figure 15.13 Light processing in the vertebrate retina In the dark, the photoreceptor cell is depolarized, releasing the neurotransmitter glutamate (Glu). Glutamate inhibits a bipolar cell type expressing an inhibitory Glu receptor and activates another bipolar cell type expressing an excitatory Glu receptor. As a consequence, an OFF ganglion cell is activated (black arrows). Upon illumination, the situation becomes reversed: the glutamate release of the hyperpolarized photoreceptor is diminished with opposite effects on the two types of bipolar cells and the activation of an ON ganglion cell (red arrows).

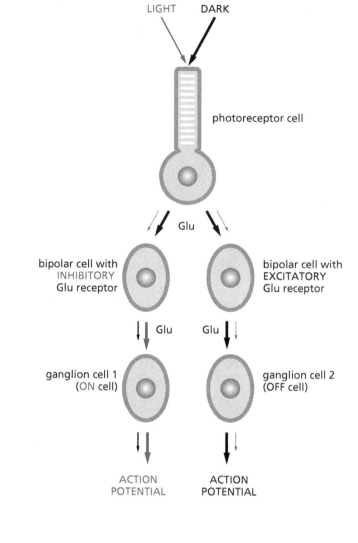

Figure 15.14 Cellular processing of light signals in vertebrate and insect eyes In the vertebrate eye, the light-induced degradation of cGMP along the retinal–transducin (G_t)–cGMP phosphodiesterase pathway leads to a closing of CNG cation channels and hyperpolarization. In the insect eye light opens TRP cation channels along the 3-OH-retinal–$G_{q,11}$–phospholipase C pathway and induces depolarization. The mechanism of TRP channel gating is still not entirely clear (shown by the question mark). Red bars show inhibition of G-proteins by cGMP phosphodiesterase, by regulator of G-protein signaling type 9 (RGS-9), and by phospholipase Cβ4, acting as GTPase-activating proteins. More details are found in the text. CNG, cyclic nucleotide gated; TRP, transient receptor potential.

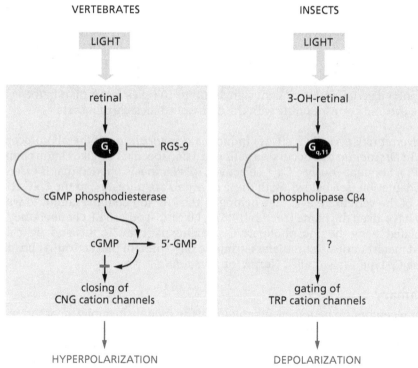

messenger cGMP. This leads to hyperpolarization due to closing of cGMP-gated cation channels. As a consequence, inhibitory neurons positioned between the sensory cell and the visual nerve are switched off, enabling the generation of an action potential in the visual nerve. In nonvertebrates (insects) a $G_{q,11}$-protein-controlled signaling cascade directly leads to gating of depolarizing cation channels in the sensory cell.

15.6 Sensory adaptation

The more sensitive a signal-processing device, the more efficient it is. That means that already low signal intensities evoke maximal activation. To respond with always-maximal sensitivity over a wide range of intensity the system must adapt. It must become insensitive for a signal just processed before in order to be able to respond maximally to a stronger signal (the apparatus is said to be "reset to zero"). Adaptation thus resembles a learning process that is indispensable for "intelligent" signal processing. We have already discussed this phenomenon for bacterial chemotaxis as an example (see Section 3.3.4).

A second condition for outstanding performance is rapid temporal discrimination or dissolution of signals. As far as visual signal processing is concerned, this is immediately conceivable: the transmission of a signal from rhodopsin to the visual nerve lasts only about 150 µs. Therefore, an image disappears as soon as we close our eyes and does not become blurred by eye movements, provided these are not too fast.

On the cellular level, adaptation and temporal discrimination are controlled by highly efficient mechanisms of negative feedback and signal extinction. These will be discussed in more detail for visual signal processing as an example.

The signaling apparatus of a photoreceptor cell adapts at four levels at least: receptor, G-protein, guanylate cyclase, and CNG ion channel. The key signal of sensory adaptation is Ca^{2+} (Figure 15.15). It enters the cell through the open

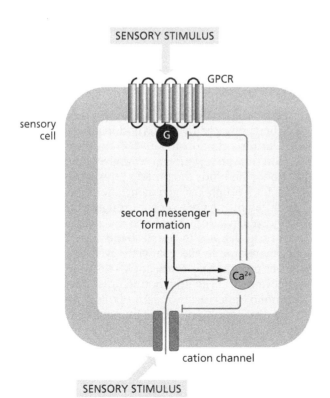

Figure 15.15 Key function of Ca^{2+} ions in sensory adaptation Ca^{2+} entering the cell through a cation channel that is gated by sensory stimuli, either directly or via a signaling cascade (involving second messengers) inhibits signal transduction at several sites. GPCR, G-protein-coupled receptor.

Figure 15.16 Adaptation of light-sensory signal processing in the vertebrate eye Adaptation is due to a modulation of guanylate cyclase and phosphodiesterase PDE6 activities that regulate the intracellular cGMP level. Guanylate cyclase is activated by two guanylate cyclase-activating proteins (GCAP). PDE6 is activated by rhodopsin via transducin (G_t) and becomes inactive when rhodopsin is inhibited by phosphorylation through rhodopsin kinase and when the interaction with transducin (G_t) is interrupted by arrestin (recruited by phosphorylated rhodopsin) and the GTPase-activating proteins PDE6 and RGS9 (as a complex with the G-protein $G\beta_5$). These reactions are under the control of Ca^{2+} ions entering the cell through CNG cation channels. In the dark, when the CNG channels are open, Ca^{2+} inactivates GCAP and activates PDE6 by inhibiting rhodopsin kinase through an activation of the kinase inhibitor recoverin. As a result rhodopsin activity is increased, the cGMP level drops, and more CNG channels are closed: the system becomes more light-sensitive (dark adaptation). Upon illumination the channels close, the Ca^{2+} influx is interrupted, and guanylate cyclase and rhodopsin kinase are activated. As a result the rhodopsin activity decreases, cGMP level increases, and more CNG channels are opened: the system becomes less light-sensitive (light adaptation). Ca^{2+} controls its influx by inhibiting the CNG channel. Red, inhibitory effects; black, stimulatory effects. CNG, cyclic nucleotide gated. For further details, see text.

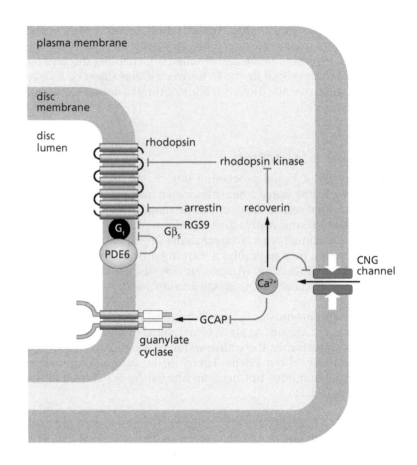

cation channels (in the dark), approaching a cytosolic concentration of 300–500 nM. Upon illumination, this value drops to 50–100 nM because the ion channels are closed whereas ion exchangers and Ca^{2+} pumps remain active. As a signal of dark adaptation, calcium sensitizes the system for light by activating rhodopsin and counteracting all events that lead to an increase of the cGMP concentration and to depolarization. The primary point of attack is the CNG channel itself becoming blocked by a Ca^{2+}–calmodulin complex. In parallel, Ca^{2+} turns down the activity of guanylate cyclase by binding to and inhibiting two guanylate cyclase-activating proteins.

Rhodopsin becomes insensitive upon illumination (Figure 15.16). The adaptive mechanism is closely related to the adaptation of β-adrenergic signal transduction (Section 5.7). When activated by light, rhodopsin, by means of the β,γ-subunit of transducin, recruits a rhodopsin kinase from the family of GPCR kinases (see Section 5.7). Upon phosphorylation at several Ser and Thr residues in the C-terminal domain, rhodopsin binds the protein arrestin that interrupts the interaction with transducin. At high Ca^{2+} concentrations (in the dark), rhodopsin kinase becomes inhibited by the Ca^{2+}-sensor protein recoverin and rhodopsin is reactivated by dephosphorylation (catalyzed by permanently active protein phosphatases). Recoverin belongs to the EF-hand superfamily of Ca^{2+}-binding proteins and is closely related to the guanylate cyclase-activating proteins (for more details about this protein type, see Section 14.6.1). The interaction of recoverin with Ca^{2+} uncovers a covalently bound myristoyl side chain normally hidden inside the protein molecule. By means of this fatty acid anchor, recoverin becomes fixed in the disc membrane in the immediate vicinity of its target, the rhodopsin/rhodopsin kinase complex (Ca^{2+}–myristoyl switch, see Section 14.6.1).

Transducin undergoes auto-inactivation due to its GTPase activity. However, for a rapid temporal discrimination of light signals this reaction, lasting 10–20 s, is by far too slow. Therefore, it is stimulated enormously by a GTPase-activating protein complex consisting of the inhibitory γ-subunit of the effector enzyme PDE6 and the protein RGS9-1. The latter, a member of the family of Regulators of G-protein Signaling (Section 4.3.2), stimulates the GTPase activity of transducin only as a complex with another protein called $G\beta_5$. It is an unusual G-protein subunit found only in nervous tissue and showing, in contrast to other Gβ-subunits, little affinity for Gγ-subunits.

In the cell membrane, the components of visual signal transduction and adaptation are assembled into multiprotein complexes that are, in addition, fixed by anchor proteins. Such an arrangement guarantees a high temporal discrimination of signals. In an especially impressive manner, this principle is realized in the ultrafast insect eye. Here a scaffolding protein, like a circuitboard, assembles the switching elements into a "transducisome" (or "signalplex") resembling a microprocessor. However, in contrast to conventional technical devices, the biological system is by no means firmly wired but rearranges its components depending on the actual requirements such as adaptation. For example, in strongly illuminated rods, transducin migrates from the outer into the inner cell segment within a few minutes, returning when the light becomes dimmed, while the inhibitor protein arrestin translocates in the opposite direction.

As far as adaptation is concerned, other sensory systems keep up with the visual complex. The sensitivity and adaptability of the sense of smell is legendary. The human nose, though by no means representing the most sensitive olfactory organ in the animal kingdom, is able to recognize only 4 mg of methyl sulfide in an air volume resembling 100 large sports halls but loses this ability quickly through adaptation (this also limits the effectiveness of perfumes). The adaptive mechanisms of olfactory neurons resemble those of photoreceptor cells: Ca^{2+}–calmodulin blocks the cAMP-controlled CNG channel, specific protein kinases phosphorylate and inactivate the odorant receptors, and a specific RGS protein (RGS2) stimulates the GTPase activity of the G-protein G_{olf}.

In the sensory cells of the inner ear, around 30 different proteins are involved in signal processing and adaptation. Again, Ca^{2+} is the primary adaptive signal that blocks the mechanosensitive TRP channels just as it inhibits the CNG channels in the eye and the nose. Moreover, for a slow adaptation to the sound volume, Ca^{2+} activates a special type of myosin, thus reducing the mechanical tension between the channel protein and anchoring structures.

Summary

Adaptation enables sensory cells to respond over a wide range of signal intensity. It is a complex process involving feedback loops at every level of signal transduction. A master regulator of adaptation is Ca^{2+}, which invades the cell through cation channels that have become gated in the course of sensory signal processing.

Further reading

Biggin PC & Sansom MSP (2003) Mechanosensitive channels: stress relief. *Curr. Biol.* 13, R183–R185.

Burns ME & Arshavsky VY (2005) Beyond counting photons: trials and trends in vertebrate visual transduction. *Neuron* 48, 387–401.

Calvert PD, Strissel KJ, Schiesser WE et al. (2006) Light-driven translocation of signaling proteins in vertebrate photoreceptors. *Trends Cell Biol.* 16, 560–568.

Chandrashekar J, Hoon MA, Ryba NJ et al. (2006) The receptors and cells for mammalian taste. *Nature* 444, 288–294.

Christensen AP & Corey DP (2007) TRP channels in mechanosensation: direct or indirect activation? *Nature Rev. Neurosci.* 8, 510–521.

Dhaka A, Visvanath V & Patapoutian A (2006) TRP ion channels and temperature sensation. *Annu. Rev. Neurosci.* 29, 135–161.

DiMarzo V, Blumberg PM & Szallasi A (2002) Endovanilloid signaling in pain. *Curr. Opin. Neurobiol.* 12, 372–379.

Dulak C & Grothe B (eds) (2004) Sensory systems. *Curr. Opin. Neurobiol.* 14, issue 4, 403–518.

Field GD & Chichilnisky EJ (2007) Information processing in the primate retina; circuitry and coding. *Annu. Rev. Neurosci.* 30, 1–30.

Filipek S, Teller DC, Palczewski K et al. (2003) The crystallographic model of rhodopsin and its use in studies of other G protein-coupled receptors. *Annu. Rev. Biophys. Biomol. Struct.* 32, 375–397.

Firestein S (2001) How the olfactory system makes sense of scents. *Nature* 413, 211–218.

Giamarchi A, Padilla F, Coste B et al. (2006) The versatile nature of the calcium-permeable cation channel TRPP. *EMBO Rep.* 7, 787–793.

Gillespie PG & Walker RG (2001) Molecular basis of mechanosensory transduction. *Nature* 413, 194–202.

Gunthorpe MJ, Benham CD, Randall A et al. (2002) The diversity in the vanilloid (TRPV) receptor family of ion channels. *Trends Pharmacol. Sci.* 23, 183–191.

Hardie RC & Raghu P (2001) Visual transduction in *Drosophila*. *Nature* 413, 186–193.

Hefti FF, Rosenthal A, Walicke PA et al. (2006) Novel class of pain drugs based on antagonism of NGF. *Trends Pharmacol. Sci.* 27, 85–91.

Julius D & Basbaum AI (2001) Molecular basis of nociception. *Nature* 413, 203–210.

Kaupp UB & Seifert R (2002) Cyclic nucleotide-gated ion channels. *Physiol. Rev.* 82, 769–824.

Lindemann B (2001) Receptors and transduction in taste. *Nature* 413, 219–225.

Lumpkin EA & Caterina MJ (2006) Mechanisms of sensory signal transduction in the skin. *Nature* 244, 858–865.

Ridge KD, Abdulaev NG, Sousa M et al. (2003) Phototransduction: crystal clear. *Trends Biochem. Sci.* 28, 479–487.

Spehr M, Schwane K, Riffell JA et al. (2006) Odorant receptors and olfactory-like signaling mechanisms in mammalian sperm. *Mol. Cell. Endocrinol.* 250, 128–136.

Svensson CI & Yaksh TL (2002) The spinal phospholipase–cyclooxygenase–prostanoid cascade in nociceptive processing. *Annu. Rev. Pharmacol. Toxicol.* 42, 553–583.

Thorne N & Amrein H (2003) Vomeronasal organ: pheromone recognition with a twist. *Curr. Biol.* 13, R220–R222.

Vosshall LB (2004) Olfaction: attracting both sperm and the nose. *Curr. Biol.* 14, R918–R920.

Signaling at Synapses: Neurotransmitters and their Receptors

This chapter provides a brief overview of neurotransmission with emphasis laid on signaling mechanisms at synapses. In fact, neuronal signaling adds some new aspects to the principles discussed in previous chapters. This holds true, in particular, for ion channel-coupled neurotransmitter receptors, which are only distantly related to the ion channel families discussed in Chapter 14 but, like those, can be traced back to bacterial precursors. For complementary information regarding further questions of development, anatomy, (electro)physiology and pathology, the reader may consult textbooks of neuroscience.

The chemical vehicles of neurotransmission are the neurotransmitters. These are defined as signaling molecules that are released by neurons to interact at specific intercellular contact sites, the synapses, in either para- or autocrine mode with membrane receptors of the post- and presynaptic cell. The concept of chemical signal transmission at synapses was put forward by Otto Loewi in Graz, Austria, in 1921 after he had succeeded in slowing down the beat of a frog heart in tissue culture not only by an electrical stimulation of the parasympathetic nervus vagus but also by a *Vagusstoff* (vagus substance) found in the culture medium of a heart electrically stimulated before. Together with Sir Henry Hallett Dale, Otto Loewi was awarded the Nobel Prize in Physiology or Medicine for 1936. Later the *Vagusstoff* was identified as acetylcholine, which still is the prototype of a neurotransmitter.

The term neurotransmitter comprises simple molecules such as acetylcholine, lipid derivatives, amino acids, and biogenic amines derived from amino acids by decarboxylation (such as dopamine, histamine, serotonin, and noradrenaline) as well as peptides and proteins. There is no clear boundary between neurotransmitters and what are called neuromodulators, even though the latter are understood commonly as signaling factors that modulate the release and effects of neurotransmitters.

16.1 Neurons and synapses

Neurons, the prototypes of excitable cells, have developed to perfection the basic cellular property of data processing, which is to "calculate" an output signal from a large number of input signals and to store acquired information in a data-processing network. The communicative cellular structures are the synapses where neurons and their target cells come up to within a distance of 20–40 nm. The concept of synaptic contacts was introduced by the Spanish neurophysiologist Santiago Ramon y Cajal (Nobel Prize in Physiology or Medicine 1906) and his British colleague Sir Charles Scott Sherrington (Nobel Prize in Physiology or Medicine 1932), two founders of modern neuroscience. As a rule, a neuron has many more synapses for reception than for transmission of signals. The receptive or input synapses are concentrated on the neurites and the cell body, while the transmitter or output synapses are found at the axons, ending in so-called presynaptic boutons. With up to 10^4 contact sites per cell, the data-processing potency of a (brain) neuron reaches an astronomical value. The total number of synapses in the adult human brain has been estimated at 10^{14}, surpassing the number of stars in the galaxy for a factor of 1000.

16.1.1 Mode of action of synapses

The input signals received by a neuron are deciphered and bundled to generate an action potential running down the axonal membrane to the presynaptic bouton, where in most neurons it triggers the release of neurotransmitters (Figure 16.1). [Note that in some tissues, such as hypothalamus, pituitary gland, and adrenal medulla, neurons release neurotransmitters as hormone-like neurosecretions into the bloodstream; that is, they function as endocrine glands.] Neurotransmitters diffuse across the synaptic cleft and interact with receptors of the postsynaptic cell (for an illustration see Figure 10.17 in Section 10.4.1). When such receptors are coupled with ion channels, they evoke a change of the postsynaptic membrane potential; when they are coupled with signal-transducing enzymes such as G-proteins, they trigger the generation of second messengers.

The distinct signaling functions of the active pre- and postsynaptic zones are mirrored by characteristic morphological features. The presynaptic bouton, primarily acting as signaling device, is filled with exocytotic vesicles containing the neurotransmitter. Those vesicles that are primed for release assemble on a protein meshwork. This **presynaptic web** lines the inner side of the presynaptic membrane and contains the protein machinery of exocytosis. The postsynaptic membrane, being a zone of signal reception, contains clusters of receptors, ion channels, scaffold proteins, and other constituents of the signal-processing protein network. Under the electron microscope, this zone is visible as **postsynaptic density**.

The combination of electrical with chemical signaling ensures that in chemical synapses the signal is transmitted only in one direction (unidirectional signaling). Nevertheless, chemical synapses also process retrograde signals, in particular to control neurotransmitter release by negative feedback. For an example,

Figure 16.1 Morphological basis of neuronal signal transmission An idealized neuron is shown (left). Its dendrites are occupied by a large number of synapses receiving input signals, the processing of which results in an action potential. The latter is propagated along the axon to the presynaptic bouton of the axonal synapse where it triggers the release of neurotransmitter resembling the output signal. The basic architecture of the presynaptic terminal of a chemical synapse is shown in the inset to the right. Secretory vesicles filled with neurotransmitter (shown in color) have been transported along cytoskeletal fibers (neurofilaments and neurotubules) from the cell body to the presynaptic bouton where those primed for release become assembled on a meshwork of proteins (presynaptic web). Upon depolarization of the presynaptic membrane (arrival of an action potential), the vesicles are emptied into the synaptic cleft. The mechanism of this exocytotic process is explained in more detail in Section 10.4.2. (Adapted in part from R.V. Krstic, Die Gewebe des Menschen und der Säugetiere. Heidelberg: Springer-Verlag, 1976.)

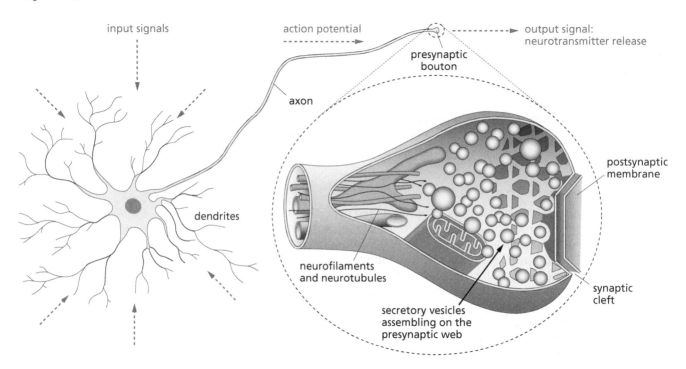

the reader may recall the autocrine self-regulation of α-adrenergic sympathetic signaling as explained in Section 5.3.3.

As with any exocytotic process, neurotransmitter release is strictly calcium-dependent. A key event of synaptic signal transmission is, therefore, the gating of voltage-dependent Ca^{2+} channels. This is brought about by depolarization, through the action potential running down the axonal membrane. In fact, in neurons the major role of action potentials is to trigger the influx of Ca^{2+} ions into the presynaptic terminals. Since Ca^{2+} influx and neurotransmitter release are stochastic rather than all-or-nothing events, synapses release small "packages" (resembling vesicles) of neurotransmitter even in the absence of an input signal, evoking postsynaptic miniature potentials that are either excitatory or inhibitory depending on the type of neurotransmitter and receptor. In other words, as long as a neuron is alive, its synapses never rest. Any excitatory or inhibitory effect of signaling is superimposed upon this apparently chaotic background activity.

For signal termination the neurotransmitter is removed rapidly from the synaptic cleft by presynaptic endocytosis, better known as recapture, or (rather an exception) by enzymatic degradation (see Sections 16.2 and 16.9).

Synapses that communicate via neurotransmitters are called chemical synapses. In vertebrate nervous systems they are the predominant type. Another type of structure is the **electrical synapse**. Here the gap between the cells is bridged by a particular type of membrane channels. Although these **gap junctions** allow ions and other low-molecular electrolytes to pass (see Sidebar 16.1), they are structurally unrelated to voltage- and receptor-dependent ion channels. In contrast to chemical synapses, electrical synapses work almost without delay, transmitting signals in *both* directions (bidirectional signaling). Their predominant function is to synchronize large collectives of cells and to trigger very rapid responses such as flight reflexes and the processing of sensory stimuli in the brain. However, as compared with chemical synapses, their capabilities to filter and to modulate signals are rather limited.

16.1.2 Axon growth: how signals regulate a developmental process

Tissue formation depends on directed growth and targeted migration of cells in combination with the induction of specific genetic programs of differentiation.

Sidebar 16.1 Intercellular communication by gap junctions Gap junctions are large membrane channels bridging the gap between adjacent cells. Each cell provides half of the channel called a **connexon**. A connexon consists of six protein subunits. These subunits, known as **connexins**, are built in the membrane by four transmembrane domains each. Human cells express about 20 connexin isoforms in a tissue-specific pattern. The central channel pore is 1.2 nm wide and penetrable for molecules up to 1000 Da, enabling an electrical and metabolic coupling of cells and an exchange of small hydrophilic signaling molecules such as cAMP, cGMP, Ca^{2+}, etc. The connexon pore opens or closes by an irislike shift of the subunits. This gating is signal-controlled and is regulated by the membrane potential and intracellular signaling reactions. Thus, protein phosphorylation may either open or close the channel depending on the protein kinase and the connexin and the intramolecular site to be phosphorylated. Signals that trigger an immediate closing are an acidification of the cytoplasm or a high Ca^{2+} concentration, as they both are typical of damaged cells. As a consequence, those cells become sealed. By means of gap junctions, tissue cells may become organized as a syncytium and respond synchronously. Well-known examples are the heart muscle, smooth muscles, and certain regions of the brain. In the course of evolution, gap junctions must have arisen independently at least twice, since the junctional proteins of vertebrates and invertebrates are entirely unrelated. Their pendants in plants are the **plasmadesmata** that connect the endoplasmic reticula of adjacent cells, penetrating the cell walls and transporting macromolecules such as proteins.

In these processes, each individual step is controlled by intercellular signals acting as morphogenetic factors. The formation of synaptic contacts provides a typical and particularly exciting example. Here a key event – the outgrowth of axons – is controlled and guided toward a target by a wide variety of factors produced by surrounding cells. We have already encountered some of these factors in previous chapters. Correct targets emit attractive signals, while wrong targets emit repulsive signals. The signal-processing apparatus of the growing neuron translates such signals primarily into cytoskeletal changes required for the protrusion or retraction of the nerve fiber. The corresponding receptors are concentrated at the tip of the developing axon, called the growth cone. This is a highly dynamic structure that continuously explores its environment in an extremely choosy fashion.

The factors dominating axon guidance are both diffusible and membrane-bound proteins exerting either paracrine long-range or juxtacrine short-range effects. Paracrine mediators include growth factors, in particular **neurotrophins** such as nerve growth factor (NGF) interacting with Trk receptor tyrosine kinases (Section 7.1.1), and glial cell-derived neurotrophic factors (Section 7.1.5) activating the Ret-type receptor Tyr kinases and interacting also with the neuronal cell adhesion molecule N-CAM. Neurotrophins are survival signals stimulating mitogenic and anti-apoptotic pathways such as the MAP kinase and P13K–PKB/Akt cascades (Section 4.6.6). Their major function is, therefore, to promote the proliferation of immature neurons and to protect growing neurons from cell death, which occurs automatically in the absence of survival factors. On the other hand, NGF-like factors also inhibit cell growth and cause cell death when they interact with the low-affinity NGF receptor p75 instead of with Trk. As explained in Section 13.2.3, p75 stimulates pro-apoptotic signaling. This situation reminds us of the principle that the meaning of a signal exclusively depends on the receiver and the context.

A family of diffusible proteins directly involved in axon guidance is the **netrins**, which (again depending on the context) exhibit either attractive or repulsive effects. Their axonal receptors are homodimeric single-pass transmembrane proteins of the immunoglobulin-like cell adhesion molecule (Ig-CAM) family (see Section 7.3). Since originally these receptors were discovered as tumor suppressor proteins deleted in large bowel cancer, they are called **DCCs** (Deleted in Colon Carcinoma). This indicates that they must have additional functions beside their role in axon guidance. Netrins interact with DCC homodimers. As heterodimers with the related Ig-CAM **Robo** (referring to the *Drosophila* mutant "roundabout"), DCCs prefer other ligands, the **slit proteins** (referring to the *Drosophila* mutant "slit"). These are large transmembrane glycoproteins exposed at the surface of axon-guiding cells. They primarily mediate repulsive juxtacrine signaling but also may be split off acting as diffusible paracrine signals. Slit proteins share this property with other axon-guiding proteins such as the **ephrins** (treated in detail in Section 7.1.4) and the semaphorins.

With more than 30 isoforms, the **semaphorins** are the largest family of axon-guiding proteins. Their major function is that of repellents that prevent axons from approaching a wrong target (because they provoke a collapse of the axonal growth cone, semaphorins are known also as collapsins). By the way, their morphogenetic effects are not restricted to the nervous system.

Like ephrins, semaphorins are expressed as either single-pass transmembrane proteins, proteins anchored in the membrane through a glycosylphosphatidyl-inositol residue, or in a diffusible form. On the tip of the growing axon are found two types of semaphorin receptors, the **neuropilins** and the **plexins**, which form homo- and heterooligomers. Both make up large families of single-pass transmembrane proteins.

A common feature of the transmembrane receptors of the axon-guiding factors is that they transduce the signal to the Rho–Rac–Cdc42 machinery as well as to the integrin complex, the two major control systems of the cytoskeleton. This occurs either through direct interactions or via cytoplasmic Tyr kinases that associate with the receptors or, as in the case of ephrin receptors, are integrated in the receptor molecule. An additional mechanism is used by plexins: their cytoplasmic region contains a GTPase-activating protein domain, which catalyzes the inactivation of Rho/Rac/Cdc42 G-proteins and thus induces a retraction of cytoskeletal filaments.

16.1.3 Birth and death of synapses

The goal of axon growth is the formation of synapses. Therefore, the signaling events involved in both axon guidance and synaptogenesis overlap to a large degree. At the cellular level the number of synapses is anything but a fixed quantity. In fact, synapses are generated and removed on demand. This birth and death of synapses starts during embryogenesis and is continued up to adulthood. For the development of the nervous system, the formation of synapses is as important as their elimination. During embryogenesis, many more neurons are established than are used later, and the same holds true for synapses. While the surplus cells die from apoptosis, unnecessary synapses are removed by ubiquitylation and proteasomal degradation of key proteins, followed by a retraction of the communicating nerve fibers. However, these developmental processes are not the end of the story, because in the adult brain synapses also come and go. Together with changes of synaptic conductivity (known as synaptic plasticity; see Section 16.4.1) and the birth and death of neurons, this dynamic process is considered to be the cellular basis of learning and memory.

Synapses develop at sites where a growing axon meets a correct target cell. Like axon guidance, this process is controlled by stimulatory and inhibitory signals that are released from the target tissue: peripheral organs for the vegetative nervous system, or other neurons for the central nervous system. Many of these factors are identical with those regulating axon growth. However, synapse formation is a matter of mutual giving and receiving because the axon is not only guided by factors of the target cell but also, vice versa, transmits signals that organize the functional architecture of the postsynaptic target membrane. This is best exemplified by the events occurring during the formation of a neuromuscular synapse.

16.1.4 Agrin and neuregulin: signals coordinating the development of neuromuscular synapses

Studies on neuromuscular synapses have deepened considerably our understanding of developmental processes in the nervous system. Here the key protein of the postsynaptic membrane is the nicotinic acetylcholine receptor. When axons of motor neurons approach muscle fibers, the receptors scattered across the cell surface assemble into large aggregates at the postsynaptic membrane zone in a process called synaptic targeting of receptors. Simultaneously, gene transcription is selectively turned on in those nuclei of the multinuclear muscle cell that are positioned in the immediate vicinity of the developing synapse. The corresponding genes encode proteins that reinforce the synaptic contact.

The major signal inducing neuromuscular synapse formation is transmitted by **agrin**. This heparan sulfate proteoglycan of about 2000 amino acids is expressed in a large number of splice variants, ensuring a high degree of specificity of neuromuscular contacts. It is released from the approaching axon and stored in the extracellular matrix filling the synaptic cleft (Figure 16.2). The agrin receptor of the muscle cell is the receptor Tyr kinase **MusK** (<u>Mu</u>scle-<u>s</u>pecific <u>K</u>inase), which is found only in skeletal muscle. Both MusK and agrin are essential proteins; mice with the corresponding gene knockouts die immediately after birth due to a failure of the respiratory system.

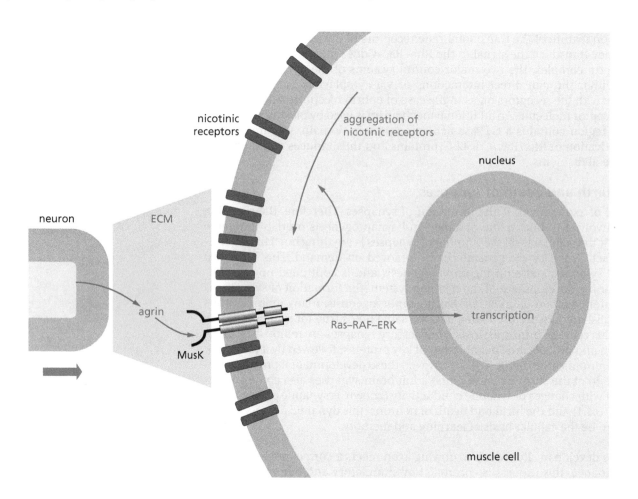

Figure 16.2 Formation of a neuromuscular synapse The tip of a nerve fiber (left) approaching the skeletal muscle cell (right) produces an extracellular matrix (ECM), which is impregnated with the protein agrin. By interacting with the receptor Tyr kinase MusK (muscle-specific kinase), agrin induces the aggregation of nicotinic receptors in the postsynaptic muscle cell membrane opposite to the nerve fiber, and via signaling cascades such as Ras–RAF–ERK, the transcription of various genes encoding synaptic proteins is induced. The red bars in the dimeric MusK protein symbolize the Tyr kinase domains.

Agrin activates MusK in the conventional way, by stimulating Tyr-directed trans-autophosphorylation within the receptor dimer (see Section 7.1). One major substrate of active MusK is the nicotinic receptor, the aggregation of which at the site of synapse formation depends on phosphorylation. This response probably requires the recruitment at the receptor of a large signal-transducing protein complex that is organized by the scaffold protein **rapsyn**. Gene knockout experiments have shown rapsyn to be as essential for life as agrin and MusK.

As is typical for a receptor, the Tyr kinase MusK activates, in addition, signaling cascades, which turn on the transcription of genes encoding synaptic proteins such as the nicotinic receptor, MusK, and acetylcholinesterase. Such genes are characterized by a synapse-specific response element called N-box that interacts with the transcription factor GABP (Guanosine- and Adenosine-Binding Protein). Being a member of the Ets family of transcription factors, GABP is a target of the Ras–RAF–ERK signaling cascade (Section 11.7.1). This cascade is activated by MusK.

Another neuronal signal cooperating with agrin is the epidermal growth factor (EGF)-like growth factor **neuregulin.** Its receptor is the receptor Tyr kinase ErbB, the EGF receptor proper, that like MusK evokes the whole program of downstream responses (including activation of the Ras–RAF–ERK cascade) that is, as described in Chapter 7, typical for this receptor family. There is every reason to assume that related events also control the formation of synapses between the vegetative nervous system and its target tissues.

16.1.5 Synaptogenesis in the central nervous system

The wiring of the human brain ranks among the most fantastic phenomena in nature: within a relatively short period, more than 1 million kilometers of axons and dendrites interconnecting 10^{11} neurons by 10^{14} synapses are established. As explained earlier, these growth processes driven by the dynamics of the actin cytoskeleton (Section 10.3) are guided by numerous diffusible and membrane-bound signaling molecules. In the central nervous system, such factors are derived from both neurons and the surrounding glial cells.

To generate an intraneuronal synapse, the dendrite of the target neuron stimulates the approaching axon to develop a presynaptic bouton equipped with the signaling machinery, while vice versa, the axon prompts the dendrite to protrude a dendritic spine, which is derived from a filopodium, and to develop postsynaptic density. This is a rapid event lasting less than an hour, whereas the maturation of the initial contact zone to a fully active synapse takes considerably more time.

For intraneuronal synapse formation, diffusible proteins seem to play a similar role as agrins and neuregulins do for neuromuscular synapses. Examples are provided by **ephrin B1**, probably acting in a proteolytically detached form (Section 7.1.4), and **neuronal activity-regulated pentraxin** (pentraxins are a family of secreted synaptic proteins). Both are required for the clustering of post-synaptic receptors such as the excitatory glutamate receptors, an effect that is important for both synapse formation and synaptic plasticity (see Section 16.4). Downstream signaling is mediated by PDZ domain interactions (see Section 7.1.4). [The acronym PDZ refers to three proteins – Postsynaptic density protein 95 and Discs large protein of *Drosophila* and Zona occludens protein of vertebrates – where these domains were found for the first time.]

Ephrin B1 also participates in the induction of presynaptic boutons, where its effect is assisted by the membrane proteins synaptic cell adhesion molecule (Syn-CAM) and neuroligin. These factors, which combine the properties of receptors with those of CAMs, transduce juxtacrine signals that play an essential role in the formation and maturation of synapses. **Syn-CAM** is one of more than 35 different N-CAMs, a neuronal subfamily of Ig-CAMs (Section 7.3). They all play an essential role in the development of the nervous system as well as in its remodeling in the course of learning and memory processes. Syn-CAMs interact homotypically. For other N-CAMs, heterophilic ligands such as proteoglycans and neurotrophic factors of glial cells exist, which in addition activate receptor Tyr kinases of the Ret type (Section 7.1.5).

Interaction of N-CAMs with each other or with their ligands has widespread consequences, reaching from direct effects on the cytoskeleton to gene activation mediated by MAP kinase cascades, NFκB modules, and nuclear receptors. For signal transduction, N-CAMs may associate, like ephrins, with proteins containing PDZ domains; like integrins, with Tyr kinases, particularly FAK and Fyn; or on the same membrane (in *cis*) with the fibroblast growth factor receptor and the receptor Tyr kinase Axl, which both belong to the immunoglobulin (Ig) superfamily. In fact, N-CAMs may be considered as Tyr kinase-associated receptors that assemble, like integrins and other receptors of this type, large signal-transducing protein complexes at their cytoplasmic domains. N-CAMs not only control the outgrowth of neurites but also play a key role in the targeted growth of axons, again cooperating with ephrins.

By alternative splicing of pre-mRNA (Section 8.9), neuronal adhesion receptors of the Ig family may be generated in an astronomically high number of variants, providing a kind of cell recognition code that ensures a highly selective interlinking of nerve cells. A striking example of biochemical combinatorics is provided by **DS-CAM**. The term stands for Down Syndrome Cell Adhesion Molecule, because the DS-CAM gene maps to a Down syndrome region of chromosome 21 (see Section 14.6.4). DS-CAM is a member of the Ig-CAM family. It comprises about 1500 amino acids and has one transmembrane domain and a large extracellular extension exposing a series of Ig- and fibronectin-like motifs. Alternative splicing of the pre-mRNA of the *Drosophila* homologue (Ds-cam) has been estimated to yield about 38,000 isoforms that differ from each other in their extracellular domains. The number of Ds-cam variants thus outdoes the total number of *Drosophila* genes by a factor of 2.5, and it is assumed that individual groups of neurons (or even single individual nerve cells) each express a unique pattern of Ds-cams. In fact, Ds-cam interactions have been found to be highly selective, allowing cells to communicate with an extreme degree of privacy. The situation is assumed to be quite similar for mammalian DS-CAMs.

Nerve fibers expressing identical DS-CAM patterns repel each other. Therefore, DS-CAMs are thought to prevent short cuts such as synaptic contacts between neurites and their cell of origin and to cause a repulsion of dendritic branches of the same neuron, thus allowing for correct outgrowth of neurite trees. To this end DS-CAMs interact with each other (homotypically). Via adaptor proteins (such as Nck in mammals and DOCK in *Drosophila*) they activate signaling cascades such as the PAK–Rac/Cdc42 sequence, which controls the architecture and function of the actin cytoskeleton (see Section 10.3). The existence of heterotypically interacting DS-CAM ligands has been postulated but not proven.

Neuronal adhesion proteins of the Ig family such as N-CAMs, DCCs, and DS-CAMs share common structural features that include a single transmembrane domain, clusters of Ig-like and fibronectin-like domains in the extracellular part, and cytoplasmic signaling domains. A prototypical structure is shown in Figure 16.3.

There are other factors involved in neuronal differentiation that do not belong to the Ig-CAM family. In addition to the ephrins, these are the **neuroligins**. In synaptogenesis they represent those dendritic signals that induce the formation of presynaptic boutons. Neuroligins are bound in the postsynaptic (dendritic) membrane, where they are fixed via PDZ domains to the scaffold protein PDZ95. Their receptors are presynaptic (axonal) transmembrane proteins, the **neurexins**, which via PDZ domains interact with proteins organizing the secretory machinery. Like DS-CAMs, neuroligins, and particularly neurexins, are expressed in a large number of splice variants. The variations concerning the extracellular domains of these isoforms represent a splice code that renders the interactions highly specific, determining, for instance, whether a synapse acquires a stimulatory or an inhibitory potential. Some neurexin variants also interact with diffusible peptide factors called neurexophilins. Neurexins are the major target of α-latrotoxin, a component of the poison of the black widow spider (*Latrodectus tredecimguttatus*).

Figure 16.3 Prototypical structure and mechanism of signaling of Ig-CAMs controlling the wiring of the nervous system The neuronal cell adhesion molecules of the immunoglobulin-like family (such as N-CAMs, DS-CAMs, Syn-CAMs, and DCCs) are characterized by series of immunoglobulin-like and fibronectin-like motifs in the extracellular domain and a single transmembrane domain. Upon *cis* and *trans* interactions with each other or other cell adhesion molecules and soluble or membrane-bound ligands, respectively, they trigger intracellular signaling cascades, probably by recruiting cytoplasmic protein kinases or by *cis*-interactions with receptor Tyr kinases.

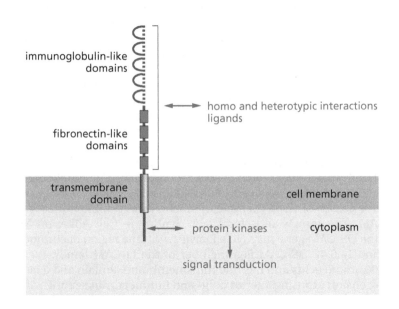

The mechanism of neuroligin–neurexin signaling closely resembles that of the ephrin system except that these proteins do not exhibit an intrinsic Tyr kinase activity. Instead, they are assumed to interact in both a direct and a retrograde fashion with cytoplasmic Ser/Thr and Tyr kinases (such as possibly Abl, see Section 7.1.1) and other signaling proteins. As in the case of ephrins, such interactions are due mainly to PDZ domains, leading to the formation of signaling multiprotein complexes.

Nerve cells express, in addition, a wide variety of other CAMs including cadherins, protocadherins, and integrins. They are important for stabilization of synaptic contacts rather than for synapse formation. In fact, CAMs that control synapse formation also balance the presynaptic with the postsynaptic functions in mature synapses and, therefore, play a fundamental role in synaptic plasticity. In other words, pre- and postsynaptic membranes are interconnected by interacting proteins, and the conventional idea of the synaptic cleft as an empty gap is not confirmed by the reality.

16.1.6 Ionotropic and metabotropic neurotransmitter receptors

The key proteins of signal processing in chemical synapses are the neurotransmitter receptors. Their functional organization is decisively determined by interacting CAMs. One distinguishes between ionotropic and metabotropic receptors. An ionotropic receptor is a part of an ion channel, whereas metabotropic receptors are coupled with trimeric G-proteins; they belong to the large family of heptahelical transmembrane proteins introduced in Chapter 4 (Figure 16.4).

Ionotropic receptors are found predominantly (but not exclusively) in the postsynaptic membrane and provide rapid paracrine signal transmission occurring within milliseconds or less. The considerably slower metabotropic receptors are expressed both pre- and postsynaptically and mediate auto- and paracrine neurotransmitter effects such as the autocrine regulation of neurotransmitter release (Figure 16.5). It should be noted that in interneuronal signaling the principle of double negation (stimulation by inhibition of an inhibitory effect) is particularly abundant. Thus, an excitatory neurotransmitter may have an inhibitory effect when (via a presynaptic receptor) it facilitates the release of an inhibitory transmitter, and an inhibitory transmitter would enhance neurotransmission when it acts on an inhibitory neuron.

As shown in Table 16.1, acetylcholine, serotonin, γ-aminobutyric acid (GABA), glutamate, ATP, and anandamide interact with both ionotropic and metabotropic receptors, whereas for the other neurotransmitters only metabotropic receptors are known, and for glycine only an ionotropic receptor has been found. The effect a neurotransmitter receptor has on postsynaptic signaling depends on the type of effector protein. Thus, ionotropic receptors

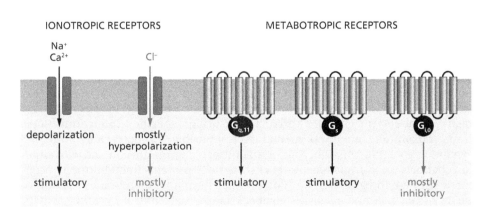

Figure 16.4 Neurotransmitter receptors: the basic types
Receptor ion channels (ionotropic receptors) and G-protein-coupled receptors (metabotropic receptors) are shown. Depending on the intracellular Cl^- concentration, an opening of chloride channels may either hyper- or depolarize the cell. $G_{i,o}$-protein-coupled receptors may exhibit either an inhibitory or (along the Gβγ–phospholipase-Cβ pathway) a stimulatory effect.

Table 16.1 Neurotransmitter receptors: an overview

| Transmitter | Ionotropic receptors | | Metabotropic receptors coupled with | | |
	Anion channels	Cation channels	G_s	$G_{q,11}$	$G_{i,0}$
acetylcholine		Nicotinic α_{1-10}		M_1, M_3, M_5	M_2, M_4
noradrenaline			β_{1-3}	$\alpha_{1A,B,D}$	$\alpha_{2A,B,C}$
dopamine			D_1, D_5		D_2, D_3, D_4
histamine			H_2	H_1	H_3, H_4
serotonin (5-OH-tryptamine)		$5\text{-}HT_3$	$5\text{-}HT_{4-7}$	$5\text{-}HT_2$	$5\text{-}HT_1$
ATP, etc.		P2X		$P2Y_{1,2,4,6,11}$	$P2Y_{12}$
opioid					MOP, DOP, KOP, NOP
anandamide (cannabinoid)		TRPV			CB_1, CB_2
glutamate		NMDA, AMPA, kainate		$mGlu_{1,5}$	$mGlu_{2-4,6-8}$
GABA	$GABA_A$				$GABA_B$
glycine	GR				
peptide transmitters			metabotropic receptors only		

coupled with cation channels and metabotropic receptors coupled with G_s- or $G_{q,11}$-proteins are stimulatory, whereas a coupling with anion channels or $G_{i,0}$-proteins mostly (but not always) has an inhibitory effect. In this chapter some emphasis is laid on ionotropic receptors and those metabotropic receptors that hitherto have not been mentioned. The function of other metabotropic receptors (for acetylcholine and noradrenaline) has been discussed in some detail in Chapter 5.

Fundamental insights into synaptic signal transduction are due to the work of Arvid Carlsson, Paul Greengard, and Eric R. Kandel, who were awarded the Nobel Prize in Physiology or Medicine for 2000.

Summary

Neurons are the prototypes of excitable cells, able to transmit signals by means of action potentials. The contact sites between neurons and their target cells are the synapses, which exist in two types. While electrical synapses are composed of transmembrane channels (gap junctions) enabling a direct exchange of molecules up to 1 kDa between adjacent cells, chemical synapses are bridged by neurotransmitters that are released from the presynaptic cell by depolarization-induced exocytosis and interact with receptors on the pre- or postsynaptic membrane. Development of neuromuscular synapses, including the expression of nicotinic acetylcholine receptors, is controlled by paracrine signaling between the approaching neuron and the muscle cell. Major signaling factors secreted by the neuron are agrin and neuregulin, which both stimulate muscle cell differentiation through receptor Tyr kinases. The development and function of intraneuronal synapses is regulated by paracrine and juxtacrine signals emitted by both pre- and postsynaptic cells. Juxtacrine signals are transmitted by cell membrane-bound proteins combining the properties of receptors with those of cell adhesion molecules (CAMs). Some of these proteins are expressed in astronomically high numbers of splice variants, ensuring specific interconnections. Neurotransmitters are local mediators responsible for para- and autocrine synaptic signal transmission. Neurotransmitter receptors are coupled with either ion channels (ionotropic receptors) or trimeric G-proteins (metabotropic receptors). They are integrated in both pre- and postsynaptic membranes as components of positive and negative feedback loops that control synaptic activity.

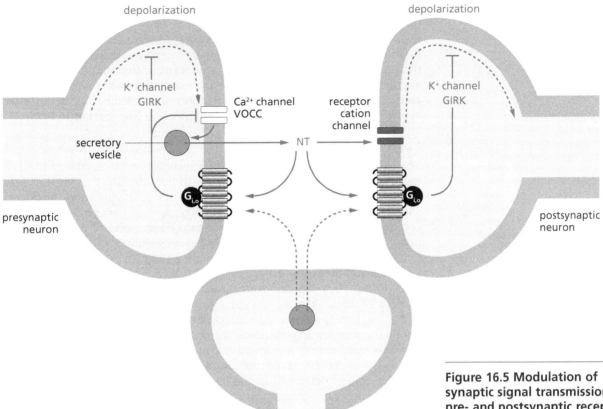

depolarization   depolarization

K⁺ channel GIRK   Ca²⁺ channel VOCC   receptor cation channel   K⁺ channel GIRK

secretory vesicle   NT

presynaptic neuron   Gᵢ,ₒ   Gᵢ,ₒ   postsynaptic neuron

associated cell

While cation channel- and G_s- or $G_{q,11}$-coupled receptors exert stimulatory effects, anion channel- and $G_{i,0}$-coupled receptors mostly transmit inhibitory signals.

16.2 Acetylcholine receptors

The classical example of a neurotransmitter interacting with both types of receptors is acetylcholine: its muscarinic receptors are of the metabotropic type, while its nicotinic receptors are of the ionotropic type. Since the muscarinic receptors have been discussed in Section 5.4, here we shall confine the discussion to the nicotinic receptors.

16.2.1 Nicotinic receptor and the family of cysteine-loop receptors: prototypes of ligand-gated ion channels

In 1906, the British physiologist J. N. Langley in Cambridge had found that nicotine caused skeletal muscle contraction and that this response could be prevented by curare, the arrow poison of native South Americans. This led Langley to postulate "receptive substances" binding pharmacologically active compounds in the corresponding target tissues. In an attempt to explain the specific effects of antigens, Paul Ehrlich in Frankfurt had put forward a similar concept of cellular receptors in his famous *Seitenketten-Theorie* some years before. Not until 20 years later was acetylcholine identified as the endogenous ligand of nicotine-binding sites at neuromuscular synapses. The "nicotinic" acetylcholine receptors, called nAchR, were found to be concentrated in the membrane of the postsynaptic muscular cell, where upon ligand binding they triggered an action potential, inducing muscle contraction. Today we know that the mammalian nicotinic receptor is expressed in more than 10 isoforms: according to their ligand-binding α-subunits (each receptor is composed of five variable subunits),

Figure 16.5 Modulation of synaptic signal transmission by pre- and postsynaptic receptors
As an example, an excitatory synapse is shown where the neurotransmitter (NT) released by the presynaptic neuron opens a postsynaptic receptor cation channel, thus depolarizing the postsynaptic cell. Simultaneously, the transmitter interacts with pre- and postsynaptic metabotropic receptors, which for the sake of clarity are assumed here to be $G_{i,0}$-coupled only. In reality, all kinds of G-protein couplings exist. Activation of these receptors leads to a gating of the hyperpolarizing G_i-protein-regulated K⁺ channel (GIRK) and to a blockade of presynaptic voltage-dependent Ca²⁺-channels (VOCC), thus impeding depolarization and neurotransmitter release by negative feedback (G_s- and $G_{q,11}$-coupled receptors may, vice versa, amplify synaptic signaling by positive feedback). Pre- and postsynaptic receptors may also interact with signal molecules released from associated cells. These cells may be neurons or glial cells, which become stimulated by the neurotransmitter or by other means. To modulate synaptic signaling, neurotransmitters and signals from associated cells may react also with pre- and postsynaptic receptor ion channels (ionotropic receptors; not shown).

these are called nAchα_1–α_{10}. Providing an impressive example of receptor multiplicity, they are tissue-specifically expressed. Only the isoform nAchRα_1 is found in skeletal muscle cells, whereas the other isoforms are restricted to individual neurons in the ganglions of the vegetative nervous system (Section 4.1) and in the central nervous system. In the brain a major function of acetylcholine is to augment neurotransmitter release by interacting with presynaptic nicotinic receptors, thus facilitating depolarization and calcium influx, an effect that explains some of the addictive effects of nicotine (see Section 16.9).

Isolation from electrical organs

The α_1-nicotinic receptor was the first receptor to be isolated and purified to homogeneity, mainly by J.-P. Changeux in the laboratory of Jacques Monod in Paris. This success was made possible by nature providing an extraordinarily rich source of the receptor in the electrical organs, which in certain fish species may come to 20% of body weight. These organs have evolved from skeletal muscles; however, in their cells, called electrocytes, nicotinic receptors serve the generation of electrical rather than of mechanical energy. To produce a high voltage, the synapses and the receptors are densely packed. So in the electrical organ of the torpedo ray (*Torpedo californica*), about 50% of the postsynaptic cell surface consists of nicotinic receptors corresponding to 15,000 receptor molecules/μm^2 and a total receptor-packed surface area of 0.07 m^2/animal (Figure 16.6). To obtain the receptor in pure form, only a 300-fold enrichment was necessary. This provides an extremely favourable starting situation as compared with the efforts necessary to isolate other receptors such as the uterine estrogen receptor (100,000-fold enrichment) or the insulin receptor of liver (500,000-fold enrichment). Another advantage consisted of the ability to label the receptor with α-**bungarotoxin**, enabling the investigator to follow up the purification procedure (for this purpose bungarotoxin had been labeled before with radioactive iodine). This peptide from the venom of the Taiwanese snake *Bungarus multicinctus* binds strongly and highly specifically to the receptor, causing a complete blockade of neuromuscular signaling. Related α-neurotoxins are also found in the venom of cobras, mambas, and coral and sea snakes (the acetylcholine receptors of these animals are insensitive to the toxins).

To isolate the nicotinic receptor, a novel approach was introduced that since then has become a standard method of protein purification. This approach is **affinity chromatography**, where a ligand (here curare) covalently bound to the chromatographic matrix selectively holds back the protein until it is eluted from the column by an excess of the same or another ligand. To prove the identity of the purified receptor, a biological assay was performed. Upon injection of the material, rabbits developed the symptoms of a severe muscular weakness resembling **myasthenia gravis**, an autoimmune disease of humans caused by an inhibitory antibody directed against muscular nicotinic receptors (it is instructive to compare this classical approach of receptor isolation with present methods as described in Section 15.1, Sidebar 15.1).

The ion channel function of the isolated receptor was demonstrated by incorporating the material into artificial lipid vesicles filled with a solution of radioactively labeled NaCl or KCl. Provided the preparation contained active receptors, acetylcholine could induce an ion flux across the vesicular membrane by gating these receptors.

Structure and properties

Nicotinic receptors are composed of five subunits, which are closely related isoforms of a basic polypeptide structure with four transmembrane helices. The subunits are arranged like a rosette and exhibit two acetylcholine-binding sites at the interface between the extracellular N-terminal extensions and the transmembrane helices of the α-subunits (Figure 16.6). At the cytoplasmic side, they are anchored at the cytoskeleton.

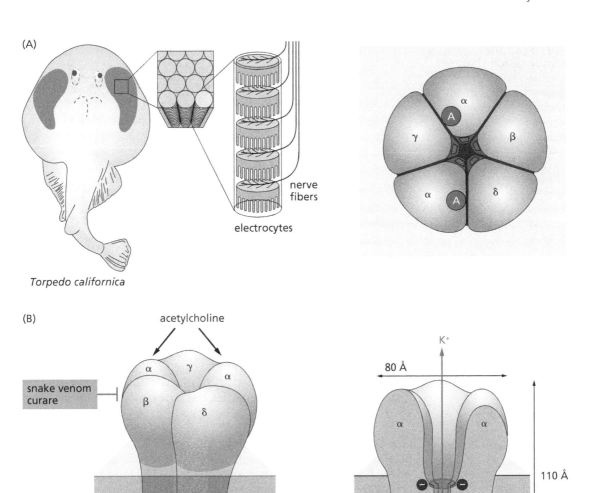

The subunits are tissue-specifically expressed in various isoforms. Depending on the isoform composition, one distinguishes between nicotinic receptors of the muscle, ganglion, and brain types. Each type comprises several receptor variants exhibiting different properties such as toxin sensitivity. α-Bungarotoxin inhibits only the muscle type, whereas the related κ-bungarotoxin inhibits only the brain type. The receptors of the electrical organ belong (as would be expected) to the muscle type, consisting of the subunits α_1, α_2, β, γ, and δ.

All nicotinic acetylcholine receptors have turned out to be nonspecific cation channels permeable for Na^+, K^+, and Ca^{2+}. Gating causes membrane depolarization and Ca^{2+} influx and facilitates the generation of action potentials. Both are essential for muscle contraction. The ion pore is formed by five helices (one per subunit) rather than by P-loops as in more specific cation channels (Section 14.1). Since P-loops control cation selectivity, their absence may explain the nonselectivity of the receptor ion channels. Novel approaches provide a clear picture of the receptor structure. A breakthrough was the X-ray crystallographic analysis of an acetylcholine-binding protein from the freshwater snail *Lymnaea stagnalis*. In its pentameric configuration this protein is very closely related to the nicotinic receptor and can be used as a template for nicotinic receptor modelling.

Figure 16.6 Nicotinic acetylcholine receptor cation channel of the skeletal muscle type (A) Left: Californian torpedo ray with the electrical organs composed of columns of electrocytes. Right: pentameric receptor with the acetylcholine-binding sites (A) seen from above. (B) Vertical view and section of the receptor with the negatively charged zones determining the cation selectivity of the central channel pore. According to the ion distribution, K^+ leaves and Na^+ and Ca^{2+} enter the cell through the channel gated by acetylcholine binding. (A, from Z.W. Hall, Molecular Neurobiology. Sinauer Publishing: Sunderland, MA, 1992.)

Figure 16.7 Gating of the nicotinic receptor cation channel For the sake of clarity, only two of the five receptor subunits are shown. In the closed conformation, hydrophobic interactions between the M2 helices hinder the ion flux. The binding of acetylcholine (ACh) induces a conformational change involving a minimal irislike twist of the subunits. As a result, the hydrophobic interactions in the pore become disrupted. (Modified from N. Unwin, *FEBS Lett.* 555:91–95, 2003.)

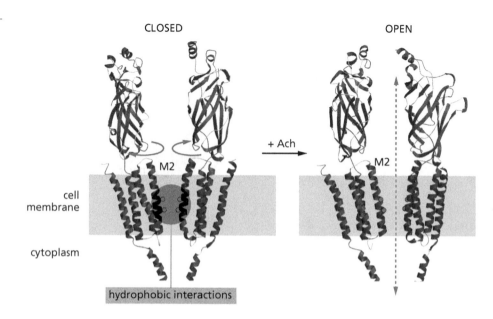

Such studies show that, in the absence of the ligand, the central ion pore of the channel is blocked by hydrophobic interactions. The binding of two acetylcholine molecules induces a minimal twist of the subunits, relieving the hydrophobic blockade and opening the ion channel like an iris aperture (Figure 16.7).

Like voltage-controlled Na^+ and Ca^{2+} channels (Sections 14.2 and 14.5.2), the nicotinic receptor cation channel exists in three states: closed, open, and desensitized (Figure 16.8). Desensitization is caused by long-lasting or high-frequency acetylcholine pulses. Hyperpolarization, Ca^{2+} ions, and phosphorylation of the cytoplasmic receptor domains by protein kinases A and C as well as by Tyr kinases of the Src family facilitate desensitization. The desensitizing protein kinases are activated by neuropeptides and the neuromodulator adenosine interacting with G_s-coupled receptors, as well as by $G_{q,11}$-coupled muscarinic acetylcholine receptors. These reactions, which provide an illustrative example of the various interactions (cross talk) between ionotropic and metabotropic

Figure 16.8 Functional states of the nicotinic receptor The receptor exists in the three states "closed", "open," and "desensitized," which are stabilized by the agents indicated.

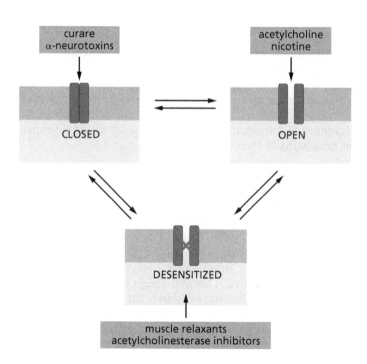

receptors, represent a kind of molecular short-term memory ensuring a continuous fine-tuning and adaptation of receptor function.

Metabolically stable synthetic derivatives of acetylcholine cause long-term desensitization of the nicotinic receptors. In the clinic, they are used to immobilize an area for surgery. Such drugs share this effect with inhibitors of acetylcholinesterase, the enzyme that degrades acetylcholine to choline and acetate in the synaptic cleft. Acetylcholinesterase inhibitors include highly toxic insecticides (like E605) and nerve gases for chemical warfare. In lower doses, such agents are also employed clinically for the treatment of myasthenia gravis. The inhibitory effect exerted by these compounds on the neuromuscular junction rivals that of curare and snake venoms.

Other receptors of the nicotinic receptor type

A structure resembling that of the nicotinic acetylcholine receptor is exhibited by the ionotropic receptors of the neurotransmitters serotonin (5-hydroxytryptamine), GABA, and glycine as well as by a receptor of zinc ions discovered recently (Figure 16.9). Together with the nicotinic receptors, these receptors constitute the family of **Cys-loop receptors** (named after a characteristic structural element; see Figure 16.9) with at least 40 members in humans. In contrast, the ionotropic glutamate and purine nucleotide receptors are members of other ion channel families. Genetic studies indicate that Cys-loop receptors join the large family of ion channels with prokaryotic ancestry.

Among the Cys-loop receptors, the **zinc receptor ZAC** (Zinc-Activated Channel) is a special case because its ligand is an ion rather than an organic molecule. Being the most abundant trace element next to iron, zinc is enriched in the ionic form (Zn^{2+}) in a particularly high concentration in certain brain areas where it is stored in secretory vesicles that become emptied upon depolarization. According to the motto "think with zinc," Zn^{2+} seems to be involved in synaptic signal transmission and other brain functions (see below). ZAC has been found as yet only in humans and dogs but not in rodents.

As shown in Figure 16.9, Cys-loop receptors transport either cations or anions. The ion selectivity is determined by rings of charged amino acid residues surrounding the pore: positive charge repels cations and attracts anions, while negative charge has the opposite effect. Cys-loop receptors are purely ligand-gated since they lack a structure acting as a voltage sensor. A multimeric structure with at least two ligand binding sites, as found for nicotinic receptors (and other Cys-loop channels), allows positive cooperativity of the subunits (Section 1.3). As repeatedly shown in previous chapters, such arrangement provides a powerful noise filter and enables rapid, switchlike channel gating.

Summary

Together with the closely related ionotropic receptors of serotonin, γ-aminobutyric acid (GABA), glycine, and zinc ions, the ionotropic (nicotinic) acetylcholine receptor constitutes the Cys-loop receptor family. These receptors consist of five subunits with four transmembrane domains each forming a ligand-operated nonspecific cation (acetylcholine, serotonin, zinc) or anion channel (GABA, glycine). Nicotinic receptors exist in various isoforms differing in subunit composition. They are particularly abundant in electrical organs, skeletal muscle, ganglions of the vegetative nervous system, and certain brain areas.

16.3 γ-Aminobutyric acid and glycine receptors: mediators of inhibitory neurotransmission

The ionotropic receptors of glycine and GABA belong to the same protein family as nicotinic receptors. In contrast to those, however, they are coupled with anion

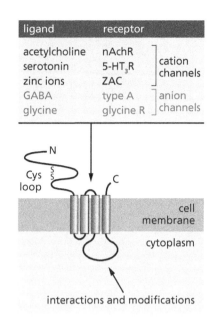

ligand	receptor	
acetylcholine	nAchR	cation channels
serotonin	5-HT$_3$R	
zinc ions	ZAC	
GABA	type A	anion channels
glycine	glycine R	

Figure 16.9 Basic structure and membrane topology of an ionotropic receptor subunit of the nicotinic receptor type A single receptor subunit is shown; each receptor is composed of five subunits. Depending on the ligand (inset), a receptor is coupled with either a cation or an anion channel. The cylinders symbolize the transmembrane domains, with the domain lining the ion pore shown in color. A characteristic structural feature of this receptor type is an extracellular N-terminal Cys loop formed by a disulfide bridge between two Cys residues. This loop stabilizes the conformation of the ligand-binding site. The large cytoplasmic loop between transmembrane helices 3 and 4 contains several Ser, Thr, and Tyr residues, the phosphorylation of which modulates the function of the ion channel. GABA, γ-amino butyric acid.

channels, mainly transporting Cl^- ions. As explained in Section 14.7, such channels may cause either hyperpolarization or depolarization, depending on the intracellular chloride concentration. In most neurons, this concentration is kept at a low level by exchanger proteins (such as, for instance, a K^+–Cl^- exchanger) in the cell membrane. This ensures a hyperpolarizing influx of chloride ions through glycine and GABA receptor channels. As a result, the generation of action potentials as well as neurotransmitter release at synapses is impeded. [In neonatal neurons, the chloride concentration may be so high that GABA acts as a depolarizing excitatory neurotransmitter. Even in the adult brain, the polarity of GABA receptors may be reversed by changing the intracellular chloride concentration and thus be adapted to the particular conditions.] In fact, GABA and glycine are the most abundant and putatively most important inhibitory neurotransmitters of the central nervous system. It has been estimated that up to 20% of all brain neurons are glycinergic or GABAergic.

16.3.1 Ionotropic GABA$_A$ receptors

One distinguishes between ionotropic GABA$_A$ and metabotropic GABA$_B$ receptors (Figure 16.10). As relatives of the nicotinic receptor, GABA$_A$ receptor anion channels differ entirely from other anion channels (Section 14.7), and within the Cys-loop family, they show the most complex architecture. Their pentameric structure arises from an assembly of different subunits expressed in seven isoforms. In the brain, dozens of individual combinations have been found that differ in their physiological and pharmacological properties. The major function of these and the glycine receptors is to dampen the excitability of neurons. Appropriately, GABA agonists mostly exhibit anticonvulsant, sedating, and narcotic effects, whereas antagonists cause hyperactivity, convulsions, and panic attacks. Both effects may be deadly.

Examples of GABA agonists are ethyl alcohol, the narcotic halothane (which also activates hyperpolarizing K_{2P} potassium channels; see Section 14.4.3), barbiturates known as sleep inducers, and benzodiazepines used as fear relievers, whereas avermectin (an insecticide) and picrotoxin (a plant poison used for fishing in southeast Asia) represent the analeptic (fear- and convulsion-inducing) GABA antagonists. In the clinic, analeptics are used against alcohol and barbiturate intoxications as well as for the treatment of psychotic diseases.

Many of the agonists and antagonists interact with specific domains of the receptor that are not identical with the GABA-binding site. Therefore, they func-

Figure 16.10 Ionotropic and metabotropic GABA receptors
(Left) The chloride channel of the ionotropic A-type receptor is gated by γ-amino butyric acid (GABA), ethyl alcohol, barbiturates, and benzodiazepines and blocked by analeptics (fear- and convulsion-inducing drugs). (Right) Metabotropic B-type receptors are exclusively coupled with $G_{i,o}$-proteins.

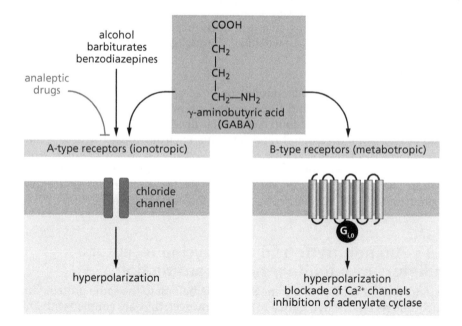

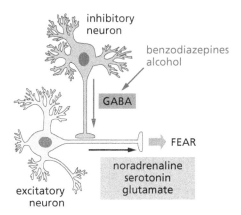

inhibitory
neuron

benzodiazepines
alcohol

GABA

excitatory
neuron

FEAR

noradrenaline
serotonin
glutamate

Figure 16.11 Fear-relieving effect of GABA neurons: an example of presynaptic modulation Shown is an excitatory (fear-inducing) brain neuron releasing, for instance, the neurotransmitters noradrenaline, serotonin, or glutamate. A presynaptically associated GABA neuron (shown in color) inhibits transmitter release by inducing hyperpolarization. This effect is augmented by alcohol or benzodiazepines.

tion as allosteric modulators of GABA binding. The most prominent example is provided by the **benzodiazepines** that bind to the α_2-subunit and enhance the interaction of the receptor with GABA. It has been speculated that these drugs mimic the effect of an endogenous ligand that remains elusive, however (the pacifying effect of mother's milk, for instance, has been suggested to be due to such a factor). Diazepam and valium, the prototypes of fear-relieving benzodiazepines or tranquilizers, rank among the most successful medical drugs. However, the hope that they are devoid of addictive effects has not been satisfied. In addition, in several cases synergistic effects with alcohol and barbiturates resulted in fatal consequences.

GABA neurons do not only inhibit the transmission of fear signals (Figure 16.11) but also dampen the excitability of neurons that are engaged in **memory fixation**. In fact, the forgetting of unnecessary information, being one of the essential conditions of successful learning, seems to be primarily under the control of GABA. As everybody knows, memory and learning capability are massively impaired in states of fear and other types of stress. Probably this is due to an activation of GABA neurons (perhaps resulting from a release of the elusive endogenous benzodiazepines?) aiming at a compensatory control of such situations. Vice versa, GABA antagonists have been found to increase learning capabilities.

In the course of learning and memory fixation, the conductivity of the synapses involved is changed. This phenomenon, called synaptic plasticity, is observed in both excitatory and inhibitory neurons and will be discussed below in more detail.

16.3.2 Metabotropic GABA_B receptors

These receptors are separated into two subfamilies B1 and B2, which both are expressed in numerous splice variants and exposed in the presynaptic as well as in the postsynaptic membrane. A peculiarity of the GABA_B receptors is that, due to special structural parameters, they are active only as B1/B2 heterodimers (see also Section 5.2).

Although characterized by seven transmembrane domains, the GABA_B receptors differ from the other G-protein-coupled receptors (GPCRs) to such an extent that they were classified as a separate subfamily that comprises, in addition, the metabotropic glutamate receptors, the T1R taste receptors, and the Ca^{2+} receptors. A characteristic structural feature of this receptor type is an extended extracellular domain that binds the ligand between two jaws like a Venus flytrap (Figure 16.12). The binding proteins of amino acids in the periplasm of Gram-negative bacteria have a similar shape and it is assumed that the extracellular domains of the GABA_B receptors and their relatives evolved by a fusion of these prokaryotic precursor proteins with the heptahelical structure of GPCRs (see also Section 3.5.7).

Figure 16.12 Metabotropic GABA$_B$ receptors Metabotropic GABA receptors heterodimerize by an interaction of the cytoplasmic C-terminal domains forming a "coiled coil" structure. The ligand is bound by extended extracellular structures reminiscent of a Venus flytrap. Quite a similar structure is exhibited by the metabotropic glutamate receptors (see Figure 16.13). For comparison, a conventional G-protein-coupled receptor (GPCR) of the rhodopsin family is shown. In contrast to the GABA$_B$ receptor, this receptor binds its ligand in the pocket formed by the seven transmembrane helices.

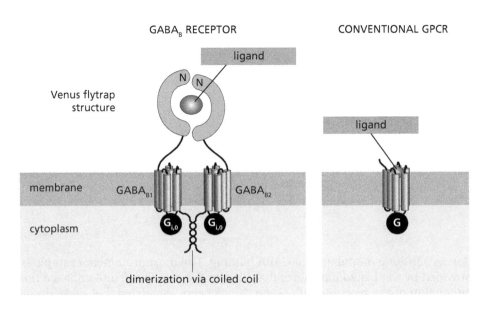

As discussed in Section 5.2, conventional GPCRs of the rhodopsin family bind their ligand in the central hollow of the rosette formed by the seven transmembrane domains. This results in deformation of the transmembrane structure leading to activation of the associated G-protein. For GABA$_B$ receptors and their relatives with the ligand-binding site positioned outside of the central hollow such a mechanism has to be ruled out, of course. Here the transmembrane signal is transduced by a mutual shift of the two monomers. Therefore, these receptors require dimerization.

Since GABA$_B$ receptors are coupled exclusively with G$_{i,0}$-proteins, they primarily inhibit synaptic signaling. Nevertheless, they may also stimulate nervous transmission if they are exposed in the presynaptic membrane of *inhibitory* neurons, thus dampening neurotransmitter release by negative feedback. The principle of such an inhibition of an inhibition resembling the arithmetical instruction "minus times minus is equal to plus," is depicted in Figure 16.5. Moreover, depending on the cell type and the receptor isoform, an activation of phospholipase Cβ signaling by the G$\beta\gamma$-subunit of the G$_{i,0}$-protein may lead to a release of diacylglycerol (DAG) and inositol 1,4,5-trisphosphate (InsP$_3$)/Ca^{2+}. This mostly stimulatory pathway plays a role in long-term potentiation of ionotropic GABA$_A$ synapses (see below).

16.3.3 Glycine receptors

For the neurotransmitter glycine, only ionotropic receptors have been found. Although their pentameric structure resembles that of GABA$_A$ receptors, it is composed of α- or α- and β-subunits only and contains at least three rather than two ligand-binding sites. These sites are blocked by strychnine, the alkaloid of the nux vomica plant (*Strychnos*) that is used as rat poison. Depending on the intracellular chloride concentration or the presynaptic receptor localization in inhibitory neurons, glycine, like GABA, may function as either an inhibitory or an excitatory transmitter. Moreover, glycine is a co-activator of excitatory ionotropic glutamate receptors of the NMDA type (see Section 16.4). At this opportunity, the reader may be reminded once more of the principle that signals do not have a meaning *per se* but trigger preformed responses of the receiver. This principle is illustrated in a particularly clear manner by glycine.

Summary

The neurotransmitters γ-aminobutyric acid (GABA) and glycine interact with anion channel-coupled ionotropic receptors. Their major effect consists of a dampening of synaptic signal transmission. They play an important role in learning processes and memory fixation. Ionotropic GABA$_A$ receptors are activated by various drugs including alcohol, barbiturates and benzodiazepines, while the glycine receptor is blocked by strychnine. For GABA, but not for glycine, metabotropic receptors coupling with G$_{i,0}$ proteins are also known. These GABA$_B$ receptors are heterodimers that differ from most other G-protein-coupled receptors by an extracellular "Venus flytrap" structure for ligand binding, which is assumed to be of prokaryotic origin.

16.4 Glutamate receptors: favorites of molecular brain research

In the brain, the excitatory neurotransmitters glutamate and aspartate are the major opponents of GABA and glycine. Like GABA, they interact with ionotropic as well as with metabotropic receptors (Figure 16.13). In mammals the ionotropic glutamate receptors are encoded by 18 genes giving rise to a large number of isoforms and splice variants. They are grouped in three families according to their exogenous ligands: **NMDA receptors** (named for the synthetic agonist N-Methyl-D-Aspartate), **AMPA receptors** (named for α-Amino-3-hydroxy-5-Methyl-4-isoxazole-Propionic Acid, another artificial agonist), and **kainate receptors** (kainic acid is a neurotoxin of a sea plant with a glutamate-like structure). NMDA receptors are expressed in seven isoforms; AMPA receptors, in four isoforms; and kainate receptors, in five isoforms (the residual two genes encode individual channel subunits).

The ionotropic glutamate receptors are a separate family that, being unrelated to other neurotransmitter receptors, is traced back to the bacterial glutamate receptor GluR0. As explained in Section 3.5.7, this receptor is thought to have evolved by a fusion of a periplasmic glutamate-binding protein with a prokaryotic K$^+$ channel of the type KcsA integrated upside-down into the membrane. This also is the basic structure of the mammalian glutamate receptor ion channels; however,

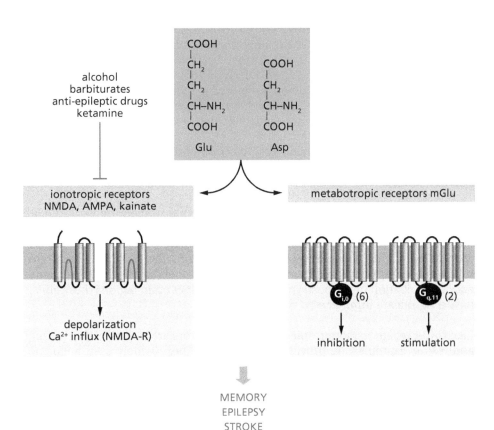

Figure 16.13 Ionotropic and metabotropic glutamate receptors For the metabotropic receptors, the number of isoforms is given in parentheses. Details are found in the text. NMDA, N-methyl-D-aspartate; AMPA, α-amino-3-hydroxy-5-methyl-4-isoxazolepropionic acid; R, receptor.

Figure 16.14 Glutamate receptors and their bacterial precursors
Shown in a schematic form are the subunits of the K$^+$ channel KcsA of *Streptomyces lividans*; the bacterial glutamate receptor GluR0 (a fusion product of a periplasmic glutamate-binding protein with a KcsA-like potassium channel integrated upside-down into the membrane); and the mammalian ionotropic glutamate receptor, exhibiting an additional transmembrane helix M4. Each ion channel consists of four subunits. In addition, the heptahelical structure of the dimeric metabotropic glutamate receptors is depicted. A characteristic feature that ionotropic glutamate receptors share with the metabotropic glutamate and GABA receptors is the Venus flytrap structure (red) derived from prokaryotic precursor proteins. The extended amino-terminal domain (ATD) of the ionotropic receptors is important for receptor oligomerization. The red balls represent the ligands. The P-loops lend some cation specifically to the channels (see text).

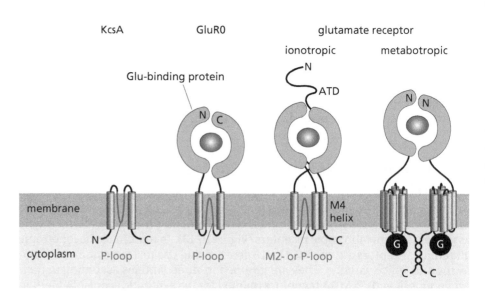

these subunits contain an additional C-terminal transmembrane helix (Figure 16.14). The ion channel is composed of four subunits, mostly different isoforms, with one ligand binding site each. Via their cytoplasmic domains, ionotropic glutamate receptors interact with a large number of intracellular proteins to form extended signal-transducing complexes. The functions of these accessory proteins are still far from being understood in all details.

The ionotropic glutamate receptors are exclusively coupled with depolarizing cation channels; AMPA and kainate receptors primarily transport Na$^+$ and K$^+$. Due to a distinct amino acid composition of the ion pore, AMPA receptors are impermeable for Ca^{2+}. In animal experiments, mutations causing Ca^{2+} permeability result in epilepsy and premature death. Analogous defects may be involved in amyotrophic lateral sclerosis, a progressive degeneration of motor neurons ending in fatal muscle weakness. On the contrary, NMDA receptors exhibit a strong preference for Ca^{2+} ions. An additional peculiarity of NMDA receptors is that, to become fully active, they require membrane depolarization and glycine or D-serine as a second ligand (D-serine is derived from enzymatic isomerization of L-serine). Membrane depolarization is required to remove Mg^{2+} ions that plug the NMDA receptor ion channel. This is achieved by the gating of AMPA receptors. Thus, for excitation of the postsynaptic cell, both receptor types share the job: AMPA receptors take care of depolarization and NMDA receptors are responsible for Ca^{2+} influx. We shall see below that, because of this teamwork, this receptor pair is especially suited to translate neuronal signals into long-lasting changes of synaptic conductivity experienced as short- and long-term memory.

Kainate receptors are found in both pre- and postsynaptic membranes. Although they share many properties with AMPA receptors, their contribution to synaptic firing is much smaller. They probably have modulatory functions: by causing partial long-lasting (also called tonic) depolarization, they facilitate neurotransmitter release by the presynaptic neuron and promote AMPA receptor-dependent depolarization in the postsynaptic cell. In distinct brain regions these effects have a significant influence on synaptic plasticity.

Like GABA receptors, ionotropic glutamate receptors are targets of many drugs and toxins. Examples are provided by phencyclidine, a widely used hallucinogenic drug ("angel dust"), and ketamine, an anesthetic in veterinary medicine, which both block NMDA receptors. The excitatory functions of glutamate receptors are also inhibited by ethyl alcohol and barbiturates. By this means, the stimulatory effects of the same drugs on opposing inhibitory GABA receptors are augmented.

In addition to the receptor ion channels, at least eight different **metabotropic glutamate receptors** (mGlu) have been identified, six of which couple with $G_{i,o}$-proteins and two with $G_{q,11}$-proteins. The $G_{q,11}$-coupled receptors are found primarily in postsynaptic membranes. By stimulating the release of DAG and $InsP_3$, they activate Ca^{2+}-dependent proteins such as PKC and calmodulin-activated kinases and, thus, complete and amplify the Ca^{2+}-mediated effects of NMDA receptor gating. In contrast, the $G_{i,o}$-coupled receptors are concentrated in presynaptic membranes and glial cells. They suppress cAMP formation, inhibit voltage-gated Ca^{2+} channels, and activate GIRK-type potassium channels and, thus control glutamatergic signal transduction by negative feedback. Structurally, the metabotropic glutamate receptors are closely related to the metabotropic GABA receptors, exhibiting a Venus flytrap structure for ligand binding. It appears as though, for the individual receptor types, this structure is derived from different prokaryotic precursor proteins.

16.4.1 Data storage by synaptic plasticity

Glutamate receptors are a preferred object of neurobiological research since they are involved directly in learning and memory fixation. According to prevailing theories, learning is associated with the generation of new and the reinforcement of existing synaptic contacts (signaling becomes strengthened in frequently used neurons) as well as with the production of new neurons. The same holds true for the reverse situation: synaptic contacts used less frequently are shut down or eliminated, in extreme cases through apoptosis of the postsynaptic cell. Both effects are realized as remembering and forgetting. In other words, like contacts between proteins, the interconnections between neurons are not irreversibly fixed but flexible and adaptable, changing in response to the actual conditions. Due to this plasticity, synapses are able to adapt to endogenous and exogenous stimuli by long-lasting alterations of conductivity. This holds true for both presynaptic functions (in particular transmitter release) and the signal-transducing network, enabling the postsynaptic cell to adjust the efficacy of its metabolic and genetic apparatus to the input signals.

Long-term potentiation and depression: a matter of protein synthesis

In brain preparations, synaptic plasticity can be measured as **long-term potentiation** and **long-term depression**. These are defined as an activity-dependent increase or decrease of synaptic conductivity that arises very rapidly and may last for many hours. In the beginning, these effects are associated with post-translational modifications of signal-processing proteins induced along signaling cascades originating from neurotransmitter receptors; they are an equivalent of short-term memory. The final establishment of long-term potentiation or depression, thought to resemble long-term memory fixation, occurs after a few hours and depends on *de novo* synthesis of synaptic proteins, requiring a modulation of both mRNA translation and gene expression. Finally, once fixed, long-term memory is independent of protein synthesis, indicating that the changes of synaptic conductivity, once established, remain stable.

Among the many transcription factors controlling genes involved in memory fixation, the cAMP response element-binding protein **CREB** plays an outstanding role. As explained in Section 8.6, its activity is controlled by various protein kinases and phosphatases acting downstream of neurotransmitter receptors. However, considering the logistic problem created by the large distances between the synapses and the nucleus, neurons cannot rely solely on transcription but must also use mRNA translation to control long-term potentiation or depression. To this end, synapses contain a stock of ribosomes and more than 100 species of stable mRNA enabling a local protein synthesis on demand, which has been shown to occur even in dendrites surgically dissected from the cell body. Such local protein synthesis allows for selective and specific modification of individual synapses for the purpose of rapid adaptation to incoming stimuli.

A master switch controlling the induction of synaptic protein synthesis is the translation factor **eIF2α**. As we have learned in Section 9.3, this G-protein is inactivated by protein phosphorylation at a strategic Ser residue catalyzed by several eIF2 kinases. As a result general mRNA translation is suppressed, whereas the synthesis of proteins involved, in particular, in stress responses is stimulated due to a special structure of the cognate mRNA species (Section 9.3.2). One of these proteins selectively produced upon eIF2 phosphorylation is the transcription factor **ATF4**. It is a master regulator of stress-sensitive genes, but at the same time it acts as an antagonist of CREB-dependent gene transcription.

In brain neurons the induction of long-term potentiation or depression correlates with a dephosphorylation of eIF2, resulting in an increase of overall translation accompanied by a suppression of ATF4 synthesis. Consequently, CREB is relieved from inhibition. Experiments with gene-manipulated animals and pharmacological inhibitors demonstrate a causal relationship between the eIF2–ATF4 complex and learning. Which of the eIF2 kinases (Section 9.3.1) and phosphatases controls this process and along which signaling pathways they are regulated is not known.

The proteins modified and produced in the course of memory fixation assemble in the **postsynaptic densities**, which (as explained in Section 16.1) are characteristic of permanently active synapses. These aggregates are visible under the electron microscope. They are composed of signal-transducing proteins, including neurotransmitter receptors and various protein kinases, and are organized and anchored to the cytoskeleton and the cell membrane by adaptor and scaffold proteins.

As an example of scaffold proteins, the **shank proteins** may be mentioned. Through their SH3 domains, ankyrin repeats, PDZ domains, and proline-rich motifs, they interact with a large number of other proteins. Moreover, by means of SAM domains (referring to a Sterile Alpha-helical Motif of yeast), they polymerize into helical fibers that interact with each other to form large sheets beneath the synaptic contact zone. These interactions are promoted by zinc ions binding to the SAM domains. Since in many neurons presynaptic vesicles have been found to contain both neurotransmitter (particularly glutamate) and Zn^{2+}, it has been speculated that Zn^{2+} may enter the postsynaptic cell through ionotropic neurotransmitter receptors and induce the organization of the shank lattice as the structural backbone of the postsynaptic density. Whether such an effect has anything to do with the above-mentioned ionotropic zinc receptors is not known.

Another synaptic scaffold protein worth mentioning is **PICK1** (Protein Interacting with C-Kinase 1). Via PDZ and other interaction domains, it organizes signal-transducing complexes that may contain glutamate, dopamine, acetylcholine, GABA and serotonin receptors, regulators of G-protein signaling that control the GTPase activity of G-proteins coupled with metabotropic neurotransmitter receptors, and the SNARE complex protein SNAP25, as well as an atypical PKC isoform. Presently, PICK1 is gaining much interest because genetic defects of this protein strongly correlate with an increased susceptibility for schizophrenia.

Role of glutamate receptors

As we have seen, learning and memory fixation correlate with specific changes of the postsynaptic proteome due to post-translational modifications and effects on mRNA translation and gene transcription. These lead to remodeling of synaptic contacts, measured as long-term potentiation and depression. The question arises how a simple electrochemical nerve impulse is interpreted by the intracellular network of signal processing in such a way that it can evoke such far-reaching changes.

Neuronal signals are transmitted by waves of depolarization and by neurotransmitters released in small packages or quanta. Such signals are frequency-modulated. Therefore, neurons must possess proteins able to read the information encoded in the frequency of the input signal and to transform this information into output signals that are perceivable for the data-processing protein network, thus changing synaptic plasticity. This condition is fulfilled perfectly by **NMDA receptors** because they are able to read frequency-modulated membrane potentials. To this end NMDA receptors have to cooperate with **AMPA receptors**, which transform the neurotransmitter signal received from the presynaptic neuron into postsynaptic voltage pulses that the NMDA receptors then translate into pulses of Ca^{2+} influx. These pulses are read by intracellular frequency detectors such as **calmodulin kinase II** (CaMKII), an outstanding monitor of oscillating calcium signals that controls the function of numerous synaptic proteins including the receptors themselves (Section 14.6.3).

How do NMDA receptors read the voltage signals emitted from AMPA receptors? The answer is found in the special type of voltage sensor with which the receptor is equipped. This sensor is a channel-blocking Mg^{2+} ion rather than a special transmembrane domain as in other voltage-controlled ion channels. To enable ligand-induced channel gating, this Mg^{2+} ion has to be removed by depolarization. The higher the frequency of the voltage pulses emitted by the AMPA receptor, the more NMDA receptors are activated. Consequently, NMDA receptor activity critically depends on the frequency with which the synapse is used. However, as we have seen, voltage alone is not sufficient to gate NMDA receptors; it must cooperate with two ligands, glutamate and glycine or D-serine. In other words, NMDA receptors are equipped with three safety locks; they represent a special type of logical AND gates.

As already mentioned, NMDA receptors predominantly allow the passage of Ca^{2+} **ions.** These are the key regulators of synaptic plasticity. For long-term potentiation, they strengthen the synaptic contact by positive feedback and render it long-lasting by activating a series of Ca^{2+}-dependent signaling proteins such as the above-mentioned CaMKII and other Ca^{2+}-activated enzymes. One of these is the neuronal NO synthase nNOS. It catalyzes the formation of nitric oxide, which functions as a retrograde messenger, stimulating the presynaptic release of glutamate (Figure 16.15; for more details see Section 16.5.1). A major task of Ca^{2+}-dependent protein kinases such as CaMKII is to increase the activity and promote *de novo* synthesis and membrane localization of AMPA and NMDA receptors. While phosphorylation promotes receptor expression required for long-term potentiation, dephosphorylation has the opposite effect, causing long-term depression. In this system a major opponent of kinases is the protein phosphatase calcineurin. Since this enzyme is Ca^{2+}-activated, Ca^{2+} is a signal for both long-term potentiation and long-term depression. Apparently, it is a matter of NMDA receptor efficacy which of the two responses gets a chance: a highly active receptor evokes a strong Ca^{2+} influx that causes long-term potentiation via CaMKII, whereas the weak Ca^{2+} signal induced by a less active receptor activates only calcineurin, thus causing long-term depression. In fact, the number and subunit composition of NMDA and particularly AMPA receptors depends on the frequency with which the synapse is used and is determined by the steady state between up- and down-regulation of the receptors. This equilibrium is highly dynamic, ensuring a permanent turnover of ionotropic receptors in the neuronal cell membrane. So the cell is able to adjust its receptor capacity very rapidly to the particular circumstances. To avoid overshooting, several negative feedback loops are built into the system. They control the Ca^{2+} influx through NMDA receptors; this influx is suppressed by dephosphorylation and N-nitrosylation (by NO) of the receptor.

Depending on the conditions, **metabotropic glutamate receptors** (as well as other neurotransmitter receptors) augment or dampen the processes leading to

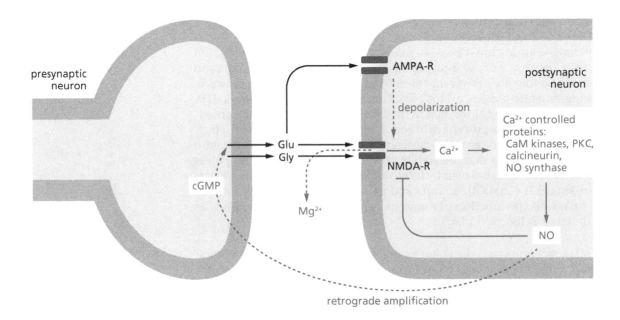

Figure 16.15 Mode of NMDA receptor activation Shown in schematic form is a central nervous glutamate-operated synapse. For gating of the postsynaptic NMDA receptor cation channel, three signals have to cooperate: these are transmitted by the two ligands glutamate (Glu) and glycine (Gly) and by membrane depolarization caused by glutamate-activated AMPA receptor ion channels. The depolarization removes Mg^{2+} that blocks the NMDA channel. The NMDA receptor preferentially transports Ca^{2+} ions, thus modulating the functions of postsynaptic Ca^{2+}-dependent proteins. One response is the production of nitrogen monoxide (NO), which exhibits a dual effect as an inhibitor of NMDA receptor activity and a retrograde amplifier of presynaptic transmitter release. The latter reaction is mediated by cyclic GMP (Section 4.6.2).

long-term potentiation or depression in that they activate G-protein-dependent signaling cascades. For instance, PKC, stimulated along $G_{q,11}$- and $G_{i,0}$-controlled pathways (Section 4.6.3), facilitates the incorporation of NMDA receptors from secretory vesicles into the plasma membrane but also promotes the formation of endocannabinoids that, as retrograde messengers, diffuse from the post- to the presynaptic cell and inhibit presynaptic glutamate release by activating $G_{i,0}$-coupled receptors (Section 16.7). Vice versa, NO synthase, CaMKs, and calcineurin are stimulated along the $G_{q,11}$–phospholipase C–InsP$_3$–Ca^{2+} pathway (Section 14.6.4).

The molecular mechanisms of long-term potentiation and depression are not restricted to glutamatergic neurons but hold true also for synapses that use other neurotransmitters. The important role played by GABAergic neurons in synaptic plasticity has been mentioned already. Studies that are more detailed have shown that here again the increase of the cytoplasmic Ca^{2+} level in the postsynaptic cell provides a key signal. This signal may be generated by GABA$_A$ receptors, provided these exhibit a depolarizing effect, thus facilitating the gating of voltage-dependent Ca^{2+} channels. However, even in hyperpolarizing GABA$_A$ neurons, a Ca^{2+} signal may arise, in this case through metabotropic receptors (of GABA or of other neurotransmitters) or NMDA receptors activated simultaneously. The downstream effects of Ca^{2+} resemble those discussed for glutamate receptors.

16.4.2 A decision between life and death of brain neurons

As we have seen, NMDA receptors are protected against accidental failure. This is for good reason, since hyperactivation would result in a flooding of neurons by Ca^{2+} ions, causing cell death. Indeed, such a fatal event is a characteristic feature of stroke, brain injury, and neurodegenerative disorders such as Alzheimer's and Huntington's disease. A permanent inactivation of glutamate receptors is, on the other hand, just as fatal because nonstimulated brain neurons also are doomed to die.

How do glutamate receptors ensure neuronal survival? Neurons belong to those cell types that automatically undergo apoptosis when survival factors such as neurotrophic growth factors are withheld; these are essential, in particular, during embryonic development, and neurotransmitters like glutamate are required throughout life. The apoptotic response is prevented by Ca^{2+} in lower concentrations entering the cytoplasm through NMDA receptors or other mem-

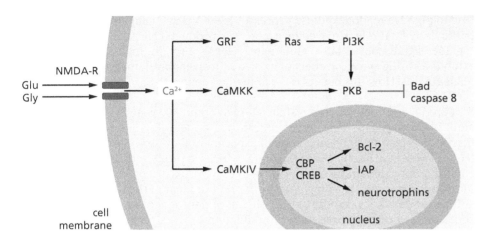

Figure 16.16 Activation of anti-apoptotic signaling by the NMDA receptor Ca^{2+} passing the cell membrane through the NMDA receptor cation channel (NMDA-R) stimulates signaling cascades that lead to inactivation of pro-apoptotic proteins (Bad, caspase 8) and to expression of genes encoding anti-apoptotic factors (Bcl2, IAP, neurotrophins). Details are found in the text. GRF, guanine nucleotide releasing factor; CaMKK, calmodulin kinase kinase; CaMKIV, calmodulin kinase IV; PI3K, phosphatidylinositol 3-kinase; PKB, protein kinase B; CREB, cAMP response element binding protein; CBP, CREB binding protein; IAP, inhibitors of apoptotic proteins.

brane channels (Figure 16.16). The signaling cascades that mediate this survival effect include the Ras–PI3K–PKB pathway, which leads to inactivation of the pro-apoptotic protein Bad (Section 4.6.6). Here Ras is activated by the Ca^{2+}-dependent guanine nucleotide releasing factor GRF that is expressed preferentially in neurons (Section 10.1.3). PKB is stimulated, in addition, by Ca^{2+}-dependent calmodulin kinase kinase (CaMKK; see Section 14.6.3).

Another survival pathway leads to the transcription of genes encoding anti-apoptotic proteins such as Bcl2, Bcl-XL, and the caspase inhibitor IAP (Section 13.2.2) as well as neurotrophic growth factors. One of the transcription factors required for this response is CREB, which (like its co-activator CBP/p300) is activated by phosphorylation catalyzed by CaMKIV, a major transducer of gene-regulatory Ca^{2+} signals (more details about calmodulin kinases are found in Section 14.6.3).

While the survival of excitatory glutamate neurons depends on a continuous normal activation of the glutamate receptors, hyperactivation causes cell death. Such a neurotoxic response is triggered by deficiencies of blood flow (ischemia) and oxygen supply (hypoxia) as well as by injury and bleeding. An infamous clinical outcome of these disorders is **stroke**. The key mediator is again Ca^{2+}, which at higher concentrations induces apoptosis rather than promoting cell survival (Section 14.6.5). Some of the relevant signaling pathways are shown in Figure 16.17.

Figure 16.17 Neuronal cell death due to hyperactivation of NMDA receptors: a molecular mechanism of stroke Ischemia and hypoxia (as well as injury) result in an increased firing of a presynaptic glutamatergic neuron (left) resulting in a hyperactivation of NMDA-R and a flooding of the postsynaptic neuron by Ca^{2+} ions (right). As a result, metabolic pathways (here exemplified by the arachidonic acid metabolism triggered by Ca^{2+}-dependent phospholipase A_2) are running out of control, generating (as byproducts) reactive oxygen species such as superoxide anion radicals. The latter react with nitric oxide, NO (produced by Ca^{2+}-dependent neuronal NO synthase, nNOS), to produce the highly cytotoxic peroxynitrite. In addition, Ca^{2+} causes cell death by damaging mitochondria and activating proteases (calpain) and endonucleases. The process is amplified by positive feedback with NO that, as a retrograde signal, augments glutamate release.

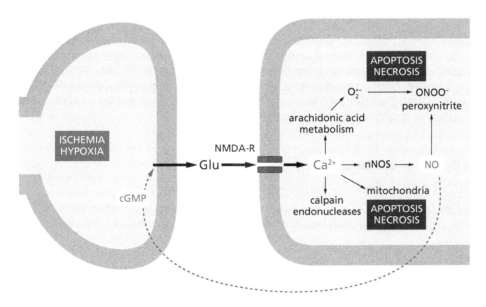

Due to positive feedback, these processes build up in a catastrophic manner. A particularly fatal role is played by excessively produced NO that, acting as a retrograde signal, potentiates glutamate release and, in high concentrations, exhibits a strong cytotoxic effect of its own by irreversibly damaging proteins and nucleic acids. How this occurs is explained in Section 16.5.

Is neuronal cell death an irreversible fate? For about 100 years the concept of the adult brain as incapable of regeneration was upheld dogmatically. At present, however, a steadily increasing body of experimental evidence indicates the existence of neuronal stem cells in the brain, which do proliferate depending on the environmental demands or in response to injury. This discovery of what is called "adult neurogenesis" is going to revolutionize our ideas of the brain's capabilities and has added a new aspect to the concept of neuronal plasticity, which now is considered to rest on three pillars: modulation of the strength of synaptic contacts, of the number of these contacts, and of the production of new neurons.

Summary

Glutamate and aspartate are the major excitatory neurotransmitters in brain. Their ionotropic receptors are coupled with semispecific cation channels, with the NMDA receptor exhibiting some preference for Ca^{2+}. The channels are composed of four subunits with three transmembrane helices and a P-loop each. The eight or more metabotropic glutamate receptors are either $G_{i,0}$- or $G_{q,11}$-coupled. Both ionotropic and metabotropic receptors are characterized by an extracellular Venus flytrap structure for ligand binding. A similar structure is found in the bacterial glutamate receptor GluR0, considered to be the evolutionary precursor of the mammalian glutamate receptors. Glutamate receptors are critically involved in learning and memory fixation. Both processes are thought to be due to a selective stimulation of existing synaptic contacts, to the formation of new synapses (regulated at both transcriptional and post-transcriptional levels), and to the formation of new neurons. Opposite effects correlate with forgetting or with a less-frequent use of synapses. Coordinated binding of glutamate and glycine, in combination with frequency-modulated depolarization waves emitted by the AMPA receptor, activate the NMDA receptor. It plays a key role in learning and memory fixation by controlling the influx of Ca^{2+} ions that (as second messengers) regulate a series of postsynaptic events leading to a strengthening of the synaptic contact. Nitric oxide generated by a Ca^{2+}-dependent synthase has an important function in this process as a retrograde messenger stimulating the presynaptic release of glutamate. The expression and activity of the ionotropic glutamate receptors (and thus the synaptic efficiency) is regulated by the Ca^{2+} influx as well as by signals interacting with metabotropic glutamate and other receptors. The continuous use of glutamate receptors keeps neurons alive by inhibiting apoptosis. However, overstimulation of these receptors causes a pathological increase of the Ca^{2+} concentration in the postsynaptic neuron. As a result, the signaling pathways of Ca^{2+}-dependent apoptosis and necrosis become activated, leading to cell death, an effect that is amplified by positive feedback through NO, which speeds up the presynaptic release of glutamate. This signal-induced neuronal degeneration is triggered by ischemia and hypoxia and plays a key role in stroke.

16.5 Two gaseous signaling molecules

16.5.1 Nitric oxide: a Janus-faced signal molecule*

As we have seen, NO is an important auto- and paracrine signaling molecule in the brain. Its biological significance was discovered, however, by studies on

---

*Janus, the Roman god of doors and gates, was depicted with two heads looking in opposite directions. The term "Janus-faced" is generally used to indicate a conflicting nature. See also Janus kinase (Section 7.2.1).

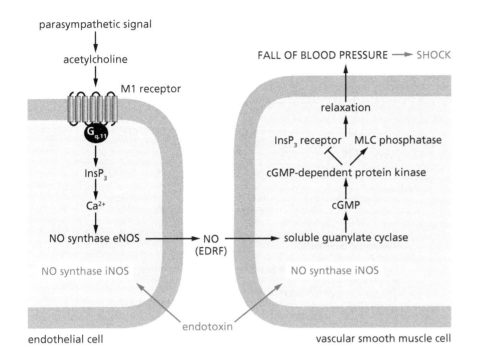

parasympathetic signal

acetylcholine

M1 receptor

$G_{q,11}$

$InsP_3$

Ca^{2+}

NO synthase eNOS

NO synthase iNOS

endothelial cell

NO
(EDRF)

endotoxin

FALL OF BLOOD PRESSURE ⟶ SHOCK

relaxation

$InsP_3$ receptor   MLC phosphatase

cGMP-dependent protein kinase

cGMP

soluble guanylate cyclase

NO synthase iNOS

vascular smooth muscle cell

Figure 16.18 Nitric oxide: the endothelium-derived relaxing factor (EDRF) In endothelial cells activated by the parasympathetic transmitter acetylcholine, Ca^{2+} stimulates the endothelial NO synthase eNOS. The NO thus generated diffuses into the surrounding vascular smooth muscle cells and stimulates the soluble guanylate cyclase. In the next step, cGMP-dependent protein kinase causes a relaxation of the muscle cell by inactivating the inositol 1,4,5-trisphosphate ($InsP_3$) receptor /Ca^{2+} channel of the endoplasmic reticulum and activating the myosin light chain (MLC) phosphatase. As a result, the blood pressure falls. In both cell types, bacterial endotoxin (interacting with the Toll-like receptor TLR4, see Section 6.3.2) induces the *de novo* synthesis of the inducible NOS isoform iNOS. This highly active enzyme dramatically amplifies the NO–cGMP signal, thus causing a catastrophic fall of blood pressure known as endotoxin shock (see Section 6.3.2). The release of cytokines additionally induced by endotoxin strengthens this fatal response.

parasympathetic blood pressure regulation. For this work, the Nobel Prize in Physiology or Medicine for 1998 was awarded to Robert F. Furchgott, Louis J. Ignarro, and Ferid Murad.

The key event of peripheral blood pressure regulation is a relaxation of vascular smooth muscle cells induced by acetylcholine and mediated by the $G_{q,11}$-coupled muscarinic M1 receptor, which activates the phospholipase Cβ–$InsP_3$–Ca^{2+} pathway (see Section 5.4). This seems to be in contradiction to the fact that Ca^{2+} triggers muscle contraction rather than relaxation. However, vascular smooth muscle cells do not express M1 receptors and thus cannot respond directly to acetylcholine. Instead, M1 receptors are found in the adjacent endothelial cells where the Ca^{2+} released by $InsP_3$ stimulates the synthesis of an **endothelium-derived relaxing factor** (EDRF) that, as a paracrine signaling molecule, triggers relaxation of the surrounding vascular smooth muscle cells. EDRF has turned out to be identical with the gas nitric oxide (Figure 16.18). This discovery was quite a surprise since at that time (in the 1980s) only one gaseous signaling molecule was known, namely, the plant hormone ethylene. The identification of EDRF as NO offered an elegant explanation of the therapeutic effects of drugs like nitroglycerol used against a constriction of coronary vessels (angina pectoris): the metabolization of such compounds results in the formation of NO.

Today it is generally accepted that NO production is not a specialty of endothelial and neuronal cells. Instead, all tissues possess the corresponding enzymatic apparatus. In other words, NO is an universal intercellular messenger. NO arises from oxidation of the guanidino group of arginine (Figure 16.19). This reaction requires nicotinamide adenine dinucleotide phosphate (NADPH) and molecular oxygen and is catalyzed by specific oxygenases, the NO synthases (NOS), with flavin mononucleotide, flavin adenine dinucleotide, and tetrahydrobiopterin as redox cofactors.

Various NOS isoforms are known, which are grouped into three families. The oxygenases of the NOS1 and NOS3 families are constitutively expressed housekeeping enzymes and are collectively called constitutive or cNOS. NOS1 is also known as neuronal or nNOS because it is particularly abundant in the central nervous system, while NOS3 has been called endothelial or eNOS because it was originally identified as the EDRF-producing enzyme. NOS1 and NOS3 need Ca^{2+}

Figure 16.19 Biosynthesis of nitric oxide NO is produced by a two-step oxidation of arginine (above). The enzymes catalyzing this reaction are shown in the box. Nitric oxide synthase (NOS) isoforms are as follows: i, inducible; c, constitutively expressed; n, neuronal; e, endothelial.

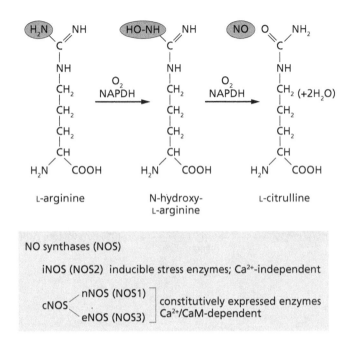

NO synthases (NOS)

iNOS (NOS2) inducible stress enzymes; Ca^{2+}-independent

cNOS $\Big\langle$ nNOS (NOS1) eNOS (NOS3) $\Big]$ constitutively expressed enzymes Ca^{2+}/CaM-dependent

and the calcium-binding protein calmodulin and are activated by any signal causing an elevation of the cytoplasmic Ca^{2+} level (such as acetylcholine in endothelial cells). The enzymes of the NOS2 family, in contrast, are Ca^{2+}-independent and become expressed in most tissues only upon demand, particularly in emergency and stress situations like infections, injury, and inflammation that require an increased output of NO. Thus they are called inducible or iNOS. Inducers of iNOS are pro-inflammatory cytokines and bacterial substances, in particular the endotoxic lipopolysaccharides. iNOS enzymes are much more active than cNOS. They produce NO in cytotoxic doses, mainly for defense purposes. However, excessive NO synthesis, such as that caused by endotoxin, may trigger a catastrophic fall of blood pressure, cumulating in shock and death.

At body temperature NO is a gas that easily penetrates lipid membranes, "soaking" the surrounding tissue cells and acting both as a signaling molecule and as a defensive compound. However, due to a metabolic lifetime of only a few seconds, its range is quite narrow unless it becomes stabilized by particular proteins (see below). Thus, in most tissues NO acts as a *local mediator*. An illustrative example is the nervous system: here NO produced by nNOS acts both as a neurotransmitter and as a neuromodulator. Its effect as a retrograde messenger in the brain has been mentioned already. In the vegetative nervous system it is responsible for so-called Non-Adrenergic, Non-Cholinergic (NANC) effects. Examples of such effects are smooth muscle relaxation in distinct organs such as gastrointestinal tract, urethra and penis (where it causes erection; see Section 4.4.2).

NO is both an endogenous messenger and a toxic gas. Therefore, its effects critically depend on its tissue concentration (Figure 16.20). At low concentrations, as generated by cNOS, NO functions as a signaling molecule that specifically interacts with a cytoplasmic receptor, **soluble guanylate cyclase** (sGC). This is a hemoprotein that binds the NO (which is a free radical) at the iron atom of the heme group (see Section 4.4.2 for more information). Upon activation by NO, the enzyme produces cGMP, which in turn stimulates protein kinases and ion channels and modulates the activity of cyclic nucleotide phosphodiesterases (Sections 4.4.2, 4.6.2, and 14.5.7). Along these signaling pathways, for instance, synaptic plasticity and blood pressure are regulated. Moreover, NO thus ensures the blood supply of the working skeletal muscle, since Ca^{2+} released from the sarcoplasmic reticulum during contraction activates cNOS, and the NO thus produced induces a dilatation of the surrounding blood vessels.

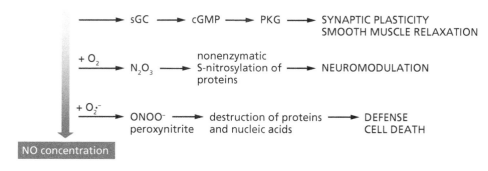

Figure 16.20 Concentration-dependent effects of NO At low and moderate concentrations, NO activates soluble guanylate cyclase (sGC) and nitrosylates proteins. At high concentrations, NO exhibits strong cytotoxic effects by its own and by combining with reactive oxygen species (here superoxide anion radical $O_2^{\bullet-}$) to highly aggressive intermediates such as peroxynitrite. PKG, cGMP-dependent protein kinase.

In higher concentrations, NO still exhibits signaling effects: it acts as a post-translational modulator by nitrosylating SH groups of proteins (see also Section 2.2). For this nonenzymatic **S-nitrosylation**, NO is mostly oxidized to the chemically more active N_2O_3. S-Nitrosylation is reversible and either inhibits or activates the functions of proteins carrying essential SH and S–S groups. Since S-nitrosylated proteins are more stable than NO, they may serve as transport vehicles, releasing NO at target sites.

As a signaling reaction, S-nitrosylation may be highly specific. A well-known example is provided by the NMDA receptor, which is inactivated by S-nitrosylation of a single SH group. Since, on the other hand, nNOS is stimulated by the Ca^{2+} influx through the NMDA receptor ion channel, S-nitrosylation probably is a means to prevent dangerous hyperactivation of the receptor (Figure 16.15). This feedback control regulates itself by S-nitrosylation of cNOS, which also becomes inhibited. Thus, in physiological concentrations NO exhibits a neuroprotective effect. Such an action is supported by an inhibition of apoptosis because S-nitrosylation inactivates the initiator caspases 3 and 9 (Section 13.2.1) as well as the pro-apoptotic protein kinase ASK1 (Section 13.2.4) but activates Ras, a central relay station of survival pathways. Moreover, NO induces the expression of heme oxygenase 1, a neuroprotective enzyme (Section 16.5.2).

At even higher concentrations, the cytotoxic effects of NO dominate. An overproduction of NO is due to an induction of iNOS expression, mostly in the course of an inflammatory reaction. As **nitrosative stress** it may have fatal consequences, since as a free radical NO inactivates essential hemoproteins such as cytochrome oxidase (thus inhibiting the mitochondrial respiratory chain) and iron-sulfur proteins. In addition, NO is transformed into highly aggressive products upon reaction with reactive oxygen species. The latter are produced by **oxidative stress**, which is another typical feature of inflammatory reactions. Particularly aggressive is peroxynitrite, which arises from a reaction between NO and superoxide anion radicals. Such agents damage proteins and nucleic acids with the result that the cell eventually may be killed. In a kind of kamikaze attack, cells of innate defense, such as granulocytes and macrophages, make use of NO toxicity for chemical warfare against microorganisms. Since collateral damage of adjacent tissue cells is unavoidable, inflammation is generally accompanied by tissue necrosis. Inflammatory reactions are typical features of **neurodegenerative disorders** such as stroke, Alzheimer's dementia, and Huntington's and Parkinson's diseases. In fact, in the morbid brain areas iNOS is strongly expressed and the NO level is pathologically high. In parallel, an overexpression of enzymes provoking oxidative stress is seen. This holds true in particular for the prostaglandin synthase cyclooxygenase 2 (COX2), which like iNOS is induced in emergency situations (see Section 4.4.5). Thus, iNOS and COX2 induction certainly are critical events in neurodegeneration, and considerable efforts are being made to develop drugs by which NO signaling and COX2 activity can be manipulated. However, these approaches are hampered by the Janus-faced role of NO as a neuroprotective and a neurotoxic signal, and the difficulty of overcoming the blood–brain barrier. This does not devalue preventive measures that generally lower the pressure of nitrosative and (even more so) of oxidative stress. Such results may be achieved by both novel drugs such as antioxidants and COX2 inhibitors and an adjustment of lifestyle.

Although the destructive role of nitrosative and oxidative stress is unquestioned, it is only one among various brain defects involved in neurodegenerative diseases. Others are a flooding of cells with calcium ions (resulting in apoptosis; see Sections 14.6.5 and 16.4.2), the generation of advanced glycation end products (Section 7.3.3), and intra- and extracellular protein deposits such as β-amyloid and Tau aggregates (Section 13.1.3). It is still not entirely clear how these processes interact with each other to bring about the overall pathological changes.

16.5.2 Carbon monoxide

For a long time carbon monoxide generated in the course of oxidative hemoglobin degradation in liver was thought to be a waste product of metabolism. However, such a concept did not really fit the observation that the corresponding enzyme, heme oxygenase, was strongly expressed in the central nervous system, a tissue normally not connected with hemoglobin metabolism. In fact, CO finally turned out to be a para- and autocrine signaling molecule sharing some functions with NO. Considering the well-known toxicity of CO, this was rather a surprising discovery.

Heme oxygenases are integrated in the endoplasmic reticulum and catalyze the opening of the hemoglobin ring to yield biliverdin, CO, and Fe^{2+}. The reaction requires molecular oxygen, NADPH as a reductant, and NADPH–cytochrome P450 reductase as a cooperating enzyme (Figure 16.21).

Heme oxygenases exist in three subfamilies: HO-1, HO-2, and a rather ill-defined HO-3. The enzymes of the HO-2 type are constitutively expressed, in particular in the liver, testes, blood vessels, and the central nervous system. HO-1 is, in contrast, an emergency enzyme expressed only in stress situations such as inflammation, heavy metal intoxication, oxygen deficiency, impaired blood supply, and heat. (Therefore, HO-1 has been identified also as heat-shock protein 32.) Thus,

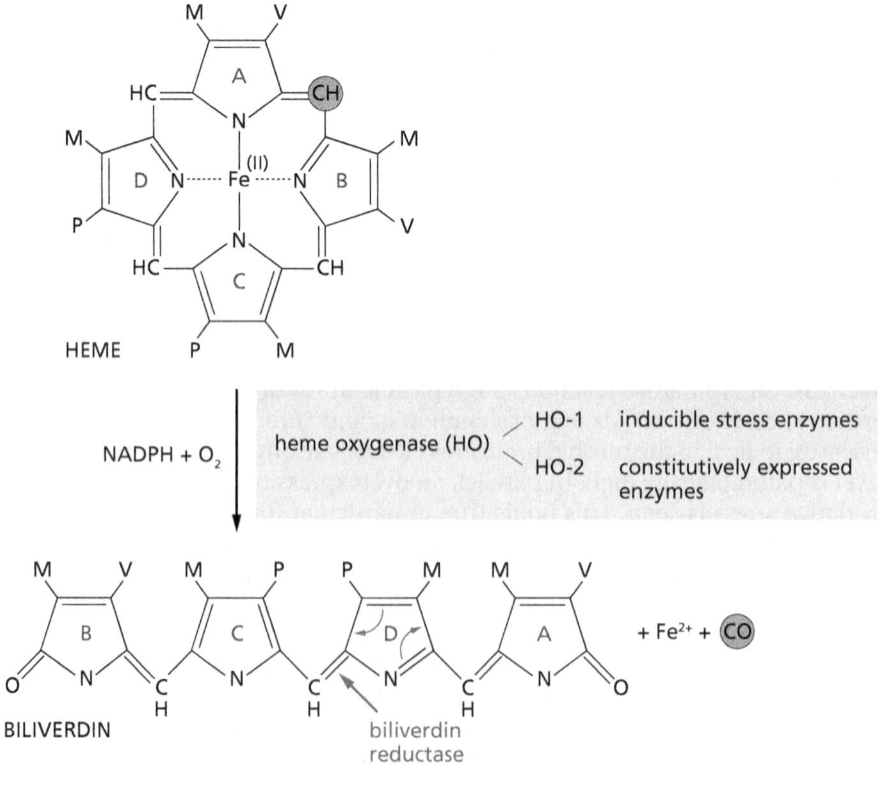

Figure 16.21 Bioformation of carbon monoxide The heme ring is oxidatively opened between the pyrrole residues A and B to yield biliverdin, iron ions, and CO (derived from the methine group in the red circle). This reaction is catalyzed by heme oxygenase expressed in two types, HO-1 and HO-2. Biliverdin is subsequently reduced between the pyrrole rings C and D by biliverdin reductase and nicotinamide adenine dinuceotide phosphate (NADPH). M, methyl; V, vinyl; P, propionyl.

heme oxygenases resemble NO synthases and cyclooxygenases (Section 4.4.5), which are also expressed both as constitutive housekeeping enzymes for everyday demands and as inducible emergency enzymes dealing with stress. Although it is a gas, CO (in contrast to NO) is not a free radical; therefore, CO is much more stable and less aggressive.

While NO may be transformed into dangerous products involved in microbial defense, the toxic effect of CO does not go beyond a blockade of Fe^{2+}-containing hemoproteins such as hemoglobin, an effect that has no consequence considering the low physiological CO concentrations. Moreover, soluble guanylate cyclase (sGC), despite being a hemoprotein, is activated rather than inhibited by CO. Although the sGC-activating efficacy of CO amounts to only 5% of that of NO, it results in comparable effects such as lowering of blood pressure (due to a relaxation of vascular smooth muscles) and inhibition of platelet aggregation. Whether CO is involved also in the regulation of synaptic plasticity, or even is a neurotransmitter of its own, remains a matter of debate.

What is the role of stress-induced CO production? It probably provides a protective mechanism against tissue damage as caused, for instance, by inflammation. Thus CO turns down the release of pro-inflammatory cytokines (such as tumor necrosis factor α and interleukin 1β) by leukocytes while at the same time inducing synthesis of the anti-inflammatory interleukin 10. Moreover, regulatory cross talk seems to exist between NO synthases and heme oxygenases, ensuring that the cytotoxic effects of NO do not go off course. In fact, NO and its aggressive metabolites belong to the most effective inducers of HO-1, whereas, vice versa, CO inhibits in many (but not all) tissues the activity and expression of iNOS.

Summary

NO gas is an abundant paracrine mediator with strictly local effects. It was discovered as an endothelium-derived relaxing factor and is generated from L-arginine by an oxidative reaction catalyzed by NO synthases. One distinguishes between constitutively expressed, Ca^{2+}-dependent NO synthases and stress-induced, Ca^{2+}-independent enzymes. Depending on the concentration, NO functions either as a signaling molecule or as a cytotoxic agent involved in defense reactions and cell death. The signaling effects are due to noncovalent activation of soluble guanylate cyclase, generating the second messenger cGMP, or to reversible post-translational modification of proteins by nitrosylation of sulfhydryl groups. Overstimulation of NO formation plays an infamous role in life-threatening pathological events such as stroke and endotoxin shock. Carbon monoxide is an endogenous messenger resembling NO. CO gas is produced by the oxidative degradation of heme catalyzed by heme oxygenase (which, like NO synthase, exists in both constitutively expressed and stress-inducible isoforms). The signaling effects of CO are known only fragmentarily. In stress situations it seems to exert a tissue-protective effect by antagonizing cytotoxic NO effects.

16.6 Receptors of purine and pyrimidine nucleotides: the ATP signal

Purine and pyrimidine nucleotides such as ATP, ADP UTP, and UDP are not only components of energy metabolism but also neurotransmitters, which are released at purinergic and pyrimidinergic synapses. Like other transmitters, they interact with both ionotropic and metabotropic receptors. Moreover, certain cell types, such as endothelial cells, have been found to contain a cell surface ATPase that directly synthesizes ATP into the extracellular space, where it acts as a paracrine signal.

The **ionotropic P2X receptors** are found in all tissues. They are highly specific for extracellular ATP. P2X receptors are coupled to nonspecific cation channels that

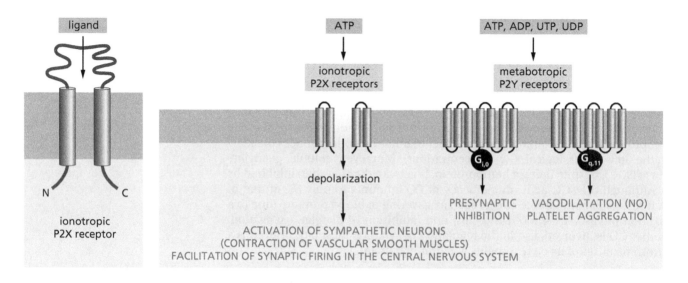

Figure 16.22 Ionotropic and metabotropic nucleotide receptors of the P2 type The inset (left) shows the subunit of the ionotropic P2X receptor cation channel with two transmembrane domains and an extended extracellular loop of about 280 amino acids that is stabilized by several disulfide bonds (not shown). Probably the ion channel consists of three subunits and is lined by all transmembrane subunits. For other details, see text.

cause depolarization when opened. The channel architecture is rather simple, resembling that of the mechanosensitive channel MscL of *Mycobacterium tuberculosis*, which probably is the evolutionary precursor of P2X receptors (Section 3.5.7). Like this channel, P2X is composed of three subunits with two transmembrane domains each (Figure 16.22). The human genome encodes seven different P2X subunits; together with various splice variants, these combine to yield a large number of receptor isoforms. Among many P2X functions, their effect on some forms of smooth muscles has been studied in more detail. Acting (like NO) as a non-adrenergic, non-cholinergic signal of the vegetative nervous system (Section 16.5.1), here ATP cooperates with the sympathetic transmitter noradrenaline to trigger muscle contraction (see Section 14.6.2). P2X receptors are quite abundant also in the central nervous system, where they enhance neurotransmitter release by facilitating presynaptic depolarization.

The **metabotropic P2Y receptors** also are widely distributed. In contrast to P2X, they are activated not only by ATP but also by ADP, UTP, and UDP; they are both purinergic and pyrimidinergic receptors. In mammals, they are expressed in at least 12 isoforms differing in ligand specificity. Some P2Y receptors expressed in presynaptic membranes are coupled with $G_{i,0}$-proteins. Among other effects, they suppress the release of their own nucleotide ligands by negative feedback, reminiscent of the regulation of α-adrenergic effects (see Section 5.3.3).

Other P2Y receptors are coupled with $G_{q,11}$-proteins; along the phospholipase $C\beta$–InsP$_3$–Ca^{2+} axis, they induce platelet aggregation (a characteristic ADP effect) and NO release.

Summary
The purine nucleotides ATP and ADP as well as the pyrimidine nucleotides UTP and UDP are paracrine mediators, acting as neurotransmitters. The corresponding receptors are the ionotropic P2X receptors, coupled with cation channels and exclusively stimulated by ATP, and the metabotropic P2Y receptors, coupled with $G_{i,0}$- and $G_{q,11}$-proteins and stimulated by each of the four nucleotides. P2X receptors play an important role in smooth muscle contraction and as enhancers of neurotransmitter release in the brain.

16.7 Cannabinoid and vanilloid receptors

The search for a cellular binding site of the marijuana ingredient tetrahydrocannabinol led to the identification of two **cannabinoid receptors, CB1 and CB2,** in the 1980s. CB1 is expressed by many tissues but particularly in brain,

Sidebar 16.2 Adenosine receptors Like nucleotides, adenosine functions as an intercellular signaling molecule. Since it is not stored in synaptic vesicles and not released by depolarization but by special membrane transporters, it is considered to be a neuromodulator rather than a neurotransmitter proper. In addition, extracellular ATP is degraded by ectonucleotidases to adenosine. In contrast to the P2 nucleotide receptors, the adenosine receptors are called P1 receptors. They exclusively belong to the G-protein-coupled metabotropic type, interacting either with $G_{i,0}$- or G_s-proteins (Figure 16.23).

A major function of adenosine is to protect tissues from damage. In fact, a strong increase of adenosine in the extracellular medium is an emergency signal indicating an acute oxygen deficiency (hypoxia, ischemia) as is characteristic for stroke, angina pectoris, and myocardial infarction. Via the $G_{i,0}$-coupled receptors A_1 and A_3, adenosine exhibits various effects that prevent (at least partially) stress-induced tissue destruction. These effects include increased blood supply (caused by vasodilatation), diminished blood clotting (due to an inhibition of platelet aggregation), and a sedating effect (resulting from presynaptic inhibition of excitatory neurotransmitter release). Vice versa, a dysregulation of the adenosine system correlates with the induction of seizures, particularly in epilepsy. Moreover, there is evidence that activation of the A_1 receptor is involved in the generation of sleep and that the stimulating effect of caffeine is due at least partially to a blockade of this receptor (see also Section 16.9.1). The strong pain indicating myocardial infarction is also triggered by adenosine, here probably through a stimulation of A_2 receptors. An overactivation of A_2 receptors is also seen in neurodegenerative disorders such as Alzheimer's, Parkinson's, and Huntington's diseases as well as in drug addiction, indicating a role of this receptor type in pathogenesis.

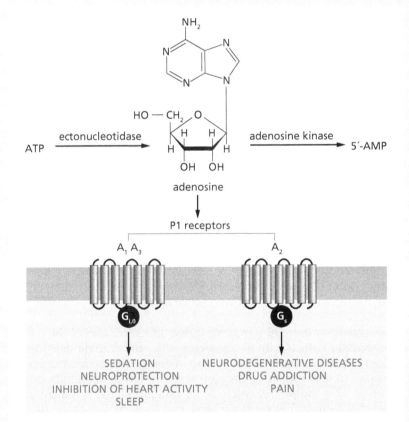

Figure 16.23 P1 adenosine receptors The isoforms are called A_1, A_2, and A_3. Extracellular adenosine is generated from ATP by ectonucleotidases (or released from the cell through special membrane transporters) and inactivated mainly by adenosine kinase.

whereas CB2 seems to be restricted to immune cells. Both receptors are coupled with $G_{i,0}$-proteins. Along this pathway, cannabinoids inhibit adenylate cyclase and voltage-dependent Ca^{2+} channels and activate GIRK channels, causing membrane hyperpolarization. Thus, they suppress neurotransmitter release, dampen neuronal excitability, and mediate inhibitory effects such as sedation and pain relief. In addition, cannabis drugs transiently induce euphoria, rendering them particularly attractive. This action is probably due to a pre- or postsynaptic inhibition of inhibitory neurons (see Section 16.9). Artificial CB1 receptor

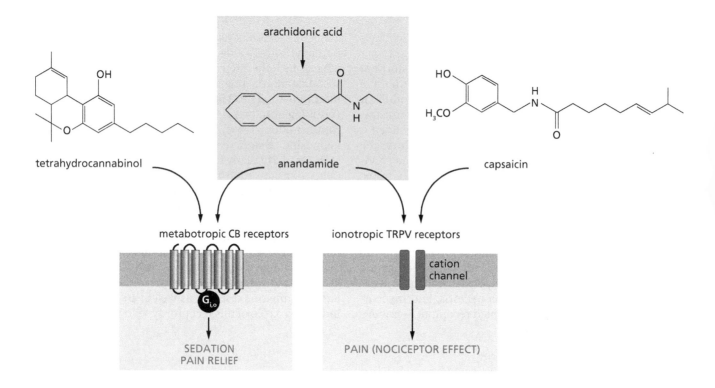

Figure 16.24 Endocannabinoid receptors The metabotropic endocannabinoid (CB) receptor with its exogenous ligand tetrahydrocannabinol is shown to the left, the ionotropic vanilloid receptor TRPV with its exogenous ligand capsaicin is shown to the right.

blockers are employed, on the other hand, as appetite suppressants in the therapy of obesity. Of course, tetrahydrocannabinol is not the actual ligand of CB receptors. Instead, endogenous cannabinoids have been identified that are derivatives of polyunsaturated fatty acids. Their prototype is arachidonylethanolamide or **anandamide** (the name refers to the Sanskrit word *ananda* for bless), which has been found in various areas of the brain but also in other tissues. The major function of endocannabinoids is the negative feedback control of synaptic activity: upon their release from depolarized postsynaptic neurons, they act as retrograde signals to inhibit presynaptic neurotransmitter release according to the mechanism shown in Figure 16.5. In certain brain regions this effect causes long-term depression of synaptic conductivity. Thus, endocannabinoids are involved in the regulation of synaptic plasticity. Endocannabinoids also play an important role in the embryonic development of the central nervous system. This may explain why excessive marijuana smoking by expectant mothers frequently causes pathological problems in their children, such as deficits of cognition, motion, and social behavior.

In addition to metabotropic CB receptors, **ionotropic receptors of the transient receptor potential (TRP) family** also interact with anandamide (Section 15.3). These TRPV or vanilloid receptors are coupled to nonspecific cation channels that actually respond to heat and acid. They are sensitized by anandamide as well as by certain exogenous compounds such as the vanillin-related capsaicin, the hot ingredient of chili pepper (for details see Section 15.3). Thus, like many other neurotransmitters, anandamide exerts a dual effect depending on the receptor type: via TRPV it enhances pain, whereas via CB it acts as a pain reliever (Figure 16.24).

Summary

Endocannabinoids such as anandamide are paracrine mediators interacting with $G_{i,0}$-coupled metabotropic cannabinoid receptors (which also bind cannabis drugs) and cation channel-coupled ionotropic vanilloid (or TRPV) receptors (which are also activated by capsaicin from chili pepper). The activation of vanilloid receptors found predominantly in sensory cells has a pain-evoking effect, whereas the cannabinoid receptors controlling neurotransmitter release mediate pain-relieving effects.

16.8 Opioid receptors

Morphine, the alkaloid of poppy isolated by Friedrich Sertürner in 1803, is the prototype of an opiate. The pain-relieving effect of these drugs has been known since early antiquity and is still unsurpassed. On the other hand, opiates are the epitome of narcotics. Rather early, the stereospecific effect of very low doses of morphine was taken as an indication of an endogenous receptor. Since 1973, three opioid receptors labeled by the Greek letters δ, κ, and μ have been identified. Since the discovery of the corresponding endogenous peptide ligands, the terms DOP, KOP, and MOP (Delta, Kappa, and Mu Opioid Peptide receptors) are now preferred. An additional opioid receptor found more recently is NOP, the nociceptive (pain-related) opioid peptide receptor. All receptors are metabotropic, that is, coupled with $G_{i,o}$-proteins (Figure 16.25); ionotropic opioid receptors have not been found.

Along the $G_{i,o}$-controlled signaling pathways, the opioid receptors trigger the usual effects such as inhibition of adenylate cyclase and voltage-dependent Ca^{2+} channels, hyperpolarization by gating the GIRK type potassium channels, and (via Gβγ-subunits) activation of phospholipase Cβ and PI3K. The endogenous ligands are small peptides: β-endorphin (31 amino acids) and enkephalins (5–8 amino acids) interact with the receptor DOP; β-endorphin also interacts with MOP; dynorphins (8–17 amino acids) bind to KOP; and nociceptin (or orphanin, 17 amino acids) binds to NOP. Morphine and heroin prefer MOP as receptor (the label μ/M refers to morphine).

The opioid peptides are released from larger precursor proteins by controlled proteolysis. They all contain the basic enkephalin sequence Tyr-Gly-Gly-Phe-Met/Leu or (as in the case of nociceptin) Phe-Gly-Gly-Phe-Met/Leu. To a certain extent, this sequence is frozen in the complex ring structure of morphine since the aromatic ring and other structural elements exactly fit the receptors' binding sites for the aromatic amino acids Phe and Tyr.

As neurotransmitters, opioid peptides inhibit interneuronal signaling post- or presynaptically according to the scheme shown in Figure 16.5. Depending on whether the opioid neuron is connected with an excitatory or an inhibitory neuron, the opioid either exerts a sedating and pain-relieving effect or causes euphoria.

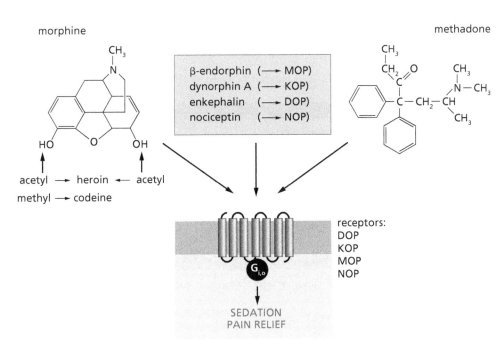

Figure 16.25 Opioid receptors All opioid receptors known are $G_{i,o}$-protein-coupled. Examples of endogenous ligands (with the preferred isoforms of opioid peptide receptors, OP, shown in parentheses are depicted in the box; examples of exogenous ligands are shown to the left and to the right.

Pain therapy with morphines is confronted by two problems: tolerance and addiction. Due to **tolerance**, the dose of the drug has to be increased permanently to overcome endogenous receptor desensitization or tachyphylaxis, while **addiction** means that a withdrawal of the drug results in severe sickness that can be prevented only by new drug administration or a time-consuming and strenuous therapy. As shown in the following section, both symptoms may be understood as the results of a dramatic and long-lasting adaptation of neuronal signal processing resembling long-term memory storage.

Summary

Endogenous opiates such as enkephalins, endorphins, dynorphins, and nociceptins are paracrine peptide mediators and neurotransmitters that interact with $G_{i,0}$-coupled metabotropic receptors (ionotropic receptors are not known) and are also activated by morphine and other opium-derived drugs. Opioid receptors play a key role in pain therapy and drug addiction.

16.9 Narcotics and drug addiction

Drug addiction is due to a misdirected adaptation of signaling events in the brain. In fact, the major targets of narcotics and psychotropic drugs are synaptic signal transmission and neuronal signal processing. Some examples are listed in Figure 16.26.

16.9.1 Nicotine, caffeine, and alcohol

First, we shall have a look at the everyday drugs nicotine, caffeine, and alcohol. The **nicotine** effects experienced by smokers are primarily due to an activation of ionotropic acetylcholine receptors in the central nervous system. These particular neurons stimulate dopaminergic and other signaling pathways of the limbic system. This so-called **rewarding center** of the brain processes emotions, in particular pleasant feelings. Adaptation due to receptor desensitization leads to tolerance and addiction, that is, to the chain-smoker.

Figure 16.26 Drug effects on the level of signal transduction: an overview

STIMULATION	
caffeine	inhibition of cAMP–phosphodiesterase stimulation of ryanodine receptors inhibition of adenosine receptors
nicotine	stimulation of nicotinic receptors
cocaine amphetamines ecstasy	inhibit the recapture of noradrenaline, serotonin, and dopamine

SEDATION, EUPHORIA	
alcohol, barbiturates	inhibition of NMDA receptor stimulation of GABA receptors
cannabis, opiates	presynaptic inhibition via $G_{i,0}$

FEAR RELIEF	
benzodiazepines	stimulation of GABA receptors

HALLUCINATIONS	
mescaline psilocybin LSD ecstasy	stimulation of serotonin, noradrenaline, and dopamine receptors

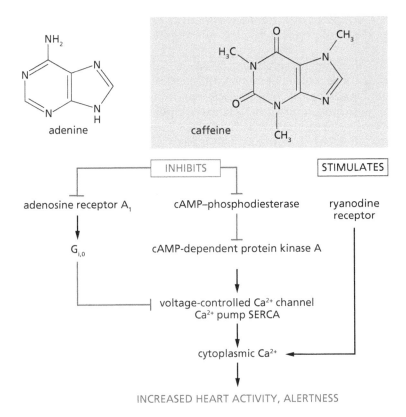

INCREASED HEART ACTIVITY, ALERTNESS

Figure 16.27 Molecular effects of caffeine Caffeine (trimethylxanthine) has an adenine-related structure explaining many of its effects that are due to an increase of the cytoplasmic cAMP and Ca^{2+} levels. Caffeine inhibits the A_1 adenosine receptor, thus relieving voltage-dependent Ca^{2+} channels (and adenylate cyclase) from $G_{i,0}$-mediated inhibition, and cAMP–phosphodiesterase, thus stabilizing cAMP and (as a consequence) activating protein kinase A. By phosphorylating phospholamban (Section 14.6.2), protein kinase A stimulates the Ca^{2+} pump SERCA of the sarcoplasmic reticulum, resulting in a storage-activated Ca^{2+} influx (Section 14.5.5). The activation of ryanodine receptors leads to an additional Ca^{2+} influx from the endoplasmic reticulum (Section 14.5.3). As a result, heart activity and synaptic signaling become stimulated, resulting in the typical caffeine effects such as increased heartbeat and alertness. For the sake of clarity, other effects of cAMP and Ca^{2+} that also are influenced by caffeine are not shown.

Because of its adenine-related structure, **caffeine** targets several cellular proteins that normally interact with adenosine and its derivatives. Caffeine inactivates the inhibitory adenosine receptor A_1 and cAMP phosphodiesterase and sensitizes ryanodine receptors, probably by mimicking the effect of cyclic ADP-ribose. Along the corresponding signaling pathways, the cytoplasmic levels of cAMP and Ca^{2+} become elevated, resulting in a reinforcement of sympathetic actions and of

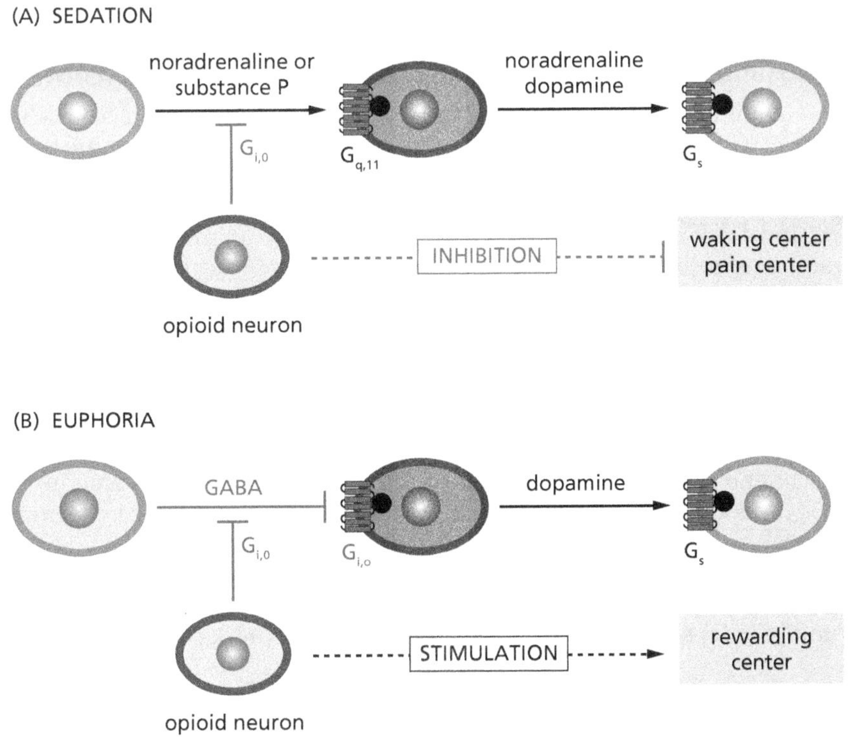

(A) SEDATION

noradrenaline or substance P

noradrenaline dopamine

$G_{i,0}$

$G_{q,11}$

G_s

opioid neuron

INHIBITION

waking center pain center

(B) EUPHORIA

GABA

dopamine

$G_{i,0}$

$G_{i,0}$

G_s

opioid neuron

STIMULATION

rewarding center

Figure 16.28 Dual effect of opioids Although the $G_{i,0}$-coupled opioid receptors primarily transmit inhibitory signals, opioids may have both sedative and stimulatory effects. (A) For sedation and pain relief, peripheral neurons are hindered by associated opioid neurons from releasing noradrenaline or substance P that, via $G_{q,11}$-coupled receptors, would trigger the release by further neurons of noradrenaline and dopamine, stimulating the waking and pain center in the brain. Substance P is a small peptide (11 amino acids) of the tachykinin family that transmits pain signals from the periphery to the spinal cord (see also Figure 14.38). It interacts with a $G_{q,11}$-coupled receptor NK1. (B) The stimulatory effect of opioids is due, at least partially, to an inactivation of GABAergic neurons that normally turn down the activity of those dopaminergic neurons, which stimulate the rewarding center in the brain. G-proteins are symbolized by black circles

central nervous processes with the well-known consequences such as increased heartbeat, desire to urinate, and alertness (Figure 16.27).

Ethyl alcohol stimulates inhibitory $GABA_A$ and blocks excitatory NMDA receptors, sharing these molecular targets with barbiturates. In principle, the physiological and psychological consequences are similar to those caused by cannabis drugs and opiates, both of which exert inhibitory effects via $G_{i,0}$-coupled metabotropic receptors. [Mice with a genetic knockout of the MOP receptor do not respond to alcohol, cannabis, and nicotine, indicating that the target neurons of such drugs release endogenous opioids that, like morphine, activate the MOP receptor.] Therefore, all these drugs (alcohol, barbiturates, cannabis, and opiates) act as sedatives and narcotics. On the other hand, they strongly induce euphoria. Indeed, this dual action renders such drugs particularly attractive. It finds an explanation in the principle of double negation; namely, an inhibitory signal has a stimulatory effect when it suppresses another inhibitory signal. This is shown for opiates in Figure 16.28.

Sidebar 16.3 Drug applications, past and present Heroin is a synthetic acetylation product of morphine. As compared with the latter, it penetrates the blood–brain barrier more rapidly. At the end of the 19th century, heroin was introduced by the Bayer Company as cough medicine. Its addictive potential was realized only 25 years later. Cough medicines developed later frequently contained the much less dangerous **codeine**, a methylated morphine derivative. The substitute drug **methadone** was developed as a pain reliever during World War II to make Germany independent from opium import. Like morphine, methadone causes addiction. However, because (as compared with heroin or morphine) it is much cheaper, exhibits a longer-lasting effect, and may be taken orally, both crimes committed in the pursuit of drug acquisition and the risk of infection (due to nonsterile hypodermic syringes) are considerably diminished.

Until the end of the 19th century, cocaine, the classical drug of native South Americans, was considered to be a harmless stimulant and was used as an ingredient of refreshing beverages such as Coca-Cola and Vin Mariani. Sigmund Freud had prescribed cocaine against morphine addiction until one of his friends fell into an incurable cocaine psychosis. When such harmful effects became known, the cocaine in Coca-Cola was replaced by caffeine.

16.9.2 Hallucinogenic and stimulatory drugs and antipsychotics

Like caffeine and alcohol, many other drugs and narcotics interact with more than one cellular protein and receptor. Typical multifunctional drugs are the **hallucinogens** depicted in Figure 16.29. Due to their chemical structure, they may fit receptors of dopamine, noradrenaline, and serotonin, neurotransmitters primarily involved in the emotional processing of sensory impressions in the limbic system. In addition, they play a key role in psychotic diseases such as schizophrenia (see below).

A major target of hallucinogens are the **serotonin** (or **5-hydroxytryptamine**) **receptors** that in mammals are expressed in no less than 14 isoforms, underscoring their importance in brain physiology. While the ionotropic serotonin receptor $5\text{-}HT_3$ is coupled with a nonspecific cation channel resembling the nicotinic acetylcholine receptor, the other receptors are metabotropic, interacting with G_s-, $G_{q,11}$-, and $G_{i,0}$-proteins. Serotonin receptors are involved in a large number of central nervous processes. Synthetic $5\text{-}HT_1$ agonists are clinically used as antimigraine drugs. They induce vascular constriction in the brain, thus preventing a migraine attack, which is accompanied by vascular dilatation. Serotonin (or tryptaminergic) signaling is also stimulated by some antidepressants.

Stimulants such as **amphetamines, cocaine,** and **ecstasy** interact with dopaminergic, adrenergic, and tryptaminergic neurons by inhibiting synaptic signal

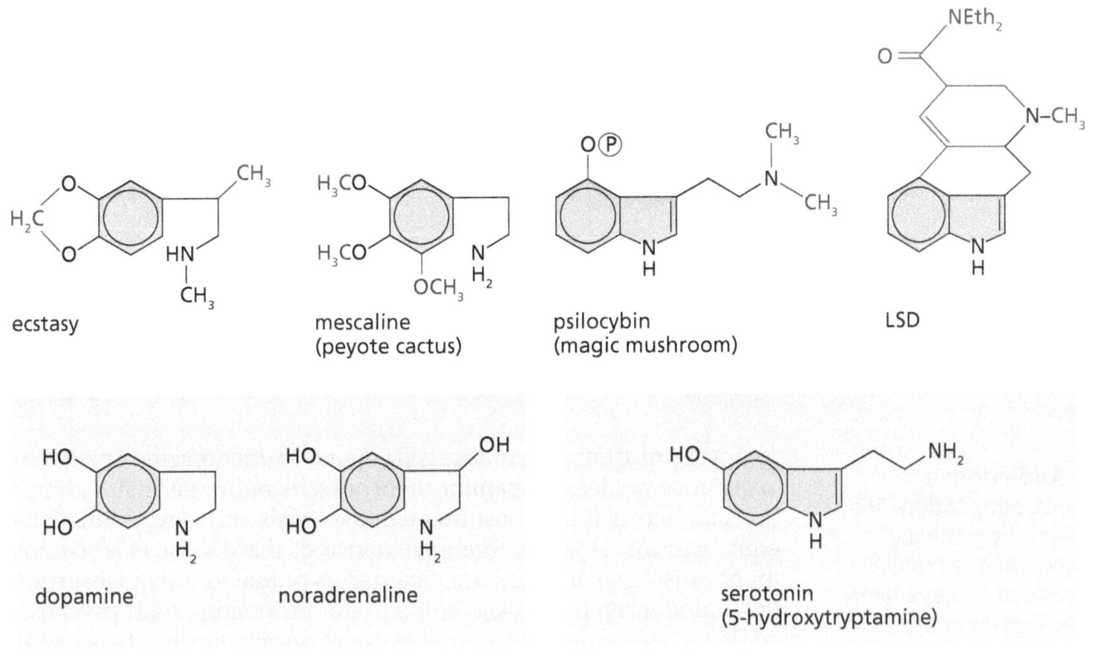

Figure 16.29 Hallucinogenic drugs The prominent hallucinogens shown have transmitter-related structures. Therefore, they are potent activators of metabotropic dopamine, noradrenaline, and serotonin receptors. These receptors dominate signal transmission in brain areas involved in the processing of emotions and sensual impressions. LSD, lysergic acid diethylamide; P, phosphate.

extinction rather than by receptor binding. To this end, they inactivate presynaptic neurotransmitter transporters, thus preventing transmitter recapture. This effect is augmented by a simultaneous stimulation of neurotransmitter release. As a result, the postsynaptic cell becomes overstimulated.

Many drugs evoke symptoms of acute psychosis. Such observations support the concept that paranoid, schizoid, and manic–depressive diseases are caused by genetically fixed or acquired disturbances of cellular signal processing. Indeed, most **antipsychotics** interact with neurotransmitter receptors, in particular with the D2 dopamine receptor, and with transport proteins catalyzing the presynaptic

Sidebar 16.4 Signal extinction by neurotransmitter transporters Synaptic signaling is terminated through a re-uptake of neurotransmitter by the pre- and postsynaptic neuron and adjacent glial cells. An exception is acetylcholine, which is hydrolyzed by acetylcholinesterase in the synaptic cleft. The re-uptake is managed by extremely efficient and specific transporter proteins within neuronal and glial cell membranes. One distinguishes between monoamine transporters for dopamine, noradrenaline, and serotonin and transporters for GABA, glutamate, glycine, and other transmitters. Most of them are characterized by 12 transmembrane domains with the C- and N-termini extending into the cytoplasm. Thus, at first glance, they seem to resemble ABC transporters (Section 3.2.1). However, neurotransmitter transporters catalyze a secondary active transport against a concentration gradient powered by a coupled ion flux (of Na^+, Cl^-, K^+, and H^+) rather than by ATP hydrolysis. Glutamate and monoamine transporters can be traced back to bacterial glutamate and leucine transporters, respectively, which have been crystallized and analyzed in detail.

Like receptors, transporters rapidly become up- and down-regulated by exo- and endocytosis depending on the situation and the demands. Moreover, their activities are controlled by post-translational modifications such as protein phosphorylation.

Release and re-uptake of neurotransmitters are interlinked. So, a blockade of transporters by cocaine, amphetamines, or ecstasy is accompanied by increased neurotransmitter secretion into the synaptic cleft. This leads to overstimulation of the particular synapses, possibly culminating in threatening psychotic symptoms. In fact, pathological malfunctions of neurotransmitter transporters belong to the molecular causes of many psychoses, and vice versa, a series of antipsychotic medical drugs, in particular antidepressants and antiepileptics, directly inhibit presynaptic re-uptake. Neurotransmitter transporters are also found in the membranes of synaptic vesicles storing transmitters. Like the transporters of the presynaptic membrane, they exhibit 12 transmembrane domains, although the types are not closely related.

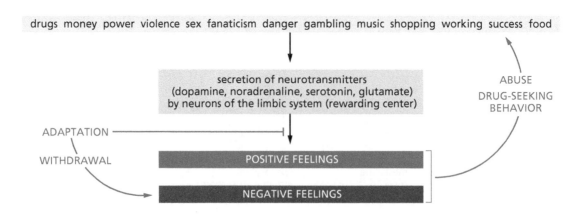

Figure 16.30 Addiction: a consequence of adaptation The brain's rewarding system may be excited by a wide variety of stimuli. Any permanent strain evokes adaptive responses of the signal-processing apparatus in the neurons involved. Such an adaptation may lead to long-term desensitization. This phenomenon yields two consequences: the dose of the stimulus has to be increased continuously in order to obtain the effect desired, and withdrawal results in pathological hyperactivity with strongly negative sensations. Both the desire for positive feelings (pleasure) and the attempt to avoid negative feelings (withdrawal symptoms) may lead to abuse and addiction, finally becoming firmly established by long-term potentiation.

re-uptake of transmitters (synaptic signal extinction; see Sidebar 16.4). Clinical experience yielded a dopamine theory of schizophrenia, maintaining a causal relationship between this most frequent psychosis and a hyperfunction of dopaminergic neurons. However, the appearance of the disease is so complex that such monocausal explanations are regarded as outdated today. Observations showing dopamine receptor blockers to be useful for treating both psychoses and opiate addiction nevertheless emphasize the close relationship between drug diseases and mental illness.

16.9.3 Drug addiction: an adaptive memory effect

Drug addiction is a phenomenon that in its complexity rivals psychotic diseases, resisting any simple explanations. Both are estimated to be about 50% genetically predisposed. The pressure to avoid withdrawal symptoms certainly is a major reason for physical dependency but is only one of several causes of addiction. As least as important is the urge to satisfy the desire for pleasure, resulting in psychological dependency. Depending on the type of drug, either component may predominate.

The major source of pleasant sensations is the previously mentioned rewarding center of the limbic system. As is well-known, this brain area is activated not only by drugs but also by a wide variety of other stimuli, the effects of which are mediated by endogenous "drugs," that is, signaling molecules such as neurotransmitters (Figure 16.30).

As is typical for cells, the neurons of the rewarding center are extremely adaptive and rapidly become desensitized upon prolonged stimulation. Consequently, the dose of the stimulus has to be increased more and more to produce the pleasant sensations desired. Thus, to understand and treat addictive diseases, one must know the cellular and molecular mechanisms of adaptation (shown as a strongly simplified scheme in Figure 16.31). Although these investigations by no means are concluded, the data available strongly indicate that addiction is due to long-term adaptation and conditioning of cellular signal processing and that drug dependency and memory fixation might be two sides of the same coin.

That adaptation occurs at both the cellular and the molecular level was shown in 1975. For these pioneering experiments, a neuroblastoma/glioma cell line was used. The cells responded to morphine treatment as expected by a diminished cAMP synthesis, because adenylate cyclase was inhibited along the MOP–$G_{i,0}$ signaling pathway. When the treatment was repeated several times, an adaptive *de novo* synthesis of adenylate cyclase was observed, settling down the cAMP level at normal values. To lower the level again, the morphine dose had to be increased continuously. In other words, single cells were rendered drug-dependent. Upon withdrawal of the drug, the cells responded by an overproduction of cAMP, that is, they exhibited some sort of withdrawal symptoms. This process is shown in Figure 16.32. It may be asked why the cells did not adapt by a down-

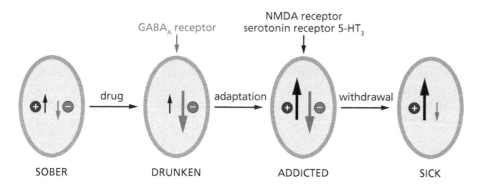

Figure 16.31 Tolerance and addiction: an adaptive process explained for alcohol as an example This strongly simplified scheme only serves to make clear drug addiction as a result of cellular and molecular adaptation. In the sober condition, excitatory (black) and inhibitory (red) signals compensate each other. Alcohol activates (among others) the ionotropic GABA$_A$ receptors, causing an increase of the inhibitory signal in the drunken state. Due to adaptation, this effect becomes matched by a corresponding increase of excitatory signaling, here exemplified by an expression of additional NMDA and serotonin 5-HT$_3$ receptors (many other possibilities are conceivable). Upon withdrawal of the drug, the inhibitory signal cannot balance out any longer the excitatory signal, resulting in cellular hyperactivity felt as withdrawal sickness.

regulation of MOP receptors. An answer may be found in a unique property of this receptor: namely, that adaptive endocytosis is induced only by the endogenous ligand and not by morphine.

Later, it was demonstrated that opiates, cocaine, and alcohol up-regulate cAMP signaling also in the limbic system. In addition to adenylate cyclase, PKA and its substrate proteins, including the transcription factor CREB, are also involved in this so-called **cAMP superactivation**, the molecular mechanism of which is not fully understood yet.

Although glutamate, serotonin, noradrenaline, and other transmitters also participate in drug dependency, research has been focused on **dopamine-induced signaling** because of its central role in the rewarding center and because it is stimulated by practically all addictive drugs, either directly or through adaptive mechanisms. Five different metabotropic dopamine receptors are known, two of which (D1 and D5) are coupled with G_s-proteins and the others with $G_{i,0}$-proteins. Drug addiction seems to be accompanied by almost irreversible changes of dopaminergic signal transmission. Such a **long-term adaptation** is particularly feared: it is the reason that, after successful withdrawal treatments, even a single contact with the drug may be sufficient to reconstitute dependency. This phenomenon is thought to be based on mechanisms resembling that of long-term memory fixation. In fact, in both cases similar changes of synaptic signal transmission caused by altered gene expression are observed. These include (apart from the above-mentioned cAMP superactivation) an overexpression of NMDA and AMPA receptors, neurotrophins, NO synthase, G-proteins, and tyrosine hydroxylase, the key enzyme of dopamine synthesis.

Recently, two transcription factors have moved into the focus of research: CREB and δFosB. **CREB**, a substrate of PKA (Section 8.6), is activated in the course of cAMP superactivation. The reader may recall the central role this factor plays in long-term potentiation of synaptic signaling (synaptic plasticity). **δFosB**, a truncated splice variant of the transcription factor FosB, is (in contrast to other Fos-type factors) metabolically extremely stable, being able to stimulate adaptive gene transcription even long after the drug has been withdrawn. The reason for δFosB stability is unknown. It might be the truncation of the C-terminal region

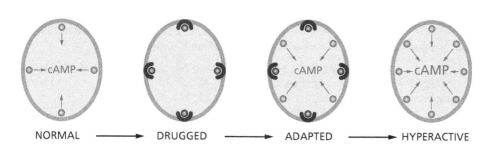

Figure 16.32 Drug-addicted cells From left to right: when stimulated by an input signal (not shown), the normal cell starts to produce cAMP due to an activated adenylate cyclase (red spheres). A drug (shown as black semicircles) blocks signal transduction and interrupts cAMP synthesis. The cell adapts to the new situation by expressing additional adenylate cyclase molecules, thus reestablishing the original state of signaling. Removal of the drug results in an overproduction of cAMP. This withdrawal symptom can be cured only by new drug administration.

and protein phosphorylation. In the dopaminergic neurons of the limbic system, the formation of the factor is induced by all sorts of drugs and other addictive stimuli but also by nonspecific stress. As indicated by experiments with dominant-negative mutants, an immediate relationship exists between δFosB expression and drug sensitivity. The role δFosB plays under normal conditions is not known. Possibly, it is used by the central nervous system to cope with chronic stress by learning an adaptive behavior. For survival it would be a tremendous advantage to remember this trained ability for life.

Summary

Nicotine interacts with ionotropic ("nicotinic") acetylcholine receptors both in the brain and in peripheral organs. Caffeine stimulates sympathetic effects as well as the central nervous system by inhibiting cAMP degradation, blocking sedative adenosine receptors, and activating ryanodine receptors (causing an increase of cytoplasmic Ca^{2+}). The sedative effects of ethyl alcohol are due to a blockade of excitatory glutamate receptors and an activation of inhibitory GABA receptors. Stimulatory effects of alcohol (and other sedative drugs such as opiates) are caused by an inhibition of inhibitory neurons. Most hallucinogenic drugs seem to activate more than one type of neurotransmitter receptor, particularly dopamine, serotonin, and noradrenaline receptors involved in the processing of sensory emotions. Stimulants such as cocaine exert their effects mainly by inhibiting the presynaptic re-uptake of dopamine, serotonin, and noradrenaline (synaptic signal extinction). The endogenous targets of many antipsychotics resemble those of hallucinogenic and stimulatory drugs. Drug addiction is a complex phenomenon based on both physiological and psychological effects. On the molecular level, continuous stimulation of a receptor-controlled signaling mechanism by a drug results in overactivation of opposing signaling. This becomes dominant upon drug withdrawal. The resulting adaptation is a long-term effect resembling long-term memory fixation. In fact, the cellular and molecular events involved in synaptic plasticity also seem to play a critical role in addiction.

Further reading

Ahern GP, Klyachko VA & Jackson MB (2002) cGMP and S-nitrosylation: two routes for modulation of neuronal excitability by NO. *Trends Neurosci.* 25, 510–517.

Albright TA, Jessell TM, Kandel ER et al. (2000) Neural science: a century of progress and the mysteries that remain. *Cell Rev. Suppl.* 100, S1–S55.

Ashcroft FM (2006) From molecule to malady. *Nature* 440, 440–447.

Auerbach A (2003) Life at the top: the transition state of AchR gating. *Sci. STKE* 2003/188/re11.

Baranano DE, Ferris CD & Snyder SH (2001) Atypical neural messengers. *Trends Neurosci.* 24, 99–106.

Baranano DE & Snyder SH (2001) Neural roles for heme oxygenase; contrast to nitric oxide synthase. *Proc. Natl Acad. Sci. U.S.A.* 98, 10996–11002.

Berg K & Dani JA (eds) (2003) Nicotinic signaling. *J. Neurobiol.* 53, No. 4 (special issue).

Bezakova G (2002) Agrin: architect of the synapse. *ELSO Gazette.* http://www.the-elso-gazette.org/magazines/issue 10/reviews/review 1.asp.

Billinton A, Ige AO, Bolam P et al. (2001) Advances in the molecular understanding of $GABA_B$ receptors. *Trends Neurosci.* 24, 277–285.

Bogdan C (2001) Nitric oxide and the regulation of gene expression. *Trends Cell Biol.* 11, 66–75.

Boison D (2006) Adenosine kinase, epilepsy and stroke: mechanisms and therapies. *Trends Pharmacol. Sci.* 27, 652–658.

Buisson B & Bertrand D (2002) Nicotine addiction: the possible role of functional upregulation. *Trends Pharmacol. Sci.* 23, 130–136.

Burnstock G (2007) Physiology and pathophysiology of purinergic neurotransmission. *Physiol. Rev.* 87, 659–797.

Busse R, Edwards G, Feletou M et al. (2002) EDHF: bringing the concepts together. *Trends Pharmacol. Sci.* 23, 374–380.

Calabrese V, Mancuso C, Calvani M et al. (2007) Nitric oxide in the central nervous system: neuroprotection versus neurotoxicity. *Nat. Rev. Neurosci.* 8, 766–775.

Carroll RC & Zukin RS (2002) NMDA-receptor trafficking and targeting: implications for synaptic transmission and plasticity. *Trends Neurosci.* 25, 571–577.

Changeux J-P (1993) Chemical signaling in the brain. *Sci. Am.* 269, 29–37.

Cryan JF & Kaupmann K (2005) Don't worry, Be happy!: a role for GABA$_B$ receptors in anxiety and depression. *Trends Pharmacol. Sci.* 26, 36–42.

Dalva MB, McClelland AC & Kayser MS (2007) Cell adhesion molecules: signalling functions at the synapse. *Nat. Rev. Neurosci.* 8, 206–220.

Davies PA, Wang W, Hales TG et al. (2003) A novel class of ligand-gated ion channels is activated by Zn^{2+}. *J. Biol. Chem.* 278, 712–717.

Davis KL, Martin E, Turku IV et al. (2001) Novel effects of nitric oxide. *Annu. Rev. Pharmacol. Toxicol.* 41, 203–236.

Dev KK & Henley JM (2006) The schizophrenic faces of PICK1. *Trends Pharmacol. Sci.* 27, 574–579.

Frerking M (2004) When astrocytes signal, kainate receptors respond. *Proc. Natl Acad. Sci. U.S.A.* 101, 2649–2650.

Freund TF, Katoma I & Piomelli D (2003) Role of endogenous cannabinoids in synaptic signaling. *Physiol. Rev.* 83, 1017–1066.

Gaiarsa JL, Caillard O & Ben-Ari Y (2002) Long-term plasticity at GABAergic and glycinergic synapses: mechanisms and functional significance. *Trends Neurosci.* 25, 564–570.

Gainetdinov RR & Caron MG (2003) Monoamine transporters: from genes to behavior. *Annu. Rev. Pharmacol. Toxicol.* 43, 261–284.

Gether U, Andersen PH, Larsson OM et al. (2006) Neurotransmitter transporters: molecular function of important drug targets. *Trends Pharmacol. Sci.* 27, 376–383.

Glebova NO & Ginty DD (2005) Growth and survival signals controlling sympathetic nervous system development. *Annu. Rev. Neurosci.* 28, 191–222.

Gundelfinger ED, Boeckers TM, Baron MM et al. (2006) A role for zinc in postsynaptic density assembly and plasticity? *Trends Biochem. Sci.* 31, 366–373.

Hara MR & Snyder SH (2007) Cell signaling and neuronal death. *Annu. Rev. Pharmacol. Toxicol.* 47, 117–141.

Hardingham GE & Bading H (2003) The Yin and Yang of NMDA receptor signaling. *Trends Neurosci.* 26, 81–88.

Harris RA & Mihic SJ (2004) Alcohol and inhibitory receptors: unexpected specificity from a nonspecific drug. *Proc. Natl Acad. Sci. U.S.A.* 101, 2–3.

Hess DT, Matsumoto A, Kim SO et al. (2005) Protein-S-nitrosylation: purview and parameters. *Nat. Rev. Mol. Cell Biol.* 6, 150–166.

Hoeffer CA & Klann E (2007) Switching gears: translational mastery of transcription during memory fixation. *Neuron* 54, 186–188.

Hogg RC, Raggenbass M & Bertrand D (2003) Nicotinic acetylcholine receptors: from structure to brain function. *Rev. Physiol. Biochem. Pharmacol.* 147, 1–46.

Jingami H, Nakanishi S & Morikawa K (2003) Structure of the metabotropic glutamate receptor. *Curr. Opin. Neurobiol.* 13, 271–278.

Kalamida D, Poulas K, Avramopoulou V et al. (2007) Muscle and neuronal nicotinic acetylcholine receptors. Structure, function and pathogenicity. *FEBS J.* 274, 3799–3845.

Kauer JA & Malenka RC (2007) Synaptic plasticity and addiction. *Nat. Rev. Neurosci.* 8, 844–858.

Kempermann G (2006) Adult neurogenesis. Oxford University Press.

Kenny PJ & Markou A (2004) The ups and downs of addiction: role of metabotropic glutamate receptors. *Trends Pharmacol. Sci.* 25, 267–272.

Lerma J (2003) Roles and rules of kainate receptors in synaptic transmission. *Nat. Rev. Neurosci.* 4, 481–495.

Li YV, Hough CJ & Sarvey JN (2003) Do we need zinc to think? *Sci. STKE* 2003/182/pe19.

Li Z & Sheng M (2003) Some assembly required: the development of neuronal synapses. *Nat. Rev. Cell Biol.* 4, 833–841.

Linden J (2001) Molecular approach to adenosine receptors: receptor-mediated mechanisms of tissue protection. *Annu. Rev. Pharmacol. Toxicol.* 41, 775–787.

Lucas KA, Pitari GM, Kazerounian S et al. (2000) Guanylyl cyclases and signaling by cyclic GMP. *Pharmacol. Rev.* 52, 375–413.

Lüscher C & Frerking M (2001) Restless AMPA receptors: implications for synaptic transmission and plasticity. *Trends Neurosci.* 24, 665–671.

Lynch JW (2004) Molecular structure and function of the glycine receptor chloride channel. *Physiol. Rev.* 84, 1051–1095.

Mackie K & Stella N (2006) Cannabinoid receptors and endocannabinoids: evidence for new players. *AAPS J.* 8, E298–E306.

Mayer ML (2006) Glutamate receptors at atomic resolution. *Nature* 440, 456–462.

Mayford M (2007) Protein kinase signaling in synaptic plasticity and memory. *Curr. Opin. Neurobiol.* 17, 313–317.

Manji HK, Gottesman II & Gould TD (2003) Signal transduction and genes – to behaviors pathways in psychiatric diseases. *Sci. STKE* 2003/207/pe49.

McClung CA, Ulery PG, Perrotti LI et al. (2004) ΔFosB: a molecular switch for long-term adaptation in the brain. *Mol. Brain Res.* 132, 146–154.

Miller FD & Kaplan DR (2003) Signaling mechanisms underlying dendrite formation. *Curr. Opin. Neurobiol.* 13, 391–398.

Mody I & Pearce RA (2004) Diversity of inhibitory neurotransmission through GABA$_A$ receptors. *Trends Neurosci.* 27, 569–574.

Münzel T, Feil R, Mülsch A et al. (2003) Physiology and pathophysiology of vascular signaling controlled by cyclic guanosine 3′,5′-cyclic monophosphate-dependent protein kinase. *Circulation* 108, 2127–2138.

Nestler EJ, Barrot M & Self DW (2001) ΔFosB: a sustained molecular switch for addiction. *Proc. Natl. Acad. Sci. U.S.A.* 98, 11042–11046.

Nestler EJ (2004) Historical review: molecular and cellular mechanisms of opiate and cocaine addiction. *Trends Pharmacool. Sci.* 25, 210–218.

Newman EA (2002) New roles for astrocytes: regulation of synaptic transmission. *Trends Neurosci.* 26, 536–542.

Passani MB, Lin JS, Hancock A et al. (2004) The histamine H_3 receptor as a novel therapeutic target for cognitive and sleep disorders. *Trends Pharmacol. Sci.* 25, 618–625.

Pilz RB & Casteel DE (2003) Regulation of gene expression by cyclic AMP. *Circ. Res.* 93, 1034–1046.

Robinson MB (2003) Signaling pathways take aim at neurotransmitter transporters. *Sci. STKE* 2003/207/pe50.

Ryter SW, Morse D & Choi AMK (2003) Carbon monoxide: to boldly go where NO has gone before. *Sci. STKE* 2004/230/re6.

Schlossmann J, Feil R & Hofmann F (2003) Signaling through NO and cGMP-dependent protein kinases. *Ann. Med.* 35, 21–27.

Seal RP & Amara SG (1999) Excitatory amino acid transporters: a family in flux. *Annu. Rev. Pharmacol. Toxicol.* 39, 431–456.

Shibasaki F, Hallin U & Uchino H (2002) Calcineurin as a multifunctional regulator. *J. Biochem.* 131, 1–15.

Sine SM & Engel AG (2006) Recent advances in Cys-loop receptor structure and function. *Nature* 440, 448–455.

Snyder SH & Paternak GW (2003) Historical review: opioid receptors. *Trends Pharmacol. Sci.* 24, 198–205.

Snyder SH (1986) Drugs and the brain. New York: Scientific American Books.

Soderling TR & Derkach VA (2000) Postsynaptic protein phosphorylation and LTP. *Trends Neurosci.* 23, 25–29.

Song I & Huganir RL (2002) Regulation of AMPA receptors during synaptic plasticity. *Trends Neurosci.* 25, 578–588.

Straiker A & Mackie K (2006) Cannabinoids, electrophysiology, and retrograde messengers: challenges for the next 5 years. *AAPS J.* 8, E272–E276.

Sutton MA & Schuman EM (2006) Dendritic protein synthesis, synaptic plasticity, and memory. *Cell* 127, 49–58.

Swope SL, Mosse SJ, Raymond LA et al. (1999) Regulation of ligand-gated ion channels by protein phosphorylation. *Adv. Second Messenger Phosphoprotein Res.* 33, 49–78.

Vial C, Roberts JA & Evans RJ (2004) Molecular properties of ATP-gated P2X receptor ion channels. *Trends Pharmacol. Sci.* 25, 487–493.

von Zastrow M, Svingos A, Haberstock-Debic H et al. (2003) Regulated endocytosis of opioid receptors: cellular mechanisms and proposed roles in physiological adaptation to opiate drugs. *Curr. Opin. Neurobiol.* 13, 348–353.

Waites CL, Craig AM & Garner CC (2005) Mechanisms of vertebrate synaptogenesis. *Annu. Rev. Neurosci.* 28, 251–274.

Waldhoer M, Bartlett SF & Whistler JL (2004) Opioid receptors. *Annu. Rev. Biochem.* 73, 953–990.

Watanabe M, Maemura K, Kanbara K et al. (2002) GABA and GABA receptors in the central nervous system and other organs. *Int. Rev. Cytol.* 213, 1–47.

Wedel BJ & Garbers DL (2001) The guanylyl cyclase family at Y2K. *Annu. Rev. Physiol.* 63, 215–233.

Putting Together the Pieces: The Approach of Systems Biology

A central theme of this book is that a signal, even when it interacts with only one receptive molecule, in a cell evokes a complex excitation pattern rather than opening a molecular one-way street. Moreover, the biological meaning of a signal, or its physiological function, cannot be found in its structure but is attributed exclusively to the recipient. In other words: meaning always depends on the recipient's inherited and learned program and the environmental context. We have tried to make this clear by introducing the metaphor of the "brain of the cell."

A signal-induced excitation pattern resembles a data-processing apparatus at work. It is generated by a system of interacting switching elements, neurons in the brain, and proteins (and nucleic acids) in the cell. Understanding the circuit pattern and dynamics of such a data-processing system (and other complex systems of interconnected entities) and predicting their behavior has become a major research project. This is the task of systems biology, a new branch of life sciences.

A system is understood as an organized assemblage composed of interacting components with identifiable function. This is identical with what has been called a network in Chapter 1. Societies, organisms, cells, proteins, and genes represent biological systems at different scales, while computers and other machines are technical systems.

In the past, the lack of suitable methods prevented any scientifically sound access to the problems of biological systems in spite of so much talking about cross talk. To circumvent this dilemma, molecular and cell biologists traditionally were satisfied by collecting and describing hard facts. This has been done, as we have seen, with tremendous success; it is tempting to make the erroneous assumption that the properties of complex systems, such as protein molecules and cells, could be explained by identifying the parts. Such a conclusion disregards Aristotle's dictum "the whole is greater than the sum of its parts" or, in modern terminology, that it is a fundamental property of complex systems to develop in an unpredictable way emergent functions that are not found in the individual components. This makes it impossible to predict something like mitosis, the behavior of a flock of birds, or a Bach cantata, even if we would possess a complete list of all biochemical reactions or all neuronal interactions. After all, mitosis, bird swarming, and the cantata are the results of a system's emergent properties. These examples clearly demonstrate that in biology the functions of the systems are the facts that finally count. Even though systems biology is still in its infancy, a brief outline of its principles, current methods, and inherent problems is given in the following.

17.1 Systems biology: origin and focus

By combining the traditional reductionist methodology with a holistic approach, systems biology aims at an understanding of complex systems in a quantitative and predictable way. Systems biology has its origin in both theoretical biology and engineering science. Already in the early 1960s, the Austrian-American biologist Ludwig von Bertalanffy formulated a "general system theory" emphasizing that all living systems share "the common properties of being composed of interlinked components, in which case they might share similarities in detailed structure and control design." Fifteen years earlier the American mathematician Norbert Wiener, the founding father of cybernetics, had developed a theory of control mechanisms stressing the importance of negative feedback for maintaining stability of technical systems.

Rapid progress in cell and molecular biology, culminating in the characterization of macromolecular structures as well as in the development of recombinant and high-throughput techniques that allow the genome-wide census of components and interactions (genomics, proteomics, transcriptomics, metabolomics, and other "omics"), has provided a basis for the experimental analysis of biological systems. This analysis may proceed in two directions: from the parts to the whole (bottom-up) or from the whole to the parts (top-down). In the **bottom-up approach** one tries to deduce the emergent properties of a system from detailed knowledge of subsystems and molecular interactions. The **top-down approach** is used to identify the molecular networks and mechanisms that are thought to underlie emergent properties of the system. This approach is feasible when one possesses a list of components generated – for instance, by an "omics" census – but lacks a detailed knowledge of their functions. Since both the bottom-up and top-down approaches necessarily complement each other, it has been proposed to combine them in what has been called a "middle-out approach." However, this is hardly more than semantics, saying what has been known for centuries: namely, that the progress of science, including biology, depends on a continuous interplay between inductive and deductive approaches.

A further point deserves to be addressed. The theories of evolution, developmental biology, genetics, and non-equilibrium thermodynamics say that life is the result of a unique historical process. This means that biological systems must have an extremely high degree of individuality in their details, a fact that from the beginning excludes any too far-reaching generalization of statements, whereas on the other hand it calls for a strict standardization of experimental approaches.

Summary

New techniques allowing a system-wide census of cellular components and the dynamics of their interactions provide the conditions for a systematic analysis of biochemical networks. This is achieved by a combination of deductive (bottom-up) and inductive (top-down) approaches.

17.2 System structure: basic network topologies and properties

Customary methods of molecular cell biology as they have been employed long before the dawn of systems biology strictly follow a bottom-up approach. A typical example is the gene knockout technique, by which one tries to find out the systemic consequences at the "top" of a single perturbation at the molecular "bottom." Modern systems biology tries to improve the bottom-up approach aiming at the generation of more quantitative data.

Top-down biology has received a strong impetus with the introduction of the "omics" based on the development of high-throughput techniques, which allow the genome-wide identification of all sorts of biomolecules, their expression patterns, and their interactions. An example is the identification of the sour-taste receptor described in Section 15.1.1. An inherent problem of top-down systems biology is that it starts with an unwieldy mass of data. Even though the data thus obtained may still be incomplete and afflicted by errors, they can be used to construct **networks**, or maps of interactions. Representing a system in a highly simplified and abstract form, albeit based on assumptions and approximations, a network may help to uncover general design principles and related control patterns.

A network consists of **nodes** that are interconnected by **links** or edges. In biological networks, nodes may symbolize proteins or genes, and links are the interactions between them. Depending on the type of interactions, one distinguishes between undirected and directed networks. Protein–protein interactions, for example, are usually represented by undirected networks, whereas transcription, signal transduction, and most metabolic reactions are targeted events depicted by directed networks. Here the nodes are connected by incoming and outgoing links.

Today, most networks still resemble street maps with the interactions described only qualitatively. Of course, for the future a quantitative assessment is indispensable: this would take into account also the strengths of interactions (resembling traffic types and densities). It is easily seen that this goal can be achieved only by a combination of top-down and bottom-up biology.

However, even the assembly of networks requires considerable effort and a combined application of different experimental techniques. The construction of protein interaction networks focusing on the assembly of protein complexes may give an idea of the problems inherent in such an approach, particularly where the faithful proof of interactions is concerned. In the beginning such networks were based on data obtained by using the yeast two-hybrid method. The value of this technique is limited, however, by false-positive results (Section 5.2). Later on, high-affinity purification followed by mass spectrometry was employed to characterize protein complexes. However, here the disadvantage is that binding partners in substoichiometric concentrations are not detected. The data sets generated with both techniques overlap only partially, but together they provide useful information regarding the global topology of protein–protein interaction maps.

17.2.1 Network parameters and nomenclature

Like a map, a network is described theoretically by a set of parameters:

- the **degree**, k (connectivity), gives the number of links connected to a node; in directed networks one distinguishes between incoming and outgoing degrees
- the **degree distribution**, $P(k)$, describes the probability that a node has k links
- the **distance** is the shortest path length between two nodes
- the **diameter** is the maximum distance between any two nodes in a network
- the **clustering coefficient** gives the percentage of existing links in the neighborhood of a node
- the **betweenness** is understood as the fraction of shortest paths between all pairs of nodes that pass through one node

Networks with a small diameter, or short paths between the nodes, are frequently termed "small-world" networks, referring to a social study where such properties were discovered for the first time. Neural and metabolic networks, for instance, may exhibit a small-world topology, indicating a minimal transition time between the different states. In such networks a local effect spreads rapidly.

Figure 17.1 Two types of networks Networks are described by the number of nodes, the number of links connected to a node (degree k), and the probability $P(k)$ of connections with a given k value (degree distribution). For random networks, P shows a bell-shaped Gaussian distribution when plotted against k, indicating that on average each node has a similar number of links (in our example, 2–4). Scale-free networks show a small number of high-degree nodes and a power-law distribution of links connected to a node that results in a straight line when P is plotted against k on a log–log scale. Regular nodes are indicated in red and high-degree nodes (hubs) are shown in gray.

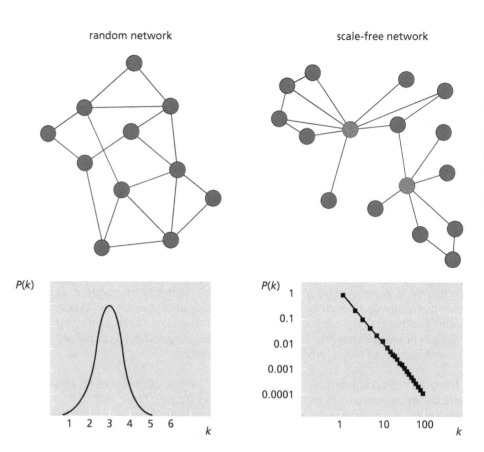

The theory predicts a few principles that dominate most networks despite their obvious diversity. A major principle concerns the quality of the nodes, which allows a distinction between two types of network organization. In a **random network**, on average each node has a similar number of links, whereas in a **scale-free network**, nodes with many connections, called hubs, are surrounded by a large number of nodes with only one or a few connections (Figure 17.1). The analysis of biological networks reveals that most of them are organized in a scale-free rather than in a random manner. For a protein network, a scale-free topology means that most proteins interact with only a few partners, whereas a few proteins exhibit a high number of interactions. The advantage of a scale-free organization is obvious: it renders the system more robust because a loss of one of the many non-hub nodes is less disruptive than in a random network.

A characteristic feature of both street maps and biological networks is a hierarchical organization: lower-degree networks are the components of higher networks. Moreover, even within a network some kind of hierarchy is found, becoming visible in the graphs as groups of highly interconnected nodes, known as clusters or modules. Nevertheless, network topology and organization resemble each other at each level.

Despite their complexity, biological networks contain a limited set of recurring regulatory elements or network motifs, representing interactions that facilitate the progress and control of biological functions, for instance by positive or negative feedback loops. Examples of such subgraphs are shown in Figure 17.2. One easily recognizes that such network motifs resemble the biochemical signaling patterns described in Section 1.9.

17.2.2 Signal-processing networks

Traditionally, signal-processing systems have been analyzed by the bottom-up approach. Only recently has the top-down approach entered the field, and the

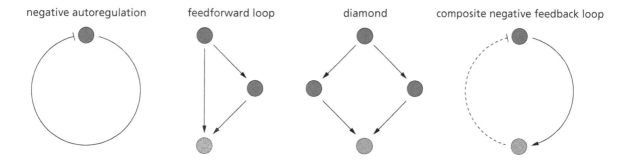

negative autoregulation   feedforward loop   diamond   composite negative feedback loop

results are still rather sparse. To handle the problem of unwieldy data sets, which are typical for this approach, one divides the network into several subnetworks. Of course, this is an artificial subdivision, since in reality the subnetworks are fully interconnected in a hierarchically organized supernetwork. One of the signaling subnetworks describes the interactions of transcription factors and genes. These **gene regulatory networks** are considered to be slow because input signals change the activity of transcription factors in a matter of seconds, whereas the output signals consisting of gene transcription, RNA processing, and accumulation of protein products may develop in hours. Despite variable quality of the data upon which the networks are based (obtained by experiments with model organisms and by large-scale analysis of transcription factor binding sites), interesting and meaningful results were obtained. These allowed predictions that afterward could be verified. An example is the global identification of targets of factors controlling embryonic stem cell maintenance, which has suggested novel pathways of stem cell self-renewal.

Much faster are **signal transduction networks**, which operate on the time scale of seconds to minutes. In these networks, signaling proteins represent the nodes and the edges reflect direct noncovalent interactions or covalent post-translational modifications such as protein phosphorylation.

The most rapid signal processing is observed in **neuronal networks**, where the response time is in the range of milliseconds. In these networks, nodes symbolize neurons and edges represent synaptic connections.

A major function of signal processing networks is to control **metabolic networks**, which combine information on proteins (enzymes) and metabolites and focus on the mass flow of chemical compounds such as amino acids, sugars, lipids, and energy required for the biochemical reactions. To examine the dynamics of these networks, a flux balance analysis is performed, based on the assumption that all metabolites exist in steady-state concentrations and that an organism optimizes the flux of metabolites to optimize biomass production. Computerized flux models have been successfully used to predict the phenotype of mutant yeast strains grown under different media conditions. These studies revealed that the majority of nonessential enzymes in yeast are required for growth under certain specialized conditions and that only a small subset of these enzymes is compensated by isoenzymes or parallel pathways.

Summary

A major goal of current systems biology is to construct maps of interactions. In such networks, components are symbolized by nodes and interactions by links. Networks are described by a set of parameters that characterize the quality of nodes and links. Most biological networks are scale-free: a limited number of nodes with many links (called hubs) are mixed with a large number of nodes with only a few links. As compared with a random network (in which, on average, each node has the same number of links), a scale-free network is more resistant to disturbances. For the sake of clarity, signal-processing networks

Figure 17.2 Recurring motifs in signaling networks A frequent network motif observed in signaling networks is negative autoregulation. Feedforward loops occur in gene regulatory and signal transduction networks, whereas a diamond relationship is observed only in signal transduction networks and can be combined in multiple layers to carry out functions on multiple input signals. Composite network motifs combine different types of interactions. Most common is a composite negative feedback loop in which one arm represents a transcriptional interaction (arrow) and the other arm represents an inhibitory protein–protein interaction (broken line). The separation of time scales between the slow transcription arm and the faster protein–protein arm helps to stabilize the dynamics of the composite loop.

may be subdivided in gene-regulatory networks, signal-transduction networks, and neuronal networks. They all communicate with metabolic networks.

17.3 The iterative cycle: laboratory experiments and model building

Scientific work may be described as an interplay between imagination and experiment. On the basis of available knowledge potentially contradictory hypotheses are formulated and experiments are designed to falsify them in an iterative process. This holds true also for systems biology , which depends on a collaboration between theoreticians and experimenters (Figure 17.3). The analysis of the system is particularly supported by mathematical models including computer simulations, which embed the working hypotheses and are designed to describe the experimental data. As a result, new predictions are made that, in turn, can be validated or falsified experimentally, leading to modified hypotheses that are examined again in the iterative cycle. This process is repeated until the theory becomes as congruent as possible with reality. Thus, by combining quantitative data generation, mathematical modeling, and computer simulations, a systematic and hypothesis-driven analysis of complex biological systems, which is not feasible with the traditional methods of molecular cell biology, becomes possible. To reach this goal, biologists must become familiar with the principles of mathematical approaches, while mathematicians must make some effort to arrive at an understanding of biological problems.

Some future aims of such approaches are a prediction of new therapeutic targets and the development of synthetic systems with desired properties obtained by artificially modifying an existing system or constructing a new system (synthetic biology).

Early investigators who applied a mathematical approach were the neurophysiologists A. L. Hodgkin and A. F. Huxley (1952), who put forward a model of the neuronal action potential, and Denis Nobel (1960), who developed the first computer model of a beating heart.

Summary

The task of systems biology is investigation of the structure and dynamics of complex networks. The goal is to explain the emergence of network properties

Figure 17.3 Systems biology approaches consist of an iterative process between laboratory experiments and mathematical modeling In a collaboration between experimenters and theoreticians, contradictory issues in biology are answered in an iterative cycle of quantitative data generation, mathematical modeling, *in silico* predictions, experimental validation, and design of new experiments. Advancement of research in computational science, analytical methods, technologies for measurements, and improved high-throughput methodologies will help to gradually transform biological research into a more systematic and hypothesis-driven science, as represented by this iterative cycle process. (Adapted from H Kitano, *Science* 295:1662–1664, 2002.)

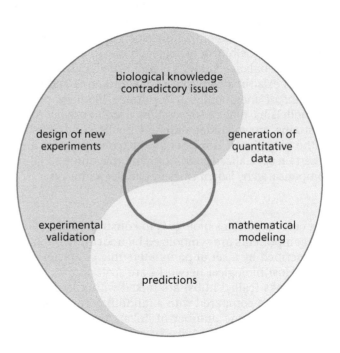

that cannot be deduced from the properties of the single components. The approach consists of combination of experimental data with mathematical models and computer simulations, requiring a collaboration of experimenters and theoreticians.

17.4 Problems of quantitative data generation and model building

Since a system is more than an assembly of its components and of the linkages between them, its properties cannot be understood solely on the basis of qualitative data. Rather, it is essential to consider the strength of the interactions and their dynamic behavior, since otherwise the construction of networks would be a descriptive rather than an explanatory venture.

To arrive at an in-depth understanding of a system, the following must be known:

- its **structure** involving the components, their mode of interaction, and mechanisms regulating these interactions

- its **dynamics**, or the behavior over time and the role of dynamic motifs such as gradual response, oscillation, adaptation, feedback control, and switchlike behavior

- its **control** by key mechanisms based on conserved design principles and dominating the state of the biological network

It must be emphasized, however, that it is still extremely difficult to measure kinetic parameters of signal-processing reactions in such a way that they would reflect the situation *in vivo*. Thus, most of the available data on signal processing are qualitative, and in many cases the temporal resolution of the experimental techniques is not high enough to obtain detailed information on process dynamics. Nevertheless, the oscillatory behavior of signaling reactions repeatedly postulated in this book is becoming evident more and more with the development of novel methods. Moreover, strong efforts are made to improve and standardize the generation of high-throughput transcriptional data (the transcriptome). This is an important prerequisite to establish mathematical models that connect signal transduction networks with gene regulatory networks and permit quantitative predictions.

Another problem of quantitative systems biology is that several key parameters of cellular signaling can be studied only in a single cell. One example is provided by switchlike responses (Figure 17.4); another by oscillatory processes, which in

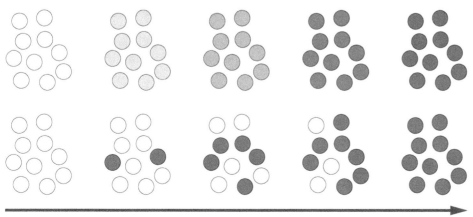

increasing stimulus

Figure 17.4 Gradual versus switchlike response Increasing stimulus concentrations may trigger a gradually increasing effect (top row), such as the expression of a distinct protein in a cell population. Alternatively, an all-or-nothing effect may be switched on in an increasing number of cells (bottom row). Upon averaging over the cell population, the two response types cannot be distinguished from each other. However, at the single cell level, the differences become evident.

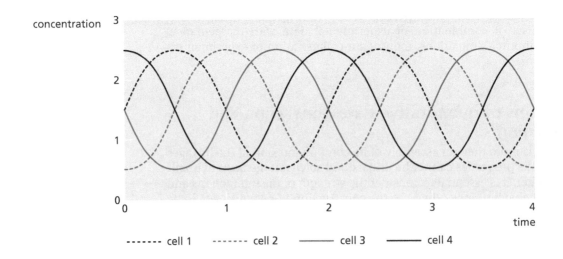

Figure 17.5 Oscillatory response
Upon averaging over the cell population, the concentration of a protein of interest appears constant if the cells are not in phase. However, detection at the single cell level reveals oscillatory behavior.

a cell population are masked unless the cells are in phase (Figure 17.5). Therefore, strong efforts are being made to improve analysis at the single cell level. At present, particular emphasis is laid on live cell imaging techniques monitoring proteins labeled with fluorescent probes such as green fluorescent protein (Section 5.2).

Finally, our knowledge of signaling pathways and gene regulatory networks is based mainly on artificial systems, in particular on cell lines that considerably deviate from normality as far as control mechanisms and genetic stability are concerned. Therefore, to yield reproducible results for model building, it is mandatory to standardize the cell type, cell culture conditions, and experimental techniques rather than to derive data from the literature.

Summary

Without quantitative data evaluation, networks remain entirely descriptive. Therefore, the progress of systems biology critically depends on the improvement and standardization of analytical methods and sample-taking as well as techniques of computerized data processing.

17.5 Mathematical modeling of a signaling pathway

As stated above, the approaches of systems biology are based on the combination of laboratory with computer work. Mathematical models of systems are abstractions aiming at the discovery of more universal or generic principles. If the data set is sparse, statistical models can be developed. Provided enough experimental data of high quality are available, the dynamic behavior and design principles of biological systems can be simulated by kinetic (reaction) models of covalent modifications, intermolecular association, and intracellular localization. Mathematical equations describe processes such as catalysis and assembly, and the parameters represent concentrations, binding affinities, and reaction rates. A **deterministic** representation considers bulk concentrations of pathway components, not individual molecules, and assumes the cell to be a well-stirred reactor. **Stochastic** models are applied for reactants that are present in small concentrations and include effects arising from random fluctuations around an average behavior.

In the majority of present-day mathematical models, changes in the concentration of reactants over time are described by coupled ordinary differential equations. This model structure is chosen to evaluate the internal structure (intermediary steps) lying between input and output of the system. Overall orientation is given by maps that qualitatively depict the connections between indi-

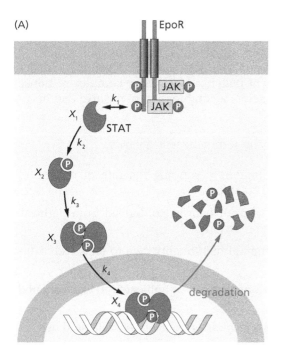
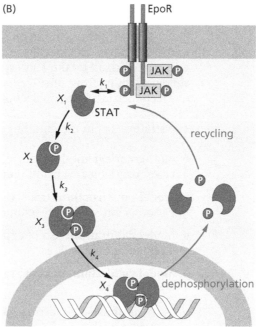

(C)

(1) unphosphorylated STAT5

$$\dot{x}_1(t) = -k_1 x_1 EpoR_A$$

(2) tyrosine-phosphorylated STAT5 monomer

$$\dot{x}_2(t) = -k_2 x_2^2 + k_1 x_1 EpoR_A$$

(3) tyrosine-phosphorylated STAT5 dimer

$$\dot{x}_3(t) = -k_3 x_3 + \tfrac{1}{2} k_2 x_2^2$$

(4) tyrosine-phosphorylated STAT5 dimer
in the nucleus

$$\dot{x}_4(t) = \qquad + k_3 x_3$$

(1) unphosphorylated STAT5

$$\dot{x}_1(t) = -k_1 x_1 EpoR_A + \boxed{2 k_4 x_3^\tau}$$

(2) tyrosine-phosphorylated STAT5 monomer

$$\dot{x}_2(t) = k_2 x_2^2 + k_1 x_1 EpoR_A$$

(3) tyrosine-phosphorylated STAT5 dimer

$$\dot{x}_3(t) = -k_3 x_3 + \tfrac{1}{2} k_2 x_2^2$$

(4) tyrosine-phosphorylated STAT5 dimer
in the nucleus

$$\dot{x}_4(t) = \boxed{-k_4 x_4^\tau} + k_3 x_3$$

Figure 17.6 Mathematical treatment of two conflicting hypotheses on signal transduction in the JAK–STAT cascade (A) Linear signal transmission by a feedforward cascade from the cell surface to the nucleus and signal termination by STAT5 degradation in the nucleus (one-way model). (B) For signal transductions by iterative cycles, STAT5 is dephosphorylated in the nucleus and returns to the cytoplasm to re-enter subsequent activation cycles (cycling model). (C) The STAT5 populations described by coupled ordinary differential equations are labeled in panels A and B by the concentrations x_1, x_2, x_3, and x_4. The dynamic parameters are indicated by rate constants k_1, k_2, k_3, and k_4 and exponent τ, the delay term introduced for the cycling model. $x(t)$ is equal to the differential quotient dx/dt. EpoR, erythropoietin receptor; P, phosphate residue; JAK, Janus kinase; STAT, signal transducer and activator of transcription.

vidual components without providing information about their dynamic behavior (Figure 17.6). Traditionally, such maps have been used to illustrate signaling pathways. They are found throughout this book. Such illustrations can be translated into a set of mathematical equations.

For this purpose, some basic facts regarding the kinetics of molecular interactions are taken into account. For instance, in a first approximation the noncovalent dimerization or oligomerization of proteins frequently observed in signaling pathways can be considered to follow the law of mass action (Sidebar 17.1), resulting in a **linear response** curve.

Enzymatic reactions such as phosphorylation/dephosphorylation or protein synthesis/degradation in the simplest case exhibit the **hyperbolic response** curves of Michaelis–Menten kinetics (Sidebar 17.1). **Sigmoidal response** curves indicating positive cooperativity finally are the result of allosteric interactions of proteins with multiple subunits or domains (Section 1.3). This enables the protein to maximally respond to small changes in ligand concentration, resulting in a switchlike behavior, while on the other hand it provides an effective noise filter. A well-known example for allosteric regulation is the binding of cyclic AMP to the inhibitory subunit of protein kinase A (Section 4.6.1). Sigmoidal kinetics are characterized by the Hill exponent (Sidebar 17.1). More complex response curves resulting from positive feedback or a combination of positive and negative feedback are found in Section 1.9.

The strategies of model building may be summarized as follows:

- **Mathematical model:** sets of equations describing the rate of change of system variables (x) by differential equations

- **Model construction:** translation of prior knowledge derived from pathway cartoons into a list of reactants and reactions

- **Model calibration:** use of experimental high-quality data to effectively constrain parameter values

- **Model validation:** making predictions that can be subjected to experimental tests

- **Systems analysis:** inferring systems properties from a set of validated mathematical models

In the following, the strategy of model building is explained with the help of a distinct signal transducing event, the JAK–STAT pathway, which is one component of the complex signaling system activated by cytokine receptors. As shown in Section 7.2.1, the transcription factors of the Signal Transducer and Activator of Transcription (STAT) family dimerize upon phosphorylation by the cytokine receptor-associated tyrosine kinase JAnus Kinase (JAK) and then migrate into the nucleus, where they bind to the promoter sequences of target genes (in our example the isoform STAT5 is activated by the erythropoietin receptor, EpoR). Signal termination may be assumed to be due either to dephosphorylation, allowing STAT to recycle into the cytoplasm (cycling model), or to ubiquitylation/proteolytic degradation of STAT (one-way model). To decide between these two possibilities, a set of coupled ordinary differential equations is formulated that describes the changes over time in the concentration of unphosphorylated as well as of Tyr-phosphorylated monomeric and dimeric STAT5, both in the cytoplasm and in the nucleus. To capture the cycling behavior, the model is extended by a delay term accounting for the time it takes the dephosphorylated STAT5 to return to the cytoplasm.

In the beginning of the operation, the ligand-induced phosphorylation of EpoR is measured and used as the input function for the mathematical model. In the next step, unknown parameters must be estimated to arrive at an optimal fit between the mathematical simulation and experimental data. This **parameter estimation** can be done by comparing time series values provided by a parameterized model with measured data via the mean square distance. When the estimated parameter values are changed, this distance increases or decreases. The parameter set that leads to a global minimum of the mean square distance is referred to as the least square estimate of the model parameters. For the example of JAK–STAT signaling, a parameter set can be estimated by fitting the cycling model only, but not the one-way model (Figure 17.7).

The dynamic behavior of most signaling pathways is determined by negative feedback loops, resulting in adaptation, or by a combination of negative and positive feedback loops, resulting in an oscillatory response or pattern formation (Section 1.9). To systematically identify parameters critically determining such dynamic behavior of a system and to predict the consequence changes have on signaling output, a **sensitivity analysis** can be performed. This mathematical

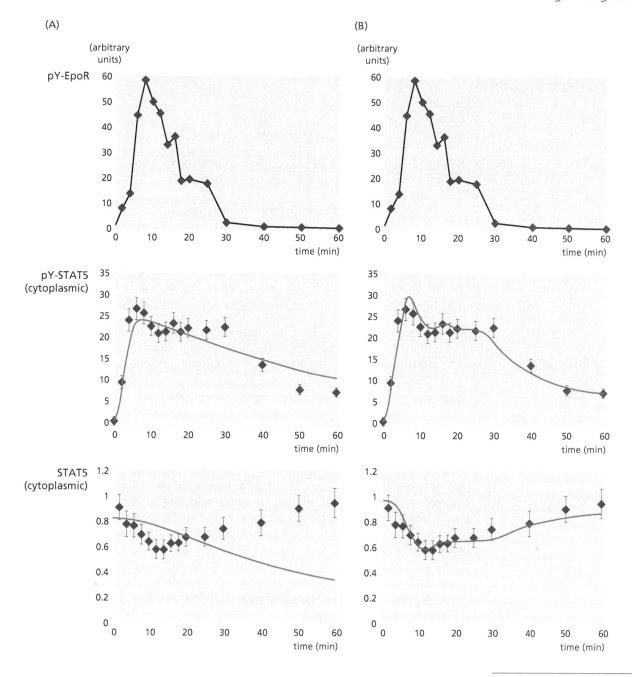

(A)

(B)

Figure 17.7 Mathematical modeling of the dynamic behavior of STAT5
The time course of ligand-induced phosphorylation of the erythropoietin receptor (EpoR) and signal transducer and activator of transcription 5 (STAT5), and the total amount of STAT5 in the cytoplasm, was determined by quantitative immuno-blotting given in arbitrary units (black diamonds). Linear interpolation of EpoR phosphorylation is indicated in gray and used as input function for the mathematical models. The mathematical fit achieved with the one-way cascade model (A) and the cycling model (B) is represented in red. (Adapted from I. Swameye, T.G. Müller, J. Timmer et al. *Proc. Natl Acad. Sci. U.S.A.* 100:1028–1033, 2003.)

approach reveals which concentrations and rate constants have the largest influence on the overall behavior of the system. Usually, sensitivity analysis can be applied only if most parameters can be identified. In the mathematical model of JAK–STAT signaling, the parameters of nuclear export and import turn out to be most sensitive to perturbation, a prediction that is confirmed experimentally. These investigations reveal rapid nucleocytoplasmic cycling as a general design principle for signaling cascades involving latent transcription factors such as STAT. This has been found to hold true also for SMAD factors (Section 6.2.2).

A related method called metabolic control analysis has been proved to be particularly useful for determining rate-limiting steps in metabolic networks. It has shown, for instance, that in such networks a single step is not rate-limiting but rather many components contribute simultaneously to the control of the network. This means that, as might have been expected, control is not the property of single components but a network property.

Sidebar 17.1 Kinetics of protein interactions

Law of mass action kinetics: When a reversible chemical reaction A + B = C + D is in equilibrium, the law of mass action defines an equilibrium constant as $K_{eq} = C_C C_D / C_A C_B$, with C_C, C_D, C_A, and C_B being the concentrations of the initial products A and B and the end products C and D. For a reversible chemical reaction, rates of the forward and backward reactions are proportional to the concentration of the reaction partners.

Michaelis–Menten kinetics: An approximation to mass action kinetics is used for enzyme-catalyzed reactions. Maximal reaction velocity is reached when the enzyme is saturated with its substrate. When the reaction rate is plotted against the substrate concentration, a hyperbolic curve is obtained. The reversible formation of an intermediary enzyme–substrate complex reaching a steady state is assumed. Limitations are that, in signal-transducing networks, intermediary complexes do not reach a steady state or reactions are tightly coupled to other processes. The Michaelis–Menten equation for an enzymatic reaction with one substrate is $v_0 = V_{max} C_S / K_M + C_S$, where C_S is the substrate concentration, K_M is the Michaelis constant, V_{max} is the maximal velocity, and v_0 is the initial rate of the enzymatic reaction. K_M is equal to $^1/_2 V_{max}$. The smaller K_M is, the higher is the enzyme–substrate affinity. For details see textbooks of biochemistry.

Hill coefficient: For enzymes consisting of several subunits that allosterically cooperate with each other, a sigmoidal curve is obtained when the reaction rate is plotted against the substrate concentration. This cooperative behavior is described by the Hill function, which is an extension of the Michaelis–Menten equation: $v_0 = V_{max} C_S^h / K_M + C_S^h$. Being proportional to the steepness of the sigmoidal function (that is, the deviation from the normal hyperbolic Michaelis–Menten function), the exponent h (Hill coefficient) is a measure for the strength of the cooperative effect.

Summary

Mathematical modeling and computer simulation of signaling pathways as components of signal-processing networks is based on experimental data and takes into account the basic laws of (bio)chemical kinetics. To arrive at a best fit between experiment and model, methods allowing an estimation of those parameters known incompletely or not at all are applied. Mathematical sensitivity analysis is performed to find out those parameters that dominate the overall behavior of a system.

17.6 Synthetic and predictive biology: the improvement of biological systems

An in-depth investigation of the nature and dynamics of biological systems is expected to have far-reaching practical consequences, for instance in agriculture, environmental protection, and medicine. This is illustrated with the help of two examples: systems robustness and drug discovery strategies.

17.6.1 Systems robustness

Perturbation insensitivity or **robustness** is a characteristic property of biological systems that is essential for performance and survival. As shown in this book, it is particularly critical for cellular signal processing. Robustness is due to the following.

(1) **Adaptation**, or the system's ability to cope with perturbations over a wide range of sensitivity. As a rule, adaptation is achieved by negative feedback control or by a combination of positive and negative feedback (Section 1.9).

(2) **Parameter insensitivity**, or the ability to cope with changes in internal kinetic parameters. This property protects the system against signaling noise. Parameter insensitivity is achieved by sigmoidal kinetics, as found for proteins consisting of allosterically interacting subunits or subdomains (Section 1.3).

(3) **Fault tolerance**, which allows the system to continue operating, albeit at lower performance, even when some of its components fail. Fault tolerance requires stability of components and processes, which mostly is due to a redundancy (backup) of regulatory proteins and pathways. Temporal and spatial separation of redundant components and pathways (modularity) prevents a failure from spreading over the whole system.

In engineering science, robustness indicates a superior design. In fact, the approaches used to achieve robustness resemble each other in living and engineered systems. Therefore, in principle, systems biologists can employ the methods used by engineers to analyze and improve the robustness of biological networks.

17.6.2 Improvement of drug development

Diseases can be understood as perturbations of biological systems ranging from the molecular up to the social level. Consequently, mathematical models integrating various signal-transducing cascades are expected to provide clinical researchers with novel mechanism-based drug discovery strategies. In fact, under the conditions mentioned above, mathematical models may predict the therapeutic and side effects of drugs in a more reliable and cost-effective way than traditional methods.

Many diseases, particularly the most complex ones such as diabetes and cancer, are caused and promoted by multiple defects of cellular signal processing. Like a signal that induces an extended excitation pattern, a defect of signaling is also expected to evoke in cells a widespread perturbation of the signal-processing network (after all, and to stress it again: most signaling cascades are complex, highly interconnected systems with multiple feedback loops rather than linear pathways). As a rule, such a network perturbation is not restricted to one cell type but afflicts surrounding tissues, as indicated, for instance, by the role the interaction between tumor and stroma cells plays in cancer development. To design new therapeutic strategies for such diseases, it appears to be mandatory, therefore, to investigate networks and systems at different levels of complexity, rather than individual components. In fact, predictive drug design is one of the major practical goals of systems biology.

For obvious reasons, advanced cancer research presently tries to cope with the effects of protein overexpression due to oncogenic defects such as somatic gene mutations. The most serious obstacles to this approach are intolerable side effects and the development of drug resistance. A prominent case illustrating this dilemma is found in Section 7.1.1. An example where mathematical modeling has led to new insights is provided by the serine/threonine kinase RAF (Section 10.1.4), which is constitutively activated in many tumors due to oncogenic mutations. Nevertheless, as compared with normal cells, the activity of the downstream kinase ERK (Extracellular signal-Regulated Kinase) is increased only moderately. Mathematical modeling has revealed that half-maximal inhibition of the mutated RAF by a potential anti-cancer drug would lead to some decrease of ERK activity but might entirely abrogate RAF and ERK activity in healthy cells, thus suggesting that the mutated RAF is a poor drug target since it has lost part of its control (of course, the prediction of the model has to be verified experimentally). An alternative strategy would be to interfere with another node (protein) of the data-processing subsystem to which RAF belongs. However, this would require a detailed knowledge of systems dynamics.

Thus, although systems biology is still in its infancy, its potential benefits are considerable, probably rendering it an indispensable tool for both basic and applied research in the future. However, there is still a long way to go, particularly as far as the highly complex systems of cellular data processing, the "brain of the cell," are concerned.

Summary

Up to a certain extent, methods used by engineers to analyze and improve a technical system can be employed also in systems biology. Thus, in both cases the insensitivity of systems to disturbances (robustness) is based on the same properties: adaptation, parameter insensitivity (for instance, noise insensitivity), and fault tolerance (achieved by redundancy). The analysis of signal-processing systems is expected to have far-reaching practical consequences such as optimized drug design. This holds true particularly for the treatment of highly complex diseases of signaling such as cancer.

Further reading

Albeck JG, MacBeath G, White FM et al. (2006) Collecting and organizing systematic sets of protein data. *Nat. Rev. Mol. Cell Biol.* 7, 803–812.

Aldridge BB, Burke JM, Lauffenburger DA et al. (2006) Physicochemical modelling of cell signalling pathways. *Nat. Cell Biol.* 8, 1195–1203.

Alon U (2003) Biological networks: the tinkerer as an engineer. *Science* 301, 1866–1867.

Alon U (ed) (2007) An Introduction to Systems Biology. Design Principles of Biological Circuits, pp 1–295. Chapman & Hall/CRC Mathematical and Computational Biological Series, Chapman & Hall.

Barabasi AL & Oltvai ZN (2004) Network biology: understanding the cell's functional organization. *Nat. Rev. Genet.* 5, 101–113.

Bothwell JFH (2006) The long past of systems biology. *New Physiol.* 170, 6–10.

Bray D (2003) Molecular networks: the top-down view. *Science* 301, 1864–1866.

Csete ME & Doyle JC (2002) Reverse engineering of biological complexity. *Science* 295, 1664–1669.

Fu P (2006) A perspective of synthetic biology: assembling building blocks for novel functions. *Biotechnol. J.* 1, 690–699.

Hornberg JJ & Westerhoff HV (2006) Oncogenes are to lose control on signaling following mutation: should we aim off target? *Mol. Biotechnol.* 34, 109–116.

Kirschner W (2005) The meaning of systems biology. *Cell* 121, 503–504.

Kitano H (2002) Systems biology: a brief overview. *Science* 295, 1662–1664.

Kollmann M, Lovdok L, Bartholome K et al. (2005) Design principles of a bacterial signalling network. *Nature* 438, 504–507.

Kremling A & Saez-Rodriguez J (2007) Systems biology—an engineering perspective. *J. Biotechnol.* 129, 329–351.

Noble D (2002) Modeling the heart—from genes to cells to the whole organ. *Science* 295, 1678–1682.

O'Malley MA & Dupré J (2005) Fundamental issues in systems biology. *BioEssays* 27, 1270–1276.

von Bertalanffy L (1968) General system theory. G. Braziller.

Westerhoff HV & Palsson BO (2004) The evolution of molecular biology into systems biology. *Nat. Biotechnol.* 22, 1249–1252.

Wiener N (1961) Cybernetics or Control and Communication in the Animal and the Machine. MIT Press.

Wolkenhauer O (2001) Systems biology: the reincarnation of systems theory applied in biology? *Briefings Bioinf.* 2, 258–270.

Wolkenhauer O, Ullah M, Wellstead P et al. (2005) The dynamic systems approach to control and regulation of intracellular networks. *FEBS Lett.* 579, 1846–1853.

Zhu X, Gerstein M & Snyder M (2007) Getting connected: analysis and principles of biological networks. *Genes Dev.* 21, 1010–1024.

Index

Note: Page references followed by the suffixes F, S and T refer to figures, sidebars and tables respectively.